Microbeam and Nanobeam Analysis

Edited by
D. Benoit, J.-F. Bresse,
L. Van't dack, H. Werner,
J. Wernisch

Mikrochimica Acta
Supplement 13

Springer-Verlag Wien GmbH

Dr. Daniele Benoit
Beauchamp, France

Dr. Jean-Francois Bresse
Department PAP/CDP/CMT, C.N.E.T. France Telecom
Bagneux, France

Luc Van't dack
Department of Chemistry, University of Antwerp (UIA),
Antwerp-Wilrijk, Belgium

Univ.-Prof. Dr. Helmut Werner
Waalre, The Netherlands

Univ.-Doz. Dr. Johann Wernisch
Institut für angewandte und technische Physik,
Technische Universität, Wien, Austria

Typesetting: Thomson Press, New Delhi, India

Graphic design: Ecke Bonk
Printed on acid-free and chlorine free bleached paper.

With 394 partly coloured Figures

Die Deutsche Bibliothek - CIP-Einheitsaufnahme
Microbeam and nanobeam analysis / D. Benoit ... (ed.). - Wien ; New York : Springer, 1996
(Mikrochimica acta : Supplementum ; 13)
ISBN 978-3-211-82874-8 ISBN 978-3-7091-6555-3 (eBook)
DOI 10.1007/978-3-7091-6555-3
NE: Benoit, Daniele [Hrsg.]; Mikrochimica acta / Supplementum

ISSN 0026-3672
ISBN 978-3-211-82874-8

Preface

This supplement of *Mikrochimica Acta* contains selected papers from the Fourth Workshop of the European Microanalysis Society (EMAS) on "Modern Developments and Applications in Microbeam Analysis" which took place in May 1995 in Saint Malo (France).

EMAS was founded in 1986 by members from almost all european countries in order to stimulate research, applications and development of all forms of microbeam methods. One important EMAS activity is the organisation of biennial workshops for demonstrating the current status and developing trends of microanalytical techniques. For this meeting, EMAS chose to highlight the following topics: Monte-Carlo simulations, transport calculations and use of soft X-rays for electron probe microanalysis (EPMA), dynamic secondary ion mass spectrometry (SIMS), detection of small quantities using different techniques: synchrotron radiation X-ray fluorescence, particle induced X-ray emission (PIXE), cathodoluminescence microscopy (CL). Two new kinds of instrumental techniques were also presented: atomic probe and scanning probe microscopy (STM). The aim of the conference is to give introductory lectures corresponding to the topics of the meeting and to have contributions in the form of poster sessions. More than 80 posters were presented. Most of them gave a short oral presentation. The poster subjects were related to the use of microanalytical techniques: EPMA with wavelength dispersive spectrometry (WDS) and energy dispersive spectrometry (EDS), Auger electron spectrometry (AES), secondary ion mass spectrometry (SIMS), scanning electron microscopy and other topographical methods like scanning tunneling microscopy (STM) or atomic force microscopy (AFM). The field of applications is very broad: metallurgy, mineralogy, semiconductor, ceramic, glass, composite, polymer and biological material.

The authors originate from almost all european countries, including eastern european ones and Russia. One invited lecture was given by a USA speaker.

This issue contains the full texts of the ten introductory lectures and 50 brief articles. These articles are the condensed scientific informations of the poster sessions. They were subjected to peer-review according to the usual procedure and some of them were rejected.

Proceedings of previous EMAS workshops were also published, either as a supplement to or in a regular issue of *Mikrochimica Acta*. We hope that these contributions to the field of microbeam techniques will be found to be useful.

September 1996

The Editors

Contents

Listed in Current Contents

Mikrochim. Acta [Suppl.] 13, 1–12 (1996)

Monte Carlo Simulation Techniques for Quantitative X-Ray Microanalysis

Ludwig Reimer

Physikalisches Institut, Universität Münster, Wilhelm-Klemm Str. 10, D-48149 Münster, Federal Republic of Germany

Abstract. Monte Carlo simulations of electron diffusion become of increasing interest for scanning electron microscopy (SEM), X-ray microanalysis (XRMA) and Auger electron spectroscopy (AES) due to the increasing speed and storage capability of modern PCs. Depth distribution functions can be calculated in less than a minute also for complex specimen structures. To apply Monte Carlo simulations in the energy range of 0.1–50 keV it is necessary to use a data base of Mott elastic cross-sections calculated by the partial-wave method. For most applications it is sufficient to consider inelastic scattering by the Bethe continous-slowing-down approximation and inner-shell ionisations with energy losses larger than 100–200 eV by the Gryzinski cross-section. In future, energy-loss functions obtained by a Kramers-Kronig analysis of experimental electron energy-loss spectra (EELS) will become of interest for a better consideration of straggling effects during the slowing-down.

Key words: Monte Carlo simulation, electron scattering, X-ray microanalysis, scanning electron microscopy.

We need for the X-ray microanalysis (XRMA) with 1–50 keV electrons the depth distribution functions $\Phi(\rho z)$ of ionizing the corresponding inner atomic shells to calculate the ratio

$$k = \frac{N_a}{N_s} = \frac{c_a}{c_s} k_Z k_A k_F \tag{1}$$

of X-ray counts from an element (a) in the specimen and a pure-element or compound standard (s) with the ZAF factors for atomic mumber, absorption and fluorescence correction, respectively. This allows to calculate the fraction c_a/c_s of mass concentrations from the measured ratio N_a/N_s of X-ray counts.

The depth distribution function is strongly influenced by the electron diffusion which can be described by a Boltzmann transport equation but not by an analytical solution. In case of plane surfaces and multilayers on substrates the transport equation can be simplified because the number of orginally 7 coordinates reduces to 3, one z coordinate (depth below surface), one angle θ_z relative to the surface normal and the electron energy E. The actual electron energy is assumed to depend only on the path length and alternatively on time. Therefore, the propagation of electrons between different equidistant layers can be treated successively by reducing the electron energy step by step [1–3]. Other methods are the solution by the so-called matrix method [4], an integral equation [5] or the radiative field similarity principle [6].

The most frequent method is the Monte Carlo simulation of a large number of electron trajectories resulting in mean values of the wanted quantities [7–27]. It shows a good flexibility for more complex specimen geometries and compositions and has been applied in SEM, XRMA and AES.

Monte Carlo Simulation

Elastic Scattering

The differential cross-sections of elastic scattering can be written as

$$\frac{\mathrm{d}\sigma_{\mathrm{el}}}{\mathrm{d}\Omega} = r(\theta)\frac{e^4 Z^2}{4(4\pi e_0)^2 m^2 v^4}\frac{1}{\sin^4(\theta/2)}. \tag{2}$$

The last two factors represent the unscreened Rutherford cross-section. The first factor $r(\theta)$ is the ratio of Mott to unscreened Rutherford cross-sections. The differential Mott cross-sections result from solving the relativistic Schrödinger equation or better the relativistic Pauli–Dirac equation [11, 25–28], especially at low electron energies, using a screened Coulomb potential and the partial-wave method. The dense packing of atoms can be considered in first-order approximation by an overlap of neighboured atomic potentials (muffin-tin model) in a WKB approximation [29] as well as the partial-wave method [26]. Figures 1a–f show for some elements and electron energies E plots of $r(\theta)$ versus the scattering angle θ. The ratio $r(\theta)$ becomes unity when the unscreened Rutherford cross-section is valid. (The decrease of $r(\theta)$ to zero for $\theta \to 0$ is a consequence of screening.) For low Z elements (B and Si in Figs. 1a, b) and large θ the $r(\theta)$ values are strongly increased at low energies. This results in an increase of the backscattering coefficient η with decreasing energy below $E \simeq 5\,\mathrm{keV}$. With increasing Z the maximum of $r(\theta)$ shifts to higher energies and very low values are observed at low energies (e.g. for Au and U in Figs. 1e, f), which results in a decrease of η with decreasing E. These strong deviations from Rutherford scattering demand to use Mott cross-sections. For example, the electron backscattering coefficients at low energies only agree with experiments when using these cross-sections in Monte Carlo simulations [16, 30, 31].

Analytical fits have been proposed for the total elastic cross-sections [32, 33] (see also discussion in [23]) and the integrated differential cross-sections [34]. However, the large available memory capacity of a modern PC allows to avoid such fits, which cannot describe exactly in all cases just the less frequent large-angle scattering processes.

Because of the concentration of elastic scattering at low scattering angles θ it will be convenient to limit individual elastic scattering processes to $10^\circ \leqslant \theta \leqslant 180^\circ$

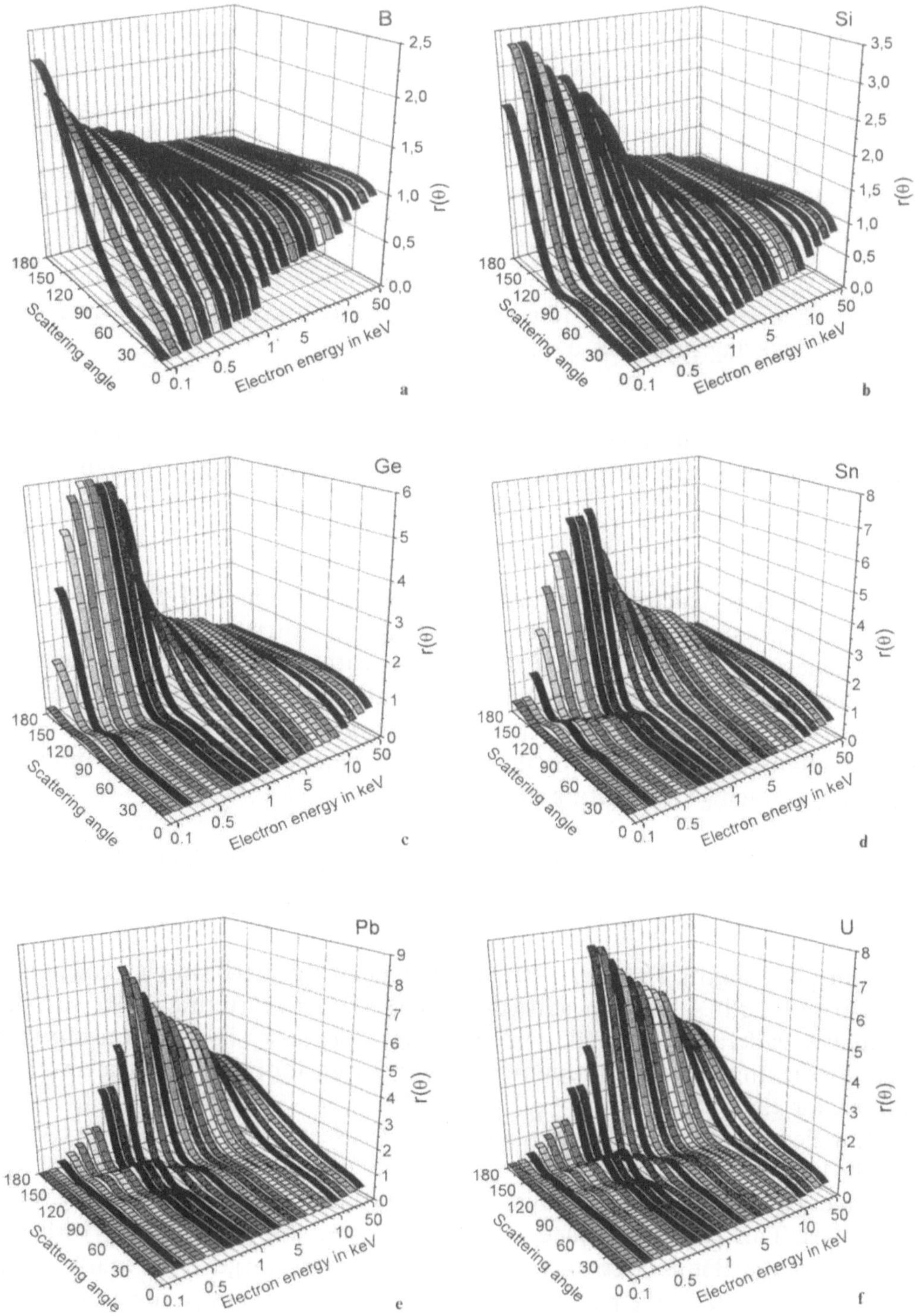

Fig. 1 a–f. Differential Mott elastic cross-sections $d\sigma_{el}/d\Omega$ for scattering angles $\theta = 10° - 180°$ plotted as ratios $r(\theta) = (d\sigma_{el}/d\Omega)/(d\sigma_{Ruth}/d\Omega)$ of Mott-to-Rutherford cross-sections for B, Si, Ge, Sn, Pb and U at electron energies $E = 0.1$–0.9 keV, 1.0–9.0 keV and 10–50 keV in Monte Carlo simulations for reducing the computation time, and to consider small-angle scattering through angles $\theta \leqslant 10^\circ$ by a mean scattering angle.

The large-angle scattering results in a partial elastic cross-section

$$\sigma_{el} = \int_{10^\circ}^{\pi} \frac{d\sigma_{el}}{d\Omega} 2\pi \sin\theta \, d\theta \tag{3}$$

and the mean-free-path length

$$\Lambda_{el} = 1/N\sigma_{el} \tag{4}$$

where $N = N_A \rho/A$ is the number of atoms per unit volume (N_A = Avogadro's number, ρ = density). With a random number $R \in [0, 1]$, the free-path between two large-angle scattering processes becomes

$$s = -\Lambda_{el} \ln R. \tag{5}$$

At the end of this trajectory, the new scattering angle can be obtained with a random number by an inversion of the normalized distribution function

$$R(\theta_{el}) = \frac{1}{\sigma_{el}} \int_{10^\circ}^{\theta_{el}} \frac{d\sigma_{el}}{d\Omega} 2\pi \sin\theta \, d\theta \tag{6}$$

that an electron is scattered through angles smaller than θ. By a numerical inversion of (6), $\cos\theta_{el}$ as a function of R is stored in a matrix for different electron energies. Due to the importance of large-angle scattering processes for electron backscattering and the low probability of scattering angles $\theta_{el} \geqslant 90^\circ$ the last interval $0.9 \leqslant R \leqslant 1$ should be stored with a ten times finer division of R-values.

The small-angle scattering through $0^\circ \leqslant \theta \leqslant 10^\circ$ can be considered by a multiple scattering approach [35] resulting in:

$$\begin{aligned} \langle\cos\theta\rangle &= 1 - \overline{\theta^2}/2 + \cdots = \exp\left[-2\pi s N \int_0^{10^\circ} \frac{d\sigma}{d\Omega}(1-\cos\theta)\sin\theta \, d\theta\right] \\ &= 1 - s/\Lambda_{tr} + \cdots \end{aligned} \tag{7}$$

(Λ_{tr} = transport length). This mean scattering angle $(\overline{\theta^2})^{1/2}$ can be introduced at half the free-path-length s between subsequent large-angle scattering processes. An individual consideration of small-angle scattering processes is only necessary for calculating the angular distribution or the lateral spread of electrons transmitting thin films, for example.

At low electron energies the partial elastic cross-section (3) becomes very small due to the low values of $r(\theta)$ and too large free path lengths are simulated which will be incorrect because the larger decrease of energy along the trajectory changes the probabilities. This can be avoided by limitating free paths to $s \leqslant s_{min}$ with e.g. $s_{min} = 1/20$ of the Bethe electron range. In this case no large-angle scattering is considered at the end of the path length but the small-angle scattering at its centre is applied as described above.

Electron Channeling Effects

The approach of Monte Carlo simulation does not consider the crystalline structure of most solids and the question arises how the neglection of electron channelling effects influences the accuracy of Monte Carlo simulations. Due to the dynamical theory of electron diffraction, electron waves in crystals propagate in a Bloch wave field with nodes and antinodes near nuclei [36]. Therefore all interaction processes concentrated at the nuclei as backscattering through large angles and the ionisation of inner shells are proportional to the orientation-sensitive amplitude (probability density) of the antinodes and the backscattering coefficient [37, 38] and X-ray emission [39] depend on the inclination of the incident electron beam to the lattice planes. This can be observed as an electron channelling pattern (ECP) when rocking the incident electron beam [40]. Also the scattering out of the Bloch wave fields is very orientation-dependent and results in the electron backscattering pattern (EBSP) [41]. Both types of pattern contain excess or defect Kikuchi bands which can be explained by a theorem of reciprocity [42]. The depth of modulation is only a few percents at high energies beyond 10 keV but can increase to the order of ten percent below 5 keV [38].

Thin amorphous or polycrystalline surface layers increase the effective aperture of the electron beam by multiple scattering and in SEM and X-ray microanalysis we work with large apertures of the order of a few tens of milliradian to get a small spot size and/or large intensity of the electron probe. This strongly blurs the modulation of ECP. The intensity of EBSP is not influenced by the probe aperture, however in SEM we work with a large detector aperture for BSE resulting in a strong averaging. This is the reason why the neglection of channelling effects in Monte Carlo simulations gives no significant discrepancies in comparison with experiments on single-crystalline materials.

Inelastic Scattering and Energy Losses

The most frequent used method to consider the decrease of mean electron energy E_m along the trajectories in a compound with weight fractions c_i is the Bethe continuous-slowing-down approximation [43] with the stopping power (mean energy loss per unit mass thickness $x = \rho s$, s = path length of trajectory):

$$S(E) = \left|\frac{dE_m}{dx}\right|_{\text{Bethe}} = \sum_i \frac{c_i N_A}{A_i} \frac{2\pi e^4}{(4\pi\varepsilon_0)^2 E} \sum_{nl} f_{nl} \ln(E/E_{nl}) \tag{8}$$

($f_{nl} \simeq Z_{nl}$ = number of electrons in the subshell nl) where the ionization energies E_{nl} of the atomic shells are considered individually or by

$$S(E)_{\text{Bethe}} \simeq \sum_i \frac{c_i N_A Z_i}{A_i} \frac{2\pi e^4}{(4\pi\varepsilon_0)^2 E} \ln\left(\frac{1.166E}{J_i}\right) \tag{9}$$

using a mean ionization potential [44]

$$J = 9.76Z + 58.5Z^{-0.19} \quad \text{or} \quad J = 11.5Z \quad \text{for } Z \leqslant 12 \tag{10}$$

in units of electron volts.

This continuous slowing down of electron energy in steps of $S(E)\rho s$ per free-path length s results in a total path length (Bethe range) where the electrons come to rest. The Bethe range for low Z material is of the same order as the practical electron range but about a factor 2–3 larger for high Z material.

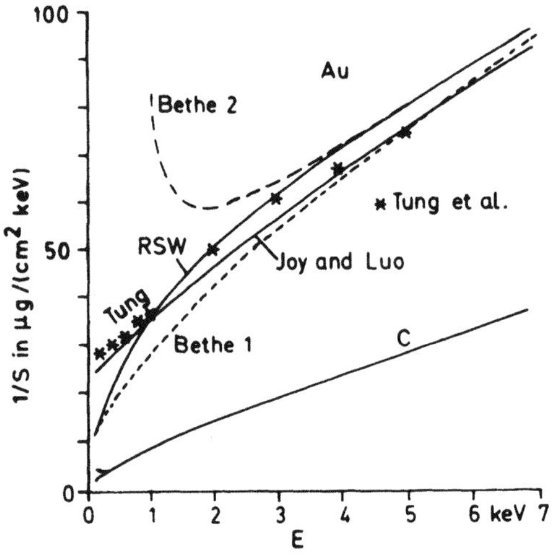

Fig. 2. Inverse stopping power $1/S$ of gold and carbon calculated from Eqs. (8) (Bethe 1) and (9) (Bethe 2) [43], Eq. (11) (Joy and Luo [45]), the Rao-Sahib and Wittry approach (RSW) [47] and calculations by Tung et al. [48]

Figure 2 shows in examples for gold and carbon different approaches of the inverse stopping power. The first formula (8) (Bethe 1) considers that atomic shells with ionization energies higher than the kinetic energy E do not contribute. The approximation (9) with a mean ionization potential J (Bethe 2) results in wrong values at low electron energies. In a proposal by Joy and Luo [45] an energy-dependent value

$$J' = J/(1 + kJ/E) \tag{11}$$

is substituted for J with k varying between 0.77 (C) and 0.85 (Au). Another proposal is to substitute Z in (9) by $N_{\mathrm{eff}}(E)$ and J by J_{eff} which can be calculated from an electron energy-loss spectrum (EELS) at high electron energies of 100–200 keV [46]. Rao-Sahib and Wittry [47] approximated $1/S$ below $E/J = 6.3$ by a parabolic curve $1/S \propto E^{0.5}$. Because this approach results in the best agreement to measurements of electron backscattering coefficients [30, 31], we prefer this simplest approach in our Monte Carlo simulations. The stars in Fig. 2 are direct calculations of $1/S$ by Tung et al. [48]. Due to the much lower value of J, $1/S$ from (9) holds down to less than 1 keV for carbon, for example.

A large part of inelastic scattering processes with energy losses W results in plasmon losses, intra- and interband transitions and inner shell excitations which are concentrated at small scattering angles of the order of $\theta_{\mathrm{E}} = W/2E$. Large-angle inelastic scattering results mainly from Compton electron–electron scattering with the maximum at the Bethe ridge at

$$\theta_{\mathrm{C}} = (W/E)^{1/2}. \tag{12}$$

Individual Compton scattering processes with an energy loss $W \geqslant 100\,\mathrm{eV}$ have been considered in [11, 18, 20], for example. A semi-classical approach is proposed by Gryzinski [49]. The differential cross-section for an energy loss W, a binding energy E_{nl} (e.g. tabulated in [50]) and the incident electron energy E can be written as

$$\left(\frac{d\sigma}{dW}\right)_{Gryz} = \frac{\pi e^4}{(4\pi\varepsilon_0)^2} \frac{E_{nl}}{EW^3} \left(\frac{E}{E+E_{nl}}\right)^{3/2} \left(1-\frac{W}{E}\right)^{E_{nl}/(E_{nl}+W)} \cdot \left\{\frac{W}{E_{nl}}\left(1-\frac{E_{nl}}{E}\right) + \frac{4}{3}\ln\left[2.7+\left(\frac{E-W}{E_{nl}}\right)^{1/2}\right]\right\}. \tag{13}$$

For the valence electrons, E_{nl} can be substituted in (13) by the Fermi energy E_F. This results in the Streitwolf formula [51] used also in [52]. For $W \gg E_F$ the cross-section

$$\frac{d\sigma}{d\Omega} = \frac{\pi e^4}{(4\pi\varepsilon_0)^2} \frac{1}{EW^2} \tag{14}$$

approximates that for scattering at a 'free' atomic electron.

With (13) the contribution of inelastic scattering with energy losses $E_{nl} + \varepsilon(\varepsilon \geqslant 100 - 200$ eV) to the stopping power becomes

$$S(E, \varepsilon)_{Gryz} = \sum_i \frac{c_i N_A}{A_i} \left(\sum_{nl} Z_{nl} \int_{E_{nl}+\varepsilon}^{E} \frac{d\sigma_{nl}}{dW} W dW\right)_i \tag{15}$$

and has to be subtracted from the Bethe stopping power $S(E)_{Bethe}$ (9). For avoiding differences between the Bethe and Gryzinski cross-sections, the Gryzinski cross-sections can be corrected either by fitting the Z_{nl} for low E_{nl} [25] or by correcting the Gryzinski cross-sections [20] to

$$\frac{d\sigma_{nl}}{dW} = \left(\frac{d\sigma_{nl}}{dW}\right)_{Gryz} \frac{S(E)_{Bethe}}{S(E, \varepsilon = 0)_{Gryz}}, \tag{16}$$

so that the sum of the remaining Bethe and Gryzinski stopping powers equalize the Bethe stopping power (9) which is in good agreement with experiments.

After passing the free-path length s, the condition

$$R \leqslant \sigma_{el}/(\sigma_{el} + \sum_{nl} \sigma_{nl}) \tag{17}$$

results in an elastic scattering process or else in an inelastic Compton scattering at the nl shell electrons, respectively, with an energy loss $W \geqslant E_{nl} + \varepsilon$ and a scattering angle θ_C (12). The trajectories of the excited fast secondary electrons (FSE) of energy ε can also be simulated until their energy is smaller than a threshold. This consideration of FSE is important e.g. for a detailed simulation of electron backscattering at thin films [10, 11] or for low-energy X-ray emission discussed below.

The scattering angles θ_{el} or θ_C and an isotropically distributed azimuth

$$\chi = 2\pi R \tag{18}$$

allow to calculate the new direction of the electron trajectory.

The Bethe continuous-slowing-down approximation and the Compton scattering cannot consider completely straggling as the statistical broadening of the energy distribution with increasing path length. Calculations of a more accurate energy distribution of backscattered or transmitted electrons, for example, need the

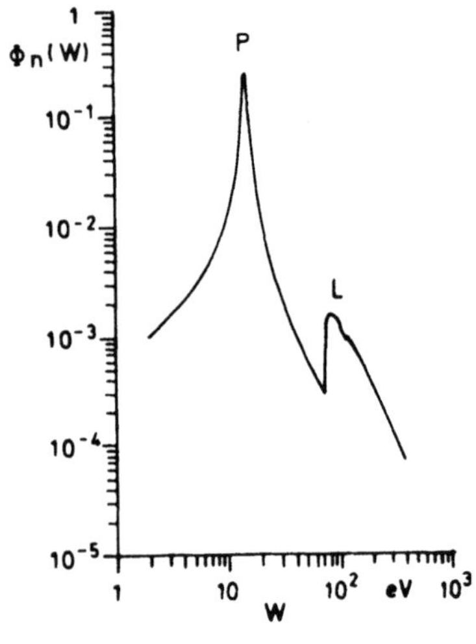

Fig. 3. Double-logarithmic plot of the analytical approach of differential inelastic cross-section $d\sigma/dW$ of aluminium with the plasmon loss (P) and L-shell ionization edge

use of differential inelastic cross-sections $d^2\sigma_{\text{inel}}/dW d\Omega$. Proposals to use Landau-distributions [12] or exponential distributions [52] of energy losses after passing a free path are only first-order approximations. The Landau theory also results in too large half-widths of the energy distribution [53].

In case of aluminium with a sharp plasmon loss, it is possible to simulate energy-loss spectra by using formula for the free-path of plasmon energy losses, the Gryzinski formula (13) for L shell ionisation and the Streitwolf formula for Compton scattering [54] or by an analytical approach $d^2\sigma_{\text{inel}}/dW d\Omega$ for the plasmon loss and the Compton scattering [55] and additional inner-shell ionization cross-sections calculated with the SIGMAK and SIGMAL programs [56] as shown in Fig. 3. This results in a good agreement with experimental energy-loss spectra of electrons transmitted through aluminium foils [23, 57]. In case of a more complex plasmon spectrum, the relation

$$\frac{d^2\sigma}{dW d\Omega} = \frac{1}{\pi^2 a_H E N_e} \cdot \frac{\text{Im}\{-1/\varepsilon(\omega)\}}{\theta^2 + \theta_E^2} \tag{19}$$

of the dielectric theory can be used, where $\varepsilon(\omega) = \varepsilon_1 - i\varepsilon_2$ is the complex dielectric constant, Im denotes the imaginary part, $W = \hbar\omega$, a_H = Bohr radius, N_e = number of electrons per unit volume. The energy-loss function $\text{Im}\{-1/\varepsilon[0]\} = \varepsilon_2/(\varepsilon_1^2 + \varepsilon_2^2)$ can be obtained from a Kramers-Kronig analysis [56] of experimental electron energy-loss spectra (EELS) at high electron energies (80–200 keV). This energy-loss function can also be used to correct the Bethe stopping power for low energies [46, 58].

Unfortunately, the EELS spectra can only be recorded with a small aperture of the order of $\alpha = 5$–10 mrad and the inelastically and multiple elastic–inelastically electrons scattered through $\theta \geqslant \alpha$ are lost. The latter can be corrected by models of the angular distribution before applying the Kramers-Kronig analysis. Nevertheles, the Compton scattering through angles $\theta \geqslant \alpha$ should be considered by the Gryzinski cross-section (13) and the Compton angle θ_C (12).

X-ray Emission

The Gryzinski cross-sections discussed above can also be used to calculate the probability of ionization of an inner shell which can result in the emission of X-ray quanta or Auger electrons. However, these processes are not frequent enough and of the order of 10^6 trajectories will be necessary to give a good statistics for X-ray depth distributions. Therefore it is much more time-consuming to calculate the probability of X-ray emission along a free-path length s between two scattering processes. This can be done by integrating the Gryzinski cross-section (13) over W and writing it as a function of the overvoltage ratio $u = E/E_{nl}$:

$$\sigma_{nl}(u) = \frac{\pi e^4 Z_{nl}}{(4\pi\varepsilon_0)^2 E_{nl}^2} \frac{1}{u} \left(\frac{u-1}{u+1}\right)^{1.5} \left\{1 + \frac{2}{3}\left(1 - \frac{1}{2u}\right) \ln[2.7 + (u-1)^{0.5}]\right\} . \quad (20)$$

Multiplying with the exponential attenuation $\exp[-(\sum_i c_i(\mu/\rho)_i)\operatorname{cosec}\psi]$ considers the X-ray absorption under an X-ray take-off angle ψ. Figure 4 shows as an example, the calculation of depth distributions $\Phi(\rho z)$ for K, Si and O Kα lines in a $KAlSi_3O_8$ (sanidine) standard with and without absorption (low for K Kα and high for O Kα).

Comparison of simulations for a pure element or compound standard and the probe allows directly to calculate the k-ratio (1) and the ZAF correction factor k_{ZA}, which no further has to be introduced as a product $k_Z k_A$ used in the conventional ZAF correction scheme. Already 1000–10000 trajectories (20–200 s computation time) are sufficient for good statistics. Such a Monte Carlo simulation can also be used in an iteration routine.

The depth distribution of X-ray emission is also needed for the calculation of the fluorescence correction by characteristic lines of other elements just beyond the ionization energy for the line of interest. The isotropic emission of the characteristic lines can simplify the integration over contributions from different parts of the specimen.

For the analysis of low Z elements with low-energy characteristic X-ray energies of only a few hundreds of electron volts, also the fast secondary electrons with energies $E \geqslant 100$–200 eV can contribute to the X-ray excitation and have to be considered in a Monte Carlo simulation of the depth distribution [59].

The Monte Carlo simulation of X-ray emission cannot only be used for a parametrization of a parabolic $\Phi(\rho z)$ model according to the PAP correction method [60] but also for multilayer systems [20, 59], implantation and diffusion depth profiles [61, 62] and X-ray linescans over more complex specimen structures [24].

An example of a linescan over a composite is shown in Fig. 5 for a 1 μm gold layer in an Al matrix scanned head-on by 15 keV electrons demonstrating the influence of electron diffusion on the line profile. Au L X-ray quanta with a higher ionization energy are generated in a smaller volume than Au M quanta. Figures 4 and 5 are calculated with the Monte Carlo Simulation program MOCASIM described in [24].

Conclusion

The presented background of a Monte Carlo simulation of electron diffusion show that elastic differential Mott cross-sections from the solution of the Schrödinger of

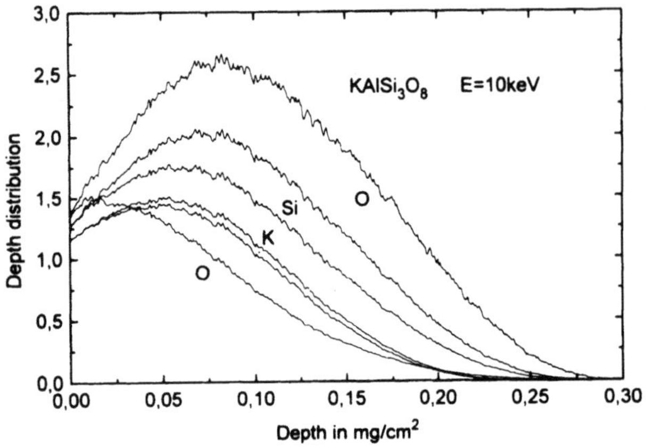

Fig. 4. Depth distribution functions $\Phi(\rho z)$ of the excitation of K, Si and O Kα X-ray lines by 10 keV electrons in a $KAlSi_3O_8$ standard without (upper) and with absorption (lower curves) 10000 trajectories, 200 s computation time)

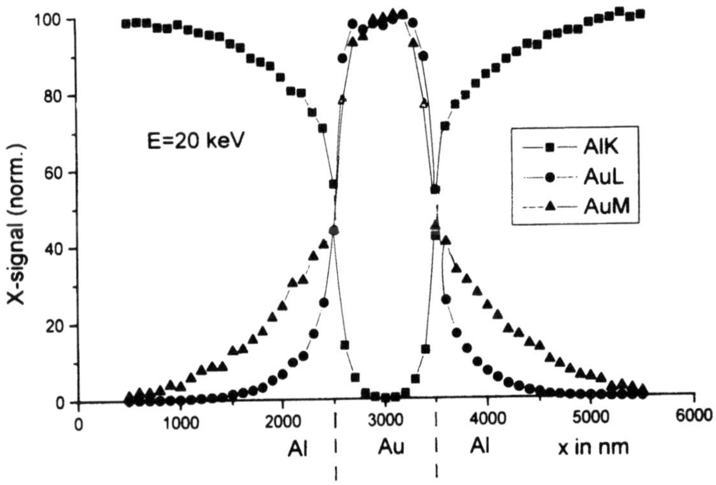

Fig. 5. Monte Carlo simulation of a linescan of Al K, Au L and Au M X-ray quanta excited by 15 keV electrons across a 1 µm Au layer in an Al matrix viewed head-on (Signal intensities arbitrary and normalized)

Pauli equation by the partial-wave method are essential. The inelastic scattering can be considered either by the Bethe continuous-slowing-down approximation corrected at low incident electron energies, an additional consideration of individual inelastic collisions with the excitation of fast secondary electrons with energies larger than 100–200 eV, or an energy-loss function $\mathrm{Im}\{-1/\varepsilon\}$ from EELS experiments when detailed energy loss spectra of backscattered or transmitted electrons shall be simulated. This allows accurate calculations in the range 1–50 keV of primary electron energies. For more complex structures the Monte Carlo simulation of X-ray depth distributions and k-factors is superior to all other methods.

References

[1] D. Liljequist, *J. Phys. D* **1977**, *10*, 1363.

[2] H. J. August, J. Wernisch, *X-Ray Spectrom.* **1991**, *20*, 131 and 141.

[3] L. Reimer, R. Böngeler, M. Kässens, F. F. Liebscher, R. Senkel, *Scanning* **1991**, *13*, 381.
[4] D. J. Fathers, P. Rez in: *Electron Beam Interactions with Solids* (D. F. Kyser, et al., eds.) SEM. Inc., AMF O'Hare, Chicago, 1984, p. 193.
[5] K. E. Hoffmann, H. Schmoranzer, in: *Electron Beam Interactions with Solids* (D. F. Kyser et al., eds.), SEM Inc., AMF O'Hare, Chicago, 1984, p. 209.
[6] I. S. Tilinin, W. S. M. Werner, *Mikrochim. Acta* **1994**, *114/115*, 485.
[7] M. Green, *Proc. Phys. Soc.* **1963**, *82*, 204.
[8] H. E. Bishop, *Proc. Phys. Soc.* **1965**, *85*, 855.
[9] L. Reimer, *Optik* **1968**, *27*, 86.
[10] E. R. Krefting, L. Reimer, in: *Quantitative Analysis with Electron Microprobes and Secondary Ion Mass Spectrometry* (E. Preuss, ed.) Kernforschungsanlage Jülich, Jül-Conf-8, 1973, p. 114.
[11] L. Reimer, E. R. Krefting, *NBS Spec. Publ.* **1975**, *460*, 45.
[12] J. Henoc, F. Maurice, *NBS Spec. Publ.* **1975**, *460*, 61.
[13] R. L. Myklebust. D. E. Newbury, H. Yakowitz, *Spec. Publ.* **1975**, *460*, 105.
[14] G. Love, M. G. C. Cox, V. D. Scott, *J. Phys. D* **1977**, *10*, 7.
[15] J. Henoc, F. Maurice, *J. Micr. Spect. Electr.* **1980**, *5*, 347.
[16] M. Kotera, K. Murata, K. Nagami, *J. Appl. Phys.* **1981**, *52*, 997 and 7403.
[17] F. Salvat, J. Parellada, *J. Phys. D* **1983**, *17*, 185.
[18] L. Reimer, D. Stelter, *Scanning* **1986**, *8*, 265.
[19] D. C. Joy, *J. Microbiology* **1987**, *147*, 51.
[20] N. Ammann, P. Karduck, *Microbeam Analysis*, San Fransisco Press, San Francisco, 1990, p. 150.
[21] M. Ichimura, M. Aramata, R. Shimizu, *J. Appl. Phys.* **1980**, *51*, 2853.
[22] K. Murata, M. Kotera, K. Nagami, *J. Appl. Phys.* **1983**, *54*, 1110.
[23] L. Reimer, R. Senkel, *Optik* **1995**, *98*, 85.
[24] L. Reimer, M. Kässens. L. Wiese, *Mikrochim. Acta* **1996**, *[Suppl.] 13*, 485.
[25] L. Reimer, B. Lödding, *Scanning* **1984**, *6*, 128.
[26] F. Salvat, R. Mayol, *Comp. Phys. Comm.* **1993**, *74*, 358.
[27] A. Jablonski, G. Gergely, *Scanning* **1989**, *11*, 29.
[28] Z. Czyzewski, D. O. MacCallum, A. Romig, D. C. Joy, *J. Appl. Phys.* **1990**, *68*, 3066.
[29] L. Reimer, K. H. Sommer, *Z. Naturforschg.* **1968**, *23a*, 1569.
[30] R. Böngeler, U. Golla, M. Kässens, L. Reimer, R. Senkel, M. Spranck, *Scanning* **1993**, *15*, 1.
[31] L. Reimer, *Image Formation in Low-Voltage Scanning Electron Microscopy, SPIE Tutorial Text, Vol. TT12*, SPIE Optical Eng. Press, Bellingham, 1993.
[32] R. Browning, *Appl. Phys. Letters* **1991**, *58*, 2845.
[33] R. Gauvin, D. Drouin, *Scanning* **1993**, *15*, 140.
[34] D. Drouin, R. Gauvin, D. C. Joy, *Scanning* **1994**, *16*, 67.
[35] H. W. Lewis, *Phys. Rev.* **1950**, *78*, 526.
[36] L. Reimer, *Transmission Electron Microscopy, Springer Ser. Opt. Sci. Vol. 36, 3rd Ed.*, Springer, Berlin Heidelberg New York Tokyo, 1993, p. 286.
[37] G. R. Booker, A. M. B. Shaw, M. J. Whelan, P. B. Hirsch *Phil. Mag.*, **1967**, *16*, 1185.
[38] L. Reimer, H. G. Badde, H. Seidel. *Z. Angew. Phys.* **1971**, *31*, 145.
[39] B. Neumann, L. Reimer, *J. Phys. D* **1980**, *13*, 1737.
[40] D. G. Coates, *Phil. Mag.* **1967**, *16*, 1179.
[41] J. A. Venables, C. J. Harland, *Phil. Mag.* **1973**, *27*, 1193.
[42] L. Reimer, *Scanning* **1979**, *2*, 3.
[43] H. Bethe, *Ann. Physik* **1930**, *5*, 325.
[44] M. J. Berger, S. M. Seltzer, *Nucl. Sci. Ser. Rep. No. 39*, NAS-NRC Publ. No. 1133, Nat. Acad. Sci., Washington, DC, 1964.
[45] D. C. Joy, S. Lou, *Scanning* **1989**, *11*, 176.
[46] S. Luo, X. Zhang, D. C. Joy, *Radiat Effects Defects Solids* **1991**, *117*, 235.

[47] T. S. Rao-Sahib, D. B. Wittry, *J. Appl. Phys.* **1974**, *45*, 5060.
[48] C. J. Tung, J. C. Ashley, R. H. Ritchie, *Surf. Sci.* **1979**, *81*, 427.
[49] M. Gryzinski, *Phys. Rev. A* **1965**, *138*, 336.
[50] K. D. Sevier, *Atom. Data Nucl. Tables* **1979**, *24*, 323.
[51] H. W. Streitwolf, *Ann. Physik* **1959**, *7*, 186.
[52] R. Shimizu, Y. Kataoka, T. Matsukawa, T. Ikuta, K. Murata, H. Hashimoto, *J. Phys. D* **1975**, *8*, 820.
[53] L. Reimer, K. Brockmann, U. Rhein, *J. Phys. D* **1978**, *11*, 2151.
[54] R. Shimizu, Y. Kataoka, T. Ikuta, T. Koshikawa, H. Hashimoto, *J. Phys. D* **1976**, *9*, 101.
[55] R. H. Ritchie, A. Howie, *Phil. Mag.* **1977**, *36*, 463.
[56] R. F. Egerton, *Electron Energy-Loss Spectroscopy in the Electron Microscope*, Plenum, New York, 1986.
[57] L. Reimer, R. Senkel, *J. Phys. D* **1992**, *25*, 1371.
[58] D. C. Joy, S. Luo, J. R. Dunlap, D. Williams, S. Cao, *Proc. Micr. Soc. America, 42nd Ann. Meeting*, 1994, p. 948.
[59] P. Karduck, N. Amman, W. P. Rehbach, *Microbeam Analysis*, San Francisco Press, San Francisco, 1990, p. 21.
[60] J. L. Pouchou, F. Pichoir, in: *Electron Probe Quantitation* (K. F. J. Heinrich, D. E. Newbury, eds.), Plenum, New York, 1990, p. 31.
[61] N. Ammann, P. Karduck, *Surf. Interface Anal.* **1994**, *22*, 54.
[62] N. Ammann, G. Gleitsmann, M. Heuken, K. Heime, P. Karduck, *Mikrochim. Acta* **1994**, *114/115*, 165.

Mikrochim. Acta [Suppl.] 13, 13–38 (1996)

Transport Equation Approach to Electron Microbeam Analysis: Fundamentals and Applications

Wolfgang S. M. Werner

Institut für Allgemeine Physik, Vienna University of Technology, Wiedner Hauptstrasse 8–10, A-1040 Vienna, Austria

Abstract. Recent developments in analytical solution approaches to the transport equation for medium and high energy electrons relevant for microbeam analysis are reviewed. The two main problems in this connection are: 1. finding an effective and accurate approximation to the elastic collision term in the Boltzmann equation and; 2. dealing with the energy dependent scattering characteristics in the case of a true slowing down problem. It is found that for smooth source angular distributions the radiative field is very weakly influenced by the details of the scattering term, allowing to replace the exact elastic collision cross section by an isotropic scattering function. This is possible without imposing any additional restrictions on the relevant scattering parameter which is determined by the relative influence of elastic and inelastic scattering. For electrons in the energy range of relevance for microbeam analysis, the scattering parameter is only weakly dependent on the energy and atomic number in the quasi-elastic regime while in the true slowing down regime the energy dependence is also very weak. This allows for certain important cases to tackle the many speed transport problem in a true slowing down situation by accounting for the relevant energy dependence in a selfconsistent way.

The results for AES/XPS which have been obtained on this basis are presented and compared with experimental results and model calculations. Among the quantities for which analytical solutions were found we note the angular/energy signal electron yield in the near peak region, the escape probability, the inelastic reflection coefficient, the backscattering factor or surface ionization, the elastic reflection coefficient etc. The agreement between the analytical results and experiment or model calculations is within 5% in most cases allowing to conclude that the proposed approach may serve as a fundament for modern microbeam analysis techniques.

Key words: electron-solid interaction, multiple scattering, Boltzmann equation, AES/XPS, quantification.

Modern microbeam analysis techniques require increasingly accurate quantification schemes. In order to achieve the required accuracy, an evaluation procedure should be based on a rigorous physical description of the phenomena involved in the signal generation. Although the basic principle of signal excitation, generation and transfer are well understood, it is difficult to describe the resulting main quantities used to assess e.g. the composition of a specimen. The main problem in this connection is the description of multiple scattering of the primary particles as well as the signal carriers.

For non-crystalline materials the particle transport may be described by a Boltzmann type kinetic equation. Several numerical procedures have been developed to find solutions of this equation. Among these we note Monte Carlo models (see [1, 2] for reviews of this field), direct integration of the transport equation [3–6] or related techniques [7]. Such numerical calculations are very useful by virtue of their general applicability but they suffer from the drawback that quite extensive computations are usually required to obtain the desired statistical accuracy and moreover it is difficult to analyze the influence of the basic physical parameters on the results.

Analytical solutions of the transport equation and their application to surface analysis have also been studied in various approximations [8–12] resulting in useful evaluation procedures for experimental spectra, in particular for electron spectroscopy (see [13] for a review). In general, the accuracy of the employed approximations cannot be easily assessed for arbitrary materials and electron energies. Moreover, the accepted formalisms in most electron beam surface analysis techniques still lack a sound theoretical foundation consistently accounting for the signal generation process.

In the present paper, a recently developed approach to overcome these problems is outlined and its application to the theory of quantitative Auger and photoelectron spectroscopy is discussed in some detail.

Analytic Solution of the Transport Equation: Generalized Radiative Field Similarity Principle

Prior to a detailed discussion of any approach it is useful to summarize and discuss the usual model assumptions of medium energy electron transport of relevance for electron microbeam analysis:

(1) The target is semiinfinite and assumed to be bounded by an ideally flat surface.
(2) It is assumed that the sample is amorphous or polycrystalline implying that coherent scattering effects (channeling, diffraction) are negligible.
(3) For emission problems, it is assumed that excitation of signal electrons is homogeneous in the analyzed volume.
(4) The scattering characteristics are independent of the depth.
(5) Elastic scattering may be described in terms of an atomic cross section.
(6) Inelastic scattering is described by the differential inverse inelastic mean free path (DIIMFP).
(7) The scattering centers are randomly distributed in space implying that multiple scattering is governed by the Poisson stochastic process.

The first assumption excludes surfaces exhibiting microroughness or self supporting thin films. Neither of these restrictions is essential however. For the techniques of main interest for the present work (AES/XPS), the analyzed depth is small ($\sim$ 10Å) compared to the thickness of any sample which can realistically be prepared and the assumption of a semiinfinite target is always reasonable. It should be emphasized that all presented results can be generalized to the case of self supporting thin films [14] in situations where this is necessary, e.g. calculation of the Auger backscattering factor for a thin self supporting target. The assumption that the surface from which the signal electrons are emitted is ideally flat is also made for simplicity only. In fact, the emission characteristics for a rough sample can be calculated once the expression for the ideally flat surface is established [15].

The next assumption, that diffraction effects can be neglected, is essential. It allows to treat the transport problem in the framework of a Boltzmann type transfer equation instead of the more general but highly more involved quantum kinetic equation. Comparison of experimental results with model calculations support the considerations in Ref. [16] from which follows that such an approach is also justified for amorphous as well as polycrystalline samples. Not that the same restriction applies to numerical solution approaches: while for non-crystalline materials a Monte Carlo simulation is applicable, electron transport in crystalline materials can only be adequately treated by e.g. a dynamical diffraction calculation [17].

Since in AES/XPS the analyzed depth is usually orders of magnitude smaller than the penetration depth of the exciting primary radiation, the excitation depth distribution function may be taken to be constant over the relevant depth range. As a direct consequence, the depth distribution of the signal electron source in AES/XPS is proportional to the depth profile of the emitting species.

The assumption of homogeneous scattering properties with depth implies two simplifications: 1. surface excitations are neglected and 2. the main parameters characterizing the electron transport should be independent of depth. Neglect of surface excitations generally only introduces significant deviations for rather low energies $\lesssim 500$ eV and oblique emission [18] while the scattering parameter relevant for XPS/AES depends only very weakly on the material and energy [19]. Therefore, this assumption does not pose a severe restriction on the applicability of the general results.

The fact that elastic scattering can be described employing an atomic cross section deserves a slightly more elaborate comment. First of all this implies that the binary encounter approximation be valid for elastic scattering in solids. This assumption is realistic in as much as the particle wavelength as well as the interaction radius are small compared to the interatomic distance. An elementary assessment shows that both conditions are satisfied for electron energies exceeding 50 eV. Secondly, the question why elastic scattering may be conceived as plane wave scattering by an atomic potential should be addressed. In this connection it should be borne in mind that the influence of elastic scattering on the transport process mainly stems from large angle deflections. The latter classically correspond to small impact parameters and quantum mechanically to a small angular momentum where the interaction mainly takes place with the (screened) Coulomb field of the nucleus. The weakly bound electrons on the other hand are mainly responsible for the details in the forward (small angle) part of the cross section which does not at all strongly affect the influence of multiple scattering on the particle flux density. Therefore, the details of the screening by the loosely bound electrons is unimportant and consequently the use of free atomic potentials to describe elastic scattering is justified.

These considerations are supported by a rather large number of experimental results on quasi-elastic electron backscattering from solid surfaces [16, 20, 21].

The inelastic electron solid interaction is mainly governed by the loosely bound electrons with binding energies of the order of $\sim e^2/a_0$ where a_0 is the Bohr radius and e is the elementary charge. In the case of interaction with delocalized conduction electrons the collisions are not well separated in space and time or, in other words, the electron keeps interacting with the solid after each loss [22]. Therefore, the notion of a cross section is not valid for the inelastic interaction. On the other hand, the electrons' energy, momentum etc. is dissipated away over the many degrees of freedom of the solid in the course of the thermalization of its own degrees of freedom. It is quite clear that this thermalization is an irreversible process. This implies that the effective correlation of the interaction between the probe electron and the solid must be small compared to the inelastic mean free path. Moreover, the interaction at a certain time must be independent of earlier collisions. This allows to describe electron transport in solids comprising both inelastic and elastic collisions in the framework of a Boltzmann type equation, in spite of the fact that the inelastic interaction occurs with delocalized electrons.

Nonetheless, the above establishes that the inelastic interaction is essentially characterized by the solid state properties of the target. This also has important practical implications. The differential inelastic inverse mean free path, i.e. the probability for energy loss per path length, is the basic quantity characterizing the interaction and is also a solid state quantity. From this it follows that the inelastic characteristics of a compound material in general cannot be conceived of as some linear combination of the characteristics of its constituents. This is in contrast to the elastic scattering case, where the interaction is described in terms of atomic quantities.

Finally, it is noted in this connection, that the characteristic length for deflection in the course of an inelastic interaction is small compared to the elastic transport mean free path. This allows to neglect deflections due to inelastic collisions.

On the basis of the above assumptions, the general equation for electron transport in the case of plane symmetry can now be formulated. Let $N(z, \mu, E)$ be the flux density of particles with energy E moving along the direction $\mu = \arccos\theta$ at the depth z. Here θ is the polar angle defined relative to the z-axis pointing inside the target. The transfer equation for the flux density may be written as [23]:

$$\mu\frac{\partial N}{\partial z} = -\left(\frac{1}{\lambda_i} + \frac{1}{\lambda_e}\right)N + \frac{1}{2\lambda_e}\int_{-1}^{1} I(\mu, \mu')N(z, \mu', E|E_0)d\mu' + \frac{1}{\lambda_i}\int_0^\infty w(E_0|\varepsilon)N(z, \mu, E+\varepsilon|E_0)d\varepsilon + q(\mu, z, E) \tag{1}$$

Here $I(\mu, \mu')$ and $w(\varepsilon)$ are the normalized differential elastic scattering cross section and DIIMFP respectively. The elastic and inelastic mean free path (EMFP and IMFP respectively) are represented by λ_e and λ_i while $q(\mu, z, E)$ is the source function. Equation 1 must be supplemented by an appropriate boundary condition depending on the nature of the problem under consideration (reflection, emission).

In general, the exact solution to Eq. (1) with an adequate boundary condition represents a formidable task and the solutions are very involved rendering them unsuitable for quantitative routine analysis. The main problem in this connection is the complex angular, energy and atomic number dependence of the scattering function $I(\mu, \mu')$ which, together with the DIIMFP determine the behaviour of the particle flux density. This problem may be overcome by means of the so called generalized radiative field similarity principle [19, 24, 25], allowing to simplify Eq. (1) using an effective approximation to $I(\mu, \mu')$. Furthermore, the solution of Eq. (1) is in general further complicated in situations where the energy dependence of the scattering characteristics is important, i.e. when describing some broad spectral feature. A self consistent one-speed variant of the general multi-speed equation Eq. (1) may lead to effective and accurate solutions in such cases [26]. These approaches will be discussed further below. In order to fully appreciate the basic idea behind these approaches, it seems useful to first investigate the elastic and inelastic scattering characteristics in some detail.

The differential cross section for elastic scattering can be calculated on the basis of a given potential employing the partial wave expansion method [27]. Usually, a Thomas–Fermi–Dirac potential is used for this purpose. Comparison of different types of potentials [2] demonstrates that this yields accurate results and moreover an accurate parameterization for elements throughout the periodic system is available for this type of potential [28]. The differential Mott cross section for elastic scattering for Si and Au calculated in this way is shown in Fig. 1 for several energies. It is seen that the presence of a pronounced forward scattering peak is common to all cross sections independent of atomic number and energy. Furthermore, a backscattering peak is seen to be present for the heavy target, and at low energies also for Si. For Au, the behaviour of the cross section between the forward and backscattering

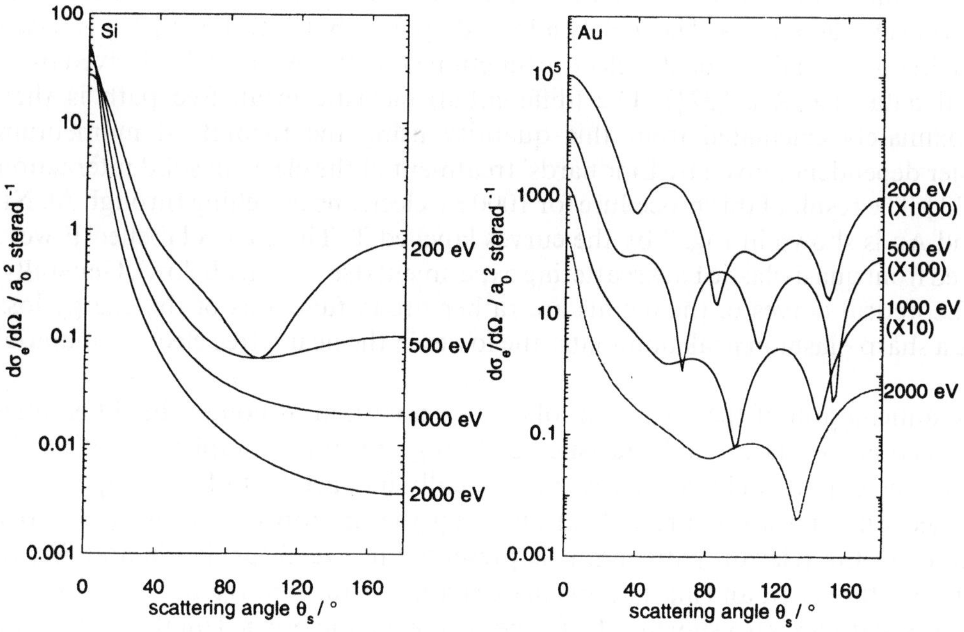

Fig. 1. Mott cross section for elastic scattering for Au and Si at different energies

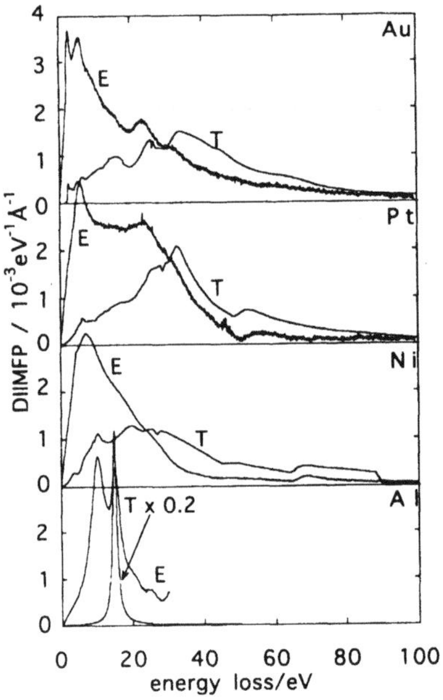

Fig. 2. Differential inverse elastic mean free path (DIIMFP) for 1000 eV electrons in Al, Ni, Pt and Au. Curves labeled E are the experimental results obtained from measurements applying the method in ref. [9] to angle integrated REELS spectra. The curves labeled T were calculated from the optical data compiled in ref. [32]

part exhibits strong oscillations. These features are often referred to as generalized Ramsauer–Townsend oscillations and can be qualitatively explained in the framework of a quasi-classical description of the scattering process [29].

As explained above, the inelastic electron solid interaction is characterized by a semimacroscopic quantity describing the probability for energy loss $\hbar w$ and momentum transfer $\hbar k$ from the electron to the solid. In the framework of dielectric response theory the ability of a probe electron to polarize the medium is given by the dielectric function $\varepsilon(\hbar w, \hbar k)$ [30] which may thus be used to assess the inelastic scattering characteristics. The widely adopted approach due to Penn [31] is based on the frequency independent dielectric function $\varepsilon(\hbar w, 0)$ which can be derived from optical data (e.g. Ref. [32]). The (differential) inelastic mean free path is then approximately calculated from this quantity using the theoretical momentum transfer dependence given by Lindhards' treatment of the electron solid interaction [30, 33]. The result of this procedure for 1000 eV electrons travelling through Al, Ni, Pt and Au is shown in Fig. 2 by the curves labelled T. The curves labelled E were derived from quasi-elastic backscattering experiments (see further below). Generally, the theoretical curves of the metals are rather broad functions of the energy loss while a sharp plasmon peak dominates the losses in the nearly free electron material Al.

In anticipation of the results to follow, we note that although the differential cross section for elastic and inelastic scattering are quite complex quantities in general, the emission characteristics are virtually independent of the shape of the cross sections and are entirely determined by the appropriate mean free paths. A survey of the relevant pathlengths is presented in Fig. 3 for Si, Cu and Au. In addition to the elastic and inelastic mean free path, the linear range and the transport mean free path are also shown. The former is the characteristic length for inelastic scattering in the continuous slowing down model. It is calculated from the stopping

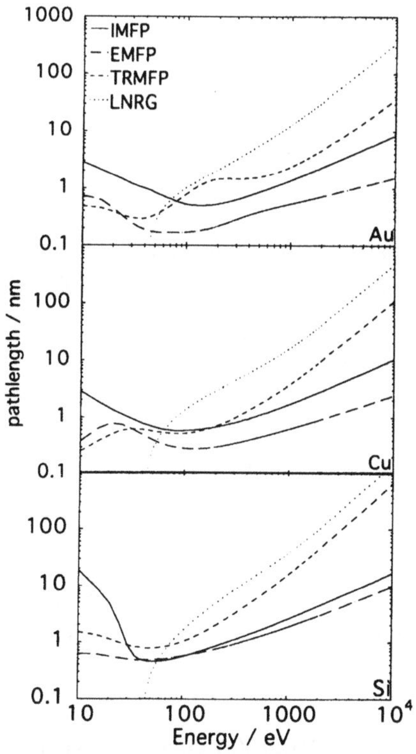

Fig. 3. Inelastic, elastic, transport mean free path and linear range for Si, Cu and Au as a function of the energy

power dE/ds via

$$R_0 = \int_{E_m}^{E_p} \frac{dE}{ds}^{-1} dE \tag{2}$$

The stopping power was calculated from optical data using Penn's algorithm [31, 33] and the cut-off energy E_m was chosen to be 50 eV, as usual. The transport mean free path characterizes the probability for large angle deflections. It was calculated from the elastic cross section [27] via the definition:

$$1/\lambda_{tr} = N_a \int_{-1}^{1} \frac{d\sigma_e}{d\mu_s}(\mu_s)(1-\mu_s)\, d\mu_s \tag{3}$$

where N_a is the atomic density of the material and $d\sigma_e/d\mu_s$ is the elastic cross section, differential with respect to the cosine of the scattering angle, μ_s.

A monotonic increase is seen for all four quantities for energies exceeding $\gtrsim 200$ eV. Generally the pathlengths are seen to decrease with increasing atomic number. In this energy range the linear range is the largest quantity followed by the transport mean free path, the inelastic and the elastic mean free path. It is important to note that the quantity

$$\sigma = R_0/\lambda_{tr} \tag{4}$$

is only very weakly energy dependent for $E > 1$ keV while it varies appreciably (but slowly) with the atomic number. (See Fig. 4). On the other hand, the ratio

$$\chi = \lambda_i/\lambda_{tr} \tag{5}$$

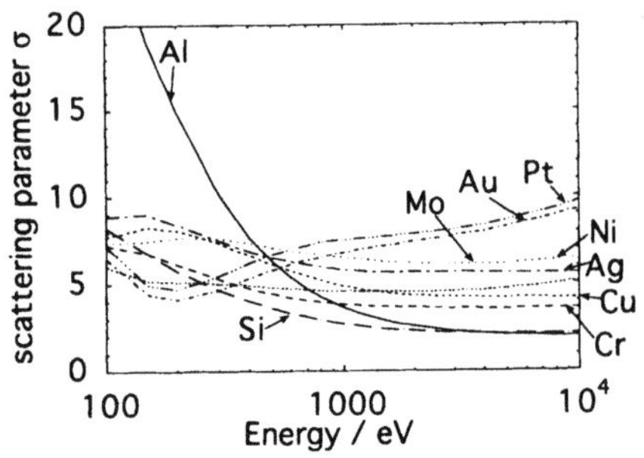

Fig. 4. Scattering parameter σ for the true slowing down regime for nine elements in the energy range 100–10000 eV

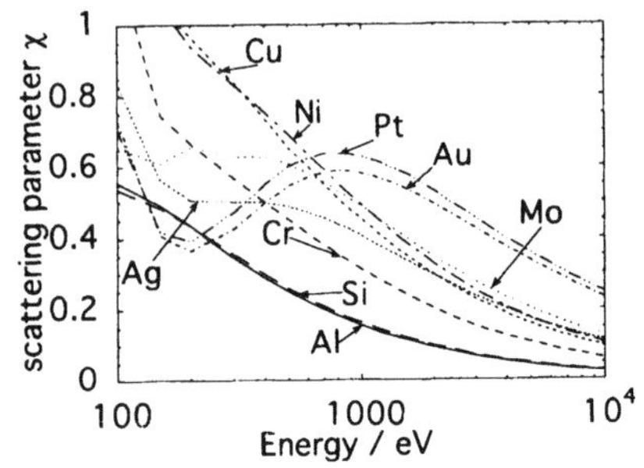

Fig. 5. Scattering parameter χ for the quasi-elastic regime for nine elements in the energy range 100–10000 eV

is seen to be of the order of unity in the energy range 200–2000 eV for all studied elements (see Fig. 5). These two parameters, hereafter referred to as scattering parameters, are of pronounced importance since they govern the scattering characteristics in the two most relevant regimes of electron transport: the quasi–elastic and the true slowing down regime. The former is of particular importance for signal electron transport in Auger electron spectroscopy (AES) and X-ray photoelectron spectroscopy (XPS) when the scattering cross sections are independent of the energy to a good approximation. The scattering parameter χ indicates the importance of elastic scattering in this case [34, 16]. For larger energy losses exceeding the quasi-elastic regime, the continuous slowing down approximation provides an adequate description of the particle transport [19] and consequently the scattering parameter σ governs the scattering characteristics. This regime is of importance to most applications of electron probe microanalysis (EPMA) and related techniques.

It should be noted at this point that convenient expressions exist to calculate the main characteristic pathlengths for electron scattering. For the inelastic mean free path, the empirical expression by Tanuma and coworkers [35] is known to be sufficiently accurate, while the quasiclassical expression for the transport mean free path derived by Tilinin [36] agrees within a few percent with values calculated on the basis of the Mott cross section [37]. For the linear range, Bethe's formula may be used at sufficiently high energies $\gtrsim 10$ keV.

As mentioned above, the most important question connected with the solution of the kinetic equation is how to effectively approximate the elastic cross section entering the collision term in the transport equation. Thus, we seek for an approximation which sufficiently simplifies the equation while retaining an accuracy as high as possible. Obviously, the choice of a given scattering cross section will significantly alter the shape of the individual trajectories in the solid. However, it is well known that under certain circumstances the exact shape of the differential elastic scattering cross section does not noticeably influence the radiative transfer [38]. Let us try to develop a feeling for the physical processes governing the characteristics of the radiative field. For this purpose, we consider a source of electrons located at a certain depth below the surface and emitting electrons with a certain angular distribution. The main question is which processes determine the amount of electrons escaping without energy loss, i.e. emerge from the sample as signal electron. Clearly, two mechanisms play a role: energy absorption and momentum relaxation. Although the origin of these mechanisms is entirely different (electron–electron vs electron–core or inelastic vs elastic processes) they exert a combined influence on the radiative field. To elucidate this we first introduce the characteristic lengths for energy loss and deflection. The former is given by the inelastic mean free path while the latter is most conveniently characterized by the pathlength for momentum transfer, or transport mean free path. The combined influence of these two competing processes is quantitatively measured by the scattering parameter which is an indication of the importance of elastic scattering. For small scattering parameters $\chi \ll 1$ the characteristic length for deflection is large compared to the inelastic loss probability and the particles' trajectories resemble straight lines. Then the characteristic length for absorption determines the escape probability. In the opposite limiting case of large scattering parameters $\chi \gg 1$, momentum relaxation dominates the transport and diffusion like trajectories will occur. The net displacement from the source is thus kept small for a given pathlength due to intense scattering and the emerging flux density of no-loss electrons is thereby reduced. Note that for isotropic source emission the momentum relaxation process is not expected to significantly modify the angular characteristics of the particle flux density.

For these two extreme cases of strong absorption ($\chi \ll 1$) and strong scattering ($\chi \gg 1$) effective approximate solutions to the transport equation are at hand. For $\chi \ll 1$ the trivial straight line approximation (SLA) holds, while for the case of weak absorption ($\chi \gg 1$) the diffusion approximation is adequate. For the intermediate case of scattering parameters of the order of unity, neither the SLA nor the diffusion or P_1-approximation provides an accurate description of the transport process. Unfortunately, the overwhelming majority of characteristic energies encountered in XPS and AES correspond to this case.

In view of the above one might expect that any approximation to the collision term in the transport equation which reduces to the SLA or diffusion equation in the limit of weak and strong scattering respectively should also work reasonably well in the intermediate case of the scattering parameter comparable to unity $\chi \sim 1$, provided the particle flux density is not highly anisotropic. This is the basic idea of the generalized radiative field similarity principle which states that the realistic cross section in the transport equation may be replaced by any approximate cross section if the following conditions are met: 1. in the case of weak absorption the usual similarity relationships should hold [38]. 2. In the opposite limiting case the SLA should be attained and 3. the source angular distribution should be sufficiently smooth. A very effective and convenient cross section obeying condition 1. and 2. is the isotropic transport cross section ($\sigma_{tr} = 1/N_a\lambda_{tr}$).

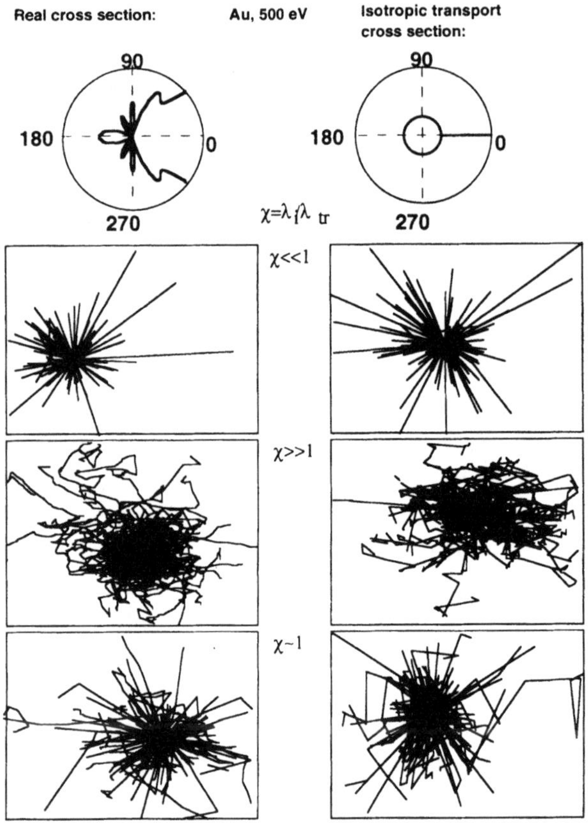

Fig. 6. Radiative fields of a single monoenergetic isotropic electron source for strong scattering ($\chi \gg 1$), weak scattering ($\chi \ll 1$) and the intermediate case. The left part of the figure corresponds to MC calculations using the realistic Mott cross section displayed in the upper panel, while the right side displays the corresponding cases for the transport cross section. Note that the individual trajectories are markedly different, while the radiative field is similar in both cases

In Fig. 6 the three regimes of electron transport mentioned above are visualized for 500 eV electrons in Au. A single monoenergetic isotropic source is located in the centre of the sets of trajectories, which are displayed until the first energy loss occurs. For the case of the intermediate scattering parameter, a realistic value of the IMFP was used while the limits of strong and weak absorption were modelled by accordingly changing the IMFP. The left part of the figure represents MC results for the realistic Mott cross section, while in the right part the isotropic transport cross section was used. It is seen that the individual trajectories markedly differ in shape (the transport approximation emphasizing large angle scattering) while the radiative field far away from the source is similar.

In order to avoid misunderstandings we emphasize that the generalized radiative similarity principle is in effect a hypothesis and it is difficult to assess the accuracy of any approach based on it for an arbitrary relationship between the relevant scattering parameters. Fortunately, for the transport approximation, which meets the similarity relationships, the accuracy of the results obtained in this way turned out to be very high. This conclusion follows from several comparisons with numerical calculations and experiments [37, 25, 16, 24].

The advantage of the transport approximation lies in the fact that the kinetic equation for isotropic scattering has been extensively studied in neutron and astrophysics [39] and several techniques have been developed to tackle boundary value problems in this approximation. In the transport approximation, the kinetic equation may be written as:

$$\mu \frac{\partial N}{\partial z} = -\left(\frac{1}{\lambda_i} + \frac{1}{\lambda_{tr}}\right) N + \frac{1}{2\lambda_{tr}} \int_{-1}^{1} N(z, \mu', E | E_0) d\mu' + \frac{1}{\lambda_i} \int_{0}^{\infty} w(E_0 | \varepsilon) N(z, \mu, \; E + \varepsilon | E_0) d\varepsilon + q(\mu, E, z) \tag{6}$$

Following the approach due to Landau [40] we introduce the relative energy loss $\Delta = (E - E_0)/E_0$ at this stage and perform a Laplace transform of Eq. (6) with respect to this variable. Designating the Laplace transform with a tilde Eq. (6) may be transformed into:

$$\mu \frac{\partial \tilde{N}}{\partial \tau} + \tilde{N} = \frac{\omega(p)}{2} \int_{-1}^{1} \tilde{N}(\tau, \mu') d\mu' + \omega(p) \tilde{q} \tag{7}$$

where the reduced depth $\tau = z/\lambda_{tr}\omega(p)$ was introduced and the complex single scattering albedo is given by:

$$\omega(p) = \chi \{\chi + \tilde{w}_\Delta(p)\}^{-1} \tag{8}$$

Here $\tilde{w}_\Delta$ (p) is the Laplace transform of the inverse inelastic mean free path differential with respect to the relative energy loss Δ. The same transformations should be applied to the boundary conditions of the considered problem. Written in this form, the transport equation is amendable to solution using e.g. Case's method of singular eigenfunctions [39]. Solutions to the emission [25, 24, 23, 41] and reflection [16, 42] problem in the quasi–elastic regime ($\Delta \lesssim e^2/a_0 E_0$) as well as the inelastic reflection problem [26] have been derived previously and will be analyzed and compared with experimental results and numerical simulations in the next section.

Application to Auger and Photoelectron Emission from Solids

In this section, the application of the above general results to describe the most important quantities of AES/XPS will be discussed. A typical XPS spectrum consists of broad spectral features of comparatively low intensity at high kinetic energies ($E_k \lesssim h\nu$) on which strong peaks are superimposed which are due to inner shell transitions and are accompanied by a background of inelastically scattered signal electrons. At low kinetic energies ($E_k \lesssim 50$ eV) a peak of true secondaries is observed. An (electron excited) Auger spectrum exhibits the characteristic Auger peaks and their background, as well as the secondary electron peak, but in addition there is a broad contribution from inelastically backscattered electrons on which the characteristic spectral features are superimposed. We will concentrate on the characteristic features first i.e. the angular and energy distribution of signal electrons with the associated inelastic background. Description of the inelastic background of reflected primaries as well as the additional excitation of Auger electrons due to backscattering will be discussed further below.

A. The Yield Equation

The formal solution of Eqs. (1) and (6) for the quasi-elastic case $\Delta E \lesssim e^2/a_0$ has been extensively discussed in Refs. [25, 24, 23]. The electron yield comprises contributions from particles having experienced a different number of n inelastic collisions. The energy distributions of these contributions (the so-called partial loss distributions) are entirely determined by the DIIMFP while the partial escape distributions contain all information about elastic scattering. By virtue of the generalized radiative field similarity principle the latter quantities can be adequately characterized by the inelastic and transport mean free path only. Weighting the loss distributions with their associated escape probabilities and adding all relevant contributions of n-fold inelastically scattered electrons, the general expression for the energy/angular distribution in the case of isotropic source emission may be written as [23, 41]

$$Y(\mu, E) = A \sum_{n=0}^{\infty} \int_0^{\infty} dE' \, L_n(E - E') f(E') \times \int_0^{\infty} dz P_n(\mu, z) c(z) \tag{9}$$

where $f(E)$ is the true intrinsic spectrum and $c(z)$ is the depth profile of the considered species. The partial loss distributions $L_n(E)$ describe the probability for the energy loss E after n inelastic collisions and are given by the recurrence relationship [23]:

$$L_0(E) = \delta(E),$$
$$L_n(E) = \int_0^E dE' L_{n-1}(E') w(E - E'). \tag{10}$$

The factor A in Eq. (9) describes the generation rate of signal electrons which for photoelectrons is given by:

$$A_{XPS} = I_x N_a \sigma_{ph} / 2\mu \tag{11}$$

In the latter expression I_x is the incident photon flux, N_a is the atomic density of the sample and σ_{ph} is the total photoelectron cross section. For AES the corresponding quantity reads:

$$A_{AES} = I_p(1 - \omega_x)(1 + r)\sigma_{ei} N_a / 2 \tag{12}$$

Here I_p is the primary current, σ_{ei} is the electron impact ionization cross section, ω_x is the fluorescence yield and r is the so-called backscattering factor. Eq. (9) was derived on the assumption that the source emission process is isotropic. The generalization of the presented formalism to anisotropic emission can be found in Refs. [24, 43].

B. The Escape Probability

The remaining problem to describe the angular/energy distribution is thus reduced to finding an appropriate expression for the partial escape distributions $P_n(z, \mu)$. Analysis of Eq. (1) shows that these quantities can be represented by [23]:

$$P_n(z, \mu) = \frac{1}{n!} \left(\frac{\partial^n}{\partial x^n} P_0(z, \mu, \omega(x)) \right) \Bigg|_{x=0} \tag{13}$$

where

$$\omega(x) = \chi[\chi + 1 - x]^{-1} \tag{14}$$

and $P_0(z, \mu)$ is the zero order partial escape distribution, being equal to the emission depth distribution function in AES/XPS [44] to within a normalization factor. Eqs. (9), (10) and (13) clearly show that the key problem in the theory of quantitative AES/XPS is an effective and accurate expression for the zero order partial escape distribution. At this point it is necessary to invoke the transport approximation in order to arrive at tractable mathematical expressions for the emission characteristics. In this approximation, the zero order escape distribution is given by [25, 41]:

$$\begin{aligned} P_0(z, \mu, \omega) = {} & \frac{\mu\omega}{4} H(\mu, \omega) \int_0^1 \frac{dv}{v - \mu} \left[\frac{g(v, \omega)}{H(v, \omega)} \right. \\ & \times \{\exp(-z/v\omega\lambda_{tr}) - \exp(-z/v_0\omega\lambda_{tr})\} \\ & \left. - \frac{g(\mu, \omega)}{H(\mu, \omega)} \{\exp(-z/\mu\omega\lambda_{tr}) - \exp(-z/v_0\omega\lambda_{tr})\} \right] \\ & + \frac{1}{2} g(\mu, \omega) \left[1 - \frac{\mu\omega}{2} \ln\left(\frac{\mu(1+\mu)}{(1-\mu)^2} \right) \right] \\ & \times \{\exp(-z/\mu\omega\lambda_{tr}) - \exp(-z/v_0\omega\lambda_{tr})\} \\ & + \frac{1}{2} H(\mu, \omega) \exp(-z/v_0\omega\lambda_{tr}) \end{aligned} \tag{15}$$

Where the definitions

$$g(v, \omega) = \left\{ \left(\frac{\omega\pi v}{2} \right)^2 + \lambda^2(v) \right\}^{-1}$$

$$\lambda(v, \omega) = 1 - (\omega v/2) \ln \frac{1 + v}{1 - v}$$

were used. The function $H(\mu, \omega)$ is the Chandrasekhar H-function for an isotropically scattering medium [45]. It may be accurately approximated by the expression [23]:

$$H(\mu, \omega) = \frac{1 + \alpha}{1 + \alpha\sqrt{1 - \omega}}$$

where $\alpha = H(\mu, 1) - 1$. The function $H(\mu, 1)$ is tabulated in the literature [45] and may be evaluated from the polynomial fit:

$$H(\mu, 1) - 1 = 0.5073\mu^3 - 1.0016\mu^2 + 2.4102\mu$$

This approximation is usually sufficiently accurate and convenient for numerical calculations. For the typical signal electron energies in AES/XPS where $\chi \lesssim 0.5$ the quantity v_0 is given to a good approximation by:

$$v_0 \approx 1 + 2\exp(-2/\omega). \tag{16}$$

For a homogeneous sample, the total escape probability $P_0(\mu)$ is found by integrating Eq. 15 over the depth. This yields the simple expression [23]:

$$P_0(\mu) = \frac{\mu\lambda_{tr}\omega H(\mu, \omega)}{2\sqrt{1 - \omega}} \tag{17}$$

The higher order escape probabilities in this case are found to be given by

$$P_n(\mu) = \kappa(\mu)^n P_0(\mu) \tag{18}$$

where

$$\kappa(\mu) = 1 - \frac{\omega/2}{1 + \mu\sqrt{3(1-\omega)}} \tag{19}$$

The yield equation can then be written as:

$$Y(\mu, E) = A \sum_{n=0}^{\infty} P_n(\mu) \int_0^{\infty} dE' L_n(E - E') f(E') \tag{20}$$

Figure 7 shows a comparison of the partial escape distributions calculated in the transport approximation (Eq. (15), solid lines) with model calculations based on the more realistic Mott cross section for elastic scattering (displayed as data points). To demonstrate the influence of elastic scattering the partial escape distributions in the straight line approximation (SLA) are also shown. The latter are given by a simple Poisson distribution

$$P_n(z, \mu) = \frac{1}{2n!}\left(\frac{z}{\mu\lambda_i}\right)^n \exp(-z/\lambda_i\mu) \tag{21}$$

It is seen that the MC results using the Mott cross section for elastic scattering and the theoretical results based on the transport approximation are in excellent

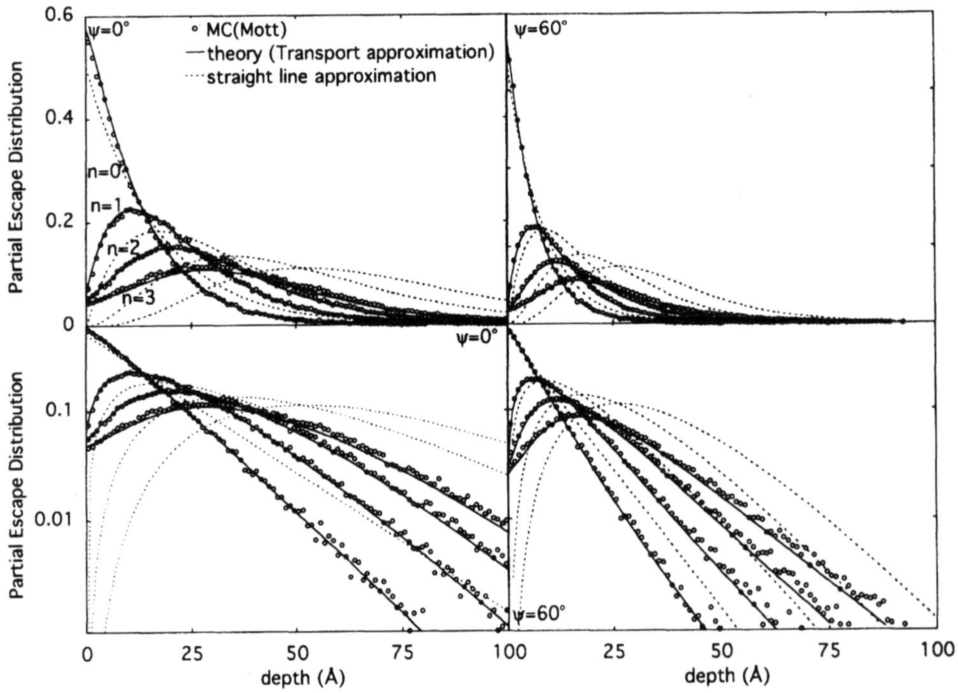

Fig. 7. Partial escape distributions for 1000 eV electrons in Ni. Data points: Monte Carlo calculations based on the Mott cross sections for elastic scattering. Solid lines: theoretical partial escape distribution in the transport approximation Eqs. (13) and (15). Dashed lines: straight line approximation results

agreement with each other. All the effects of elastic scattering are accurately described within the framework of the transport approximation. Among them we point out the nonvanishing contribution to the escape probability of the electrons generated at the surface for $n \geqslant 1$, the shift of the maximum position of the PED towards the surface and the emission angle independence of the slope of the PED at large depths. The latter effect has been extensively discussed previously [46, 47]. The tendency of the maximum position to be closer to the surface than that predicted by the straight line approximation can be easily understood. Elastic scattering generally reduces the mean displacement of an electron from the point of generation. Consequently the most probable escape depth becomes smaller for any n when elastic scattering is taken into account.

This is in contrast to the distributions calculated in the straight line approximation. They agree reasonably with the MC data for $n = 0$ only at shallow depths where the elastic scattering effect is negligible. The discrepancies are especially clearly seen for $n \geqslant 1$. Formula (21) fails to correctly predict the maximum positions and the behaviour of the escape distributions at large depths. From this it follows that the shape of the inelastic background in the near peak region of the spectrum cannot be properly described when elastic scattering is neglected.

C. *Line Shape Analysis and Background Subtraction*

Next, the application of the presented formalism to calculation of the characteristic spectral features of signal electrons will be illustrated. As a first example, a homogeneous sample will be considered. In this case the line's shape can be calculated with Eqs. (17–20). An example of this procedure is shown in Fig. (8) for a Lorentzian true spectrum with a FWHM of 1 eV for 1000 eV electrons emitted from an Ag substrate for two different geometries. Here the universal differential inverse IMFP proposed by Tougaard [13] and the IMFP given by the formula of Tanuma, Powell and Penn [35] were used in the calculations. The total spectrum, comprising a broad inelastic background as well as the true initial Lorentzian spectrum are shown as solid lines. The partial contributions of electrons which have been inelastically scattered 0–n

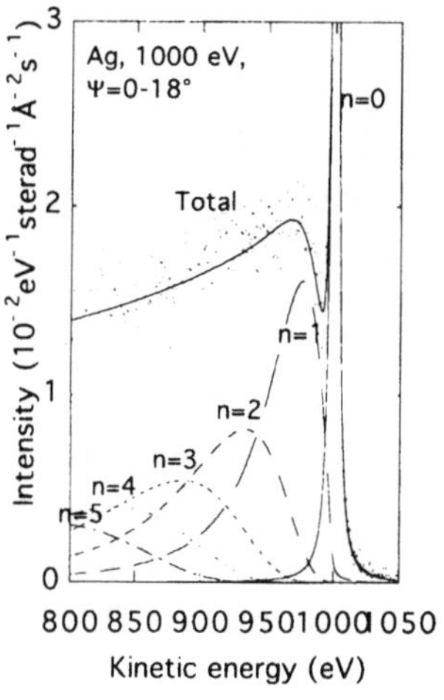

Fig. 8. Model energy spectrum of a 1000 eV signal electron source in Ag. Data points: Monte Carlo results. Solid line: spectrum with inelastic background calculated with Eq. [20]. Dashed lines: contributions of electrons having undergone n inelastic collisions. Note that the ordinate scale is in absolute units. No normalization of the data has been performed

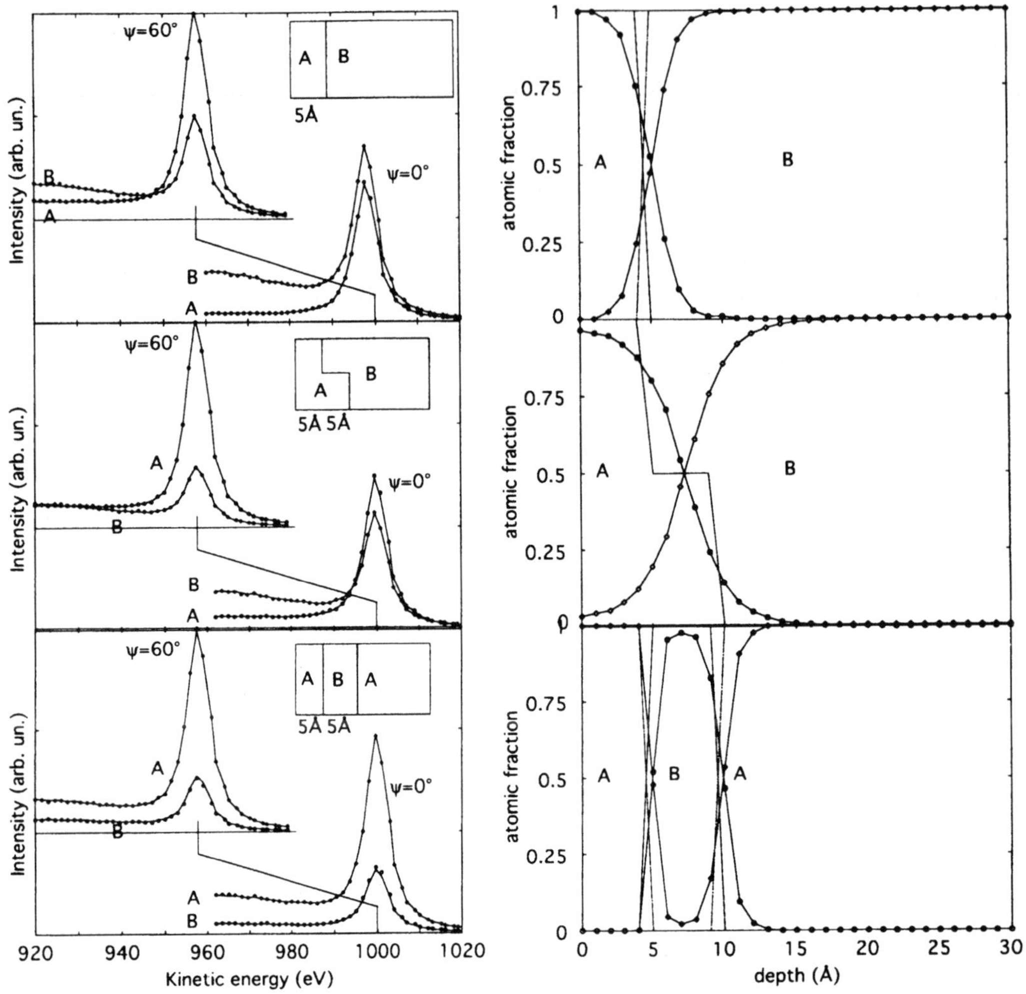

Fig. 9. (Left hand side). Energy spectra for three model multilayered samples at two different emission angles calculated with the Monte Carlo technique. The solid lines are the response of the reconstructed true intrinsic spectrum and depth profile (cf. Fig. 10) found by the evaluation procedure

Fig. 10. (Right hand side). Reconstructed depth profile based on the data shown in Fig. 9

times are also shown as dashed lines. Note that the background over the first 100 eV is almost entirely made up of (0–3)-times inelastically scattered electrons. The figure also shows MC results based on the Mott cross section for elastic scattering. The comparison yields an almost perfect quantitative agreement in absolute units.

Application of the general expression Eq. (9) for a sample with an arbitrary depth profile is illustrated in Fig. 9 for three different model depth profiles, indicated in the inset. The scattering properties of elements A and B were assumed to be independent of depth and corresponding to ~ 1000 eV electron traversing an Au target. In the analytic calculations (represented by the solid lines) elastic scattering was accounted for in the transport approximation using Tilinin's formula for the transport mean free path [36]. The inelastic mean free path was evaluated by the formula of Tanuma and coworkers [35]. For the differential inelastic mean free path, Tougaards universal cross section was used [13]. These results are compared with Monte Carlo calculations using the more realistic Mott cross section for elastic scattering [27] while the differential inelastic mean free path was derived from optical data [32] by means of a dielectric response calculation (cf. Fig. 2).

At this point, it should be noted that not only the relative intensities, but also the peak shape contains information about the specimen's depth profile. Supposing that the substrate composition is known, it is relatively easy to make qualitative inferences about the depth profile on the basis of the presented spectra. For instance, in Fig. 9a, the observation that peak A increases for a grazing geometry while peak B decreases, is consistent with the lower background level for peak A and with the assumption that element A is located at the specimen surface. The same qualitative assessment may be made on considering the spectra in Fig. 9 B. The angular dependence in the spectra for the last depth profile (5 Å A/5 Å B/A substrate) is extremely moderate and at first sight the spectra seem to contain little information about the depth profile. In order to extract more detailed information from these simulation data, the Monte Carlo results were used to quantitatively reconstruct the model depth profile employing an optimization scheme on the basis of Eq. (9) [41]. These reconstructions are shown as solid lines in Fig. 10. The spectra corresponding to these reconstructed depth profiles are shown as solid lines in Fig. 9. From this comparison it may be concluded that the yield equation Eq. (9) together with the expression for the partial escape distributions Eqs. (13) and (15) adequately describe the influence of the particle transport on the emission process in AES/XPS.

As one of the main applications of the presented theory we will now consider the problem of separation of the extrinsic and intrinsic features of a given spectrum, i.e. the extraction of the true intrinsic spectrum from the observable spectrum containing features due to the particle transport. The usual procedure in finding a deconvolution scheme for distributions of the type Eq. (20) is to perform a Laplace transform on the energy variable, which is trivial observing the convolution theorem. The next step consists of making use of some kind of recursion property of the quantities P_n in such a way that the inverse transform poses no major difficulties, eventually yielding the true intrinsic spectrum f.

We start with the usual step of performing the Laplace transform of Eq. (20) on the energy E. Dropping the variables for clarity and denoting the quantities in Laplace space with a tilde ($\sim$) we may write:

$$\frac{\tilde{Y}}{A\lambda_{\mathrm{tr}}} = \sum_{n=0}^{\infty} \tilde{w}^n \tilde{f} P_n \tag{22}$$

Introducing the reduced quantities $y = Y/P_0 A\lambda_{\mathrm{tr}}$, $p_n = P_n/P_0$ and rearranging we find:

$$\frac{\tilde{y}}{\tilde{f}} = 1 + \sum_{n=1}^{\infty} p_n \tilde{w}^n \tag{23}$$

Now we multiply Eq. (23) by $(1 - p_1 \tilde{w})$, effectively increasing the exponent on the right-hand side by one:

$$(1 - p_1 \tilde{w})\frac{\tilde{y}}{\tilde{f}} = 1 + \sum_{n=2}^{\infty} (p_n - p_1 p_{n-1})\tilde{w}^n \tag{24}$$

and note at this point that for isotropic source emission we have $p_n = \kappa^n$ (see Eq. (18)). Consequently the sum in Eq. (24) vanishes. Therefore, by going back to energy space, we immediately find the deconvolution formula valid for an isotropic source distribution:

$$f(E) = y(\mu, E) - \kappa(\mu) \int_E^{\infty} dE' y(\mu, E') w(E' - E) \tag{25}$$

The general case of non-isotropic source emission is dealt with in Ref. [43].

The raw LMM Auger spectrum of a polycrystalline Ni sample irradiated by 8 keV electrons is shown in Fig. 11a. The smooth solid line is a polynomial fit to the background of inelastically scattered electrons, which has to be removed prior to application of Eq. 25. The upper curve of Fig. 11b shows the spectrum corrected for this background and for the analyzer transmission employing the elastic peak technique [48]. The lower solid curve is obtained by application of Eq. 25 and represents the true intrinsic spectrum. The dotted line is the background of inelastically scattered signal electrons. It can be seen that this background is consistently removed over a large energy range. At energies $\lesssim 600$ eV the spectrum drops below zero. This can be attributed to a slight uncertainty in connection with the fitting procedure of the inelastically reflected electrons (see Fig. 11a).

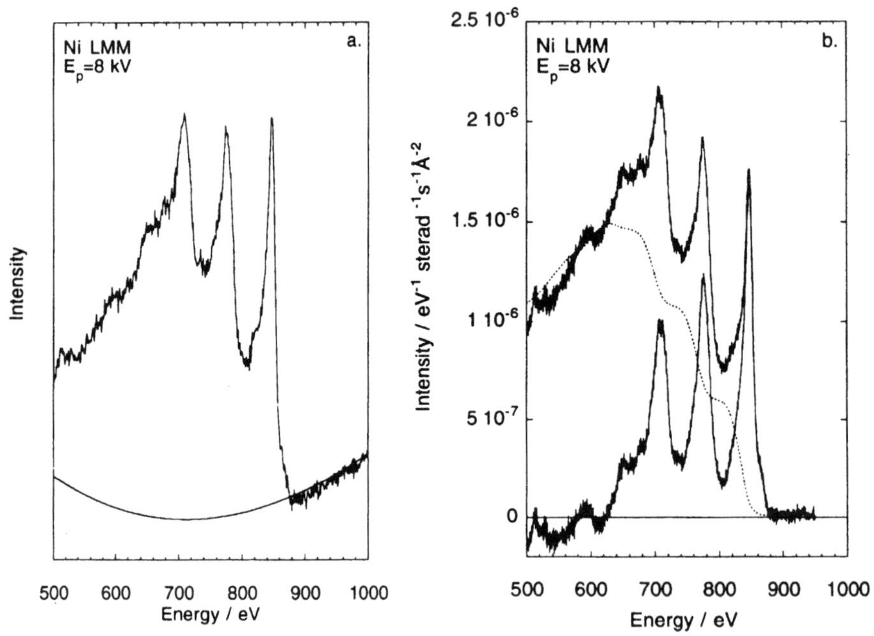

Fig. 11. Illustration of the background subtraction procedure (Eq. 25) to an experimental Ni LMM spectrum obtained by 8 keV irradiation of a polycrystalline Ni sample. **a** Raw data and polynomial background of inelastically reflected electrons. **b** Spectrum corrected for the inelastic background and the analyzer transmission (upper curve) and the reconstructed true intrinsic spectrum (lower curve). The dotted line represent the contribution from inelastically scattered signal electrons

D. Inelastic Electron Backscattering

The inelastic reflection coefficient is of importance for the calculation of the broad background on which the peaks in an Auger spectrum are superimposed, the Auger backscattering factor and to interpret the contrast in electron microscopy when no energy filtering is applied. To describe an essentially broad spectral feature like the backscattering coefficient, the energy dependence of the scattering characteristics in Eq. (7) must be taken into account. A method to tackle this problem was proposed by Tilinin and successfully applied to calculate the backscattering coefficient [26]: it has been shown above that the scattering parameters χ and σ depend only very weakly on the energy (see Figs. 4 and 5). Therefore, to a first approximation, only the energy dependence in the derivative term of Eq. 7 has to be taken into account: $\lambda_{tr} = \lambda_{tr}(E)$. Subsequently an effective mean energy $\langle E_{eff} \rangle$ is used and the resulting one speed equation can be solved. The effective energy can then be found selfconsistently on the basis of the derived energy distribution. The transport mean free path used should then be evaluated at this energy.

Furthermore, the isotropization of the incident particle flux proceeds very rapidly compared to the energy dissipation process for large incident energies allowing to use the transport approximation in spite of the sharply peaked initial angular distribution. The energy loss process on the other hand may be described in the continuous slowing down approximation since after multiple inelastic collisions taking place along the highly randomized electron trajectories the discrete energy losses are sufficiently smeared out. This implies that the Laplace transform of the DIIMFP may be written as [40, 26] $\tilde{w}(p) = p$ and the scattering albedo becomes:

$$\omega(p) = \frac{\sigma}{\sigma + p}$$

Thus the quantity σ is the relevant scattering parameter in the true slowing down regime.

Using the outlined approach, highly accurate and simple expressions have been found for the inelastic reflection coefficient, X-ray yield and Auger backscattering factor of homogeneous samples [26, 49] (see also [19]).

It should be noted that for the integral properties of homogeneous targets the solutions do not depend on λ_{tr} explicitly and therefore the selfconsistent procedure to calculate $\lambda_{tr}(\langle E_{eff} \rangle)$ is only necessary to verify the general procedure. Recent results on depth dependent characteristics in the true slowing down regime indicate that such a selfconsistent procedure is effective in such a case [50].

Medium Energy Electron Reflection from Non-Crystalline Solids

Quasi-elastic electron reflection from non-crystalline solids is both of fundamental as well as practical importance for electron microbeam analysis. From a practical point of view the interest mainly originates from elemental contrast measurements in the electron microscope. Moreover, electron reflection experiments are of great fundamental benefit to verify the model assumptions concerning electron transport and allow to assess the basic quantities for the phenomenon like the differential elastic scattering cross section, the IMFP and the DIIMFP. In the present context this subject is of particular interest since it may serve to elucidate the requirement of a smooth initial source distribution of the generalized radiative field similarity principle.

Consider a monoenergetic well collimated beam of electrons incident on a homogeneous target. For medium energies, the scattering parameter χ is of the order of unity implying that neither the single deflection model nor the diffusion approximation is adequate to describe the phenomenon. On the contrary, the contribution of single and multiple scattered particles is expected to be of the same order of magnitude. For a well collimated primary beam, representing a sharp initial angular distribution, the third requirement of the generalized radiative field similarity principle is not fulfilled and in consequence replacement of the realistic cross section by an approximate cross section (e.g. isotropic) is not allowed. To put it more precisely, the elastic reflection coefficient is expected to be strongly influenced by the exact shape of the elastic cross section. This is in perfect agreement with experimental observations [16, 20, 21, 51].

In order to overcome this problem it may be noted that after several elastic collisions, the incoming flux becomes increasingly randomized. The extent of this isotropization is determined by the scattering parameter, but also by the shape of the scattering cross section and the reflection geometry. It turns out that for medium energies $200 \lesssim E \lesssim 2000$ eV the particle flux density (in particular its backward part) after two elastic processes is generally sufficiently smooth for the third condition of radiative field similarity to hold. Therefore, the transport approximation can be applied to describe the contribution of multiple scattered electrons (more than two elastic collisions), taking the flux density of the particles scattered twice as source function. The solution of the reflection problem in this so-called double scattering transport approximation has been derived in Ref. [16]. The differential reflection coefficient $R_0(\Omega, \Omega_0)$, i.e. the ratio between the amount of incoming and outgoing particles in the directions Ω and Ω_0 per unit area is found to be given by [16]:

$$R_0(\vec{\Omega}|\vec{\Omega}_0)=(-\mu)n_0\lambda_t\left\{\frac{1}{\mu_0-\mu}\frac{d\sigma_{el}(\vec{\Omega}_0|\vec{\Omega})}{d\vec{\Omega}}\right.$$
$$+n_0\lambda_t^{tr}\left[\int_{4\pi} D(\mu,\mu_0|\mu')\frac{d\sigma_{el}(\vec{\Omega}_0|\vec{\Omega}')}{d\vec{\Omega}'}\frac{d\sigma_{el}(\vec{\Omega}'|\vec{\Omega})}{d\vec{\Omega}}d\vec{\Omega}\right.$$
$$+\frac{1}{2\pi}\int_{4\pi}\int_{4\pi} B(\mu,\mu_0|\mu',\mu'')$$
$$\left.\left.\times\frac{d\sigma_{el}(\vec{\Omega}_0|\vec{\Omega}'')}{d\vec{\Omega}''}\frac{d\sigma_{el}(\vec{\Omega}''|\vec{\Omega}')}{d\vec{\Omega}'}d\vec{\Omega}'\,d\vec{\Omega}''\right]\right\} \tag{26}$$

where n_0 is the atomic density of the target, λ_t and λ_t^{tr} are the total mean free path and the total mean free path in the transport approximation:

$$\lambda_t=\left(\frac{1}{\lambda_e}+\frac{1}{\lambda_i}\right)^{-1} \tag{27}$$
$$\lambda_t^{tr}=\left(\frac{1}{\lambda_{tr}}+\frac{1}{\lambda_i}\right)^{-1}$$

and λ_e is the elastic mean free path. Expressions for the coefficients $D(\mu, \mu_0|\mu')$ and $B(\mu, \mu_0|\mu', \mu'')$ can be found in the appendix of Ref. [16]. The first term in formula (26) represents the contribution of electrons which are backscattered in a single collision, the second term accounts for those particles which have experienced a deflection at a large or small angle and an additional large angle scattering. The third term is the background of diffusively scattered particles.

The angular distribution of elastically scattered electrons Eq. (26) is the theoretical foundation of the elastic peak electron spectroscopy technique (EPES) and is the basis to calculate the higher order reflection coefficients, i.e. the contributions to the quasi-elastic reflection coefficient of electrons having experienced a certain number of inelastic collisions. These distributions together describe the angular/energy distribution of the reflection coefficient in the near peak energy range used in reflection electron energy loss spectroscopy (REELS) in complete analogy to the yield equation for emission Eq. (20).

Figure 12 shows experimental angular distributions of elastically backscattered electrons for polycrystalline Al, and Pt as a function of the scattering angle θ_s. The polar angle of incidence θ_0 was chosen to be 75° (defined relative to the outward surface normal). The theoretical angular distributions given by Eq. (26) are also shown for comparison. The experimental data obtained in relative units were normalized to the theoretical results in order to facilitate comparison.

For the Al target a weakly pronounced minimum is seen at $\theta_s = 90°$ for $E_0 = 300\,\text{eV}$. As the primary energy increases this minimum becomes less pronounced and shifts slightly towards the greater scattering angle $\sim 100°$. At 1000 eV the minimum vanishes and a monotonic behaviour of the angular distribution is observed. For Pt the strongest deviations from theory are also seen at small scattering angles $\theta_s \lesssim 60°$ where the experimental angular distribution shows

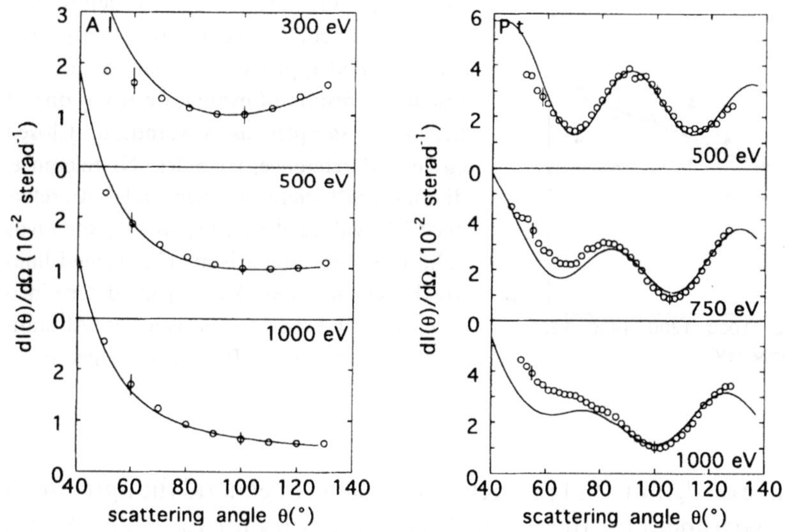

Fig. 12. Comparison of measured angular distributions of electrons reflected from an Al and Pt target at different primary energies with theory. The data points represent the experimental results, the solid lines are the double scattering/transport approximation theory (Eq. 26). The experimental data were normalized to theory at 120°

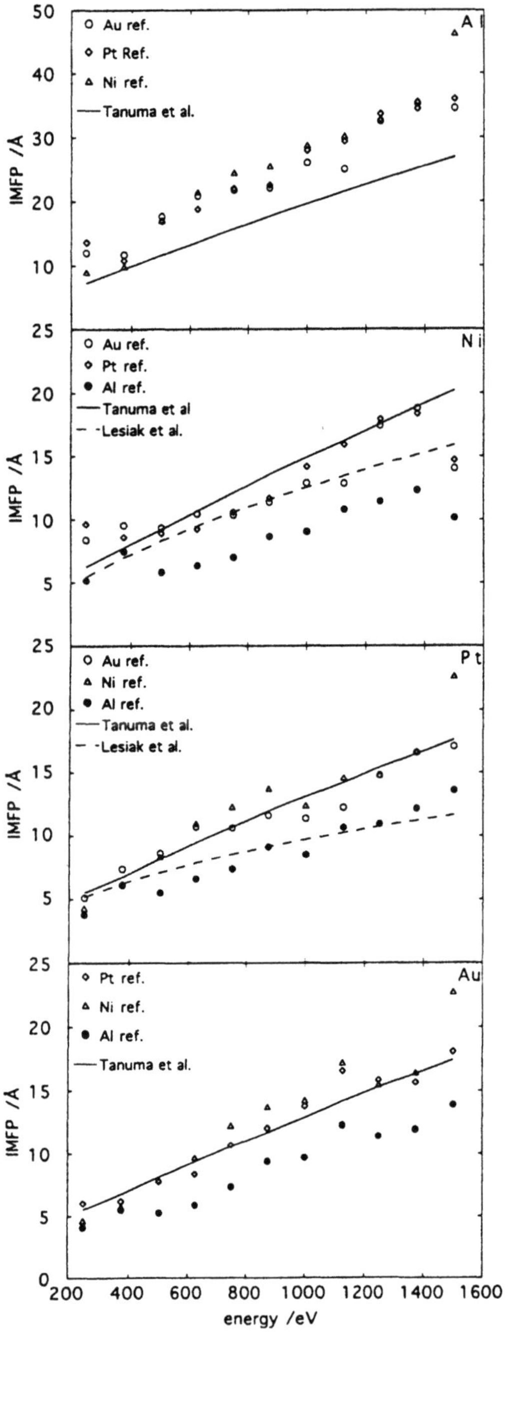

Fig. 13. IMPF values for Al, Ni, Pt and Au in the energy range 250–1500 eV, as determined from experimental reflection coefficients using the presented approach. The data points represent the results of evaluation using one of the other samples as a standard. Closed circles: Al reference; triangles: Ni reference: diamonds: Pt reference; open circles: Au reference. The values of ref. [35] are also shown as solid lines for comparison. The dashed lines are the results of Lesiak et al. [52], determined from similar experiments using an Al standard: **a**, Al; **b**, **c**, **d** Ni, Pt and Au respectively

a smoother dependence on θ_s than that predicted by theory. In general the agreement between theory and experiment for these targets is satisfactory for $E_0 \gtrsim 500$ eV and scattering angles $\theta_s \gtrsim 50$–$60°$. It is believed that the discrepancies at smaller angles are due to surface excitations [16].

The elastic peak measurements can be used to derive IMFP values from the reflection coefficient. To this end, Eq. (26) was used to calculate the reflection coefficient for the considered geometry and element as well as for a reference material. If the IMFP of the standard is assumed to be known, the experimental and theoretical ratios of the reflection coefficient can be equated and numerically solved for the IMFP of the material of interest.

Figure 13 displays a survey of this procedure for polycrystalline Al, Ni, Pt and Au. These measurements were carried out at normal electron incidence while the emission angle was chosen to be 35° relative to the surface normal. Open circles represent the evaluations of a certain datum against the Au standard, while diamonds, triangles and filled circles are used to represent the evaluations using the Pt, Ni and Al standard respectively. Within the experimental error of 10%, the results for Al, using the other materials as reference, are clearly consistent. For the other materials, on the other hand, a comparable consistency is only seen if any material except Al is used as standard, while the use of the Al reference leads to markedly lower IMFP values. Also shown in this figure are the results of [35] based on optical data and Penn's formalism to calculate the IMFP. Comparison of our results with the IMFP values found by Lesiak and coworkers [52], who used a CMA and employed the MC technique to relate the measured reflection coefficient to the IMFP, is also shown in this figure as a dashed line. These authors used Al as reference.

The systematic inconsistency of the results for Al seen in this figure is of practical importance since Al is frequently used as a standard in these types of measurements [52]. The present results seem to indicate that Al is not well suited for this purpose. This is not quite true however. Assessment of the reflection electron energy loss spectra [53] (see below) reveals that the inelastic excitation properties of all the samples are dominated by surface excitations, not accounted for by the present theory. The extent of this effect is extremely pronounced for Al and is much weaker (but certainly not negligible) for the other samples which exhibit a comparable influence of this effect. This explains why the results for Ni, Pt and Au are consistent since the influence of surface excitations will cancel out to a large degree whereas it is expected to lead to strong deviations for Al.

In conclusion of this section we will consider the reconstruction of the DIIMFP from measurements of the energy distribution of quasi elastically reflected electrons. The curves labelled T in Fig. 2 were obtained by applying the method of Tougaard and Chorkendorff [9] to experimental REELS spectra. The theoretical curves labelled T which were derived from optical data deviate significantly from the former. This is believed to be caused by two effects: surface excitations [18] and the influence of the elastic scattering cross section on the REELS energy distribution [53], which is not accounted for by the procedure of Ref. [9]. The latter may be understood by considering Fig. 14. This figure shows a comparison of experimental REELS spectra (lower panel) with Monte Carlo results (upper panel) of 300 eV electrons impinging on an Au target at 15° grazing incidence for several geometries. It is seen that the shape of the REELS spectra depends significantly on the geometry. For example, curve I for 300 eV electrons decreases by a factor ~ 4 when going from 300 to 150 eV, while curve IV exhibits a very flat energy dependence. In curve III the low energy losses are also seen to dominate, whereas curve II is again quite flat.

An explanation of the above effects is easily found by inspection of the portion of the elastic cross section associated with a particular spectrum, shown in the inset. It

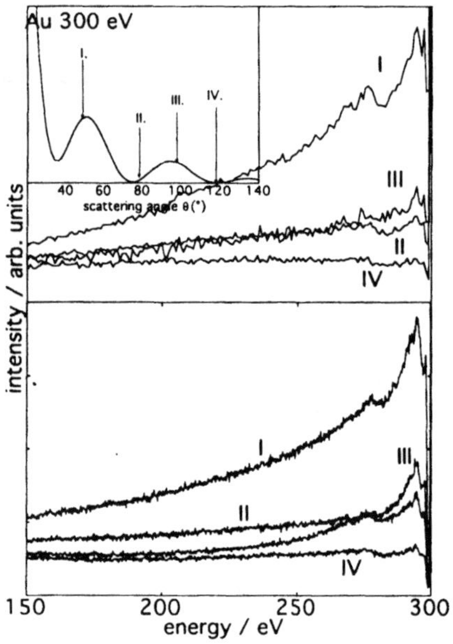

Fig. 14. Experimental REELS spectra (lower part) compared with Monte Carlo results (upper part) for 300 eV electrons in Au. The inset shows the part of the elastic scattering cross section corresponding to the experimental geometry

is seen that the spectra with strong low loss contributions are those recorded at a geometry for which the elastic cross section is relatively high. In particular, for Au these spectra were recorded at a maximum of the single scattering probability but also for the other cases for which the cross section varies smoothly with the scattering angle a similar correlation between the scattering probability and shape of the energy spectrum is clearly seen. For those geometries for which the single scattering reflection probability is large, multiple elastic scattering processes will contribute less than in the opposite case. This implies that the pathlength travelled by an electron in the solid will be less in the former case. Moreover the trajectory of an electron in such a geometry will be located more in the vicinity of the surface than for a particle which is reflected after multiple elastic collisions owing to a smaller single scattering probability. Therefore it is also to be expected that for geometries corresponding to a minimum in the elastic cross section larger losses contribute stronger and surface excitations will be less pronounced and vice versa for a maximum. This is in perfect agreement with the experimental observations. This influence of the shape of the elastic cross section on the energy distribution of the reflection spectrum emphasizes the importance of the third requirement of the generalized radiative field similarity principle: neglect of the exact shape of the scattering cross section in any approach is only allowed if the source angular distribution is sufficiently smooth.

Summary and Conclusions

The general approach to tackle electron transport problems in non-crystalline materials on the basis of transport theory has been outlined. The concept of similarity of the radiative field has been shown to be of foremost importance for analytical treatment of the electron transport process, since it allows to significantly simplify the collision term in the kinetic equation. In this way sufficiently simple and accurate expressions for the basic quantities in microbeam analysis have been derived.

For the quasi-elastic case, mainly important to AES/XPS and characterized by the scattering parameter $\chi = \lambda_i/\lambda_{tr}$, the analytical results obtained so far have been presented and discussed in some detail. Comparison with experimental data and model calculations show that the developed approach meets the modern requirements of microbeam analysis in terms of accuracy and effectiveness.

In the true slowing down regime of energy losses large compared to the considered characteristic energy, the electron transport is governed by the scattering parameter $\sigma = R_0/\lambda_{tr}$ since the energy dissipation process is mainly determined by the continuous loss along the trajectory of the multiple scattered electron. Here an additional problem is encountered in connection with solution of the kinetic equation: the energy dependence of the elastic and inelastic scattering characteristics. Fortunately the energy dependence of the corresponding scattering parameter is extremely weak and consequently only the energy dependence of the transport mean free path turns out to be essential. Thus a realistic solution of the multi-speed equation is provided by a self consistent solution of an appropriate one-speed variant. Simple expressions for quantities like the inelastic reflection coefficient, the Auger backscattering factor and the X-ray yield have been found previously by means of this approach. These theoretical results describe experimental data with high accuracy and prompt for transport theoretical calculations of other quantities relevant for EPMA and related techniques.

In view of the considerable success of analytic approaches for quantitative microbeam analysis the question whether numerical calculations like Monte Carlo simulations are expected to become of secondary importance in the future deserves to be addressed. In this connection, it should be kept in mind, that the presented results, although general with respect to the scattering characteristics, were derived for situations which should be regarded as idealized with respect to the boundary conditions.

In particular, analysis of samples with a compositional variation along all spatial coordinates is by no means straightforward using transport theory. For the quasi-elastic regime this problem is less serious owing to the weak energy and element dependence of the scattering parameter χ, while in the true slowing down regime accounting for the dependence of σ on the spatial coordinates is only possible with considerable effort. It can be concluded that in such cases simulation codes which are able to account for the most general boundary conditions, like the SESAME code recently developed at the CEC in Vienna [54] are of pronounced importance. In particular so since this program incorporates all scattering, excitations and transfer mechanisms of relevance for EPMA, electron spectroscopy etc. On the other hand, analytic solutions for simple cases like homogeneous elemental samples are invaluable for testing purposes of such simulation tools and because they provide physical insight concerning the role of certain parameters on the signal generation process. In this connection, the importance of the concept of radiative field similarity cannot be overemphasized.

References

[1] R. Shimizu, Z.-J. Ding, *Rep. Prog. Phys.* **1992**, 487.
[2] A. Jablonski, *Surf. Interface Anal.* **1989**, *14*, 659.
[3] D. J. Fathers, P. Rez, *Secondary Electron Microscopy*, AMF O'Hare, 1979, p. 55.
[4] D. J. Fathers, P. Rez, *Secondary Electron Microscopy*, AMF O'Hare, 1982, p. 193.

[5] J. P. Spencer, C. J. Humphreys, *Phil. Mag.* **1980**, *A42*, 433.
[6] K. E. Hoffmann, H. Schmoranzer, *Secondary Electron Microscopy*, AMF O'Hare, 1982, p. 209.
[7] R. A. Dashen, *Phys. Rev.* **1964**, *134*, A1025.
[8] S. Tougaard, P. Sigmund, *Phys. Rev.* **1982**, *B25*, 4452.
[9] S. Tougaard, I. Chorkendorff, *Phys. Rev.* **1987**, *B35*, 6570.
[10] A. L. Tofterup, *Surf. Sci.* **1986**, *167*, 70.
[11] V. M. Dwyer, J. A. D. Matthew, *Surf. Sci.* **1984**, *143*, 57.
[12] V. M. Dwyer, J. A. D. Matthew, *Surf. Sci.* **1988**, *193*, 549.
[13] S. Tougaard, *Surf. Interface Anal.* **1988**, *11*, 453.
[14] I. S. Tilinin, *Phys. Chem. Mech. Surfaces* **1989**, *5*, 476.
[15] W. S. M. Werner, *Surf. Interface Anal.* **1995**, *23*, 696.
[16] W. S. M. Werner, I. S. Tilinin, M. Hayek, *Phys. Rev.* **1994**, *B50*, 4819.
[17] J. B. Pendry, *Low Energy Electron Diffraction*, Academic Press, London, 1974.
[18] F. Yubero, S. Tougaard, *Phys. Rev.* **1992**, *B46*, 2486.
[19] I. S. Tilinin, W. S. M. Werner, *Mikrochim Acta* **1994**, *114/115*, 485.
[20] R. Schmid, *Ph.D. Thesis*, University of Tübingen, 1982.
[21] I. M. Brohnstein, V. P. Pronin, *Sov. Phys. Solid State.* **1975**, *17*, 1363.
[22] U. Fano, J. A. Stephens, *Phys. Rev.* **1986**, *B34*, 438.
[23] I. S. Tilinin, W. S. M. Werner, *Surf. Sci.* **1993**, *290*, 119.
[24] W. S. M. Werner, I. S. Tilinin, *J. Appl. Phys.* **1993**, *70/71*, 29.
[25] I. S. Tilinin, W. S. M. Werner, *Phys. Rev.* **1992**, *B46*, 13739.
[26] I. S. Tilinin, *Sov. Phys. JETP.* **1982**, *55*, 751.
[27] A. C. Yates, *Comp. Phys. Comm.* **1971**, *2*, 175.
[28] R. A. Bonham, T. G. Strand, *J. Chem. Phys.* **1963**, *39*, 2200.
[29] W. F. Egelhoff, Jr., *Phys. Rev. Lett.* **1993**, *71*, 2883.
[30] J. Lindhard, K. Dan, *Vidensk Selsk. Mat. -Fys. Medd.* **1954**, *8*, 28.
[31] D. R. Penn, *Phys. Rev.* **1985**, *B35*, 482.
[32] E. D. Palik (ed), *Handbook of Optical Constants of Solids*, Academic Press, New York, 1985.
[33] Ding Ze-jun, R. Shimizu, *Surf. Sci.* **1988**, *179*, 539.
[34] W. S. M. Werner, I. S. Tilinin, *Prog. Surf. Sci.* **1994**, *46*, 241.
[35] S. Tanuma, C. J. Powell, D. R. Penn, *Surf. Interface Anal.* **1988**, *11*, 577.
[36] I. S. Tilinin, *Sov. Phys. JETP.* **1988**, *67*, 1570.
[37] W. S. M. Werner, I. S. Tilinin, *Surf. Sci.* **1992**, *268*, L319.
[38] V. V. Sobolev, *A Treatise on Radiative Transfer*, van Nostrand, Princeton, 1963.
[39] K. M. Case, P. F. Zweifel, *Linear Transport Theory*, Addison-Wesley, Reading, 1967.
[40] L. Landau, *J. Phys. (Moscow)* **1944**, *8*, 201.
[41] W. S. M. Werner, I. S. Tilinin, H. Beilschmid, M. Hayek, *Surf. Interface Anal.* **1994**, *21*, 537.
[42] H. Beilschmidt, W. S. M. Werner, I. S. Tilinin, *Surf. Interface Anal.* **1994**, *22*, 120
[43] W. S. M. Werner, *Phys. Rev.* **1955**, *15*, 2964.
[44] A. Jablonski, H. Ebel, *Surf. Interface Anal.* **1988**, *11*, 627.
[45] S. Chandrasekhar, *Radiative Transfer*, Dover Publications, New York, 1960.
[46] W. S. M. Werner, W. H. Gries, H. Störi, *J. Vac. Sci. Technol.* **1991**, *A9*, 21.
[47] W. S. M. Werner, W. H. Gries, H. Störi, *Surf. Interface Anal.* **1991**, *17*, 693.
[48] H. Wagner, C. Schiebl, W. S. M. Werner, *Mikrochim. Acta* **1996**, *[Suppl.] 13*, 623.
[49] I. S. Tilinin, *Poverkhnost* **1988**, *8*, 5. (in Russian).
[50] W. S. M. Werner, 1996, to be published **1996**
[51] J. S. Schilling, M. B. Webb, *Phys. Rev.* **1970**, *B2*, 1665.
[52] B. Lesiak, A. Jablonski, Z. Prussak, P. Mrozek., *Surf. Sci.* **1989**, *223*, 213.
[53] W. S. M. Werner, M. Hayek, *Surf. Interface Anal.* **1994**, *22*, 79.
[54] H. Wagner, A. Pfeiffer, C. Schiebl, W. S. M. Werner, *Mikrochim. Acta* **1996**, *[Suppl.] 13*.

Mikrochim. Acta [Suppl.] 13, 39–60 (1996)

Use of Soft X-Rays in Microanalysis

Jean-Louis Pouchou

O.N.E.R.A., Department of Materials, 29 Avenue Division Leclerc, F-92320, Châtillon, France

Abstract. The use of soft X-rays in microanalysis is illustrated by several new practical examples, concerning the K lines of the ultra-light elements, and the L lines of medium-Z elements that can be used for near-surface analysis and in situations where eliminating the secondary fluorescence emission is of interest. Some of the specific problems of the soft X-ray field are examined: surface contamination, coating of insulating specimens, line interference, background subtraction, line distortion, mass absorption coefficients.

Key words: soft X-rays, low-energy lines, X-ray microanalysis, electron microprobe, analytical scanning electron microscopy.

In microanalysis, the soft X-ray range can be defined as extending from ~1 keV down to ~100 eV. This corresponds to photons that could not be detected in the early times of the technique. The wavelength dispersive spectrometers (WDS) of the first generation microprobes could only detect photons above ~1.5 keV, i.e. the Al Kα line. Similarly, the first generation of energy dispersive spectrometers (EDS) could not detect photons below the Na Kα line (~1 keV). Even when they became detectable, these photons could not be used efficiently for meaningful quantitative analysis because, until the eighties, the approximations in the existing absorption correction models were a limiting factor, as well as the poor knowledge of the values of the mass absorption coefficients, and also the lack of attention paid to the so-called chemical effects (line shift and change of shape due to the chemical bonding).

Nowadays, lines with energies down to ~100 eV (the Be K line) can be detected by WDS. The use of multilayer structures as monochromators has strongly improved the X-ray collection efficiency in the low-energy range. However, even in the best excitation conditions, the peak-to-background ratio is not very high (less than 100 for the lightest pure elements), so that the minimum detectable concentration is not very low (a few hundred to a few thousand μg/g, depending on the situation), and the experiments always require more thoughtfulness than at higher energy.

In the case of EDS spectrometry, the new detectors including high-quality semiconductor crystals, improved electronics and ultra-thin atmospheric windows enable measuring elements down to boron (B K line at ~185 eV). However, the spectral resolution of these detectors is poor compared to WDS spectrometers (e.g. ~70 eV FWHM for the C K line, i.e. at least ten times more than a WD spectrometer with a lead stearate pseudo-crystal), and the peak-to-background ratio is consequently typically ten times smaller. Hence, one should be aware of the fact that, in spite of the improvements of EDS detectors and software, their potential for fine and accurate analysis will always remain limited in the field of low-energy photons, where spectral interference of elements frequently occurs.

For the ultra-light elements (beryllium to fluorine) there is obviously no other alternative than using soft X-rays. But for other elements, one can wonder if there is some interest in using the low-energy lines (i.e. L lines for medium-Z elements or M lines for heavier elements), since in most current applications these elements can be analyzed without major problem by using the high-energy radiation (i.e. K lines for medium-Z elements or L lines for heavier elements). The aim of this paper is not to give an extensive review of the use of soft X-rays, but to illustrate with practical examples typical situations where the analyst can take advantage of using soft X-rays rather than conventional higher energy lines, and to draw reader's attention to some particular problems with these soft X-rays.

When to Use Soft X-rays?

– To analyze light and ultra-light elements (third and second lines of the periodic table), the soft K lines are the only characteristic lines available.

– To perform analyses with an improved sensitivity to the near-surface region, one has to operate at low accelerating voltage of the electron beam, and to measure the L lines of the medium-Z elements instead of the K lines, or the M lines of the heavier elements instead of the L lines.

– To minimize the effect of secondary emission by fluorescence in multiphase or stratified specimens, one can take advantage of the low fluorescence contribution in the soft X-ray lines.

Use of Soft X-Rays for Surface Analysis

Depth and Efficiency of Excitation

More and more applications of X-ray microanalysis are now concerning surface layers or more generally stratified specimens. When the structure of the stratified specimen is a priori known, measuring the specimen with a high depth resolution is not absolutely necessary. But if the near-surface structure has to be investigated, an improved depth resolution is needed to enable the formulation of reasonable hypotheses. The only way for improving the depth resolution and keeping at the same time a sufficient excitation efficiency is to operate at low accelerating voltage with low-energy characteristic lines. Table 1 illustrates this point in the case of copper. It is interesting to compare the situation obtained at 10 kV for the Cu Kα line (critical energy 8.979 keV) with the situation at 5 kV for the Cu Lα line (critical energy 0.933 keV): these operating conditions give approximately the same mean ionization depth ($\sim 30\ \mu g/cm^2$) and the same maximum ionization depth ($\sim 115\ \mu g/cm^2$), but the peak count rate is ~ 6 times higher and the peak-to-background ratio ~ 8 times higher for the low-energy line. Hence, for the same excited depth, the limit of detection is ~ 7 times better with the Lα line than with Kα. If a smaller analyzed

Table 1. Comparison of the ionization depths[a], peak intensities and peak-to-background ratios for Cu Kα (LiF monochromator) and Cu Lα (TAP monochromator) in copper

Cu Kα	Accelerating voltage (kV)	9.5	10	11	12	13	14	15	
	Overvoltage ratio	1.058	1.114	1.225	1.336	1.448	1.559	1.671	
	Mean ion. depth (μg/cm²)	17	31	53	74	93	113	133	
	Max. ion. depth (μg/cm²)	65	115	200	270	350	425	510	
	Peak (c/s/nA)	0.7	10.4	28	60.4	103.6	154.2	212.1	
	Peak/background ratio	4	27	53	76	116	128	136	
Cu Lα	Accelerating voltage (kV)	1.5	2	2.5	3	3.5	4	4.5	5
	Overvoltage ratio	1.608	2.144	2.680	3.215	3.751	4.287	4.823	5.359
	Mean ion. depth (μg/cm²)	3.7	6.6	10	14	18	22	27	33
	Max. ion. depth (μg/cm²)	14	25	37	49	65	81	98	115
	Peak (c/s/nA)	3.4	9.5	17	27	36	45	57	68
	Peak/background ratio	100	131	148	169	175	176	195	208

[a] Note: the depth from which 90% of the intensity is generated is approximately twice the mean ionization depth.

depth is required, Table 1 shows that a mean ionization depth less than 4 μg/cm² (~4 nm) can be obtained at 1.5 kV in copper with the Cu Lα line. In spite of the low peak counting rate, such an extreme operating condition can be employed in practice, because of the high peak-to-background ratio (~100).

Example of Surface Analysis

The practical interest of soft X-rays can be illustrated by the concrete example of a specimen provided by the N.I.S.T. in view of a round robin test. This specimen was described as a pure chromium layer on a GaAs substrate.

A series of measurements at variable voltage with the usual Kα lines of Cr, Ga and As have been done. Processing the data with the automatic iteration in the program *Strata* seems to give a satisfactory result: as Fig. 1 shows, the computed *k*-ratio curves fit correctly the experimental data (only 0.26% residual absolute deviation) for a Cr layer thickness of 82.7 μg/cm² (i.e. 115 nm assuming 7.19 for the density), and for a substrate composition equal to Ga 49.5 – As 50.5 at.%.

It is almost impossible to get more information with the K lines of the elements declared. However, if one observes the specimen more carefully, the presence of oxygen can be detected. One can measure, at 10 kV for example, a relative intensity greater than 5% of that of an $Y_3Fe_5O_{12}$ conductive oxygen standard. Such a single oxygen measurement does not tell us whether oxygen is incorporated inside the Cr layer, or is in an oxide layer at the surface, or both. To know more, one has to measure the oxygen at several voltages (in particular at lower voltage) and also to generate a Cr signal from a reduced depth, i.e. by use of the Cr Lα line at low voltage. The results are displayed on Fig. 2: at 2.5 kV, the oxygen signal increases strongly, indicating that there is certainly some oxide at the surface. However, the assumption that all the oxygen is in a surface oxide layer does not produce a satisfactory fit of the oxygen data from 2.5 to 15 kV. To get a good global agreement, one has to postulate that some oxygen is also present inside the Cr layer. This is confirmed by measuring

Table 2. Ionization depths, peak intensities and peak-to-background ratios for Cr $L\alpha$ lines in chromium, for the low voltages used in Fig. 2, (W/Si monochromator, with $2d = 5.9$ nm)

Cr $L\alpha$	Accelerating voltage (kV)	1.28	1.50	1.80	3.50	5.12
	Overvoltage ratio	2.230	2.613	3.136	6.098	8.920
	Mean ionization depth (nm)	5	6.6	9	27	50
	Max. ionization depth (nm)	18	24	32	93	167
	Peak (c/s/nA)	4.5	6.9	10.2	30	47
	Peak/background ratio	15	17	18	25	29

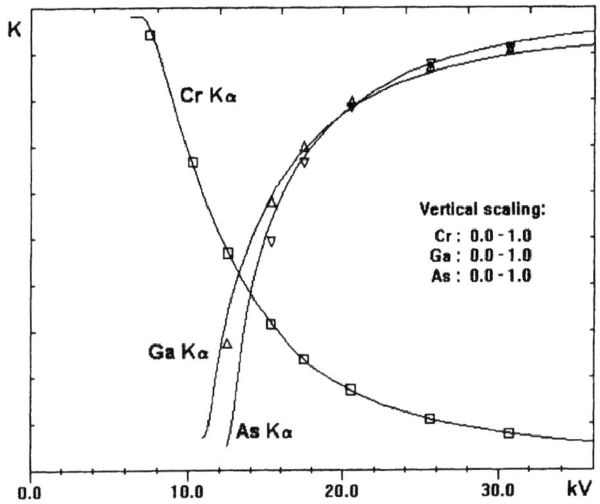

Fig. 1. Comparison of the computed *k*-ratio curves with the experimental data for a Cr layer on a GaAs substrate. K lines for all elements

the Cr $L\alpha$ line, between ~1.3 kV and ~5 kV. This permits to scan progressively the layer, from near-surface to deeper regions (Table 2). On Fig. 2, the observed decrease of the Cr $L\alpha$ data at very low voltage confirms the assumption of a surface oxide layer, while the level of the "plateau", near 4 kV, agrees with the hypothesis of incorporated oxygen. Finally, the curves of Fig. 2 are obtained for the following specimen description: a GaAs substrate, covered by a 87 μg/cm² layer (i.e. 121 nm assuming 7.19 for the density) of Cr containing 3.2 at.% oxygen, and with a surface oxide layer of 1.25 μg/cm² (i.e. 25 nm assuming Cr_2O_3 with a density equal to 5).

Use of Soft X-Rays to Minimize Fluorescence Effects

In the Cr $K\alpha$ *k*-ratio curve of the previous example (Fig. 1), it can be noted that the value of the narrow plateau computed at very low overvoltage for a pure chromium layer is lower than 100%. This is the effect of the secondary emission by fluorescence, the contribution of which to the total emission is less for the thin Cr film than for a bulk reference.

Fluorescence is probably the most worrying effect for high-energy lines, because it may produce not only quantitative errors, but also wrong qualitative interpretation of the measurements, since the volume concerned by fluorescence is much greater than the primary volume excited by the incident electrons, so that a radiation

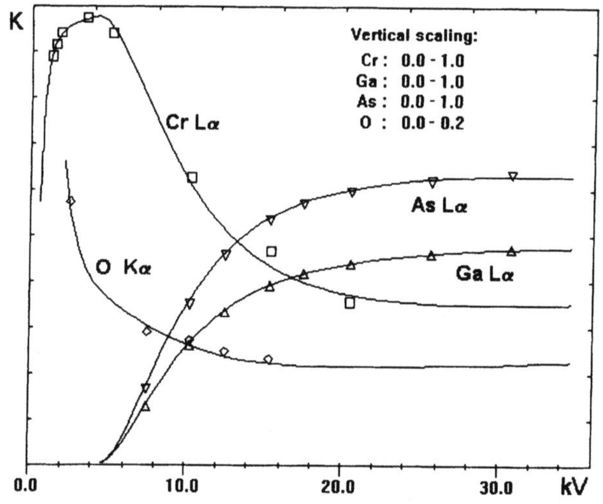

Fig. 2. Comparison of the computed k-ratio curves with the experimental data for a Cr layer on a GaAs substrate. L lines for all elements, except oxygen

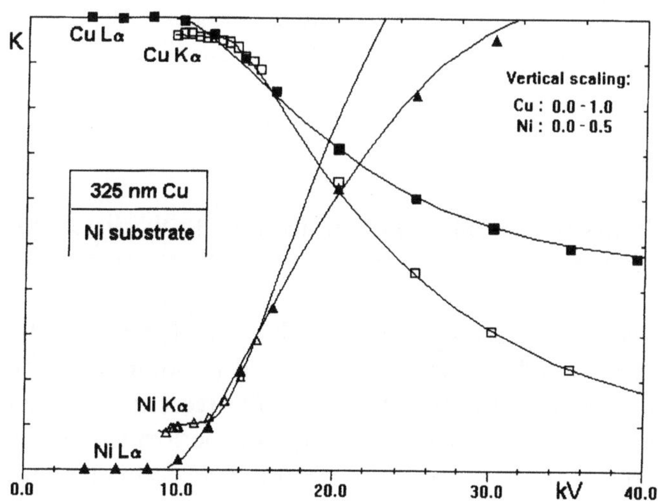

Fig. 3. k-ratio vs. accelerating voltage for a 325 nm Cu layer on Ni substrate. The K lines (open symbols) show the effect of fluorescence, negligible for the L lines (close symbols). Curves modeled with *Strata*

can be excited far away from the point of electron impact. This paragraph illustrates this point, and shows that the problem can be eliminated with soft X rays.

Fluorescence in Stratified Specimens

The importance of the fluorescence for high energy lines can be appreciated easily on Fig. 3, in the simple case of a Cu layer (325 nm) on a Ni substrate. In that case, the Ni Kα line is excited both by the continuum and by the Cu Kβ line, whereas the Cu Kα line is excited by the continuum only. Provided that the surface cleanliness is the same for the Cu layer and for the Cu standard (the best is to evaporate simultaneously the Cu layer onto the bulk Cu standard), one can measure at low overvoltage that the Cu Kα plateau is not at 100%, but near 96.5% at 10 kV. At the same time, the Ni Kα k-ratios are not equal to zero below the threshold for primary substrate excitation ($\sim$12 kV), but are near 4.5% at 10 kV. The interesting point to observe on Fig. 3 is that the use of the Lα lines, for which the fluorescence yield is much lower,

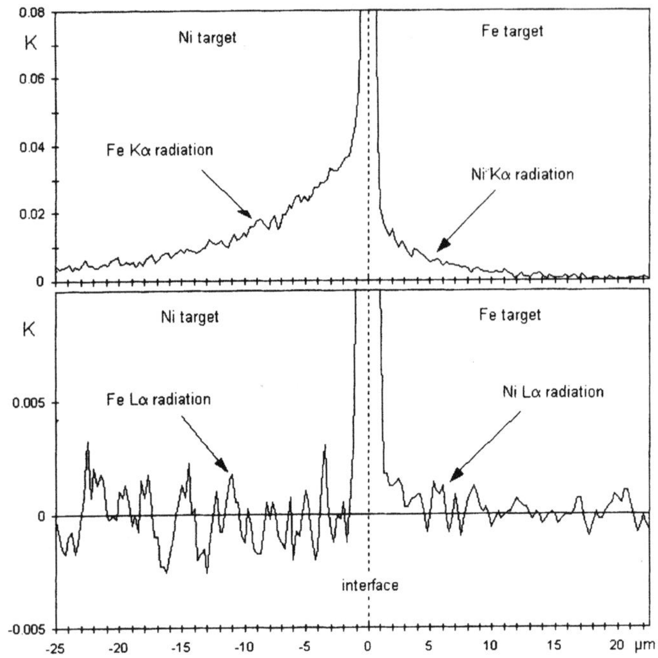

Fig. 4. Fe and Ni *k*-ratios measured with K and L linss on each side of a Ni–Fe interface. The fluorescence is eliminated by use of L lines

eliminates completely (in the limits of the experimental accuracy) this disturbing effect: for example, the Ni Lα *k*-ratios mesasured from 4 kV to 8 kV are ranging between 0.01% and 0.02%.

This example of stratified specimens corresponds to a well-defined geometry, for which calculations have been developed [1] and incorporated in some available computer programs (e.g. *Strata, Multifilm*). But the problem of fluorescence may also be crucial in many other current situations where the analysis is performed not far from an interface (e.g. diffusion near an interface, analysis of particles smaller than a few tens of μm).

Lateral Fluorescence Near Interfaces

The example of Fig. 4 illustrates the lateral fluorescence in the vicinity of an interface, in the case of a Ni-Fe couple on which line scans across the interface have been performed. To avoid absorption artifacts, the plane of the interface was located in the vertical plane defined by the electron beam and the direction of the spectrometer. Both the Kα and the Lα lines of Fe and Ni have been measured. To obtain primary excitation volumes having approximately the same dimensions, the voltage was set to 15 kV for the K lines and 12 kV for the L lines. The difference in the behavior of both elements, on each side of the interface, is due to the fact that Ni K is only excited by the continuum, whereas Fe K is excited by continuum and by the Ni Kα and Kβ lines. Figure 4 shows clearly that the fluorescence is much stronger on the Ni side (for the Fe Kα radiation) than on the Fe side (for the Ni Kα radiation): respectively 3.3% and 1% at 3 μm of the interface, 1.4% and 0.2% at 10 μm, 0.7% and 0.03% at 20 μm.

On the other hand, when the soft Lα lines are used (bottom of Fig. 4), one can observe that the contribution of the fluorescence is almost undetectable. Hence, this shows again that the analyst can take advantage of using soft lines instead of the usual ones in order to eliminate strong fluorescence effects and to determine more easily if an element is actually present or not in a given phase, particle, or layer.

Particular Problems of Line Shape Distortion

Since the effort made in the beginning of the eighties to quantify the ultra-light elements, it has been clearly shown that the characteristic emission bands of these elements were suffering the so-called "chemical effects", i.e. the spectral distribution of the intensity was dependent on the nature of the compound. Generally speaking, the stronger the chemical bonding, the narrower the line of the ultra-light element. Although some basic work is still needed in this field [2], the experience has shown that in most situations (except the Ni and Co borides [3]) the ultra-light elements can be properly analyzed quantitatively by WDS [4], provided that peak areas are used to form the relative intensities, either directly [5] or through the so-called APF (area-to-peak factors) [6]. Recent WDS software, such as the programs distributed by *CAMECA* and *SAMx*, include automatic area acquisition by scanning the spectrometers. In addition, the latter also enables to combine conventional peak measurements with APF factors.

For other soft X-rays, such as the Lα lines of the 3d transition elements, other phenomena may appear, i.e. important changes in the Lα generation and in the self-absorption, depending on the chemical state of the transition atom [7]. Understanding these phenomena is necessary to avoid serious errors in using these soft lines. This is the objective of this paragraph.

In the examples of the previous paragraphs, there was no danger in using the Lα lines of transition elements Cr, Fe and Ni, because these elements were pure (or almost pure) in the specimens as well as in the standards. But when such a transition atom is chemically bonded with another element in a compound, the partial filling of the 3d electron band modifies both the generated Lα intensity (because Lα results from a 3d–2p transition, so that its intensity increases with the 3d population), and also its self-absorption (because the filling of the 3d band attenuates the anomalous Lα absorption resonance corresponding to the possible 2p–3d promotion of electrons by the most energetic photons of the Lα line). This paragraph aims to clarify some observations about the Ni Lα line, which could be interpreted as chemical effects (i.e. the image of strong changes in the electron distribution of the 3d band), but which actually result mainly from an absorption effect, the attenuation of the Ni Lα resonant absorption when the nickel is alloyed.

If one compares the spectra acquired at 30 kV on pure Ni and on the NiAl compound, in the region of the Ni L lines (Fig. 5), it seems that the Ni Lα line of the compound is shifted towards the high energies by ~0.6 eV. Incidentally, it can be observed that the high-energy tail of the Ni Lα line appears weaker in the compound than in the pure metal. Actually, the apparent shift is not due to a chemical effect, as demonstrated by Fig. 6: if the same spectra are recorded at 3 kV, where the absorption is very weak, the Lα line has almost the same position and the same shape in NiAl as in pure Ni (only a slight difference in the high- and low-enery tails can be observed). This demonstrates clearly that the effect observed at high voltage is not

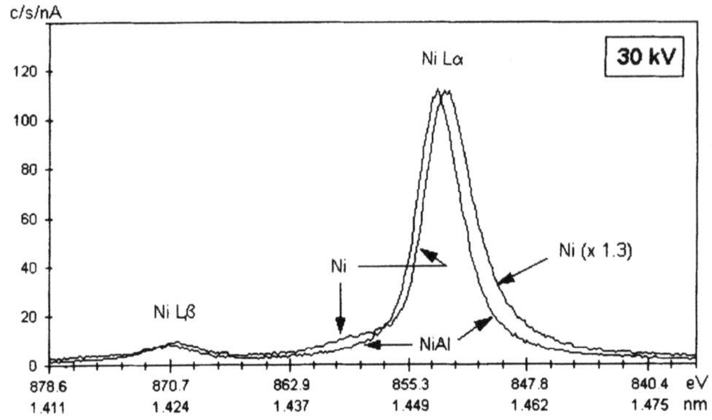

Fig. 5. Ni L spectra from Ni and NiAl at 30 kV. WDS spectrometer with TAP monochromator

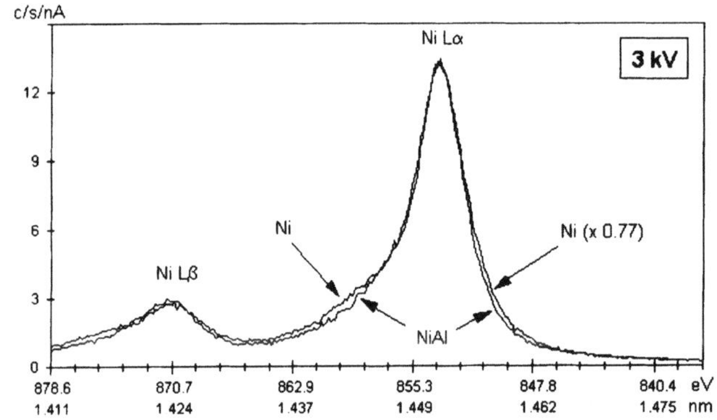

Fig. 6. Ni L spectra from Ni and NiAl at 3 kV. WDS spectrometer with TAP monochromator

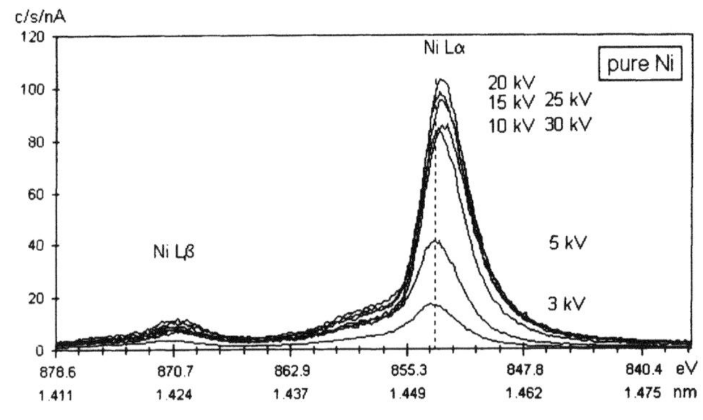

Fig. 7. Evolution of the Ni L spectrum with accelerating voltage in pure Ni

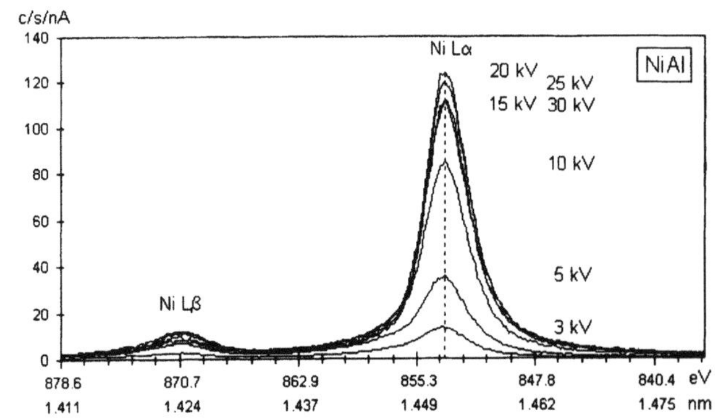

Fig. 8. Evolution of the Ni L spectrum with accelerating voltage in NiAl

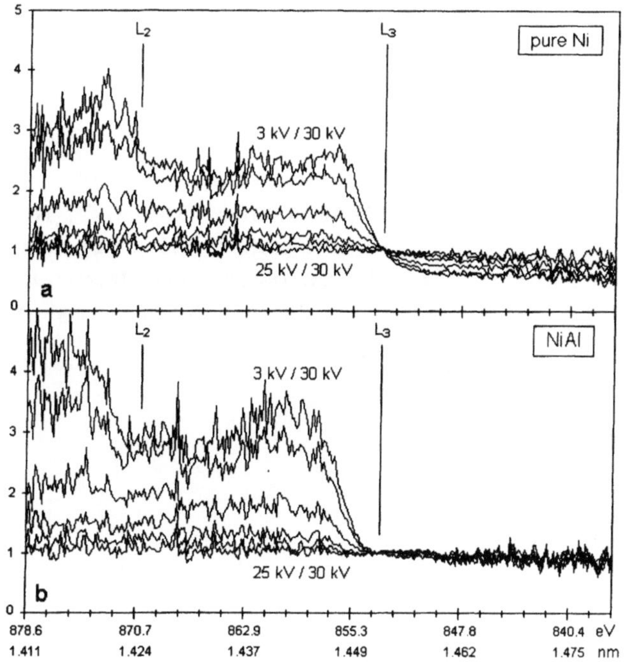

Fig. 9. Ratio of Ni L spectra to the 30 kV spectrum. **a** Case of pure Ni, **b** case of NiAl

a real chemical effect, but an absorption effect due to the attenuation of the resonant self-absorption in NiAl compared to pure Ni.

If one compares on Fig. 7 (case of Ni) and Fig. 8 (case of NiAl) the variations of the NiLα spectra with the accelerating voltage, one can see that the depression in the Ni spectrum near the L_3 edge (854 eV according to [8]) almost disappears in the NiAl spectrum. Similar measurements on the Ni_3Al compound (not presented here) show an intermediate behavior between Ni and NiAl.

By ratioing the spectra recorded at every voltage to the 30 kV spectrum (taken as reference because it has the best counting statistics), one obtains the graphs of Fig. 9a and 9b, which show clearly the actual positions of the L_2 and L_3 absorption edges (the positions marked on the graph are from [8]). By dividing the 3-to-30 kV NiAl ratio (which represents the maximum absorption effect) by the corresponding ratio for pure Ni, one gets a graph (Fig. 10) which enhances the difference in the behavior of pure and bonded Ni atoms, and shows the energy of the resonant absorption for the Ni Lα line and incidentally for the Ni Lβ line.

The fact that the 3d–2p transition rate and the L self-absorption coefficient vary with the chemical state in the 3d transition elements is a strong limitation for the use of their L lines in routine applications of quantitative microanalysis. However, these lines can be used efficiently in particular cases of thin film analysis: when the transition element is almost pure in a given layer, or, in the opposed situation, when a standard with a similar composition is available (for example a thick film, with a composition previously measured with the K lines). Whatever the situation, the effective value of the self-absorption coefficient of the L line in the transition element is needed.

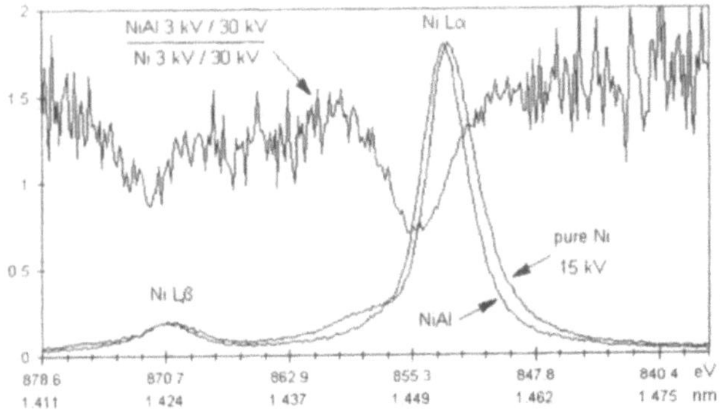

Fig. 10. Ratio of (3 kV/30 kV) NiAl curve (Fig. 9b) to the corresponding Ni curve (Fig. 9a). The observed resonant absorption is on the high-energy side of Ni L lines

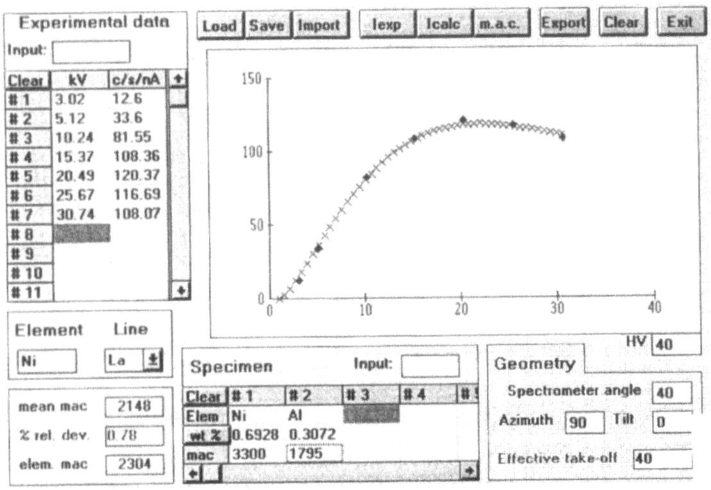

Fig. 11. Determination of the mass absorption of Ni Lα by Ni atoms in NiAl (*XMAC* program)

The Determination of Soft X-Ray Absorption Coefficients

Accurate $\phi(\rho z)$ models have been developed during the last decade for quantitative analysis, but none of them is able to produce good results without a set of reliable values for the X-ray mass absorption coefficients.

Frequently, in the soft X-ray field, one has the problem that an analyzed line is close to an absorption edge, or suffers some anomalies as in the previous example. Consequently, the absorption coefficient may be quite uncertain. An efficient method for determining an absorption coefficient is to measure and process the variation of the line intensity emerging from a known composition as a function of the accelerating voltage. The advantage and the principle of the method, initially based on the $PAP\phi(\rho z)$ model, have already been described [9]. Recently, a new software (*XMAC*) based on the *XPP* model has been developed by the author for the *SAMx* user's group.

An application of this program, in the case of the Ni Lα line in NiAl, is shown on Fig. 11. In this typical example, the absorption coefficient of Ni Lα in Al is supposed to be known exactly (1795 cm^2/g according to [10]). The self-absorption of Ni Lα by Ni atoms bonded with Al has to be determined. It should be noted that the value of 3300 displayed in the "Specimen" box is the indicative value of the self-absorption

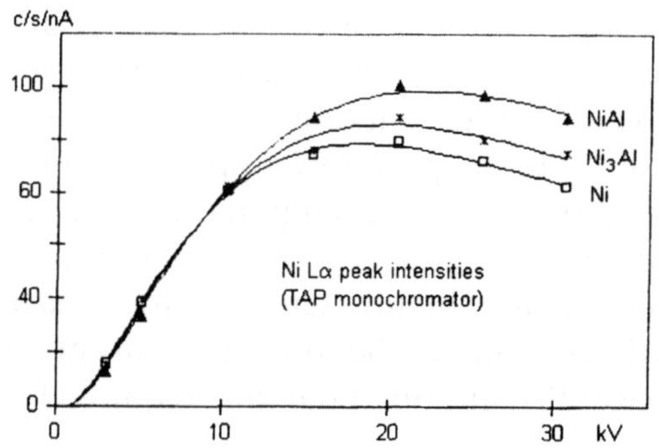

Fig. 12. Variation of the Ni Lα peak intensity vs. accelerating voltage emerging from pure Ni, Ni_3Al and NiAl

proposed by the program for pure Ni atoms. For bonded Ni atoms, one expects a lower value. The emerging intensity is measured at the Ni Lα peak maximum from ~3 kV to ~30 kV. An automatic iterative procedure searches the value of the mass absorption coefficient for which the computed intensity variation (the series of crosses) fits at best the experimental data. The resulting absorption coefficient of Ni Lα in Ni bonded with one Al atom is displayed at the lower-left corner of Fig. 11. This result (2304 cm^2/g) is significantly lower than the value 3300 for pure Ni. It should be mentioned that, if area measurements are performed instead of peak measurements, the resulting value of the effective self-absorption coefficient is slightly higher (2575 cm^2/g), because of the greater contribution of the resonant absorption). In the case of the Ni_3Al compound, a similar measurement of the peak intensity versus the accelerating voltage has given 2850 cm^2/g, i.e. an intermediate value between pure Ni and NiAl. Hence, the attenuation of the resonant absorption near the Ni L_3 energy, when the concentration of Al atoms increases, is responsible for the apparent shift of the Ni Lα line, because the absorption is stronger on the high-energy side than on the low-energy side of the line.

Finally, if the absolute emerging intensity curves for Ni, Ni_3Al and NiAl are plotted on the same graph (Fig. 12), one can paradoxically observe at medium and high accelerating voltages that, the lower the concentration of nickel, the higher the emerging intensity of its Lα line! This illustrates that in the field of soft X-rays, mainly for the transition elements, the analyst should always keep in mind that unusual effects may appear, and that he should approach very carefully any new problem.

The Problem of Surface Contamination

When soft X-rays in general and ultra-light elements in particular have to be analyzed, the polishing and rinsing of the specimens and standards have to be done with a lot of care. Any significant deterioration of the target under the beam should also be avoided. The most frequent problem is the carbon contamination due to the cracking of the residual organic molecules of the vacuum, and also the contamination coming from the target itself.

Case of Electron Microprobes

In the electron microprobes, the problem is solved by combining an oxygen- (or air-) jet with a N_2-cooled trap. Such a system is able, generally after typically 2 or

3 minutes, to eliminate the absorbed carbon from the region of the surface bombarded by the electrons. The most recent microprobe automation programs (*XMAS* for example) allow the user to define a decontamination time prior to the measurement at each point. Future versions should probably apply statistical criteria to the variation of the carbon signal versus time at each point, in order to determine and manage automatically the decontamination process. One should also be aware that some materials (e.g. zirconium) are highly sensitive to surface oxidation and may react under the beam in the presence of the oxygen- or air-jet. In such cases, an hydrogen-jet can be used successfully, although the time required for decontamination is longer [11].

Case of Scanning Electron Microscopes

Most often, the scanning electron microscopes don't have any anticontamination device. The main reason is that, contrary to the electron microprobes, these instruments have to be operated in widely variable conditions of working distance, stage position and tilt, so that the attachment of a fine capillary tube near the specimen surface, very close to the beam, is not easy. The consequence is that the carbon contamination is important, even in the microscopes equipped with a turbo-molecular pump, since most of the contamination comes from the primary rotary pump.

On Figs. 13 and 14, the EDS spectra recorded with a *Kevex Super Quantum* detector illustrate how the carbon signal measured on a Fe_3C specimen varies with the counting conditions. All the spectra correspond to a total counting time of 1000

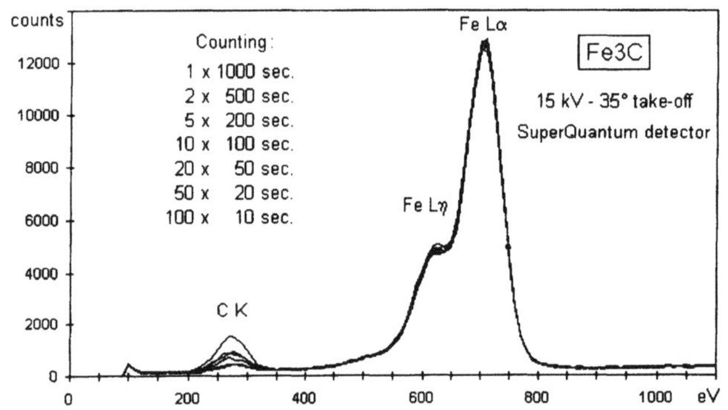

Fig. 13. Low-energy portion of Fe_3C energy dispersive spectra for variable counting time per point

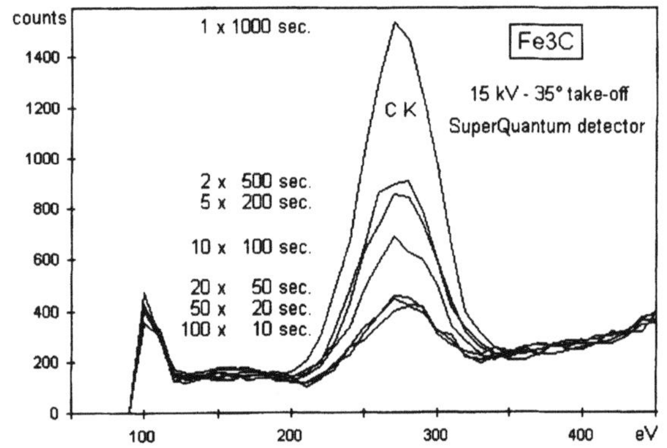

Fig. 14. Enlarged view of the carbon spectra of Fig. 13

Table 3. Influence of carbon contamination on the result of standardless carbon analysis in Fe_3C (6.68 wt.% nominal). *Quantex*$^+$program with *XPP* model

Time per point (seconds)	10	20	50	100	200	500	1000
wt. % carbon (*XPP* procedure)	1.56	6.35	6.47	10.84	13.68	14.52	22.77

seconds, achieved either by counting once on a single point, or by summing N spectra acquired on N different points during 1000/N seconds. The first observation from Fig. 13 is that the carbon peak has a low intensity compared to the Fe L lines. This is due to the strong absorption of the C K line both in Fe and in the detector window (the main constituent of the *SuperQuantum* window is boron, which absorbs the C K line with a maximum efficiency). This low intensity justifies that long counting times have to be used to get a statistically well-defined peak for ultra-light elements. Zooming over the C K peak (Fig. 14) shows the effect of the contamination. For individual counting periods of 10, 20 and 50 seconds per point, the C K spectra are almost identical. But for longer periods, the "apparent" carbon signal increases strongly (more than threefold for 1000 s.). In the present example, processing the spectra with the standardless *Quantex+/XPP* procedure [12] leads to a carbon content in good agreement with Fe_3C (see Table 3) only for the short counting periods (less or equal 50 s). By applying *Strata* to the high carbon signal obtained for 1000 seconds per point, one obtains $\sim 15\,\mu g/cm^2$ carbon (i.e. ~ 75 nm) due to the contamination on the specimen surface.

It should not be forgotten that the contamination by carbon is a problem not only in the case of the carbon measurement, but also for the measurement of other soft X-rays (mainly the nitrogen line, which is the most strongly absorbed by carbon). For example, to give an order of magnitude, one can calculate with *Strata* that, at 10 kV and 35° take-off, the presence of 75 nm carbon at the surface of a boron nitride produces a relative loss of intensity of 11% for the B K line and 46% for the N K line. At 5 kV, where the losses of direct ionization becomes significant, these values are respectively 23% and 54%. Obviously, this is a strong source of error in EDS standardless analysis, since there is in that case no check of the result, because of the necessary normalization of the concentrations.

The Problem of Coating the Insulating Targets

In many cases, ultra-light elements have to be analyzed in insulating specimens (oxides, nitrides, ...). The conventional procedure consists in depositing simultaneously the same conductive coating on the specimen and on the standards, considering that the influence of the coating on the intensity loss for a given X-ray line is the same for all targets. To minimize the loss of intensity due to the coating, most people evaporate a light element, carbon, except when this element is present inside the specimen.

The analysis of insulating specimens is actually a serious source of problems. The main question is to know if the electrical field which subsists inside the specimen [13], in spite of the conductive coating, produces significant changes in the $\phi(\rho z)$

functions, able to modify strongly the X-ray emission, or not. To try to answer this question, it is essential to eliminate all the sources of errors, which would lead to erroneous conclusions. This paragraph comments a few aspects that need to be examined with care.

Influence of the Coating on the Emerging Intensity

The first point is that it may be a wrong assumption to consider that the influence of a given coating is the same for a given line, whatever the nature of the target. Actually, this is completely false in all the situations where the mass absorption coefficient of the line of interest is very high in the target and lower in the coating, for example when boron is analyzed in a Si-rich target coated with carbon or gold (μ/ρ B K in C ~ 6750; (μ/ρ B K in Au ~ 10400; μ/ρ B K in Si ~ 80000). In such situations, the emission of the coated target can be higher than that of the same, but uncoated, target (hypothetically conductive). For example, in the case of a borosilicate glass supposed to contain 10 wt. % B_2O_3 in SiO_2, one can calculate with *Strata* the values reported in Table 4. They show that there is a range of coating thickness (up to ~150 nm for C and ~15 nm for Au, i.e. up to ~30 µg/cm^2 for both coating materials) that produce an emission enhancement.

This surprising effect can be simply explained on the basis of the $\phi(\rho z)$ model. The intensity emerging from an homogeneous specimen is proportional to the integral of $\phi(\rho z)\cdot\exp(-\chi\cdot\rho z)\cdot d\rho z$, from $\rho z = 0$ to infinity. For the coated specimen, with the approximation that the $\phi(\rho z)$ function is not significantly modified, the emerging intensity is the integral of $\phi(\rho z)\cdot\exp[-\chi\cdot(\rho z - \rho z_c)]\cdot d\rho z$ from $\rho z = \rho z_c$ to infinity, multiplied by the factor $\exp(-\chi_c\cdot\rho z_c)$ to account for the absorption by the coating (mass thickness ρz_c and absorption factor χ_c for the line of interest). In the case of the borosilicate glass, the absorption of the B K line is very strong in the target, so that only the initial part of the $\phi(\rho z)$ function at 10 kV has actually some effect on the result (Fig. 15). For a simple calculation "by hand", one can take the approximation that in this region the slope S of $\phi(\rho z)$ is constant, so that the interesting part of $\phi(\rho z)$ can be expressed as $\phi(0) + S\cdot\rho z$. Here, $\phi(0)$ is so-called surface ionization.

With this approximation, the calculation of the intensity for the uncoated specimen gives

$$I_{\text{uncoated}} \sim [\phi(0) + S/\chi]/\chi$$

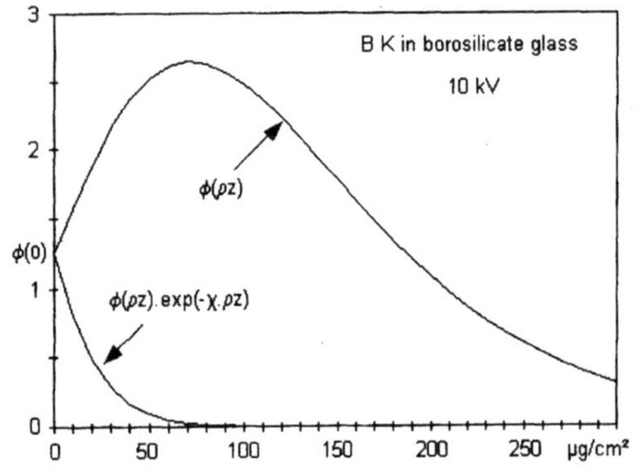

Fig. 15. Depth distributions of the generated and emerging B K intensities for a borosilicate glass at 10 kV. (*XPP* model, 40° take-off angle)

Table 4. Computed relative change in the B K intensity emerging from a borosilicate glass coated with carbon or gold, compared to the uncoated target (supposed to be conductive). 10 kV, 40° take-off angle, *Strata* with *XPP* model

C coating thickness (nm)	5	10	20	40	80	120	160
Relative change of intensity (%)	+0.5	+0.9	+1.7	+3	+3.9	+2.7	−0.4
Au coating thickness (nm)	1	2	4	8	12	16	
Relative change of intensity (%)	+3.3	+6	+9.9	+11	+6.6	−1.9	

Table 5. Result of the analysis of an olivine coated with carbon (4.5 μg/cm²). *Strata* program. All elements analyzed

Element	O	Mg	Si	Fe	Total
wt. %	48.1	29.1	19.3	9.1	105.6

For the coated specimen, one obtains

$$I_{\text{coated}} \sim \exp(-\chi_c \cdot \rho z_c) \cdot [\phi(0) + S(1/\chi + \rho z_c)]/\chi$$

Hence, if the quantify $1 + S \cdot \rho z_c/[\phi(0) + S/\chi]$ is greater than $1/\exp(-\chi_c \cdot \rho z_c)$, then I_{coated} is greater than I_{uncoated}. This condition can be realized only by the combination of a high absorption factor χ in the specimen, a lower absorption factor χ_c in the coating, and a moderate thickness of the coating. The higher effect reported in Table 4 for the gold coating than for the carbon coating results from the higher electron scattering power of gold, which increases the slope S.

Nature of the Coating

It is not sure that coating the insulating specimens (and mainly oxides) with carbon is the best procedure to adopt. We had already briefly mentioned [14], that we were obtaining better quantitative analyses for a SiO_2 specimen after coating it with aluminium than with carbon. New comparisons have been made, in the case of a couple of olivine specimens (having slightly different MgO and FeO fractions), the one coated with evaporated carbon, the other one with sputtered gold. The oxygen is analyzed with respect to a conductive $Y_3Fe_5O_{12}$ standard (uncoated), the other elements with pure standards. The program *Strata* is employed for the data reduction.

For the C-coated specimen (4.5 μg/cm² carbon, i.e. 22.5 nm assuming that the density equals 2), all the elements being analyzed at 10, 15, 20 and 25 kV, one obtains before normalization the unsatisfactory results of Table 5, with the sum of the concentrations close to 105%. It can be mentioned that a bad sum is also found if, instead of processing all the voltages together, one considers only the data at the lowest or at the highest voltage.

If oxygen is not measured, but obtained by stoichiometry, one gets the results of Table 6, showing that only the oxygen result changes significantly, and that the measurement of the other elements leads to a correct sum, close to 100%.

Table 6. Same as Table 5, but with oxygen determined by stoichiometry

Element	O	Mg	Si	Fe	Total
wt. %	43.6	28.7	19.4	9.1	100.8

Table 7. Results of the results of an olivine coated with gold (17.2 μg/cm²). *Strata* program. All elements analyzed

Element	O	Mg	Si	Fe	Total
wt. %	43.7	26.2	19.1	11.9	100.9

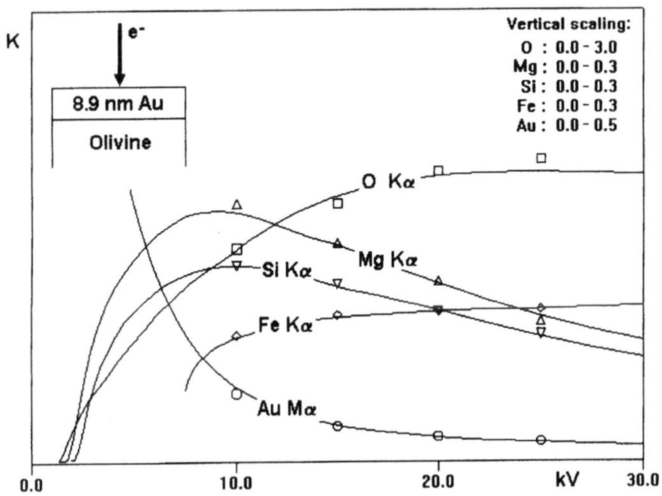

Fig. 16. Display of *Strata* relative intensity curves vs. accelerating voltage for a gold-coated olivine (~8.9 nm gold)

On the opposite, for the Au-coated specimen (17.2 μg/cm² Au, i.e. 8.9 nm assuming the nominal density), the results of Table 7 show that there is no problem with the sum when all the elements are analyzed. The *k*-ratio curves drawn for the specimen description resulting from the iterative procedure of *Strata* fit satisfactorily the experimental data (Fig. 16).

One of the reasons that could explain the excess of oxygen found for a carbon-coated specimen is that some oxygen could be incorporated into the coating during the carbon evaporation. This possibility has been checked by evaporating a similar carbon film on one half of a clean nickel substrate (the other half was hidden during the evaporation in order to deduce the natural amount of oxygen at the surface of nickel). An oxygen concentration of ~4% was actually found in the carbon film. However, such a concentration is quite insufficient to explain the excess of oxygen reported in Table 5: when 4% oxygen is incorporated into the 22.5 nm carbon layer, the result given by *Strata* for the oxygen in olivine decreases only slightly (47.9% instead of 48.1%). Hence, one has to admit that an evaporated carbon film with a thickness in the range of ~20 nm might not be conductive enough to ensure an efficient electrical grounding of the specimen surface. On the opposite, we have the feeling that a thin gold coating (a few nanometers thick) made by sputtering (to avoid

Table 8. Some first order lines interfering with ultra-light elements

Analyzed line	Energy (eV)	Interfering lines
Be K	109	Sr Mζ
B K	183	Nb Mζ
C K	277	K L*l*, η Mo Mζ
N K	392	Sc Lα, β Ti L*l* Sn Mζ
O K	525	V Lα, β Cr Lη Mn L*l*

the formation of islets and to obtain a better adherence to the substrate) might be the best choice for the analysis of insulating materials.

Line Interference in WDS Spectrometry

It is frequent in the soft X-ray region that the characteristic line of interest interferes with some lines of other elements. It is easy to eliminate the interference with high-order lines, by selecting an appropriate band width for the pulse height analyzer (PHA). It can be noted too that there is an advantage in using multilayer monochromators, which are optimized to strongly attenuate the 3n orders of diffraction. Hence, the problem of interference actually subsists for the first order lines. Table 8 gives a few examples of such lines interfering with the ultra-light elements. Fig. 17 shows the example of the interference of the oxygen K line with the chromium L lines.

The most recent computer programs for electron microprobes include the automatic processing of the interference. To make this possible, one should perform a preliminary measurement of an "interference standard" that emits the same interfering line as in the specimen (same wavelength and shape). With this assumption, two different approaches can be used:

Peak Acquisition

If the line of interest is not subject to detectable shape alterations by chemical effects, the intensity I_{peak} of this line can be measured at peak maximum. The intensity I_{int} of the "interference standard" is measured at the same position. In the *CAMECA* or in the *SAMx* microprobe automation and data processing software, the interference is taken into account at each loop of the iteration procedure which leads from the experimental relative intensities to "corrected" mass concentrations. At every loop, the concentration C_{int} of the interfering element (measured with another non-interfering line) is evaluated. Using this particular concentration and the concentrations of the other elements presents in the specimen, the contribution $C_{int} * I_{int} * ZAF$ of the interfering line to the line of interest in the specimen is computed by applying the atomic number, absorption and fluorescence corrections through the correction procedure (*PAP*, *XPP* or other).

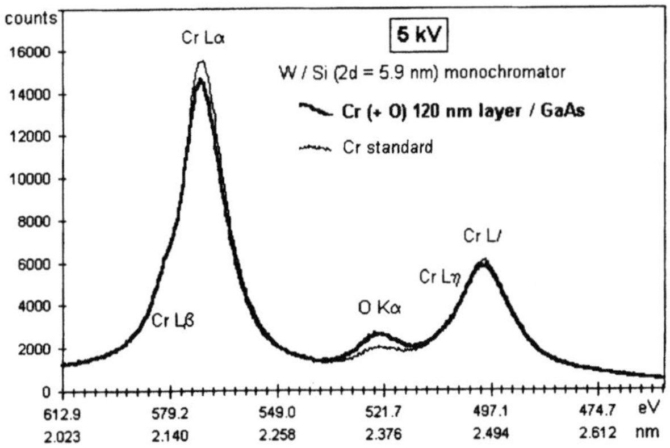

Fig. 17. Interference of the oxygen K line with the Cr L lines (W/Si multilayer monochromator 2d = 5.9 nm)

Integral Acquisition

In the software produced by *SAMx*, an integral mode is additionally available. In this mode, the line of interest, as well as the interfering line, are measured by scanning the spectrometer over a wide spectral region. This allows to apply the interference processing to those situations where the line of interest is shifted and/or distorted by chemical effects. Obviously, the "interfering standard" is also acquired in integral mode. Then, as in the previous mode, its intensity is incorporated in the correction procedure. It should be pointed out that in the present version of the software, the absorption coefficient applied to all the interfering band is that of the interfering line. This can be a limitation if a strong absorption jump is present in the integration band. For such cases, the software should be further improved by allocating the proper absorption coefficient to each channel of the spectrum.

Case of Stratified Specimens

In the case of stratified specimens, there is presently no program able to do the same as for bulk specimens (except in the case of a surface layer thicker than the analyzed depth, where one of the programs mentioned previously may be used). In the example of the Cr layer on GaAs, previously presented on Fig. 2 and Fig. 17, the oxygen contamination had to be determined. The only way to get the oxygen peak coming from the specimen was to deduce from the specimen spectrum the spectrum of a pure chromium standard, after scaling it down to superimpose the Cr Lα lines. The general problem with the analysis of oxygen is that it is almost impossible to get a standard with no oxygen or oxide at the surface. In the present case (Fig. 17), the spectrum of the standard was exhibiting an oxygen peak that could correspond to ~1 nm Cr_2O_3 oxide at the surface.

Background Subtraction in Low-Energy EDS Spectra

Bremsstrahlung and Actual Background

To succeed in performing quantitative analysis with soft X rays, one has firstly to overcome the problem of the background subtraction. Two different methods are available in the commercial programs: the quickest one is to apply over the spectrum a "top hat" numerical filter; the other one is to model the background theoretically. In the case of the *Kevex* system used in our laboratory, the filtering the low energy region numerically leads at the end to poor quantitative results. For example, in the case of SiC standardless analysis, the carbon concentration is overestimated by more than 20% relative.

If the theoretical background modeling is applied, the *Quantex +/XPP* procedure actually computes the emerging intensity of the "bremsstrahlung" radiation, taking into account its absorption inside the specimen and through the *Super Quantum* detector window. The problem is that in the low-energy region, even when the "optimize" option of the background computation is selected [15], the computed "bremsstrahlung" does not fit the actual background (Fig. 18), which contains other contributions, i.e. mainly the electronic noise of the detector, and a "spur" [16] with a tail extending to ~500 eV due to incomplete charge collection. This "spur" depends on the shape and on the intensity of the spectrum. It is obvious on the example of Fig. 18 that below the C K line, the predicted "bremsstrahlung" intensity is almost zero (because of the absorption), whereas the actual background represents at least 10% of the peak height. To overcome the problem, the simplest way is to use a standard emitting a spectrum similar to the specimen's one, but without any line in the region of interest. In the present example, the best choice is to record a spectrum on pure Si. Obviously, this spectrum (Fig. 19) exhibits the same problem as the SiC spectrum: its computed "bremsstrahlung" is lower than its actual background. Additionally, the Si reference spectrum exhibits small carbon and oxygen peaks corresponding respectively to carbon contamination and native oxide. After subtracting the "bremsstrahlung" from the Si spectrum and erasing all the channels above ~600 eV, one gets experimentally the dotted region (Fig. 19), consisting

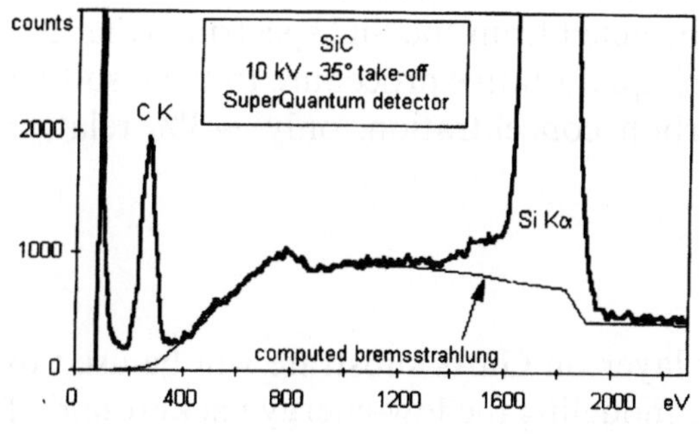

Fig. 18. Low-energy region of an EDS spectrum from SiC and from the computed bremsstrahlung

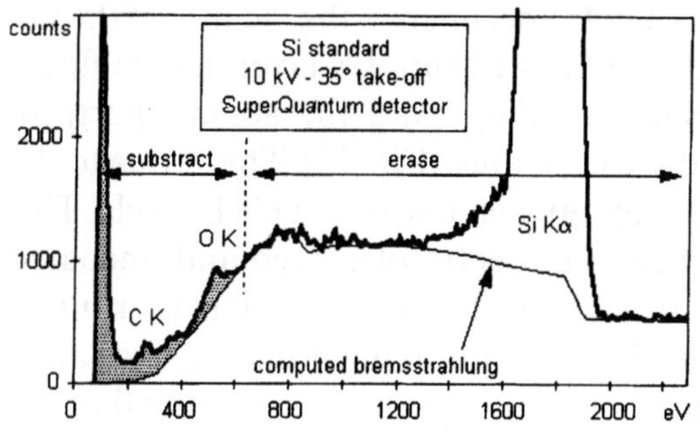

Fig. 19. Low-energy region of an EDS spectrum from a Si standard. The dotted region is the "spur"

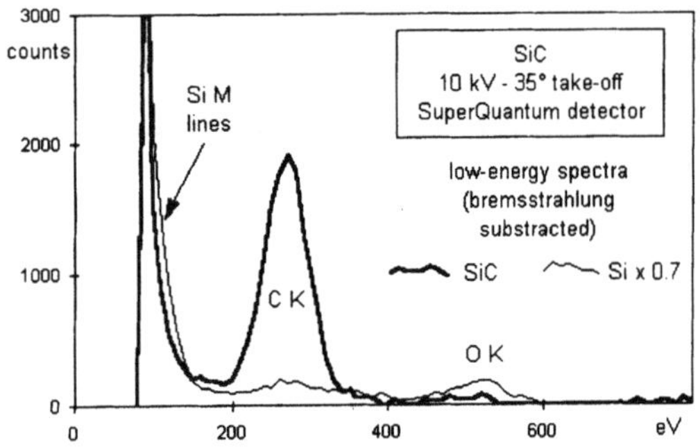

Fig. 20. "Spur" contribution (obtained from Fig. 19), to be subtracted from SiC spectrum

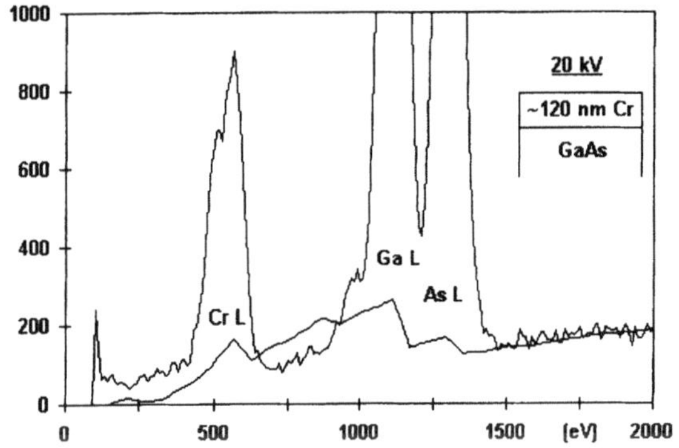

Fig. 21. Comparison of the actual low-energy EDS background for a Cr/GaAs stratified specimen with the background modeled in *Quantex*$^{+}$*XPP* assuming an homogeneous specimen

mainly of the "spur", plus the electronic noise, the surface imperfections and, in this particular case, some contribution of the Si L lines. The final step of the processing is to deduce from the SiC the proper fraction of this dotted region. Since the "spur" is the main constituent, one can take the ratio of the total number of counts in the SiC spectrum to the Si spectrum, i.e. ~70%. In this way, 70% of the carbon contamination present on the Si standard is also subtracted from the SiC spectrum (Fig. 20). Applying the standardless *Quantex* +/*XPP* quantitative procedure to the resulting SiC spectrum leads to a satisfactory carbon concentration, only ~3% relative higher than the nominal one.

Background in Stratified Specimens

In this paragraph, the example of the Cr layer on GaAs substrate will be used to illustrate the problems that may appear in modeling the low-energy background of stratified specimens. First, it should be noted that there is no problem above ~600 eV to model the background of a bulk GaAs specimen in the vicinity of the Ga and As L lines. But when AsGa is coated with a ~120 nm Cr layer, a significant disagreement between the computed "bremsstrahlung" and the actual spectrum appears in the region between the Ga L and the Cr L lines (Fig. 21). This is due to the strong absorption of the photons having an energy just above the Cr L levels. This absorption is clearly visible on Fig. 22, where the spectra of coated and uncoated GaAs are compared. At the present time, the practical problem is that no routine EDS procedure takes properly into account this absorption phenomena, since all of them perform the computation of the background absorption by assuming that the

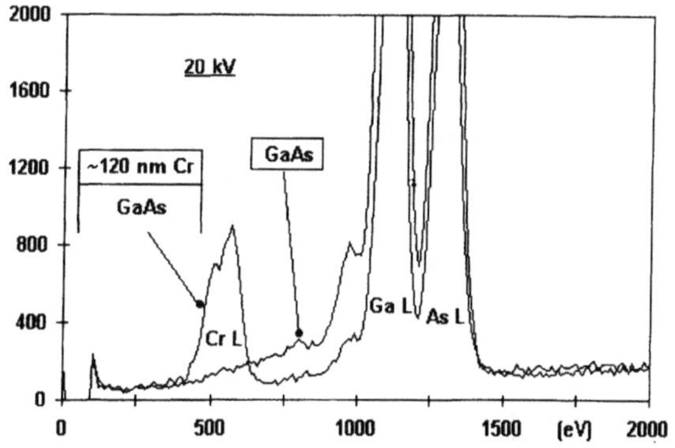

Fig. 22. Comparison of the backgrounds of uncoated and Cr-coated GaAs

elements are homogeneously distributed inside the specimen. Hence, the analyst has to be conscious of the problem that, in the present example, subtracting the background of Fig. 21 would lead to analyze Ga by default compared to As. The other alternative, which would lead to analyze Ga by excess, would be to subtract a linear background by interpolating from the low-energy side of the Ga line to the high-energy side of the As line. By doing so, the "true" value would be somewhere between the two results.

Regarding the background subtraction below the Cr L lines, it can be seen on Fig. 21 that taking a linear interpolation would be better than taking the theoretical "bremsstrahlung".

Conclusions

In the field of material science, and also in other fields (geology, biology,...), there is a growing interest for the microanalysis of ultra-light elements and for the measurement of soft X rays in general. An increasing number of problems require to adopt a "stratified specimen approach", either because of the actual structure of the specimen, or because it has to be coated to avoid charging effects. For applications to layered specimens, the use of soft X rays enables to obtain a reduced depth of ionization. For the microanalysis of multiphase specimens, it also permits to eliminate most of the secondary emission by fluorescence that occurs when lines of higher energy are employed.

A lot of difficulties subsist (contamination, coating, background subtraction, line interference, line distortion, absorption, ...), so that any experiment requires a cautious approach and a real understanding of the physical processes involved.

Acknowledgements. The author acknowledges D. Boivin and Y. Pioche for their contribution to the measurements, most of them specially made for this paper. The development of the program *Strata* and the implementation of the model *XPP* into the *Quantex* software were done in cooperation with SAMx, under contract *ONERA/SAMx* # 6361.

References

[1] J. L. Pouchou, F. Pichoir, *Electron Probe Quantitation*, Plenum, New York, 1991, p. 31.
[2] M. Fialin, G. Remond, C. Bonnelle, *Microbeam Analysis*, VCH, New York, 1994, p. 237.
[3] J. L. Pouchou, F. Pichoir, *J. Microsc. Spectrosc. Electron.* **1986**, *11*, 229.

[4] G. F. Bastin, H. J. M. Heijligers, *Scanning* **1990**, *12*, 225.
[5] J. L. Pouchou, F. Pichoir, *La Recherche Aérospatiale* (English edition), **1984**, *3*, 13.
[6] G. F. Bastin, H. J. M. Heijligers, *X-Ray Spectrometry* **1986**, *15*, 135.
[7] J. L. Pouchou, F. Pichoir, *J. Microsc. Spectrosc. Electron* **1985**, *10*, 291.
[8] J. A. Bearden, *X-Ray Wavelengths*, U.S. Atomic Energy Commission Publ., 1964.
[9] J. L. Pouchou, F. Pichoir, *Microbeam Analysis*, San Francisco Press, San Francisco, p. 319.
[10] K. F. J. Heinrich, *Proc. ICXOM 11*, Univ. Western Ontario Publ., 1987, p. 67.
[11] P. Willich, *Personal Communication*, 1992.
[12] J. L. Pouchou, *Mikrochim. Acta* **1994**, *114/115*, 33.
[13] J. Cazaux, *J. Appl. Phys.* **1986**, *59*, 1418.
[14] J. L. Pouchou, F. Pichoir, *Scanning Microscopy* **1993**, [*Suppl.* 7], 167.
[15] J. L. Pouchou, F. Pichoir, D. Boivin, *Microbeam Analysis*, San Francisco Press, San Francisco, 1990, p. 120.
[16] G. Love, V. D. Scott, A. O. Sandborg, *Microbeam Analysis*, San Francisco Press, San Francisco, 1984, p. 191.

Mikrochim. Acta [Suppl.] 13, 61–86 (1996)

Intensity Measurement of Wavelength Dispersive X-Ray Emission Bands: Applications to the Soft X-Ray Region

G. Remond[1,*], C. Gilles[1], M. Fialin[2], O. Rouer[3], R. Marinenko[4], R. Myklebust[4], and D. Newbury[4]

[1] BRGM, DR/PCI, BP 6009, 45060 Orleans Cedex 2, France, [2] Université P. et M. Curie, Paris, France, [3] CNRS-CRSCM, Orleans, France, and [4] NIST, Gaithersburg, MD, U.S.A.

Abstract. Line shapes of atomic lines and soft X-ray emission bands are described using pseudo-Voigt profiles accounting for the presence of non-diagram features (low and high energy satellites) and instrumental distortions. Peak shape changes of soft X-ray emission bands as a function of the excitation conditions and the matrix composition are related with self-absorption phenomena. Implications of spectral deconvolution of WDS X-ray intensity measurement for quantitative analysis of complex spectra and for studying valence states are reviewed.

Key words: line shape, satellites, instrumental distortion, valence states, self-absorption, spectral deconvolution.

Modern computer controlled electron probe micro analysers (EPMA), more frequently equipped with wavelength dispersive spectrometers (WDS) are used for repetitive analyses on a routine basis in a variety of metallic and non-metallic compounds. Recently, soft X-ray emission bands ($E < 1\,keV$) became detectable using layered-multi-structures (LMS) as monochromators and the nature and thickness of the superimposed light and heavy metals can be selected in order to optimise the detection sensitivity for specific soft X-ray emission bands. In practice, for the analysis of minerals containing many elements with very different atomic numbers, it is usual to use both *atomic X-ray lines* resulting from radiative transitions involving core electrons and soft *X-ray emission bands* resulting from radiative transitions involving valence electrons. The first reason for adding soft X-ray emission bands in an analytical procedure for routine analyses with the EPMA is to obtain information on the chemical environment of the analysed elements since the production of soft X-ray emission bands involves valence hole

* To whom correspondence should be addressed

states and these X-ray bands have their shape depending upon the electronic structure of the element within the matrix. The second reason for investigating the possible use of soft X-ray emission bands for the analysis of elements with intermediate atomic numbers is justified by the escape depth of the analysed X-ray photons. When analysing low atomic number elements such as carbon and oxygen the escape depth for the soft Kα emission bands of these elements is small and the analyses are sensitive to the surface composition of the specimens. In order to measure X-ray intensities originating from similar depths below the specimen surface for all elements it is tempting to measure the L emission spectra of elements of medium atomic number rather than the K X-ray spectra usually used for routine analyses. For many years it has been well accepted to use the peak height intensity measurement procedure to determine the intensity of a WDS X-ray peak measured with the EPMA. This procedure assumes that the shape of the analysed X-ray peak is symmetrical, i.e., the peak height remains proportional to the peak area. An X-ray line (or diagram line) resulting from a transition between two levels in the energy-level diagram is frequently accompanied by non-diagram lines (or "satellites") whose energies do not correspond to the difference of two energy levels. These "satellites" may occur either on the high or the low energy side of the diagram line and may contribute to the asymmetry of the energy distribution of the generated X-ray emission band. The observed line shape of an X-ray peak is the convolution product of the intrinsic (or natural) shape with the response function of the spectrometer and thus the width of the measured WDS emission peaks or bands is much broader than that of the natural width. In addition, peak shape alterations may result from instrumental distortions.

Although the peak height measurement procedure has been proved to be a sufficient application for quantitative analysis of mineral materials, this procedure is questionable when soft X-ray emission bands are concerned and peak area measurement should be preferred. Direct on-line peak integration has been proposed (Fialin et al. [1]). However, peak integration by summing the number of counts as a function of the channel number does not account for the origins of the mechanisms leading to the observed peak shape. Peak area measurement based on a least-squares fitting technique of the experimental data to an analytical description of the emission band can be used. For this purpose, the line shape of X-ray peaks measured with the usual monochromators of commercial EPMA has been investigated by Remond and Campbell [2] and by Remond et al. [3]. The final objectives of spectral deconvolution of WDS X-ray spectra are to i) generate synthetic spectra to be used as reference data, ii) resolve peak interferences and iii) use a unique approach for WDS and EDS spectra. *It is the purpose of this paper to illustrate the merits and limitations of spectral deconvolution of WDS X-ray spectra accounting for the presence of "satellites" and instrumental distortions both effects existing in different proportions as a function of the analysed X-ray emission lines.* The problems encountered for the case of atomic X-ray lines will be briefly mentioned to examine the more complex situation when soft K X-ray emission bands of low atomic number elements and soft L X-ray emission bands of the first transition-series elements are analysed simultaneously. The possibility of studying chemical environment and oxidation states from soft X-ray emission bands will be discussed in the case of iron bearing minerals. Changes in the O Kα and Fe Lα, β bands will be examined, since X-ray spectra changes as a function of the matrix composition and the excitation conditions.

1. Experimental

1.1. Instrumentation

The computer and microprocessor controlled microprobes (Cameca Camebax-Microbeam and SX 50 types respectively) equipped with WDS's and installed at the joint Bureau de Recherches Géologiques et Mineres (BRGM) – Centre National de la Recherche Scientifique (CNRS) laboratory were used. LiF and PET monochromators were used for the analysis of the high energy atomic X-ray emission lines. For the examples of soft X-ray emission bands, the Lα, β spectra of transition elements were analysed with a TAP monochromator. The O Kα emission band was measured with a W/Si LMS as monochromator. A Cameca microprobe installed in the Cameca application laboratory was also used. This EPMA is equipped with a WDS with improved mechanical displacement of the monochromator in order to scan high Bragg angles making it possible to measure the O Kα emission band with a TAP monochromator. For the case of the analysis of the soft Lα, β and O Kα spectra, the measurements were also repeated with the JEOL microprobe (8600 type) installed in the Surface and Microanalysis Science Division at the National Institute of Standards and Technology (NIST). All X-ray spectra were digitally recorded by stepping the monochromator. Each analysed wavelength domain contained 1024 channels.

1.2. Analytical Description and Processing of WDS X-Ray Peaks

The line shape of an observed WDS X-ray peak is intermediate between Gaussian and Lorentzian distributions. It is common to use a summation of analytical functions for describing the envelope of the observed X-ray peaks. A summation of two Gaussian distributions has already been used to describe K and L X-ray peaks as illustrated by Salem et al. [4] and by Salem and Scott [5]. The effectiveness of using a hybrid Gaussian and Lorentzian summation to describe any X-ray line shape has been demonstrated by Doyle and Shafroth [6], by Huang and Lim [7] and by Remond et al. [2, 3] among others.

The observed peak profile, P(λ), was described by a pseudo Voigt-function, i.e., a linear combination of a Gaussian and a Lorentzian distribution according to:

$$P(\lambda) = Cg\ G(\lambda) + C_1\ L(\lambda) \tag{1}$$

where $O < Cg < 1$ and $C_1 = 1 - Cg$ are the proportions of the Gaussian G(λ) and Lorentzian L(λ) of same width and centred at the same position.

The digitally recorded X-ray spectra were deconvoluted into pseudo-Voigt profiles, using Eq. [1] as fitting function. For describing asymmetrical peak shapes, Remond et al. [2, 3] showed that adding Gaussian or pseudo-Voigt offsets to the major pseudo-Voigt diagram line profile was a good enough approximation to describe a complex peak containing several discrete spectroscopic features, e.g., diagram and non-diagram contribution and possible instrumental effects. The continuous X-ray emission distribution was assumed to vary linearly below the analysed peaks.

For energy dispersive spectra (EDS) processing, it is usual to model the continuous emission distribution and to subtract the calculated distribution from the experimental data. Such a procedure allows us to account for the presence of absorption edges lying within the energy domain of the analysed characteristic X-ray emission peak. In order to correct the WDS spectra in the soft X-ray region for absorption effects, the WDS spectra were transformed into EDS spectra. For this purpose, the NIST-NIH Desktop Spectrum Analyser (DTSA) software [8] was used. The merit of the procedure used in the DTSA program for transferring a WDS spectrum into the energy space is that the width of each energy channel is constant while this quantity varies from a channel to another when the wavelength scale of a WDS spectrum is simply expressed in terms of $1/\lambda$. Let γ_i, be the intrinsic width of the energy distribution of the *generated* emission band within the target, Γ_i, the width of the energy distribution of the photons *emitted* at the specimen surface and Γ_{obs} the width of the *measured* spectrum. In absence of instrumental distortion and broadening ($\Gamma_{obs} \cong \Gamma_i$), and if Γ_1 is in the same order of magnitude of the intrinsic width γ_i, the experimental data can be, in a first approximation, corrected for absorption by multiplying the number of counts $N^{obs}(j)$ in each channel (j), by the absorption correction factor, $[1/f(\chi_j)]$ corresponding to the photon energy at each successive channel contained within the digitally measured spectrum. The absorption correction factor can be calculated applying a ZAF procedure at each channel energy within the analysed energy domain. In practice, such a calculation is difficult to apply and an easier approach was used consisting in deriving the absorption correction factors from the ratio of the intensities of two continuous emission distributions successively calculated when accounting or not for the presence of absorption jumps. The EDS background subtracted spectra derived from the experimental WDS data were also deconvoluted into several components according to the least squares-fitting technique using pseudo-Voigt profiles as fitting functions.

2. Atomic X-Ray Line Shapes

2.1. Asymmetry Resulting from Spectroscopic Effects

Changes in the peak shape as a function of the analysed radiative transitions are illustrated in Fig. 1 for the case of the S Kα, Nb Lα and Au Mα atomic emission lines measured with a PET monochromator [3]. The analysed emission lines have similar energies, i.e., 2308 eV, 2166 eV and 2123 eV respectively. The Bragg angles for the first order reflection with the PET crystal are located midway within the Bragg angle domain imposed by the mechanical limitations of the monochromator displacement within the spectrometer. As discussed by Remond et al. [3], for such conditions, distortions induced by instrumental factors are negligible and the asymmetry of the analysed peaks shown in Fig. 1 must be related with spectroscopic effects.

An X-ray emission peak consists of the normal (or diagram) line and non-diagram lines. The satellite lines have been intensively studied by Parratt [9] for the case of Kα lines and by Randall and Parratt [10] for the case of Lα lines. The origins of satellite lines or bands with very weak intensities and occurring either on the high or low energy side of the parent peak have been recently reviewed by Bonnelle [11].

The first class of satellites consists of weak features occurring on the high energy (short wavelength) side of the parent peak as shown in Fig. 1. The positions of the high energy offsets added to the diagram lines were found in good agreement with the calculated energy shifts resulting from multiple ionisation mechanisms. For the case of the Kα lines, when 1s and 2p vacancies are created simultaneously, the inner vacancy de-excites in presence of an outer hole which produces a change in the electrostatic potential distribution leading to shifts in energy as shown in Fig. 1a for the case of the S Kα emission peak. The energy separation distances of the satellite lines from their Kα parent line were found to linearly increase when the atomic number of the emitter also increased. As shown in Fig. 1b, c for the case of the Nb Lα and Au Mα emission lines, high energy offsets were also added to the pseudo-Voigt profiles of the diagram peaks. For L emission peaks of elements $36 < Z < 56$, the high energy satellite positions were consistent with calculated energy shifts resulting from outer vacancies. These additional vacancy states arise primarily from Coster-Kronig transitions, i.e. an Auger mechanism within the L subshells. These Coster-Kronig transitions are not permitted for all elements and for elements $36 < Z < 56$,

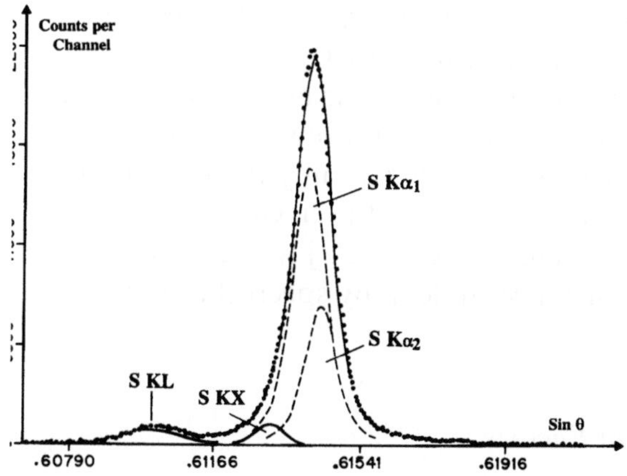

a

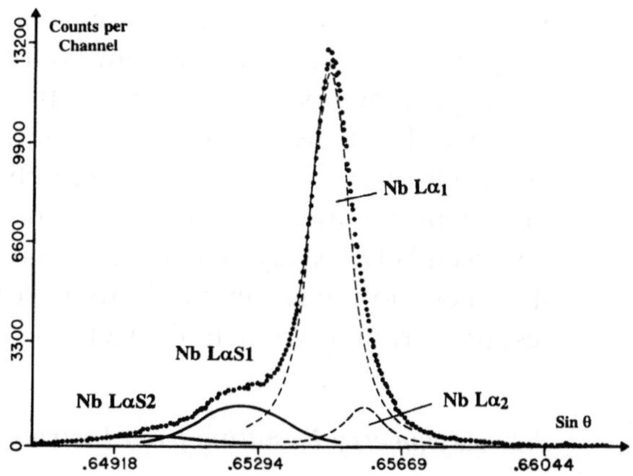

b

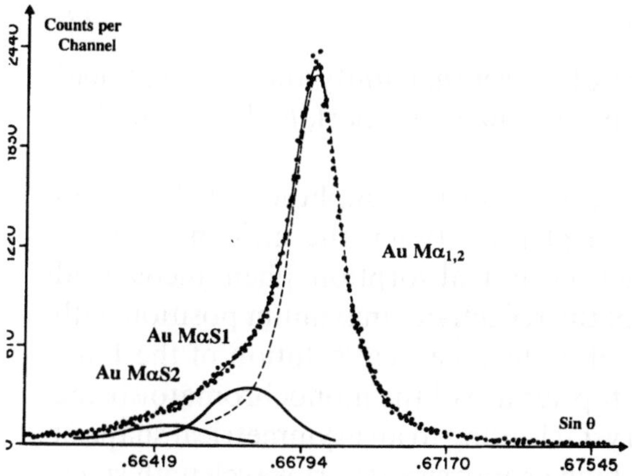

c

Fig. 1. Results of fitting to the S Kα (**a**), Nb Lα (**b**) and Au Mα (**c**) peaks. All have similar energies and were analysed with a PET monochromator (from Remond and Campbell [2])

the Lα satellite to diagram intensity ratio exhibits a sharp decrease for $Z = 49$. For the case of Lα emission lines of heavy elements such as Au, Pt, etc, with a LiF monochromator, it was not possible to measure the contribution of high energy satellites resulting from multiple ionisations because they occur within an energy domain where the high energy side of the peaks are distorted by instrumental effects as will be discussed below. A Coster-Kronig mechanism within the M subshells has also been associated with the high energy satellites as illustrated in Fig. 1c for the case of the Au Mα emission line.

The second class of satellites consists of features occurring on the low energy side of the peaks. They result from the simultaneous emission of an X-ray photon and a plasmon or an electron. When an X-ray photon and an outer electron are ejected simultaneously, the resulting low energy satellites are frequently referred as "radiative Auger emission" (RAE). According to the plasmon oscillation theory, during the emission process, the transition valence electron excites a plasmon in the valence band. The transition energy is thus shared between the plasmon and the emitting photon which will be deprived on an energy equal to the plasmon energy. Such a mechanism has been shown by Srivastava et al. [12, 13] to explain the low energy satellite to the Kβ line of transition elements and to the L X-ray spectra of rare-earth elements. Low energy satellites were measured with the EPMA by Remond et al. [14] to the $L\beta_2$, $L\gamma_1$ and $L\gamma_2$ lines of rare-earth elements (REE).

Depending upon the origins of the non-diagram lines, their energy separation distances from their parent diagram lines may range from a few eV to of a few hundreds of eV. They consist of closely spaced structures but only the *envelope* of these features are measured with the EPMA leading to the asymmetry of the observed WDS X-ray peaks as illustrated by Remond et al. [2, 3] who also showed that these non-diagram bands are in some instances hidden by spectral distortions resulting from instrumental effects.

2.2. Asymmetry Resulting from Instrumental Factors

For X-ray frequencies, the refractive index n* is a complex number:

$$n^* = n - i\beta \tag{2}$$

The real part of the refractive index, n, is slightly lower than unity and the decrement d equal to $1 - n$ characterises the dispersion. The imaginary part, β of the refractive index is the extinction coefficient.

The extinction coefficient results in a decrease of the amplitude of the waves passing through the crystal and changes of phase between the incident and the successively reflected waves. These refraction and absorption phenomena lead simultaneously to: i) a small angular shift of the reflectivity maximum position with respect to that derived from the Bragg relation (geometrical solution of the Laue conditions), and ii) a Bragg angle near the top surface of the monochromator being slightly smaller than that deeper into the crystal leading to an asymmetrical shape of the diffracted X-ray peak with a tail on its high energy side (short wavelength side).

The wavelength dispersive spectrometer (WDS) used with the EPMA are operated according to the symmetrical Bragg case, i.e., the incident and reflected rays are located on the same side of the monochromator surface, the reflecting planes (hkl) being parallel to its surface. More often, the monochromator mounting is of the Johann type. According to this mounting, a flat crystal is cylindrically bent to twice the focal circle radius. The response function (or diffraction pattern) from a curved crystal can be derived from that for a flat crystal as discussed by Cauchois and Bonnelle [15] and by Berreman [16, 17] using the dynamical theory of X-ray diffraction. For a short radius of curvature (as for the case of most commercial EPMA), the angle of reflection will change appreciably as a function of depth below the crystal surface leading to an instrumental distortion resulting from an approximately exponentially decreasing tail occurring on the high energy side of the peak. This distortion is particularly marked for low incident Bragg angles and becomes less pronounced for Bragg angles greater than about 30°. The effect of instrumental distortion on the line shape of an X-ray peak is shown in Fig. 2 and Fig. 3 comparing the first (Fig. 2) and second (Fig. 3) order reflections of the Au $L\alpha$ emission analysed with a LiF monochromator. In order to avoid high energy satellite bands resulting from the contribution of Coster-Kronig transitions the incident energy was adjusted to 12.5 keV, a value just above the L_3 excitation threshold (11.9 keV) and below the excitation threshold of the L_1 subshell (14.3 keV). With the LiF (200) monochromator the Bragg angle for the Au $L\alpha$ peak is 18°5 and the asymmetry due to the

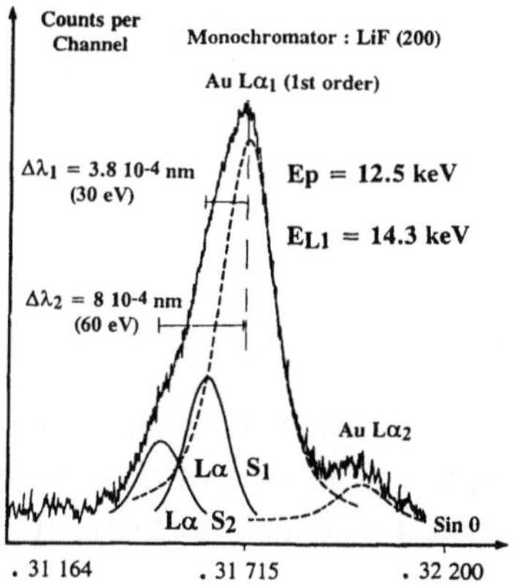

Fig. 2. First order reflection of the Au $L\alpha_{1,2}$ emission band analysed with a LiF monochromator and a 12.5 keV incident energy just above the Au L_3 excitation and threshold and below the L_1 excitation threshold (from Remond et al. [3])

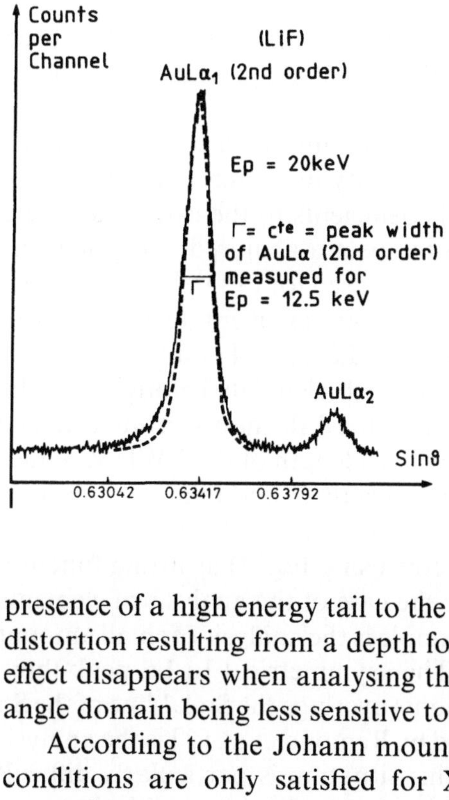

Fig. 3. Second order reflection of the Au $L\alpha_{1,2}$ emission bands analysed with a LiF monochromator, and a 20 keV incident energy (from Remond et al. [3])

presence of a high energy tail to the Au Lα peak in Fig. 2 is mostly an instrumental distortion resulting from a depth focusing defect within the monochromator. This effect disappears when analysing the second order reflection occurring in a Bragg angle domain being less sensitive to focusing defect as shown in Fig. 3.

According to the Johann mounting of the bent monochromator, the focusing conditions are only satisfied for X-ray beams incident at the "centre" of the monochromator, i.e., the point where the focal circle is tangent to the crystal surface. A deviation from the focusing conditions will increase the farther the incident beam is from the "centre" of the monochromator. Away from the "centre" of the crystal, the small distance between its surface and the focal circle will give rise to a focusing defect resulting in a broadening and a decrease in intensity of the observed X-ray peak. An example of spectral distortion resulting from focusing defect along the focal circle has been given by Remond et al. [3] comparing the Au Lα peak shape changes as a function of the dimensions of the active area of the LiF monochromator.

The focusing defect along the focal circle will combine with peak shape alterations resulting from the focusing defect in the depth direction of the crystal. In addition some instrumental distortions may occur at some specific wavelengths. Anomalous intensity of the continuous emission has been attributed to multiple reflections on differently oriented crystallographic planes below the monochromator surface (Self et al. [18]). If an X-ray beam makes an angle θ with the (hkl) planes and an angle θ' with the (h'k'l') planes, the same wavelength will be diffracted when:

$$2d(hkl)\sin\theta = 2d(h'k'l')\sin\theta' \qquad (3)$$

In such a situation, diffraction by the (h'k'l') planes will cause a decrease in the intensity of the primary beam to be diffracted by the (hkl) planes. The positions of the intensity distribution of the X-ray emission analysed with a LiF monochromator were calculated by Self et al. [18].

2.3. Implications in Quantitative Analysis of Complex X-Ray Spectra

The use of spectral deconvolution procedures for quantitative analyses has been illustrated by Remond et al. [3, 14, 19]. Some of the potential applications are listed below.

REE bearing compounds exhibit numerous peak interferences between the L emission spectra of the elements. To obtain intensity data free of interferences, a common approach is to apply predetermined coefficients to the raw data. These overlap correction factors can only be applied to spectra acquired with the same instrumental and experimental conditions than those used for obtaining the correction factors. Peak deconvolution based on a least-squares fitting technique is of a more general application. For this purpose, Remond et al. [14] built a data base describing the L X-ray emission lines of the rare-earth elements belonging to the lanthanide element series ($57 \leqslant Z \leqslant 71$) analysed with a LiF monochromator. The data base includes the variations as a function of wavelength of the FWHM values for the Lα, Lβ_1, Lβ_3 and L1 peaks, the separation distances and relative intensities for the low energy satellites associated with the parent diagram lines.

Spectral deconvolution of digital WDS spectra using Eq. (1) as fitting function has also been applied by Remond et al. [3] for the case of the analysis of elements present at low concentration levels. The first example is the analysis of small amount of indium (In) in cassiterite (SnO_2) minerals. The asymmetrical In Lα peak (high energy satellites resulting from multiple ionisations) overlaps the Sn L1 line of the Sn major constituent. The second example reported by Remond et al. [3] is the analysis of gold at trace levels using the Au Lα line. This situation is particularly difficult because two spectral distortions occur simultaneously, making impossible the use of the usual peak height intensity measurement. The presence of a "hole" in the continuous emission distribution is very close to the maximum position of the Au Lα peak and, modifies the shape and position of that peak when analysing gold present at low concentration levels with respect to the Au Lα emission measured at the surface of a pure gold specimen as shown in Fig. 4. The effect of the line shape changes resulting from a combination of instrumental distortions on the accuracy of the analysis of gold at trace levels has been discussed by Remond et al. [3, 19]. These authors showed that spectral deconvolution must be applied rather than direct on-line peak integration when the line shape is affected by instrumental distortions.

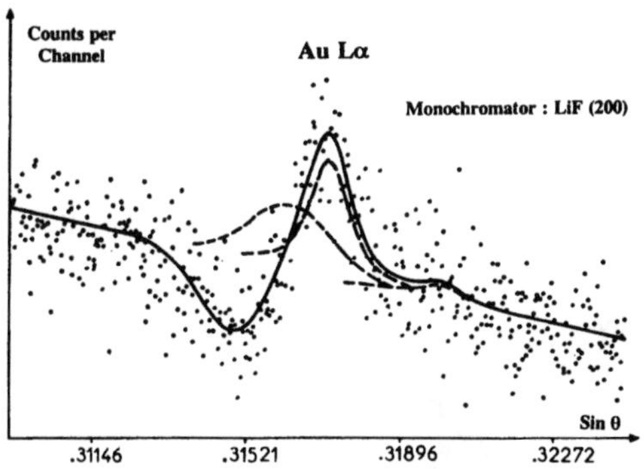

Fig. 4. Au Lα emission peak (LiF monochromator) measured at the surface of gold ion implanted arsenopyrite (AsFeS) specimen. Result of fit including the pseudo-Voigt profile for the Au Lα diagram peak, a Gaussian offset representing the focusing defect due to the monochromator, and a Gaussian offset of negative amplitude representing the "hole" in the background (from Remond et al. [3])

3. Soft X-Ray Emission Bands

3.1. K Emission Bands of Low Atomic Number Elements

All considerations mentioned above for the case of atomic lines also apply to the soft X-ray emission bands. High energy satellites resulting from multiple ionisations are observed when analysing K emission bands of low atomic number elements. Such high energy satellites have been reported by Fisher [20], Fisher and Baun [21] and by Fialin et al. [22] for the case of the F Kα emission band for LiF, topaz ($Al_2SiO_4(F,OH)_2$), and fluorite (CaF_2) specimens. The high energy satellite of 3.4 eV distance from the F Kα peak for the CaF_2 specimen is attributed to the radiative de-excitation of F^- ions in the presence of one vacancy in the L shells, created at the same time as the initial K hole. The atom is thus doubly ionised, leading to the $K\alpha_{3,4}$ satellite. A weaker satellite distant by 8.8 eV from the parent peak is also observed and is due to the presence of two additional vacancies on the L levels and is generally labelled as $K\alpha_{5,6}$. These two high energy satellites are frequently referred to as shake-off peaks. The distance of the high energy satellites from the Kα parent peak varies as a function of the F bearing compound.

The Kα intensity emitted from low Z elements in solids depends on the property of the valence band, i.e. the density of occupied valence states and the energy width of the band. These parameters influence not only both the peak position and shape but also the rate of photon generation (Fialin et al. [23]). The complex shape of the K soft X-ray emission band is shown in Fig. 5 for the case of the O Kα band measured at the surface of CuO and Cu_2O specimens analysed with a TAP monochromator. For the case of the CuO spectrum, a low energy feature centered at 5.2 eV from the main peak is related to the valence band and is not observed for the case of the Cu_2O specimen. High satellites are observed in both spectra but only the peak labelled $K\alpha_{3,4}$ on the Cu_2O spectrum results from shake-off mechanisms while the weaker band labelled $K\alpha''$ results from instrumental defects. The $K\alpha''$ feature results from interactions of the emitted X-ray photons with the monochromator as demonstrated by Bodeur and Barchewitz [24] and illustrated by Fialin et al. [23] for the case of the O Kα emission band analysed with a TAP and a KAP monochromator respectively. An intense high energy band is observed on the O Kα band when analysed with a KAP monochromator while this feature is strongly attenuated when a TAP monochromator is used. These features result from reflectivity anomalies of monochromators containing oxygen under Bragg conditions in the region of the O K ab-

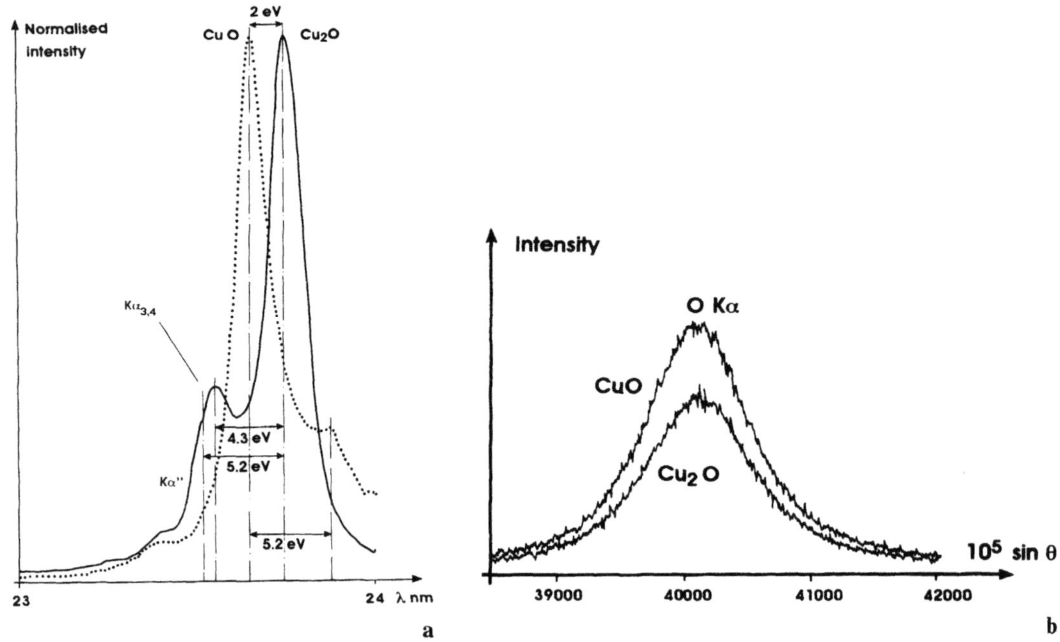

Fig. 5. O Kα emission band from a CuO and a Cu_2O specimen measured with a TAP monochromator (**a**) and with a W/Si LMS monochromator (**b**)

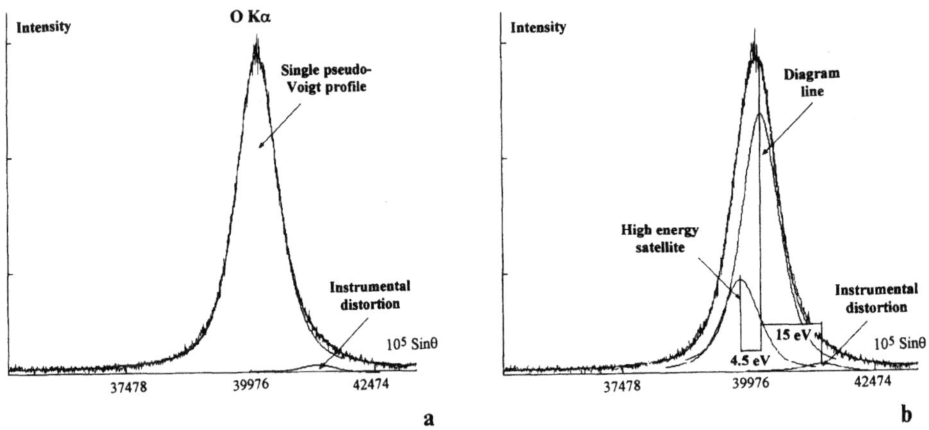

Fig. 6. a Result of fit of the O Kα emission band measured with a W/Si multilayer structure at the surface of a Cu_2O specimen to a single pseudo-Voigt profile (diagram peak) and a low energy pseudo-Voigt offset (instrumental distortion) and (**b**): spectral decomposition of the O Kα band for the Cu_2O specimen analysed with the W/Si monochromator accounting for the high energy feature distant by 4.3 eV from the diagram peak as shown on the O Kα sectrum analysed with the TAP monochromator (see Fig. 5a)

sorption edge. One advantage of the energy resolution of the TAP monochromator is to resolve the fine structure of the soft O Kα emission band. However, because of the low detection efficiency of the TAP monochromator in the soft energy region, the use of LMS is often preferred allowing us short data acquisition time for on-line peak integration as discussed by Fialin et al. [1]. The O Kα emission bands for the CuO and Cu_2O specimens analysed with a W/Si LMS monochromator are shown in Fig. 5b. Spectral deconvolution did not reveal significant changes for the width and the maximum position of the O Kα emission band for the CuO and Cu_2O specimens. The full-width at half-maximum (FWHM) was 13 eV, a value twice the distance between the main peak and the high energy satellite observed on the O Kα band for the Cu_2O specimen (see Fig. 6a). By varying the Cg value in the equation (1) it was not possible to correctly describe the O Kα band with a single pseudo-Voigt profile and an offset was added to the low energy side (long wavelength side) of the band. This offset, distant by 15 eV from the major peak probably, expresses an instrumental distortion resulting from slight imperfections of the monochromator such as surface roughness or thickness variations of the superimposed layers. This effect has been already observed by Remond et al. [25] studying the O Kα emission band measured at the surface of a variety of oxides analysed with several W/Si monochromators. Using some W/Si monochromators, the O Kα band was symmetrical and for some measurements performed with other W/Si monochromators, a low energy offset was added to correctly describe the peak shape.

The O Kα band for the Cu_2O specimen analysed with the W/Si monochromator was deconvoluted into the sum of pseudo-Voigt profiles accounting for the fine structure shown with the TAP monochromator (see Fig. 5a). The Kα″ high energy satellite resulting from the TAP monochromator was neglected in the data processing of the O Kα band analysed with the LMS monochromator. A high energy pseudo-Voigt profile was added to the main profile. The energy separation (4.3 eV) and the intensity ratio of the two pseudo-Voigt distributions (estimated from the analogue recording of the spectrum with the TAP monochromator), were kept as coupled variables in the fitting procedure. A low energy feature representing the W/Si instrumental distortion was also added into the fitting routine. Result of fit is shown in Fig. 6b. Adding a low energy satellite tends to shift by 1.3 eV the maximum position of the O Kα diagram peak with respect to the observed peak maximum. Spectral deconvolution of a broad peak analysed with a LMS monochromator into several spectroscopic features is only feasible when some parameters can be predetermined and used as coupled variables in the fitting procedure, i.e., when a physical parameter data base is available to predict the position and relative intensities of the diagram and non-diagram components or when the emission band can be analysed by two different monochromators. This situation has been illustrated by Remond et al. [26, 27] for the case of the O Kα band interfering with the Zn Lα, β second reflection order. The Zn Lα, β emission bands are accompanied with high energy satellites whose amplitudes with respect to those of the diagram peaks vary as a function of Zn bearing specimen composition. The least squares fitting procedure can only be applied to the spectral deconvolution of the overlapping O Kα/Zn Lα, β bands if the Zn Lα, β for the same compound has been independently analysed with a TAP monochromator, as illustrated by Remond et al. [26].

3.2. *L Emission Spectra of Transition Elements*

As reviewed by Fabian [28], the shape of the Lα, β spectra of the first transition-series elements change as a function of the excitation conditions. The three following

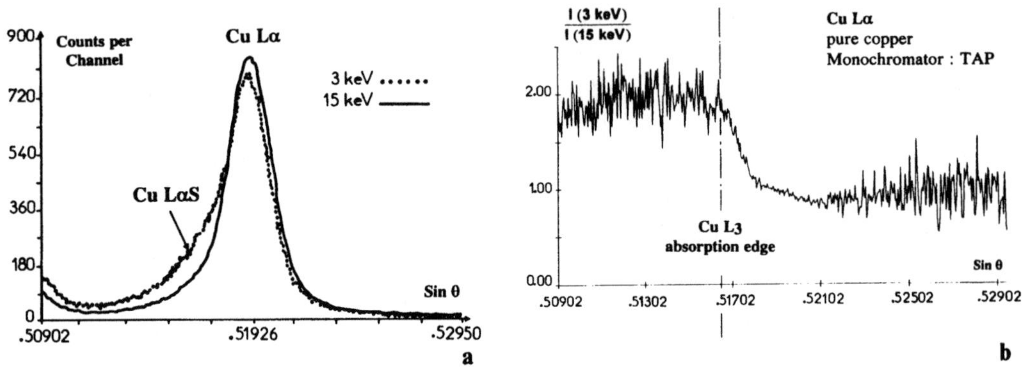

Fig. 7. Cu Lα, β emission bands for a pure copper specimen analysed with a 3 keV and a 15 keV incident energy successively (**a**). The intensities of both spectra measured with a TAP monochromator are normalised and the relative intensities I(3 keV)/I(15 keV) are plotted as a function of wavelength (**b**)

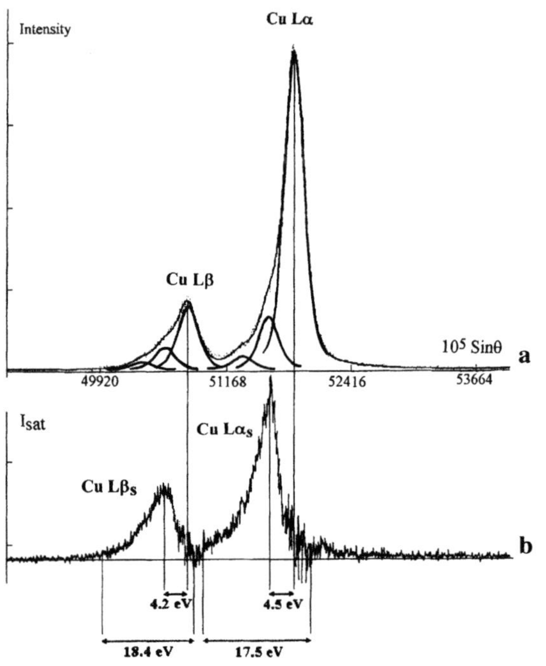

Fig. 8. Spectral decomposition into pseudo-Voigt profiles of the Cu Lα, β spectrum measured with a 5 keV incident energy (**a**). The asymmetrical high energy features (**b**) to the Cu Lα and Cu Lβ diagram peaks result from multiple ionisations

excitation regions must be considered when studying the shape of soft X-ray emission bands: i) the threshold excitation region when the incident energy lies between the L_3 and L_2 levels, a condition reducing the multiple vacancy satellites, ii) the satellite region when the incident energy increases from the L_3 threshold up to about three times that value leading to a progressively distorted spectrum and iii) the self-absorption region when the incident energy is greater than about 3 or 4 times the L_3 excitation threshold, the satellite structure vanishes due to self-absorption.

The effect of line shape changes of soft X-ray emission bands as a function of the incident energy is shown in Fig. 7 for the case of the Cu, Lα, β band analysed with a TAP monochromator and measured with a 3 keV and a 15 keV incident energy successively. The experimental spectra were processed using pseudo-Voigt profiles (Eq. (1) as fitting function). The best description of the high energy side of the Lα peak assumed to be free of non-diagram emission and of instrumental distortions was obtained for a Cg Gaussian to Lorentzian ratio in Eq. (1) equal to 0.6. As shown in Fig. 8 the residuals between the experimental spectrum measured for a 5 keV energy and the theoretical pseudo-Voigt profiles of the Cu Lα and Cu Lβ diagram peaks showed high energy satellites labelled Cu Lα_S and Cu Lβ_S. These satellites bands are asymmetrical. As shown in Fig. 8 the Cu, Lα, β of a pure copper specimen analysed with a 5 keV incident energy is correctly described by adding two offsets on the high energy side of each pseudo-Voigt profile representing the main Cu Lα and Cu Lβ diagram peak respectively. In order to reduce the number of variables in the fitting procedure the pseudo-Voigt profiles of the stellite bands were kept with the same width and same Cg values as those used in the pseudo-Voigt profiles describing the diagram bands. These high energy offsets are the envelope of satellites lines which were measured by Bonnelle [29] to be distant by 1 eV, 2.2 eV and 3.9 eV successively from the main Cu Lα peak and by 2.2 eV and 4.2 eV from the Cu Lβ peak for pure metallic copper analysed with a 2.5 keV incident energy.

Results of fit of the Cu Lα, β emission spectrum of the pure metallic specimen are given in Table 1 showing that both the intensity ratios between the high energy non-diagram bands to the diagram peaks and the Lβ/Lα intensity ratios decreased when the incident energy was increased from 5 keV to 15 keV. The FWHM values we measured with a TAP monochromator are in the same order of magnitude as those reported by Bonnelle [29]. Using a mica crystal as monochromator, Bonnelle [29] showed that the instrumental broadening of the peaks was negligible and the widths were measured 3.8 eV and 3.2 eV when the incident energy was decreased from 2.5 keV to 1 keV. Comparing emission and absorption spectra Bonnelle [29] reported that the changes in the widths of the Cu Lα and Cu Lβ diagram peaks as a function of the excitation conditions, resulted from the changes in the intensity ratios between the high energy non-diagram bands and the diagram emission peaks. Simultaneously, when a 1 keV energy was used, Bonnelle [29] also reported that the satellite lines were no more observable and that the Cu Lα peak position was moved by 0.3 eV towards the lower energies with respect to that observed for the case of a 2.5 keV energy.

In the present study, the shape of the Cu Lα, β spectra was also found to vary when the incident energy was increased from 5 to 15 keV. These two values are much higher than the L_1 excitation threshold and Coster-Kronig transitions are thus

Table 1. Results of fit of Cu Lα, β emission spectra for a pure copper specimen measured with a 5 keV and 15 keV incident energy successively

Ep keV	FWHM eV	I[Lα_S]/ I[Lα + Lα_S]	I[Lβ_S]/ I[Lβ + Lβ_S]	I[Lβ + β_S]/ I[Lα + Lα_S]	I[Lβ]/ I[Lα + Lβ]
5%	4.4%	17.5%	32%	24.4%	20%
15%	4.2%	5.9%	21.7%	12.8%	11%

permitted. The decrease of the satellite to diagram line intensity ratios shown in Table 1 will be related with self-absorption effects.

According to Fabian [28], a self-absorption spectrum is the ratio of two spectra measured for high and low incident energies. The two Cu Lα, β spectrum acquired with a 3 keV and 15 keV incident energy successively were normalised. The intensity ratio I (3 keV)/I (15 keV) as a function of wavelength in Fig. 7b, shows a jump around the Cu L_3 edge. The L_3 absorption edge for copper lies between the Cu Lα emission peak and the maximum intensity of the high energy satellite band. A differential absorption by the matrix will affect the generated Cu Lα diagram and non-diagram bands. Increasing the absorption path-length by increasing the incident energy will lead to higher absorption effect for the satellite photons than for the Cu Lα photons and the measured satellite to diagram intensity ratio will decrease when the incident energy is increased leading to line shape changes as we observed.

The effect of self-absorption on peak shifts was shown by comparing experimental data with those corrected for absorption using calculated absorption factors. Based on the consistency between the FWHM values we measured in the present study and those reported by Bonnelle [29], it is reasonable to assume that the observed width Γ_{obs} of the measured X-ray emission bands is close to the intrinsic width. The WDS X-ray spectra were transformed in energy dispersive spectra. The energies of the maximum positions for the Cu Lα, β bands derived from the initial WDS data were found 0.8 eV lower than the nominal values stored into the data base of the DTSA program. The EDS spectrum was corrected for that shift so that the Cu Lα, β energy positions corresponded to the theoretical values. The continuous emission distribution underlying the Cu Lα, β emission bands was modelled accounting for the presence of the L_2 and L_3 edges. These calculations were repeated by omitting the presence of these absorption jumps and the two distributions were normalised and divided each other in order to obtain the absorption correction factor $1/f(\chi_j)$ at each channel (j) of the energy spectrum as illustrated in Fig. 9. The absorption correction curve was applied to the energy Cu Lα, β spectrum leading to an absorption free spectrum as shown in Fig. 10. As expected, the relative Lβ/Lα intensities and the maximum positions were different between the raw spectrum and

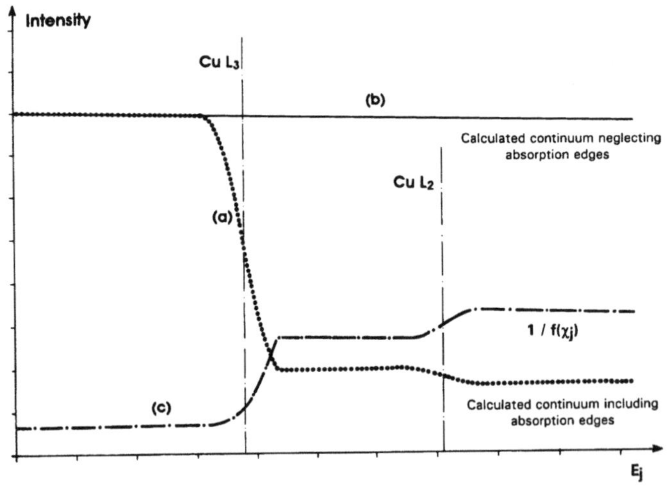

Fig. 9. Continuous emission distribution as a function of photon energies within the energy region containing the Cu Lα, β emissions for a pure copper: *a* calculations accounting for the absorption jumps at the L_2 and L_3 edges, *b* calculations neglecting the absorption edges and, *c* absorption correction factors for the photon energies Ej given by the ratio of the two calculated *a* and *b* continuous emission distributions

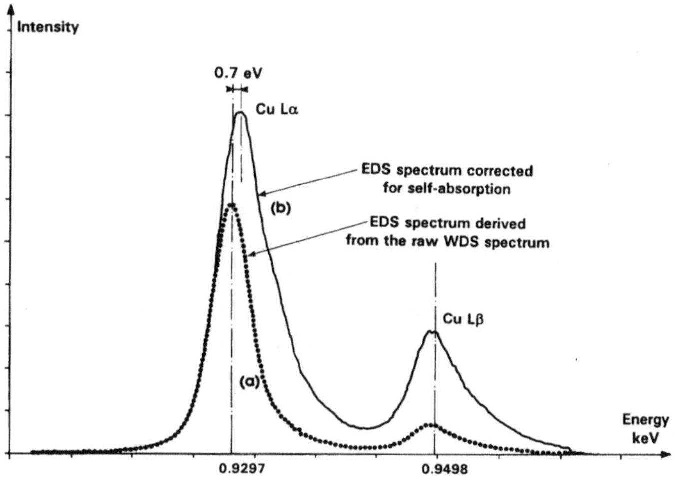

Fig. 10. Effect of self-absorption on the Cu Lα, β emission spectrum of pure copper: *a* The measured WDS spectrum is transformed into an energy dispersive spectrum (EDS) of a constant energy width between two successive channels and *b* Cu Lα, β EDS spectrum corrected for self-absorption by applying the 1/f(χj) correction factors at each channel j of the experimental spectrum

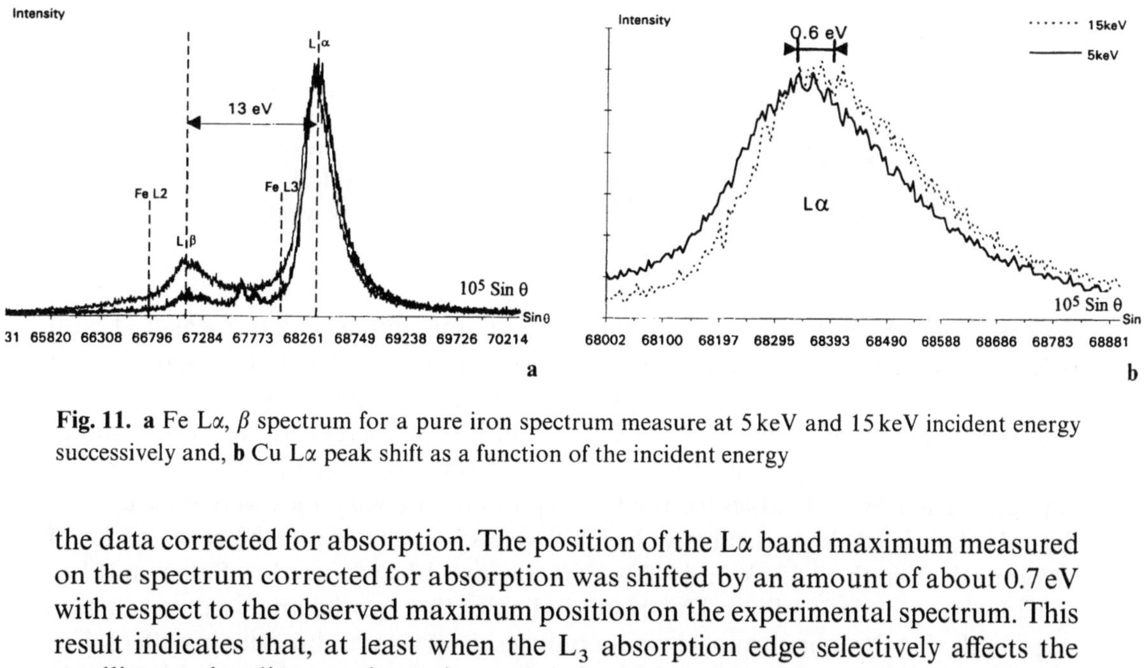

Fig. 11. **a** Fe Lα, β spectrum for a pure iron spectrum measure at 5 keV and 15 keV incident energy successively and, **b** Cu Lα peak shift as a function of the incident energy

the data corrected for absorption. The position of the Lα band maximum measured on the spectrum corrected for absorption was shifted by an amount of about 0.7 eV with respect to the observed maximum position on the experimental spectrum. This result indicates that, at least when the L_3 absorption edge selectively affects the satellite to the diagram intensity ratio, the absorption effect tends to move the maximum position towards lower energies (longer wavelengths). The 0.7 eV shift in energy of the maximum position of the Cu Lα emission peak resulting from absorption effect is consistent with that we observed on the initial WDS spectrum with respect to the theoretical value indicating that the deviation between the measured and nominal positions resulted from absorption effect rather than from an inaccurate calibration of the WDS spectrometer.

The shape of the Fe Lα, β WDS emission spectrum for a pure metallic iron specimen measured with a 5 keV and 15 keV incident energy successively are shown in Fig. 11a. The two Lα, β peaks are asymmetrical and are separated by an intense continuum intensity distribution. The measured peak positions were shifted by 0.6 eV towards the shorter wavelength when the incident energy was decreased from

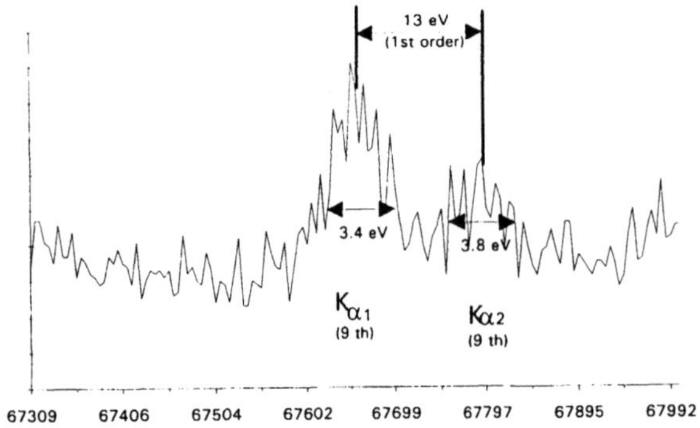

Fig. 12. Ninth Fe $K\alpha_{1,2}$ reflection order occurring between the Fe $L\alpha$ and Fe $L\beta$ emission bands for a pure iron specimen analysed with a TAP monochromator and a 15 keV incident energy

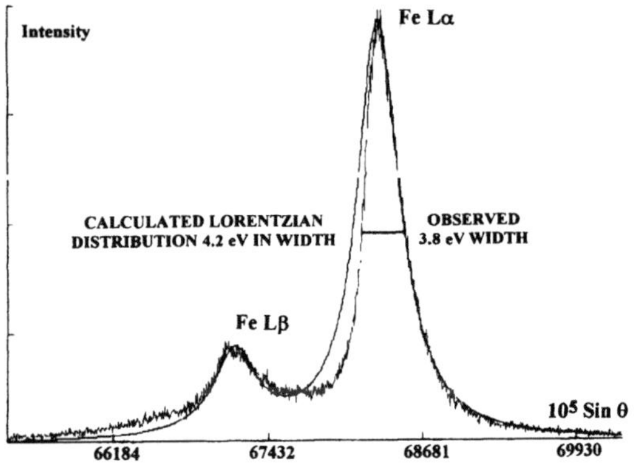

Fig. 13. Lorentzian shape of the long wavelength side (low energy side) of the Fe $L\alpha$, β emission bands. The assumption of a Lorentzian shape leads to an excess of width on the high energy side of the calculated peaks with respect to the observed peak widths

15 keV to 5 keV (Fig. 11b). From the Fe $L\alpha$, β emission spectrum measured with a 15 keV incident energy, the measured FWHM value was 3.84 eV. For that incident energy the Fe $K\alpha_{1,2}$ ninth order reflection are also detected (Fig. 12). The energy difference between the Fe $K\alpha_1$ and Fe $K\alpha_2$ multiple order reflections correspond to a 13 eV incident energy difference between the fundamental Fe $K\alpha_1$ and Fe $K\alpha_2$ energies. The FWHM values for the Fe $K\alpha_1$ and the Fe $K\alpha_2$ lines were measured to be 3.4 eV and 3.8 eV respectively (Fig. 12). These values are in the same order of magnitude than the 2.45 eV and 3.20 eV intrinsic widths reported by Salem and Lee [30] for these emission lines. This result indicates that for the analysed wavelength region, the TAP monochromator only induces a very small broadening effect. For this reason, the observed shape of the analysed emission bands is probably close to that of the intrinsic emission bands and pure Lorentzian distributions were tentatively used as fitting functions for describing the measured Fe $L\alpha$, β emission spectra. As shown in Fig. 13 the low energy sides of the $L\alpha$ and $L\beta$ bands are correctly described by a Lorentzian function having a 4.2 eV width assuming that the low energy part of the diagram bands is not distorted by non-diagram features. As shown in Fig. 13 such a Lorentzian band profile leads to an excess of width on the high energy side of the calculated peak with respect to the 3.8 eV FWHM value measured on the experimental spectrum. The observed 3.8 eV width of the Fe $L\alpha$ emission band is in a very good agreement with the value of 3.7 eV reported by Bonnelle [29]. In the present study, no secondary emission features were observed on the high energy side of the $L\alpha$ and $L\beta$ major emission bands as it was reported by Bonnelle [29] studying

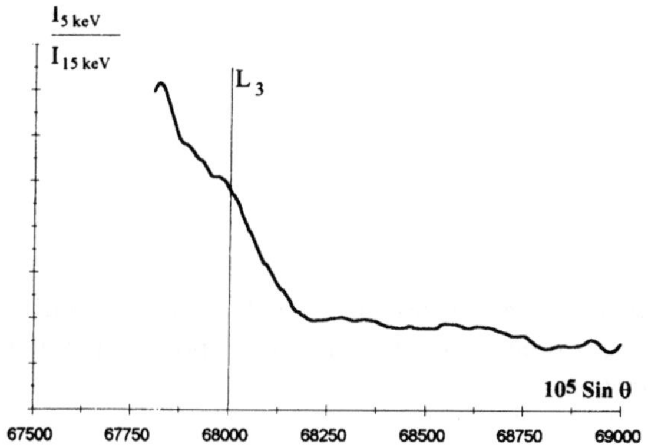

Fig. 14. Intensity ratio as a function of wavelengths of the Fe Lα spectrum measured at 5 keV and 15 keV incident energy respectively

the Fe Lα, β emission of pure iron with a 2.5 keV and a 1.38 keV incident energy successively. This author showed the presence of a secondary emission located 2.7 eV higher than the energy of the main emission Lα peak but, this secondary emission was not observed for the case of the Fe Lβ peak. According to Bonnelle [29], the intensity of the high energy emission band relative to that of the Lα intensity rapidly decreased as a function of the self-absorption phenomena, i.e., when the incident energy was increased. In addition, according to Bonnelle [29], the high energy secondary emission to the main Fe Lα band was not associated with a satellite emission but was attributed to the *characteristic emission* of the pure metallic iron specimen. However, the presence of high energy satellites resulting from multiple ionisations cannot be excluded because for the atomic number of iron, Coster-Kronig transitions are permitted.

The departure from the Lorentzian profile on the high energy side of the Lα peak with respect to that of the low energy part of the band may result from an asymmetrical intrinsic line shape since for a soft X-ray emission band, the energy distribution of the outer hole is not a Lorentzian distribution. In addition to the spectroscopic origin for the asymmetry of the high energy side of the Fe Lα, β diagram bands a variation of the self absorption effect within the analysed wavelength (energy) domain may also be invoked. This assumption is supported by the results in Fig. 14 illustrating the variations of the intensity ratio I(5 keV)/I(15 keV) obtained after the two Fe Lα, β spectra measured with a 5 keV and 15 keV incident energy have been normalised.

For the reasons mentioned above it is not possible to describe both sides of the measured Fe Lα, i.e. peak shape uncorrected for self-absorption, using a Lorentzian distribution. In order to obtain a satisfactorily fit of the experimental data a sum of the pseudo-Voigt profiles ($Cg = 0.6$ in Eq. (1)) was used. As shown in Fig. 15 a satisfactorily description was obtained but, obviously such approach only provides a mathematical solution to the description of the complex Fe Lα, β emission spectrum.

4. Study of the Valence States: Application to Iron Bearing Minerals

Synthetic $Fe_{0.94}O$ (referred below as FeO), natural hematite (Fe_2O_3) and natural magnetite (Fe_3O_4) specimens were analysed. The Fe Lα, β emission bands were analysed with several TAP monochromators.

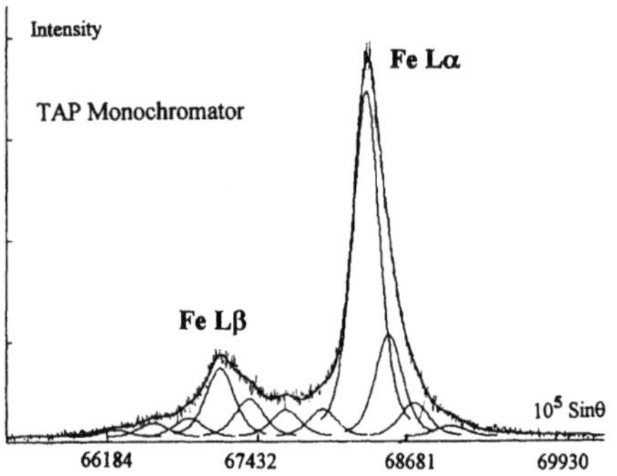

Fig. 15. Empirical description of the Fe Lα, β spectrum using a summation of pseudo-Voigt profiles as fitting functions

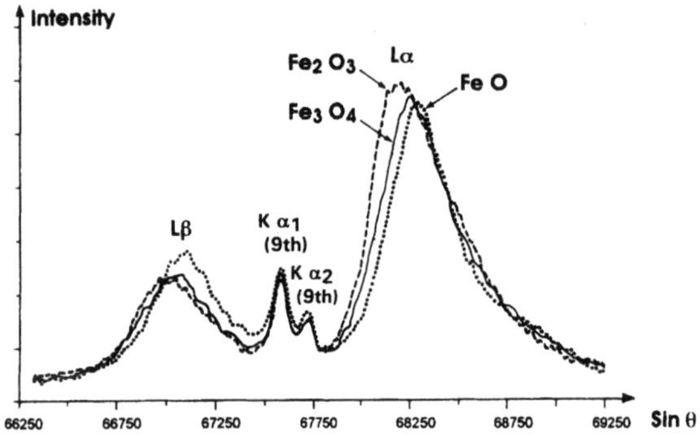

Fig. 16. Fe Lα, β spectrum measured with a TAP monochromator for an FeO, Fe_2O_3 and an Fe_3O_4 specimen respectively analysed with a 15 keV incident energy

4.1. Fe Lα, β Emission Bands of Iron Oxides

The Fe Lα, β spectra were measured with a 8 keV and 15 keV incident energies and a 50 nA beam current. The counting time was one second per channel. For a 15 keV incident energy, the ninth order reflection of the Kα_1 and Kα_2 lines were detected between the Fe Lα and Fe Lβ emission bands. The shape of the Lα, β emission bands was reproducible for all TAP monochromators used.

As shown in Fig. 16, the Fe Lα, β spectra measured at the surface of the Fe_3O_4 specimen is intermediate between the spectra of the Fe_2O_3 and FeO specimens. For easier comparisons of the Fe Lα, β shape and peak position changes as a function a the analysed specimens, the spectra were smoothed using spline functions before to be displayed. We tried to resolve the Fe Lα, β spectrum of the Fe_3O_4 specimen which contain both Fe^{2+} and Fe^{3+} ions as the sum of the Fe Lα, β spectra of the Fe^{2+} and Fe^{3+} bearing iron oxide specimens respectively. For this purpose, the Fe Lα, β spectra for the Fe_2O_3 and FeO specimens were decomposed according to the procedure described above. These parameters were used as coupled variables for the spectral deconvolution of the Fe Lα, β spectra of the Fe_3O_4 specimen as shown in Fig. 17. The Fe Lα, β proportion of Fe_2O_3/FeO spectra in the Fe_3O_4 spectra derived from the fitting procedure applied to spectra acquire for several different experimental conditions are shown in Table 2.

The differences in the calculated FeO/Fe_2O_3 ratios shown in Table 2, may result from either i) a differential absorption effect of the Fe Lα, β bands associated with the FeO and Fe_2O_3 components or from, ii) beam damages occurring during the digital recording of the spectra.

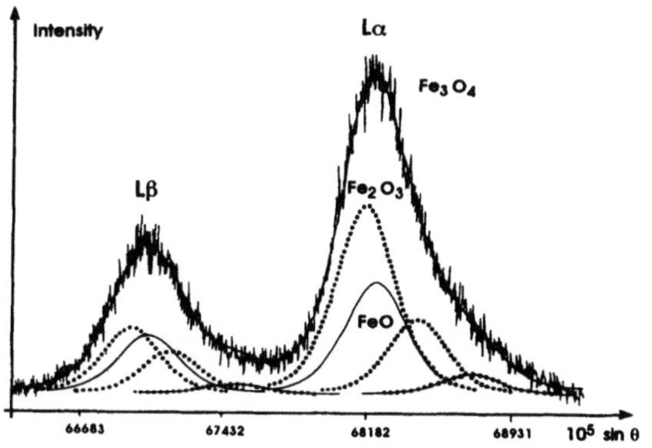

Fig. 17. Spectral decomposition of the FeLα, β spectrum of the Fe_3O_4 specimen into the sum of the Fe Lα, β spectrum of an FeO and an Fe_2O_3 specimen respectively

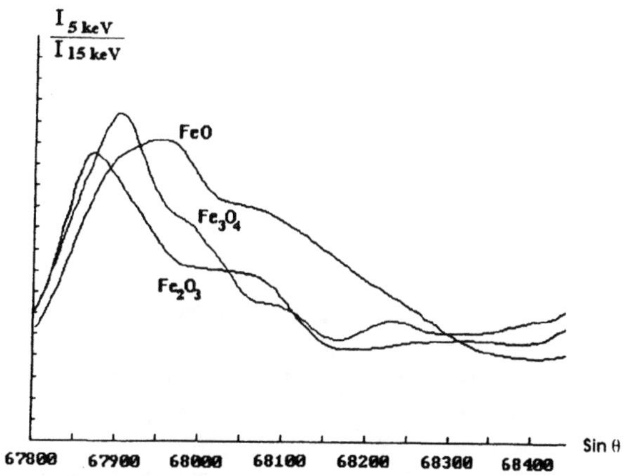

Fig. 18. Self-absorption curves for the three iron oxide specimens resulting from the intensity ratio of the Fe Lα, β spectra measured at 5 keV and 15 keV incident energy respectively

Table 2. Results of the spectral deconvolution of the Fe Lα, β spectra measured at the surface of an Fe_3O_4 specimen as the sum of the Fe Lα, β spectra of an FeO and Fe_2O_3 spectra

Ep keV	i_b nA	FeO/Fe_2O_3 %
8	50	55.4
15	50	57.6
8	90	59.1
15	90	78.5

The effect of differential absorption on the peak position and shape changes as a function of the matrix composition is shown in Fig. 18 illustrating the self-absorption curves for the three iron oxide specimens obtained by dividing the normalised Fe Lα, β spectra measured at 5 keV and 15 keV successively. As shown in Fig. 18 the absorption curve for the Fe_3O_4 specimen lies between the curves for the FeO and Fe_2O_3 specimens. The shape of the self-absorption curve derived from

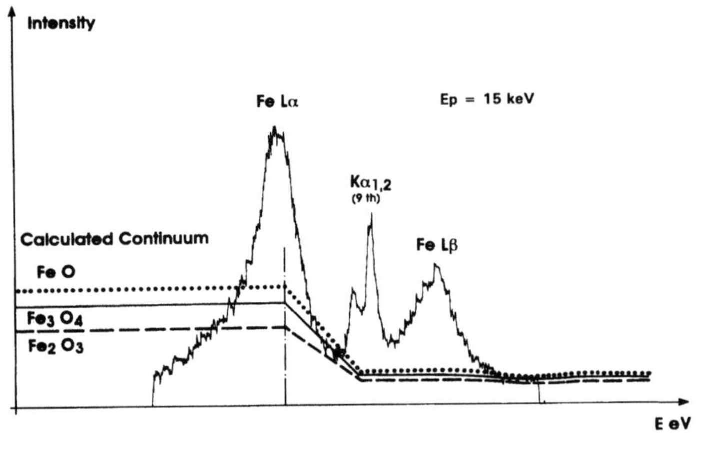

Fig. 19. Calculated energy dispersive continuous distributions for FeO, Fe_2O_3 and Fe_3O_4

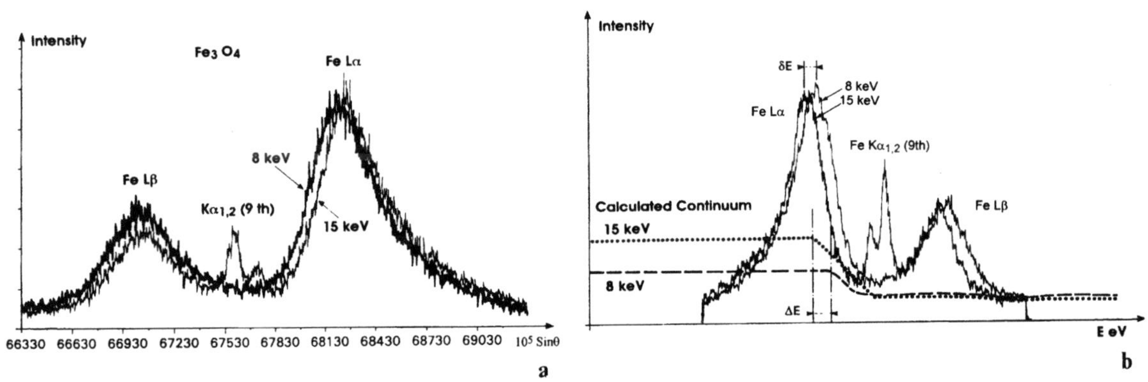

Fig. 20. **a** Fe Lα, β WDS spectrum for the Fe_3O_4 specimen showing a shift towards the long wavelength side (low energy side) when the incident energy is increased from 8 keV to 15 keV and, **b** Fe Lα, β energy spectrum derived from WDS experimental data and calculated continuous emission distributions. The δE peak shift results from the shift ΔE of the continuous distributions, i.e.; a change of self-absorption as a function of the incident energy

measurements carried out at two different excitation conditions are consistent with the absorption curves obtained from the continuous emission distribution calculated from the energy spectrum as shown in Fig. 19 for the case of the iron oxide specimens. Differential absorption effects are also responsible for the observed peak shifts of the Fe Lα, β spectra measured at the surface of the FeO, Fe_2O_3 and Fe_3O_4 specimens successively. This result is supported by the changes in the continuous emission distribution as illustrated in Fig. 20a, b for the case of the Fe_3O_4 specimen analysed with a 8 keV and 15 keV successively. The energy Fe Lα, β spectra were re-scaled by adding offsets in order that the peak positions correspond to the theoretical values. The offsets were found different when the spectra measured at 8 keV and 15 keV were re-scaled indicating for peak shifts as a function the incident energy, consistent with the shift observed on the initial WDS spectra. The continuous emission intensity distribution for the two incident energy values were cal-

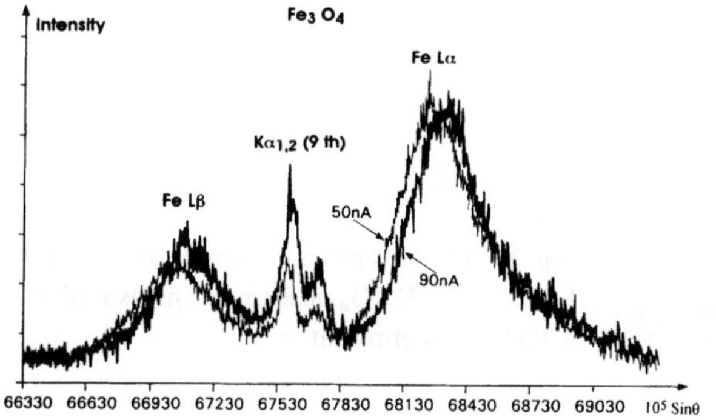

Fig. 21. Shift towards the long wavelengths (low energies) of the Fe Lα, β spectrum for an Fe_3O_4 specimen (15 keV incident energy) when the beam current is increased from 50 nA to 90 nA

culated and scale to the high energy side of the spectra. As shown in Fig. 20b the decrease in the continuous emission intensity occurs at a higher energy value for the case of the spectrum measured with a 8 keV energy than that observed on the spectrum that was measured with a 15 keV incident energy. The energy shift of the two continuous emission distribution is approximately equal to the energy difference between the observed peak positions measured at two different incident energies. The same results were obtained when comparing the continuous emission distributions for the case of the FeO and Fe_2O_3 specimens respectively.

As shown in Fig. 21 for the case of the Fe_3O_4 specimen measured with a 15 keV energy, a shift towards the lower energies (longer wavelengths) of the Fe Lα, β spectrum is also observed when the beam current is increased from 50 nA to 90 nA. This shift may result from a reduction effect of the iron oxide consistent with the results in Table 2 indicating for a higher proportion of the FeO spectrum than that of the Fe_2O_3 spectrum for describing the Fe_3O_4 spectrum when the beam current was increased from 50 nA to 90 nA. The effect of beam damage leading to peak shifts was less marked when the spectra were measured with a 8 keV incident energy.

Both effects, beam damages and absorption phenomena tend to induce similar peak shifts between the Fe Lα, β spectra measured at the surface of the three iron oxide specimens. For a given incident energy, the shifts of the Fe Lα, β spectra of the three iron oxides can be related with absorption effects. An increase of the absorption effect resulting from matrix composition changes or resulting from an increase of the absorption path-length when increasing the incident energy tend to move the maximum peak position towards the lower energies.

4.2. *O Kα Emission Band*

The O Kα emission spectrum measured at the surface of the three iron oxide specimens with a 8 keV incident energy are shown in Fig. 22. As for the case of the Fe Lα, β emission spectra for these specimen the O Kα peak position of the Fe_3O_4 specimen is intermediate between the peak positions observed for the FeO and Fe_2O_3 specimen respectively. The emission spectra of the FeO and Fe_2O_3 were processed using a set of pseudo-Voigt profiles. The weak feature on the high energy side of the O Kα band is attributed to an instrumental defect resulting from anomalous reflectivity within the TAP monochromator. The results of fit of the FeO and Fe_2O_3 spectra were used as coupled variables to decompose the Fe_3O_4

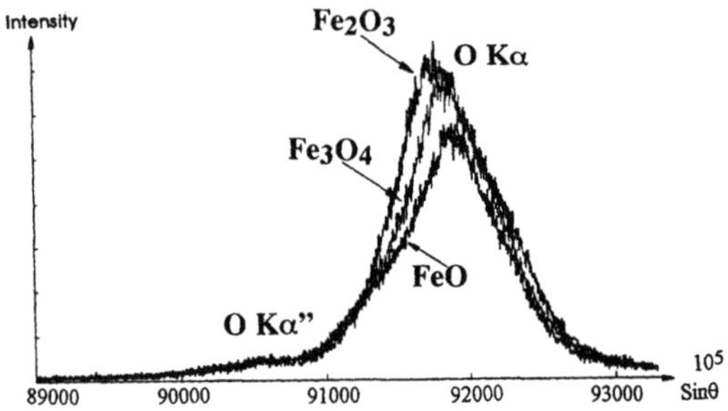

Fig. 22. O Kα emission bands for the FeO, Fe_2O_3 and Fe_3O_4 specimens measured with a TAP monochromator

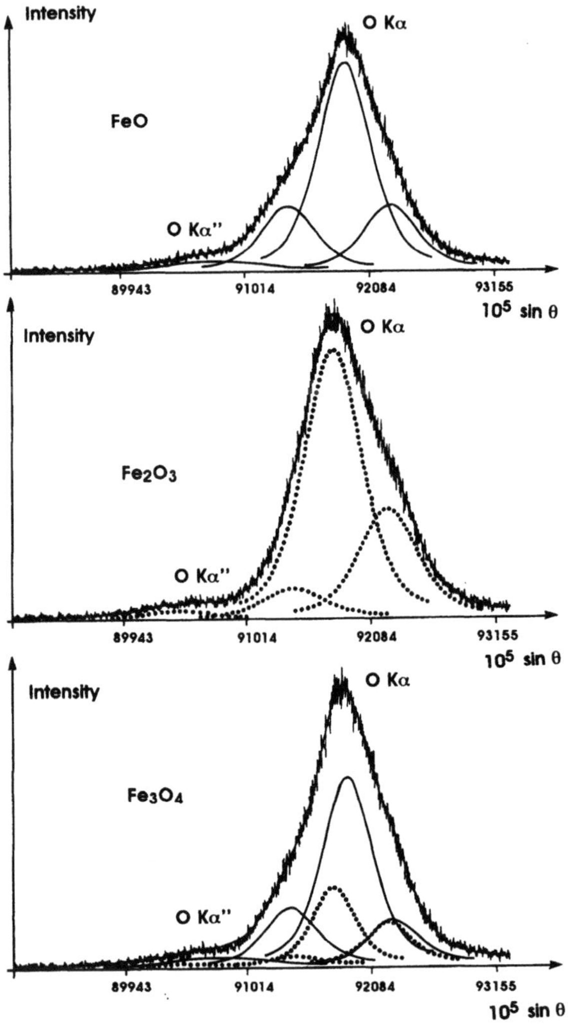

Fig. 23. Spectral deconvolution of the Fe Lα, β spectrum of the Fe_3O_4 specimen as the sum of the spectrum of FeO and Fe_2O_3 specimens respectively

spectrum as the sum of the two previous spectra. Results of the spectral deconvolution of the O Kα emission for the Fe_3O_4 specimen are shown in Fig. 23 for the case of measurements performed with a 8 keV energy. The experiments were repeated when the incident energy was increased to 15 keV. The proportion of the FeO spectrum to that of the Fe_2O_3 spectrum was 41% and 34% successively when the Fe_3O_4 specimen was measured with a 8 keV and 15 keV energy successively. The calculated

Table 3. O Kα spectral deconvolution of the Fe_3O_4 spectrum as the sum of FeO and Fe_2O_3 spectra

Ep keV	FeO/Fe_2O_3 %	Calculated O Kα k-ratios corresponding to the FeO and Fe_2O_3 concentrations in Fe_3O_4	
		wt% O associated with Fe^{2+}	wt% O associated with Fe^{3+}
5	41%	8.2%	21.1%
15	34%	7.2%	23.1%

O Kα intensities of the FeO and Fe_2O_3 contributions to the Fe_3O_4 spectrum were compared to the intensities measured at the surface of the FeO and Fe_2O_3 specimens respectively. The resulting O Kα k-ratios for the oxygen concentration participating to the FeO and Fe_2O_3 components are shown in Table 3 (the theoretical O concentrations associated with the Fe^{2+} and Fe^{3+} weight concentrations in Fe_3O_4 are 6.9% and 20.7% respectively.) The O Kα peak shifts we observed in Fig. 22 as a function of the matrix composition are more probably significant of a chemical environment effect rather than an effect of beam damage during the acquisition time of the O Kα band recording. This statement is supported by comparing the O Kα. emission of the Fe_3O_4 specimen measured with a 8 keV energy before and after the analysed area was exposed to a 25 keV irradiation for 10 minutes. No peak shift was observed but, only a decrease of the O Kα intensity resulted from the high energy pre-irradiation.

5. Discussion and Conclusion

The description of an X-ray peak using pseudo-Voigt profiles (Eq. (1)) has been proved to be a sufficient approximation for describing atomic X-ray lines. These profiles can be used in a least squares fitting procedure in order to resolve spectral interferences or to process X-ray spectra exhibiting asymmetrical shape dominated by instrumental distortions.

A detailed examination of changes in the shape of a soft X-ray emission band as a function of the electronic structure of the compound may provide information on chemical bonding. However, such interpretation can only be obtained after the shape of the measured emission band (position, width and relative intensities of the spectroscopic features) has been corrected for self-absorption effect, as it has been illustrated in the present study for the case of Lα, β emission spectrum of first series transition elements.

Many attempts have been made to determine oxidation states from the study of peak position, peak width and relative intensity changes of soft X-ray emission bands. For this purpose, Albee and Chodos [31] proposed to relate the iron oxidations states with changes in the Lβ/Lα intensity ratios. However, the observed variations reflected the absorption effect as a function of the matrix composition. Recently, Hofer et al. [32] investigated the possible use of peak shift and broadening for determining the Fe oxidation states in iron oxide bearing minerals. As shown in the present study, the modifications of the profile of soft X-ray emission bands of transition elements can be explained in terms of absorption phenomena. This relationship between the observed X-ray band profiles and absorption phenomena is consistent with the results from Armstrong and Carpenter [33] who related similar Cu Lα, β spectrum alterations with absorption effects when studying Cu/Au alloys (no bonding differences) and copper oxide compounds. The use of emission band profile changes for studying valence states is thus questionable. The change in the Fe^{2+}/Fe^{3+} ratio in a variety of iron oxide bearing minerals modifies the shape and position of the soft Fe Lα, β X-ray emission band but these changes are only indirectly related to the valence states because, varying the Fe^{2+}/Fe^{3+} ratio is equivalent to change the weight Fe concentrations and consequently to modify the absorption phenomena. In addition, when studying peak shape changes, beam damages cannot be neglected because of possible reduction effect induced by the incident electrons.

Armstrong and Carpenter [33] investigated the O Kα emission band in a variety of silicate minerals analysed with LMS monochromators. Owing to the energy resolution of these monochromators, the peak shift and peak broadening are small and peak deconvolution techniques are difficult to apply. The use of a TAP is preferred for studying the O Kα emission band but the low efficiency of that monochromator within the soft X-ray region makes difficult its use for routine analysis (long data acquisition time and subsequent risks of beam damage). More investigations are necessary in order to try to relate the O Kα emission band changes with the chemical environment because the shape of a soft X-ray emission band depends on all the chemical and crystallographic properties of the mineral as illustrated in Fig. 24 comparing the line shape of the O Kα emission bands measured at the surface of an andradite garnet (Ca_3 Fe_2 $(SiO_4)_3$) specimen and of the Fe_2O_3 hematite crystal, both specimen containing only Fe^{3+} ions.

A possible approach for determining the Fe^{2+}/Fe^{3+} ratio is to obtain an accurate quantitative analysis of the oxygen concentration. At the present time, there is still a lot of uncertainties in quantitative analyses using soft X-ray emissions. These difficulties result both from the intensity measurement of the various components describing the shape of the emission band and from the prediction of the mechanisms involved in the production of low energy X-ray photons.

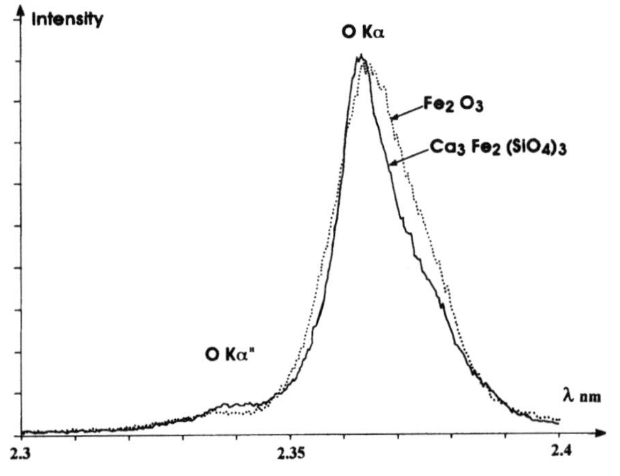

Fig. 24. Changes of the shape of the O Kα emission band for a Fe_2O_3 and a andradite garnet (Ca_3 Fe_2 $(SiO_4)_3$) containing only Fe^{3+} ions

Direct on-line peak integration allows us to measure the total number of photons regardless of their origins (diagram, non-diagram emissions and instrumental distortions). The peak integration method avoids the need for recalibration of the spectrometer in presence of chemical effects. For obtaining a reliable X-ray intensity of a soft X-ray emission band it is necessary to distinguish the effects being matrix dependent from those being related to instrumental effects. In addition to the focusing defects occurring into the depth or along the focal circle of the monochromator and to possible anomalous reflections ("hole" in the continuum or $K\alpha''$ satellite to the O $K\alpha$ band), peak shape alterations may result from the crystallographic orientation of the specimen as shown by Bastin and Heijligers [34] studying boride specimens. For all of these reasons, spectral deconvolution using least squares fitting technique to an analytical description of the peak shape is recommended for quantitative spectrum processing.

In quantitative analysis of insulating materials using soft X-ray emission bands, the crystallographic and dielectric properties of the materials must be considered for choosing the reference compounds. It has been shown that accurate quantitative analysis can be obtained if a relationship exists between both the position and width of the peak of the O $K\alpha$ emission measured at the surface of the unknowns and the reference materials as discussed by Fialin et al. [23]. For wide band gap insulating materials, the radiative transitions to the 1s vacancy from states located in the band gap may represent a contribution of several percent of the total emitted intensity. These bandgap emissions are expected to have higher radiative probabilities than those involving transitions from the valence band (Fialin et al. [35]). In addition, these bandgap emissions involve electrons which do not originate from the density of valence states on the emitting atom site but are electrons trapped in pre-existing point defects or induced defects by the electron bombardment. The presence of point defects in strongly insulating materials are responsible for electrostatic phenomena as reviewed by Cazaux [36] and by Cazaux and Le Gressus [37] and illustrated by Remond et al. [38] for the case of quantitative analysis of minerals using soft X-ray emission bands.

Acknowledgements. For C. Gilles and G. Remond, this is the BRGM contribution number 95017; this work was financially supported by a BRGM research project. G. Remond wishes to thank NIST for supporting his collaborative research with the Surface and Microanalysis Science Division.

References

[1] M. Fialin, J. Henoc, G. Remond, *Scanning Micros.* [*Suppl.*] **1993**, *7*, 153.
[2] G. Remond, J. L. Campbell, in: Microbeam Analysis (J. R. Michael, P. Ingram, eds) San Francisco Press, San Francisco, 1990, pp. 59–66.
[3] G. Remond, J. L. Campbell, R. H. Packwood, M. Fialin, *Scanning Microsc.* **1993**, [*Suppl.*], 89
[4] S. I. Salem, G. M. Hockney, P. L. Lee, *Physical Rev. A* **1976**, *13*, 330.
[5] S. I. Salem, B. L. Scott, *Physical Rev. A* **1974**, *29*, 690.
[6] B. L. Doyle, S. M. Shafroth, *Physical Rev. A* **1979**, *19*, 1433.
[7] T. C. Huang, G. Lim, *Adv. X-Ray Anal.* **1986**, *29*, 461.
[8] C. E. Fiori, C. R. Swyt, R. L. Myklebust, *Desktop Spectrum Analyser*, National Institute of Standards and Technology, Gaithersburg, MD, US Patent 529913.

[9] L. G. Parratt, *Physical Rev.* **1936**, *50*, 1.
[10] C. A. Randall, L. G. Parratt, *Physical Rev.* **1940**, *57*, 786.
[11] C. Bonnelle, *Annual Report C*, The Royal Society of Chemistry, London, 1987, pp. 201–272.
[12] K. S. Srivastava, S. Singh, A. K. Srivastava, R. S. Nayal, A. Chaubey, P. Gupta, *Physical Rev. A* **1982**, *25*, 2838.
[13] K. S. Srivastava, A. K. Srivastava, K. S. M. Husain, S. Singh, *Ind. J. Pure Appl. Phys.* **1983**, *21*, 256.
[14] G. Remond, Ph. Coutures, C. Gilles, D. Massiot, *Scann. Microsc.* **1983**, *3*, 1059.
[15] Y. Cauchois, C. Bonnelle, in: *Atomic Inner-Shell Processes, Tome II: Experimental Approaches and Applications* (B. Crasemann, ed.), Academic Press, 1975, pp. 84–121.
[16] D. W. J. Berreman, *Physical Rev. B* **1976**, *10*, 4313.
[17] D. W. J. Berreman, *Physical Rev. B* **1979**, *19*, 560.
[18] P. G. Self, K. Norrisk, A. R. Milnes, J. Graham, B. Robinson, *X-Ray Spectrom.* **1990**, 59.
[19] G. Remond, R. H. Packwood, C. Gilles, *Analyst* **1995**, 1247.
[20] D. W. Fischer, *J. Chem. Phys.* **1965**, *42*, 3814.
[21] D. W. Fischer, W. L. Baun, *Adv. X-Ray Anal.* **1967a**, *11*, 230.
[22] M. Fialin, G. Remond, *Microbeam Anal.* **1993**, *2*, 179.
[23] M. Fialin, G. Remond, C. Bonnelle, *Microbeam Anal.* **1995**, 211.
[24] S. Bodeur, R. Barchewitz, *Solid State Commun.* **1984**, *49*, 11.
[25] G. Remond, M. Fialin, R. H. Packwood, C. Gilles, O. Rouer, *Microbeam Anal.* **1995**, submitted.
[26] G. Remond, M. Fialin, R. H. Packwood, C. Gilles, O. Rouer, *Microbeam Anal.*, Proc. 28th Annual MAS Meeting, New Orleans LA, July 31-August 5, 1994 (J. Friel ed.), pp. 235–236.
[27] G. Remond, J. Cazaux, M. Fialin, C. Gilles, RH Packwood, O. Rouer, *Microbeam Anal.* **1995**, submitted.
[28] D. Fabian, *CRC Critical Review in Solid State Sciences*, CRC, Cleveland, Ohio, 1971, pp. 255–316.
[29] C. B. Bonnelle, *Thèse*, Masson, Paris, 1966.
[30] S. I. Salem, P. L. Lee, in: *CRC Handbook of Chemistry and Physics, 1979* (R.C. Weast, M. J. Astle eds.), CRC, Boca Raton, FL, E 207.
[31] A. L. Albee, A. C. Chodos, *Am. Mineral.* **1970**, *55*, 491.
[32] H. E. Hofer, G. P. Brey, B. Schulz-Dobrick, R. Oberhansil, *Eur. J. Mineral.* **1994**, *6*, 407.
[33] J. T. Armstrong, *Microbeam Analysis, Proc. 28th Annual MAS Meeting, New Orleans, LA, USA, July 31-August 5, 1994* (J. Friel, ed.), pp. 53–54.
[34] G. F. Bastin, HTM Heijligers, in: *Electron Probe Quantitation* (K. F. J. Heinrich, D. Newburry, eds.) Plenum, 1991, pp. 145–162.
[35] M. Fialin, C. Bonnelle, G. Remond, P. Blanc, *29th Annual MAS Meeting*, 1995, *Microbeam Anal.* p. 199.
[36] J. Cazaux, *J. Microsc. Spectrosc. Electron.* **1986**, *11*, 215.
[37] J. Cazaux, C. Le Gressus, *Scann. Microsc.* **1992**, *5*, 17.
[38] G. Remond, J. Cazaux, R. Chapoulie, P. Ermakoff, M. Fialin, R. H. Packwood, *Microbeam Anal.* **1995**, submitted.

Mikrochim. Acta [Suppl.] 13, 87–115 (1996)

Synchrotron Radiation Induced X-Ray Microfluorescence Analysis

K. Janssens[1,*], L. Vincze[1], B. Vekemans, A. Aerts[1], F. Adams[1], K. W. Jones[2], and A. Knöchel[2]

[1] Department of Chemistry, University of Antwerp (UIA), B-2610 Antwerp, Belgium
[2] Department of Applied Sciences, Brookhaven National Laboratories, Upton 11973, NY, USA
[3] Department of Chemistry, Universität Hamburg, D-20146 Hamburg, Federal Republic of Germany

Abstract. μ-XRF is the microscopic equivalent of the well-established multi-element analytical technique. In this paper, after comparing the interaction of X-ray photons, electrons and protons with matter and an introduction to synchrotron rings and microfocussing of X-rays, the instrumentation for μ-XRF is discussed, both for laboratory source and synchrotron based setups and the analytical characteristics of μ-XRF are contrasted to that of other microanalytical techniques. Also, this issue of quantification of μ-XRF data is addressed; the applicability of the method in archeological and geological analysis is illustrated.

Key words: synchrotron radiation, microscopic X-ray fluorescence analysis, X-ray fluorescence analysis, trace elements, procelain analysis, fluid inclusions, archeological glass.

---

* To whom correspondence should be addressed

1. Introduction

One of the most widely applied analytical technologies that makes use of X-rays, both as a means of excitation and as analytical signal is X-ray fluorescence analysis (XRF) [1–3]. Although the basis of qualitative and quantitative X-ray spectrometry was already established by Mosely in 1913, only with the introduction of the first commercial X-ray spectrometry instrument in 1948, XRF started towards becoming a standard multi-element analytical technique present in many research and industrial laboratories. The worldwide number of XRF instruments today is approximately 18000 units. This equipment is mostly used for routine bulk elemental analysis of a variety of materials (e.g., steel, polymers, ceramics, raw materials in the microelectronics industry...) and is often used for quality control of start- or finished products. For a number of years, however, XRF had been considered as a very useful, but not a very modern technique. Developments in instrumentation (X-ray optics, X-ray sources) and methodology have changed this picture in the last 5 years and two important variants of the technique have evolved [4]. Both variants are based on the confinement of the interaction volume of the primary X-ray beam with the material being analysed.

In total reflection XRF (TXRF) [5], by irradiating an (optically flat) material with a parallel X-ray beam below the angle of total reflection, the in-depth penetration of the primary X-rays can be confined to a few tens of nanometers below the surface. As a result, ultra-trace analysis of liquids which have been pipetted onto a flat support and allowed to dry has become possible on a routine basis. Typical detection limits for TXRF are at or below the 10 ng level for a 1000 s counting time in conventional instrumentation [6].

The second important variant of the bulk XRF technique is μ-XRF and will be discussed in this paper. Instead of a broad beam of radiation with a foot-print of, e.g., a few cm^2 on the sample, a photon beam of a few microns in diameter is used to irradiate a specific location on the sample surface [7, 8]. The element-specific radiation emitted by the sample atoms in that spot is used to determine the local composition. X-ray microbeams of sufficiently small size ($\leqslant 10\,\mu m$ diameter) and of sufficient intensity to allow energy-dispersive (ED) data collection rates in the 1000–10000 cps range can be obtained via the use of highly intense X-ray sources and/or by means of focussing of the primary X-rays.

At synchrotron facilities, i.e., extremely intense sources of X-rays, around the world, several types of μ-SRXRF(synchrotron radiation induced μ-XRF) spectrometers have been installed [7, 9–11] and are planned [12–15]. In these instruments, the white synchrotron beam can be simply collimated down to micrometer dimensions by means of suitable apertures [7, 9]. However, also focussed, monochromatic microbeams can be used for μ-XRF purposes [8, 10, 11].

When a more conventional type of X-ray source (such as an X-ray tube) is employed, the use of fine collimators usually results in too low an X-ray flux that finally impinges on the sample to be useful in ED-XRF experiments. However, as a result of recent developments in focussing optics for hard X-rays [16–18], laboratory scale μ-XRF experiments using relatively simple equipment also have become feasible [19–21]. By means of glass capillary focussing devices, in which, through repeated total reflection on the inner walls of the glass tube, X-ray beams are 'squeezed' to micron-sized dimensions, cost-effective focussing of radiation generated in an X-ray tube can be accomplished.

In this paper, after a brief review of a few fundamental concepts regarding X-ray/matter interactions, synchrotron radiation and micro-focussing X-ray optics, the analytical advantages and limitations of μ-SRXRF (microscopic synchrotron radiation induced XRF) are compared to those of similar microanalytical techniques such as EPXMA (electron probe X-ray microanalysis) and μ-PIXE (proton induced X-ray emission). Also, the advantages associated with using synchrotron light as opposed to conventional (tube-generated) X-rays for microanalytical purposes are highlighted. Next, the problems associated with quantification of μ-(SR)XRF data are discussed in various experimental situations. Finally, in order to illustrate the potential of the method, a number of applications from the archeometrical and geological field will be described.

2. Theoretical Background

The most striking feature of μ-(SR)XRF is that it is a micro-analytical technique with trace-level sensitivity. It is therefore relevant to consider the factors determining the sensitivity of this technique and its closest analogues. For all X-ray emission techniques that employ some form of photon counting in the detection chain (mostly when energy-dispersive detectors are used), the lowest detectable concentration level c_{MDL} (MDL = minimum detection limit) for a given chemical element can be estimated from the net photopeak intensity N and the corresponding background intensity B which are collected from a sample containing a concentration c of a given element, as:

$$c_{\mathrm{MDL}} = c\frac{k\sqrt{B}}{N} = \frac{k\sqrt{\sigma_{\mathrm{B}}}}{\sigma_{\mathrm{N}}}\frac{1}{\sqrt{It}} \tag{1}$$

where k usually is a number in the range 1–5 (usually set to 3); $\sigma_{\mathrm{N}} = N/cIt$ is the net X-ray yield for this element and $\sigma_{\mathrm{B}} = B/It$ the corresponding background yield. The sensitivity of any X-ray emission technique therefore can be increased by

(i) decreasing the background σ_{B}
(ii) increasing the elemental yield σ_{N} and/or
(iii) increasing the beam intensity I and/or the collection time t.

It is useful to consider the relative and absolute magnitudes of the above mentioned quantities when comparing the analytical characteristics of various X-ray emission techniques. Since in the present context different microanalytical techniques are compared, also the achievable beam size S and the beam flux density $F = I/S$ (expressed, e.g., in photons, electrons or protons per second and per μm^2) are important figures-of-merit.

Interaction of Photons, Electrons and Protons with Matter

In Table 1 the penetration depth of X-ray photons, electrons and protons of energies giving rise to appreciable characteristic X-ray emission are compared. For photons and electrons, this is typically the 0–100 keV range; for protons, useful energies are in

Table 1. Penetration depth of electrons, protons and X-ray photons in various materials

Ele-ment	Atomic number	Atomic mass	Density (g/cm^3)	μ_L (cm^{-1})	Penetration range (μm) X-rays[a] (20 keV)	Electrons[b] (20 kV)	Protons[b] (2.5 MeV)
C	6	12.0	1.9	0.79	12000	5	55
Si	14	28.1	2.3	10.0	1000	5	68
Fe	26	55.9	7.8	200	50	1.6	27
Ag	47	107.9	10.5	189	50	1.7	28
Pb	82	207.2	11.3	971	10	2.0	37

μ_L = linear mass absorption coefficient; [a] 1/e range; [b] Bethe range.

the range 1–4 MeV. The much larger penetrative power of photons in light/medium heavy matrices ($Z < 20$) relative to charged particles is immediately clear; protons take an intermediate position between electron and photons. This difference becomes also clear when, e.g., trajectories of photons and electron of the same energy are compared. The penetration range for charged particles is determined by the gradual energy loss that occurs as a result of many inelastic scattering interactions with the sample material (quantitatively described by the stopping power function $S(E)$) and by the curvature of the trajectories (resulting from many elastic collisions between projectile and sample atoms). On the other hand, the most probable interaction mechanism for X-ray photons in the range 1–50 keV (for most sample materials) is the photo-electric effect. E.g., for 10 keV photons in Al, the probabilities for photo-ionisation, Compton (i.e., inelastic) and Rayleigh (elastic) scattering are resp. 97.5%, 0.4% and 2.1%. Since the photo-electric effect causes the primary photon to be annihilated, the majority of the primary photons will only be subject to a single (scattering or photo-electric) interaction; only a very small fraction will first be (in)elastically scattered by a sample atom before ionising a second atom. (An even smaller fraction is subject to double scattering or higher order combinations). The consequence of this different behaviour is that samples irradiated with monochromatic X-ray photons of energy E_0, next to the characteristic radiation of the sample atoms, only will emit scattered photons in a limited energy window from a few keV below upto the primary energy E_0. A typical feature are the Compton and Rayleigh scatter peaks that appear in XRF spectra (see Fig. 1b), caused by the primary photons that were scattered towards the detector during resp. a single inelastic and elastic scattering interaction in the sample. In principle, the rest of the spectrum shown in Fig. 1b should be background free; in reality, processes such as incomplete charge collection and Compton scatter inside the Si crystal of the detector give rise to a background continuum that extends from the position of the principal (photo)peaks towards lower energy.

In contrast, the irradiation of a material with charged particles of energy E_0 will result in a *bremsstrahlung* continuum (primarily due to decelerating electrons) spanning the entire energy range from 0 to E_0. This continuum is especially noticable in EPXMA (electron probe X-ray microanalysis) spectra where the continuum is generated at relatively shallow depths [23] (typically $\leqslant 5$ μm; see Table 1 and

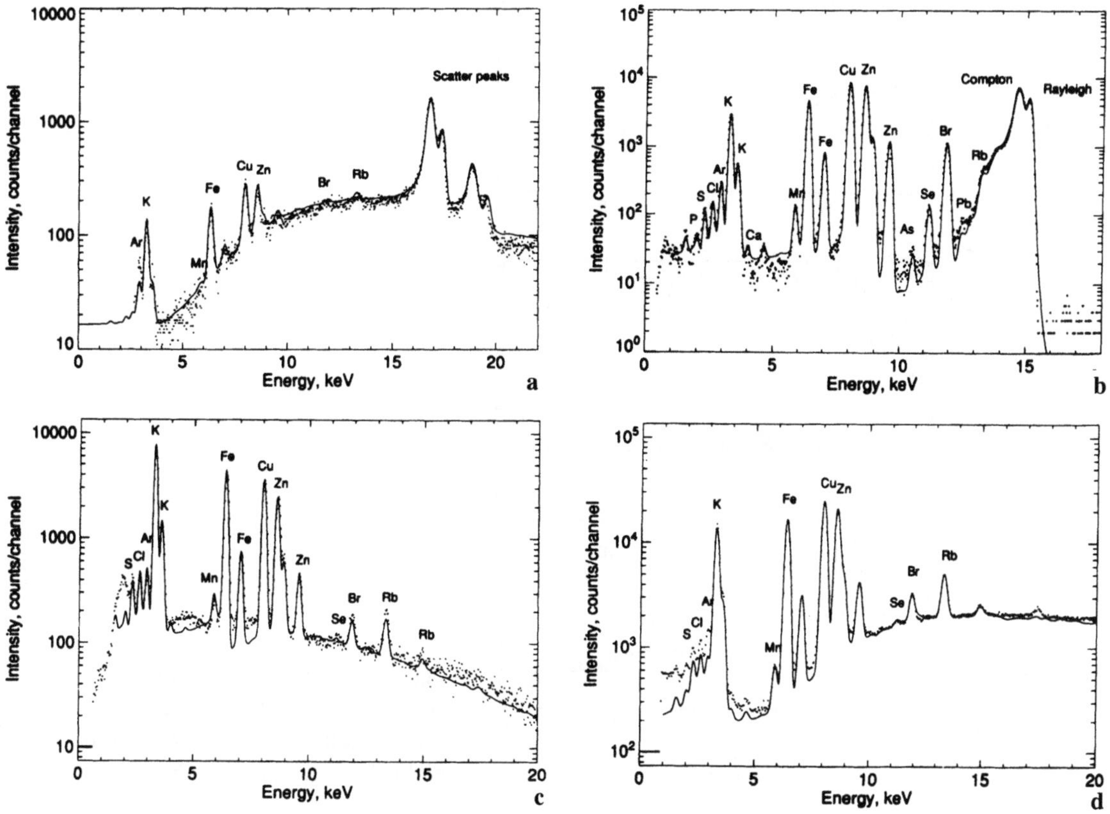

Fig. 1. ED-XRF spectra of NIST SRM 1577a Bovine Liver obtained using **a** a conventional Mo X-ray tube using a 200 × 200 μm^2 collimator (300 s counting time), **b** a 2 × 0.5 mm^2 monochromatic 15 keV SR beam at the NSLS X26A station (300 s), **c** a 8 × 8 μm^2 polychromatic SR beam (at the same station, 220 s) and **d** a 50 × 50 μm^2 white SR beam at the Hasylab L SR-XRF station (900 s). Dots: experimental spectra, lines: predicted spectral distributions (see text)

Fig. 2a). In case of proton induced X-ray emission (PIXE), the abundance of the continuum is supressed since it is caused by secondary electrons of relatively low energy generated by the primary protons [24] (Fig. 2b). When a sample is irradiated with a polychromatic X-ray beam, also in XRF spectra, a scatter continuum spanning the entire energy range can be observed (Figs. 1a, b, d and 2c).

In Fig. 3, the cross-sections for characteristic X-ray production by X-ray photons, electrons and protons are compared [25]. Two important observations can be made: (i) for the three projectile types, the overall probability for ionisation is comparable; (ii) whereas for the charged particles, the ionisation cross-sections decrease with increasing atomic number of the target atoms, the photo-ionisation process favours heavier over lighter elements.

Considering Eq. (1), it can therefore be concluded that for equal beam intensity I and collection time t, the lower background levels that are recorded with monochromatic photon-induced X-ray emission, together with the comparable or higher ionisation cross sections, will result in lower minimum detection limits (MDL's) for μ-XRF than for EPXMA and μ-PIXE in the case of heavy elements ($Z > 25$).

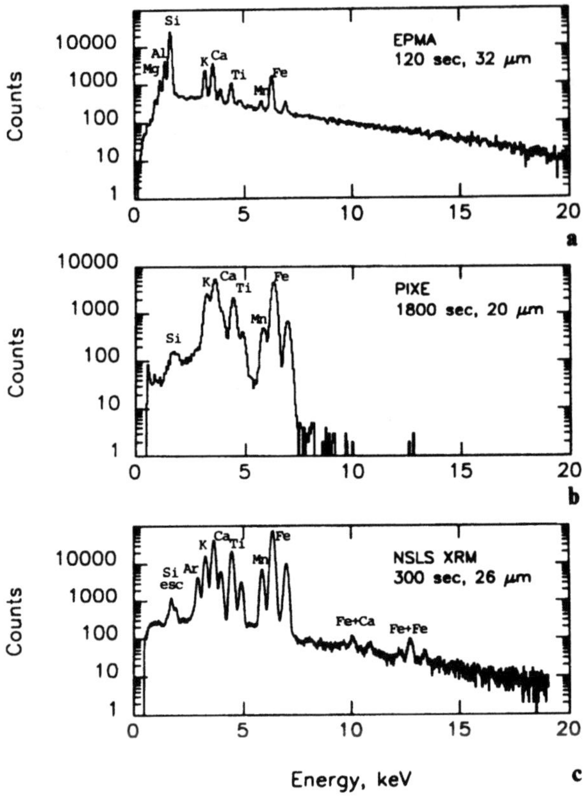

Fig. 2. **a** EPXMA, **b** μ-PIXE, **c** white beam SR-XRF (NSLS X26A station) spectra collected by irradiating NIST SRM K961 glass microspheres

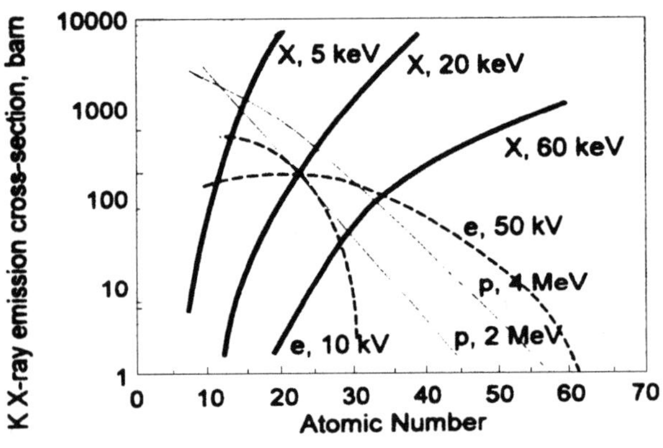

Fig. 3. Cross sections for *K*-shell X-ray production by X-rays, protons and electrons as a function of target atomic number for various projectile energies

Unfortunately, where in conventional electron microprobes routinely currents of, e.g., 1 nA (i.e., about 6.10^9 elections/s are focussed in an area smaller than 0.001 μm^2, and whereas in μ-PIXE apparatus, usually 10–1000 pA are concentrated into a 1–10 μm^2 spot (resulting in fluxdensities in the range of 10^6–10^9 protons/s/μm^2), equivalent photon flux densities cannot be straightforwardly achieved using conventional X-ray sources. The reasons for this are twofold:

(a) the spot emitting X-rays on the anode of a conventional X-ray tube is comparatively large (1 to a few tens of mm^2) while the generated photon beam is emitted into a broad cone and
(b) it is not possible to focus X-rays as easily as charged particles or as radiation of longer wavelengths.

The problem is even agravated by the fact that X-ray tubes (based on electron induced X-ray emission) produce a continuum of X-rays rather than monochromatic radiation. Typical primary flux densities on a sample surface achievable with a standard X-ray tube are situated in the range of 10^4–10^5 photons/s/μm^2 (Cu-tube, 30 mA, 45 kV, sample at 5 cm from the anode spot) when only the principal anode line intensity is considered.

Therefore, in order to be able to take advantage of the inherently better peak-to-background ratio of photon-induced X-ray emission relative to excitation by charged particles, either much more brilliant X-ray sources (see below) than conventional tubes must be used or the radiation from an X-ray tube must be focussed into a microspot on the sample or both.

Tube-Excited vs. Synchrotron Radiation

In comparing the qualities of various X-ray sources, it is relevant to consider, next to the flux F (i.e., the total number of emitted photons/energy interval/s) of an X-ray source, also its brightness B (i.e., the emitted flux per unit solid angle) and brilliance b (being the brightness per unit source size). When an X-ray microbeam is created through collimation of a broad X-ray beam (e.g., by means of a pin-hole), the flux F_{ph} arriving at the sample will be proportional to the source brightness B; in case a focussed X-ray beam is employed, generated by some optical device that forms a (demagnified) image of the X-ray source onto the sample, the flux density at the sample surface is proportional to the source brilliance b. In Fig. 4, the source brilliance of Cu and Mo X-ray tubes is compared to that of various synchrotron sources. Synchrotron radiation is generated when a beam of relativistic electrons (or positrons) is forced to change its direction inside a storage ring (see Fig. 5). When the speed of the electrons is very close to the speed of light, the radiation will be emitted in a narrow cone (a few mrad wide), centered around the velocity vector of the electrons at the time of the acceleration. As a result, synchrotron sources by their very nature are much more brilliant sources of X-rays than X-ray tubes and are therefore much more suitable for generating micro-beams with. In Fig. 5, a schematic layout of a first generation synchrotron is shown; the actual storage ring consists of a number of straight sections in which the electrons do not experience any acceleration; at the intersection of two straight sections, the electrons pass through a bending magnet. Only when the electrons pass through these magnetic fields is synchrotron radiation produced; the radiation is emitted tangentially to the curved path of the electrons and is employed at experimental setups at the end of the beamlines. As can be seen in Fig. 4, bending magnets produce a polychromatic spectrum of X-ray photons extending from the infra red up to the hard X-ray region. The average photon energy in the spectrum is called the critical energy E_G and depends on the energy E_{mach} of the circulating electrons (called the machine energy)

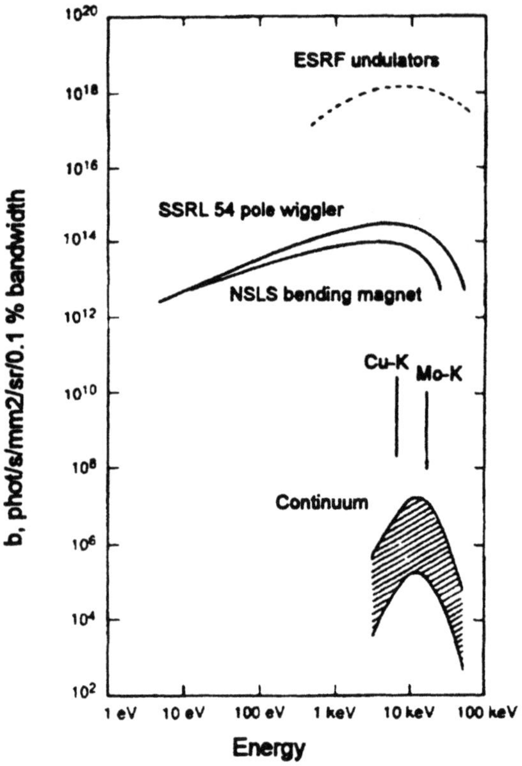

Fig. 4. Source brilliance of various conventional and synchrotron sources

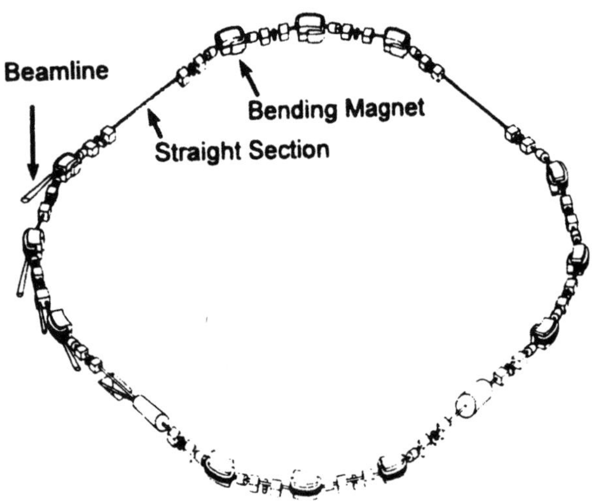

Fig. 5. Schematic drawing of a storage ring

and the field strength B_M of the bending magnets [26]:

$$E_C[keV] = 0.667\ B_M[Tesla]\ E^2[GeV] \quad (2)$$

For instance, the critical energy of the ESRF (European Synchrotron Radiation Facility, Grenoble, France) bending magnets ($E_{mach} = 6\,\text{GeV}$, $B_M = 0.8\,\text{T}$) is 19.2 keV. In second and third generation synchrotrons, other magnetic structures, called 'insertion devices' have been introduced into the straight sections of the ring, forcing the electrons into multiple oscillatory trajectories, thus forming even more brillant X-ray sources (see Fig. 4). 'Wigglers' are insertions devices which produce a polychromatic spectrum similar to bending magnets but 10–100 times more intense; 'undulators' generate spectral distributions with very sharp maxima at certain energies, called undulator harmonics. Within certain limits, the energy of the harmonics is tunable; undulators are therefore excellent sources of (quasi) monochromatic X-rays. (In Fig. 4, the brilliance of the first harmonic of the ESRF undulators is plotted for variable harmonic energy.) Since the electrons that circulate in the storage ring are not uniformly distributed along their trajectory but are concentrated in 'bunches', synchrotrons are pulsed X-ray sources with typical pulse durations in the 10^2–10^3 ps range and a period in the ns-μs range.

Another difference with conventional X-ray sources is that, as a result of the slow decay of the electron current inside the storage ring, the output of a synchrotron source is not constant with time. Injections of new electrons once or twice a day are therefore necessary; in the 12–24 h period in between injections, the synchrotron light is available for experiments. Typical electron currents used in storage rings are currently in the 100–250 mA range.

For μ-SRXRF, a particularly important property of synchrotron radiation is its high degree of linear polarisation when the radiation is sampled in the ring plane. The total emitted radiation intensity can be considered to consist of two components $I_{\parallel}$, and $I_{\perp}$ corresponding to photons whose electrical field vectors are resp. situated parallel and perpendicular to the storage ring plane. The degree of linear polarisation P is defined as:

$$P = \frac{I_{\parallel} - I_{\perp}}{I_{\parallel} + I_{\perp}} \tag{3}$$

Due to this polarisation, scattering interactions that occur with atoms of a sample placed in the ring plane will cause photons to be scattered preferentially out of this plane. Therefore, when SR-XRF experiments are done, as shown in Fig. 6, with the sample e.g. at 45° to the primary beam and with the Si(Li) detector at 90° to the

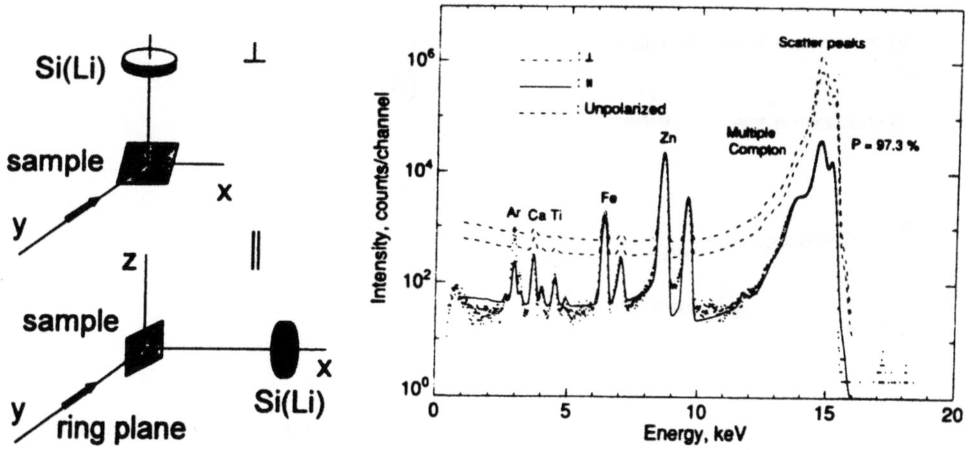

Fig. 6. Experimental arrangements for SR-XRF leading to (∥) minimal and (⊥) maximal scatter background. The background in either geometry using unpolarised radiation is also indicated

primary beam, but also *in the storage ring plane*, the resulting spectra will show a markedly reduced scatter background level compared to when the Si(Li) would not be positioned in this plane. The degree of linear polarisation P of microbeams obtained through collimation of the white synchrotron spectrum generated in a bending magnet can be as high as 99%. To reach such high levels, however, the optical axis of the SR-XRF spectrometer must be carefully aligned in height to that of the storage ring plane. In practice, values in the range 90–99% are routinely achievable. Compared to the unpolarised case, however, even a polarisation value of 90% yields a scatter background reduction of approximately 10 times in SR-XRF spectra.

Microfocussing of X-Ray Beams

As stated earlier, it is not straightforward to focus X-ray beams down to micrometer sized cross sections. Unlike radiation in or near the visible part of the electromagnetic spectrum, X-rays are (almost) not refracted at the boundary of two materials but simply penetrate from one medium into the other. As synchrotron sources, a wide range of optical devices are being used for X-ray focussing [26, 27]. Within the scope of this paper, only a few will be touched upon.

An overview of focussing devices used for generating X-ray microbeams is given in Figs. 7–9. In many cases, these devices employ the phenomenon of total external reflection: this occurs when X-ray photons of energy E impinge on an optically flat

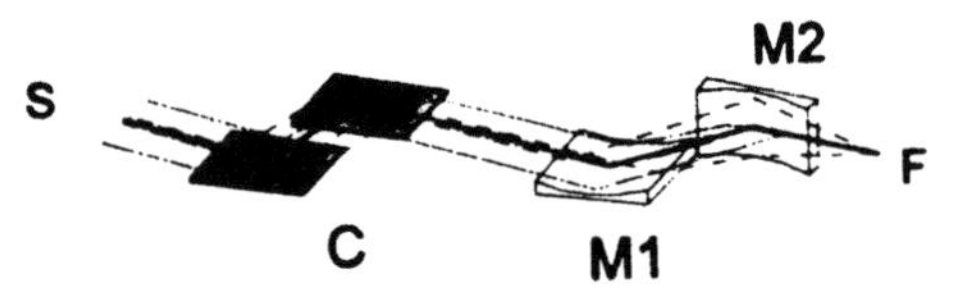

Fig. 7. Monochromatisation (*C*) and two-dimensional focussing of hard X-rays by elliptical mirrors (*M1*, *M2*) in Kirkpatrick-Baez geometry (*S* source, *F* focus)

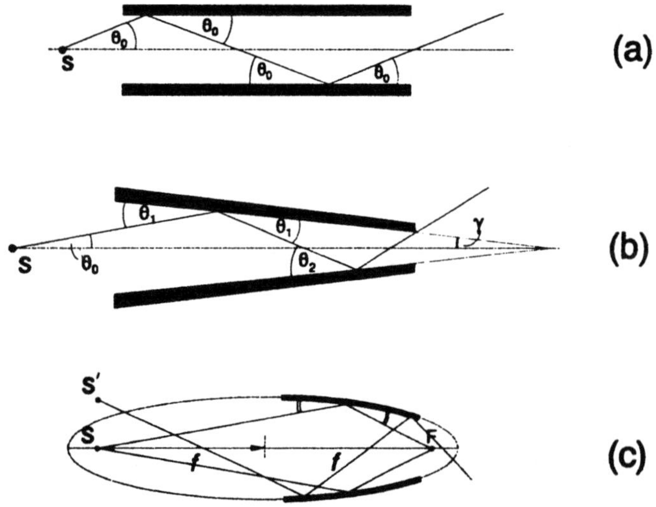

Fig. 8. Principle of capillary X-ray concentrators: **a** straight, **b** conical, and **c** ellipsoidal capillaries

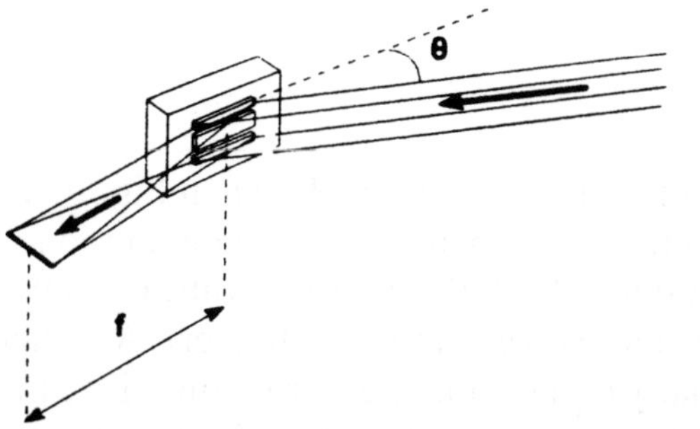

Fig. 9. Principle of operation of 1-dimensional Bragg-Fresnel lenses; f focal distance of the lens, θ Bragg angle

boundary (e.g., air/quartz or vacuum/platinum) below the critical angle of total reflection $\theta_C(E)$ for this interface. Under conditions of total reflection, the photons only very shallowly penetrate into the material and most of the beam intensity (up to 99% for air/quartz) is reflected back. By fashioning, e.g., Pt or Rh coated, elliptical mirrors, the total reflection principle can be employed for focussing X-rays that originate from one focal point of the ellipse into the other (see also Fig. 8c) [27]. By using two elliptical mirrors in Kirkpatrick-Baez geometry, two-dimensional focussing can be achieved (see Fig. 7); an alternative is to employ a single ellipsoidally shaped mirror. Next to totally reflecting mirrors, also Bragg reflection has been employed for creating X-ray microbeams for μ-SRXRF purposes; by means of an ellipsoidally double bent Si(111)-wafer, simultaneous monochromatisation and focussing of 15 keV bending magnet radiation to a spot of 10–20 μm has been accomplished [10].

In glass capillaries, multiple total reflection of X-rays on the inner walls takes place (Fig. 8). Straight capillaries, similar to optical waveguides simply 'transport' X-rays from one end of the glass tube to the other (see Fig. 8a). When these glass tubes are fashioned to have conical inner shapes, under certain conditions, X-ray beams can be 'squeezed' down to the final inner diameter of the conical tube, as shown in Fig. 8b [28]. A recent development is the possibility to manufacture ellipsoidally shaped capillaries which operate in a similar manner as the above mentioned elliptical mirrors (Fig. 8c). However, whereas elliptical mirrors are an established type of focussing device for synchrotron beams, the use of capillaries for focussing synchrotron light is still in the evaluation phase [29]. On the other hand, when tube generated X-rays are concerned, glass capillaries are successfully and routinely being employed for generating microbeams with cross sections in the 5–20 μm range, as part of laboratory scale μ-XRF spectrometers [16, 17, 19–21].

Another type of optical device that in recent years has been gaining attention for obtaining (sub)micron sized spots of hard X-rays at synchrotron facilities are Bragg-Fresnel lenses [30]; their principle of operation is illustrated in Fig. 9 for the case of one-dimensional focussing. Bragg-Fresnel lenses work similarly as Fresnel zone plates but employ Bragg-reflection rather than absorption and transmission of the X-rays in alternate zones. By means of the one-dimensional lens shown in Fig. 9, in the focal plane, a central line of high intensity is created in an area which is depleted of X-rays. Using circular Bragg-Fresnel lenses, at the ESRF, microspots with an area below 1 μm^2 have recently been obtained with a monochromatic photon flux of about 10^9 ph/s. Whether this type of device will be useful for SR-XRF experiments in the near future is currently being evaluated [30, 31].

3. Instrumentation for μ-XRF

Laboratory-Scale μ-XRF

A schematic layout of a laboratory scale μ-XRF spectrometer is shown in Fig. 10. In principle, the basic layout is very similar to the experimental arrangement used for EPXMA and μ-PIXE, with the exception that in μ-XRF the detector is always at 90° to the primary beam to minimise the scatter background (see higher). Another difference is that μ-XRF spectrometers usually do not operate in vacuum; since the highest sensitivity is obtained for the heavier elements (see Fig. 3), usually air absorption of the fluorescent radiation between specimen and detector is not a limiting factor. Accordingly, the building blocks of the setup shown in Fig. 10 can simply be put on a support table which is attached to the X-ray source being used. Figure 11 shows a picture of the μ-XRF spectrometer in our laboratory: a glass capillary is used for focussing radiation generated by a rotating anode source onto the sample which is mounted onto an XYZθ-stage. The sample is viewed with an optical microscope using long working distance lenses. Elemental maps can be recorded by moving the sample through the microbeam and collecting an ED-XRF spectrum at each location. Since the efficiency with which conical and ellipsoidal capillaries focus X-rays decreases with increasing photon energy, often Cr and/or Cu tubes are employed as X-ray sources in these instruments. With these tubes, beam sizes down to 5–10 μm can be employed with good sensitivity [20]. When Mo or Rh anodes are used, the smallest beam size offering a sufficient photon flux is in the range 20–40 μm.

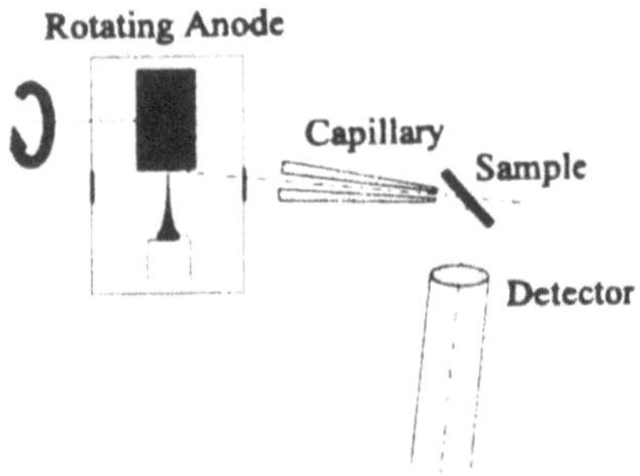

Fig. 10. Schematic layout of a laboratory scale μ-XRF spectrometer

Fig. 11. Photograph of the laboratory scale μ-XRF spectrometer of the University of Antwerp

Fig. 12. Photograph of the Hasylab Line L μ-SRXRF spectrometer. Primary beam (from right) is collimated by means of a straight, 10 μm diameter capillary. Fluorescent X-rays are detected at 90° in the ring plane by a (collimated) HP-Ge detector; samples are at 45° to the beam and detector axis. Sample viewing is done by a long-distance microscope

Synchrotron Radiation Induced μ-XRF

When polychromatic synchrotron light is employed as means of excitation, more or less the same layout is employed with the exception that the entire arrangement (collimator or focussing device, sample stage, microscope, detector) is mounted onto a table which can be adjusted in height (with an accuracy of ca. 100 μm) to the center of the synchrotron beam in order to maximise the degree of polarisation P of the primary beam (see above). A photograph of the Hasylab μ-SRXRF spectrometer is shown in Fig. 12. In this setup, straight glass capillaries with a bore hole of 30, 20 and 10 μm diameter are used as collimators for the primary beam. Since the high intensity of the primary beam, for some samples, may give rise to intense matrix lines, between detector and sample, an absorber (e.g., one or more thin polymer or Al foils) may be placed. Similarly, the primary beam may be filtered to reduce the overall intensity or to influence the shape of the polychromatic excitation spectrum (if any). The current generation of collimated-beam SR-XRF spectrometers employs beam sizes in the range 5–10 μm. When a focussed (monochromatic) microbeam is used, the arrangement is more elaborate and also includes a monochromator and focussing optics in between the ring and the sample (see refs. 10–12 for details); at SR-XRF stations installed at second generation synchrotrons, focussed monochromatic beam size in the 5–30 μm range have been reported [10–12].

Analytical Characteristics of μ-(SR)XRF vs. Other Microanalytical Techniques

From the NIST SRM 1577 a Bovine Liver spectra in Fig. 1, an idea can be obtained of the difference in sensitivity between on the one hand monochromatic and polychromatic SR-XRF and on the other hand conventional and synchrotron based (μ-)XRF. In Fig. 13, the corresponding MDL values for the different modes of excitation are compared. Clearly the high degree of polarisation of synchrotron radiation significantly lowers the scatter background level; however, the most

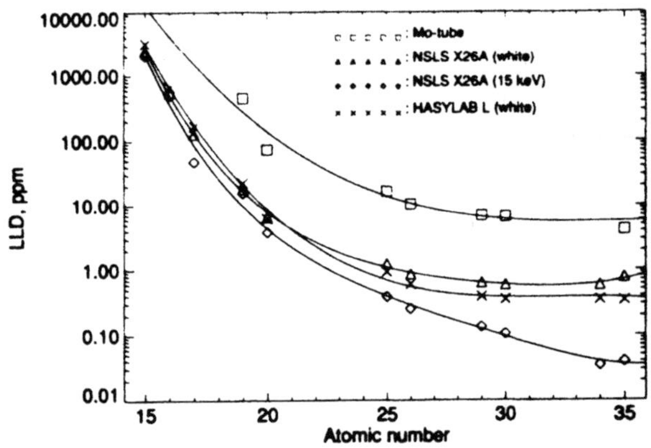

Fig. 13. MDL values corresponding to the spectra shown in Fig. 1: **a** conventional XRF (Mo-anode), **b** monochromatic 15 keV SR-XRF, **c** polychromatic μ-SRXRF at NSLS station X26A and **d** at Hasylab station L

Table 2. Analytical characteristics of various microanalytical techniques compared to those of μ-(SR)XRF

	Projectile	Quantum energy (keV)	Destructive?	Beam size (μm)	Sampling depth (μm)	MDL (ppm)	Quant. accuracy (%)
EPXMA	e^-	5–100	no	<0.1	1–10	>100	b.t. 20%
μ-PIXE	p^+, α	1–4 10^3	yes/no	0.2–5	5–100	5–10	b.t. 20%
μ-SRXRF	X	2–80	no	3–20	10–1000	1–100	b.t. 10%
μ-XRF	X	2–60	no	5–60	10–1000	10–100	b.t. 10%
SIMS	M^+, N^-	10–30	yes	0.5–1	0.010	<1	w.t. 10%

b.t. = better than; w.t. = worse than.

important improvement in peak-to-background ratio is obtained when monochromatic instead of polychromatic excitation is employed. In Table 2, next to the sensitivity, also other analytically important characteristics of μ-(SR)XRF are compared to those of more conventional microanalytical techniques. Probably the most advantageous characteristic of μ-(SR)XRF is its ability of yielding reliable quantitative information on the microscopic distribution of trace elements (see next section); the weak spot of current-day SR-XRF instruments (as installed at second generation synchrotrons) is the limited lateral resolution that they offer [8]. The lateral resolution is primarily defined by the beam size, but is also negatively influenced by the fact that the penetration and sampling depths are equal or larger than the spot size (cfr. Table 1 and Table 2). The next generation of instruments, which are planned to be installed at third generation rings such as the ESRF [13–15] will probably employ beams with cross-section down to 1 μm.

4. Quantitative Microanalysis Using μ-(SR)XRF

Quantification of Conventional XRF Data

One of the advantages of XRF is its quantitative reliability. This advantageous property is a direct consequence of the simple and well-defined way in which characteristic radiation is generated as a result of photon bombardment of a sample material. Since in most cases the cross-section for photo-ionisation is much larger than those for scattering interactions, the majority of the primary photons are annihilated in photo-ionisation processes which directly result in the production of characteristic radiation (next to Auger electrons). Therefore, in (μ-)XRF, for homogeneous samples of thickness T, the observed characteristic intensity I_j of element j (with energy E_j) for monochromatic excitation with energy E_0 is proportional to [1–3]:

$$I_j(E_0) \propto \int_0^T e^{-[\mu_0 \rho\ z \csc(\phi) + \mu_j \rho\ z \csc(\psi)]} c_j I_0(E_0) \tau_j(E_0) \rho \mathrm{d}z$$

$$\propto c_j I_0(E_0) \tau_j(E_0) \left(\frac{1 - \mathrm{e}^{-\chi \rho T}}{\chi} \right) \tag{4}$$

where $X = \mu_0 \csc(\phi) + \mu_j \csc(\psi)$; ϕ and ψ are resp. the incident and take-off angles of primary and fluorescent X-rays, ρ is the density of the material, $\tau_j(E_0)$ is the photo-electric cross-section of element j at energy E_0 and μ_0, μ_j are the mass absorption coefficients of the material at the primary energy E_0 and the fluorescent energy E_j, c_j is the concentration of element j and I_0 is the primary beam intensity. (For clarity, in Eq. (4) contributions to I_j due to higher order phenomena such as secondary fluorescence were omitted.)

On the other hand, in EPXMA and PIXE, the production of element-specific radiation is due to (slowly decelerating) electrons or protons, whose energy may vary between its original value E_0 and the absorption edge energy of the atom to be ionised. As a result, the relation between I_j, c_j, τ_j and T becomes much more complicated for charge-particle induced X-ray emission, as an integration of the corresponding photo-electric cross-sections along the deceleration path of the primary particles is involved [23, 24].

In XRF quantification, Eq. (4) or its equivalent (essentially, $I_j = f(c_j)$) is used in reverse order (i.e., $c_j = f^{-1}(I_j)$) by the so called 'fundamental parameter methods' [2, 32]. This approach to quantification has the advantage of being general in nature (i.e., applicable to any type of sample matrix and/or excitation conditions) but is limited in accuracy as a result of the uncertainty in the fundamental atomic constants employed (photo-electric cross-sections, mass-absorption coefficients, fluorescence yields, transition probabilities etc.) to the level of about 5% [32]. More accurate quantification schemes are empirical in nature (the 'alpha coefficient methods') but mostly employ calibration relations which are only valid in a narrow matrix composition range and usually require an extensive set of calibration standards with compositions very similar to the unknown samples (see, e.g., ref. 33).

In view of the greater flexibility of the fundamental parameter method, this approach is the most often used for quantification of SR-XRF and μ-XRF data.

However, in order to be able to use calibration relations such as Eq. (4), also the sample thickness T and the composition of the so-called 'dark matrix' must be known [32]. The latter is that fraction of the sample material (usually consisting of low-Z elements such as C, N, O) which does not yield measurable characteristic X-ray lines. However, since it can significantly contribute to the self-absorption of the primary and fluorescent radiation, its presence must be corrected for. When these quantities are not *a priori* known, in conventional XRF, usually the sample mass ρT and its mean atomic number $\langle Z \rangle$ are estimated from resp. the absolute and relative intensities of the Rayleigh and Compton scatter peak intensities I_R and I_C; for a given experimental setup, empirical or theoretically calculated calibration curves between on the one hand ρT and the total scattering intensity $I_R + I_C$ and on the other hand between $\langle Z \rangle$ and the Compton-to-Rayleigh ratio I_C/I_R can be established. As an example, in Fig. 14, the calibration curves corresponding to irradiation using monochromatic 15 keV synchrotron radiation are shown [35] for the NSLS SR-XRF setup. (Complicating factors such as the scatter contribution due to the air medium are not discussed here – see Ref. 35 for details.)

Quantification Problems Associated with SR-XRF Data

When trying to apply a similar approach to synchrotron radiation induced XRF data, several difficulties are encountered:

(a) when polychromatic radiation is employed, as a result of the absence of sharp maxima in the excitation spectrum, no separate Compton and Rayleigh scatter peaks can be distinguished. It is therefore not possible to estimate $\langle Z \rangle$ from the scatter background intensity; in this case, the composition of the 'dark matrix' must be determined by another method such as EPXMA.

(b) since the scatter intensities I_R and I_C depend on the degree of linear polarisation P of the primary radiation, the latter must be known during each measurement so that the sample mass ρT can be accurately estimated. As can be seen from Fig. 14a, small errors in the determination of P can lead to significant over- or underestimation of the sample mass [35]. (An additional difficulty in this respect is that for polychromatic excitation, P is not constant over the energy range spanned by the excitation spectrum [36].)

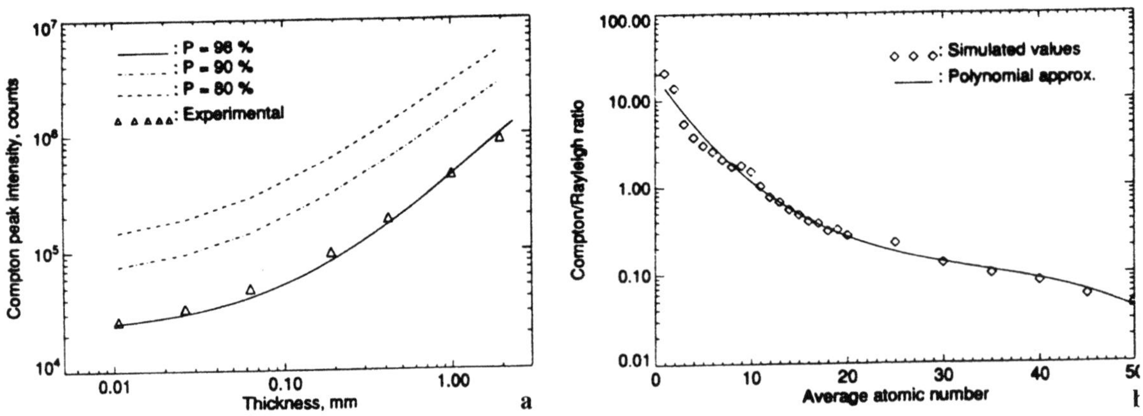

Fig. 14. Operational calibration curves relating (**a**) the sample mass ρT to the scatter intensity I_c for different values of P and for polypropylene foils, (**b**) the sample mean atomic number $\langle Z \rangle$ to the Compton/Rayleigh ratio I_c/I_R

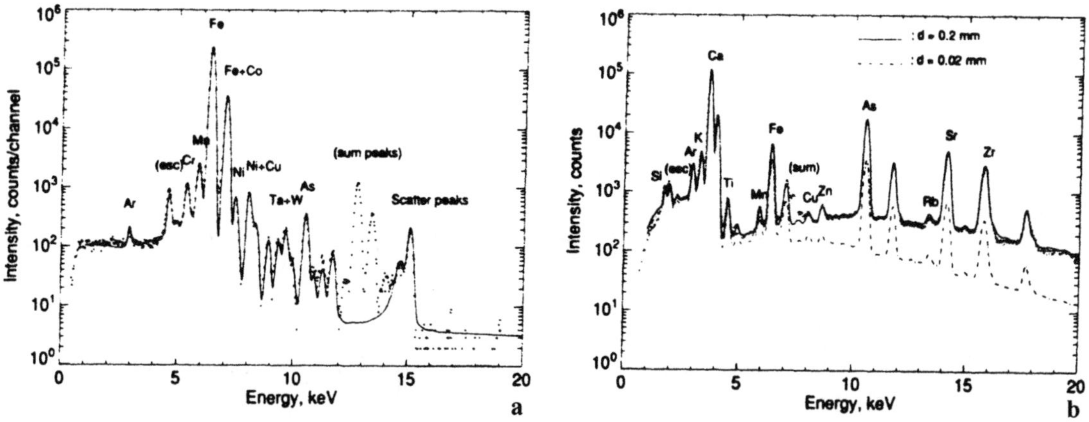

Fig. 15. Experimental (dots) and simulated (lines) ED-XRF distributions of NIST SRM 662 AISI 94B17 Steel using monochromatic 15 keV excitation (**a**) and NIST SRM 620 Soda Lime Glass using white beam excitation (**b**). Both experimental spectra were recorded at the NSLS X26A SR-XRF station. **b** also shows the simulated spectrum corresponding to 1/10th of the actual sample thickness, indicating the difference in sampling depth among, e.g., Ca and Sr

In order to gain a better insight in how experimental parameters such as the excitation spectrum, its degree of polarisation, the detection geometry, the sample composition and thickness influence the characteristic and scatter intensities in SR-XRF spectra and thus determine the sensitivity and quantitative accuracy of the method, a detailed Monte Carlo simulation model was developed to model the interactions between a beam of polarised synchrotron radiation and a given sample material [34–37]. The code takes as input parameters the excitation conditions, the detection geometry and the characteristics of the sample being irradiated and generates the equivalent of experimentally collected ED-XRF spectra. Using this code, it is possible to systematically study the influence of all above-mentioned parameters and evaluate their significance. One can obtain an idea of the predictive power of this Monte Carlo model from Figs. 1 and 15 in which experimental and predicted XRF spectra are compared, corresponding to different experimental conditions and sample materials. In view of the extensive similarity of the experimental with the simulated spectral data, it has been possible to use the simulation model for quantification purposes in various ways. One approach is to simulate ED-XRF spectra corresponding to (unavailable or unmeasured) standards, which then can be used in conventional empirical or fundamental parameter based calibration schemes [38]. Another way is to vary the input parameters of the simulation program (primarily the sample characteristics) until the resulting simulated spectral distribution shows the highest overlap with the experimental spectrum that one wants to quantify [35, 36]. In Table 3, some quantitative results obtained with the latter approach are summarised for different excitation conditions and sample types using experimental data collected at the Hasylab SR-XRF station [39]. In general, root-mean-square deviations between certified and calculated concentrations in the range of 5–15% are observed; in view of the above-mentioned specific problems that are associated with the quantification of SR-XRF data, we consider these to be quite acceptable results [39].

Table 3. Results of quantitative analysis of various types of materials by means of the Monte Carlo model. All results in μg/g (ppm) or in weight percentage (%)

Matrix	Biological materials				Metallic materials				Geological materials			
Material	NIST SRM 1571		NIST SRM 1577a		NIST SRM 1206		NIST SRM 1208		KL2		ATHO	
Element	Cert. (μg/g)	Calc. (μg/g)	Cert. (μg/g)	Calc. (μg/g)	Cert. (%)	Calc. (%)	Cert. (%)	Calc. (%)	Cert.[a] (μg/g)	Calc. (μg/g)	Cert.[a] (μg/g)	Calc. (μg/g)
K	1.47%	1.41%	0.97%	0.92%					0.81%	0.77%	2.26%	2.21%
Ca	2.09%	2.1%	0.012	⟨MDL					7.96%	10.5%	1.25%	1.14%
Ti					2.94	2.74	0.46	0.52	1.52%	1.96%	1620	1550
Cr					19.7	18.4	17.5	17.3	315	280	4.0	3.6
Mn	91	87	10.3	7.9					1290	1310	0.077	–
Fe	300	258	270	261	0.46	0.41	19.2	19.16	8.79%	(ref)	2.59%	(ref)
Co					11.55	11.9	0.82	–				
Ni					53.3	53.4	51.9	50.9	112	117	7	–
Cu	12	14.8	193	204							13	23
Zn	25	22.7	130	134							140	152
As	14	14.5	0.055	<MDL								
Br	(10)	9.2		9								
Rb	12	12.7	18.3	19.6					9.36	5.0	67	65
Sr	37	35.2							363	342	105	94
Y									24.4	27.0	213	105
Zr									163	151	562	613
Nb							5.38	5.30	17	14		
Mo					10.3	10.4	3.24	3.35				
Ba									124	102	740	556
La									13.7	7.8	53.2	50.0
Ce									33.7	27.0	116	112
Nd									22.3	26.9		
Pb	45	48.9	0.34	<MDL								

[a] The geological materials were irradiated as fused glass; the certified concentrations refer to the parent rocks.

In order to arrive at these results, however, it is mandatory, as a part of each measurement run, to irradiate a number of reference samples. For verifying the correctness of the expected (polychromatic) excitation spectrum and of the detector efficiency calibration, the simplest and most rapid method is to measure a multi-element thin standard, prior to or in between unknown samples. Very useful standards in this respect are NIST SRM 1832 and 1833. Secondly, to establish the average degree of polarisation of the synchrotron beam that impinges on the sample, it is convenient to measure a thick (e.g., 1 mm) polymer foil with known trace constituents. Via the Monte Carlo simulation model, it is then fairly easy (a) to estimate the average value of P in the beam and (b) to verify that the primary excitation spectrum which is assumed in the quantitative calculations agrees with the one employed in the experiments.

Quantification of μ-XRF Line-Scan and Image Data

It should be noted, that the above-described calibration method and the results listed in Table 3 are valid for homogeneous samples. In μ-XRF (either with or without the use of polarised synchrotron radiation) obviously the majority of the investigated samples will be heterogeneous in nature. We have therefore recently started to extend the above-mentioned Monte Carlo simulation code to allow it to model the interactions of an X-ray microbeam of given characteristics (such as its spatial and angular intensity and energy distribution) with samples composed of a number of distinct phases [40]. Examples of heterogeneous sample types are, e.g., particles on a filter backing, fluid inclusions in a geological material, multilayered samples and interfaces between two minerals inside a thin rock section. Most quantification errors in this type of samples result from the incomplete and/or inaccurate correction of the complex self-absorption of the primary and the fluorescent radiation inside the sample. As an example, Fig. 16 shows experimental and predicted net intensity profiles obtained by scanning an $8 \times 8\,\mu m^2$ white synchrotron beam over a 32 μm NIST K961 glass microsphere along directions in and perpendicular to the beam-sample-detector plane (resp. horizontal and vertical scans). Whereas in the horizontal scan, the fluorescent radiation needs to pass through the entire particle on its way to the detector, in the vertical scan it can escape sideways. Accordingly, the vertical scans are much more symmetric than the horizontal scans. The solid lines in Fig. 16 also show that the Monte Carlo model is capable of predicting the self-absorption in the glass sphere in a correct way. Since the horizontal profiles differ among elements (different fluorescent energies), this implies that the XRF spectra corresponding to various impact points of the microbeam on the microsphere will also be different. This is illustrated in Fig. 17 which shows two SR-XRF spectra collected during the horizontal scan, corresponding to locations -10 and $+10\,\mu m$ left and right from the geometrical center of the sphere. It is clear from this example that the sample topology can cause significant systematic errors during the quantification step if too simple calibration models are employed for treating μ-XRF data derived from heterogeneous samples. We are currently investigating the possibility to deal with this type of problems in a general way [41].

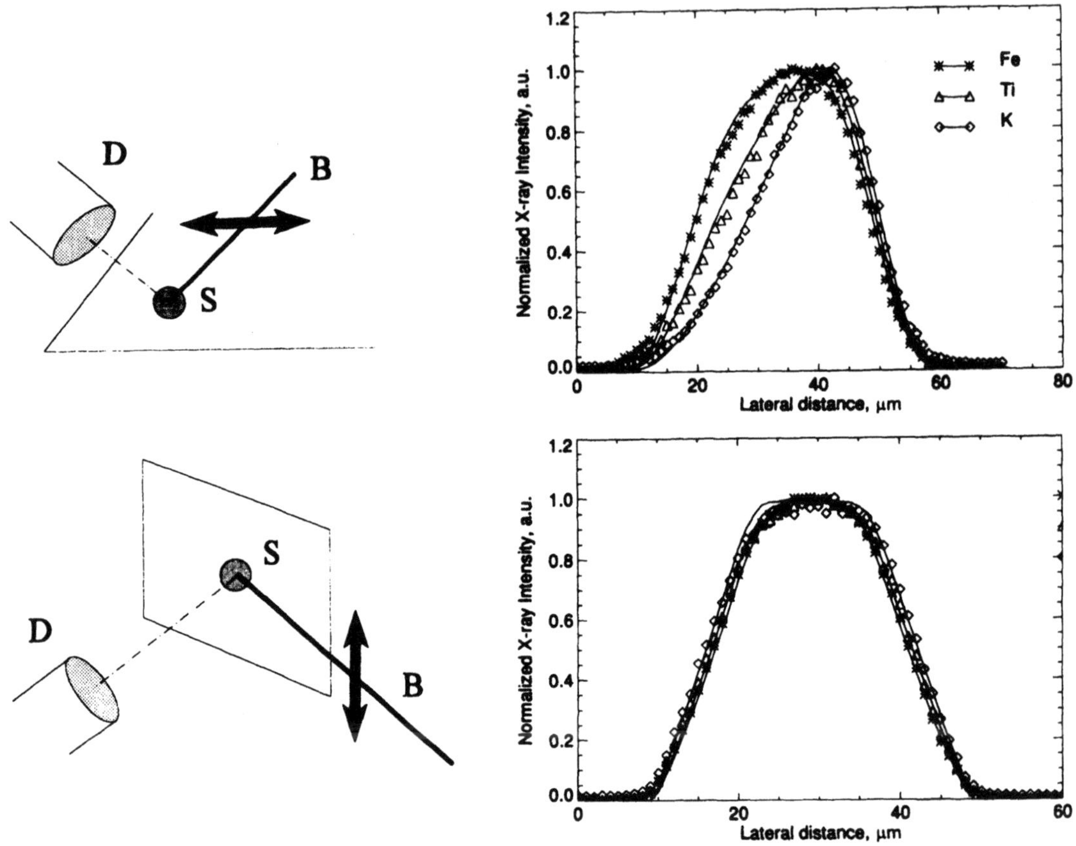

Fig. 16. Experimental (symbols) and simulated (lines) net intensity profiles resulting from horizontally (top panel) and vertically (bottom panel) scanning a $8 \times 8\,\mu m^2$ white SR beam over a $32\,\mu m$ diameter K961 glass microsphere. *B* beam, *S* sample, *D* detector. The asymmetric shapes obtained during the horizontal scan result from the extensive self-absorption in the particle

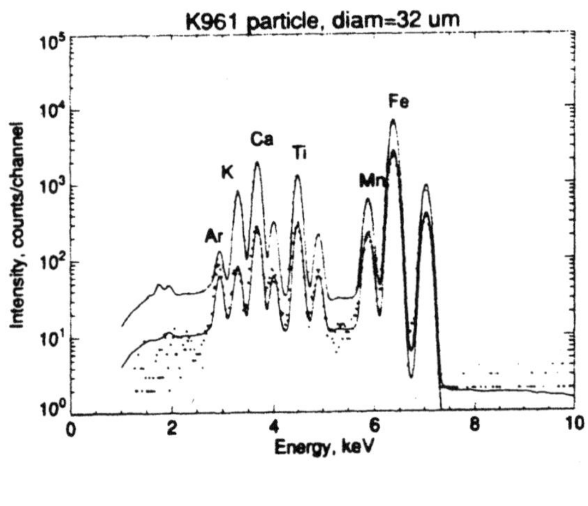

Fig. 17. Experimental (dots) and simulated (lines) SR-XRF spectra collected during the horizontal scan over the K961 microsphere of Fig. 16, at $10\,\mu m$ to the left (lower curve) and right (upper curve) of the geometric centre of the particle. The difference, both in absolute and relative intensity (e.g., Ca relative to Fe) among the two spectra is evident

5. Applications

In this section, a few examples of applications of laboratory scale μ-XRF and μ-SRXRF are briefly discussed. The first SR-XRF example deals with the analysis of corrosion layers in Roman glass objects and is illustrative for the way SR-XRF can provide complementary information on trace-elements when used together with EPXMA. In the second example, the possibilities and problems regarding the non-destructive analysis of fluid inclusions in quartz are outlined. This type of investigation employs the unique blend of characteristics of μ-SRXRF for obtaining laterally resolved data on trace elements from deep inside materials.

Finally, the analysis of Chinese porcelains for dating/authentification purposes using laboratory μ-XRF is briefly described.

Corrosion Analysis of Roman Glass Objects

Glass objects such as cups, goblets, vials etc. which are excavated from Archeological sites dating from Medieval or Ancient times frequently feature a multi-coloured, flaky, irridiscent surface layer. This corrosion layer is formed as a result of the interaction of the glass with the surrounding (moist) soil in which it has been buried for extended periods of time. Recently, in our laboratory, the corrosion layers on a series of about 90 glass objects found at *Khirbet Qumran* (situated on the North Western shore of the Dead Sea, Jordan) were studied using various micro-analytical techniques [42]. Figure 18 shows an electron micrograph of a cross-section of a fragment from a corroded glass flask; next to the original glass (central layer, thickness about 800 μm), on both sides and inside the original glass body, corrosion layers of about 100–200 μm thickness have formed during the period of about 1900 years the object was buried. On top of the original surface, deposits of Ca and K salts (mostly sulphates) can be seen. As is indicated in Table 4, the major composition of the three layers is quite different. Before investigating the corrosion layers them-

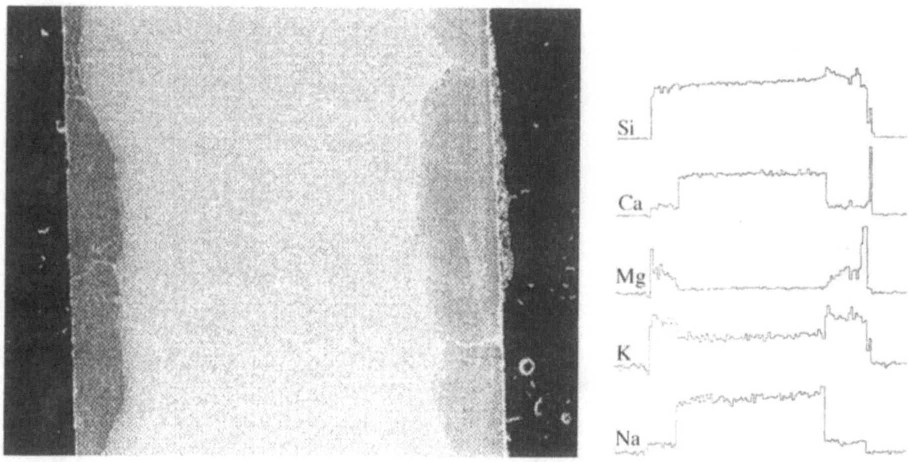

Fig. 18. Left panel: Backscattered electron micrograph of a cross section of Qumran Fragment #22; the original glass can clearly be distinguished from the corroded layer. (Total bottom width of image is 1.3 mm.) Right panel: Major element line profiles across the horizontal line in the micrograph obtained through EPXMA (15 kV, 1 nA)

Table 4. Quantitative EPXMA analysis results (in % w/w) for some NIST glass standards and major element composition for original glass, corroded layer and salt crust of Qumran fragment #22

	NIST SRM 1830		NIST SRM 620		Qumran object #22		
	Certified	Measured	Certified	Measured	Bulk	Leached layer	Crust
Na	10.20	10.59	10.68	10.59	11.66	0.27	0.01
Mg	2.35	2.17	2.23	1.92	0.10	4.24	4.75
Al	0.06	0.13	0.95	1.02	1.33	2.17	2.39
Si	34.16	34.72	33.70	34.09	32.88	38.08	20.16
P					0.04		0.20
S	0.10	0.10	0.11	0.11	0.06	0.33	0.21
Cl					0.87	1.24	0.46
K	0.03	0.02	0.34	0.33	0.75	2.32	1.74
Ca	6.12	5.83	5.08	5.00	5.95	0.64	26.68
Ti					0.06	0.08	0.35
Mn					0.32	0.14	0.26
Fe	0.11	0.09	0.03	0.01	0.33	0.62	2.08
Cu					0.10	0.12	0.01
O	46.73	47.02	46.70	46.92	45.62	49.74	40.71

selves, first the composition of the original (unaffected) glass in all fragments was determined. For determining the matrix composition, EPXMA was employed; as can be seen from Table 4, a standardless ZAF quantification approach provides sufficient accuracy for this type of material. Based on their major element composition, hierarchical cluster analysis was used to separate the 90 fragments into 2 groups, a large one with, on average, a CaO content of $8.4 \pm 0.5\%$ and a smaller one containing less CaO ($5.9 \pm 0.8\%$) (see Table 5). However, the distinction between the two groups is not very clear: only one element (Ca) significantly differs (at the 1 s

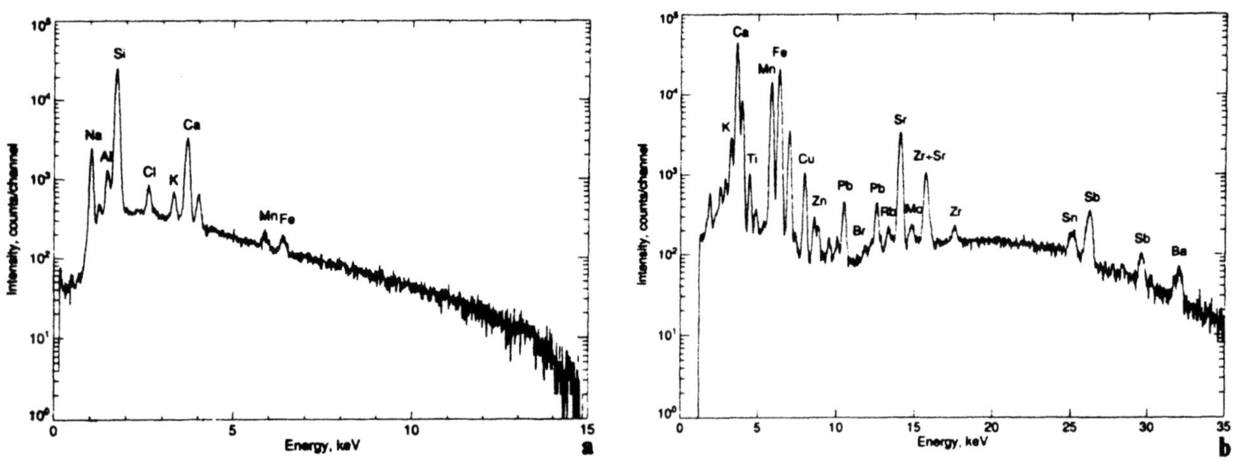

Fig. 19. EPXMA (**a**) and white beam SR-XRF (**b**) spectra of Qumran glass fragment #11 (SR-XRF spectrum collected at the NSLS X26A station). The complementary sensitivity of the two techniques is evident

Table 5. Individual and average compositions of fragments belonging to the two groups of Qumran objects

	High Ca group					Low Ca group			
	#4	#13	#41	#75	Average	#12	#43	#73	Average
				Major composition (% w/w)					
Na_2O	16.4	16.3	16.6	16.0	16.5 ± 0.5	17.3	17.2	17.1	17.0 ± 0.5
MgO	0.5	0.3	0.2	0.2	0.2 ± 0.1	<0.1	<0.1	<0.1	0.1 ± 0.1
Al_2O_3	2.5	2.5	2.5	2.5	2.5 ± 0.1	2.5	2.2	2.5	2.4 ± 0.3
SiO_2	68.2	69.7	70.0	70.1	69.6 ± 0.7	71.8	71.5	72.1	71.6 ± 0.8
P_2O_5	0.1	0.1	0.1	0.1	0.1 ± 0.04	<0.1	0.1	<0.1	<0.1
SO_3	0.1	0.2	0.1	0.1	0.1 ± 0.1	0.2	0.1	0.1	0.2 ± 0.1
Cl	0.7	0.8	0.9	0.8	0.8 ± 0.1	1.2	1.1	1.2	1.1 ± 0.1
K_2O	0.9	0.8	0.8	0.9	0.8 ± 0.1	0.6	0.6	0.6	0.6 ± 0.1
CaO	9.2	8.2	7.9	8.4	8.4 ± 0.5	5.1	6.2	5.0	5.9 ± 0.8
TiO_2	0.1	0.1	<0.1	<0.1	0.1 ± 0.02	<0.1	<0.1	<0.1	<0.1
MnO	0.5	0.5	0.4	0.4	0.4 ± 0.1	0.9	0.7	1.0	0.8 ± 0.3
Fe_2O_3	0.6	0.6	0.5	0.6	0.5 ± 0.1	0.3	0.3	0.3	0.4 ± 0.1
				Trace composition (μg/g)					
Cr_2O_3	26	29	24	14	22 ± 5	38	69	9	30 ± 20
NiO	10	7	7	7	8 ± 1	14	6	10	9 ± 4
CuO	126	124	136	131	143 ± 36	10	24	31	13 ± 9
ZnO	29	28	24	26	32 ± 21	20	33	16	19 ± 7
Rb_2O	11	12	13	11	12 ± 2	12	13	11	12 ± 2
SrO	571	613	624	495	595 ± 99	502	557	561	540 ± 38
Y_2O_3	9	10	10	8	9 ± 2	6	4	8	7 ± 2
ZrO_2	96	100	96	76	86 ± 13	71	71	68	71 ± 9
Mo_2O_3	4	1	3	3	3 ± 2	3	1	4	2 ± 2
SnO_2	165	100	100	113	117 ± 33	42	125	51	52 ± 29
Sb_2O_5	360	318	497	291	281 ± 127	3	50	71	17 ± 26
BaO	212	179	192	182	165 ± 56	122	51	166	129 ± 56
Ta_2O_5	16	16	18	18	18 ± 3	2	11	2	3 ± 3
PbO	113	152	143	104	128 ± 26	9	26	50	16 ± 14

level) among the two groups; within each group, a relatively large variation of the CaO content can be noticed. A much better distinction between the two groups was found when SR-XRF data are considered. EPXMA and SR-XRF spectra of the same fragment are shown in Fig. 19. Instead of obtaining information on the major constituents (Na, Mg, Si, Al, K, Ca) only, in the SR-XRF spectra, clear signals of trace elements such as Cu, Zn, Sr, Rb, Pb, Mo, Zr, Sn, Sb and Ba can be distinguished. By using the trace element information, the existence of the two above-mentioned groups of fragments could be confirmed. In Table 5, the major and trace composition of a few representative members of both above-mentioned groups are listed, together with the group averages. Clearly, a large difference in the

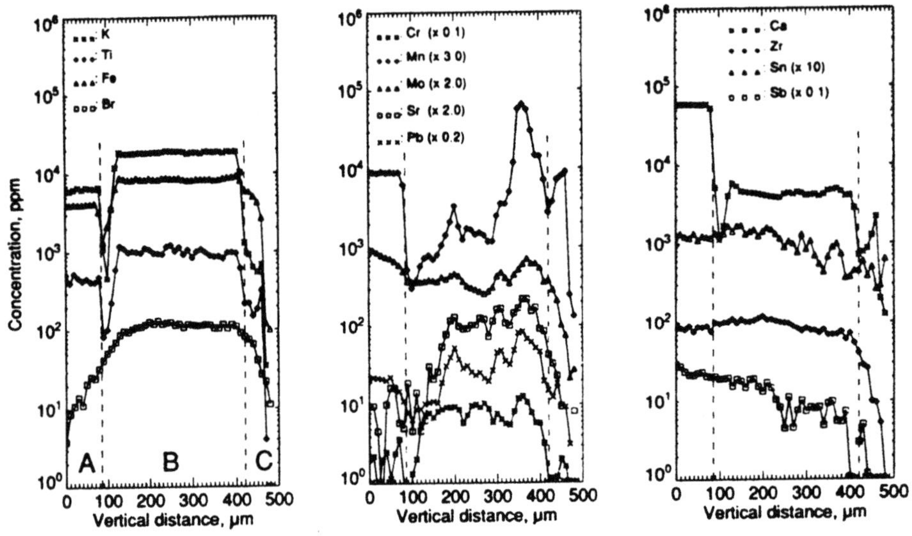

Fig. 20. Quantitative SR-XRF line profiles across the corrosion layer of Qumran fragment #11; *A* original glass; *B* corrosion layer; *C* resin, used for embedding. (Data collected at the NSLS X26A station)

concentration of elements such as Cu, Sn, Sb and Pb exists among the two groups. Considering the fact that in Ancient times, glass objects were rare and valuable items, the fact that most fragments appear to have identical major and trace element composition is remarkable and points to the common origin of all objects. Either all these objects were manufactured locally at the Qumran site from one single stock pile of glass through remelting or they were acquired in large numbers from elsewhere. Both hypotheses, however, seem to suggest the extensive use of glass receptacles at the Qumran Site. This tentative conclusion may support the hypothesis [43] that Khirbet Qumran in Ancient times was a centre of the perfume trade in the Middle East. In Fig. 20, the quantitative line profiles of a number of major and trace elements across the corrosion layer of fragment #11, as determined using SR-XRF, are shown. Most trace elements appear to be enriched in the corrosion layer, with the exception of Rb and Zr which remain at the same level as in the bulk glass. Some of the enriched elements (Ti, Br) behave similar to K and Fe and feature a uniform concentration level throughout the layer; others, however, show asymmetric distributions or are concentrated at the boundary between original glass and corrosion layer (Cu, Bi), or near the surface of the corroded coating (Cr, Mn, Zn, Sr, Mo, Pb). A few elements shown a monotonous decrease in concentration from their bulk value towards the surface (Sn, Sb). These profiles, together with those observed for other fragments, indicate that the corroded layer is not formed as the result of a simple, monotonous leaching process where the major cations (Ca, Na) are replaced by others (Mg, K). Rather, complex migration of major and trace elements in and out of the glass during subsequent periods of time appears to have taken place. To be able to interpret these profiles, more systematic research on the migration of trace elements in glass during moisture-induced corrosion is necessary on model-glass materials.

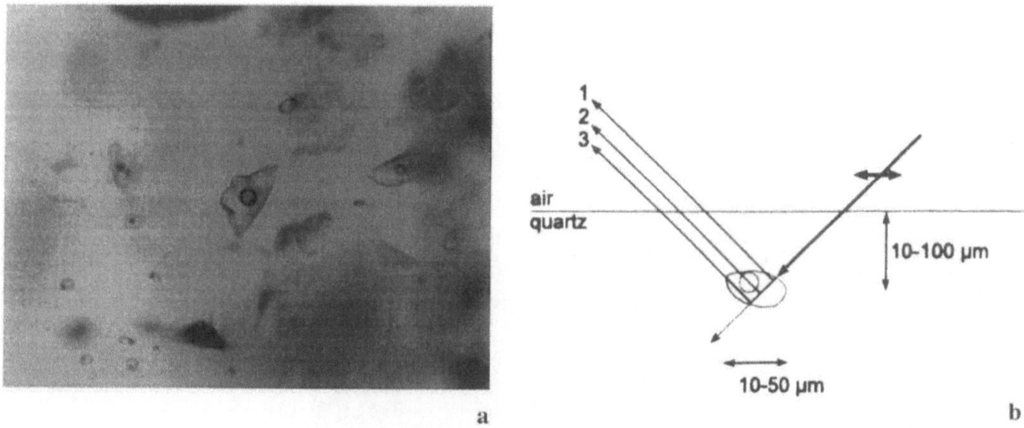

Fig. 21. **a** Photograph of a ca. 50 × 70 μm² fluid inclusion in quartz; **b** schematic drawing of the geometry during X-ray microbeam irradiation. The pathways 1–3 experience different absorption inside and outside the inclusion

Non-Destructive Analysis of Fluid Inclusions in Quartz

Fluid inclusions occur in certain types of minerals (e.g., quartz, fluorite, ⋯) and are formed during the genesis of the minerals: upon cooling of the parent hydrothermal fluid and crystallisation of the minerals, substances which cannot be incorporated into the forming mineral lattice may be concentrated into inclusions [45]. A picture of a typical fluid inclusion in quartz is shown in Fig. 21a. Next to a fluid of high salinity (NaCl, KCl, and/or $CaCl_2$ usually are the major dissolved salts) which is enriched in trace elements such as Br, Sr and Rb, also a CO_2 bubble under high pressure may be present. Next to the fluid and gaseous phases, also microcrystals of the above-mentioned salts may have formed inside the inclusion. Since the major and trace composition of the fluid inside the inclusion contains valuable information about the composition of the original hydrothermal fluid the minerals crystallised from, the analysis of fluid inclusions is of major geological interest [45]. For analysis, usually 1 mm thick sections of the minerals are employed (see Fig. 21). Whereas the major salt concentration inside the fluid can be identified by microthermometry, it is fairly difficult to obtain information on the trace element content. One frequently used approach is to open up inclusions situated at shallow depths by focussed laser bombardment in a mass spectrometer: upon opening, the fluid is released into the mass spectrometer vacuum and a fraction can be (post-) ionised by the laser pulse [46]. Since this method is destructive in nature and involves a rather violent (and not reproducible) way of opening the inclusions, it is difficult to obtain quantitative results with it. By using μ-SRXRF, however, it is possible to obtain data from the inclusion without destroying them [46–48]. Because of the high penetrative power of both the primary and the fluorescent radiation in a matrix such as quartz, trace element signals from inclusions which are situated up to 50–100 μm below the sample surface can still be used for this purpose. μ-SRXRF currently appears to be the only trace-level method with which this is possible. As an example, Fig. 22 shows X-ray maps obtained by scanning a 8 × 8 μm² X-ray beam over the fluid inclusion shown in Fig. 21a. The average depth of the inclusion in 17 μm; its approximate

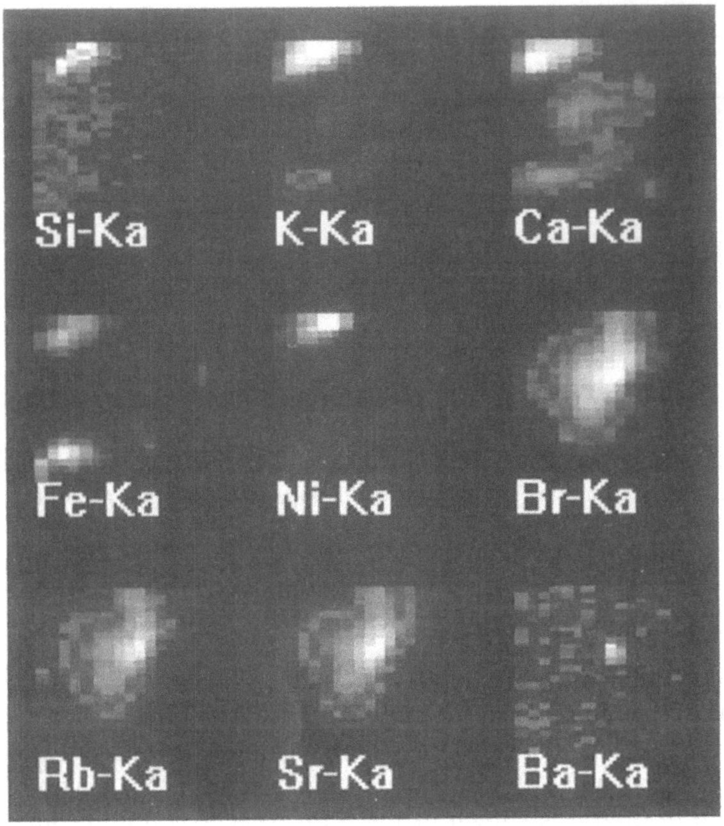

Fig. 22. SR-XRF maps of the fluid inclusion shown in Fig. 21 obtained with a $8 \times 8\,\mu m^2$ white synchrotron beam at the NSLS X26A station

dimensions are $70 \times 50\,\mu m$. The X-ray maps clearly show the enrichment of elements such as K, Ca, Rb and Sr inside the inclusion. A problem currently being studied in this respect is the quantitative analysis of this type of heterogeneous samples: not only must the depth of the center of the inclusion be known with sufficient accuracy [48], also the three-dimensional shape of the bubble can significantly influence the collec/ted X-ray intensities, as was already illustrated previously with the glass microsphere example of Fig. 16. Whereas e.g., the full shape of the inclusion can be discerned in the Br, Rb and Sr signal maps (of which the sampling depth in quartz is larger than $100\,\mu m$), only the parts of the inclusion that are closest to the surface are visible in the K and Ca images (for both signals, the sampling depths is about $10\,\mu m$). A systematic investigation into these matters using synthetic fluid inclusion samples [49, 50] with known trace element content is currently underway.

Dating/Authentification of Chinese Porcelain

Collectors and musea throughout the world are faced with the problem of determining the authenticity of Chinese porcelain of different dynasties [51]. Next to subjective methods based on the stylistic attributes of the objects, also a number of objective methods have been used. One of them is based on the Mn/Co ratio in

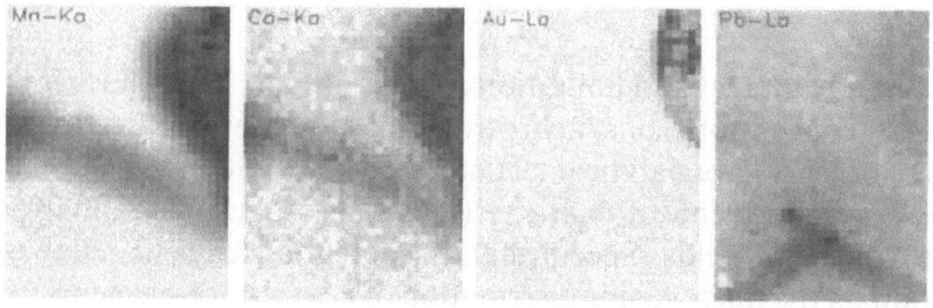

Fig. 23. μ-XRF maps of a 3 × 4 mm² area of the inner surface of an 18th Century Chinese porcelain cup. The Au and Pb signals originate resp. from gold and red decorative lines painted on the blue and white porcelain

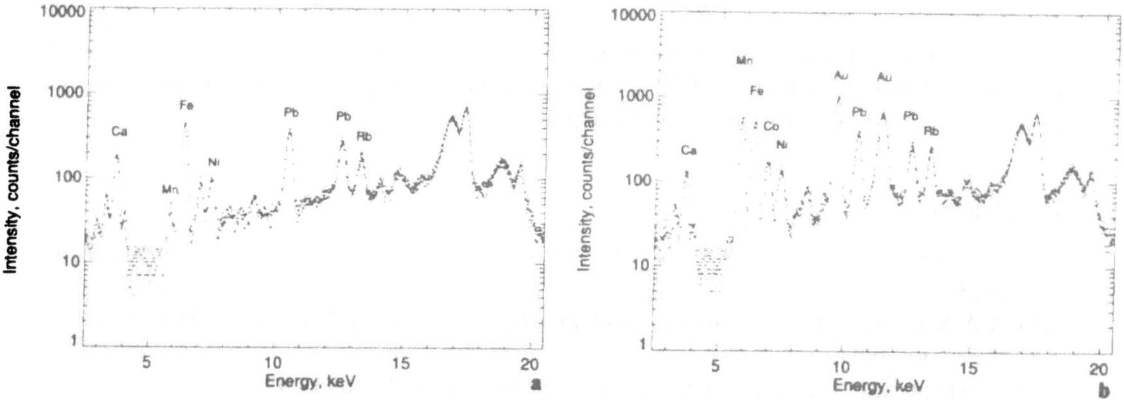

Fig. 24. μ-XRF spectra obtained from the white (**a**) and blue (**b**) areas of the porcelain using a 40 × 40 μm² Mo-Kα beam

blue-and-white porcelains which can be used for classifying objects in groups corresponding to different dynasties [52]. Other methods use the Zn/Rb ratio, the absolute amount of Zn or the Nb/Zr ratio. Recently, in our laboratory, multivariate analysis of XRF data was used for age determination of different Chinese porcelains from the 18th, 19th and 20th century [51]. Although classical bulk XRF can and thus far has been used for this type of work [51–53], in practice these methods sometimes fail because differently coloured parts of the porcelain surface are irradiated simultaneously. On the other hand, since most of these analyses need to be done in a non-destructive way and on large objects, using EPXMA or other microscopical vacuum techniques in many cases is not possible. Using the setup shown in Fig. 11, it proved to be straightforward to collect X-ray spectra from different parts of a fragment of an 18th century Chinese porcelain cup. As an illustration, in Fig. 23, the Mn, Co, Pb and Au maps of a 3 × 4 mm² area (part of a floral motif on the cup), obtained by using a focussed Mo-K_α beam of 40 μm cross section are shown. Figure 24 shows spectra corresponding to the white and blue colored areas visible in Fig. 23; the strong Au signal visible in Fig. 24b originates from a decorative pattern of lines which has been painted on top of the blue colored flower leaves. A fairly high Pb signal is present throughout the images; red decorative lines painted across the floral design also are rich in Pb.

6. Summary

In this tutorial paper, the advantages and limitations of microscopic X-ray fluorescence analysis (using either conventional or synchrotron radiation) were outlined and compared with those of other microanalytical techniques. It can be concluded that μ-(SR)XRF combines a unique blend of characteristics that makes it a valuable analysis method in a variety of fields. Specifically, its potential to yield reliable quantitative information on the microscopic distribution of heavy trace-elements in a non-destructive way is unsurpassed. The method's weak point, as implemented in current instruments is its limited lateral resolution. However, both the sensitivity and/or the lateral resolution of μ-SRXRF will significantly improve when within a few years, third generation synchrotron sources will be available for material characterisation experiments that employ microscopic X-ray fluorescence.

Acknowledgements. The authors would like to express their thanks to M. Rivers, S. Bajt, S. Sutton, F. Lechtenberg, S. Garbe, M. Radtke, M. Haller, P. K. Jochum, J. Hertogen and N. De Biscop for helping with experiments and for providing sample materials.

References

[1] E. P. Bertin, *Principles and Practice of X-ray Spectrometry Analysis, 2nd Ed.*, Plenum, New York, 1975.
[2] A. A. Markowics, R. E. Van Grieken (eds.), *The Handbook of X-ray Spectrometry*, Marcel Dekker, New York, 1993.
[3] R. Jenkins, *An Introduction to X-ray Spectrometry*, Heyden, London, 1976.
[4] K. H. Janssens, F. C. Adams, *J. Anal. At. Spectrom.* **1989**, *4*, 123.
[5] *Spectrochim. Acta* **1991**, *46*, 1313.
[6] H. Schwenke, J. Knoth, in: *Handbook on X-ray Spectrometry* (R. E. Van Grieken, A. A. Markowics, eds.), Marcel Dekker, New York, 1993, pp. 453–490.
[7] K. W. Jones, B. M. Gordon, *Anal. Chem.* **1989**, *61*, 341A.
[8] K. H. Janssens, F. Adams, K. W. Jones, *Anal. Chim. Acta* **1993**, *283*, 98.
[9] M. Bavdas, A. Knöchel, P. Ketelsen, W. Petersen, N. Gurker, M. H. Salehi, T. Dietrich, *Nucl. Instr. Meth.* **1988**, *A266*, 308.
[10] F. Van Langevelde, D. K. Bowen, G. H. J. Tros, R. D. Vis, A. Huizing, D. K. G. de Boer, *Nucl. Instr. Meth.* **1990**, *A292*, 719.
[11] Y. Goshi Y., S. Aoki, A. Iida, S. Hayakawa, H. Yamaji, K. Sakurai, *Jpn. J. Appl. Phys.* **1987**, *26*, L1260.
[12] A. C. Thompson, J. H. Underwood, Y. Wu, R. D. Giauque, K. W. Jones, M. L. Rivers, *Nucl. Instr. Meth.* **1988**, *A266*, 318.
[13] J. V. Smith, *Conf. Ser. Inst. Phys.* **1992**, *160*, 605.
[14] Y. Goshi, *Personal Communication.*
[15] K. Janssens, F. Adams, A. Snigirev, P. Engström, C. Riekel, in: *Proc. 24th Microbeam Analysis Meeting, New Orleans, Louisiana, July 31st-August 5th, 1994* (J. Friel, ed.), VCH, New York, 1994, pp. 23–24.
[16] A. Rindby, *Nucl. Instr. Meth.* **1986**, *A249*, 536.
[17] P. Engström, S. Larsson, A. Rindby, A. Buttkewitz, S. Garbe, G. Gaul, A. Knöchel, F. Lechtenberg, *Nucl. Instr. Meth.* **1991**, *A302*, 547.
[18] D. J. Thiel, D. Bilderback, A. Lewis, *Nucl. Instr. Meth.* **1992**, *A317*, 597.
[19] K. Janssens, L. Vincze, J. Rubio, F. Adams, *J. Anal. At. Spectrom.* **1994**, *9*, 151.
[20] B. A. Carpenter, *Adv. X-Ray Anal.* **1989**, *32*, 115.

[21] A. Rindby, P. Vogils, G. Nillson, B. Stocklassa, *Adv. X-ray Anal.* **1992**, *35*, 1247.

[22] R. Jenkins, *Spectrochim. Acta* **1982**, *37B*, 207.

[23] S. J. B. Reed, *Electron Microprobe Analysis*, Cambridge University Press, Cambridge, 1975.

[24] S. Johansson, J. Campbell, *PIXE, A Novel Technique for Elemental Analysis*, Wiley, Chichester, 1988.

[25] R. Vis, in: *Applications of Synchrotron Radiation* (C. Catlow, G. Greaves, eds), Blackie, London, 1990, pp. 311–332.

[26] G. Margaritondo, *Introduction to Synchrotron Radiation*, Oxford University Press, New York, 1988.

[27] B. W. Batterman, D. H. Bilderback, in: *Handbook on Synchrotron Radiation, Vol. 3*, North-Holland, Amsterdam, 1991, pp. 105–154.

[28] L. Vincze, K. Janssens, F. Adams, A. Rindby, *X-ray Spectrom* **1995**, *24*, 27.

[29] L. Vincze, K. Janssens, F. Adams, S. Larsson, A. Rinbdy, P. Engström, *Proc. 25th Meeting of the Microbeam Analysis Society, Breckenridge, Colorado, 16–11 August 1995* (E. Etz, ed.), VCH, New York, 1995, pp. 43–44.

[30] D. Erko, P. Chevallier, in: *X-Ray Microscopy III* (A. Michette et al., eds.), Springer, Berlin Heidelberg New York Tokyo, 1992, pp. 105–107

[31] P. Chevallier, P. Dhez, F. Legrand, A. Erko, Yu. Agafonov, L. Panchenko, A Yakshin, *J. Trace Microprobe Techn.*, **1996**, *14*.

[32] F. He, P. Van Espen, *Anal. Chem* **1991**, *61*, 2237.

[33] J. W. Criss, L. S. Birks, *Anal. Chem.* **1968**, *40*, 1080.

[34] K. K. Nielsson, *Anal. Chem.* **1977**, *49*, 641.

[35] L. Vincze, K. Janssens, F. Adams, M. Rivers, K. Jones, *Spectrochim. Acta* **1995**, *50B*, 127.

[36] L. Vincze, K. Janssens, F. Adams, K. Jones, *Spectrochim. Acta B* **1995**, *SOB*, 1481.

[37] L. Vincze, K. Janssens, F. Adams, *Spectrochim. Acta B* **1993**, *48B*, 553.

[38] K. Janssens, L. Vincze, F. Adams, P. Van Espen, *X-Ray Spectrom.* **1993**, *22*, 234.

[39] L. Vincze, K. Janssens, F. Adams, H. Amort, M. Radtke, S. Garbe, F. Lechtenberg, M. Haller, A. Knöchel, K. P. Jochum, *Hasylab Jahresbericht 1994*, Hasylab, Hamburg, 1995, pp. 917–918.

[40] K. Janssens, L. Vincze, F. Adams, K. Jones, *Proc. 25th Microbeam Analysis Society, Breckenridge, CO, 6–11 August 1995*, VCH, New York, 1995, pp. 68–69.

[41] L. Vincze, K. Janssens, F. Adams, *Spectrochim. Acta* **1996**, submitted.

[42] A. Aerts, K. Janssens, F. Adams, H. Wouters, in: *Proc. 24th Microbeam Analysis Society, New Orleans, LA, July 31st August 5th 1994* (J. Friel, ed.), VCH, New York, 1994, pp. 45–46.

[43] P. Donceel-Voûte, *Archeologia* **1994**, *298*, 26.

[44] K. Janssens, A. Aerts, I. De Raedt, L. Vincze, F. Adams, *μ-SRXRF Analysis of Medieval and Roman Glass*, Hasylab Berichte 1994, Hamburg, 1995, pp. 980–990.

[45] E. Roedder, *Geochim. Cosmochim. Acta*, **1990**, *54*, 495.

[46] A. H. Rankin, M. H. Ramsey, B. Coles, F. Van Langevelde, C. R. Thomas, *Geochim. Cosmochim. Acta*, **1992**, *56*, 67.

[47] J. D. Frantz, H. K. Mao, Y. Zhang, Y. Wu, A. Thompson, J. Underwood, R. Giauque, K. Jones, M. Rivers, Chem. Geology **1988**, *69*, 235.

[48] D. A. Vanko, S. R. Sutton, M. L. Rivers, R. J. Bodnar, *Chem. Geology* **1993**, *109*, 125.

[49] R. J. Bodnar, S. M. Sterner, in: *Hydrothermal Experimental Techniques* (G. C. Ulmer, H. L. Barnes, eds.), Wiley, New York, NY, 1987, pp. 423–457.

[50] D. A. Vanko, R. J. Bodnar, M. A. Sterner, *Geochim. Cosmochim. Acta* **1988**, *52*, 2451.

[51] J. Swerts, A. Aerts, N. De Biscop, F. Adams, P. Van Espen, *Chemom. Intell. Lab. Systems.* **1994**, *22*, 97.

[52] C. T. Yap, S. M. Tang, *Archeometry* **1984**, *26*, 78.

[53] C. T. Yap, S. M. Tang, *Archeometry* **1985**, *27*, 61.

Mikrochim. Acta [Suppl.] 13, 117–133 (1996)

Particle-Induced X-Ray Emission – A Quantitative Technique Suitable for Microanalysis

Klas G. Malmqvist

Department of Nuclear Physics, Lund University and Institute of Technology, Sölvegatan 14, S-223 62 Lund, Sweden

Abstract. The characteristic features of elemental analysis using the particle induced X-ray emission technique is discussed with emphasis on its implementation in nuclear microprobes. An account of the various benefits of using micro PIXE is given and the advantages of using other simultaneous techniques of ion beam analysis is stressed.

Key words: PIXE, microanalysis, trace elements, quantitative.

The development of particle-induced X-ray emission (PIXE) has mostly been carried out in nuclear physics laboratories where accelerators are available. Because of the versatility, high sensitivity and multi-elemental capability of PIXE analysis an increased interest has led to the establishment of dedicated PIXE facilities. A further very important advantage is that PIXE can be used as a microbeam technique, which makes imaging and elemental mapping with a spatial resolution of about 1 μm possible. PIXE can be combined with other ion beam-based phenomena such as particle elastic scattering and nuclear reaction analysis (NRA) for determination of light elements and ionoluminescence (IL) for characterisation of chemical state. The usefulness of PIXE will be illustrated by a number of typical applications in biology, medicine, geology, air pollution research, archaeology and the arts in which PIXE will be applied either as a single technique or in combination with others.

PIXE

Particle induced X-ray emission has been developed over a period of two decades. The growth in activity has been rapid, partly because there are many small accelerators that have become obsolete for nuclear physics research but which are ideal for use in MeV ion beam analysis (IBA). The number of acelerators and even more so the manpower are limited and most of the suitable accelerators are in operation at academic institutions, where often formal problems restrict the use of these machines for applied work, especially on a commercial basis. However, these restrictions are less severe for PIXE than for some other IBA methods. The optimum particle energy is 2–3 MeV and hence a suitable accelerator can be rather small, at costs no higher than for some other complex analytical equipment. Hence, any research organisation of a reasonable size can set up a PIXE laboratory with a dedicated accelerator and this also has happened in recent years. PIXE is developing into an independent and well-established analytical method. The reason for this is essentially that PIXE has some unique and advantageous characteristics which make such an investment worthwhile.

Advantages of PIXE

Some of the advantages that can be offered by PIXE analysis are as follows:

– *High sensitivity*: In routine analysis the detection limits are of the order of 0.1–1 ppm. The mass of a typical sample is 0.1–1 mg, which means that the detection limits in absolute terms are 1–100 pg. With even smaller samples the detection limits become correspondingly smaller. An especially important example, which will be discussed in detail below, is the use of a finely focused beam that selects only a small part of the sample for analysis. With a spatial resolution of 1 μm the minimum detectable amounts can be as low as 10^{-17}–10^{-16} g.

– *Versatility and direct analysis*: Any solid object can be irradiated directly provided it can be mounted in the irradiation chamber. For the analysis of larger objects the particle beam can be extracted out through a thin window and the object irradiated in air.

– *Good spatial resolution*: The particle beam can be focused to a very small cross-section and it can be deflected in much the same way as the electron beam in scanning electron microscopy. This allows for very sensitive analysis with good spatial resolution, micro PIXE. When combined with simultaneous nuclear techniques and various imaging methods a *nuclear microprobe* [1] is obtained.

– *Non-destructive analysis*: Some delicate objects such as organic tissue or printed documents can be affected by heating and radiation damage, but nearly always these effects can be avoided by maintaining a sufficiently low current density and total charge. In the microprobe mode the conditions of non-destructive analysis are harder but still possible to meet or circumvent [2].

– *Multi-element analysis*: In conventional PIXE analysis all elements from aluminium to uranium can be measured. The important feature is that there are only small variations in sensitivity between different elements. The light elements can also be determined if a special detector for low-energy X-rays is used. However, a better alternative is to combine PIXE with simultaneous nuclear methods.

None of these characteristics is unique to PIXE. There are analytical methods that are more sensitive and there are other multi-elemental and non-destructive methods, but it can be claimed that no other method combines so many advantages: Hence, PIXE should primarily be used for analytical problems that require most of these characteristics and not when other, less complicated and less expensive methods, can be applied.

Basic Principles of PIXE

As the acronym PIXE indicates, it is an X-ray method with the characteristic feature that the excitation is performed by charged, heavy particles, usually protons or α-particles. This method of excitation has some advantages over the more common method of excitation using X-rays (X-ray fluorescence) or electrons (the electron microprobe). Compared with traditional tube-excited X-ray excitation, the

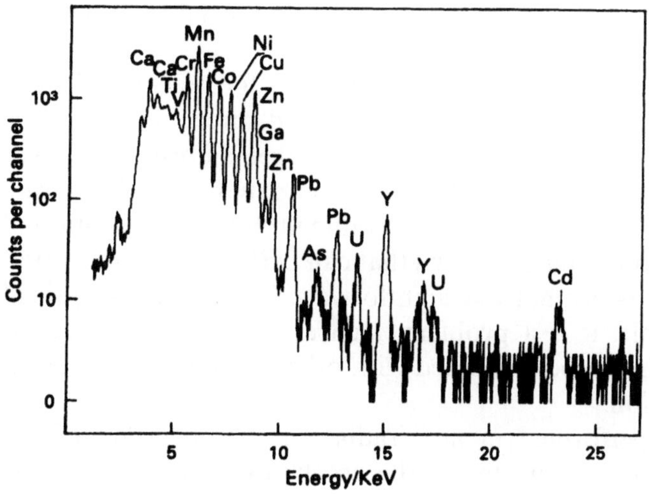

Fig. 1. X-ray spectrum from PIXE analysis of a thin specimen with 100 ng of each of a number of elements (adopted from [3])

product of the beam intensity and cross-section of X-ray production is larger which accounts for the higher absolute sensitivity of PIXE. Compared with electron excitation the heavy particles are less easily scattered than electrons and hence, the intensity of the primary Bremsstrahlung produced in the sample is much lower.

A typical PIXE spectrum for a sample containing about 100 ng each of a number of elements is shown in Fig. 1 (adopted from [3]) showing well-resolved K X-ray peaks for the light and medium-heavy elements and L X-ray peaks for Pb and U. At low X-ray energies the background is produced by secondary electrons and the high-energy tail by the particle beam directly. As the cross-section for X-ray production of different elements is well known, the minimum detection limits can be calculated. PIXE has its maximum sensitivity for atomic number (Z) $20 < Z < 30$ and $Z \approx 80$ which are regions of great interest in many applications. In low-Z matrices the minimum detection limits are of the order of 0.1 ppm for most elements of interest. It should be emphasized that a considerably worse sensitivity is obtained if the matrix contains appreciable amounts of heavier elements, for instance archaeological and geological samples. Despite lower sensitivity, PIXE has been shown to be a useful method even in these fields, especially for geology, which has turned out to be a major field of applications for PIXE.

Experimental Procedure

Comprehensive accounts of the PIXE technique and all its applications can be found in the proceedings of the series of International PIXE Conferences [4–6], arranged every three years and in the conferences on the use of nuclear microprobes, every second year [7, 8]. For a detailed description of the experimental procedures the reader is referred to two textbooks on PIXE [9, 10].

As mentioned above one advantage of PIXE is its versatility. Many different types of specimens can be analysed without any extensive preparation. PIXE analysis of a thin specimen is particularly advantageous because the projectiles lose only a small part of their energy when passing through the sample and that the absorption of the X-rays in the sample is also relatively insignificant. The sample preprocessing in micro PIXE analysis is a tedious procedure crucial for obtaining accurate analytical results [11].

Sometimes, sample preparation is either not possible (e.g., for precious objects) or is difficult to perform (e.g., for refractory minerals), and direct irradiation is carried out. In this alternative (thick-target analysis) the target is normally thicker than the range of the particles. The main difference, compared with thin-target analysis, is that the calculation of the results is more difficult. If the composition of the matrix is known there are computer programs that can perform this calculation [12]. For organic material this is normally straightforward but for the analysis of geological material, for example, it is a problem. Hence, it is an advantage for accurate quantification if PIXE can be combined with other methods for determination of the matrix composition.

PIXE analysis is usually performed in an irradiation chamber under high vacuum connected directly to the beam line of the accelerator. This means that there are practical limits to the size of the objects that can be analysed. Sometimes, if large or sensitive objects are to be analysed, such as paintings, books, archaeological artifacts etc., there is the possibility of bringing the particle beam out through a thin window at the end of the beam line, so called external-beam PIXE [13]. As the charged particles lose energy rather rapidly in air the object to be analysed must be placed very close (preferably within a few mm) to the exit window. An additional advantage is that the ambient air has a cooling effect, so that delicate specimens such as paper or parchment are not so easily damaged by the beam. The extra background radiation production can be reduced and the cooling effects enhanced if the sample is instead flushed by helium gas.

Microbeam Systems

To fully exploit the potential of PIXE the development of high resolution micro PIXE is a natural consequence. Special experimental requirements must be met in a nuclear microprobe facility. The ion beam from the accelerator is focused through a primary collimator, the object aperture. In the focusing of the beam on to the target the large mass and high energy of the ions, makes it impossible to use normal solenoid lenses as in electron microscopes. Instead, normally a set of strongly focusing quadrupole magnets (normally 2 to 4) are used. In Fig. 2 the principles of a NMP compared to an electron microscope column are shown. Due to field impurities, lens aberrations occur and limit the lateral resolution of a nuclear microprobe for micro PIXE analysis to approximately 10^{-6} m. Hitherto, the best obtained value is 3.10^{-7} m [14]. Using a deflection system, either before or after the focusing system, the ion beam can be scanned in two dimensions over the sample in much the same way as the electron beam in a SEM.

The specimen chamber of a NMP is normally equipped with an xyz-translator or goniometer system for mounting the samples. Focusing and positioning require access to optical microscopes and fibre optics illumination to view the sample surface. Inside the chamber a variety of detectors simultaneously produce many signals for imaging and analysis. By tradition, the energy-dispersive X-ray detection (micro PIXE) has been the most important. Surface barrier detectors determine the intensity and energy distribution of scattered primary ions (micro RBS), charged

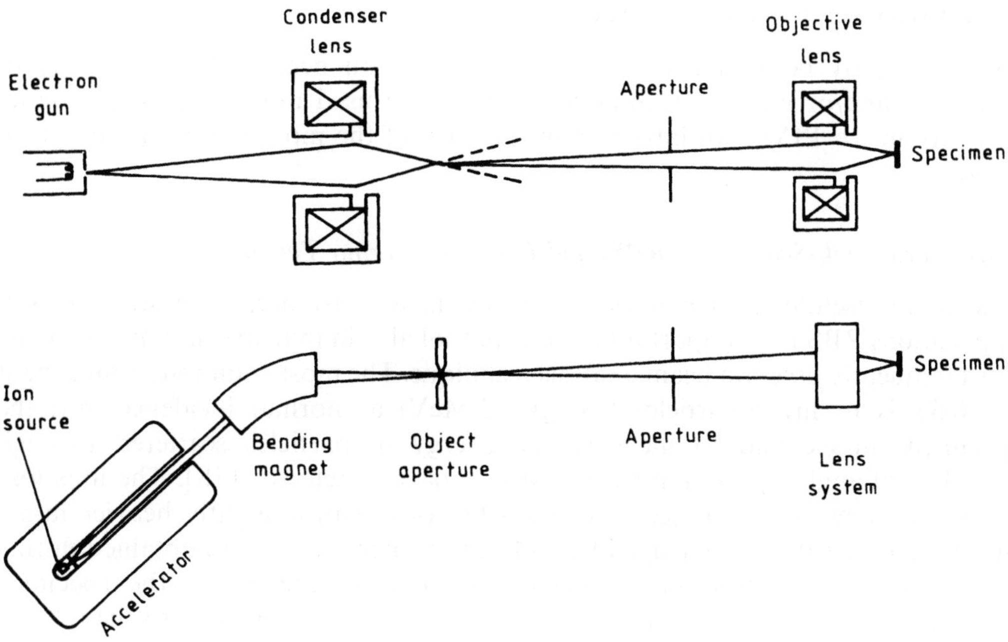

Fig. 2. Schematic experimental arrangements for an electron microscope (top) and a nuclear microprobe (bottom)

particles from nuclear reactions and the transmitted primary ions. It is also feasible to extract the ion beam through a thin foil and bombard the specimen under atmospheric pressure. Due to scatter in the foil and with gas molecules a practical lateral resolution will then be limited to approximately 100 μm.

Imaging of the specimen can be performed using the intensity of detected X-rays or other radiation components to produce, for instance, elemental maps. To fully utilise this possibility the data acquisition systems of most nuclear microprobes are designed to collect data event-by-event [15]. Storing the events consecutively allows a computer to sort the events and produce such two-dimensional element on- or off-line. The maps can be grey scale or colour coded and are intuitive when interpreting data. In analysis of organic materials it is important to reduce losses and to keep track of those which do occur. The organic part of the matrix is mainly affected and biological specimens are normally analysed for heavier elements, for instance metals and the losses will mainly create problems in measuring the concentration of elements in the tissue. Systematic studies of this phenomenon have revealed that it is mainly dose dependent and only to a lesser extent affected by temperature [16]. It has been demonstrated that a single cell may undergo a linear shrinkage of more than 10% under ion bombardment in a nuclear microprobe [17]. Since most of the ion beam analysis methods are quantitative, it is essential that the matrix effects are fully considered. The event-by-event mode of collection allows monitoring of possible element losses during irradiation. The use of high intensity MeV focused ion beams will cause inevitable losses, in particular of light elements like hydrogen and oxygen.

Complementary Analytical Methods

As has already been discussed the use of MeV ion beams offers a variety of analytical and/or imaging techniques that can be used more or less simultaneous with micro PIXE. An introduction to some of the more important are given below.

Rutherford Back-Scattering (RBS) and Other Scattering Techniques

Due to its usefulness in materials analysis in, for instance, solid state physics applications RBS is an important analytical tool also in industry and probably the most frequently used ion beam analysis technique. The most common arrangement for RBS is to fire α-particles (energy < 2 MeV) at normal incidence into the specimen surface and to measure the energy of particles scattered through a backward angle by using a silicon surface barrier detector [18]. The ions lose most energy when the projectile collides head-on with a slightly heavier target nucleus and has its path changed by 180°. In an energy spectrum obtained during RBS analysis of a thin target, each nuclear mass present in the specimen corresponds to a peak. The areas of the peaks will be proportional to the relative concentrations of the nuclei and the peak shape and centroid depend on the elemental depth distribution. Using computer codes the obtained energy spectrum can be used for a simulation of the depth distribution of an element. Since the mass resolution is best for low-Z elements the main use of RBS in combination with PIXE is to determine the concentrations of bulk elements to facilite quantitative PIXE analysis. For MeV protons as projectiles the classical Rutherford conditions are not fulfilled and other kinds of scattering take place as well and the technique is then normally called particle elastic scattering (PESA). This technique is often used simultaneously with micro PIXE [19].

Nuclear Reaction Analysis (NRA)

In matter, the projectiles in an MeV ion beam normally interact with orbital electrons by Coulomb force. However, some rare events of close encounters between incident ions and the nuclei will also occur. The detection and identification of products from the induced nuclear reactions form the basis of NRA. The cross sections for nuclear reactions are determined by complex quantum mechanical rules and depend on the incident energy and on the internal properties of the nucleus. The Coulomb barrier will keep a charged projectile from reacting strongly with a nucleus unless the projectile energy is high enough. With ions of modest energy it is clear that nuclear reactions will be important only for low-Z elements (rule of thumb: $Z < 20$). The usual mode of operation is to chose an ion velocity to match the energy of a resonance of a specific selected nuclear reaction and hence NRA is normally not multielemental. By varying the initial ion energy the resonance energy will be reached at different depth in the sample and the depth distribution of that particular nucleus can be determined. The ability to determine specific low-Z elements in high-Z matrices is a major feature of NRA. Normally NRA relies on the measurement of the energy of various reaction products (charges particles, gamma rays). In a recent development of the technique the fast timing properties of special detector/electronics arrangements are used to combine charged particle and photon detection in coincidence experiments (photon-tagged NRA, pNRA) to significantly reduce the background level in the energy measurement thus significantly reducing the detection limits [20].

One special type of NRA analysis, PIGE, is based on emission of gamma rays from thick targets [21]. Instead of studying specific resonant reactions with a selected value of the ion velocity a standard value is chosen that is suitable both for PIXE and PIGE and assuming homogeneous depth distribution of elements the concentration of several high elements can be deduced simultaneously with the heavier ones from PIXE.

Ionoluminescence (IL)

Normal ion beam analysis is limited to determining elemental composition while the chemical state often is very important. The electronic state of the elements or the crystal state can be revealed if the UV, visible or IR light is measured during bombardment with MeV ions. This IL carries chemical information as to the nature of chemical bonds, the band structure of semiconductors and impurity and defect distributions in crystalline matter [22]. Ionoluminescence resembles cathodoluminescence and photoluminescence but the mode of excitation affects the characteristics of the light emission.

In a nuclear microprobe the irradiated area is a few μm^2 and by recording the signals emitted it is straightforward to form an image. In panchromatic mode all light is collected and the intensity is used for imaging while when a monochromator is used a single spectral line can be used for imaging the presence of, for instance, a rare earth element (REE) in a geological substrate (see also below sect. *Geological applications*). The sensitivity for qualitative analysis can for some REE's be very high (MDL: 10–100 μg/g). This can be combined with quantitative trace element analysis by micro PIXE analysis. In Fig. 3a an IL spectrum for analysis of a synthetic zircon doped with gadolinium is shown. The corresponding PIXE spectrum is shown in Fig. 3b.

Scanning Transmission Ion Microscopy (STIM)

The STIM technique makes use of the principle that the bombarding ions lose energy as they pass a thin sample depending on the local sample areal mass density and composition. The intensity of the ion beam can be as low as few 1000 ions per second. By measuring the energy loss using a surface barrier detector or a photo diode at 180° relative to the beam direction a map showing the areal mass density of the sample can be produced. This can be used for orientation or to normalise elemental maps obtained from micro PIXE [23]. In Fig. 4 an example of STIM imaging of a thin section of a brain tumour is shown clearly showing the individual cells in four different magnifications [24]. The individual cells and the cell nuclei are clearly visible. Since only a few ions through each pixel are needed to determine the average energy loss, the imaging process is fast and genuinely non-destructive. After STIM analysis, the slits are opened up to give an analytically useful beam current (100 pA or more), and if effects of beam damage are critical, STIM analysis can be

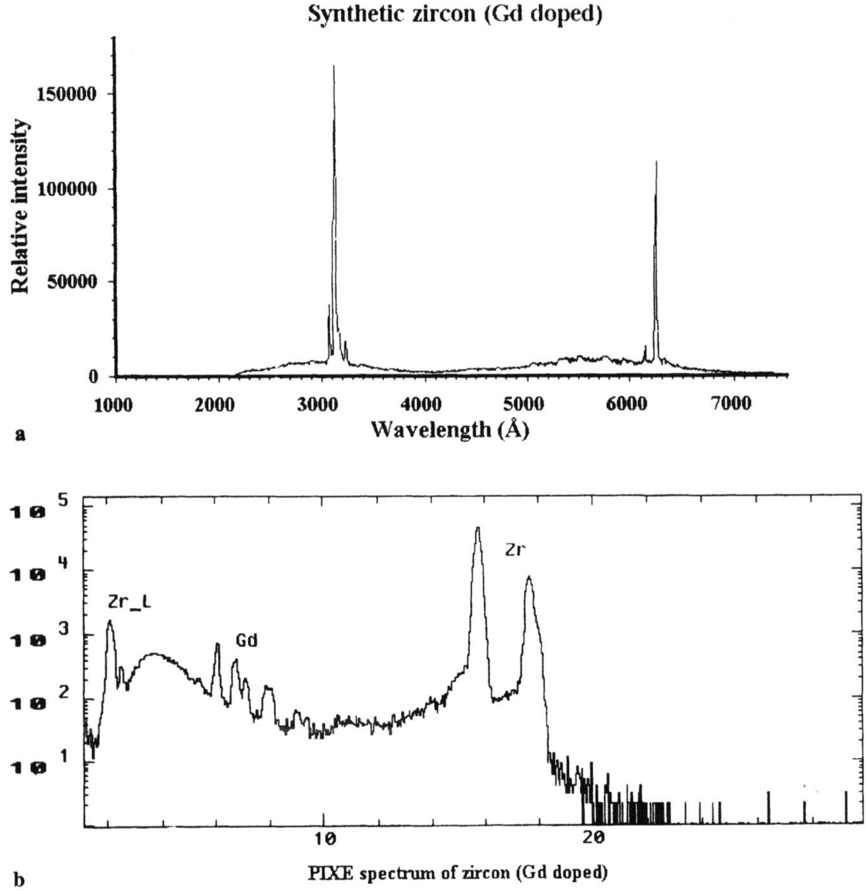

Fig. 3. a IL spectrum from proton irradiation of a synthetic zircon doped with gadolinium. **b** The PIXE spectrum corresponding to **a**. The abscissa shows the X-ray energy. The current was 2 nA and the irradiation time 10 min (Courtesy of Yang and Elfman)

performed again after the elemental analysis to compare the mass map with the initial one.

In another mode, off-axis STIM, the structural determination is carried out simultaneously with, for instance, micro PIXE. Using a beam diameter of a few micrometers and a high enough current to obtain acceptable X-ray pulse statistics, the particle detector cannot be exposed to the direct beam but only at an angle relative to the beam (off-axis). With a suitable angle and collimation the detector will not be saturated.

The combination with crystal channeling or with computerised tomographic reconstruction further expands the usefulness of STIM in materials characterisation. The transmittance of ions through specific crystal axes makes it possible to image crystal defects (channeling contrast microscopy) and evaluate crystal quality without any significant radiation damage effects [25]. Since the projectiles lose energy in a well-defined manner and follow almost straight paths in the specimen they can be used for three-dimensional imaging by acquiring two-dimensional images in varying directions and use computer-based reconstructions. Such ion

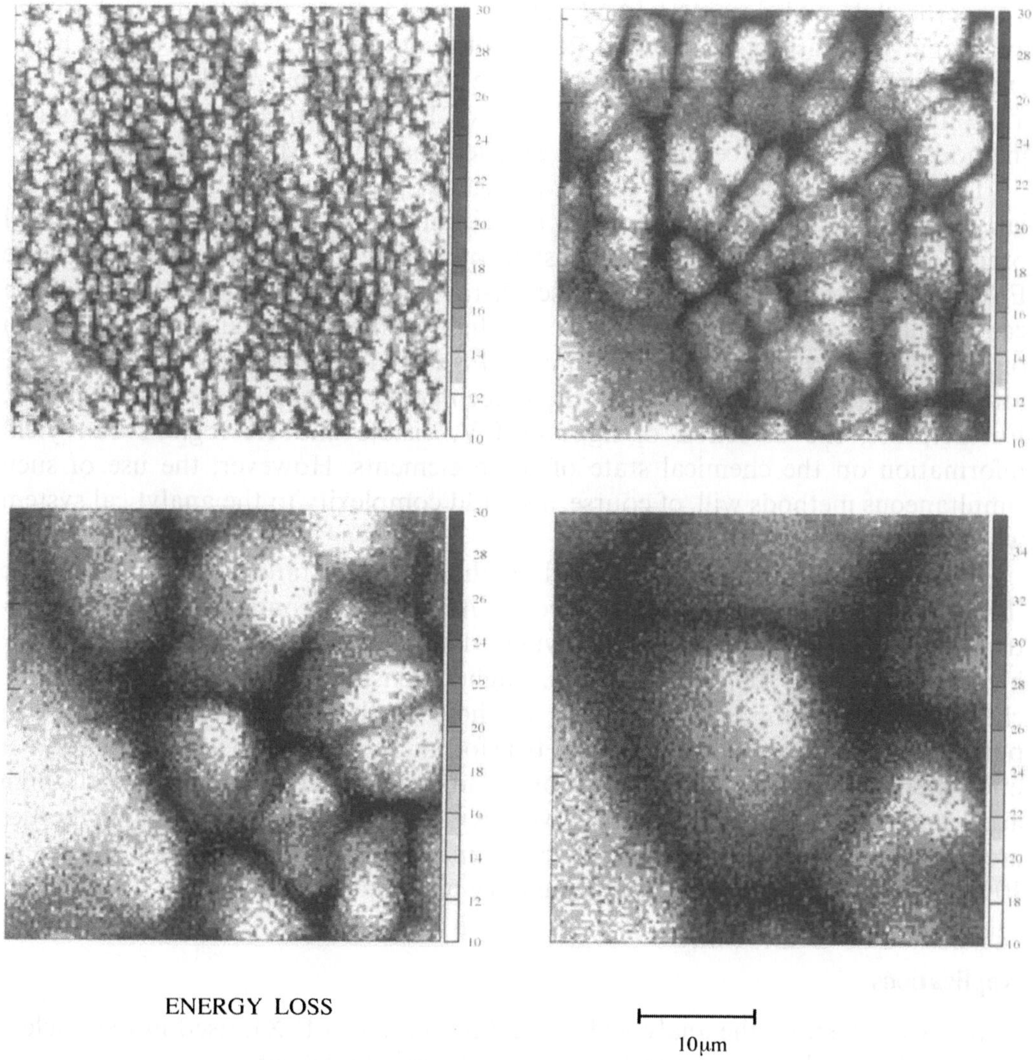

Fig. 4. STIM images of tissue from a brain tumour. The maps show the energy loss in grey scale (arbitrary units) and the magnification increases gradually from left to right and from top to bottom. The scale bar refers to the highest magnification (Courtesy of Kristiansson and Sjöland)

microtomography is especially useful when brittle or hard microsamples, not suited for physical sectioning, need to be characterised. By combining micro PIXE measurements with STIM measurements and ion microtomography a powerful combination for three dimensional imaging of elemental concentrations in microstructures can be obtained [26].

Comparison with Other Microprobe Techniques

Comparing an analytical technique to alternative methods is always a hard venture. It is particularly difficult to produce a fair and objective evaluation of the properties of a well known technique in relation to others in which only a superficial knowledge is at hand. The discussion here will be limited to the microanalytical aspects of micro PIXE although the traditional broad beam excitation PIXE also has an important role to play in analytical chemistry. For a detailed investigation of pros and cons of PIXE relative to other analytical methods reference can be made to an excellent paper by Maenhaut [27].

In the earlier discussion on the advantageous properties of PIXE one particularly important factor is the possibility of a simultaneous combination with several other ion beam techniques. In contrast to other X-ray methods, for instance the EMP or the synchrotron radiation based X-ray microprobe (SXRF), this approach can extend the analytical range to the very light elements. The superior resolution of EMP and the superior detection limits of SXRF do not necessarily compensate for this advantage since each of these techniques display some weak points. Furthermore, the detection of light in UV, visible and IR regions can yield information on the chemical state of some elements. However, the use of such simultaneous methods will, of course, also add complexity to the analytical system which is always a disadvantage.

There are several microanalytical methods, for instance, laser microprobe (LAMMA), Auger electron spectroscopy (AES) and secondary ion mass spectrometry (SIMS), that can yield analytical data superior to that of micro PIXE: higher lateral and depth resolution, higher sensitivity, isotopic information, chemical information etc.. But just as these methods serve many analytical purposes in their own right the same is true for micro PIXE with its combination of micrometre resolution, μg/g detection limits, quantitative analysis with high accuracy and precision, non-destructiveness etc. When several of these properties are relevant for the solution of a particular analytical problem the choice of micro PIXE with its simultaneous complementary techniques can be very successful.

Applications

In order to illustrate the analytical capabilities of micro PIXE used in the nuclear microprobe, a selection of applications will be briefly described.

Trace Elements in Biomedicine

With detection limits of the order of 0.1 μg/g some of the most important trace elements such as iron, copper and zinc can easily be detected by PIXE in an organic matrix. In addition to other elements, both essential and toxic can be determined if the sensitivity is optimised. The possibility of performing such a multi-elemental survey is very valuable as one often does not know beforehand which elements are of importance in an investigation. A further reason for multi-elemental studies is that the function of the trace elements often is inter-correlated. Since trace elements play such important roles in biomedicine many types of biomedical tissues have been analysed using PIXE. The potential of micro PIXE to quantify trace elements even inside single cells opens up new possibilities in this field.

In biological microanalysis sample preparation plays a paramount role for accurate interpretation of obtained data. The electron microprobe community has developed a variety of sample preparation techniques for biomedical specimens.

These techniques have been transferred to the NMP, but since PIXE analysis gives lower detection limits than EMP, it has yet to be investigated whether in vivo gradients of the trace elements remain unchanged during the sampling, sectioning and drying of the specimen. It is possible to preserve intracellular gradients of very mobile elements such as electrolytes (e.g., calcium and potassium), which indicates that the normally less mobile trace elements should be stable under preparation. Since staining causes severe alterations in the concentrations of elements stained tissue sections cannot be used. The development of rapid STIM imaging has circumvented the orientation problems on freeze dried cryo-sections. As was shown in Fig. 4 the image resolution is fully adequate to allow for identification of intra and extra cellular structures.

In our laboratory we have used the NMP extensively in the fields of dermatology and neurobiology. Calcium homeostasis disorder in rat brain tissue following ischemia and epileptic seizures has been investigated. The detected (total) calcium concentrations were in the range of 1–50 µmol/g and would have been difficult to quantify using other microprobe techniques. By analysing brain tissue sections dissected at intervals after an epileptic seizure, a net calcium increase in brain regions known to be prone to these attacks could be observed [28]. The multi-element capability of PIXE was useful since a simultaneous decrease in the potassium level could also be observed. In studies of human skin both normal and pathological material has been investigated. An interesting finding in analysis of epidermis of normal looking skin collected from psoriatic patients is the abnormal distribution of iron which implies that abnormal metabolism of this element is one of many dysfunctional features of this state.

In Melbourne the NMP has been used for medical studies around AIDS and other diseases. They measured elements in human skin fibroblast cells cultured in vitro and taken from patients suffering from Menkes' disease, known to be linked with impaired copper metabolism. By scanning the cells with a 3-µm proton beam detailed information on several elemental distributions was obtained. Noticeable was a sixfold increase in intracellular copper concentrations in Menkes' cells compared with normals [29].

Trace Elements in Minerals

Geochemistry plays an important role in geology both in basic studies and in investigation of economically interesting mineralogical deposits. The concentration of the interesting elements in such deposits are often very low and, therefore, a sensitive analytical method such as PIXE is required. In geochemistry one of the standard analytical techniques, the electron microprobe (EMP), has several limitations, for instance, low sensitivity, that have led to the increasing use of micro PIXE in the NMP in this field, especially for localised and quantitative analysis of trace elements. Geological materials are normally complex, heterogeneous structures with mineral grains and inclusions. Elemental, isotopic and chemical characterisation of the various compartments have long been important for understanding the basic processes of formation as well as for investigating the economical potential of deposits. In many instances, there are significant advantages to use trace elements instead of major elements when constructing models of the genesis of a geological system. The increasing use of SIMS is also due to the interest in analysing trace elements, although normal SIMS analysis is of limited use for quantitative analysis.

In the CSIRO Laboratory in North Ryde, Australia, an ion beam analysis laboratory has been running micro PIXE for more than a decade, primarily serving the mineral industry with analyses for applied geology [30]. At the ion beam laboratories in Heidelberg, Germany, and Guelph, Canada, the PIXE programmes are also more or less dedicated to geology including the characterisation of extraterrestial material.

The standard samples used for EMP analysis also lend themselves to micro PIXE analysis: approximately 30 μm thick polished sections mounted on a glass backing. Since the ion beam in the NMP penetrates deeper than the EMP, the analytical properties will be somewhat different. There is a potential risk of penetrating into an underlying region of completely different composition, and hence, the results of the analysis of such standard samples have to be evaluated with care. However, this capability to see "below the surface" could also be used in investigating unopened liquid inclusions, which cannot be analysed with the traditional EMP technique. The depth resolution available when combining micro PIXE with micro RBS makes information on the underlying structures obtainable. The distribution of elements in mineral grains can be used to facilitate the understanding and modelling of the formation and transformation of the mineral.

Ore genesis is one field in the earth sciences in which PIXE has been applied. It is especially suited for the study of trace element partition between co-existing minerals, which gives information about the chemical and physical conditions (e.g. temperature) during the ore-forming processes. Diamond genesis can be studied though the analysis of small grains, which have been trapped during the growth of the diamond [31]. As these inclusions have been shielded from any outside influence they carry important information on the physical and chemical parameters of the environment during the formation of the diamonds.

As was discussed above (sect. ionoluminescence) the IL technique is particularly useful for REE determination. Since a variety of REE elements are present in many geological structures the combined use of micro PIXE and IL in the nuclear microprobe will significantly enhance the potential of the NMP for geosciences. The detection limits for REE using IL (sub μg/g), are a few orders of magnitude better than for micro PIXE [32]. One must keep in mind, however, that the quantitative capability of IL in its present from is very limited and hence the simultaneous use of micro PIXE and IL is very advantageous. The related technique of cathodoluminescence lacks the availability of a simultaneous trace element technique.

Another way of using the IL technique in geochemistry is the production of maps that will reveal the internal structure which has been overgrown and is invisible to SEM imaging. Due to deep penetration of the ions in transparent minerals light will be emitted and detected from fairly "deep" layers. The possibility of using a broad ion beam excitation combined with image grabbing with a sensitive CCD camera is very promising in the routine use of the NMP in mineralogy. The images can be readily obtained and be used for detailed orientation on the specimen to facilitate a detailed analysis using micro PIXE. In

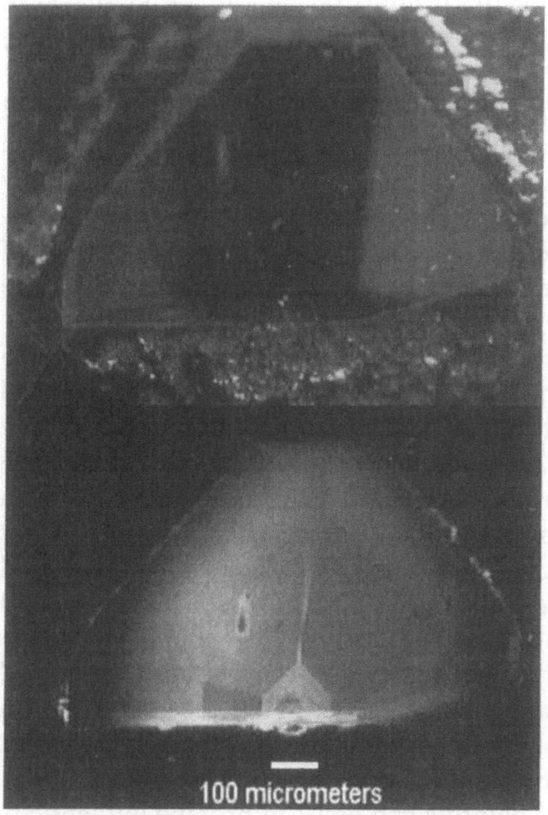

Fig. 5. Images of a gadolinium doped zircon crystal in incident light (top) and with IL imaging (bottom) (cf. Fig. 3). The concentration of gadolinium is approximately 1100 μg/g by weight. (Courtesy of Yang and Elfman)

Fig. 5 a synthetic zircon crystal doped with gadolinium is shown in incident light (top) and with IL imaging (bottom).

It is interesting to demonstrate that in geochemistry PIXE can play an important role even at broad beam excitation. A special method of mineral exploration is based on a phenomenon called "geogas". It has been found that there is a slow flow of gas through the cracks and fissures of the bedrock. This gas probably consists of air that has been dissolved in rain water which trickles down through the ground to the groundwater table. When the air is liberated it slowly diffuses upwards through the bedrock. It has been found that small amounts of dust are carried by the gas to the surface where it can be collected. This phenomenon is interesting because a relation has been found between the elemental composition of the dust and the existence of minerals in the ground. This is the basis of a new method of mineral exploration which has turned out to be successful [33]. Because the amount of dust is small, less than 1 ng for most elements, PIXE appears to be one of the few methods that can be used for a reasonably rapid and efficient analysis.

Arts and Archaeology

When analysing archaeological artefacts a major concern is the non-destructiveness of the analysis. Many objects are irreplaceable and it is therefore necessary to use an analytical method which avoids observable damage to the object. The nuclear microprobe in combination with PIXE fulfils this condition. Similarly to capillary tube XRF microprobes [34], PIXE can be easily used for non-destructive analysis of sub mm^2 regions and for element mapping of larger areas.

In arts and archaeology, interest in and the use of microanalysis have increased significantly over the last decade. The well known Louvre museum in Paris has since several years an ion beam analysis laboratory in operation for many applications in art science and archaeometry [35]. To further extend the laboratory's analytical capability a nuclear microprobe has also been installed. Because of sample size and lowered sample deterioration external beam microprobes are particularly useful in this field.

While metallic objects, for instance of bronze and gold, are less sensitive to radiation damage, the corrosion of non-noble metals represents a severe analytical problem. The shallow analytical depth of the NMP requires the removal of any such layers to obtain analytical results representative of the bulk material. In some cases, removal of this patina may be prohibitive. Even if this is a serious problem that has to be addressed the situation is normally even worse for most other analytical techniques. Large-scale studies have been performed on copper-based alloys, mainly antique bronzes. In such studies, great care has to be taken to avoid erroneous analytical data due to surface effects caused by corrosion [36].

Paintings can be characterized by the pigments used by the artist as these varied at different periods. It is known that some pigments were introduced at a certain time, sometimes rather late. For example, the pigment called Naples yellow was not used before 1640 and the white pigment titanium oxide was not used before 1920. Hence, an analysis of the pigments in a painting can assist in proving authenticity or reveal forgeries.

An external PIXE beam is a very efficient tool for a rapid and non-destructive analysis of paintings and the interest in this type of analysis has increased in recent years [37]. In a study by Neelmeijer et al. a combination of three ion beam technique was applied in an external "milliprobe" system. PIXE, RBS and PIGE (see Fig. 6) were all used simultaneously during bombardment with 1-mm diameter 4 MeV proton beam in air of oil paintings as well as book pages and ancient glass [38]. Paper and related material have low-Z matrices but are sensitive to radiation damage and heating effects. In a specially designed nuclear "milliprobe" at Crocker

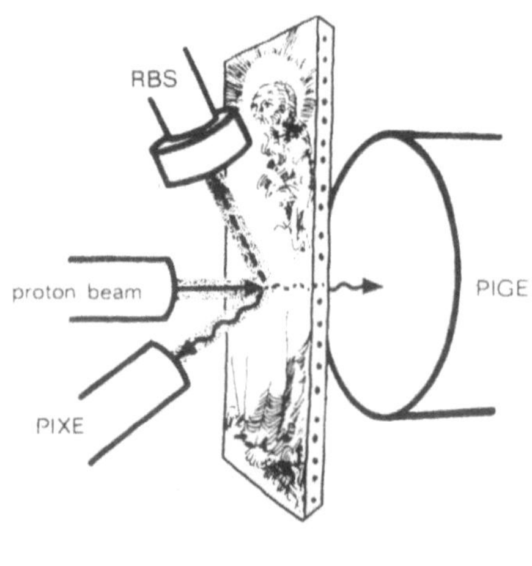

Fig. 6. Principal arrangement for a "milliprobe" external beam setup for arts analysis. Three different IBA techniques are employed simultaneously (adopted from [38])

National Laboratory at Davis, California, multi-element analysis is performed in air [39]. The Gutenberg Bible and the Vinland map are examples of well-known works which have been analysed non-destructively at this facility. However, by using the high lateral resolution of an external nuclear microprobe, it is possible to extract new information on material and printing or writing procedures, identifying faded letters and characterising inks.

In our laboratory at Lund an external scanning nuclear microprobe combined with sophisticated multivariate statistical methods was used to identify faded letters in ancient Greek handwriting on papyrus [40]. By moving the specimen on an x-y scanning table in air during the analysis, two-dimensional element maps were produced. The hypothesis was that invisible letters could be identified by detecting traces of elements from the ink. It turned out, however, that this simple approach did not yield good image contrast. After applying chemometric methods and compiling maps of several elements by multivariate statistical methods, the contrast in the maps obtained was significantly enhanced, and the letters could be identified.

Environmental Research

One of the first applications of PIXE was the study of air pollution. In such studies it is desirable to collect the aerosol with size fractionation and good time resolution. This often results in samples of very small mass. Multi-elemental PIXE analysis is ideal, and macro PIXE has, therefore, been extensively used in atmospheric research where it is now a well-established method. One type of investigation is concerned with special air pollution situations, for instance, dust storms originating in the east Asian deserts and passing over the Pacific, volcanic emissions, urban aerosol in big cities, Arctic haze (an aged anthropogenic aerosol of continental origin) and identification of sources of emission from various industries. In another type of investigation the ambient air is monitored for extended periods (often several years) at selected measuring stations. A good example of this is the network operated in the USA by the PIXE group at the University of California, Davis [41]. At more than 70 stations all over the USA samplers of different types collect the aerosol for analysis by various methods including PIXE for the elemental analysis.

Even if broad beam excitation using PIXE has been shown to contribute significantly to resolve problems in atmospheric chemistry the use of high lateral resolution of micro PIXE has the potential to cast even more insight into the problems of how individual aerosol particles are generated and transformed. In the '90s several studies including the NMP and other microanalytical methods, e.g. EMP and LAMMA have been carried out [42]. By characterisation of a large number of individual particles rather than averaging over a large particle population with many different histories an improved understanding of the particle formation processes is obtained.

Conclusions

From the discussions above it is clear that PIXE has an important analytical role to play. In particular, when the high lateral resolution of the nuclear microprobe is used for micro PIXE and complementary ion beam techniques PIXE's full potential in many areas of application is released. Few, if any, other micro probe techniques combine so many advantages as micro PIXE.

References

[1] M. B. H. Breese, G. W. Grime, F. Watt, *Annu. Rev. Nucl. Part. Sci.* **1992**, *42*, 1.

[2] B. J. Kirby, G. J. F. Legge, *Nucl. Instr. Meth.* **1991**, *B54*, 98.

[3] S. A. E. Johansson, *Analyst* **1992**, *117*, 259.

[4] H. Van Rinsvelt, S. Bauman, J. W. Nelson, J. W. Winchester (eds.), *Nucl. Instr. Meth.* **1987**, *B22*, 1–475.

[5] R. D. Vis (ed.) *Nucl. Instr. Meth.* **1990**, *B49*, 1–589.

[6] M. Uda (ed.) *Nucl. Instr. Meth.* **1993**, *B75*, 1–589.

[7] U. Lindh (ed.) *Nucl. Instr. Meth.* **1993**, *B77*, 1–556.

[8] F. Yang, J. Tang, J. Zhu (eds.) *Nucl. Instr. Meth.* **1995**, *B*, 104.

[9] S. A. E. Johansson, J. L. Campbell, *PIXE: A Novel Technique for Elemental Analysis*, Wiley, Chichester, 1988.

[10] S. A. E. Johansson, J. L. Campbell, K. G. Malmqvist (eds.) *Particle-Induced X-ray Emission Spectrometry (PIXE), Chemical Analysis, Vol. 133*, Wiley, New York, 1995.

[11] B. J. Kirby, G. J. F. Legge, *Nucl. Instr. Meth.* **1993**, *B77*, 268.

[12] J. L. Campbell, W. Maenhaut, E. Bombelka, E. Clayton, K. G. Malmqvist, J. A. Maxwell, J. Pallon, J. Vandenhaute, *Nucl. Instr. Meth.* **1986**, *B14*, 282.

[13] J. Räisänen, *Biol. Trace Elem. Res.* **1987**, *12*, 55.

[14] G. W. Grime, M. Dawson, M. Marsh, I. C. McArthur, F. Watt, *Nucl. Instr. Meth.* **1991**, *B54*, 52.

[15] N. E. G. Lövestam, *Nucl. Instr. Meth.* **1989**, *B36*, 455.

[16] K. Themner, *Nucl. Instr. Meth.* **1991**, *B54*, 115.

[17] M. Cholewa, G. J. F. Legge, H. Weigold, G. Holan, C. Birch, *Nucl. Instr. Meth.* **1993**, *B77*, 282.

[18] W. -K. Chu, J. Mayer, M. -A. Nicolet, *Back Scattering Spectrometry*, Academic Press, New York, 1978.

[19] M. Hult, B. Bengtsson, N. P. -O. Larsson, C. Yang, *Scanning Microsc.* **1992**, *6*, 581.

[20] P. Kristiansson, E. Swietlicki, *Nucl. Instr. Meth.* **1990**, *B49*, 98.

[21] L. -E. Carlsson, K. R. Akselsson, *Nucl. Instr. Meth.* **1981**, *181*, 531.

[22] C. Yang, N. P. -O. Larsson, E. Swietlicki, K. G. Malmqvist. D. N. Jamieson, C. G. Ryan, *Nucl. Instr. Meth.* **1993**, *B77*, 188.

[23] J. Pallon, U. A. S. Tapper, *Nucl. Instr. Meth.* **1991**, *B56/57*, 1270.

[24] Kristiansson, Sjöland, *Private Communication*, Lund, 1955.

[25] J. C. McCallum, C. D. McKenzie, M. A. Lucas, K. G. Rossiter, K. T. Short, J. S. Williams, *Appl. Phys. Lett.* **1983**, *42*, 827.

[26] R. Schofield, *Nucl. Instr. Meth.* **1995**, *B104*, 212.

[27] W. Maenhaut, *Nucl. Instr. Meth.*, **1990**, *B49*, 580.

[28] K. Inamura, M. Martins, K. Themner, U. A. S. Tapper, J. Pallon, G. Lövestam, K. Malmqvist, B. Siesjö, *Brain Research* **1990**, *514*, 49.

[29] G. L. Allan, J. Camakaris, G. J. F. Legge, *Nucl Instr Meth.* **1991**, *B54*, 175.

[30] C. G. Ryan, W. L. Griffin, *Nucl. Instr. Meth*, **1993**, *B77*, 381.

[31] W. L. Griffin, C. G. Ryan, D. R. Cousens, S. H. Sie, G. F. Suter, *Nucl. Instr. Meth.* **1990**, *B49*, 318.

[32] C. Yang, N. P. -O. Homman, K. G. Malmqvist, L. Johansson, N. M. Halden, V. Barbin, *Scanning Microsc. Int.* **1995**, *9(1)*.

[33] K. Kristiansson, L. Malmqvist, *Geoexploration* **1987**, *24*, 517.

[34] A. Rindby, *X-ray Spectrometry* **1993**, *22*, 187.

[35] M. Menu, *Nucl. Instr. Meth.* **1991**, *B56/57*, 597.

[36] S. J. Fleming, C. P. Swann, M. Jaksic, *Nucl. Instr. Meth.* **1992**, *B64*, 499.
[37] M. Menu, *Nucl. Instr. Meth.* **1993**, *B75*, 469.
[38] Ch. Neelmeijer, W. Wagner, H. P. Schramm, *Deutsche Gesellschaft für Zerstörungsfreie Prüfung e.V.*, Berlin, 1994, p. 296.
[39] B. H. Kusko, T. A. Cahill, R. A. Eldred, R. N. Schwab, *Nucl. Instr. Meth.* **1985**, *B3*, 689.
[40] N. E. G. Lövestam, E. Swietlicki, *Nucl. Instr. Meth.* **1989**, *B43*, 104.
[41] T. A. Cahill, *Nucl. Instr. Meth.* **1993**, *B75*, 217.
[42] P. Artaxo, M. L. Rabello, F. Watt, G. Grime, E. Swietlicki, *Nucl. Instr. Meth.* **1993**, *B75*, 521.

Mikrochim. Acta [Suppl.] 13, 135–166 (1996)

Cathodoluminescence Microscopy and Spectroscopy of Semiconductors and Wide Bandgap Insulating Materials

J. F. Bresse[1], G. Remond[2,*], and B. Akamatsu[1]

[1] France Telecom, CNET/PAB, Laboratoire de Bagneux, 196 avenue H. Ravéra, BP 107, 92225 Bagneux Cedex, France

[2] BRGM, Direction de la Recherche, Départment Géomateriaux et Géoprocédés, avenue de Concyr, BP 6009, Orléans Cedex 2, France

Abstract. Cathodoluminescence (CL) corresponds to the emission of photons observed from semiconductors and wide bandgap insulating crystals exposed to an electron bombardment. The radiative de-excitation of the created hole-electron pairs may occur according to an intrinsic mechanism (direct band to band recombination) or may be an extrinsic mechanism involving charge recombination through energy levels associated with impurities or point defects. For low bandgap materials, local analysis of semiconductor physical properties such as surface (interface) recombination velocity and minority carrier diffusion length are studied by means of CL measurements at different incident electron energies. For wide bandgap insulators the complementarity of CL and X-ray spectrometries is illustrated for the case of the localisation of trace elements and for the case of charging phenomena related to the presence of pre-existing point defects and induced defects during the electron bombardment. Examples of CL microscopy and spectroscopy applied to the microcharacterisation of semiconductors (GaAs, InP and related materials) and of synthetic and natural mineral materials (ZnS, SiO_2, rare-earth bearing crystals), are presented.

Key words: cathodoluminescence, intrinsic luminescence, extrinsic luminescence, liquid helium, impurities, lattice defects, semiconductors, insulators, carrier recombination, minority carriers diffusion length.

Cathodoluminescence (CL) corresponds to the emission of photons at the surface of semiconductor and insulating crystals exposed to an electron bombardment. The created carriers, i.e. holes or electrons, diffuse into the material until they recombine. The radiative de-excitation may occur according to an intrinsic mechanism (direct band to band recombination) or may be an extrinsic mechanism involving charge recombination through energy levels being associated with impurities or point defects.

* To whom correspondence should be addressed

For the low bandgap materials, typically lower than 3 eV, the energy levels, associated with a doping impurity, are either located near the conduction band (CB), i.e. donor levels or near the valence band (VB), i.e. acceptor levels, and the energy of the emitted photons extends from the visible to the infrared part of the spectrum. CL allows local analysis of semiconductor physical properties. Main parameters, such as surface (interface) recombination velocity and minority carrier diffusion length can be evaluated by means of the signal measurement at various incident energies. In order to take into account hole-electron generation, diffusion phenomena and sample geometry, various analytical expressions and models will be presented.

For silicates and oxides, the bandgap being greater than a few eV (as an example, about 5 eV for diamond, 10 eV for quartz and 12 eV for LiF), the mechanisms of charge recombination lead to photon energy ranging from the UV to the visible region. In addition, some impurities play the role of luminescent centers, as for the case of rare-earth elements: the de-excitation process leads to a CL emission spectrum exhibiting spectroscopic features being characteristic of this centre. Applications of CL properties in geosciences have been recently reviewed by Remond et al. [1] including an abundant bibliography on instrumentation, experimentation, specific applications of CL emission as a signature of composition and genesis of minerals. Some of these results will be summarised in the present paper and updated results will be presented emphasising the complementarity of CL and X-ray spectrometries to the microcharacterisation of insulating materials.

Examples will illustrate the CL emission microscopy and spectroscopy contribution to the microcharacterisation of semiconductor devices (GaAs, InP, and related compounds), and of mineral materials (ZnS, SiO_2, natural and synthetic rare-earth bearing crystals).

1. Physical Phenomena in Crystals

1.1. Intrinsic Luminescence

The intrinsic luminescence emission results from direct band to band recombination of electrons and holes according to the decay routes marked (1) and (2) in Fig. 1. The maximum of the intrinsic emission band occurs at energy $h\nu = E_g$ where E_g is the absorption edge of the material. The term near bandgap radiation is also used because excitons and shallow recombination centres may contribute to this emission. The term intrinsic luminescence is also used to designate the emission characteristic of the crystal lattice of minerals. The term "intrinsic" is improperly used because most minerals, silicates, oxides and phosphates among other are wide

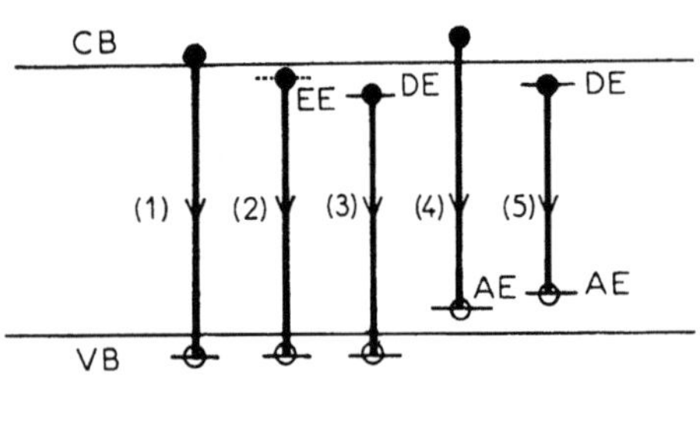

Fig. 1. Schematic diagram of radiative transitions between the conduction band (*CB*), the valence band (*VB*), the excitonic level (*EE*), the donor level (*DE*) and the acceptor level (*AE*)

bandgap insulating crystals and the actual intrinsic emission cannot be measured experimentally.

As an order of magnitude the bandgap is typically about 10 eV in most silicate minerals. Thus, for all types of crystal, we will use intrinsic luminescence to designate emission involving lattice defects as charge recombination centres. Because of the large band gap of the insulating materials, the intrinsic emission can only take place in the presence of major lattice perturbations.

In the case of semiconductors, the bandgap energy is varying from 0.1 to 3 eV. Usually, the intrinsic luminescence can be detected using photomultipliers or photodiodes. If the created holes and electrons are free, they can move independently or form a bound state, the exciton. The excitonic binding energy is slightly different from the bandgap energy. The excitonic transition corresponds to route (2) in Fig. 1.

1.2. Extrinsic Luminescence

Extrinsic luminescence arises from electron transitions on shallow or deep energy states located within the forbidden bandgap of the energy diagram of the crystal. For both semiconductors and insulating materials, the concept of the band theory is used for illustrating the extrinsic luminescence mechanisms involving the ionisation of ionic impurities or charge recombination through energy levels associated with impurities or defects in the crystal. These imperfections responsible for the extrinsic luminescence are called activators. When two activators must be simultaneously present to provoke the emission, the secondary impurity is generally called co-activator. Extrinsic luminescence results from charge recombination according to the decay routes marked (3) and (4) in Fig. 1. The recombination centres are donor-acceptor pairs associated with point defects resulting from the presence of impurities or lattice defects. The recombination levels are called luminescence centres when they lead to radiative transitions or killer centres when they lead to non-radiative transitions. The materials exhibiting luminescence involving electron displacement through the conduction band before to recombine with a hole in the valence band are photoconductor luminescent crystal. Luminescence in non-photoconductor materials results the de-excitation of excited levels of luminescent ions without any necessity to raise the electron into the conduction band. The emission mechanism is that of an atomic emitter in which excitation and emission occur within the energy levels of the luminescent ion as illustrated in Fig. 2 for the case of the characteristic luminescence emission of Cr^{3+} ions in an Al_2O_3 crystal or a quasi-atomic emitter as illustrated by the decay route (5) in Fig. 1 involving a donor-acceptor pair for a semiconductor material.

1.3. Fluorescence and Phosphorescence

The recombination centres are luminescent centres responsible for a fluorescence emission when the charges spontaneously recombine through these levels. The levels exhibiting a large capture probability for conduction electrons are electron traps responsible for the phosphorescence mechanism. According to this mechanism, the excited electron is trapped on the metastable level which does not allow further

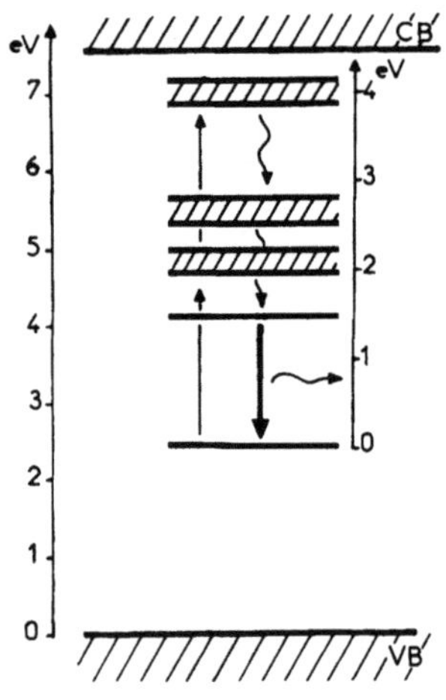

Fig. 2. Ion activator energy levels in a band scheme for the characteristic luminescence emission of Cr^{3+} ions in an Al_2O_3 host crystal. Excitation and emission occur as a result of transitions between the ion activator levels without participation of conduction and valence bands [2]

transition unless it receives an additional amount of energy sufficient to reach again an excited level with the possibility of recombination with the emission of a luminescent photon.

1.4. Influence of the Temperature

The ionisation energy of impurity which gives the donor or the acceptor level is usually in the range of 5–10 meV for donors and 25–30 meV for acceptors. The excitonic binding energy is ranging from 0.1 to 2 meV, depending on the material and the impurity for the bound excitons.

The thermal distribution of the carriers which is given by kT (26 meV at 300 K) may give a broad spectrum for the near band edge emission which includes all the intrinsic luminescence and the extrinsic luminescence through donors or acceptors.

In order to separate the different emission lines, the sample temperature has to be reduced either to liquid nitrogen (separation of acceptor level from the bandgap energy) or to a temperature near the liquid helium (separation of all the donors levels and some excitonic levels from the bandgap energy).

1.5. Deep Level Emission

Deep level emission corresponds to a broad spectrum due to the coupling to the lattice of the impurity or defect giving the deep level. The emission corresponds either to the ground state or to a state arising after an energy loss of one or several phonons.

The line shape is governed by the probability of the emissive transition:

$$P_{\mathrm{om}} = \exp(-s)\frac{s^{\mathrm{m}}}{m!}$$

where s is the Huang-Rhys coupling factor, and m the number of phonons involved in the transition.

2. The Cathodoluminescence Process

2.1. Radiative and Non-Radiative Process

Cathodoluminescence (CL) designates the luminescence emission induced by electron bombardment of the specimen and is the sum of all luminescence emissions induced by all forms of energies resulting from the dissipation of the incident energy within the specimen. CL emission is thus the sum of different luminescence mechanisms resulting from:

i) electron excitation (cathodoluminescence in the strict sense),
ii) X-ray photons generated within the specimen (roentgenoluminescence or radioluminescence),
iii) electrical field created by trapped charges (electroluminescence) and,
iv) thermal effects leading to competitive phenomena, thermoluminescence emission and thermal quenching.

As a function of the incident irradiation dose and the electrical properties of the specimen, each of the available energies released within the crystal are able to:

i) modify the relative emission intensities in the different emission bands,
ii) modify the relative probabilities of the radiative and non-radiative transitions and,
iii) modify the filling of electron traps.

2.2. Cathodoluminescence Emission Versus X-Ray Emission

Figure 3 illustrates the analogy between radiative and non-radiative transitions involving core electrons with the production of X-ray photons and Auger electrons and, radiative and non-radiative transitions involving electron traps and luminescent centres located within the forbidden bandgap of the energy diagram leading to the emissions of luminescence photons and exo-electrons. As a consequence of multiple mechanisms of energy transfer of the incident electron energy to the luminescent centres, the luminescence spectrum of a crystal containing several luminescent ions and lattice defects is not the sum of the luminescence spectra of the same crystal containing separately each individual point defects, as it is the case of the X-ray emission spectrum of each element contained within the compound. The interpretation of a cathodoluminesence spectrum is not directly related to the nature of impurities responsible of the extrinsic emission because of its dependence on all crystallographic properties of the material rather than only on the qualitative nature of the activators of the luminescence.

3. Cathodoluminescence Intensity

According to Yacobi and Holt [3], the observed CL intensity can be expressed as:

$$I_{CL} = \int_V f_D f_A f_R \, \eta \, \Delta n(r) \, d^3 r$$

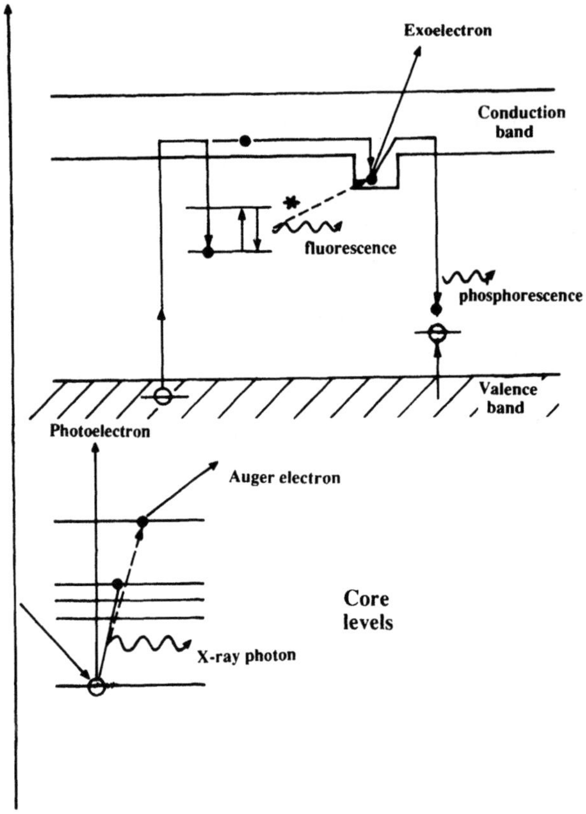

Fig. 3. Analogies between radiative and non-radiative transitions resulting from hole-electron transition on core level (X-ray and Auger electron emissions) or on point defect levels (luminescence and exoelectron emissions) [1]

where f_D is a constant factor accounting for instrumental factors (detection solid angle, collection and transmission efficiency of the detection system...), f_A is the absorption factor, f_R is a factor accounting for internal reflection loss of CL intensity, η is the radiative recombination efficiency (ratio of the radiative recombination rate to the total recombination rate or ratio of the total lifetime, τ, to the radiative lifetime, τ_r, of minority carriers), $\Delta n(r)$ is the stationary density of minority carriers (electrons in p-type material) after the effect of diffusion and recombination of the hole-electron created pairs.

3.1. Photon Generation

The photon generation is mainly governed by the stationary excess minority carrier density, $\Delta n(r)$, which is the solution of the continuity equation written in p-type semiconductor as:

$$D\nabla^2(\Delta n(r)) - \frac{\Delta n(r)}{\tau} + g(r) = 0$$

D, diffusion constant of carriers, $g(r)$, generation rate of excess carriers per unit volume and per second.

The solution of the continuity equation is depending on the expression of the generation function. Various expressions of this generation function will be mentioned in section 7.4.

For an ideal point-source, the distribution of the excess carrier density is given by:

$$\Delta n(r) = \frac{GI_b}{qD} \frac{\exp(-r/L)}{r}$$

G is the generation factor given by:

$$G = \frac{E_0}{E_p}(1 - \bar{k}\eta)$$

E_0, beam energy, I_b, beam intensity, E_p, hole-electron pair energy creation ($\approx E_g$), $\bar{k}$, ratio of the mean energy of backscattered electrons to the electron beam energy, η, backscattered electron coefficient, L, diffusion length of the minority carriers.

The non-radiative surface recombination, s, is taken into account by:

$$s\Delta n(r) = D\left|\frac{\delta\Delta n}{\delta z}\right|_{z=0}$$

The one-dimensional solution of the continuity equation gives:

$$\Delta n(z) = \frac{\tau}{2q(\pi d_s^2)L} \int_0^R \left\{\left[-\exp\frac{|z - z_s|}{L}\right] + \frac{S-1}{S+1}\exp\left[-\frac{z+z_s}{L}\right]\right\} g(z_s)dz_s$$

$S = s\ L/D$, normalised value of the surface recombination, $g(z_s)$, source depth distribution of created carriers per depth, d_s, radius of the electron beam spot size, z_s depth of the point source, R electron range.

3.2. CL Photon Absorption

The absorption loss factor f_A arises from a decrease in intensity of the form $\exp(-\alpha d)$, where α is the absorption coefficient factor, d, the length of the photon path in the interior of the material.

$$f_A = \frac{I_{CL}\ (\text{abs})}{I_{CL}\ (\text{gen})} = \frac{\int_0^\infty I_{CL}(z)\exp(-\alpha z)}{\int_0^\infty I_{CL}(z)dz}$$

when $I_{CL}(z)$ can be approximated by C $\exp(-z/L)$:

$$f_A = \frac{1}{1 + \alpha L}$$

The factor f_R accounts for the refraction and total internal reflection. The reflection factor, R, when the sample is in vacuum, is:

$$R = \left(\frac{n-1}{n+1}\right)^2$$

Although the luminescence is generated isotropically, only a small fraction inside the cone of semi-angle θ_C will escape (θ_C corresponds to the angle for total reflection).

$$f_R = (1 - R)\frac{1}{4\pi}\sum_0^{\theta c} 2\pi \sin\theta \, d\,\theta$$

$$f_R = (1 - R)\frac{n - \sqrt{n^2 - 1}}{2n}$$

For example, for SiO_2, $f_R = 4.5 \cdot 10^{-2}$, for CdS, $f_R = 3.4 \cdot 10^{-2}$, for GaAs, $f_R = 1.4 \cdot 10^{-2}$.

It is important to know what happens to the major part of CL that is totally reflected. Is it lost by absorption or does it emerge after repeated reflections? Precautions must be taken to prevent the detection of multiple reflected light [4].

f_A and f_R factors may depend on the depth. Rough surfaces and multilayers structures may lead to local variations in optical absorption and reflected losses.

3.3. CL Intensity as Function of the Excitation Conditions

In general, the dependence of CL intensity on the electron beam conditions is of the form:

$$I_{CL} = f(I_b)(E_0 - E_d)^m$$

where E_d is a "dead voltage" corresponding to a luminescent dead layer. $1 \leqslant m \leqslant 2$

This relationship has been observed for phosphors. For semiconductors, the CL intensity is proportional to the excess carrier generation rate g ($g \propto I_b\, E_0$).

For constant E_0, the dependence of the CL intensity is related to the dominant type of transition. When $\tau_r \ll \tau_{nr}$ (τ_{nr} non-radiative lifetime), the CL intensity should be proportional to the beam current only. For $\tau_{nr} \ll \tau_r$, two cases have to be considered. For the low excitation case (i.e. low I_b), the CL intensity should be proportional to N_0, equilibrium carrier concentration, and I_b, whereas at higher excitation levels, the CL intensity should be proportional to I_b^2 and independent of N_0. Experimental results are in agreement with this predictions [5]. But it has been also pointed out that the CL intensity may depend on g^m where $m \geqslant 1$ and in some cases not remain constant due to change in the occupation of recombination centres [6].

Prolonged electron beam irradiation may also cause significant changes in the CL intensity. Electron irradiation can lead to quenching or enhancement of CL signal. For example, in non-implanted silicon, an electron beam annealing has induced an enhancement of the infrared CL intensity [7].

Prolonged electron irradiation often leads to surface contamination. The deposition of a film which consists primarily of polymerised hydrocarbons is caused by electron bombardment of the absorbed molecules of the pump oil. Replacement of the diffusion pump by a turbomolecular pump may alleviate this problem significantly.

The analysis of non-conducting samples require special care due to specimen charging. A conductive thin coating will ensure the electron conduction, but it will also reduce the emitted luminescence intensity. The electric field distribution inside the coated sample may also alter significantly the electron trajectories inside the

Table 1. Available equipments for CL analysis

Light collector	Transfer of light	Monochromator	Detector
Lenses, spherical mirror (built in EPMA) parabolic or ellipsoidal mirror (SEM attachment)	Lenses, mirrors, light pipe or optical fibre	Filters, prisms, gratings interferometer or by-pass filter	Photomultiplier tube, photodiode, linear photodiodes array (LPDA), CCD camera

material and the CL generation mechanism. This may modify also the depth from which the light is emitted.

4. Instrumentation

As summarised in Table 1, a variety of components are available to develop specific attachments to EPMA, SEM and optical microscopes for the detection and the analysis of CL emission. The optimised CL equipment is depending on the type of microscope and other criteria such as spectral resolution, sensitivity, acquisition time, financial budget...

4.1. Cathodoluminescence Microscopy

Electron probe microanalysers (EPMA) are widely used for CL microscopy. For this purpose, a fixed electron beam is unfocused in order to irradiate an area of a few hundreds of microns in diameter at the specimen surface and the CL emission is observed and recorded on a photographic plate in front of the eyepiece of the optical microscope attached to the EPMA. In order to make CL microscopy easy for mineralogical and petrological applications, different CL devices have been developed allowing the simultaneous observation of CL and reflected or transmitted light with a conventional optical microscope. For this purpose, CL microscopes consisting of a thermoelectronic electron gun similar to that of the EPMA have been developed by Le Poole et al. [8] and more recently by Ramseyer et al. [9]. Commercial CL microscopes are also available. These equipments consisting of a cold cathode electron gun attached to the specimen stage of an optical microscope are derived from the equipment initially designed by Sippel [10]. Using the scanning beam facility of the scanning electron microscope (SEM) or EPMA equipments, polychromatic and monochromatic CL images can be combined with X-ray or secondary and backscattered electron images. The polychromatic CL images simply obtained with a photomultiplier tube attached to the electron column of the SEM, relates the local changes of intensity and colour of the CL emission with the distribution of impurities in the specimen. In order to improve the collection efficiency of the emitted CL photons with the SEM, several light collectors have been installed in the SEM chamber as discussed by Horl and Roschger [11], Trigg [12] and Yacobi and Holt [3] among others. An integrated single detector for simultaneous detection of CL and backscattered electrons has been proposed by Autatra et al. [13]. However, the more commonly used light collector is a parabolic mirror placed immediately above the specimen surface. The specimen is placed at the focus point of the mirror and the light is reflected through a silica lens towards the detector placed outside the SEM column. The detector can be a PMT, a silicon photodiode or a CCD camera.

4.2. Cathodoluminescence Spectroscopy

The two basic parts of a CL spectrometer are the grating or prism monochromator and the detector assembly. As for the case of X-ray spectrometry, CL spectra can be obtained using either a sequential or a simultaneous detection mode. A sequential CL spectrometer has been developed by Semo [14]. The system consists of a flint prism, a condenser lens and a PMT as detector. The analysed wavelength is sequentially selected by moving a slit in the focal plane of the optical condenser placed behind the monochromator. The second type of CL spectrometer is an optical multichannel analyser (OMA) for simultaneous detection as described by Lohnert et al. [15]. For the use of an OMA system, the CL flux collected by the optical collector is focused on the entrance slit of a grating monochromator with an intensified linear photodiode array (LPDA) or a CCD camera as a detector.

5. Experimentation

5.1. Case of Minerals

5.1.1. Specificity of each type of apparatus. Fixed or scanning beam irradiation are two complementary techniques for CL microscopy. Colour CL images using either the unfocused beam of the EPMA or the CL device attached to an optical microscope can be used only for qualitative observation.

Commercial photographs films have a broad response function allowing the recording of CL images with a broad colour domain of the visible spectrum. Polychromatic CL images with the SEM generally reveal the same structure as observed by using CL recording with static beam irradiation [1].

However, the PMT used as a CL detector with the SEM has a sharper response function to photon energies than a colour photographic plate. It is common to use PMT having the same photocathode as for secondary electrons detector. Most of the photomultipliers have a photocathode of the S20 type with a maximum sensitivity in the blue region and a very low efficiency in the red part of the visible spectrum. Thus, a high contrast in polychromatic CL imaging may occur when the spectral CL emission bands have a maximum intensity at different wavelengths.

5.1.2. Effect of electron beam heating. The increase of temperature (ΔT) due to the electron beam heating is given by [16]:

$$\Delta T = (1-\eta)\frac{3}{2\pi}\frac{P_0}{kR}$$

P_0, beam power (W)
k, thermal conductivity ($W \cdot cm^{-1} \cdot K^{-1}$)
R, electron range (cm)

For the case of glass or pure SiO_2, the thermal conductivity is less than $10^{-2}\, W \cdot cm^{-1} \cdot K^{-1}$, that gives for a beam energy of 20 keV, a beam current of 100 nA (usual electron beam conditions for CL in EPMA), an increase of temperature of 250 °C. Using a CL spectrometer attached on a SEM equipped with a parabolic mirror, a beam current of 10 nA can be used, reducing by a factor 10 the increase of temperature.

5.1.3. CL induced by internal electrical fields. For a specimen bombarded with an incident electron beam, the charge conservation at the electron beam spot on the specimen surface is:

$$I_b = I_r + (dQ/dt) + I_s$$

where I_b is the number of incident electrons, I_r is the number of electrons escaping the specimen surface, I_s is the specimen current and dQ/dt is the charge trapped during the interval time dt. The trapped charges are responsible for the development of strong internal electric fields as reviewed by Cazaux [18]. In order to explain the macroscopic electrostatic phenomena, microscopic effects resulting from energy levels acting as electron traps in the energy diagram of the crystal must be invoked as discussed by Cazaux and Le Gressus [19]. The energy levels involved in electrostatic charging phenomena are also responsible for the luminescence emissions. Due to the internal electrical field inside the insulator, a mechanism analog to that illustrated in Fig. 4 for the case of electroluminescence can take place. This mechanism can be combined with the cathodoluminescence emission in the sense strict.

5.1.4. Response function of the CL assembly system. Significant changes in the CL intensity, shape and wavelength position of the emission bands, may occur when comparing raw CL spectra with data corrected for the response function of the spectrometer. Broad emission bands undergo larger correction effects than narrower

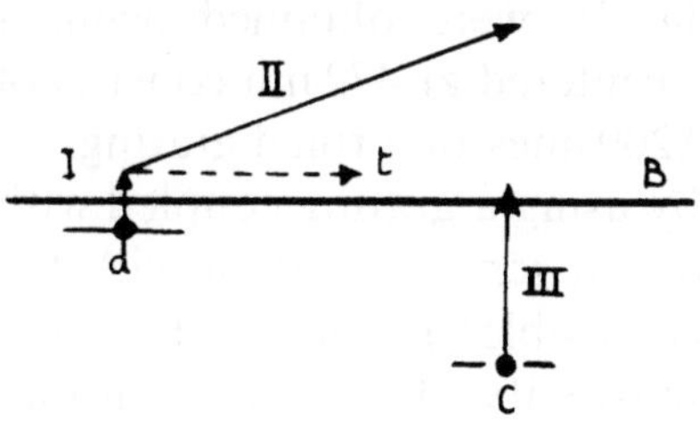

Fig. 4. Schematic picture of excitation mechanism for electroluminescence due to accelerated electron impacts. *B* conduction band, *d* shallow donor level, *C* luminescence centre, *I* ionisation of donor by electric field, *II* acceleration of some electrons raised into conduction band (t-path of electron remaining at thermal velocity), *III* ejection of electron initially in centre due to impact of an accelerated electron (reprinted from D. Curie [17])

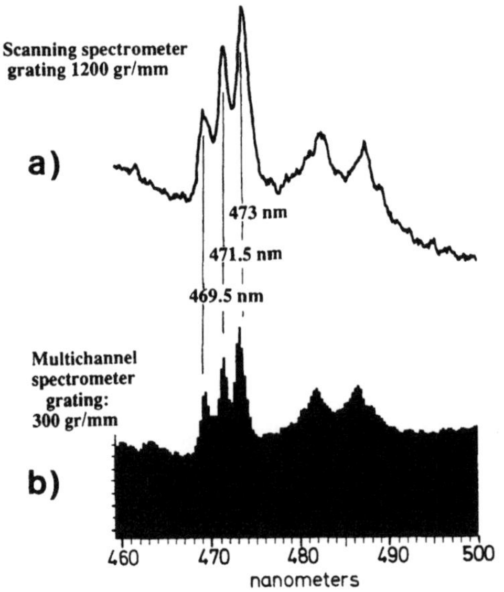

Fig. 5. Comparative wavelength resolution obtained using a scanning spectrometer (**a**) and a multichannel spectrometer (**b**) [1]

bands. Thus, the presence of narrow CL peaks can be used for comparing the respective merits of the scanning and multichannel CL spectrometers. Using a scanning monochromator, the spectral resolution can be improved by reducing the width of the entrance slits of the spectrometer. However, increasing the wavelength resolution by adjusting the slit width is obviously accompanied by an important decrease in the measured intensity. For a spectrometer of focal length 300 mm, the spectral resolution with a grating of 300 lines/mm is usually 5 nm/mm of the entrance slit width. In the case of photodiode array based spectrometer, the absence of mechanical part makes the reproducibility and accuracy of wavelength reading greater than those of any mechanically scanned spectrometer. A photodiode array has a constant length, and the theoretical spectral resolution is only governed by the reciprocal linear dispersion (nm/mm) of the grating and by the width of a diode. For practical reasons (aperture response function of the array), the spectral resolution is worse by a factor 2 of the theoretical one. For an array of 512 diodes and a grating of 300 lines/mm, the spectral resolution is 1.2 nm. The spectral resolution obtained with the 300 lines/mm grating coupled to the photodiode array was sufficient to identify all the features found using the scanning grating monochromator coupled with a photomultiplier (Fig. 5a and 5b). Data in Fig. 5b were obtained using a 300 lines/mm ruled grating and show that the peak centered at 472 nm consists of several features similar to Fig. 5a, obtained with a 1200 lines/mm ruled grating.

It is also noticeable that the gain in resolution by using a grating coupled with a photodiode array is accompanied by a lower loss in sensitivity than with the scanning grating monochromator. Furthermore, it is possible to adjust the power of the intensified photodiode array to increase the sensitivity of detection without increasing the electron beam current, i.e. limiting the heating and the beam damage.

The overall efficiency of CL detection is the net result of many instrumental factors (solid angle of light collector, transmission of the fibre optics and monochromator devices, sensitivity of detectors, ...) depending on which assembly is used as mentioned in Table 1.

5.1.5. Possible artefacts. Artefacts may arise from spurious luminescence signals induced by backscattered electrons reaching some luminescent components within the specimen chamber. When the front window of the PMT is placed in vacuum of the SEM, backscattered electrons induce the luminescence of the glass window [20]. In the case of specimen containing several phases of different atomic number, the observed CL image will be the sum of the true image plus a contribution proportional to the backscattered electrons coefficient. It is thus important to verify that any window which can be exposed to backscattered electrons within the specimen chamber consists of non-luminescent materials.

Another source of spurious light is the electron gun filament. This emission can be reflected by the sample surface which is metal coated. Fortunately, due to the presence of apertures in the electron column, this intensity is usually very low as compared to the emission of minerals. The use of an electron beam blanking with magnetic coils or electrostatic plates and of a synchronous detection will alleviate this problem.

The presence of a window between the SEM chamber and the fibre optic coupling may affect the transmission spectrum of the light, especially in the blue region. The response function of the CL assembly has to be obtained.

5.2. Case of Semiconductors

5.2.1. Effect of the electron beam heating. As expressed previously, the stationary increase of temperature is mainly given by the thermal conductivity of the semiconductor.

For direct bandgap semiconductors, the usual electron beam conditions are: electron beam energy, 5–20 keV, electron beam current, 1–10 nA. For example, in GaAs, these electron beam conditions give an increase of temperature varying from 0.6 K (20 keV, 1 nA) to 20 K (5keV, 10 nA).

When the sample temperature is maintained near liquid helium temperature, the heating of the sample has to be taken into account for the spectral line shape.

5.2.2. Effect of electron beam injection. The electron beam conditions have to be chosen in order to ensure the low injection conditions, i.e. the created minority carriers density has to be lower than the majority carriers density, that is, the doping level.

For GaAs, *n*-type $N_d = 10^{17}\,\mathrm{cm}^{-3}$, the estimated beam current for low injection conditions with a beam energy of 20 keV is about 1 nA. For GaAs, *p*-type, $N_a = 3\cdot 10^{16}\,\mathrm{cm}^{-3}$, the estimated low beam injection conditions are 0.2 nA, for 20 keV beam energy.

5.3. Artificial Contrast in Backscattered Electrons Detection by Solid State Detectors Due to CL Emission

Reciprocally to the spurious CL signals induced by backscattered electrons impinging the detector entrance window, the visible CL photons enter the backscattered

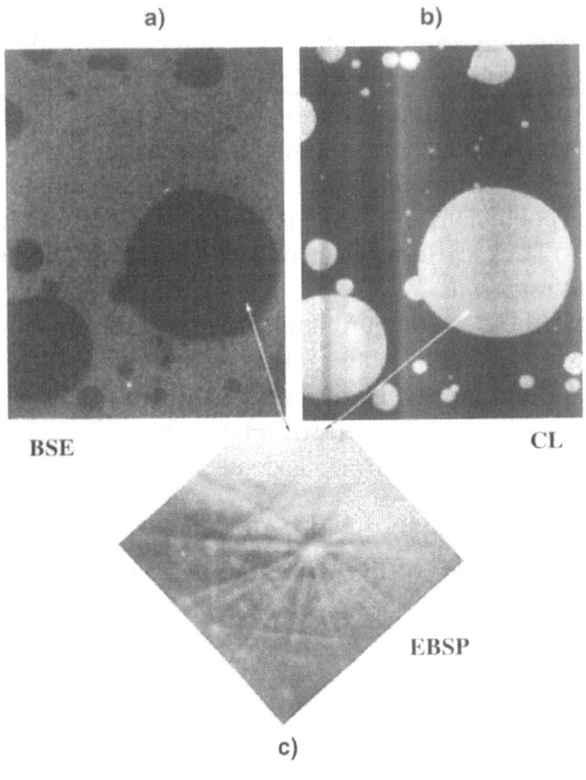

Fig. 6. Effect of CL emission inducing an artificial contrast on the backscattered electrons image using solid-state detectors for a YAG glass specimen containing crystalline structures. **a** Backscattered electrons (*BSE*) image. **b** CL image. **c** Backscattered electrons pattern (*EBSP*)

electron detector leading to a simultaneous detection of the electrons and photons emitted at the incident electron beam spot. For composition mapping based on the change in the backscattered electron yield as a function of the local change in the mean atomic number of the specimen, an array of solid-state detectors (*p–n* junctions) are usually used. These silicon based semiconductor detectors (1.1 eV bandgap) are sensitive to the visible light and both signals resulting from the absorption of the backscattered electrons and luminous photons combine to give the contrast of the backscattered electrons (BSE) image. As shown in Fig. 6a, the BSE image for the case of a 7 keV incident energy, reveals contrasts which can obviously be simply related to composition heterogeneities while in fact the specimen was expected to be chemically homogeneous. The specimen is a glass with the composition of an yttrium-aluminium garnet (YAG) prepared according to a procedure described by Coutures et al. [21]. RPE and RMN analyses [22] indicated for the presence of crystalline structures within the bulk glassy specimens. The presence of spherical particles was revealed by their intense cathodoluminescence emission exhibited during EPMA analyses as shown on the CL image in Fig. 6b. The luminescent areas corresponds to crystalline materials as shown on the backscattered electron pattern (EBSP) in Fig. 6c. Quantitative X-ray analysis with the EPMA did not show significant composition changes sufficient to explain the contrast shown on the BSE image (Fig. 6a) between the YAG glass and the small crystallised particles emitting the intense CL intensity. The same contrasts than those illustrated in Fig. 6 were observed when the incident electron energy was varied from 5 to

20 keV. Bulk charging phenomena may selectively occur within the glass and the crystal modifying the depth distribution of the electrons, leading to a change of the backscattered electron yield as a function of the crystallographic properties of the analysed areas. Using an EDS, the continuous X-ray emission distribution was measured at the surface of the uncoated glass and crystalline particles. For the case of a 5 keV incident energy, the Duane-Hunt limit (maximum energy of the continuous energy) was equal to the incident energy indicating that no retarding potential developed at the surface of the glass area. For the case of the crystalline particles, the Duane-Hunt limit was measured about 100 eV lower than the incident energy indicating for a surface or bulk small charging effect. When the specimen surface is carbon coated as for the case of the experiments illustrated in Fig. 6, no surface charging may occur but, bulk charging can take place leading to a change in the backscattered electron yield. If such phenomena exist, the contrast of the BSE image should change when the incident energy value is also changed while the contrasts we observed remained independent of the excitation conditions. Rather than bulk charging effects, the role of the CL properties of the glass and crystals of similar composition must be invoked to explain the contrasts shown on the BSE image. Because of the energy difference for the CL photons (a few eV) and the backscattered electrons (in the range of a few keV) the two ionising incident energy interact the solid-state detector at different depths with respect to the depletion area of the *p–n* junction. The difference in the penetration depth for the visible and X-ray photons may result in the two currents of opposite signs at the output of the detector. The current associated with the incident backscattered should be deprived by the current induced by the absorption of the luminous photons. The higher the CL intensity the lower will be the net current at the detector output so that for a constant mean atomic number the luminescent areas will appear darker on the BSE image. More investigations involving several types of electron detector (*p–n* junctions versus Schottky barriers) are necessary to support this tentative explanation of the contrasts shown in Fig. 6 but these observations indicate that all physical properties of the specimen and not only the chemical composition (mean atomic number) must be considered for a reliable interpretation of BSE images.

6. Examples of Application in Synthetic or Natural Insulating Crystals

6.1. CL as an Analytical Tool for the Identification and Location of Trace Elements

6.1.1. Strategy for trace element analysis. Cathodoluminescence microscopy is frequently used to reveal heterogeneities which are not observed by optical microscopy or scanning electron imaging as reported by Long and Agrell [23] and by Remond [1, 24, 25] among others for the case of minerals. These authors also showed that correlations existed between the CL images and the distribution of some trace elements revealed by X-ray imaging with the EPMA or SEM. An example of correlations between the CL image and the distribution of impurities is illustrated in Fig. 7 for the case of a natural cassiterite (SnO_2). Based on these positive correlations between the CL emission and the presence of impurities, it was thus tempting to use CL spectrometry to complement X-ray spectrometry for the identification and the localisation of impurities within minerals. However, it must be kept in mind that the

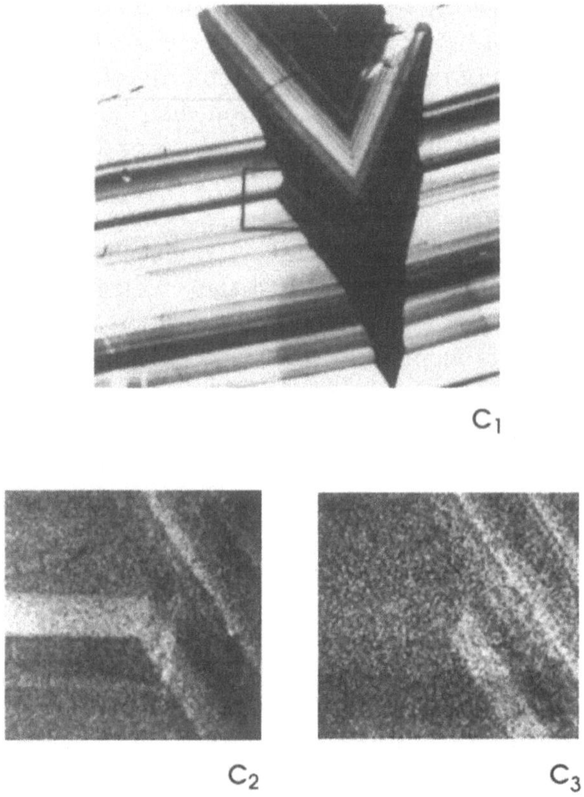

Fig. 7. Cassiterite SnO_2 sample. C_1 polychromatic CL image. C_2 and C_3 Fe and Ti distribution of the marked zone by an EPMA

detected impurities within the different CL areas may not be the unique impurities responsible for the observed luminescence emissions. In the absence of any general framework for relating CL spectra to crystal impurities, empirical approaches are used. The first step consists of the use of non-destructive analytical techniques for identifying the impurities present in the compound. The second step in establishing a relationship between the CL emission and the impurities contained in the host crystal lattice is to identify among the detected impurities those which can play a role in the observed CL emission. For this purpose, synthetic doped crystals are prepared and their CL emission spectra compared to those of the minerals. This approach was used by Schvoerer et al. [26] for identifying small among of Ti and Cr ions in an α alumina single crystal. The CL emission spectrum of this specimen was characterised by two broad bands centred at 350 nm and 522 nm respectively and by multiple narrow peaks near 694 nm. Based on the abundant literature on the role of Cr^{3+} ions in alumina, the presence of this impurity within the analysed crystal was evident. From the CL spectra of Ti doped alumina crystals, Schvoerer et al. [26] associated the CL bands at 350 nm and 522 nm with the presence of that impurity. However a complete understanding of the CL spectra can only be achieved by varying both the quantitative composition of the synthetic compounds since Schvoerer et al. [26] showed that the relative intensity of the emission bands changed as a function of the two added impurity concentrations. This dependence of the CL spectra on the concentrations of all activators and co-activators make difficult the development of a sufficient number of synthetic crystals when a variety of trace elements can be found simultaneously in various crystallographic sites as it is the case for minerals. The identification of the impurities responsible for the CL emissions become difficult when several activators are present because the CL properties will change as a function of the relative concentrations of the activators and co-activators as it is the case of ZnS crystals as discussed among others by Curie [17]. The complexity in obtaining reference CL emission spectra obtained from synthetic doped materials has been illustrated by Rouer et al. [27] for the case of zircon ($ZrSiO_4$) crystals. Studying synthetic crystals doped with phosphorus and a rare-earth element (REE), Rouer et al. [27] showed that a positive correlation existed between the CL image and the distribution of P while the doping REE element was not detected with the EPMA. This example illustrate how false interpretation of the nature of the luminescent activators based only on the comparisons of CL images and element mapping obtained with the EPMA or secondary ion mass spectrometry (SIMS).

6.1.2. Application to rare-earth bearing crystals. Impurities with incomplete electronic shells may exhibit characteristic luminescence spectra making easier the use of CL spectrometry for analytical purposes. Such a situation will be illustrated for the case of zircons crystals ($ZrSiO_4$) containing rare-earth elements (REE) at low concentration levels. Ohnensteter et al. [28] distinguished two populations of natural zircon crystals based on their blue or yellow-green CL color respectively as observed by means of an optical microscope equipped with a cold cathode electron gun. The spectra in Fig. 8 are representative of the two populations of natural zircons. For the case of the zircon crystals exhibiting a blue hue, narrow peaks are superimposed on the broad emission bands. These narrow peaks consist of many features and are associated with rare-earth luminescent ions. For a complete identification of the CL emission peaks, synthetic zircon crystals ($ZrSiO_4$), undoped and doped with Y^{3+}, La^{3+}, Ce^{3+}, Pr^{3+}, Nd^{3+}, Sm^{3+}, Eu^{3+}, Gd^{3+}, Tb^{3+}, Dy^{3+}, Ho^{3+}, Er^{3+}, Tm^{3+}, Yb^{3+}, Lu^{3+}, as well as coupled REE^{3+}-doped zircons, were growth from a flux consisting of a mixture of Li_2MoO_4 and MoO_3 heated to 1125 °C and then cooled to 750 °C [29]. The cathodoluminescence spectra of these zircons were analysed, at room and near liquid nitrogen temperature, but the experimental data were not corrected for the response function of the spectrometer used [30].

Common to all specimens, undoped or REE doped crystals, the CL emission spectra displayed broad emission bands ranging from ~ 230 nm to ~ 500 nm which

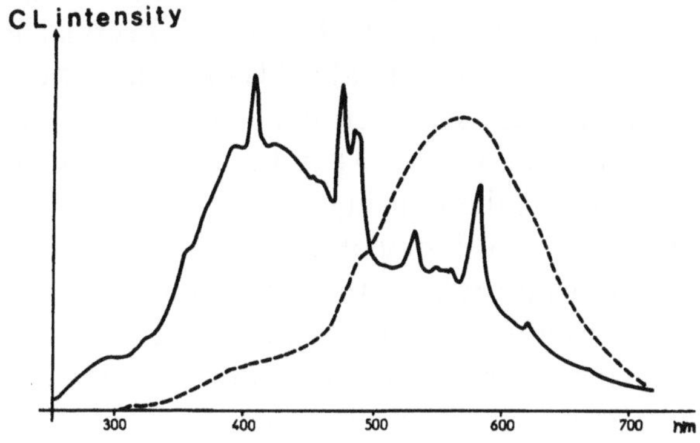

Fig. 8. Representative CL spectra of two populations of natural zircons ($Zr(SiO_4)$ exhibiting a blue (full line) and a yellow-green (dashed line) emission [1]

are intrinsic in origins and associated with lattice defects. However, the nature of the luminescent centres responsible for the zircon "intrinsic" emission band is still in debate and their possible origins have been reviewed by Cesbron et al. [30].

The transitions corresponding to the characteristic emission lines of Dy^{3+}, Tb^{3+}, Er^{3+} and Sm^{3+} have been indexed [30] by comparison with the thermoluminescence (TL) spectra of zircon doped with the same elements [31, 32].

However, the quantitative interpretation of the CL spectra of REE doped zircons is not a general rule because of the CL intensities of the intrinsic and extrinsic features depend upon numerous experimental factors such as the electron beam energy and the beam current density. In addition quantitative analysis of CL emission spectra must be corrected for the instrumental response function of the equipment and for the anisotropy of emissions as a function of the crystal orientation as shown by Cesbron et al. [30].

6.2. CL Property Changes Related to Electrostatic Charging Phenomena in Strongly Insulating Crystals

Time dependence of CL intensity is frequently observed as illustrated among other by Remond et al. [1, 33] for the case of natural quartz (SiO_2) crystals. These authors showed that the total CL intensity observed at the specimen surface rapidly increased and then decreased to almost complete extinction when the time of electron bombardment was prolonged. The time dependence of the CL intensity was found to vary as a function of the incident electron dose and to vary from a specimen to another. In fact, the variations as a function of time of the total CL intensity corresponds to a change in the colour of the CL emission during the irradiation time as illustrated by Ramseyer et al. [34] who imaged the short-lived and the long-lived quartz luminescence. Using a photodiode array as detector. Remond et al. [1] measured sequentially CL spectra at the surface of a natural quartz specimen with a 20 s acquisition time per spectrum and a constant incident fluence. The first spectrum showed a broad continuous emission distribution with a maximum intensity near 630 nm (1.97 eV). After 200 s irradiation, the intensity of the CL emission was increased and the peak in the red part of the spectrum became narrower. After about 500 s irradiation, a very intense broad CL emission band was revealed in the blue-green region of the spectrum. As discussed by Luff and Townsend [35] and by Mc Keever [36] for the case of synthetic amorphous and crystalline quartz specimen, the origin of the luminescence is very complex and no unique explanation has emerged from theoretical and experimental studies. The complexity of the emission is particularly true for the case of CL since all effects such as thermal increase and the development of internal electrical fields may contribute to the emission. In addition, CL emission changes as a function of irradiation time may result from beam damage leading to ionic diffusion [33] and to the creation of dangling non-bridging oxygen ions generated by the ionising radiation.

In wide bandgap insulating crystals, the development of surface and bulk charging phenomena probably have an important role on the variations of the CL emission distribution. The CL intensity is a complex function of the incident energy, the dead voltage and the incident dose, i.e. the beam current, the beam diameter and the irradiation time which are also the parameters involved in the development of

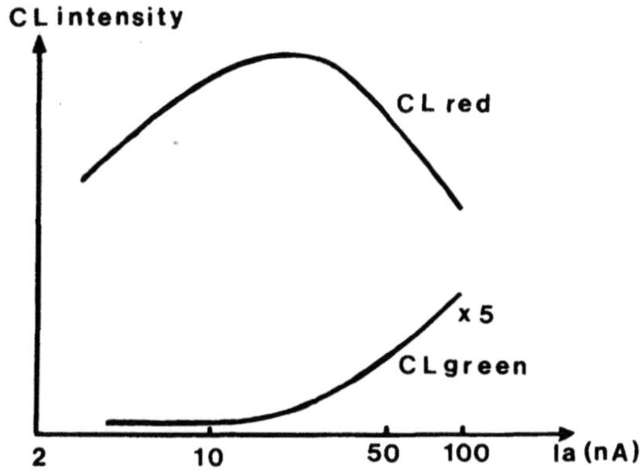

Fig. 9. Evolution of the two CL emissions (red and green) as a function of the electron beam current, measured as absorbed current, I_a, for a ZnS natural mineral (electron beam energy 20 keV, electron beam spot size, 20 μm) [25]

internal electric fields in strongly insulating materials. Studying natural ZnS crystals containing Cu and Ga as major impurities, Remond et al. [25] showed that the specimen exhibited a red CL colour when the stationary beam irradiated an area several hundreds of microns in diameter and that a blue emission was observed when the beam was focused on the specimen surface. CL spectroscopy showed that the spectrum consisted of two bands in the red and blue parts of the visible spectrum and that the relative intensities of the two bands varied as a function of the electron density. Results in Fig. 9 illustrate the variations of the intensity of the blue and red CL bands respectively as a function of the beam current. These variations are consistent with the data reported by Mattler and Ceva [37] studying the electroluminescence properties of ZnS crystals. These authors showed that the electroluminescence of crystals doped with several impurities, the different emission bands exhibited different variations as a function of the applied voltage. A possible contribution of luminescence induced by internal electric fields induced during the electron irradiation has also been postulated by Remond et al. [1, 38, 39] who showed that the CL emission may propagate at long distance from the incident beam spot and that the luminescent features were very similar to those observed on the secondary electron image by Vigouroux et al. [40, 41] showing the remote electron cascade multiplication induced by an electric field within an insulating material.

Recently, Remond et al. [39, 42] illustrated the effect of charge trapping mechanisms during the electron bombardment by comparing the CL emission properties and the X-ray emission spectra of ZnS crystals as a function of the experimental conditions. Two synthetic ZnS crystals were analysed. The crystals were grown by chemical vapour deposition using Zn vapour and H_2S in an argon carrier gas at near 700 °C. The two synthetic crystals were manufactured by Morton International (Woburn, MA, USA). The specimens are polycrystalline cubic ZnS. One specimen is water clear with grain size 20–22 μm and a resistivity greater than 10^{13} Ω.cm. The second specimen is yellow coloured with grain size 2–10 μm and a resistivity of about 10^{12} Ω.cm. As received the specimens have their surfaces optically polished according an unknown polishing procedure.

The CL spectra measured at room temperature of the yellow and clear synthetic ZnS crystals are shown in Fig. 10. The narrow CL peak occurring at 338 nm (3.67 eV) is associated with the intrinsic emission resulting from the direct recombination of electron-hole pairs in the conduction and valence bands. The intensity of the

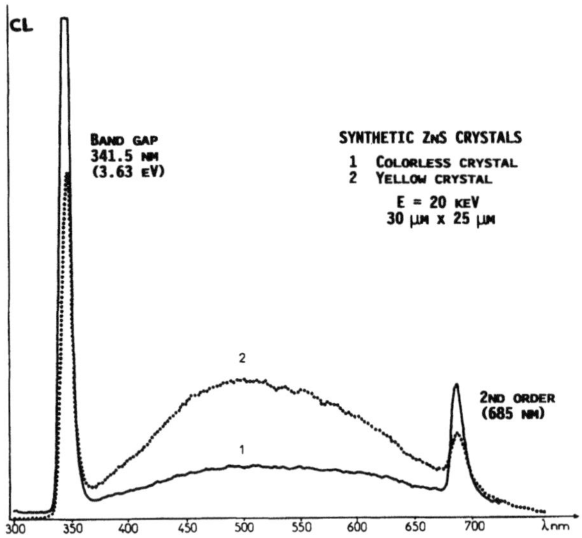

Fig. 10. CL spectra of two synthetic ZnS crystals, one giving a clear luminescence, the other giving a yellow luminescence

intrinsic emission peak was found to be much higher for the case of the clear ZnS specimen than for the yellow crystals. A second narrow CL feature is also observed on the CL emission spectra of both the clear and yellow synthetic ZnS crystals. The peak occurring at 685 nm may correspond to the second order reflection of the fundamental intrinsic emission peak measured at 338 nm. The experimental data in Fig. 10 are not corrected for both the response function of the spectrometer and for the matrix effect and it is not possible to compare the absolute CL yields for the 338 nm and 685 nm peaks respectively. The instrumental and matrix corrected CL spectra obtained by Datta et al. [43] showed that the intensity of the 685 nm peak was much higher than that occurring at 338 nm. A possible effect of multiple reflections by the grating cannot be completely rejected but according to Datta et al. [43], the 685 nm peak should be associated with a defect resulting from the presence of tin at trace levels within the ZnS specimens. This result clearly illustrates the need of correcting the raw data for the response function of the equipment for obtaining reliable data. In addition to the CL emission peaks at 338 nm and 685 nm, a broad extrinsic emission band in the visible part of the analysed wavelength range was also detected for both crystals. The intensity of the visible emission band, relative to that of the intrinsic emission peak was higher for the yellow than for the clear crystal. These differences may result from a higher density of defects within the yellow coloured specimen than that existing within the clear crystal. Consequently, the probability of direct charge recombination through the bandgap of the crystal is lower for the case of the yellow specimen than for the clear one. The development of internal electric fields resulting from charge trapping mechanisms may be a second possible explanation to the change in the intensity ratio between the intrinsic and extrinsic emission bands. In order to support this assumption, the CL intensity at the bandgap emission was studied as a function of irradiation time at the surface the clear and yellow ZnS crystals.

As shown by Remond et al. [38, 39], the intensity of the intrinsic CL peak at 338 nm measured at the surface of the yellow coloured specimen, remained approxi-

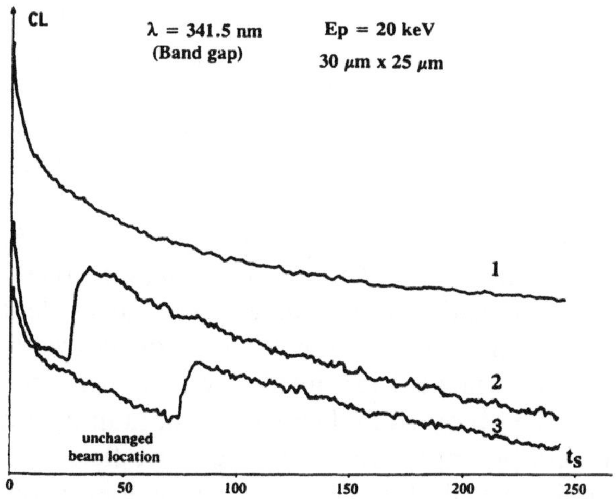

Fig. 11. Time dependence of the CL emission in the two synthetic ZnS minerals giving a clear and a yellow luminescence (see text)

mately constant as a function of the bombardment time. When the specimen was moved away from the irradiated area, the CL intensity increased but remained constant as a function of time. The CL intensity at 338 nm measured at the surface of the "as received" clear ZnS crystal exhibited a very rapid decrease as a function of irradiation time. Again changing the analysed specimen position under the electron beam led to an increase of the CL intensity immediately followed by a decrease of the CL intensity.

The two synthetic ZnS specimens were mechanically polished on a metallic polishing disk using diamond abrasives dispersed in silicone oil according to a procedure already discussed by Remond et al. [38, 39]. As shown in Fig. 11, the intensity of the intrinsic emission measured at surface of the clear polished ZnS specimen decreased rapidly for the first ten seconds of irradiation before to reach asymptotically a constant value (curve 1 in Fig. 11). For the case of the yellow polished crystal, the intensity of the intrinsic CL emission peak as a function of the electron bombardment time exhibited a rapid decrease followed by a sharp increase after $\sim$ 30 s irradiation before showing a continuous decrease. No further jump in the CL intensity was observed even when the irradiation was extended up 4 min (curve 2 in Fig. 11). The electron beam spot was then moved $\sim$ 100 μm away from its initial position and the CL intensity behaved the same variation as a function of time as was observed at the previously analysed area (curve 3 in Fig. 11).

The two ZnS specimens were analysed by energy dispersive X-ray spectrometry (EDS). Comparisons between observed and calculated continuous emission distributions were used to reveal charging effects resulting from charge trapping mechanisms involving point defects as discussed by Remond et al. [38, 39, 42]. The uncoated surface were analysed with a 5 keV incident energy and a 10^{-9} A beam current. For the "as received" yellow crystal, the Duane-Hunt limit was equal to the incident energy while after the surface was mechanically repolished, the Duane-Hunt limit was equal to 3 keV indicating that the maximum kinetic energy of the electrons into the specimen was reduced by 2 keV with respect to the incident value. However, the continuous emission was correctly described assuming that the incident energy was 3 keV instead of 5 keV. The consistency between the observed and calculated continuous emission distributions indicated that a 2 keV retarding potential was established at the surface of the repolished specimen. The continuous emission distributions measured at the surface of the clear and yellow repolished specimens were compared. The incident energy was 5 keV. A beam current of $3 \cdot 10^{-10}$ A the Duane-Hunt limit occurred at 4 keV. In order to observe the same Duane-Hunt limit on the X-ray spectra of the clear crystal the beam current was increased to $6 \cdot 10^{-9}$ A. The continuous emission distribution of the yellow ZnS specimen was correctly described assuming the existence of a surface reading potential reducing the incident energy down to 4 keV. The continuous emission distribution was no more correctly described for the case of the clear specimen when a 4 keV incident energy was used for calculations. The deviation between the observed and calculated data indicates that in addition to the surface retarding potential, an internal electric field also existed modifying the energy and spatial distribution of the electrons within the insulator.

Based on the modifications of the X-ray emission observed as a function of the electrical properties of the two synthetic ZnS specimens and as a function of the specimen preparation procedures, the differences in the CL emission properties of both specimens must be associated with charging phenomena. These effects depends on the existence of pre-existing defects and induced defects during the specimen preparation and analysis. However, the presence of point defects acting as electron traps resulting from small amount of impurities is not sufficient to explain the development of the internal electric fields and the density of defects at the grains boundaries of the polycrystalline materials must be considered as discussed by Remond et al. [42]. The difference in grain size of the two synthetic ZnS specimens is also responsible for the differences in their electric and mechanical properties consistent with the differences in microscopic roughness as observed by Remond et al. [44] at the surface of the mechanically polished specimens. As discussed by Myhajlenko et al. [45], the microroughness of the surface relates specifically to scattering phenomena associated with spatial fluctuations in the dielectric constant modifying the yield of the CL photon emission and the contrast of CL images.

Surface and bulk charging of insulators during electron bombardment are complex phenomena which tend to modify the shape and the in-depth position of the ionisation distribution with respect to those found in an electrical conductor. However, the charging phenomena are difficult to predict and to model because they depend on the spatial distribution of the trapped charges are generally unknown. As a consequence, the choice of standards for quantitative X-ray spectrometry becomes crucial, more particularly in the soft X-ray region and the electrical properties of both the unknowns and the standards must be considered and not just their chemical composition. For this purpose, CL studies may provide a guide for choosing the suitable standards for X-ray analyses.

7. Examples of Applications in Semiconductor Materials

The combined CL spectroscopy and microscopy is a powerful tool for the study of semiconductor material and devices. Zones of different compositions can be visualised using the near band edge emission. The use of the liquid helium temperature allows to separate the transition lines of different impurity contents or doping levels. Furthermore CL can be used to image these different zones.

All the experiments shown below have been performed at low temperature in a JEOL 840 modified scanning microscope. The helium cooled Oxford collection system has been adapted to the microscope [46]. The CL spectra and images have been obtained with a beam energy varying from 5 to 20 keV and a beam current ranging from 1 nA for spectra to 10 nA for images in order to ensure the low injections conditions. The spectra have been obtained with a wavelength resolution of 2.5 nm or less.

7.1. Composition Mapping

$Ga_{1-x}Al_xAs$ layers have been grown on a GaAs (100) substrate by molecular beam epitaxy. During the growth, the layers have been doped by introducing silicon. The net electron doping measured by Hall effect is $3 \cdot 10^{17}\,cm^{-3}$. The layers are around 1 μm thick. During the growth, defects have appeared. Due to their shape, they are called "oval" defects. Their density is as high as $10^{14}\,cm^{-2}$. Their origin seems to be related to the presence of impurities in the Ga source (oxide) used for the growth.

Figure 12a and 12b show cathodoluminescence spectra recorded outside (Fig. 12a) or inside (Fig. 12b) an oval defect. The 665 nm emission line corresponds to the composition of the $Ga_{0.7}Al_{0.3}As$ layer whereas the 700 nm emission in the defect corresponds to the composition of $Ga_{0.82}Al_{0.18}As$ from which the defect is mainly constituted.

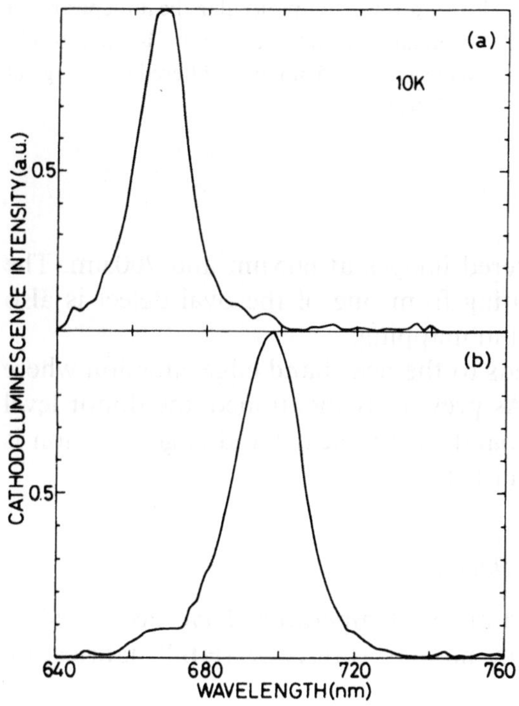

Fig. 12. Cathodoluminescence spectra ($T = 10$ K) recorded outside (**a**) and inside (**b**) an oval defect of a GaAs Si doped layer grown by molecular beam epitaxy

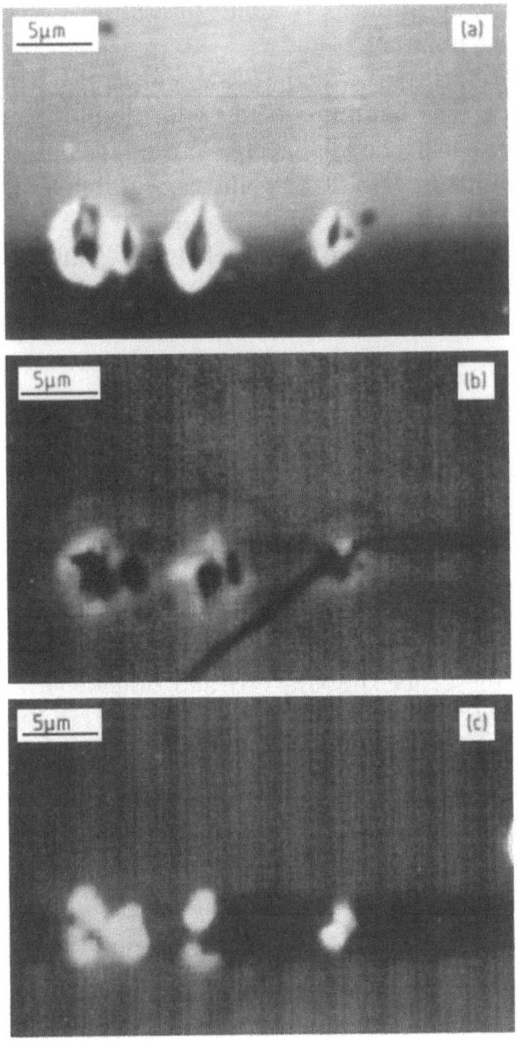

Fig. 13. Images obtained from a typical group of oval defects ($T = 10\,K$) of a GaAs Si doped layer grown by molecular beam epitaxy. **a** Secondary electron emission, **b** filtered CL image at $\lambda = 665\,nm$ and **c** filtered CL image at $\lambda = 700\,nm$

Figure 13b and 13c show the CL filtered images at 665 nm and 700 nm. The presence of non-radiative defects originating from one of the oval defect is also visualised by a black contrast on the 665 nm mapping.

The observed emission line corresponds to the near band edge emission where the D-A type transition may dominate. As previously mentioned, the donor level follows the evolution of the conduction band, so the near band edge emission is characteristic of the material composition [47].

7.2. Doped or Implanted Area Imaging in Devices

CL spectroscopy and imaging performed at low temperature have been used to qualify the GaAlAs/GaAs heterojunction bipolar transistor (HBT) fabrication technology [48].

Table 2. Layer structure of the heterojunction bipolar transistor

Layer	Material	Thickness (μm)	Net doping (cm^{-3})
Emitter cap	GaAs n+	0.15	$4 \cdot 10^{18}$
Emitter	$Ga_{0.7}Al_{0.3}As$	0.35	$2 \cdot 10^{17}$
Base	GaAs p+	0.15	$2 \cdot 10^{18}$
Collector	GaAs n	0.45	$2 \cdot 10^{16}$
Sub-collector	GaAs n+	0.5	$2 \cdot 10^{18}$
Substrate	GaAs semi-insulating		undoped

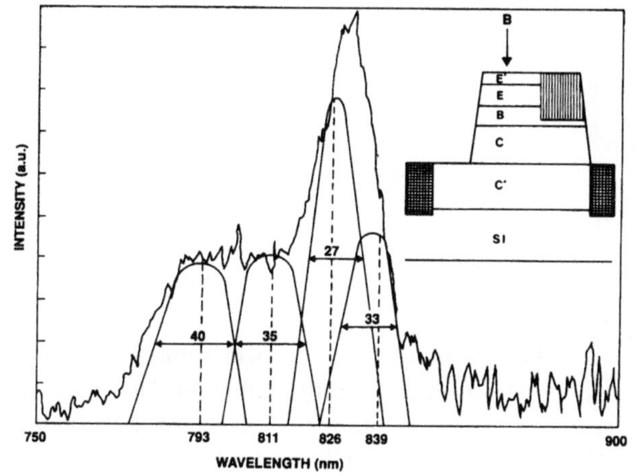

Fig. 14. Cathodoluminescence spectrum ($T = 10$ K) recorded on the whole structure of a heterojunction bipolar transistor. Electron beam conditions: energy 15 keV, intensity 2 nA. The FWHM are in meV

For the HBT fabrication, an epitaxial structure of layers of GaAs, $Ga_{0.7}Al_{0.3}As$, *p*-type and *n*-type has been grown on a GaAs substrate. Table 2 gives the composition, the doping type, the dopant concentration and the layer thickness of each layer. The *n*-type dopant is Si whereas the *p*-type dopant is Be. In order to define the active part of the HBT, several technological steps are processed: a mesa is defined by etching all the layers until to reach the collector layer. A Mg *p*-type implantation is performed to contact the base layer. Boron and proton are co-implanted outside the transistor active area to achieve a good device insulation. The *n*-type AuGeNi and *p*-type AuMn ohmic contacts are then deposited.

Figure 14 shows the CL spectrum obtained when the electron beam is located at the top of the mesa structure, outside the *p*-type implanted zone, obtained with a beam energy of 15 keV. With this beam energy, the electron range is around 1.5 μm and the incident electrons can excite each layer, even the sub-collector layer. The obtained spectrum is a convolution of 4 peaks corresponding to all layers excepted the emitter layer which is a GaAlAs layer. The position and the FWHM of each peak may be attributed to the different layers as compared to data given for the same doping type and doping level.

CL images can be used to qualify the protection of the HBT during the B-H implantation step used for device insulation. CL images allow, by filtering the emission at a specific wavelength to visualise the protected part from the

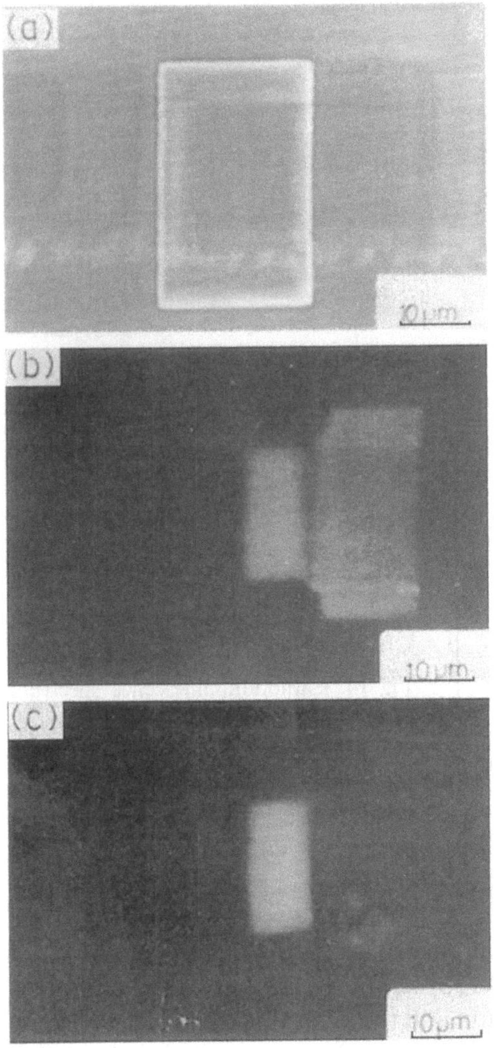

Fig. 15. Images of a heterojunction bipolar transistor structure after definition of the mesa and the B–H implantation for insulation. **a** Secondary electron image, **b** CL image ($T = 10$ K) at 811 nm: emission from the sub-collector layer on the right part, combined emission from the emitter cap layer and the sub-collector layer at the top of the mesa: **c** CL image ($T = 10$ K) at 835 nm: combined emission from the collector layer and the base layer at the top of the mesa

implantation at different levels of the structure. Figure 15b shows the CL image for the 811 nm emission. This visualises the CL emission from the sub-collector layer on the right part and on the mesa part of the emitter combined with the sub-collector. The CL image of Fig. 15c, obtained with the 835 nm emission, shows the combined emission from the collector and the base layer in the unimplanted part of the structure.

7.3. Evolution of the CL Signal as a Function of the Impurity Content

During the growth of GaAs layers by molecular beam epitaxy, the impurity content of dopant can be introduced in a controlled manner by varying the silicon cell temperature, giving evaporation as for the two major elements Ga and As.

One µm thick GaAs:Si layers have been grown on a GaAs substrate with different impurity content given by the silicon cell temperature. The total impurity content can be measured quantitatively by SIMS using implanted standards. The total net doping level (N_D–N_A) can be measured by Hall effect which corresponds to silicon in electrically active sites. Silicon is an amphoteric impurity and can be introduced in donor sites or in acceptor sites. Silicon is first introduced in donor sites and then in acceptor sites. Silicon may also create complexes involving either Ga or As in antisite or other silicon atoms.

For each layer, the CL intensity has been measured in the same electron beam conditions with the same adjustment of the assembly collection and selection in wavelength. The CL spectra taken for each layer allow to measure the intensity of the near band edge (NBE) emission and the deep level (DL) emission.

When the net doping level increases, the NBE emission increases, then goes to a maximum for a threshold doping value of $3 \cdot 10^{17}$ cm^{-3} and decreases. Above this threshold doping level, the DL emission which is negligible, increases strongly when the NBE emission decreases. The increase of the NBE intensity with N_D–N_A corresponds, as described in 3.3, to the reduction of the radiative lifetime due to the increase of the silicon in donor sites.

When we represent the results as a function of the inverse of the silicon cell temperature (Fig. 16), we can follow the different quantities: net doping level (N_D–N_A), the difference between the total incorporated impurity and the impurity in active sites (Δ[SIMS–(N_D–N_A)], the NBE emission, the DL emission, and the electron acoustic signal (EAI). We see that (N_D–N_A) goes to a maximum and decreases, when the silicon is introduced first in donor sites and then in acceptor sites. The NBE emission goes to a maximum and decreases to 0 whereas the DL emission start and increases strongly, corresponding to the creation of complexes. The Δ[SIMS– (N_D–N_A)] represents the introduction of silicon in non-electrically active sites.

The electron acoustic signal follows the evolution of the inverse of the square root of the thermal conductivity [49]. In semiconductors, the thermal conductivity has two contributions: one is the electronic part mainly given by the electrical conductivity, due to the impurities in electrically active sites, the other is the lattice part, mainly governed by the mean free path of phonons as the inverse of the cubic root of the total impurities introduced. The electronic part increases more quickly than the lattice part. But when silicon is introduced in acceptor sites, the net doping is reduced. So, at this point the thermal conductivity is mainly governed by the lattice conductivity and the thermal conductivity is strongly reduced due to the introduction of silicon in non-electrically active sites as measured by SIMS. As shown in Fig. 16, after the threshold doping, the electron acoustic signal increases strongly.

7.4. Electrical and Optical Parameters Evaluation

The characterisation of semiconductor materials is necessary to improve the technology of optoelectronic devices. One of the most important parameters to be measured is the diffusion length L of minority carriers created by an electron beam. CL and electron beam induced current (EBIC) are the two most convenient methods for the evaluation of L. EBIC is chosen in most cases, provided the presence of a *p–n* junction and ohmic contacts in the sample [50]. In other cases, CL can be used under the only condition that a clean flat surface can be prepared. However, it is difficult to

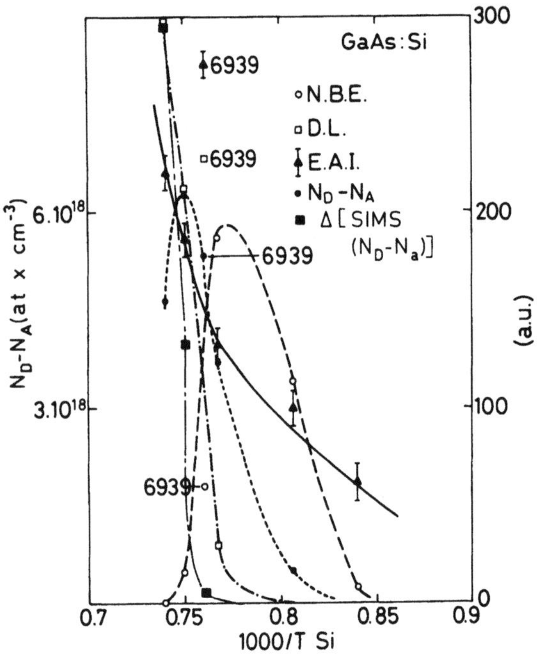

Fig. 16. Variations with the inverse of the silicon cell temperature of: the net doping level (N_D–N_A) (left scale), the difference between the total impurity concentration and the net doping level (Δ[SIMS–(N_D–N_A)] (left scale), the near band edge CL signal (*NBE*) (right scale), the deep level CL signal (*DL*) (right scale), the electron acoustic signal (*EAI*) (right scale)

obtain quantitative results, especially because the optical signal goes through a collection system, the response efficiency of which can be hardly determined.

The evaluation of the diffusion length L of a sample can be obtained by comparing the experimental curve giving the CL intensity versus beam voltage to theoretical ones [51]. These curves are calculated with the known physical data of the sample on the one hand and the two adjustable parameters L and s, the surface recombination velocity on the other hand. The abacus obtained allows the simultaneous evaluation of both parameters.

To get these theoretical curves, one may use the so-called point-source model and the calculations by Hackett [52]. Under low injection conditions, the CL intensity can be expressed as:

$$I_{CL} \propto \sum_{z=0}^{z=\infty} \Delta p(z) \exp(-\alpha z)$$

α, is the mean optical self-absorption coefficient, and is experimentally evaluated [51]. It includes both the self-absorption and the reabsorbed recombination radiation (RRR) effect [53, 54]. $\Delta p(z)$ is the concentration of the excess minority carriers in an elementary slice parallel to the surface, at depth z. This concentration depends on the pair generation function $g(x, y, z)$ at any point of the sample and on the diffusion of the minority carriers.

The choice of the generation function is very important to obtain a good accuracy for semiconductor parameter measurements. Various analytical generation functions have been published. Everhart and Hoff [55] have proposed a "depth dose function" in the form of a simple polynomial expression; Wu and Wittry [56] a Gaussian function; Donolato [57], and Oelgart and Scholz [58] a modified Gaussion function; Konnikov et al. [59] have presented a double Gaussian form and Closas and Fernandez Rubio [60] a Gaussian function plus the Everhart's polynome.

Another approach is based on the electron trajectory simulation by means of a Monte Carlo calculation; it can provide the three dimensional distribution of the electron energy loss in each elementary volume of the generation volume. With this method, Akamatsu and Hénoc [61, 62] have found a three dimensional analytical expression of the generation function. It involves three Gaussian functions: the first one corresponds to the interactions with the unscattered beam, the second one expresses the energy losses of the single scattered electrons and the third one, the energy losses due to the multiple scattering process. The variables X, Y, Z are normalised by a third of the Grün range [63]:

$$R_g(\mu m) = 4.57 \times 10^{-2} \times E^{1.73}(\mathrm{keV})/\rho(\mathrm{g.cm}^{-3})$$

Three additional parameters can be evaluated by means of CL, namely, the internal quantum efficiency, η, the apparent optical absorption coefficient, α, and the dead-layer depth z_m at the sample surface (see ref. in [64]).

Semiconductor optoelectronic devices need low-dimension structures. This leads Akamatsu and Hénoc [64] to study the CL characterisation of an epitaxial surface layer. The authors have presented a semi-quantitative method for the simultaneous evaluation of the semiconductor layer parameters, L, α, and the reduced surface and interface recombination velocities $S_{\mathrm{surface,\ interface}}$. The η can be deduced from these measurements. This method requires:

1. The quantitative CL signal calculation. It involves (i) a very simple "depth-dose" formulation describing the electron beam energy loss distribution as a function of the depth z, and (ii) the "image source and sink" model and the diffusion equation formulation of Berz and Kuiken [65].
2. The use of a reference sample, which allows (i) a semi-quantitative calculation of the CL signal as a function of the beam energy and (ii) the evaluation of the four previously listed parameters. In many cases, however, s is infinite and only three parameters remain to be determined.

A computational fitting technique for simultaneous extraction of five parameters related to the above mentioned ones has been published by Chan et al. [66]. Chim et al. [67] have presented an analysis of "modelling techniques for the quantification of some electron beam induced phenomena".

Conclusions

CL emission is an efficient method for spatially resolved analysis of point defects in minerals and dopant impurities in semiconductors.

Since the physical and chemical defects are a signature of the crystal growth conditions, CL emission is a signature of the genetic conditions. CL microscopy has been naturally used in the field of mineralogy and petrology for revealing growth zoning or identifying minerals from different sequences of crystallisation but having similar composition which could not be distinguished by means of optical microscopy.

The development of CL in semiconductors has followed the use of photoluminescence for the identification of the composition and the presence of dopants.

The avalaible modern instrumentation allows the use of CL emission analysis to complement other spectroscopies for spatially resolved studies. CL spectroscopy can be associated with photoluminescence and optical absorption measurements for the study of recombination centres. For deep levels or electron traps, complementary experiments can be made using infrared absorption, thermoluminescence or electron paramagnetic resonance.

In minerals, the CL emission bands may be intrinsic or extrinsic in origin, depending upon the nature of the point defect in the crystal. The former are associated with impurities, the latter resulting from lattice defects originally existing or induced by the electron bombardment. This may contribute to the CL property change during electron irradiation. An accurate knowledge of all factors involved in the generation, emission and detection process is needed for a reliable use of CL emission as a signature of a mineral with respect to its geological environment.

For semiconductors, combined CL spectroscopy and imaging has proved to be a very powerful tool to visualise layers or zones of different compositions or with different impurity contents in optoelectronic devices. Since CL emission is depending on electrical parameters, such as diffusion length and lifetime of minority carriers and surface recombination velocity, the modelling of the generation function allows the determination of these parameters. CL can be used as a microcharacterisation technique to analyse the effect of the different technological processes (wet or dry etching for the definition of the structures, ion implantation for the delimitation of the active part of the devices···) on the electrical parameters L and s, and eventually on the optical absorption coefficient.

Provided that a rigorous experimental procedure has been established, CL microscopy and spectroscopy should be recognised as a reliable method. In light of the recent availability of commercial CL systems, the development of CL for quantitative applications will undoubtedly appear in the near future.

Acknowledgements. The authors wish to thank the CNET MBE epitaxy group and the HBT ground for furnishing the samples and A. C. Papadopoulo for the CL expertise. For G. Remond, this is the BRGM contribution number 95020; this work was financially supported by a BRGM research program.

References

[1] G. Remond, F. Cesbron, R. Chapoulie, D. Ohnenstetter, C. Roques-Carmes, M. Schvoerer *Scanning Microsc.* **1992**, *6*, 23.

[2] A. S. Marfunin, *Spectroscopy, Luminescence and Radiation Centers in Minerals*, Springer, Berlin Heidelberg New York, 1979, p. 1.

[3] B. G. Yacobi, D. B. Holt *Cathodoluminescence Microscopy of Inorganic Solids*, Plenum, New York, 1990, p. 65.

[4] C. A. Warwick, *J. Appl. Phys.* **1986**, *59*, 4182.

[5] D. F. Kyser, D. B. Wittry, *The Electron Microprobe*, Wiley, New York, 1964, p. 127.

[6] D. B. Wittry, D. F. Kyser, *J. Appl. Phys.* **1967**, *38*, 375.

[7] S. Myhajlenko, W. Ke, B. Hamilton, *J. Appl. Phys.* **1983**, *54*, 862.

[8] J. B. Le Poole, A. B. Bok, W. J. Boogerd, *An Electron Luminescence Microscope*, Geologie en Minjbouw, 1968, p. 47 and 443.
[9] K. Ramseyer, J. Mullis, *Am. Miner.* **1989**, *75*, 791.
[10] R. F. Sippel, Proceedings of the Second Lunar Conference, *Vol. 1* (A. A. Levenson, ed.) MIT, Cambridge, 1965, p. 247.
[11] E. M. Höri, P. Roschger, *Scanning Electron Microsc.* **1980**, *I*, 285.
[12] A. D. Trigg, *Scanning Electron Microsc.* **1985**, *III*, 1011.
[13] R. Autrata, J. Jirak, J. Spinka, O. Hutar, *Scanning Microsc.* **1992**, *6*, 69.
[14] J. Semo, *Rev. Phys. Appl.* **1975**, *9*, 355.
[15] K. Löhnert, M. Hastenrath, E. Kubalek, *Scanning Electron Microsc.* **1979**, *I*, 229.
[16] L. Reimer, *Scanning Electron Microsc.*, Springer, Berlin Heidelberg New York, 1985, p. 117.
[17] D. Curie, *Luminescence in Crystals*, Methuen London, 1963, p. 1.
[18] J. Cazaux, *J. Microsc. Spectrosc. Electron.* **1986**, *11*, 293.
[19] J. Cazaux, C. Le Gressus, *Scanning Microsc.* **1991**, *5*, 17.
[20] G. Remond, *Proceedings 6th Internat. Conf. on X-ray Optics and Microanalysis* (G. Shinoda, K. Kohra, T. Ichinokawa, eds., University of Tokyo Press, Tokyo, 1972, p. 611.
[21] J. P. Coutures, J. C. Rifflet, D. Billard, Ph. Coutures, *Proceedings 6th European Symposium on Materials Sciences Under Microgravity Conditions, Dec. 1986, Bordeaux, France*, ESA Special publication, 1987, 25b, p. 427.
[22] J. P. Coutures, D. Massiot, C. Bessada, P. Echegut, J. C. Rifflet, F. Tautelle, *C. R. Acad Sci.* **1990**, *310*, 1041.
[23] J. V. P. Long, S. O. Agrell, *Mineral Mag.* **1965**, *34*, 318.
[24] G. Remond, S. Kimoto, H. Okuzumi, *Scanning Electron Microsc.* **1970**, 33
[25] G. Remond, *J. Luminescence* **1977**, *15*, 121.
[26] M. Schvoerer, P. Guilbert, D. Piponnier, F. Bechtel, *PACT15*, Edition du Conseil de l'Europe, Strasbourg, France, 1986, p. 93.
[27] O. Rouer, C. Gilles, G. Remond, C. Yang, K. Malmkvist, F. Cesbron, *EMAS 95', 14–19 May 1995*, St Malo, France, Abstract 399.
[28] D. Ohnenstetter, F. Cesbron, G. Remond, R. Caruba, J. M. Claude, *C. R. Acad. Sci. Paris*, **1991**, *313*, 641.
[29] F. Cesbron, D. Ohnenstetter, P. Blanc, O. Rouer, M. C. Sichère, *C. R. Acad. Sci. Paris*, **1993**, *316*, 1231.
[30] F. Cesbron, P. Blanc, D. Ohnenstetter, G. Remond, O. Rouer, *Scanning Microsc.* **1995**, [*Suppl. 9*], accepted.
[31] P. Iacconi, R. Caruba, *Phys. Chem. Mineral.* **1984**, *11*, 195.
[32] Y. Kirsh, P. D. Townsend, *J. Phys. C. Solid State Phys.* **1974**, *20*, 967.
[33] G. Remond, C. Legressus, H. Okuzumi, *Scanning Electron Microsc.* **1979**, *I*, 237.
[34] K. Ramseyer, J. Mullis, *Am. Minerals* **1990**, *75*, 791.
[35] B. J. Luff, P. D. Townsend, *J. Phys. Condens. Matter* **1990**, *2*, 8089.
[36] S. W. S. Keever, *Radiation Protection Dosimetry* **1984**, *8*, 81.
[37] J. Mattler, T. Ceva, *Luminescence in Crystals* (D. Curie, ed.), Methuen, 1961, p. 244.
[38] G. Remond. M. Fialin, R. H. Packwood, C. Gilles, O. Rouer, *Microbeam Anal.* **1995** submitted.
[39] G. Remond, J. Cazaux, M. Fialin, C. Gilles, R. H. Packwood, O. Rouer, *Microbeam Anal.* **1995** submitted.
[40] J. P. Vigouroux, J. P. Duraud, A. Le Moel. *J. Appl. Phys.* **1985**, *57*, 5139.
[41] J. P. Vigouroux, J. P. Duraud, C. Le Gressus, G. Petite, P. Agostini, C. Boiziau, *Scanning Electron Microsc.* **1985**, *I*, 179.
[42] G. Remond, J. Cazaux, M. Fialin, *Proceedings ICEM 13*, Editions de Physique, Paris, 1994, p. 785.
[43] S. Datta, B. G. Yacobi, D. B. Holt, *J. Mater. Sci.* **1977**, *12*, 2411.

[44] G. Remond, J. Cazaux, R. Chapoulie, P. Ermakoff, M. Fialin, R. H. Packwood, *MicrobeamAnal.* **1995**, submitted.

[45] S. Myhajlenko, R. A. Puechner, J. L. Edwards, D. B. Davito, *Scanning Microsc.* **1995**, [*Suppl. 9*], accepted.

[46] J. F. Bresse, A. C. Papadopoulo, P. Hénoc, *Scanning Microsc.* **1987**, [*Suppl. 1*] 205.

[47] A. C. Papadopoulo, J. F. Bresse, *Appl. Phys. Lett.* **1988**, *52*, 224.

[48] A. C. Papadopoulo, C. Dubon-Chevalier, J. F. Bresse, *Scanning Microsc.* **1992**, *6*, 97.

[49] J. F. Bresse, A. C. Papadopoulo, *J. Appl. Phys.* **1988**, *64*, 98.

[50] B. Akamatsu, P. Hénoc, A. C. Papadopoulo, *Scanning Electron Microsc.* **1983**, *IV*, 1579.

[51] C. de Meerschman, B. Sieber, J. L. Farvacque, Y. Druelle, *Micros. Microanal. Microstruct.* **1992**, *3*, 483.

[52] W. H. Hackett Jr, *J. Appl. Phys.* **1972**, *43*, 1649.

[53] O. Von Roos, *J. Appl. Phys.* **1983**, *54*, 1390.

[54] B. Akamatsu, J. Hénoc, P. Hénoc. *J. Appl. Phys.* **1981**, *52*, 7245.

[55] T. E. Everhart, P. H. Hoff, *J. Appl. Phys.* **1971**, *42*, 5837.

[56] C. J. Wu, D. B. Wittry, *J. Appl. Phys.* **1978**, *49*, 2827.

[57] C. Donolato, *Phys. Stat. Sol.* (*a*) **1981**, *65*, 649.

[58] G. Oelgart, H. Scholz, *Phys. Stat. Sol.* (*a*) **1983**, *75*, 547.

[59] S. G. Konnikov, V. A. Solov'ev, V. E. Umanski, V. M. Chystiakov. *Sov. Phys. Semicond.* **1987**, *21*, 1229.

[60] L. Closas, J. Fernandez Rubio, *Phys. Stat. Sol.* (*a*) **1988**, *105*, 541.

[61] B. Akamatasu, P. Hénoc, R. B. Martins, *J. Microsc. Spectrosc. Elect.* **1989**, *14*, 12a.

[62] R. B. Martins, P. Hénoc, B. Akamatsu, J. F. Palmier, *J. Appl. Phys.* **1991**, *70*, 554.

[63] A. E. Grün, *Z. Naturforsch.* **1957**, *12a*, 89.

[64] B. Akamatasu, P. Hénoc, *Proceedings of the 13th International Congress on Electron Microscopy, Electron Microscopy 1994*, Les Editions de Physique, France, 2B, p. 1133.

[65] F. Berz, H. K. Kuiken, *Solid-State Electron.* **1976**, *19*, 437.

[66] S. H. Chan, Kin Leong Pey, J. C. H. Phang, *IEEE Trans. El. Dev.* **1993**, *40*, 1417.

[67] W. K. Chim, D. S. H. Chan, T. S. Low, J. C. H. Phang, K. S. Sim, K. L. Pey, *Scanning Microsc.* **1992**, *6*, 961.

Mikrochim. Acta [Suppl.] 13, 167–182 (1996)

Time-of-Flight Secondary Ion Mass Spectrometry (ToF-SIMS)

Patrick Bertrand* and Lu-Tao Weng

Unité de Physico-Chimie ed de Physique des Matériaux, Université Catholique de Louvain, 1 Place Croix du Sud, B-1348 Louvain-la-Neuve, Belgium

Abstract. Secondary ion mass spectrometry in the "static mode" is becoming a key technique for the surface characterization of organic materials. This is due to the very specific chemical information derived from characteristic molecular secondary ions. The present expansion of this technique is related to the development of high performance time-of-flight mass spectrometers. Indeed they combine high mass resolution allowing to resolve mass interferences between isobaric molecular secondary ions, unlimited mass range, high transmission allowing to reduce the total ion fluence per spectrum ($< 10^{12}$ ions/cm^2) and the molecular imaging capabilities in microscope and/or microprobe modes.

The aim of this paper is to present a general view on the basic principles and experimental developments of ToF-SIMS. The present understanding of the secondary molecular ion formation mechanisms in organic materials (polymers) will be reviewed and illustrated. Different applications coming mainly from our recent work will be treated with special emphasis on imaging. They concern surface composition and morphology of polymer blends, contamination and additives at polymer surfaces, plasma treated polymer surfaces and carbon fibers in polymer matrix.

Key words: surface analysis, ToF-SIMS, surface imaging, polymers

The SIMS technique is known as a destructive method because it was mainly used in the past in the dynamic mode in order to obtain depth concentration profiles. However, when applied in the static mode, the surface damage can be negligible and the emitted secondary molecular ions are very characteristic of the unperturbed surface chemical composition. This method is then very complementary to XPS to obtain chemical information on the surface. The static mode requires that the total ion fluence per analysis is kept limited to a value lower than 10^{12} ions/cm^2. In these conditions, there is only one primary ion impact for 10^3 surface atoms. This low fluence was previously obtained by rastering nA primary ion beam on large surface

---

* To whom correspondence should be addressed

area ($1 \times 1\,cm^2$) combined with a secondary ion mass separation and detection using a quadrupole spectrometer [1]. This system has many limitations: poor sensitivity and low mass range (< 1000 Da) due to the low and mass dependent transmission of the quadrupole spectrometer, poor mass resolution allowing only isotope separation, low lateral resolution due to the rastering process. These limitations are overcome with the development of the time-of-flight mass spectrometer combined with the use of finely focused metal ion sources (Ga^+). Here, the "static conditions" are directly related to the method itself. Indeed pulsed ion beam is used, so that the AC current is only a small fraction of the DC current. The high transmission of the secondary ion optics associated with a multichannel ion detection leads to a very high sensitivity that allows to keep the total fluence per spectrum well below the static limit even for small irradiated area. By scanning the beam over the surface, secondary ion images can be recorded while preserving the static conditions. This allows to produce chemical mapping of the surfaces.

In this review, we will mainly focus on the applications of ToF-SIMS to the surface analyses of polymers and organic materials. Indeed the detection of large secondary molecular ions characteristic of the original surface opens new ways for their characterization.

Polymers are more and more used in connection with other materials in different applications where the control of their surface properties is crucial. Interfaces involving polymers are of course met in composite materials (polymer matrix with fibers), in layered structures like metallized polymers or protective polymer coatings on metals. In all these applications, it is important to be able to link the physico-chemical properties of polymer surfaces to their molecular structure (composition, functional groups, side group orientation, crystallinity..). For this purpose, surface analysis methods are required and ToF-SIMS brings new possibilities in complement to the electronic spectrometries more commonly used.

Basic Principles of Sputtering and Production of Secondary Ions

When a low energy ion (keV regime) interacts with a solid, it penetrates into the solid up to its implantation while losing its kinetic energy. In this energy regime, its stopping power or energy loss per unit length in the solid (dE/dx) is dominated by the elastic collisions between the ion and the targets atoms nuclei. As a result, energetic recoil target atoms are produced and, in turn, they initiate other collisions, producing new recoil atoms so that collision cascades are developed in a region around the primary ion impact. This mechanism is believed to be the major contribution to sputtering in inorganic materials [2]. Indeed the collision cascade can transfer energy and momentum back to the surface allowing the emission of secondary particles if the transferred recoil energy exceeds the surface binding energy. In an insulator, the energy deposited in the electronic system may also contribute to the sputtering, mainly because the energy deposited in excitation and ionization cannot be easily dissipated as in a conductor [2]. In the SIMS technique, only the charged (positive or negative) fraction of the secondary particles flux is detected. This fraction is very small (10^{-3}) and this introduces many difficulties in the quantitative interpretation of the results due to the associated matrix effects [2]. Our understanding of the ionization process is not yet complete and different models have been proposed but no one has up to now an universal range [2].

In the case of ToF-SIMS, we are mainly interested in the detection of large molecular ions from which a chemical information about the surface can be extracted. This requires that the observed molecular ions reflect the surface composition and are not the result of a recombination process in the vacuum between smaller fragments originating from different surface areas. Different situations have to be considered. The intact desorption of molecules adsorbed on the surfaces are observed in SIMS [3]. Their parent ion is usually formed by protonation $(M+H)^+$ or deprotonation $(M-H)^-$ of the original molecule M. A fragmentation of the molecule is also observed. This emission mechanism should be different from that of molecular emission from bulk polymer. Indeed in the latter case, the emission process requires the breaking of strong covalent bonds while maintaining intact the structure of the ejected fragment. One bond breaking is needed for the emission of a side group or a chain end group, but at least two bonds of the backbone have to be broken for the emission of a characteristic fragment and even more if the polymer is crosslinked. A direct desorption process is however also possible if organic additives or oligomers are present at the surface. The extreme situation is the intact desorption of an entire macrochain and this can be produced when very thin polymer layers are deposited on metal (silver) substrate. The charge is obtained by cationization with a metal ion $(M+Me)^+$. An example of such ToF-SIMS spectrum is given in Fig. 1. It

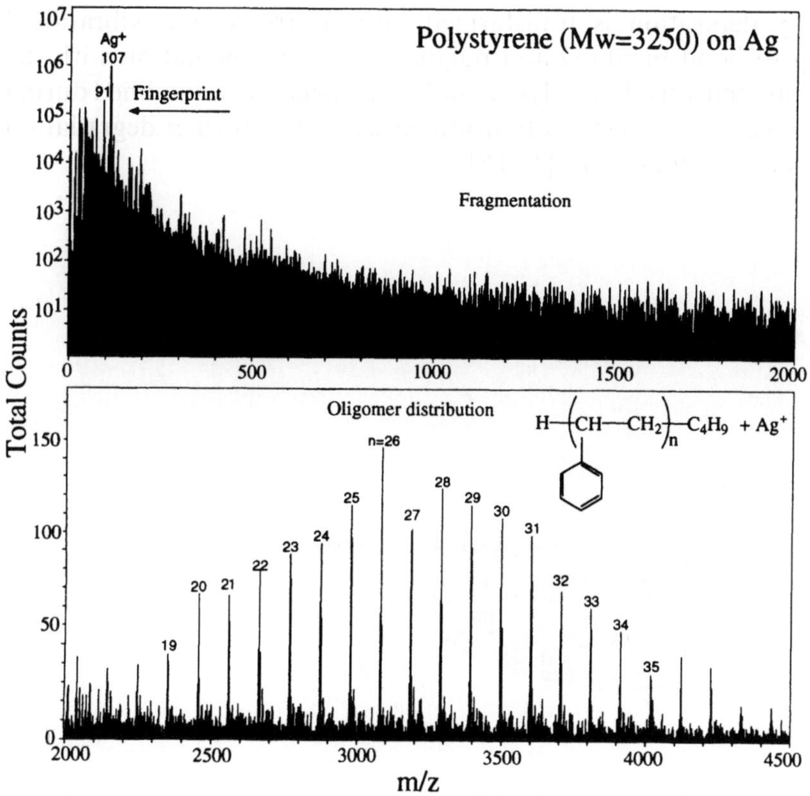

Fig. 1. ToF-SIMS spectrum of polystyrene deposited on silver

was obtained with a thin layer of a low molecular weight ($M_W = 3250$) polystyrene (PS) deposited on silver. Three different regions can be distinguished in the spectrum. The low mass range (< 500 Da) is called fingerprint region in which fragments of the macrochains are observed up to the fragmentation of the monomer repeat unit itself. In the fragmentation region (500–2000 Da), the large fragments of macrochains cationized with Ag are observed. The higher mass range (>2000 Da) is called oligomer distribution region. In this part, the intact macrochains cationized with Ag^+ are detected and the spectrum exhibits the molecular mass distribution [4]. Peaks are separated by $\Delta m = 104$ Da corresponding to the PS monomer repeat unit mass. The exact mass determination of these ions allows also to know the mass and hence the composition of the chain end groups [4]. As it will be shown later, the fingerprint region contains also specific information about the structure and chemical compositions of the polymer. With bulk polymer samples, only the fingerprint region is observed in the SIMS spectra and higher mass peaks, if detected, correspond usually to molecular additives and/or contamination.

The exact mechanisms of the molecular ion emission from polymers are not yet fully understood and cannot be explained only on the basis of the collision cascades. This constitutes a complex and challenging problem which needs to be urgently addressed theoretically. Several models have been invoked to explain the origin of secondary molecular ions [5, 6]. In the precursor model, large molecular ions are believed to be produced only at some distance from the primary ion impact point, where the deposited energy from the collision cascade is limited [7] (see in Fig. 2). However, the inelastic contribution to the ion sputtering is unclear. Part of the primary ion energy deposition is transformed into electronic and vibrational excitations leading to bond breaking and polymer chain fragmentation with the production of radicals and ions. The polymer molecular structure is modified during SIMS analysis and there is a direct relationship between the polymer degradation and the production of secondary ions [8–16].

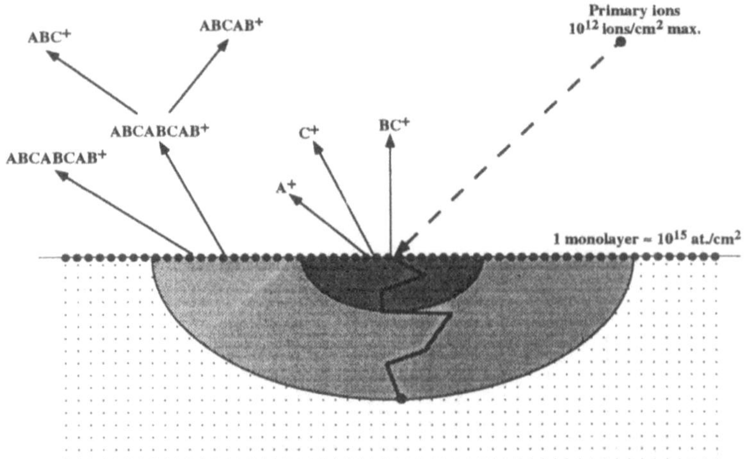

Fig. 2. Schematic showing the principle of static SIMS, large molecular ions are produced at some distance from primary ion impact

Instrumentation

Time-of-Flight Mass Spectrometer

Basic principles: The basic principle of the time-of-flight mass spectrometry is to obtain a time separation of ions with the same energy but a different mass. To apply this principle in SIMS, the first requirement is to produce short pulses of secondary ions. This is done by bombarding the sample with a pulsed ion beam. Typically the primary pulse width Δt_{beam} is of the order of a few ns and can still be reduced below 1 ns by electrodynamical bunching [17, 18].

The charged secondary particles produced by the sputtering/desorption process are then accelerated by a constant voltage V_{acc} so that they acquire a constant kinetic energy E if their initial kinetic energy is negligible. Then, their velocity in a field free drift region depends only on their mass:

$$qV_{\text{acc.}} = E = 1/2\, mv^2 \tag{1}$$

If they travel a flight path L in the drift region before reaching the detector, their time of flight t is proportional to the square root of their mass:

$$t(m) = L/v = Lm^{1/2}/(2qV_{\text{acc.}})^{1/2} \tag{2}$$

Time delay between the beam pulsing (start pulse) and the pulse produced on the detector by the first arriving secondary ion (stop pulse) is measured and converted into the amplitude of the detector response pulse (single stop time to digital converter, TDC). The pulses are sorted in order to produce the mass histogram in the multichannel analyzer. The detection system has a very high sensitivity thanks to the quasi-parallel detection of the secondary ions. Since lighter elements will always reach the detector in the first position, the number of secondary ion per primary ion pulse has to be kept lower than one in order to preserve a stochastic nature for the mass distribution of ions reaching the detector. This limits the number of ions per pulse to very low values ($\leq 10^3$ ions/pulse if one assumes that the ion yield is $\leq 10^{-3}$). If higher primary ion pulses are used or when samples with higher secondary ion yields are analyzed, a multistop TDC is required.

If one considers the mass resolution that can be achieved by such a system, a complication arises from the non nul initial kinetic energy distribution of the secondary ions. Indeed, by differentiating Eq. (2) one obtains:

$$\Delta t = \Delta m(\delta t/\delta m)_{E,L} + \Delta E(\delta t/\delta E)_{m,L} + \Delta L(\delta t/\delta L)_{E,m},$$

and the mass resolution $(\Delta m/m)^{-1}$ is given by

$$\Delta m/m = 2\Delta t/t + \Delta E/E - 2\Delta L/L \tag{3}$$

In order to improve the mass resolution, the energy term in Eq. (3) has to be compensated by the flight path term. This can be obtained by a deflection of the ions in appropriate electric fields, so that longer flight paths are produced for ions with higher energies ($\Delta E/E \cong 2\Delta L/L$). The mass resolution then reduces to

$$\Delta m/m \cong 2\Delta t/t \text{ where } \Delta t = (\Delta t^2_{\text{beam}} + \Delta t^2_{\text{detect}} + \Delta t^2_{\text{MCA}})^{1/2}. \tag{4}$$

Δt_{beam} is the primary beam pulse duration, Δt_{detect} is the response time of the detector and Δt_{MCA} is the time resolution of the detection system. In practice, the time resolution is limited by the primary pulse duration ($\Delta t_{\text{beam}} \sim\ \leq 10$ ns).

This discussion does not take into account the effect of the initial angular divergence of the emitted secondary ions accepted by the spectrometer and which contribute also to degrade the mass resolution.

Different energy-flight path compensation systems have been proposed. The most familiar are the reflectron spectrometer in which the secondary ions are $\sim 180°$ deflected in an appropriate retarding field [19] and the TRIFT™ spectrometer which uses three electrostatic hemispherical analyzers and

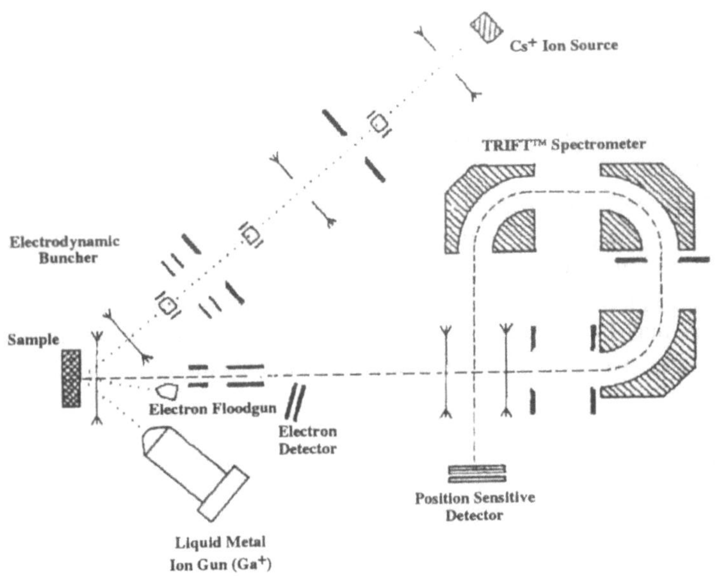

Fig. 3. Schematic diagram of TRIFT^{TM} spectrometer [17]

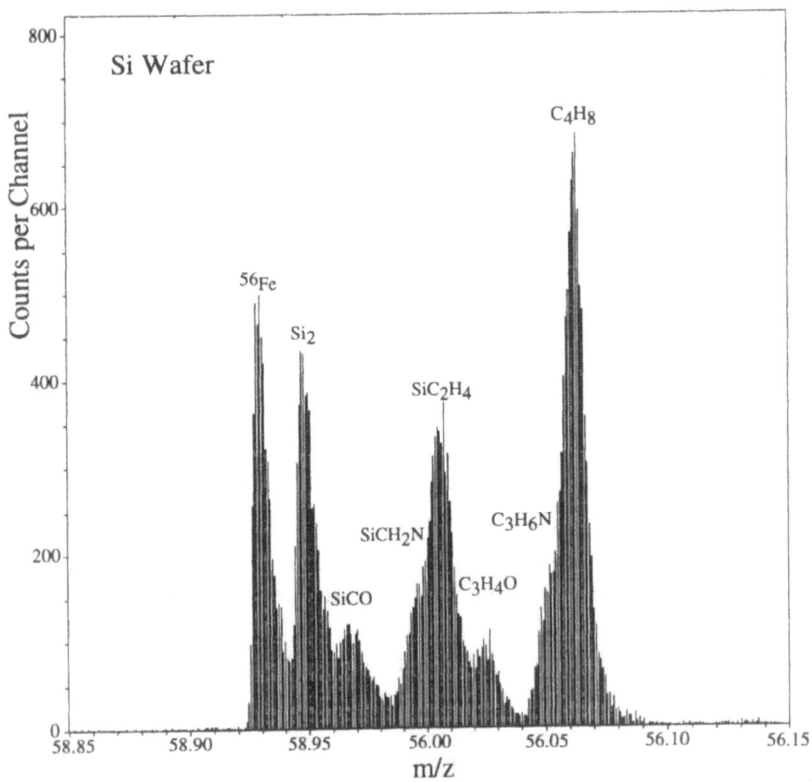

Fig. 4. Partial ToF-SIMS spectrum obtained with a Si wafer (mass 56)

a 270° deflection (schematically represented in Fig. 3) [17]. A very high mass resolution can be obtained ($m/\Delta m > 10^4$ at 28 Da) on flat and conductive surface like silicon wafer. This allows to separate the different contributions appearing in the peak at a given nominal mass (molecular or atomic ions with different chemical composition and having mass differences of only a few 10^{-2} Da, called isobaric ions). Figure 4 shows an example of such discrimination obtained at the mass 56 Da on a Si wafer. The mass resolution is degraded on rough and insulating samples. Nevertheless, in practice, it stays high enough to separate between pure hydrocarbon and oxygen containing molecular ions. The separation between hydrocarbon and nitrogen containing ions is more arduous.

The TRIFT™ spectrometer has a stigmatic secondary optics so that direct secondary ion images are formed at the detector position. These images can be directly recorded by the use of a position sensitive detector constituted by a dual microchannel plate followed by a resistive anode encoder (RAE). A lateral resolution of a few μm is obtained in this microscope direct imaging mode. A better lateral resolution (< 1 μm) can be obtained in the microprobe mode in which a finely focused metal ion beam (25 keV Ga^+ beam has a spot diameter of ≤ 0.1 μm) is rastered across the surface and combined

Table 1. ToF-SIMS typical working conditions

Ion bombardment	Spectrum acquisition	Image recording
Ion	a) $^{69}Ga^+$(15 keV) b) $^{133}Cs^+$(12 keV)	$^{69}Ga^+$(25 keV)
Beam diameter (μm)	a) 0.5 b) 30	0.06
DC current (pA)	500	60
(ions/s)	3.1 10^9	3.8 10^8
Pulse duration (ns)	5	25
Repetition rate (kHz)	5	10
AC current (ions/s)	7.8 10^4	9.4 10^4
(ions/pulse)	15.6	9.4
Rastered area (μm²)	100 × 100	100 × 100
AC current density (ions/cm²·s)	7.8 10^8	9.4 10^8
ToF spectrometer		
Flight time (μs)	2 for $m = 1$ Da 63 for $m = 1\,000$ Da	
Flight path (m)	2	
Data acquisition		
Mass range (Da)	1–5000	1–1000
Acquisition time (s)	300/spectrum	1800/image
Data memory	1.6 10^6 channels/spectrum dwell time: 156 ps/chan	256 × 256 pixels/image
Ion fluence (ions/cm²)	2.4 10^{11}/spectrum 2.4 10^7 ions/spectrum	1.7 10^{12}/image 2.5 10^3 ions/pixel

with a synchronous detection of the secondary ions. This better lateral resolution is however obtained at the expense of the mass resolution.

The extraction and focusing of the secondary ions are done by immersion and transfer lenses that control the direct image magnification and limit with the apertures the angular acceptance of the spectrometer. This leads to an overall secondary ion transmission greater than 40% [17].

For electrical insulating samples, charge compensation is done by pulsing low energy electrons during the off cycles of the ion pulses [20], with an adjustable timing and repetition rate. The microscope imaging mode allows to control the neutralization settings by looking for an uniform ion emission from the irradiated zone. Typical working conditions are summarized in Table 1.

Specificity of ToF-SIMS for Polymer Surface Analysis

In order to illustrate the sensitivity of ToF-SIMS to polymer molecular structures, a few examples are given. The first consists of three different polyolefins (polyethylene (PE), polypropylene (PP) and polyisobutylene (PIB)) samples. The molecular structure of these aliphatic polymers differs only by the presence of one (PP) or two (PIB) methyl pendant groups along the same hydrocarbon backbone. XPS is unable to discriminate between these three polymers. Their ToF-SIMS spectra are shown in Fig. 5 [21]. It is seen that their fragmentation patterns are different with regard to the relative intensities of the different carbon clusters (C_xH_i). The maximum intensity is observed for the deprotonated monomer ion $(M-H)^-$ at 27 Da for PE, 41 Da for PP and 55 Da for PIB. The pendant methyl group contributes to the high relative intensities at masses 55 and 69 amu for PP and PIB, and the double pendant methyl group of PIB accounts for the peaks at 83 and 97 amu. Their molecular structures have been previously discussed [22, 23]. Another significant difference between these

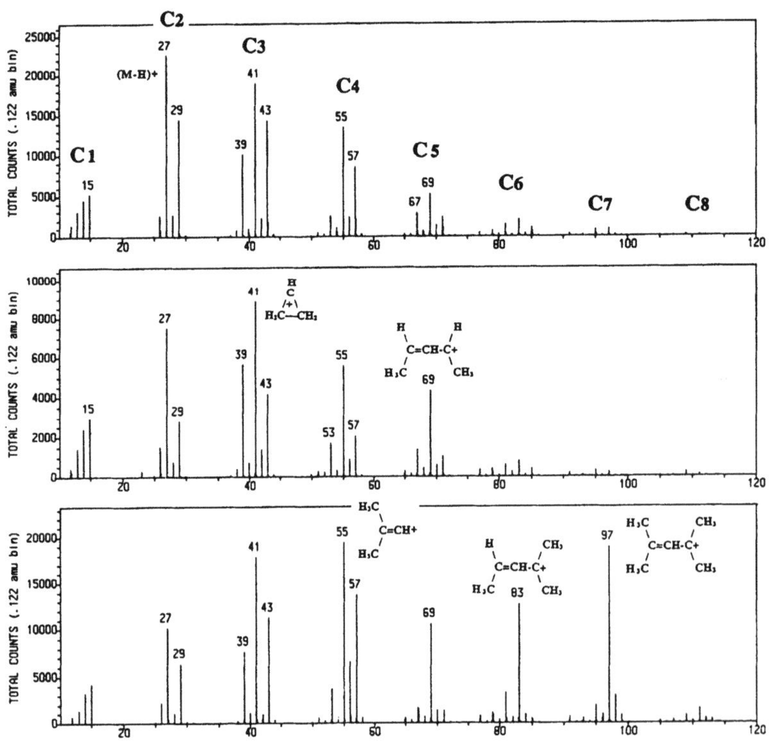

Fig. 5. ToF-SIMS spectra of polyethylene, polypropylene and polyisobutylene

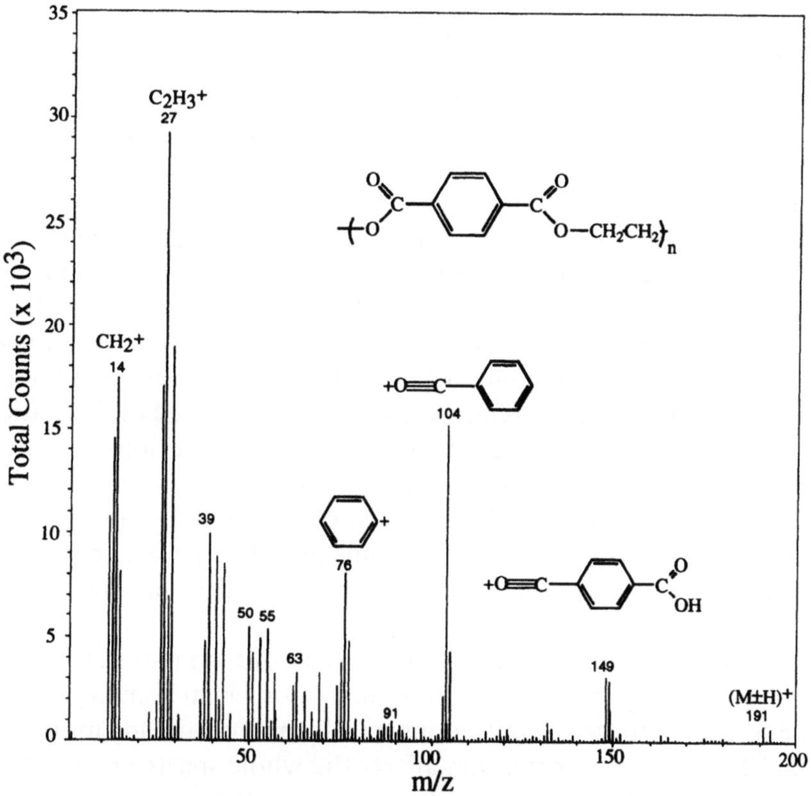

Fig. 6. Positive ToF-SIMS spectrum of PET

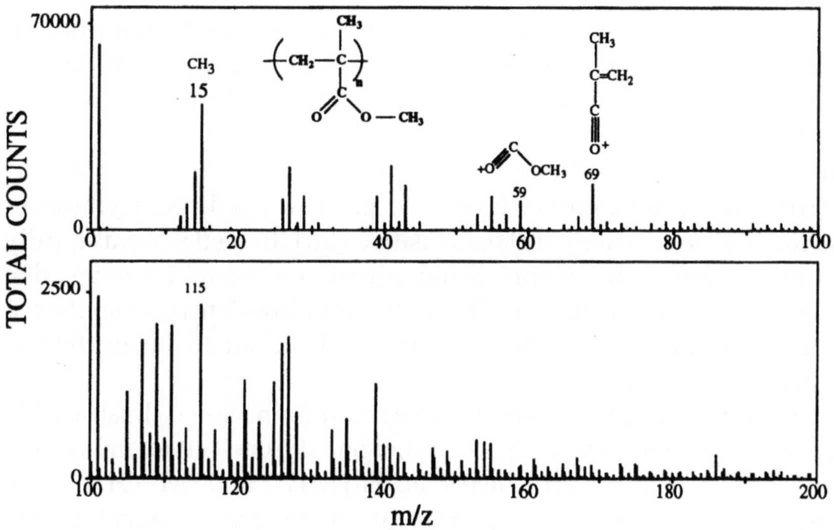

Fig. 7. Positive ToF-SIMS spectrum of PMMA [25]

spectra is the contribution of the C_8 cluster which increases with the branching level. The explanation for this effect is related to the higher stability of branched ions as compared to the linear ones [22].

If a polymer contains unsaturation, aromaticity and/or functional groups, its ToF-SIMS spectrum gives a direct signature of them. Figures 6 and 7 show the

ToF-SIMS spectra of poly(ethylene-terephthalate) (PET) and poly(methyl methacrylate) (PMMA). Both contain an ester group, one in the backbone (PET) and the other in the side group (PMMA). Moreover PET contains phenyl rings. In the mass range (0–200 Da), the positive secondary ion spectra are interpreted on the basis of the fragmentation scheme of the monomer repeat unit [24]. For PET, the ions at 191 D, 193 D correspond to $(M \pm H)^+$ where M is the monomer repeat unit. The major peaks correspond respectively to $C_6H_4^+$ (Φ) (76 D), Φ-CO^+ (104 D) and CO-Φ-$COOH^+$ (149 D). The positive ion spectrum of PMMA [25] is dominated by the ions CH_3^+ (15 D), $C_3H_5^+$ (41 D), $OCOCH_3^+$ (59 D) and $C_4H_5O^+/C_5H_9^+$ (69 D). At higher mass, a series of characteristic ions appears also at 105–111, 121, 135 D and beyond. These examples prove the fingerprint nature of the characteristic fragmentation pattern observed in the static SIMS spectra of polymers. A higher intensity is observed for the ions able to stabilize charge by inductive and/or resonant effects. Some fragments can rearrange in a complex way in order to fulfill this requirement. The relative intensity of the different peaks in the spectra may also be influenced by the experimental conditions (nature and energy of the primary ions, total ion fluence per spectrum) [11].

Concerning the sample degradation during the analysis and the requirement of "static conditions", studies of polymer modifications under ion beam bombardment have shown that the surfaces undergo a continuous transformation already initiated at very low fluences [21]. This transformation affects the whole spectrum and not only the highly characteristic SIMS peaks which are however the most sensitive to degradation. No plateau regime is seen in the peak intensity, their decrease preceded or not by a maximum is commonly observed as a function of the ion fluence. This means that it does not exist a real 'static' regime for polymers and that ion formation is intrinsically related to the degradation process. Nevertheless, reproducible molecular information can be obtained from these polymers when analyzed with ion fluence $< 10^{+13}$ ions/cm^2.

Results obtained on Langmuir Blodgett layers have indicated that molecular ions are emitted from the topmost surface layer. Therefore, when detecting molecular fragments, the method is extremely surface sensitive [26]. Indeed it seems hardly plausible that large molecular fragments could propagate from the inner side towards the surface without fragmentation. Due to the very low density of sputtered particles, recombination reactions in the vacuum are believed to be negligible, especially in negative mode [27].

The sensitivity to molecular additives and surface contamination is illustrated in the spectrum (see in Fig. 8) obtained with a commercial styrene-butadiene rubber copolymer. Parent ions of palmitic and stearic acids [MH^+ and $(M\text{–}OH)^+$] are directly observed so as the fingerprint fragmentation pattern of polydimethyl siloxane (PDMS) at 28, 73, 131, 147, 191, 207, 221 Da. PDMS is a very common surface contaminant owing to its very low surface free energy. As a consequence, a semi-quantitative interpretation of the SIMS intensity in terms of the copolymer composition, requires a sample washing (extraction in toluene) before analysis. After such a cleaning procedure, linear relationships have been obtained between the SIMS intensity of the deprotonated ions, $(M\text{–}H)^+$, of each monomer repeat unit and the copolymer bulk composition [28].

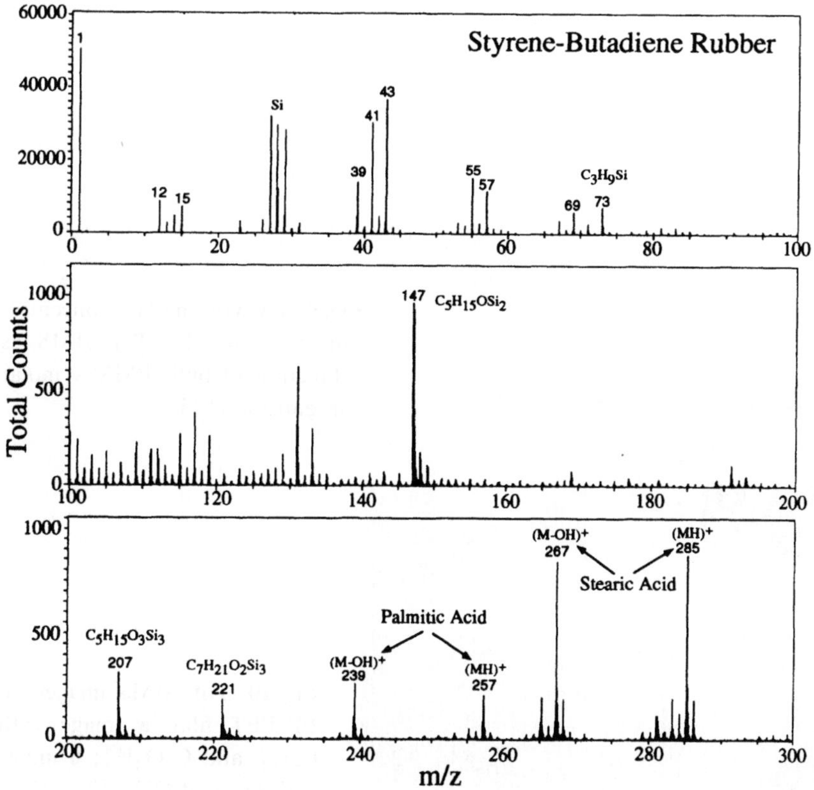

Fig. 8. Positive ToF-SIMS spectrum showing the presence of additives and PDMS contamination [28]

Examples of Applications

Surface Composition and Morphology of Polymer Blends

Polymer blending is an important way to create new materials with synergetic properties. Since most polymers are immiscible at a molecular scale, a two-phase system results from the mixture of two polymers. The blending properties depend on the morphology which is influenced by the elaboration parameters. If bulk morphology can be controlled, this is not the case for the surface morphology. However the control of surface heterogeneities is crucial if one wants to tailor properties like adhesion, biocompatibility, printability... For that purpose, it is important to characterize the blend surface in terms of chemical composition and morphology. In this field, ToF-SIMS can help a lot thanks to its molecular sensitivity and its imaging capabilities. This will be illustrated by two examples, the first concerns the surface composition of PMMA-PC (polycarbonate) blends [25] and the other the surface morphology of PET-PP blends [29]. Both systems are mixtures of incompatible polymers and their bulk characterization reveals biphasic systems (nodules of one polymer dispersed in the matrix of the major component). For PC-PMMA, blends of different composition were prepared by solution casting onto silicon wafers. ToF-SIMS showed that the negative mode was more useful because PMMA and PC have, in this mode, intense characteristic fragments which are not overlapping in the

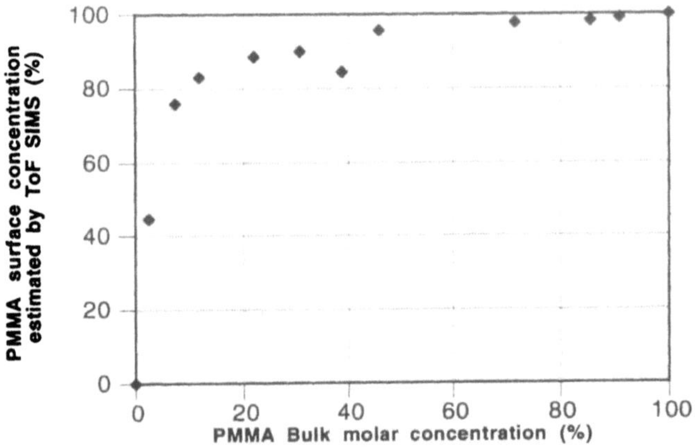

Fig. 9. PMMA molar concentration estimated by ToF-SIMS as a function of bulk PMMA molar concentration [25]

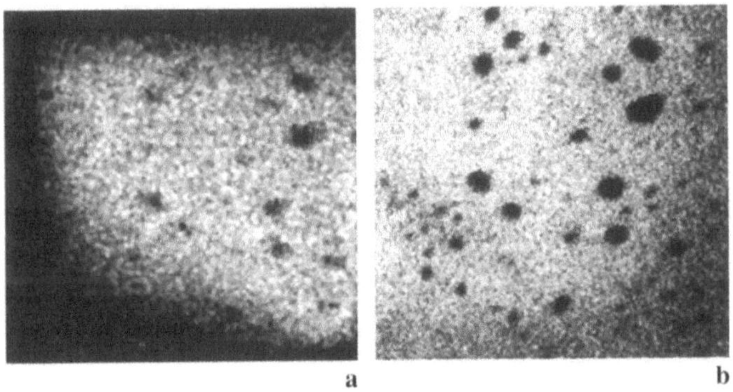

Fig. 10. ToF-SIMS images on PP/PET film, **a** image with $C_6H_4^+$ and $C_8O_3H_5^+$; **b** image with O^- and OH^- [29]. Field of view: 130 × 130 µm²

blends (among others $C_4H_5O_2^-$ at 85 Da for PMMA and $C_{15}H_{15}O^-$ at 211 Da for PC). The PMMA surface concentration X (shown in Fig. 9) was estimated from the relative intensities of peaks of 85 and 211 Da using:

$$\frac{I_{85}}{I_{85}+I_{211}}=\frac{X\cdot I_{85}^{PMMA}+(1-X)\cdot I_{85}^{PC}}{X\cdot I_{85}^{PMMA}+(1-X)\cdot I_{85}^{PC}+(1-X)\cdot I_{211}^{PC}+X\cdot I_{211}^{PMMA}}$$

where I_{85} and I_{211} are respectively the relative intensities of the peaks at 85 and 211 Da obtained with a specific blend while I_i^x is the relative intensity of the peak i obtained with pure polymer x. These results, confirmed by a more quantitative technique XPS, showed clearly a sharp PMMA segregation at surface. This can be explained by the difference in surface tension between the two polymers. On the other hand, the good agreement between XPS and ToF-SIMS results strongly suggests that for this system, matrix effects are negligible in SIMS except when the surface concentration of both polymers is comparable. In addition, ToF-SIMS allowed to give information which was not available from XPS, namely the presence of the two polymers over all the concentration range.

The second example deals with blends of 20% PP and 80% PET, prepared by compression molding [29]. The surface morphology was studied by ToF-SIMS imaging, AFM in the lateral force mode (LFM) and SEM. ToF-SIMS images were recorded with specific ions of PET:$C_6H_4^+(\Phi)$ (76 D), Φ-CO^+ (104 D) and CO-

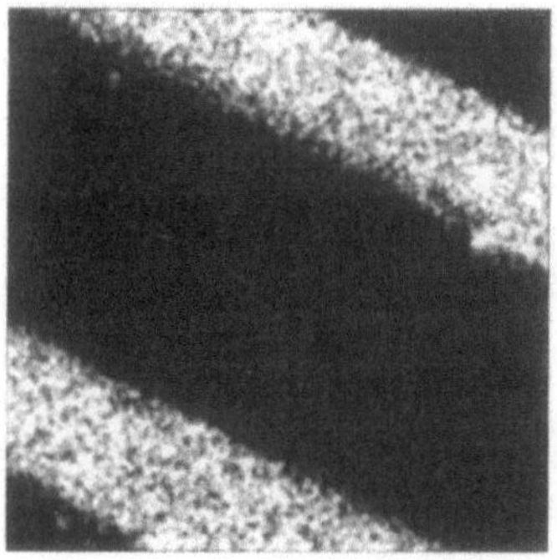

Fig. 11. ToF-SIMS image obtained with characteristic ions of protein: NH^-, CN^-, S^- and CNO^-. Field of view: $200 \times 200\,\mu m^2$

Φ-$COOH^+$ (149 D) in the positive mode and O^- and OH^- in the negative mode. The images, shown in Fig. 10, reveal a continuous PET phase with dispersed nodules (1 to 100 μm size) in which no selected ions are present. Images recorded with C^- and CH^- ions (both are present in PET and PP) were uniform, indicating that the contrast observed with the PET ions is due to chemical heterogeneities (PP nodules) and not to surface topography. A same morphology was observed with LFM and SEM. In this system the bulk morphology, as revealed by SEM, is preserved up to the extreme surface (ToF-SIMS and LFM). The lateral resolution is of course, much better in LFM (down to atomic scale) than in ToF-SIMS) ($\sim 1\,\mu m$). However ToF-SIMS allows to map the chemical surface composition.

Plasma Treated Polymer Surfaces

Plasma treatment of polymers surfaces is extensively used for modifying the surface properties, without modifying their bulk properties. The main goal is to enhance the adhesion properties [30, 31], in particular to improve biocompatibility [32, 33] and to enhance metal adhesion [34, 35]. Plasma processing of polymers can produce a broad variety of properties thanks to variable operating conditions, plasma chamber designs and especially gas composition. In order to characterize the surface modifications produced by plasma treatments, XPS has been extensively used. ToF-SIMS appears to be an ideal complementary technique to XPS owing to its molecular information, its high sensitivity and its selectivity to the uppermost surface layers [36–40]. An example is given in the field of polymer biomaterials used as a support for cell culture [41]. When cells are in contact with a surface, their behaviour depends on different factors such as the physico-chemical properties of the surface, the composition of the cell culture media, the kind of reconditioning and the rugosity of the surface. The effects of heterogeneous surface properties of polymer materials on mammalian cell behaviour are studied. PS was chosen as substrate because untreated, it is known to lead to a weak cellular adhesion. RF plasma oxygen discharge was carried out on PS samples partially covered with a photosensitive resin, used as a mask, in order to create only modified tracks at the surface. In order to optimize cell contrast, the obtained surfaces were reconditioned by a solution containing both a surfactant (Pluronic F68: cell adhesion inhibitor) and a protein (Fibronectin: cell adhesion promotor) after dissolution of the remaining resin. The surface modifications induced by the treatment were characterized by

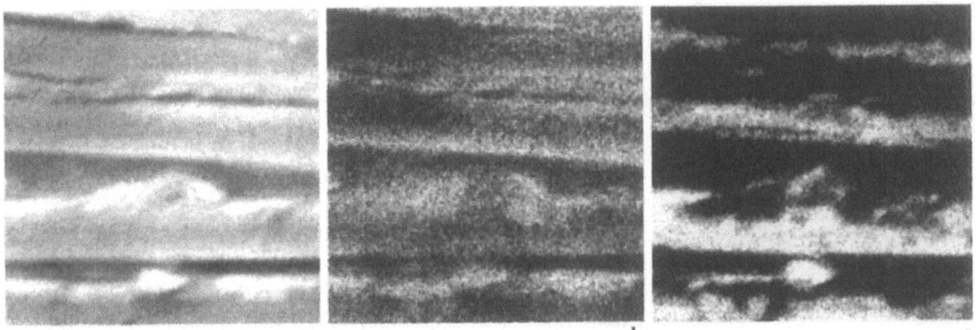

Fig. 12. ToF-SIMS negative ion images of a peeled surface of a carbon fiber-polymer composite **a** total ion image, **b** O^- and OH^- ion image (C fibers) and **c** C_2^- and C_2H^- ion image (PS) [43]. Field of view: $40 \times 40\,\mu m^2$

ToF-SIMS, XPS and Wilhelmy Plate (Dynamic Contact Angle). The plasma discharge has been found to create a large hydrophobic contrast between the exposed (unmasked) and unexposed (masked) regions. Indeed, if the unexposed regions remained hydrophobic as the native PS, the exposed regions were found much more hydrophilic after the plasma discharge. Moreover, after reconditioning, ToF SIMS images (Fig. 11) reveal that the protein adsorbs preferentially on the hydrophilic (unmasked, plasma treated) regions. Cell cultures on those treated samples showed that cell (PC12) adhesion occurs almost exclusively on the hydrophilic tracks which were covered by the protein [41].

Carbon Fibers in a Polymer Matrix

The mechanical properties of carbon fiber composites depend not only on the fiber and matrix properties but also on their interface properties. Indeed an interphase region is created which controls the stress transfer from the polymer to the fiber. In order to improve the properties of these composites, it is important to understand and to characterize the physico-chemical interactions occuring at the interface. Here also ToF-SIMS can be applied. Composites with carbon fibers and PS matrix were prepared by impregnating a carbon fiber bunch with a PS solution. The fibers were previously washed in order to eliminate their sizings [42]. The impregnated samples were then cleaved along the mean fiber chain axis and the two faces (fiber and polymer) were analyzed by ToF-SIMS imaging in order to study the decohesion and rupture mode. The results [43] are presented in Fig. 12. Characteristic ions from the polymer (C_2^- and C_2H^- for PS) and from the fibers (O^- and OH^-) are used for the imaging. The surface topography is given by the total secondary ion image which allows to vizualize individual fibers (7 μm mean diameter). From the complementarity in the contrast between the chemically resolved images, one can observe that fiber regions are covered by the polymer.

Conclusions and Perspectives

As shown by the different examples, ToF-SIMS is becoming a mature technique to look at surface chemistry. It is particularly well adapted when molecular information is required. Moreover it can also be used for chemical surface imaging. However, from a fundamental point of view, much work is still needed to understand the exact mechanisms of secondary molecular ion emission and to improve the quantitative interpretation of the spectra. Postionization by pulsed UV laser opens new perspectives in quantification (SNMS mode) and in improving the detection limit mainly for elemental species.

Acknowledgements. We gratefully acknowledge the contribution of many researchers who have participated in the overall research program involving ToF-SIMS in our group, in particular, Mr. A. Delcorte, D. Leonard, J. B. Lhoest, C. Poleunis. The ToF-SIMS system was acquired thanks to the financial support of the FRFC, the Loterie Nationale and the Région Wallonne of Belgium.

References

[1] J. C. Vickerman, *Analyst* **1994**, *119*, 513.
[2] A. Beninghoven, F. G. Rüdenauer, H. W. Werner, *Secondary Ion Mass Spectrometry, Basic Concepts, Instrumental Aspects, Applications and Trends*, Wiley, Chichester, 1987.
[3] A. Beninghoven, *Angew. Chem. Int. Ed. Engl.*, **1994**, *33*, 1023.
[4] I. V. Bletsos, D. M. Hercules, D. van Leyen, B. Hagenhoff, E. Niehuis, A. Benninghoven, *Anal. Chem.* **1991**, *63*, 1959.
[5] G. J. Leggett, J. C. Vickerman, *Int. J. Mass Spectrom. Ion Proc.* **1992**, *122*, 281.
[6] S. J. Pachuta, R. J. Cooks, *Chem. Rev.* **1987**, *87*, 647.
[7] A. Benninghoven, in: *Ion Formation from Organic Solids* (A. Benninghoven, ed.), *Springer Series in Chemical Physics 25*, Springer, Berlin Heidelberg New York, 1982, p. 77.
[8] A. Brown, J. A. van den Berg, J. C. Vickerman, *Spectrochim. Acta* **1985**, *40B*, 871.
[9] J. A. Van den Berg, *Vacuum* **1986**, *36*, 981.
[10] D. Briggs, A. B. Wootton, *Surf. Interface Anal.* **1982**, *4*, 109.
[11] D. Briggs, M. J. Hearn, *Vacuum*, **1986**, *36*, 1005.
[12] G. J. Leggett, J. C. Vickerman, *Anal. Chem.* **1991**, *63*, 561.
[13] G. J. Leggett, J. C. Vickerman, *Appl. Surf. Sci.* **1992**, *55*, 105.
[14] G. Marletta, *Nucl. Instr. Meth.* **1990**, *B46*, 295.
[15] S. Pignataro, *Surf. Interface Anal.* **1992**, *19*, 275.
[16] A. Licciardello, S. Pignataro, A. Leute, A. Benninghoven, *Surf. Interface Anal.* **1993**, *20*, 541.
[17] B. W. Schueler, *Microsc. Microanal. Microstruct.* **1992**, *3*, 119.
[18] R. W. Odom, in: *Microscopic and Spectroscopic Imaging of the Chemical State* (M. D. Morris, ed.) Marcel Dekker, New York, 1993, p. 345.
[19] J. Schwieters, H. G. Cramer, T. Heller, U. Jürgens, E. Niehuis, A. Benninghoven, *J. Vac. Sci. Technol. A* **1991**, *9*, 2864.
[20] B. Hagenhoff, D. van Leyen, E. Niehuis, A. Benninghoven, *J. Vac. Sci. Technol. A*, **1989**, *7*, 3056.
[21] A. Delcorte, L. T. Weng, P. Bertrand, *Nucl. Instr. Meth. B.* **1995**, *100*, 213.
[22] W. J. Van Ooij, R. H. G. Brinkhuis, *Surf. Interface Anal.* **1988**, *11*, 430.
[23] D. Briggs, *Surf. Interface Anal.* **1990**, *15*, 734.
[24] D. Briggs, A. Brown, J. C. Vickerman, *Handbook of Static Secondary Ion Mass Spectrometry (SIMS)*, Wiley, Chichester, 1989, p. 44.
[25] J. B. Lhoest, P. Bertrand, L. T. Weng, J. L. Dewez, *Macromolecules* **1995**, *28*, 4631.
[26] M. J. Hearn, D. Briggs, S. C. Yoon, B. D. Ratner, *Surf. Interface Anal.* **1987**, *10*, 384.
[27] M. L. Yu, *Appl. Surf. Sci.* **1982**, *11/12*, 196.
[28] L. T. Weng, P. Bertrand, W. Lauer, R. Zimmer, S. Bussetti, *Surf. Interface Anal.* **1996**, *23*, in press.
[29] B. Nysten, G. Verfaillie, E. Ferain, R. Legras, J. B. Lhoest, C. Poleunis, P. Bertrand, *Microsc. Microanal. Microstruct.* **1994**, *5*, 373.

[30] L. Mascia, G. E. Carr, P. Kember, *Plastics and Rubber Processing and Applications*, **1988**, *9*, 133.
[31] F. Aréfi, V. André, P. M. Rahmati, J. Amouroux, *Pure Appl. Chem.*, **1992**, *66*, 715.
[32] W. G. Pitt, J. E. Lakenan, A. Brent Strong, *J. Appl. Polym. Sci.* **1993**, *48*, 845.
[33] Y. Tamada, Y. Ikada, *Polymer* **1993**, *34*, 2208.
[34] C. M. Gilbert Carlsson, K. S. Johansson, *Surf. Interface Anal.* **1993**, *20*, 441.
[35] S. Nowak, M. Collaud, G. Dietler, P. Schmutz, L. Schlapbach, *Surf. Interface Anal.* **1993**, *20*, 416.
[36] R. D. Short, A. P. Ameen, S. T. Jackson, D. J. Pawson, L. O'Toole, A. J. Ward, *Vacuum* **1993**, *44*, 1143 and references therein.
[37] R. W. Linton, M. P. Mawn, A. M. Belu, J. M. DeSimone, M. O. Hunt, Jr., Y. Z. Meneceloglu, H. G. Cramer, A. Benninghoven, *Surf. Interface Anal.* **1993**, *20*, 991.
[38] D. Briggs, in: *Practical Surface Analysis* Vol II (D. Briggs, M. P. Seah, eds.) Wiley, Chichester, 1993, p. 367 and references therein.
[39] A. Benninghoven, *Surf. Sci.* **1994**, *299/300*, 246 and references therein.
[40] P. C. Zalm, *Vacuum* **1994**, *45*, 753 and references therein
[41] J. B. Lhoest, E. Detrait, P. van den Bosch de Aguilar, P. Bertrand, *ACS Colloid and Surface Science Symposium*, Salt Lake City, USA, June 11–14, 1995.
[42] L. T. Weng, C. Poleunis, P. Bertrand, V. Carlier, M. Sclavons, P. Franquinet R. Legras, *J. Adhesion. Sci. Technol.* **1995**, *9*, 859.
[43] C. Poleunis, I. Fallais, V. Carlier, P. Bertrand, R. Legras, *Proc. 10th International Conference on Secondary Ion Mass Spectrometry SIMS X*, Münster-Germany, October 1–6, Wiley 1995.

Mikrochim. Acta [Suppl.] 13, 183–194 (1996)

Three-Dimensional Nanoanalysis with the Tomographic Atom-Probe

Didier Blavette*, Annabelle Bigot, Cristelle Schmuck, Frédéric Danoix, and Pierre Auger

Laboratoire de Microscopie Ionique, URA CNRS 808, Faculté des Sciences et Techniques, Université de Rouen, F-76821 Mont-Saint-Aignan Cedex, France

Abstract. The tomographic atom-probe is a nanoanalytical instrument which provides three-dimensional maps of chemical heterogeneities in a metallic material on a subnanometric scale. The basic principle of this new generation of apparatus relies on the field evaporation and ionisation of material atoms and the localization of generated ions on a position-sensitive multidetector. Chemical species are identified by time-of-flight mass spectrometry. Local chemical composition in a small selected region of the reconstructed volume ($10 \times 10 \times 1000\,nm^3$) may be therefore quantitatively measured. Illustrations of performance of this new generation of apparatus are given (GP zones in Al Cu, precipitation and ordering in two-phase γ/γ' NiCrAl single crystal alloys).

Key words: 3D atom-probe, microanalysis, nickel, aluminium, precipitation.

The FIM atom-probe was invented by E. W. Müller 25 years ago. The new instrument that will be described in this article, namely, the tomographic atom-probe (TAP), constitutes an important extension of this apparatus. While the conventional instrument only provides one dimensional information in the form of a set of depth concentration profiles, one for each element, the TAP makes it possible to image and analyse a small volume of material ($10 \times 10 \times 100\,nm^3$) on a subnanometric scale in the three dimensions of real space.

Among the various microanalytical techniques which are used in material science, the conventional atom-probe [1] is probably the instrument which provides the highest spatial resolution. The lateral resolution at the specimen surface is a few tenth of a nanometer. Depth profiles can be obtained with a resolution of a single atomic layer. Another advantage of atom-probe techniques is their quantitativity: no cross-section or ionisation efficiency has to be taken into account in the calculation of concentrations. Under proper experimental conditions, concentrations

---

* To whom correspondence should be addressed

are reliable and are directly proportional to the numbers of ions of each species collected from the specimen. The principle is based on the field evaporation of surface atoms and the identification of ions by time-of-flight mass spectrometry. The high electric field required is obtained by applying a high voltage to the sample prepared in the form of a sharply pointed needle.

Over the past twenty years, the instrument has been intensively used to study the microchemistry of a wide range of metallic alloys. Thanks to its high spatial resolution, the atom-probe is a powerful tool for the investigation of numerous phenomena in materials, including precipitation or decomposition processes, interfacial reactions, intergranular segregation and ordering phenomena [2, 3].

The first attempts to combine both chemical and spatial information in the atom-probe techniques were carried out by Panitz et al. [4] with the introduction of the imaging atom-probe. With such an instrument, a 2D-cartography of chemical species at the specimen surface could be obtained with a resolution close to 0.5 nm. However, time-gated images produced could be obtained only for a selected type of atom (a given time-of-flight). As a result, no composition data could be derived from the distribution maps provided by the imaging atom-probe. This instrument has nevertheless been successfully used in various areas in material science including surface segregation and precipitation in metallic alloys.

A decisive advance was achieved with the implementation of position-sensitive detectors on the instrument [5]. Individual atoms field evaporated from the tip surface could be identified by time-of-flight mass spectrometry and at the same time the position of impacts could be localized. 2D images of the surface could be obtained for every chemical species. The material being depth analysed, atomic layer by atomic layer, a three-dimensional atomic map of the element distributions within the probed volume could be reconstructed.

In contrast to the 2D detector which was used in the first position sensitive atom probe (a wedge and strip detector) [5] or to the CCD camera based systems [6], the 2D multidetector that was designed and constructed in our laboratory at the University of Rouen, can give the ion positions for multiple events. The spatial resolution (<0.5 nm) is limited essentially by the trajectory aberrations of evaporated ions in the close vicinity of the surface. Multiple time-of-flight events as well as simultaneous impacts can be localized. This is a decisive advantage as compared to previous designs, if one considers the reliability of position and composition data. Taking into account these multiple events is of utmost importance as up to 10% of impacts may be multiple under standard experimental conditions.

In this paper, the principle and applications of the TAP in physical metallurgy are described and discussed.

Principle of the Tomographic Atom-Probe

A schematic drawing of the tomographic atom-probe is depicted in Fig. 1. The sample, prepared in the form of a very sharp tip ($R \sim 10$–50 nm), is subjected to a high positive voltage ($V_0 \sim 5$–20 kV). Surface atoms are field evaporated and ionized under high vacuum conditions (10^{-10} Torr) by means of high voltage pulses ($V_p \sim 1$–5 kV). The tip is maintained at low temperatures (<80 K) in order to ensure

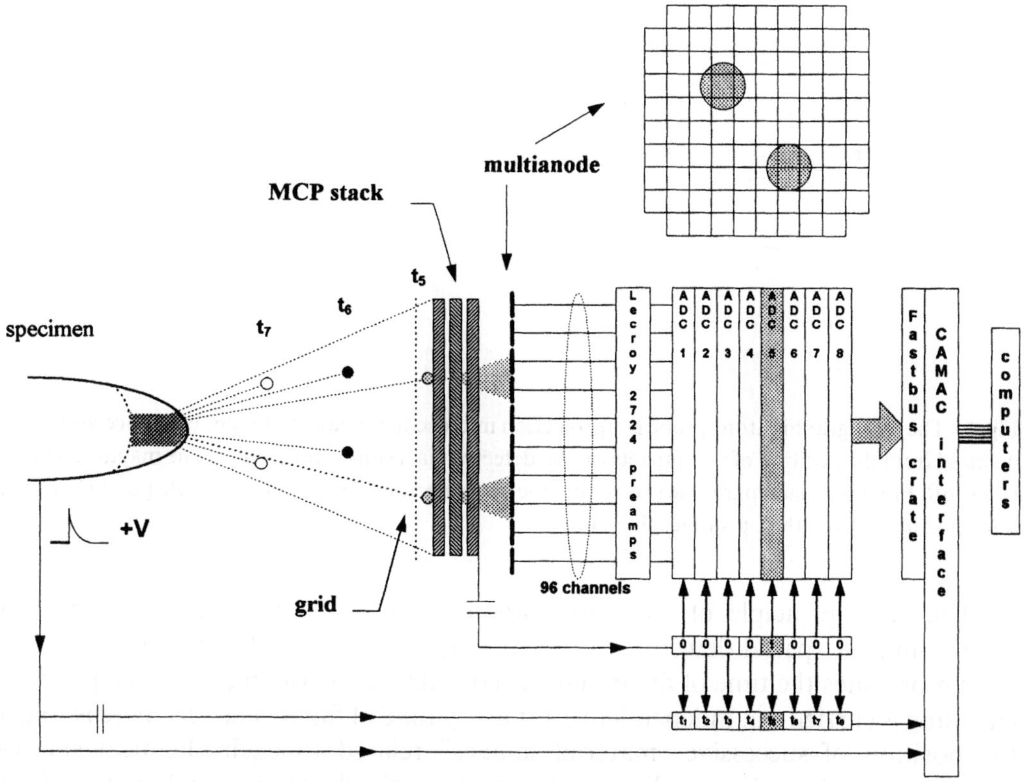

Fig. 1. Schematic view of the tomographic atom-probe. Field evaporated ions originating from the specimen surface (a tip) are identified by time-of-flight mass spectrometry. The position of atoms is deduced from the coordinate of ion-impacts on the detector. For each impact, an electron shower is produced by the microchannel plates (MCP). The measurement of charges that are received on a 10 × 10 anode array provides the position of the spot centre. For each flight time ($t_1, t_2 \ldots t_8$), the charge quantities collected on the 100 anode-array are measured independently by a set of eight 100 channel-analogic to digital converter (ADC 1 for t_1, ADC2 for t_2 .... ADC8). Up to 8 isochronal charge maps can be stored for each evaporation pulse (~100 pulses/s)

reliable composition information [3]. The ions originating from a selected region of the tip (~100 nm^2) are received on a position-sensitive detector placed at a given distance ($L \cong 0.45$ m) from the specimen. The detection signal provides the flight time (t) of field evaporated ions. The mass to charge ratio (m/n) is deduced from the flight time and the total applied voltage ($V_0 + V_p$). The conservation of kinetic energy imposes that:

$$\tfrac{1}{2}\mathrm{m}v^2 = ne(V_0 + V_p) \text{ with } v = L/t$$
$$\text{so that: } m/n = 2e(V_0 + V_p)t^2/L^2$$

For chemical species with close m/n ratios (1 a.m.u. for instance), the flight times may be as close as a few tens of nanoseconds. This is the reason for which CCD cameras [6] are unable to treat multiple events properly (long readout time response).

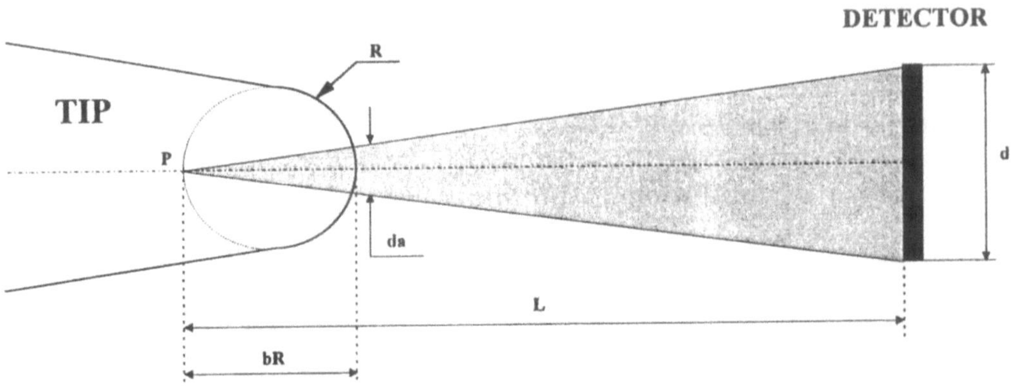

Fig. 2. The tomographic atom-probe is a projection microscope. The coordinates of surface atoms are deduced from the position of ion impacts on the detector via a simple projection. The magnification is $G = L/bR$. For a pure stereographic projection, the ion trajectories intercept at a single point (P) which is located at $2R$ from the tip surface ($b = 2$)

The basic principle of the tomographic atom-probe reposes on the use of a multi-impact spatial detector. For each evaporation pulse (~ 100 pulses/s), this system provides the time of flight and the coordinates of ion impacts. The position encoding is carried out independently for every time of flight. The idea here is to get the position of successive "material images" related to each chemical species ($m/n = \text{Const}$). Once the coordinates of impacts on the detector (x_a, y_a) are known, it becomes easy to calculate the position of atoms at the tip surface. A simple projection is involved. For a small acceptance angle (d_a/L small), ion trajectories can be considered as originating from a single origin (Fig. 2). The magnification is proportional to the flight path (L) over tip radius (R) ratio:

$$G \approx \frac{L}{bR} \quad \text{and} \quad x = x_a/G; \qquad y = y_a/G.$$

Where b (~ 2) is a constant related to the projection features.

For $R = 50$ nm and $L = 1$ m, $G \sim 10^7$. As a 0.1 nm precision at the tip surface is aimed, impacts must be localized with a 1 mm accuracy. The position sensitive detector that was designed is composed of two chevron microchannel plates (MCP charge multiplier) associated with a collector array (10×10 anodes). The collector is composed of one hundred square anodes (1 cm^2) deposited on a glass support. The detector size (10 cm $\times$ 10 cm) leads, for $G \sim 10^7$, to an analysed area close to 10 nm $\times$ 10 nm. The size of each anode corresponds to 1 nm on the tip surface. The geometry of this detector is similar to the four-sector detection system that we experimented a few years ago [7].

Coordinates of impacts are deduced from the charge spread over the anode array [8]. Each ion impact on the MCP's generates a charge cloud which irradiates several anodes. By measuring the electronic charge on each individual anode, the position of ion impacts can be calculated. The principle is depicted in Fig. 3. When the right side charge (Q_r) is equal to the left side charge (Q_l) for instance, the electron cloud barycenter is located at $x = 0$ (with respect to the local xOy reference). In

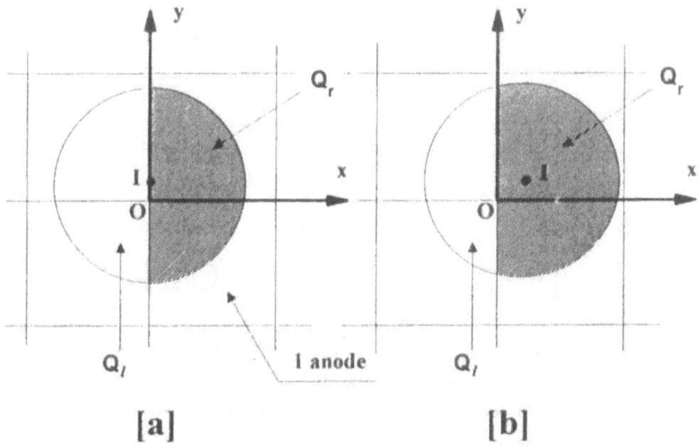

Fig. 3. The coordinates (x_a, y_a) of the ion impact (I) may be deduced from the elemental electronic charges collected on the multi-anode. The electron spot (dashed lines) produced by the ion impacts on the MCP must irradiate several anodes. When the right side charge (Q_r) is equal to the left side charge (Q_l), the abscissa of I with respect to the Oy axis is $x=0$ **a**. For each value of the reduced charge $q_x = Q_r/Q_t (Q_t = Q_r + Q_l)$, one can associate and calculate a single value of x. In **b**, $q_x > 1/2$ and $x > 0$

proportion as the impact shifts towards the right side, it is clear that the reduced charge $q_x = Q_r/Q_t$ (with $Q_t = Q_r + Q_l$) will increase and reach a value which is related to the position of the ion impact along the Ox axis. The determination of the y coordinate follows exactly the same scheme. Experiments have demonstrated that the spatial resolution of the detector is better than 0.3 mm (0.03 nm at the specimen surface). The precise procedure to determine the transfer functions $x(q_x)$ and $y(q_y)$ as well as the technical details related to the multidetector are described elsewhere [9].

In order to get the positions of each chemical species, these charge measurements are carried out independently for every time of flight. As ions with close m/n ratios may be separated by less than 100 ns, this imposes the use of fast electronics. In our detection system, up to 8 isochronal charge maps may be recorded. Under standard detection rate conditions ($\leqslant 1$ ion/pulse in average), this number of channels is shown to be sufficient. As a result, up to 800 electronic charges (8×100) may be measured on each evaporation pulse. For a pulse repetition of 100 Hz, the number of collected ions per hour is close to 10^5. Reconstructed images are typically composed of a few 10^5 atoms. This data set is subsequently treated by a graphic work station by using advanced 3D image processing softwares.

By field evaporation of surface atoms, the material can be investigated in depth, layer by layer. For each evaporated layer, the related 2D element cartography can be calculated and a three-dimensional reconstruction of the probed volume can finally be achieved by combining successive 2D maps (Fig. 4). The probed depth (z), related to the third dimension, is deduced from the number of evaporated layers (N): $z = Ne$ with e the thickness of layers.

As the evaporated volume related to each layer is known ($v = nv_a/Q$, with n the number of detected ions in the layer, Q the detection efficiency ($\sim 60\%$) and v_a the mean atomic volume), it is straightforward to calculate the thickness of evaporated

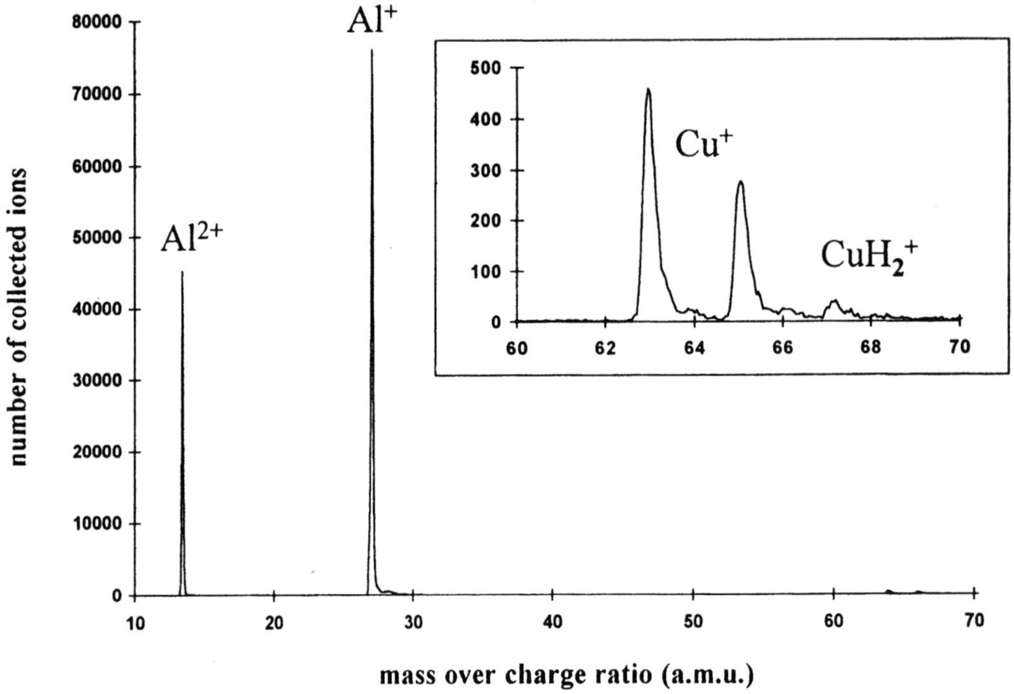

Fig. 4. Mass spectrum related to the analysis of an Al-Cu alloy

layers (e):

$$e = v/d_a^2$$

d_a is the size of the probed area (Fig. 3):

$d_a = d/G$ with d, the detector size (here, $d = 10\,\text{cm}$) and G, the magnification.

This qualitative description is, in fact, very simplified. For instance, the curvature of the tip surface from which ions originate, must be taken into account in the calculation of the depth coordinate (z) of atoms. More details on the exact reconstruction process may be found elsewhere [11].

Applications to Material Science

GP Zones in Al Cu Alloys

It is well known that the first stage of decomposition in supersaturated Al-Cu solid solutions is the formation of GP zones. Extensive X-ray diffraction, electron microscopy, as well as atom-probe field ion microscopy studies have shown that these zones are copper-rich plate-like clusters which form on the {200} planes of the aluminium matrix. But the question of their exact structure is still controversial. Recently, HREM observations of samples aged 10 h at 100 °C and 20 h at 130 °C clearly revealed that a majority of GP zones are copper rich monolayers [12–14]. However, the composition of these zones as well as the copper content of the matrix in their close vicinity have not been precisely determined yet. The T.A.P. can help to solve these points.

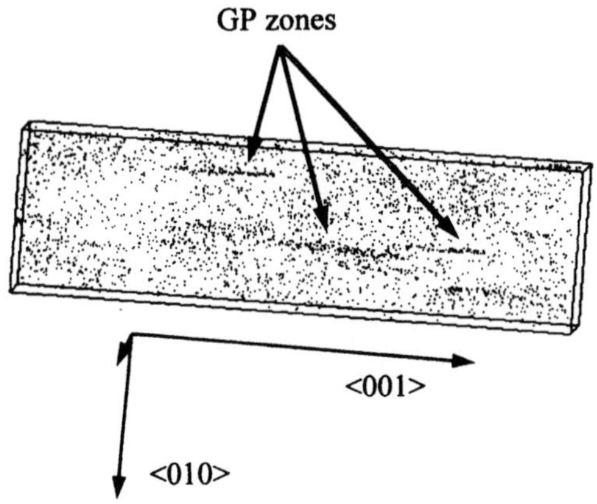

Fig. 5. 3D map of Cu atoms in Al-Cu 4 wt% (only copper atoms are represented for clarity). The analysis direction is the ⟨001⟩ direction of the Al-matrix. Cu-enriched GP zones appear as thin platelets spreading parallel to the (020) Al planes. The represented volume is $16 \times 16 \times 50\,nm^3$

Samples of an Al 4 wt % Cu alloy, aged 30 h at 100 °C, were therefore studied with the T.A.P.. Samples are ⟨001⟩ Al oriented. Analyses were performed at 35 K with a pulse to standing voltage ratio $V_p/V_0 = 19\%$, under UHV conditions ($P < 5.10^{-10}$ torr).

As can be seen on the mass spectrum Fig. 4, the aluminium ions collected are singly and doubly charged, whereas Cu ions are only singly charged. Hydride ions are also formed, especially $(Cu\,H_2)^+$. The mass resolution is close to $M/\Delta M = 100$, with ΔM the full width at half maximum (FWHM).

A 3-dimensional reconstruction of the analysed volume is shown Fig. 5. Several GP zones are observed appearing as thin copper enriched platelets. These zones are a few nanometer wide and their habit plane is here nearly parallel to the analysis direction (⟨001⟩). The analysed volume contains only one family of GP zones among the three possible variants corresponding to each base plane of the cubic cell {100}. It is possible that the strain field around a GP zone favours the formation of other GP zones parallel to it [15]. Moreover, field ion microscopy studies conducted on a similar specimen indicate that groups of GP zones with the same habit plane are often observed.

The apparent thickness of the GP zones on this side view image appears greater than one or two atomic planes. In addition, the solid solution still contains a non negligible amount of copper which is not homogeneously distributed: depleted areas can be seen on both sides of the GP zones. This observation is confirmed by the selected area composition profile obtained across a GP zone in a direction perpendicular to it (Fig. 6). The method consists in counting the Al and Cu atoms present in a selected volume (here a flat box) moved step by step throughout the volume.

The composition of the reconstructed GP zones reaches 30 at% Cu (maximum value), the full width at half of the maximum of the peak is about 6 Å: zones therefore appear to be artifically thickened. Two phenomena can account for this:

– the lateral resolution of the TAP is about 5 Å. As a result, the positions of copper atoms cannot be determined precisely enough to reconstruct a monolayer parallel to the direction of analysis.

– Under the high electric field conditions at the surface of the specimen, the copper clusters develop a radius of curvature much smaller than that of the matrix. A so called 'local magnification' effect appears greater for the GP zones than for the

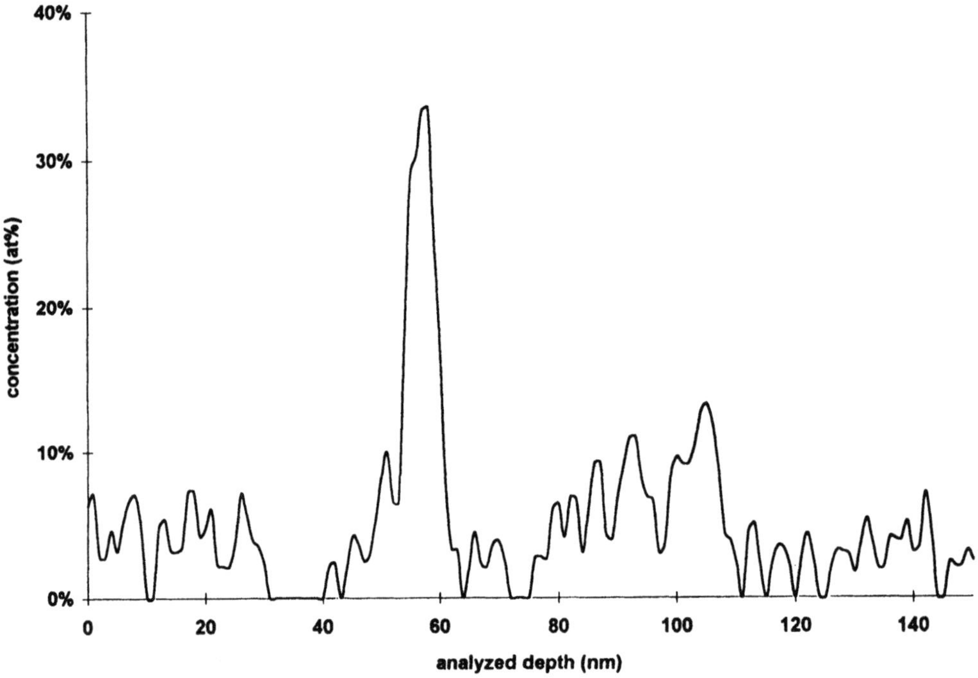

Fig. 6. Cu concentration profile thorugh a GP zone in a direction $[010]_{Al}$ normal with respect to the platelet (see Fig. 5). Compositions are measured over 50 atoms, leading to a mean standard deviation of about 5 at %

solid solution. As a consequence, copper atoms spread over three or four (020) lattice spacings, which causes an apparent dilution of the Cu concentration of the cluster. Assuming that all the Cu atoms originate from a monolayer, Cu content is estimated to range from 50 to 100 at%. For a bilayer zone, the Cu concentration evidently yields half of these values.

This first example clearly shows that analysing and imaging GP zones is a challenge. Because of the small size of these clusters – one or two atomic layers of a low index plane – the instrument is here used in the close vicinity of its ultimate lateral resolution ($\sim$0.5 nm). An absolute depth resolution of a single lattice spacing could be, in principle, achieved along the analysis direction. More precise measurements of the GP zone thickness and Cu content could therefore be obtained for the GP zones which are perpendicular to the analysis direction (one of the three variants). This work, which appears to be experimentally difficult is currently in progress. However, these preliminary reconstructions, although not perfect, have shown that the TAP is able to bring important new information on both GP zones and the surrounding matrix compositions.

Precipitation of Ordered γ′ Particles in Single Crystal Model Ni Cr Al Alloys

A 3D cartography of aluminium and chromium in a single crystal nickel base superalloy (Ni-23.4 at% Cr-10 at% Al) is displayed in Fig. 7. After homogenization at 1300 °C and water quenching, the model alloy was aged at 600 °C for 1 h in order

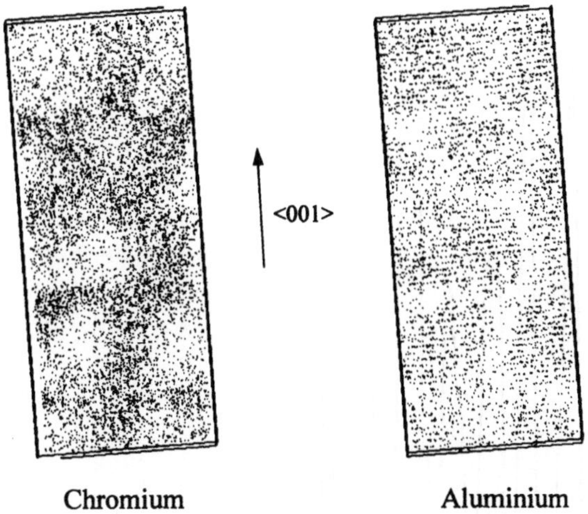

Fig. 7. Three-dimensional reconstruction of a small volume ($10 \times 4 \times 25\,nm^3$) of a nickel base superalloy containing small aluminium-enriched γ' particles ($\phi \sim 5$ nm) finely dispersed in a chromium-rich γ solid solution. Only aluminium and chromium atoms are represented. Nickel atoms are not represented for clarity. The analyzed volume is observed along a direction which is parallel to the (002) superstructure planes of the $L1_2$-type ordered γ' phase. Aluminium rich planes alternate with aluminium-depleted planes with a periodicity equal to half the lattice parameter ($a = 0.36$ nm). This clearly shows the ordered structure of precipitates

to form γ' particles in the material. The analyzed samples were $\langle 001 \rangle$ oriented. This crystallographic orientation corresponds to one of the superstructure direction of the ordered ($L1_2$) structure of γ' precipitates (Ni_3 Al).

From such a 3D image, a lot of information may be derived. This image clearly exhibits the presence of small – the particle size is close to 5 nm – aluminium enriched γ' precipitates dispersed in a chromium-rich γ solid solution. (The calculated position of each individual atom is given with a precision of less than half of the lattice parameter in the analysis direction, and a few Ångstroms in the perpendicular directions).

Local compositions can be extracted from a small selected region of the observed image. The measured compositions of both γ and γ' phases are provided in Table 1. The main features of these phases clearly appears: γ is a chromium-enriched phase while γ' is Al rich. The γ' volume fraction, as deduced from these data and using the well known lever rule, is estimated to be approximately 35%.

With the help of a graphic work station, the reconstructed volume can be oriented in a well defined direction. The view angle in Fig. 7 was selected in such a way to be perpendicular to the $\langle 001 \rangle$ direction of the crystal, one of the superstructure directions of the ordered Ni_3Al-type precipitates. Thus, the basic stacking sequence of (002) planes in the ordered phase ($L1_2$) is clearly shown.

Table 1. Measured composition of the γ chromium-rich matrix and γ' aluminium-rich precipitates in the model nickel base alloy NiCrAl. Composition data are estimated from $N = 1400$ atoms for the γ' particle, and $N = 2100$ atoms for the γ matrix

Composition (at%) $\pm 2\sigma$	Al	Cr	Ni
nominal	10	23.4	66.6
γ matrix	5 ± 1	31 ± 2	64 ± 2
γ' particle	20 ± 2	11 ± 2	69 ± 3

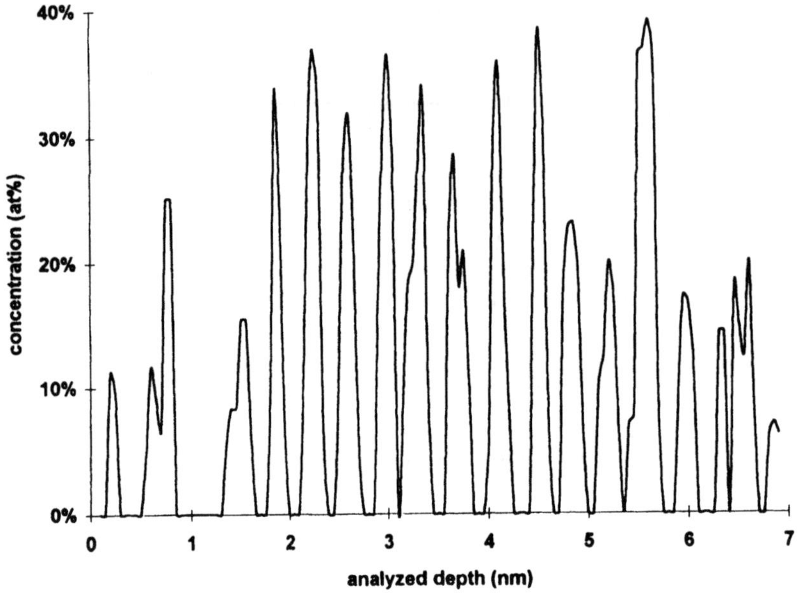

Fig. 8. Al concentration profile through a γ' precipitate. The analysis direction is $\langle 001 \rangle$, perpendicular to superstructure planes displayed in Fig. 7. Compositions are measured over 25 atoms, leading to a mean standard deviation of less than 10 at %

Aluminium rich planes are observed periodically with a spacing close to the lattice parameter ($a \cong 0.36$ nm). This example demonstrates that the tomographic atom-probe is able to reveal very fine scale features on a scale close to that of the unit cell of the material. This also brings the proof that the 3D reconstruction algorithm described previously provides reliable depth position data.

From an experiment such as that described previously, information related to the long range order in the γ' phase may be obtained. A depth profile related to Al concentration modulations through ordered particles in the $\langle 001 \rangle$ direction is displayed in Fig. 8. From such a profile, the mean composition of both types of planes (Al-rich mixed planes-Ni-rich planes) may be derived. As a result, occupation frequencies of the various elements on both types of sites (Al, Ni) may be estimated. Similarly, the local order which occurs, for instance, near interfaces or antiphase boundaries may be studied on a very fine scale. However, as for conventional atom-probes, care must be taken in the interpretation, because preferential retention effects may occur and data may be biased [16].

As already demonstrated, the tomographic atom-probe provides quantitative composition data [10]. Scatter on composition data (Table 1) is essentially controlled by statistical fluctuations due to sampling errors. They are given by the standard deviation $\sigma(\Delta C \sim 2\sigma)$ with $\sigma = \sqrt{C(1-C)/N}$; N is the number of ions over which concentrations (C) are calculated.

The typical number of collected ions related to the investigation of one sample is $N = 500\,000$. Thus, the minimum concentration which could be theoretically detected is a few ppm ($1/N = 2$ ppm). However, because of the background noise which occurs in mass spectra (Fig. 4), the actual detectability which can be reached with the tomographic atom-probe is, in reality, close to 100 ppm.

Conclusion

The major advantage of the tomographic atom-probe (TAP) is that it provides a three-dimensional nanoanalytical image with a very high spatial resolution. The reconstructed images we showed demonstrate that each atom is positioned on a near atomic-scale. The instrument has been proved to be able to image atomic planes of a material directly in real space. As compared to other position-sensitive detectors which may be implemented on atom-probes, the multidetection system that we designed has the great advantage to give the position of most of detected ions including simultaneous impacts and multiple time-of-flight events. This enables quantitative composition data as well as reliable positions to be obtained [10].

Due to the principle of the TAP multidetector (a hundred independent collectors), the tomographic atom-probe may be considered as being almost equivalent to one hundred conventional atom-probes working in parallel. A one-day experiment with the TAP is equivalent to a few week investigation with the conventional atom-probe. More, the instrument has the advantage of giving three-dimensional information: topological data and very fine scale microstructural features are rendered and can be related directly to local compositions. This is a decisive advantage as compared to conventional atom-probes. With the emergence of 3D atom-probes, a new experimental approach of the early stages of phase separation in metallic alloys is opened. One of the main promising projection comes from the direct comparison of 3D reconstructed images with those generated by Monte-Carlo simulations [17]. Solid state decomposition and ordering processes in model Ni Cr Al alloys are currently investigated by both approaches.

Acknowledgements. The development of the tomographic atom probe has been financially supported by the DRET, EDF, the CNRS, the MRT and the Région de Haute-Normandie that we wish to thank. The authors wish to thank Professor B. Jouffrey and Doctor M. Karlik (Ecole Centrale, Paris) for fruitful discussions and for supplying single crystal Al-Cu alloys. Many thanks also go to Doctor P. Caron (ONERA, Chatillon, France) for supplying single crystal nickel based alloys.

References

[1] E. W. Müller, J. A. Panitz, S. B. Mc Lane, *Rev. Sci. Instr.* **1968**, *39*, 83.
[2] D. Blavette, *Surf. Sci.* **1992**, *266*, 299.
[3] M. K. Miller, G. D. W. Smith, *Mat. Res. Soc.* Pittsburgh, PA, 1989.
[4] J. A. Panitz, *J. Vac. Sci. Technol.* **1974**, *11*, 206.
[5] A. Cerezo, T. J. Godfrey, G. D. W. Smith, *Rev. Sci. Instr.* **1988**, *59*, 862.
[6] M. K. Miller, *Surf. Sci.* **1992**, *266*, 494.
[7] D. Blavette, A. Bostel, J. M. Sarrau, *J. Physique* **1986**, *47*, 473.
[8] R. W. Wijnaendts Van Resandt, H. C. Harink, L. Los, *J. Phys. E: Scient. Instr.* **1976**, *9*, 503.
[9] D. Blavette, B. Deconihout, A. Bostel, J. M. Sarrau, M. Bouet, A. Menand, *Rev. Sci. Instr. 1993*, *64*, 2911–2919.
[10] D. Blavette, A. Bostel, J. M. Sarrau, B. Deconihout, A. Menand, *Nature* **1993**, *363*, 432.
[11] P. Bas, A. Bostel, B. Deconihout, D. Blavette, *Appl. Surf. Sci.* **1995**, *87/88*, 298.
[12] B. Jouffrey, D. Dorignac, *J. Phys.* **1992**, *I*, 1067.
[13] B. Jouffrey, M. Karlik, *Microsc. Microanal. Microstruc.* **1992**, *3*, 243.

[14] M. Karlik, *Thèse Ecole Centrale*, 1994, Paris.
[15] V. A. Phillips, *Acta Metall.* **1945**, *23*, 751.
[16] D. Blavette, A. Bostel, *Acta Metall.* **1984**, *32*, 811.
[17] T. A. Abinandanan, F. Haider, G. Martin, *Solid Phase Transformations*, (W. C. Johnson, J. M. Howe, D. E. Laughlin, W. A. Soffa, eds.) The Minerals, Metals and Materials Society, 1994.

Mikrochim. Acta [Suppl.] 13, 195–207 (1996)

Microanalysis at Atomic Resolution

Stephen J. Pennycook**, David E. Jesson, Nigel D. Browning*, and Matthew F. Chisholm

Solid State Division, Oak Ridge National Laboratory, P.O. Box 2008, Oak Ridge, TN 37831-6030, U.S.A.

Abstract. In this paper we discuss how the dedicated scanning transmission electron microscope can provide a microanalysis of crystalline materials at atomic resolution. The method requires the establishment of incoherent conditions for a reference imaging signal as well as the spectroscopic signal. The image can then be used to focus and locate the probe to atomic precision for microanalysis. The Z-contrast image provides the most convenient incoherent reference image, and X-ray and electron energy loss data may be acquired simultaneously. In zone axis crystals, strong columnar channelling delays the onset of beam broadening for several hundred Ångstroms, so that atomic resolution microanalysis may be achieved in materials specimens of significant thickness. This combination of signals provides a powerful means for studying interface structure and bonding, and avoids relying on preconceived model structures.

Key words: electron microscope, Z-contrast, spatially resolved analysis, microprobe, high resolution imaging.

Now that the dedicated scanning transmission electron microscope (STEM) allows atomic sized probes to be routinely achieved, the potential to analyse chosen atomic columns for their composition and bonding has become a reality. Studies of materials can finally be carried out at the fundamental atomic scale. For example, impurity segregation at dislocation cores in grain boundaries can be investigated atomic column by atomic column. Such capabilities are expected to have a far-reaching impact throughout materials science [1].

In this paper we discuss the conditions under which such microanalysis is possible, including considerations of probe formation, beam broadening, and localization of spectroscopic information. The first critical requirement is the ability to form an incoherent image of the material at atomic resolution, so that the probe size and position may be accurately controlled. The Z-contrast image formed from elastically scattered electrons with a high angle annular detector is convenient

* Present address: Department of Physics (M/C 273), University of Illinois at Chicago, 845 West Taylor Street, Chicago, IL 60607-7059, U.S.A.

** To whom correspondence should be addressed

for this purpose, and does not interfere with the simultaneous acquisition of X-ray or energy loss data. Secondly, incoherent conditions for the analytical signal must be established, so that it too will show an incoherent image in direct correspondence to the Z-contrast image, although typically of much weaker intensity. In this way, the Z-contrast image can provide a high intensity reference image corresponding to the much weaker inelastic image, and can therefore be used to focus the probe to an optimum size and to locate it over particular columns of interest. A point microanalysis can therefore be achieved from specific atomic columns, in principle from a single column, provided beam damage is not too rapid.

It is important to appreciate from the outset that atomic resolution microanalysis is only possible at present on the dedicated STEM, although there appears no reason in principle why this should not be achievable in a conventional TEM equipped with a field emission gun. The instrumental requirements are discussed further in the following section, while in section 3 we discuss the propagation of a coherent probe through a zone axis crystal. It is the remarkable behaviour of such a probe that makes atomic resolution microanalysis possible in specimens up to several hundred Ångstroms in thickness. In section 4 we discuss the limits on spatial resolution imposed by the delocalization of the inelastic scattering probability. Generally, losses in excess of several hundred eV are required for atomic resolution to be possible, although some improvement is possible by using an annular collector aperture to exclude low-angle scattering from the spectrometer.

Finally, we present an experimental demonstration of atomic resolution microanalysis by mapping the Co L loss intensity across an atomically abrupt $CoSi_2$/Si(111) epitaxial interface. This intensity drops dramatically on moving from the last plane of the silicide to the first plane of the silicon, a distance of only 2.7 Å.

Instrumental Requirements for Incoherent Imaging

As realized first by Crewe et al. [2–4], the high brightness of the field emission gun is essential in achieving an atomic sized probe with sufficient current for image formation. To achieve an electron-optically limited probe, a high source demagnification is used, so that the geometrical size of the source in the focal plane of the objective lens is much less than the actual probe size obtained by balancing the effects of spherical aberration and defocus. Under these conditions the probe is coherent, with an optimum size given by

$$d_{\mathrm{opt}} = 0.43\, C_s^{1/4} \lambda^{3/4} \tag{1}$$

where C_s is the objective lens spherical aberration coefficient, and λ the electron wavelength. This optimum probe is obtained with an optimum objective aperture

$$\alpha_{\mathrm{opt}} = \left(\frac{4\lambda}{C_s}\right)^{1/4} \tag{2}$$

and an optimum defocus of

$$\Delta f_{\mathrm{opt}} = -(C_s\lambda)^{1/2}. \tag{3}$$

These conditions were first identified by Scherzer [5], and later analysed in more detail by Crewe and coworkers [6, 7]. Note that they differ from the more familiar

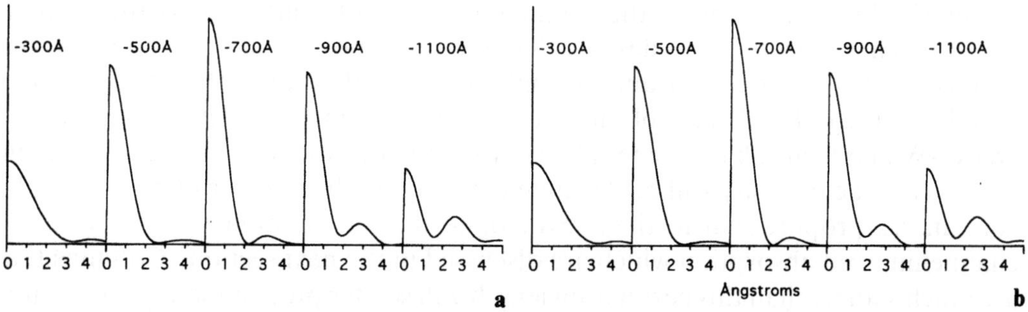

Fig. 1. Probe intensity profiles as a function of defocus for **a** the 100 kV HB 501 UX and **b** the 300 kV HB 603 at Oak Ridge National Laboratory

Scherzer conditions for bright field phase contrast imaging. In particular, the optimum objective aperture tends to be smaller, since there is no need to include particular low order diffracted beams, and d_{opt}, which is often taken as a convenient measure of microscope resolution, is significantly improved over the corresponding expression for coherent imaging,

$$d_{\text{coh}} = 0.66\, C_s^{1/4} \lambda^{3/4}. \tag{4}$$

The higher resolution is a direct result of squaring the probe amplitude profile $P(\mathbf{R})$, on which the resolution of a *coherent* image depends, to form the intensity profile $P^2(\mathbf{R})$, which controls the resolution of an incoherent image or analysis. Probe profiles are straightforward to calculate [8], and Fig. 1 shows profiles as a function of defocus for the two STEM instruments at Oak Ridge National Laboratory, a 100 kV VG Microscopes HB501UX, with an objective lens C_s of 1.3 mm giving an optimum probe size of 2.2 Å, and a 300 kV VG Microscopes HB603, with a 1 mm C_s giving a 1.3 Å probe. These profiles show behaviour typical of coherent probes, broad at low defocus values, becoming narrower at higher defoucs but also acquiring significant subsidiary maxima. The Scherzer optimum probe provides the narrowest central peak without substantial intensity appearing in subsidiary maxima. This represents the most local probe for imaging and microanalysis.

The conditions required for incoherent imaging with elastically (including thermally) scattered electrons, and inelastically scattered electrons, have been reviewed recently [9] and will only be briefly mentioned here. Incoherent conditions apply when either *all* the scattering is detected corresponding to some elastic or inelastic interaction, or when a representative *fraction* of the total scattering is detected.

In the case of elastically scattered electrons incoherent imaging is achieved by collecting a small representative fraction of the total scattering through the use of a high angle annular detector. By ensuring that the inner angle θ_i of the detector is large compared to the probe forming aperture, incoherent imaging of crystal columns is achieved down to spacings of

$$\Delta R = 1.22\, \lambda / \theta_i. \tag{5}$$

This holds even if the scattering reaching the detector is predominantly coherent [11], since the detector average over the diffraction pattern. In most cases however, thermal diffuse scattering is the dominant means of scattering to the annular detector. Thermal vibrations destroy the long-range positional correlations within each column, so that each atom sees only a few neighbours within its column as correlated [12]. The length of such a "packet" depends on detector angle and Debye-Waller factor, but is generally less than the thickness of a typical specimen. In this case the intensity will be dependent simply on the number of packets in the column, i.e. proportional to thickness t (ignoring for the moment effects due to channelling and absorption which are discussed in the next section). Columns that are much shorter than this correlation length will scatter predominantly coherently, so that the intensity will be proportional to t^2 [11].

With X-ray microanalysis, clearly we again collect a small but representative fraction of the total emission, but for electron energy loss spectroscopy (EELS), the strongly forward-peaked scattering makes it experimentally feasible to collect most of the scattering with modern aberration corrected spectrometers. Incoherent imaging conditions are again fulfilled [10], and the inelastic image is given by a convolution of the probe intensity profile with an inelastic object function centered on the atomic sites. However, because of the low characteristic scattering angles associated with low energy edges, the corresponding object functions may be substantially broader than the probe profile, so that to achieve atomic resolution a certain minimum energy loss is necessary as discussed in the section after next.

It is clear that the conditions for incoherent imaging with elastically and inelastically scattered electrons are very complementary, so that both may be achieved simultaneously with the STEM geometry. The intensity in the electron optically limited probe is however significantly reduced from the conditions generally used for microanalysis, so that use of an efficient parallel detection system for EELS is essential. The Oak Ridge system uses a McMullan detector [13], based on a Wright Instruments multi-phase-pinned CCD camera, which provides exceptional sensitivity. Indeed, calculations indicate that single atom detection is feasible with this system [14], despite the low beam currents, due to the fact we are illuminating only a few tens of atoms in one atomic column. A single atom therefore represents a much higher concentration than in situations where beam broadening is important, such as glasses or biological specimens [15]. Due to the likelihood of beam damage affecting the spectral fine structure, an important advantage of the McMullan system is its ability to record a sequence of spectra in rapid succession. Comparing the first spectrum with the last enables an assessment to be made of the importance of beam damage, and if there is none, the spectra may be summed to improve statistics. At present, only the 100 kV STEM is equipped with this EELS capability.

Although no X-ray microanalysis has yet been demonstrated at atomic resolution, the only limitation would appear to be the need for longer counting times with the increased possibility of beam damage. Although the EELS fine structure is the most sensitive indicator of beam damage, amorphization of a column is quite apparent from the Z-contrast image alone, and it would be reasonable to assume that if the image were not changed at the end of the analysis, then at least the major constituents of the column were not displaced during the analysis. X-ray analysis requires the use of the virtual objective aperture, and due to the high demagnification of the illumination system, smaller apertures than usual

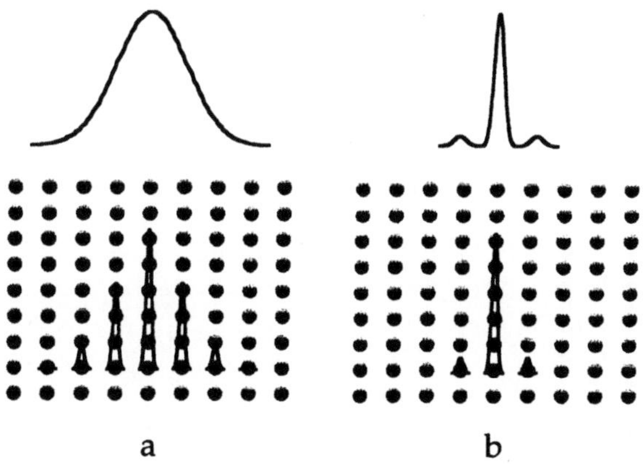

Fig. 2. Schematic indicating the columnar channelling of a coherent probe in a zone axis crystal with **a** a broad incident probe, **b** a probe size less than the column separation, giving atomic resolution analysis

are required. The optimum angle can be adjusted by varying the condenser and objective lens strengths; extra demagnification can be achieved if necessary by increasing the strength of the gun lens. In this way, Z-contrast images can be obtained of equal quality to those obtained with the real objective aperture.

Probe-Channelling

No consideration has so far been given to the effects of dynamical diffraction on the conditions needed to establish incoherent imaging. The reason for this is that dynamical diffraction manifests itself in a particularly simple way when a coherent probe is incident on a zone axis crystal. It has the effect of focussing the current onto the atomic columns, while maintaining the overall envelope of the incident probe, as indicated schematically in Fig. 2. For an incident probe size below the column spacing, most of the current can be channelled along a single atomic column, which is critical to achieve Z-contrast imaging and microanalysis at atomic resolution. Incoherent imaging is no longer restricted to weak phase objects, but applies also to crystals of substantial thickness.

This behaviour was apparent in early calculations of STEM probe propagation [16, 17], and its explanation lies in the effects of quantum mechanical superposition [18–20]. The STEM probe is a coherent superposition of plane waves emanating from the objective aperture, and is best thought of as a spherical wave converging on the specimen surface. Inside the crystal, the probe becomes a coherent superposition of fast electron Bloch states, and it is clear that if the amplitude and phase of a state is strongly dependent on incident angle, then superposition of states over a wide angular range will lead to destructive interference. The only states that survive this angular integration are those that are tightly bound to the atomic columns, since they have an amplitude and phase in reciprocal space that is slowly varying. In Si and Ge for example, only 1s states are important, and the probe therefore takes on the form of these states as it propagates into the crystal, becoming concentrated into narrow spikes around each atomic column.

With only 1s states contributing to the probe amplitude, then it is also clear that interference between different Bloch states becomes a second order effect, so that the usual oscillatory depth dependence of diffraction intensities will be suppressed to a large degree. For the same reason, beam broadening will also be strongly

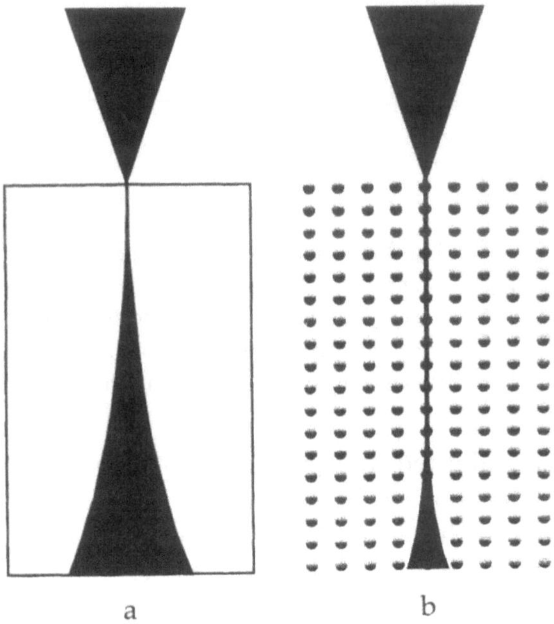

Fig. 3. Schematic indicating the reduction in beam broadening as a result of columnar channelling; **a** normal beam broadening effect in a non-channelling material, **b** delayed broadening in a zone-axis crystal

suppressed. The channelled flux propagates along the columns in the *s* states until it is scattered out to plane wave states by thermal vibrations, whereupon beam broadening is initiated in the usual way (Fig. 3). The penetration length depends on the *s* state absorption coefficient, which, since *s* states are in close proximity to the atomic nuclei, is quite strong and very sensitive on atomic number. At 100 kV the flux can penetrate hundreds of Ångstroms in Si, but only a few tens of Ångstroms along heavy columns. At 300 kV penetration lengths are considerably enhanced. Such effects are well described in the Bloch wave description through an Einstein model of thermal vibrations, and the effect can be incorporated quantitatively in the appropriate cross section for imaging or microanalysis. If $\sigma_i(\mathbf{R})$ is the appropriate atomic cross section for the elastic or inelastic scattering process being detected, the columnar cross section is given by

$$O(\mathbf{R}) = \sum_i \sigma_i(\mathbf{R}) \varepsilon^{1s^2} b^{1s^2}(\mathbf{R}) \left(\frac{1 - e^{-2\mu^{1s}t}}{2\mu^{1s}t} \right) \tag{6}$$

where the sum is over all atoms *i* in the column, b^{1s} is the 1s Bloch state of excitation ε^{1s} and absorption coefficient μ^{1s}, calculated for exact zone axis incidence. The angular dependence of these quantities is included by defining an effective probe intensity profile,

$$P^2_{\text{eff}}(\mathbf{R}) = \left| \frac{1}{\varepsilon^{1s}(0)} \int_{\substack{\text{objective}\\ \text{aperture}}} \varepsilon^{1s}(\mathbf{K}) e^{i[\mathbf{K}\cdot(\mathbf{R}-\mathbf{R}_0)+\gamma(\mathbf{K})]} d\mathbf{K} \right|^2 \tag{7}$$

where $\mathbf{K}$ is the transverse component of the incident wave vector and

$$\gamma = \frac{\pi}{\lambda} \left(\Delta f \theta^2 + \frac{1}{2} C_s \theta^4 \right) \tag{8}$$

is the transfer function phase factor for a defocus Δf and electron wavelength λ. If the Bloch wave excitation ε^{1s} was constant over the aperture, this would be identical to the incident probe profile. Typically, ε^{1s} falls to about half its axial value at the edge of the objective aperture, which has the effect of broadening the effective probe by about 10% [19]. It can be appreciated that the constancy of ε^{1s} over the objective aperture is critical to forming a probe inside the crystal. ε^{2s} for example changes sign across the aperture, which is the reason that the 2*s* Bloch state, although highly excited on axis, does not contribute significantly to the probe intensity inside the crystal. The Z-contrast, EELS or X-ray image is then given by the convolution of the effective probe with the columnar cross section, which can be considered as the object function for incoherent imaging:

$$I(\mathbf{R}_0) = \int P_{\text{eff}}^2(\mathbf{R}-\mathbf{R}_0)O(\mathbf{R})d\mathbf{R}. \tag{9}$$

Localization of Inelastic Scattering

Object functions for elastic scattering to the high angle detector are highly localized about the atomic columns, but this is not necessarily the case for inelastic scattering. Plasmons and low loss excitations may be quite delocalized, and would not carry atomic resolution chemical information. (Note that atomic resolution images in plasmon losses can be obtained through elastic scattering mechanisms, a situation usually referred to as preservation of contrast. This is unrelated to the present discussion which is concerned with the localization of the microanalytical information.) A good description of the object function for inelastic processes is therefore desirable. The object function for a transition from an initial state $|0\rangle$ to a final state $|n\rangle$ may be written in terms of the matrix elements ρ_{no} as

$$O(\mathbf{R}) = \left(\frac{e^2}{\pi\hbar v}\right)^2 \left| \int \frac{\rho_{no}}{q^2}(\mathbf{q})e^{-i\mathbf{K}\cdot\mathbf{R}}d\mathbf{K} \right|^2, \tag{10}$$

[10, 21] where the momentum transfer $\mathbf{q} = (\mathbf{K}, q_{z\min})$, with $q_{z\min}$ corresponding to the minimum momentum transfer at zero scattering angle. The object function can also be written in terms of a form factor in reciprocal space

$$O(\mathbf{R}) = \frac{1}{(2\pi)^2}\int f'(\mathbf{K})e^{i\mathbf{K}\cdot\mathbf{R}}d\mathbf{K}. \tag{11}$$

where

$$f'(\mathbf{K}) = \left(\frac{2e^2}{\hbar v}\right)^2 \int \frac{\rho_{no}(\mathbf{q}')}{q'^2}\frac{\rho_{no}^*(\mathbf{q}'+\mathbf{K})}{|\mathbf{q}'+\mathbf{K}|^2}d\mathbf{K}' \tag{12}$$

Although this form factor has a similar angular dependence to the more familiar generalized oscillator strength, it is not identical. For high resolution microanalysis we will primarily be concerned with inner shell excitations, for which a hydrogenic model has been developed by Maslen and Rossouw [22–26]. In this model, an inelastic excitation with scattering vector $\mathbf{q}$ for the fast electron is accompanied by ejection of a secondary electron with wavevector κ. As the secondary electron is not detected, κ must be integrated coherently over all secondary electron emission

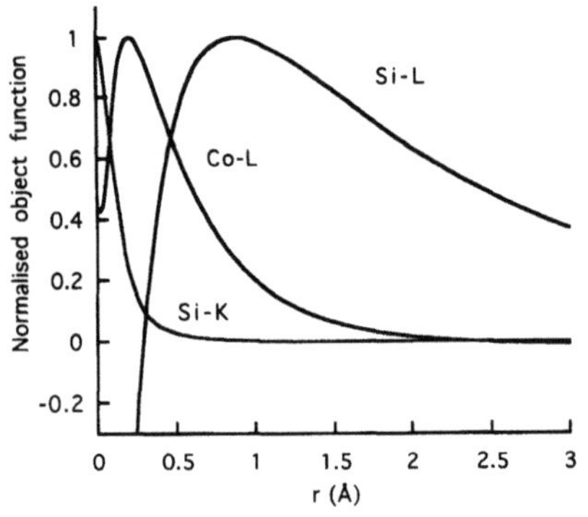

Fig. 4. Object functions for the Si K, Si-L and Co-L losses, calculated on the hydrogenic model for a 50 eV ejected secondary electron and assuming all scattered electrons are collected. Plots have been normalised to their peak intensity to emphasize differences in width. Note the object function for Si-L is negative at $r=0$

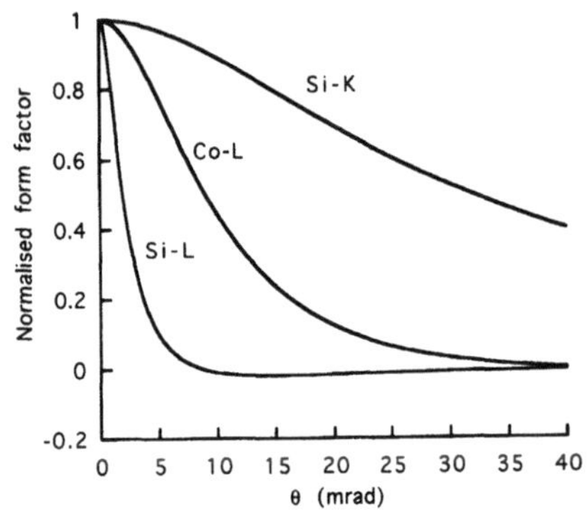

Fig. 5. Corresponding object functions in reciprocal space

directions, and the matrix product $\rho_{no}(\mathbf{q}')\rho_{no}(\mathbf{q}'+\mathbf{K})$ replaced with

$$\int \rho_{no}(\mathbf{q}', \kappa)\rho^*_{no}(\mathbf{q}'+\mathbf{K}, \kappa)d\kappa. \tag{13}$$

Convenient analytical expressions have been given for these integrals [27, 28]. Figure 4 shows object functions calculated using the hydrogenic model assuming collection of all the scattering for edge energies appropriate for the Si-K loss (1840 eV), Si-L loss (100 eV) and the Co-L loss (786 eV). Although the hydrogenic model was developed for K shells, we believe it to be the only model for inner shell excitations available at this time, and is likely to provide a more accurate description than simple estimates based on characteristic scattering angles [29–31]. The full width half maxima are 0.24 Å for the Si K loss, 4.9 Å for the Si L loss and 2.0 Å for the Co L loss. Clearly one could not expect atomic resolution from the Si L loss, although it should be possible with the other two edges. Note the sensitive dependence of the object function width on atomic number, the Co L object function being almost an order of magnitude wider than that of the Si K edge, with less than a factor of two difference in energy.

The corresponding functions in reciprocal space, $f'(\mathbf{K})$, are shown in Fig. 5. Clearly, for our detection angle of 30 mrad, almost all of the scattering is indeed detected for the Co-L and Si-L edges, so that the conditions for incoherent imaging are fulfilled. This is not the case for the Si K loss, but provided the detector angle is much larger than the probe forming aperture, incoherent imaging still applies, and the real space object function is now given by restricting the integration in Eq. 11.

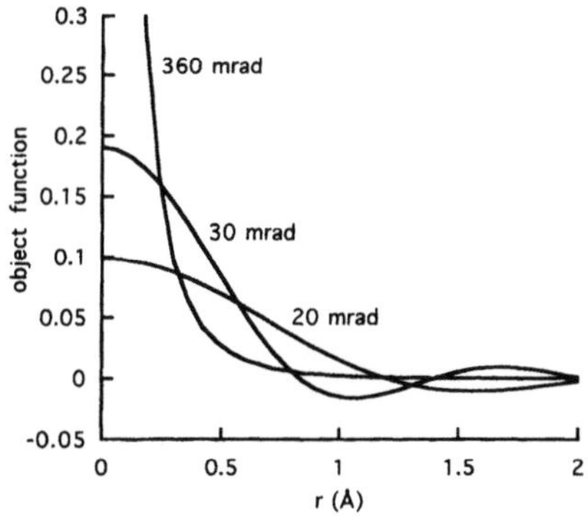

Fig. 6. Reduction and broadening of the Si-K object function due to the use of a finite collection aperture

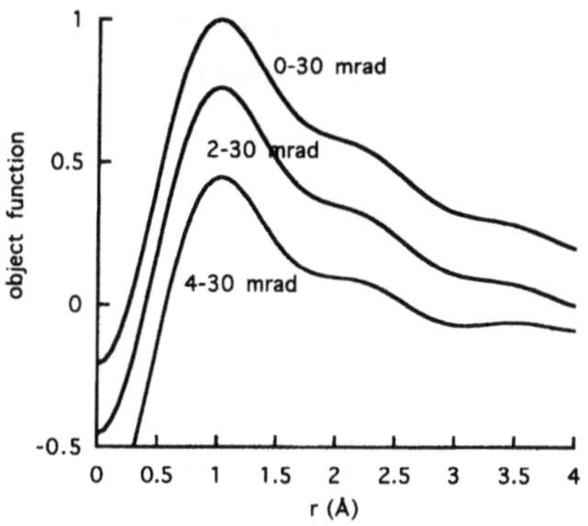

Fig. 7. Enhanced spatial resolution expected with the Si-L loss using an annular collection aperture. Note the significant enhancement of the negative central part of the object function, which will cause loss in signal intensity

to the collector aperture

$$O(\mathbf{R}) = \frac{1}{(2\pi)^2} \int_{\text{detector}} f'(\mathbf{K}) e^{i\mathbf{K}\cdot\mathbf{R}} d\mathbf{K}. \tag{14}$$

The effect of a finite collector aperture on the object function for the Si K loss is shown in Fig. 6. Clearly, provided we ensure that our collection angle is significantly greater than the objective aperture (semiangle ~10 mrad), then the use of a finite collector aperture will not cause a loss in resolution.

Note that all object functions show significant tails in real space, unlike the Gaussian forms often assumed. This will result in a non-local component to the image. It would be expected that this component could be removed by the use of an annular collector aperture to restrict low angle scattering from entering the spectrometer. Figure 7 shows the signficant resolution improvement to be expected with the Si L loss, a FWHM of 3.6 Å and 2.8 Å being achieved for an annular collection apertures of radius 2–30 mrad and 4–30 mrad respectively, which may be sufficient to allow atomic resolution to be achieved in Si. Larger inner collection angles cause substantial loss in intensity.

Atomic Resolution Spectroscopy at a $CoSi_2$/Si(111) Interface

The $CoSi_2$/Si(111) epitaxial interface is an ideal test specimen for demonstrating atomic resolution since it can be grown atomically flat and abrupt as seen in the

Fig. 8. Z-contrast image of an epitaxial $CoSi_2$/Si(111) interface showing the last layer of Si to be in a twinned orientation Numbered planes correspond to the energy loss spectra in Fig. 9

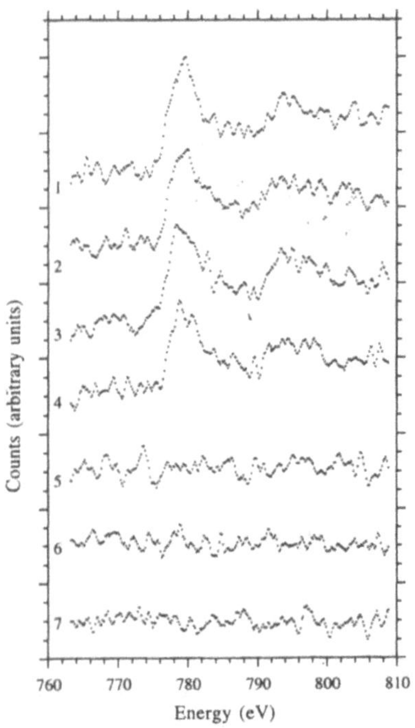

Fig. 9. Co L spectra recorded plane by plane across the interface shown in Fig. 8, using the Z-contrast image to maintain the probe over the desired plane. The Co edge disappears between planes 4 and 5, a distance of 2.7 Å, demonstrating atomic resolution microanalysis

Z-contrast image in Fig. 8. This specimen was grown by high dose ion implantation and annealing, and shows an unexpected interface structure in that the last plane of the Si is seen to be in a twinned orientation [32]. The image was taken with the 100 kV STEM, and from the profiles of Fig. 2a would not be expected to resolve the 1.36 Å separation of the dumbbells, but instead each is imaged as a single elongated feature. In the $CoSi_2$, the image is dominated by the high scattering power of the Co columns, and the Si columns are unresolved, forming a high background intensity to the image. Note that, due to the twin, the separation of the last plane of the Si and the first plane of the $CoSi_2$ is only 2.7 Å.

Given the object function shown in Fig. 4 we would anticipate achieving atomic resolution using the Co L-loss. The spectra shown in Fig. 9 were obtained by

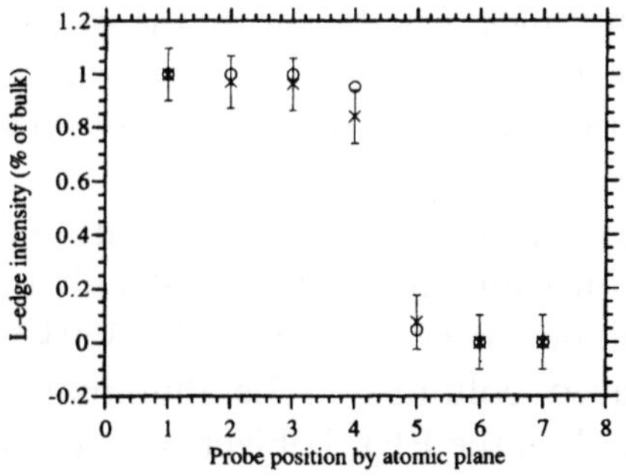

Fig. 10. Profile showing integrated Co edge intensity plane by plane across the interface (crosses). The sharp drop in Co intensity on moving a single plane across the interface demonstrates atomic resolution. Circles show the theoretical profile for a localized object function

centering a line scan over each plane in turn using the Z-contrast signal displayed on an oscilloscope, then acquiring a Co-L edge for 5 seconds [33]. In this way, the possibility of beam damage is minimised and any specimen drift is visible by observing the line scan. The use of a large collection aperture does degrade the energy resolution of the fine structure, the width of the L_3 edge being about 5 eV. Thus precision in spatial information must be traded for precision in spectral detail due to averaging over a large range of momentum transfers. It is quite apparent that the Co edge drops dramatically in moving a single atomic plane across the interface. The specimen thickness in this region was estimated at 100–200 Å, so that no significant beam broadening should occur in the Si, but some is to be expected for the heavier Co columns. A quantitative profile was obtained by least squares fitting to reference spectra taken several units cells from the interface, and resulted in the profile shown in Fig. 10. The inelastic intensity drops from 86% to 7% in a distance of 2.7 Å, demonstrating that atomic resolution is indeed achieved. Also shown is the expected profile for a 2.2 Å probe and a perfectly localized inelastic scattering potential. The observed profile is clearly somewhat broadened as would be expected from the significant width of the object function. The asymmetry in the experimental profile is consistent with the anticipated onset of beam broadening in the silicide.

Conclusions

Microanalysis at atomic or near-atomic resolution is now experimentally feasible [33–36], opening up new areas of materials science to fundamental study, such as the mechanisms of segregation and embrittlement. The simultaneous acquisition of imaging and spectroscopic data is particularly powerful, due to the complementary nature of their information. For example, in a recent study of tilt grain boundaries in $SrTiO_3$, Z-contrast imaging and EELS were combined to deduce structure models directly from the experimental data [37–39]. Heavy columns were located by a maximum entropy analysis of the Z-contrast image to an accuracy of ~0.2 Å, while the oxygen coordination was determined from the EELS fine structure. The importance of this capability was the finding of an unexpected structure, involving two half columns (or equivalently a single zig-zag column) at the core of each grain boundary dislocation. Such cores appeared in all structural units found, and appear to be a very effective way to avoid self-ion repulsion in ionic materials. Also, such columns represent attractive sites for the segregation of dopant atoms, which can now be investigated experimentally by these techniques. Furthermore, the model structures found in this way represent the perfect starting point for theoretical studies, by ab initio methods for example. Both theory and experiment may now study the same systems in parallel.

In the future we can also expect to see X-ray analysis carried out at atomic resolution, adding complementary information on elements that are difficult to detect by EELS. We can also expect these techniques to be used with the 300 kV STEM, its smaller probe allowing smaller spaced columns to be illuminated individually, in intermetallics for example. The width of the EELS object function is practically insensitive to accelerating voltage, so that real advantages are to be anticipated for materials analysis.

Acknowledgments. We are grateful to J. T. Luck, T. C. Estes and S. L. Carney for technical assistance. This research was sponsored by the Division of Materials Sciences, U.S. Department of Energy, under contract No. DE-AC05-84OR21400 with Martin Marietta Energy Systems, Inc.

References

[1] L. M. Brown, *Nature* **1993**, *366*, 721.
[2] A. V. Crewe, J. Wall, J. Langmore, *Science* **1970**, *168*, 1338.
[3] A. V. Crewe, J. Wall, *J. Molec. Biol.* **1970**, *48*, 375.
[4] A. V. Crewe, J. P. Langmore, M. S. Isaacson, in: *Physical Aspects of Electron Microscopy and Microbeam Analysis* (B. M. Siegel, D. R. Beaman, eds.) Wiley, New York, 1975, p. 47.
[5] O. Scherzer, *J. Appl. Phys.* **1949**, *20*, 20.
[6] A. V. Crewe, *Rep, Prog. Phys.* **1980**, *43*, 621.
[7] A. V. Crewe, D. B. Salzman, *Ultramicroscopy* **1982**, *9*, 373.
[8] C. Colliex and C. Mory, in: *Quantitative Electron Microscopy* (J. N. Chapman, A. J. Craven, eds.) Scottish Universities Summer School in Physics, Edinburgh, Scotland, 1984, p. 149.
[9] S. J. Pennycook, D. E. Jesson, N. D. Browning, *Nucl. Instr. Meth. B*, **1995**, *96*, 575.
[10] R. H. Ritchie, A. Howie, *Phil. Mag. A* **1988**, *58*, 753.
[11] D. E. Jesson, S. J. Pennycook, *Proc. R. Soc. Lond. A*, **1993**, *441*, 261.
[12] D. E. Jesson, S. J. Pennycook, *Proc. R. Soc. Lond. A*, **1995**, *449*, 273.
[13] D. McMullan, J. M. Rodenburg, Y. Murooka, A. J. McGibbon, *Inst. Phys. Conf. Ser.* **1990**, *98*, 55.
[14] N. D. Browning, S. J. Pennycook, *Microbeam Anal.* **1993**, *2*, 81.
[15] R. D. Leapman, in: *Transmission Electron Energy Loss Spectrometry in Material Science* (M. M. Disko, C. C. Ahn, B. Fultz, eds.) The Minerals, Metals and Materials Society, 1992, p. 47.
[16] J. Fertig, H. Rose, *Ultramicroscopy* **1977**, *2*, 269.
[17] R. F. Loane, E. J. Kirkland, J. Silcox, *Acta Cryst.* **1988**, *A44*, 912.
[18] S. J. Pennycook, D. E. Jesson, *Phys. Rev. Lett.* **1990**, *64*, 938.
[19] S. J. Pennycook, D. E. Jesson, *Ultramicroscopy* **1991**, *37*, 14.
[20] S. J. Pennycook, D. E. Jesson, *Acta Metall. Mat.* **1992**, *40*, S149.
[21] M. J. Whelan, *J. Appl. Phys.* **1965**, *36*, 2099.
[22] V. M. Maslen, C. J. Rossouw, *Phil. Mag. A* **1983**, *47*, 119.
[23] V. M. Maslen, C. J. Rossouw, *Phil. Mag. A* **1984**, *49*, 735.
[24] C. J. Rossouw, V. W. Maslen, *Phil. Mag. A* **1984**, *49*, 749.
[25] L. J. Allen, C. J. Rossouw, *Phys. Rev. B* **1990**, *42*, 11644.
[26] L. J. Allen, *Ultramicroscopy* **1993**, *48*, 97.
[27] V. W. Maslen, *J. Phys. B* **1983**, *16*, 2065.

[28] C. J. Rossouw, V. W. Maslen, *Ultramicroscopy* **1987**, *21*, 277.
[29] A. J. Bourdillon, *Phil. Mag. A* **1984**, *50*, 839.
[30] S. J. Pennycook, *Contemp. Phys.* **1982**, *23*, 371.
[31] S. J. Pennycook, *Ultramicroscopy* **1988**, *26*, 239.
[32] M. F. Chisholm, S. J. Pennycook, R. Jebasinski, S. Mantl, *Appl. Phys. Lett.* **1994**, *64*, 2409.
[33] N. D. Browning, M. F. Chisholm, S. J. Pennycook, *Nature* **1993**, *366*, 143.
[34] P. E. Batson, *Nature*, **1993**, *366*, 727.
[35] D. A. Muller, Y. Tzou, R. Raj, J. Silcox, *Nature*, **1993**, *366*, 725.
[36] D. A. Muller, J. Silcox, *Ultramicroscopy*, **1995**, *59*, 195.
[37] M. M. McGibbon et al., *Science*, **1995**, *266*, 102.
[38] N. D. Browning et al., *Interface Sci.*, **1995**, *2*, 397.
[39] M. M. McGibbon et al., *Proc. MRS* **1995**, *357*, 115.

Mikrochim. Acta [Suppl.] 13, 209–216 (1996)

Composition of Vanadium Carbides Formed by Solidification in Fe-V-C-M Alloys: Influence of Additions (M = Al, Cu, Mo)

Leïla Adnane[1], Rafika Kesri[1], and Sylvaine Hamar-Thibault[2,*]

[1] Laboratoire de Métallurgie Structurale, Institut de Chimie, U.S.T.H.B., BP 32, El Alia, Bab Ezzouar, Alger, Algérie

[2] Laboratoire de Thermodynamique et Physico-Chimie Métallurgique, E.N.S.E.E.G., I.N.P.G., BP 75, F-38402 St Martin d'Hères, France

Abstract. The solidification behavior of Fe-V-C-M alloys was studied by differential thermal analysis. The composition of cubic vanadium carbides VC_x was analysed by quantitative electron probe microanalysis. The addition of a quaternary element (Mo, Al, Cu) induced changes in the crystallization path, the morphology of the carbides (Al) and the composition of the vanadium carbides (Mo, Al). Moreover, the stoichiometry (x) of the carbides depends on the nature of the addition.

Key words: Fe-V-M-C alloys, vanadium carbides, microprobe analysis, carbon level.

Vanadium is a common alloy element in steels and cast irons which is used to improve mechanical properties, therefore, the Fe-V-C ternary system was investigated.

The liquids projection in the Fe-V rich region has been determined experimentally [1] and calculated [2–4]. Two invariant equilibria were reported:

– a ternary eutectic: 17.1 at% C and 2.75 at% V at 1120 °C:

$$Liq \Rightarrow \gamma + VC_{x=0.89} + M_3C \text{ (point E)},$$

– a ternary quasi-peritectic: 10.8 at% V and 9 at% C at 1370 °C:

$$Liq + \alpha \Rightarrow \gamma + VC_x \text{ (point P)}.$$

VC_x carbides formed in the binary V-C system, show a large deviation from stoichiometry ($0.59 < x < 0.89$) which is due to vacancies in the carbon sublattice [5–6]. As shown in a previous paper [7], the carbon level in vanadium carbides

* To whom correspondence should be addressed

formed from the melt in Fe-V-C ternary alloys decreases continuously, as the vanadium level increases.

The objective of this work is to perform quantitative electron probe microanalysis on carbides formed from the melt in quaternary systems, in order to obtain data about the influence of additions on the stoichiometry of vanadium carbides and about the composition of cementite.

Experimental

Series of alloys, with varying vanadium content (5, 8, 15 and 20 at%) and different additions were prepared. The amount of addition was below 4 at%. Two types of additions were tested:

- carbide formers such as chromium and molybdenum,
- elements which do not form stable carbides (Al, Si, Ni and Cu).

In particular, Mo, Cu and Al were investigated.

The alloys were prepared in an arc furnace using a non-consumable electrode in an argon atmosphere. The melting behaviour and the solidification path were determined by differential thermal analysis (DTA) during slow cooling (600 °C/h).

Quantitative electron probe microanalysis (CAMECA SX50) was used to determine accurately the composition of the phases and particularly the carbon level in carbides. The CKα characteristic X-ray radiation was analysed with a PC_1 monochromator which consists of layers of W-Si (d spacing $\approx$ 30 Å). A chromium carbide with known composition (Cr_7C_3 with 8.71 wt% C) was used as standard for C analysis. The spectrometer was scanned over the range 39.80 Å to 49.00 Å. A linear background is used, but a window around the second order VLα radiation ($\approx$ 48 Å) is used to eliminate the contribution of this radiation to the background. No important shift or alteration of the carbon peak was observed, as shown previously [8]. The CKα line is shifted only slightly for vanadium carbides of different composition (between 44.41 to 44.48 Å) and the standard (44.39 Å). The area/peak factor according to Bastin [9] for the carbides can be assumed as constant. Due to the small size of the carbides and the difficulty to obtain a smooth surface, experimental errors remain larger than those due to the alteration of the CKα peak.

At least 20 measurements were performed on each type of carbides in the different alloys. The quantitative analyses of the metallic elements were performed with an LiF monochromator at 15 keV (electron current: 80nA) using standards of elemental Fe, V and M. Atomic number, absorption and fluorescence effects were corrected with a computer program according to the PAP model [10]. The total error on weight concentration remained below 2%.

Results and Discussion

Liquid-Solid Equilibrium and Morphology of Vanadium Carbides

In Fe-V8%-C alloys, a small amount of primary VC_x carbide is formed. The binary eutectic between austenite(γ) and VC_x begins to solidify at 1320 °C and evolves along the monovariant line PE. The solidification path stops on the binary line and does not reach the ternary point E. The eutectic vanadium carbides formed in ternary alloys are fibrous and the primary carbides are non-facetted.

Small amounts of additions (Al, Mo or Cu) to the ternary Fe-V8%-C alloy do not significantly change the solidification path. The solidification begins at 1320 °C with primary VC_x carbides and continues with the formation of the binary eutectic γ(Fe)/VC_x. When a carbide-forming element is added, the solidification ends with the

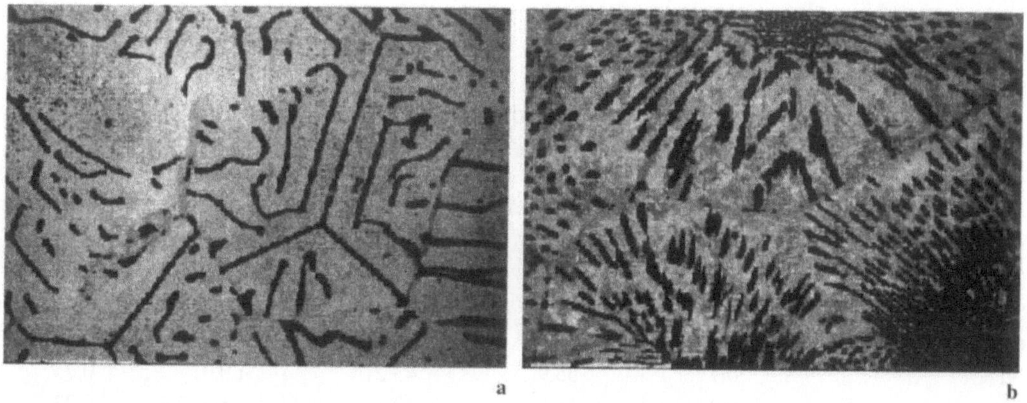

Fig. 1. Morphology of carbides formed from the melt in two alloys $FeV_8C_3M_2$, (M = Al and Cu) – backscattered electron image. **a** Al addition, **b** Cu addition

formation of a second binary eutectic between austenite and a M-rich carbide (M_7C_3 or M_2C).

The influence of the addition on the morphology of the vanadium carbides is complex. With Cr, Cu or Ni, no significant change of the morphology is observed. The vanadium carbides keep their regular and fibrous structure (Fig. 1b). On the other hand the morphology is strongly affected by Al or Si additions. In these two latter cases, the eutectic vanadium carbides are plate-like and the primary carbides facetted (Fig. 1a).

In alloys with a low vanadium level (< 5 at% V) and a high carbon content (> 12 at% C), the solidification begins with the austenite, then the binary eutectic γ(Fe)/VC_x is formed, and finally ends with the ternary eutectic γ(Fe)/VC_x/M_3C (point E). The amount of cementite formed at the end of the solidification is very small and often difficult to observe. It is frequently confused with the eutectoid cementite.

In the case of small amounts of additions (2%) of Cu, Al or Si, the solidification begins with a small amount of VC_x carbides. The end of the solidification is not changed. At a higher level of additions (4%), the solidification begins with the formation of a large amount of VC_x carbides and ends with the formation of two binary eutectics: γ(Fe)/M_3C and γ(Fe)/graphite.

Si and Cu have a graphitizing effect during the solidification of cast iron and reduce the tendency to form the metastable eutectic between austenite and cementite [10]. This effect compensates the opposing effect of carbide forming elements such as V.

In Fe-V20%-C-M alloys, a small amount of primary VC_x carbide is formed, then the binary eutectic ferrite(α)/VC_x crystallizes. Small amounts of additions do not significantly change the solidification path.

Carbide Forming Elements and Stoichiometry of Vanadium Carbides

In ternary Fe-V8%-C alloy, VC_x carbides have a carbon to metal (Fe + V) ratio (x) of around 0.80. As the eutectic solidification γ(Fe)/VC_x, evolves along the eutectic

Table 1. Microprobe analysis (in at %) of vanadium carbides formed in FeV_8C_3M alloys (M = 2%) after DTA experiments [$x = C/(V + Fe + M)$]

M addition	Composition of vanadium carbides				
	Fe	V	C	M	x
–	0.9	54.2	44.9	–	0.80
Al	0.9	54.1	44.4	<0.01	0.82
Si	0.9	55.1	44.3	<0.01	0.82
Cu	0.7	55.5	43.8	0.01	0.79
Ni	0.7	55.9	43.2	<0.01	0.75
Cr	0.9	54.8	42.1	1.7–2.4	0.73
Mo	2.2	46.2	41.4	9.8	0.71

line PE, the composition of vanadium carbides changes. The carbon level increases and a final value of 0.84 is determined in the external part of the eutectic cell.

The composition of vanadium carbides formed in quaternary Fe-V8%-M-C alloys after DTA experiments is reported in Table 1.

Chromium. Addition of chromium to the ternary Fe-V8%-C alloy does not significantly change the solidification path. The solidification begins with primary VC_x carbides, and then two binary eutectics γ(Fe)/VC_x (1297 °C) and austenite/M_7C_3 (1284 °C) are formed. The primary VC_x carbides are globular and often surrounded by a shell of M_7C_3 carbide. In the center of the eutectic cell, the vanadium carbides have a high carbon level and about 2 at% chromium. In these VC_x carbides, the carbon level ($\approx$ 42 at% C) remains constant but is slightly lower that in the carbides formed in the ternary Fe-V-C alloy.

The hexagonal M_7C_3 carbides have a constant carbon level ($\approx$ 30 at% C). Small amounts of vanadium are detected and the composition is: $Fe_{49}V_9Cr_{14}C_{27}$. The interface between the two carbides is straight from the morphological view point. The composition changes also abruptly at the interface.

Molybdenum. The addition of molybdenum leads to more complex phenomena. As shown in Fig. 2a, the solidification path of quaternary Fe-V8%-Mo-C alloys is: primary VC_x and two binary eutectics γ/VC_x (1297 °C) and γ/Mo_2C (1120 °C).

In the backscattered electron image (BEI), three zones are observed:

- vanadium carbides (black) in the center of the eutectic cell. These carbides keep their regular fibrous morphology,
- molybdenum carbides (white) in the external part of the cell,
- a transition zone, which appears with a gray contrast, where a continuously varying Mo/V ratio was qualitatively observed.

More accurate microanalysis of the carbides (Fig. 2b) showed:

- continuous substitution of the metallic elements from the vanadium carbides

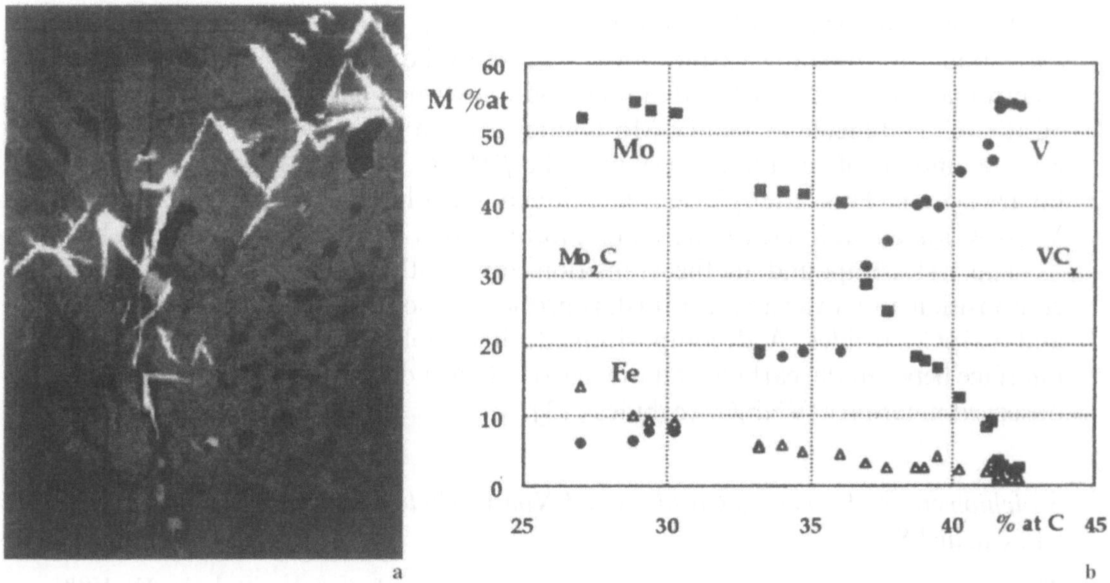

Fig. 2. Morphology of carbides formed in a quaternary alloy FeV_8C_3Mo alloy: **a** backscattered electron image and **b** composition of the carbides as a function of carbon content

with a small amount of Mo ($Mo/V \approx 0.05$) to molybdenum carbides with $V/Mo \approx 0.10$,

– continuous variation of the carbon level (that is new and important).

In the center of the eutectic cell, the darkest vanadium carbides have a high carbon level (> 42 at% C) as in the ternary Fe-V-C system. They belong to the cubic VC_x carbides, while the molybdenum carbides have a low carbon level (< 30 at% C) and belong to the hexagonal Mo_2C carbides. The solubility of V in Mo_2C is limited (< 8 at% V).

Between these two domains of compositions corresponding to well known carbides, two other domains are distinguished. The first one corresponds to a phase which appears as well characterized small plates. Microprobe analysis gives a carbon level of around 33–35 at%, with large amounts of both Mo (≈ 40 at%) and V (≈ 20 at%). Due to the small amount of this phase, the structural determination is difficult. The second domain shows a large variation in composition. As the Mo level decreases, the V and C levels increase simultaneously. The existence of a large concentration gradient inside one carbide showed that volume diffusion of metallic-species is slow even at high temperature. Moreover, the morphology of the carbide-matrix interface becomes facetted.

Isothermal sections, in the ternary V-Mo-C system [11], show a continuous solubility between V_2C and Mo_2C. No ternary compound has been reported. In the binary Mo-C diagram, four intermediate phases were characterized [12]. Three phases are of hexagonal type: αMo_2C (≈ 26–33 at% C) which is the stable phase at room temperature, βMo_2C (≈ 33–36 at% C), and ηMo_2C (≈ 37–39 at% C). The cubic carbide αMoC (≈ 40–43 at% C) is isostructural with the VC_x carbide.

The whitish phase (BEI) observed at the end of the eutectic solidification corresponds to the stable carbide αMo_2C. Due to the small amount, the structural determination of the gray facetted phase with intermediate carbon content ($\approx$ 33–36 at% C) is difficult. A binary Mo-C intermediate phase could be stabilized by a large amount of vanadium: certainly, the βMo_2C with a carbon level as in the binary system. This latter phase is mentioned in isothermal sections of the ternary V-Mo-C system at intermediate temperatures with a V/Mo ratio of about 0.2.

The last compound in the transition zone with a continuous variation of composition and a variable contrast, is probably a solid solution between the VC_x and αMoC carbides. A decomposition during cooling can induce this facetted interface between the carbide and the matrix. Such a decomposition was previously observed in ternary $(VNb)C_x$ carbides [13].

Stoichiometry of Vanadium Carbides and Non-Carbide Formers Additions (Al, Cu, and Si)

The compositions of the vanadium carbides formed from the melt in Fe-V8%-M2%-C alloys are reported in Table 1. The solubility of these non-carbide forming elements in vanadium carbides is below the detection of the measurements (<0.01%).

Accurate carbon analyses show that the carbon level depends on the nature of the addition. Cu and Al have no special influence on the carbon to metal ratio ($x = C/[Fe + V + M]$) which remains around 0.80–0.82. Depending on the amount, Si additions promote a significant increase of x (0.80 to 0.90) while x decreases to 0.75 for Ni additions.

In high-vanadium alloys, neither the solidification path nor the morphology of the carbides are changed, whereas the non-stoichiometry of the vanadium carbides is directly related to the nature of the additions. Cu has no special influence on the carbon to metal ratio which remains around 0.72–0.74 as in the ternary system. On the contrary, Al additions promote a significant decrease of x (0.66 to 0.56) bringing VC at the limit of the range of homogeneity range.

The influence of the cooling rate on the stoichiometry of the vanadium carbides depends on the nature of the additions. In the case of non-carbide forming additions, the carbon to metallic elements ratio increases while in the case of carbides forming additions this ratio decreases with cooling rate.

Influence of Additions on the Solid-State Transformations in the Matrix

Two solid-state transformations can be expected in the matrix:

- magnetic transition in high vanadium alloys: according to the results obtained by DTA a magnetic transformation occurred in the matrix of high vanadium alloys.
- cementite formation.

In the case of the ternary alloy FeV_8C_3 austenite can be decomposed progressively to three decomposition morphologies of cementite: proeutectoid grain boundary allotropics, Widmanstätten plates and pearlite lamellae. Microprobe analysis shows evidence of vanadium in allotropic grain boundary cementite, denoted M_3C.

Table 2. Microprobe analysis (in at %) of proeutectoid cementite and matrix of FeV_8C_3M alloys (M = 2%)

M addition	Composition of proeutectoid cementite				Composition of the matrix			
	Fe	V	C	M	Fe	V	C	M
–	73.8	1.9	24.3	–	92.3	0.9	6.1	–
Al	77.4	0.7	21.8	0.08	87–92.0	1.2–2.0	5.2–7.4	1.2–4.0
Cu	75.3	0.7	23.5	1.2	90.5–96	1.2–2.0	0.9–5.2	1.6–2.2
	70.3	3.2	24.8	1.7				
Si					87.8	1.7–3.1	5.5–7.3	4.3
Ni	77.4	0.6	20.3	1.2	91.5	1.3	2.3–6.6	3.7–4.1
Cr					89.0	3.7	4.2	3.1
Mo	75.2	0.35	22.5	1.9	89.5	0.2–1.6	5.8–8.1	0.2–1.2

Deliberate additions in this ternary alloy have resulted in an increase of the matrix content of the element added while, generally the cementite composition does not change (Table 2). Addition of copper resulted in the precipitation of particles in the matrix and in the proeutectoid cementite. This precipitation is very important in the proeutectoid cementite which formed near vanadium carbides. Indeed, copper does not dissolve in VC_x and thus is segregated in the zone surrounding the carbides. Observations and qualitative microanalyses of this type of cementite in Fe-V5%-C3%-Cu4% allowed the identification of large amounts of small precipitates of copper inside the cementite. In alloys with higher vanadium content, the copper precipitation is very difficult to put into evidence due to a smaller size of the precipitates. In the matrix, the precipitates are smaller and coherent as in the Fe-Cu binary system.

This precipitation occurred in the binary alloys Fe-Cu and is accentuated by the presence of carbon. Wasynczuk et al. [4] concluded that copper is partitioned during the growth of the proeutectoid cementite/austenite interphase boundaries. Fourlaris and Baker [15] have also shown that copper precipitates in the cementite lamellae of pearlite in high carbon steels. Khalid and Edmonds [16] have observed copper precipitation in the three morphologies of the cementite phase.

Silicium, aluminium, nickel, chromium and molybdenum are dissolved in the matrix and less in the proeutectoid cementite, if forms (Table 2).

Conclusion

Accurate microanalysis of vanadium carbides formed from the melt in Fe-V-C-M quaternary alloys was performed. The composition of vanadium carbides depends on the nature of the addition. We have tested two types, in the first case those elements which do not form carbides (Si, Al, Cu, Ni). We have noted that all of these were present in the matrix and less in the proeutectoid cementite if it forms and not at all in the vanadium carbides. Silicium, aluminium and nickel are dissolved in these phases, but not copper which precipitates in them.

In the second case, the added elements form carbides and then mixed carbides due to the presence of vanadium. Compositions of phases formed in Fe-V-Mo-C alloys show three zones: vanadium carbides (V, Mo)C_x in the center of the eutectic cell, these carbides kept their regular fibrous morphology; molybdenum carbides (Mo, V)$_2$C in the external part of the cell, and a transition zone with a variable Mo/V ratio. Addition of chromium leads to formation of supplementary carbide having M_7C_3 formula.

References

[1] R. Kesri, M. Durand-Charre, *Mater. Sci. Techn.* **1988**, *4*, 692.
[2] W. Huang, *Z. Metallkde.* **1991**, *82*, 391.
[3] J. F. Smith, *Bulletin of Alloy Phase Diagrams* **1984**, *5*, 184.
[4] V. Raghavan, *Bulletin of Alloy Phase Diagrams* **1984**, *5*, 293.
[5] C. H. de Novion, R. Lorenzelli, P. Costa, *C. R. Acad. Sc. Paris* **1966**, *263*, 775.
[6] J. D. Venables, D. Kahn, R. G. Lye, *Phil. Mag.* **1968**, *18*, 177.
[7] R. Kesri, S. Hamar-Thibault, *Z. Metallkde.* **1989**, *80*, 502.
[8] S. E. Amara, R. Kesri, N. Valignat, S. Hamar-Thibault, *Mikrochim. Acta* **1984**, *114/115*, 157.
[9] G. L. Bastin, H. J. M. Heijligers, *X-Ray Spectrometry* **1986**, *15*, 135.
[10] J. L. Pouchou, F. Pichoir, *J. Microsc. Spectrosc. Electron.* **1986**, *11*, 229.
[11] H. Holleck, *Binäre und Ternäre Carbide und Nitride Übergangsmetalle and ihre Phasenbeziehungen*, Kernforschungszentrum, Karlsruhe, 1981.
[12] A. Massalky, *Binary Alloy Phase Diagrams* **1990**, *12*, 863.
[13] L. Adnane, R. Kesri, S. Hamar-Thibault, *J. All. Comp.* **1992**, *178*, 71.
[14] J. A. Wasynczuk, R. M. Fisher, G. Thomas, *Met. Trans.* **1986**, *17A*, 2163.
[15] G. N. Fourlaris, A. J. Baker, *Met. Trans.* **1986**, *17A*, 204.
[16] F. A. Khalid, D. V. Edmonds, *Electron Microscopy* **1992**, *2*, 275.

Mikrochim. Acta [Suppl.] 13, 217–224 (1996)

Electron Transmission Coefficient for Oblique Angle of Incidence

Michael Andrae*, Peter Klein, Kurt Röhrbacher, and Johann Wernisch

Institute for Applied and Technical Physics, Technical University of Vienna, Wiedner Hauptstrasse 8–10, A-1040 Wien, Austria

Abstract. Starting from a scattering model the transmission coefficient of electrons for oblique angle of incidence is derived. The decrease of the transmitted fraction is described by the backscattering of electrons and the energy loss caused by the decreasing mean free path. The angular- and energy-distribution of electrons is replaced by the most probable angle and mean energy. A comparison with other models and measurements valid for normal incidence is performed. The altered transmission characteristics by tilting the sample are verified by comparison with results from Monte Carlo simulations and results from Neubert and Rogaschewski, which are based on measurements. An analytical expression for the transmission law is obtained.

Key words: transmission coefficient, tilted specimen.

The exact knowledge of the transmission characteristics is important to understand how electrons penetrate into the sample and therefore necessary for EPMA. Unfortunately, hardly any experimental material regarding the transmission coefficient for tilted specimens has been published. Moreover transmission coefficients were only published for Al, Cu, Ag and Au. For normal incidence Cosslett and Thomas [1] made detailed measurements for Cu and Au with various acceleration energies. Young [2] published measurements for Al_2O_3, whereas measurements for Al and Au were performed by Seidel [3].

Transmission coefficients for tilted specimens with EPMA-relevant-energies are available from Oppel [4] and from Neubert and Rogaschewski [5]. These investigations, however, only refer to Al and Au. This is why we used the Monte Carlo program "mcsing" published by Hénoc and Maurice [6] to investigate the variation of the transmission for various atomic numbers and tilt angles.

It is impossible to extend existing models conceived for normal incidence to the application to tilted specimens. In spite of its simple mathematical form, the model of Kanaya and Okayama [7] shows good agreement with the experimental findings for Al, Cu and Au (see Figs. 5–7). Neubert and Rogaschewski [5] adopted the equation

* To whom correspondence should be addressed

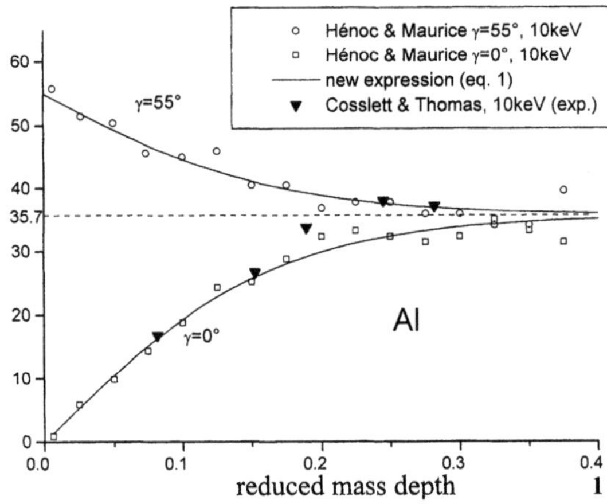

1

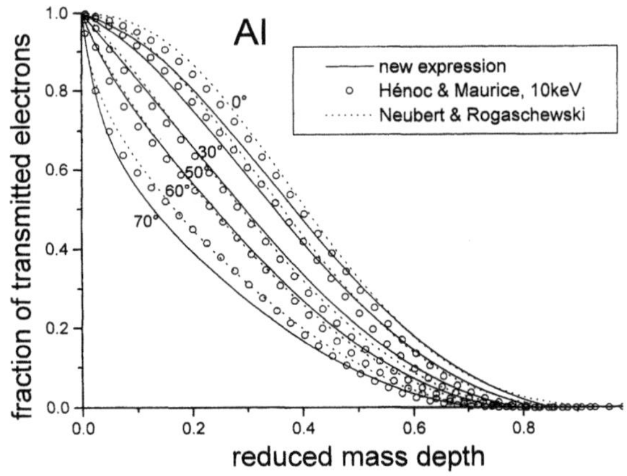

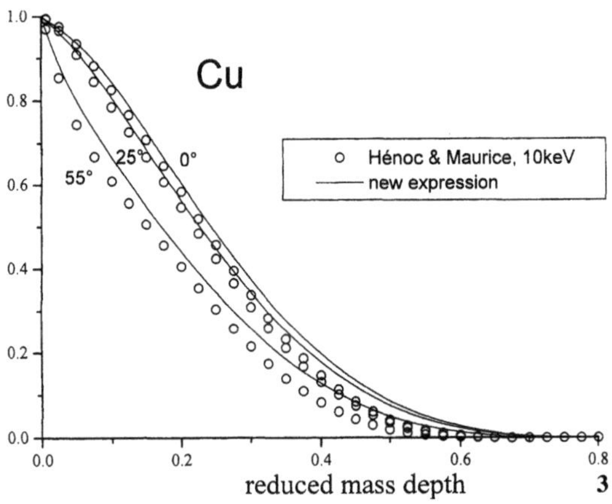

3

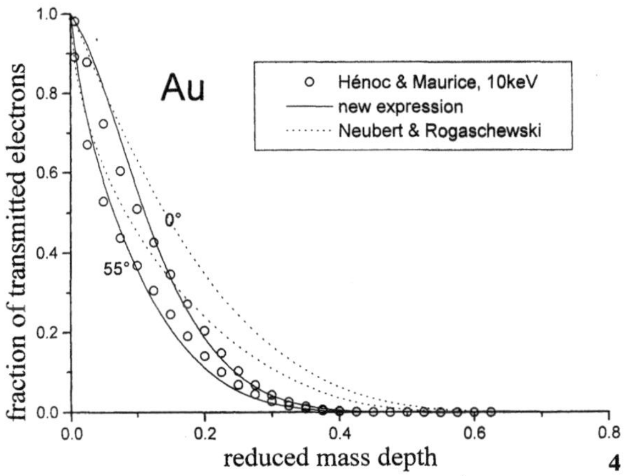

4

Fig. 1. Most probable angle versus reduced mass depth for aluminium with angle of incidence $\gamma = 0°, 55°$

Fig. 2. Fraction of transmitted electrons versus reduced mass depth for aluminium with $\gamma = 0°, 30°, 50°, 60°, 70°$

Fig. 3. Fraction of transmitted electrons versus reduced mass depth for copper with $\gamma = 0°, 25°, 55°$

Fig. 4. Fraction of transmitted electrons versus reduced mass depth for gold $\gamma = 0°, 55°$

of Vyatskin et al. [8, 9] for Al and Au to their experimental results at different angles of incidence and incident energies between 20 and 60 keV. No general formula valid for all atomic numbers and tilt angles is given. The most popular description for the transmission law at zero degrees is the model of Zeller and Ruste [10], which uses a scattering model by taking both the elastic and the inelastic collisions into account. The incident electron beam is assumed to be composed of a direct beam $T_1(\rho z)$ and an indirect beam $I(\rho z)$ representing the fraction of diffused electrons. $I(\rho z)$ originates from $T_1(\rho z)$, the part of electrons which by definition suffered no interference in the sample. We tried to adapt the model for the application to tilted specimens by changing the geometry for the calculation of the total cross sections and the geometry of the incident beam. It is impossible to describe our Monte Carlo results with Zeller-Ruste's model. The division into a direct and an indirect beam, which works at zero degrees, does not seem to be a good description for oblique incidence.

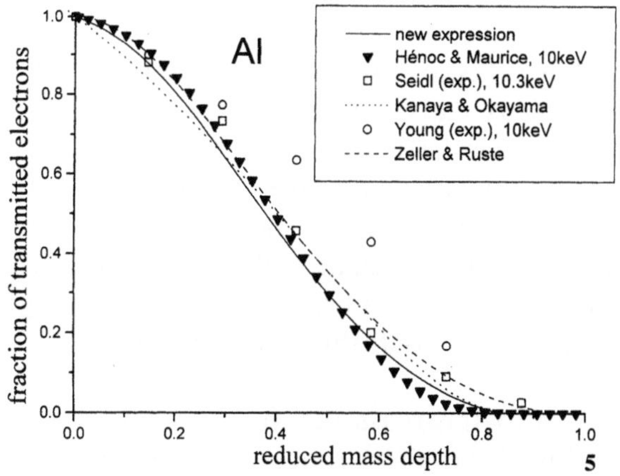

5

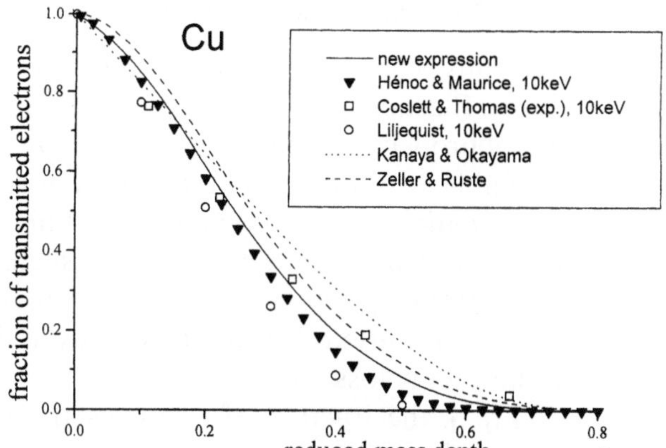

6

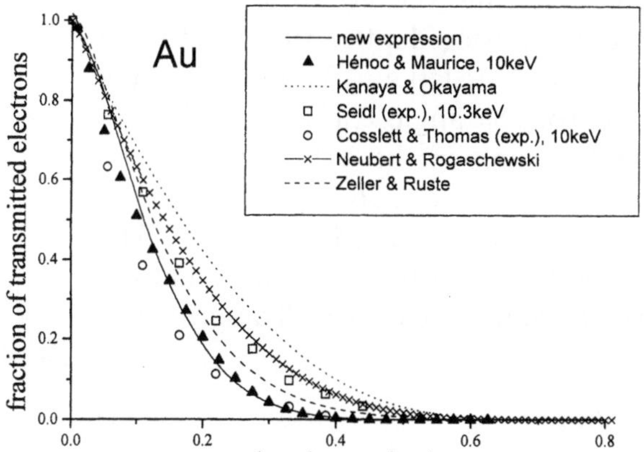

7

Fig. 5. Fraction of transmitted electrons versus reduced mass depth for aluminium at normal incidence
Fig. 6. Fraction of transmitted electrons versus reduced mass depth for copper at normal incidence
Fig. 7. Fraction of transmitted electrons versus reduced mass depth for gold at normal incidence

Several assumptions lead to errors which are compensated by adapting the parameters $\langle\cos\theta\rangle$ and $\langle\cos\theta^{-1}\rangle$ to the experimental results.

$$\langle\cos\theta\rangle = \left(\frac{\rho z_m}{\rho x_m}\right), \text{ where}$$

ρz_m total electron range,
ρx_m prolongated path of electron due to scattering events,
$\langle\cos\theta\rangle$ average $\cos\theta$ – value for all electrons between the sample surface and the electron range ρz_m.

Nevertheless it cannot be explained why according to Zeller and Ruste $\langle\cos(\theta)\rangle^{-1}$ should have a dependence on the atomic number and $\langle\cos(\theta)\rangle$ not.

Therefore a new description of the transmission law seems to be necessary, which will be used for an analysis program for oblique angle of incidence based on a multiple reflection model [11].

Most Probable Angle and Mean Energy

Electrons penetrating into the sample undergo both elastic and inelastic scattering events. Elastic collisions, interactions with the nucleus, cause changes in the direction of propagation between 0° and 180°. The energy loss of an elastic interaction is negligible. In our treatment the Rutherford cross section is used because of its simple mathematical form. The energy loss is mainly caused by inelastic collisions, interactions between electron and electron shell, with only small angular deflections and bremsstrahlung processes. The beam broadens due to numerous scattering acts. The loss of transmitted electrons caused by ionisations and bremsstrahlung corresponds to the energy loss.

$T(\rho z)$, the fraction of electrons transmitted into the mass depth ρz, can be described by the angular and energy distribution. Near the surface the angular distribution is similar to a δ-function and due to scattering events the beam broadens until the distribution reaches a gaussian form at a diffusion depth corresponding to a value for the most probable angle of about 38° (see Cosslett and Thomas [1]). We substituted the angular distribution by the most probable angle. Closer investigation shows that the exact value of the most probable angle in the diffusion depth θ_D depends on the atomic number. We obtained a value of about 36° for Al and found that the maximum of the angular distribution in Au is reached at $\approx 40.5°$. The diffusion angle has a linear dependence on the atomic number. We found that the variation of the most probable angle with the mass depth ρz can be well described by a tanh-law:

$$\theta_{MP} = \gamma + (\theta_D - \gamma)^* \tanh(b^* x) \tag{1}$$

with

$$\theta_D = 34.74611 + 0.07472 \cdot Z$$
$$b = 22.32154 \cdot \sin(0.018957 \cdot Z)$$

where

θ_{MP} is the most probable angle [degrees],

γ – angle of incidence (= deviation from the normal incidence) [degrees],

θ_D – diffusion angle [degrees],

$x = \frac{\rho z}{\rho z_m}$ $x \in [0, 1]$ – reduced mass depth,

ρz_m – total range $\left[\text{g/cm}^2\right]$,

ρz – mass depth $\left[\text{g/cm}^2\right]$,

and Z – atomic number.

This expression, obtained by fitting to Monte Carlo data, shows good agreement with the measurements performed by Cosslett and Thomas [1] but tends to reach a different saturation value. The parameter b in Eq. (1) determines the onset of diffusion. Since Monte Carlo simulations show that the diffusion depth is nearly independent of the angle of incidence γ the parameter b in Eq. (1) only depends on the atomic number. Figure 1 shows a comparison, relative to Al, among the Monte Carlo results for 0° and 55°, our equation and the results of Cosslett and Thomas for 0°.

In our calculation the energy distribution at a certain mass depth ρz is replaced by its mean energy. The average energy is supposed to follow a square law. We replaced Thomson-Whiddington's energy-law

$$E^2(\rho z) = E_0^2 \cdot \left[1 - \frac{\rho z}{\rho z_m} \right]$$

with
E_0 energy of incident electrons,
E energy at depth ρz,
by the following expression:

$$E^2(\rho z) = \begin{cases} E_0^2 \cdot \left[1 - \frac{MPP(\rho z)}{\rho z_m} \right] & MPP(\rho z) \leqslant \rho z_m \\ 0 & MPP(\rho z) > \rho z_m \end{cases} \tag{2}$$

where MPP represents the most probable path of an electron at a mass depth ρz.

$$MPP(\rho z) = \int_0^{\rho z} \frac{d\rho z}{\cos \theta_{MP}(\rho z)}$$

θ_{MP} most probable angle.

Eq. (2) is analogous to the energy-mass depth relation of Thomson and Whiddington for zero degrees, but implies the dependence on the angle of incidence, which was deduced from Monte Carlo simulations. Measurements by Cosslett and Thomas [1] prove the square correlation between mean energy and mass depth. Nevertheless it should be mentioned that the Thomson-Whiddington law runs out of validity at the end of deceleration, which is corrected for by the function $f(W_P)$ (Eq. 6).

Calculation of the Transmission Law

The incident electrons at a mass depth ρz can be scattered back into the semi-space ($90° < \theta < 180°$) on their way through the infinitesimally thick layer $d\rho z$. The number of scattering centres is given by

$$N \cdot d\rho z \cdot \frac{1}{\cos[\theta_{MP}(\rho z)]}$$

N number of atoms per unit mass depth $\left[\frac{\mathrm{cm}^2}{\mathrm{g}}\right]$

The total cross section can be calculated easily by integrating the elastic Rutherford cross section $\sigma(\theta)$ over the semi space:

$$\sigma(\theta) = \frac{e^4 \cdot Z^2}{16 \cdot E^2} \cdot \frac{1}{\sin^4 \frac{\theta}{2}} \tag{3}$$

It was already mentioned above that inelastic scattering events do not contribute to large angle back-scattering and therefore can be neglected for the calculation of the total backscattering cross-section.

$$\sigma_{\mathrm{TOT}} \propto \iint_{\text{semi-space}} \frac{\sin\theta' \cdot d\theta' \cdot d\varphi'}{\sin^4\left[\frac{\theta'}{2}\right]}$$

$$= \int_{\varphi=0}^{2\pi} \int_{\theta=\pi/2}^{\pi} \frac{4 \cdot \sin\theta \cdot d\theta \cdot d\varphi}{[1 - \sin\theta \cdot \sin\theta_{MP} \cdot \cos\varphi + \cos\theta \cdot \cos\theta_{MP}]^2} = \frac{4\pi}{\cos^2\theta_{MP}}$$

ρz, θ, φ spherical coordinates with surface normal as $z-$axis,
$\rho z'$, θ', φ' spherical coordinates with tilted $z-\text{axis}(\theta_{MP})$.

Thus we obtain for the total cross section:

$$\sigma_{\mathrm{TOT}} = \frac{e^4 \cdot Z^2 \cdot \pi}{4 \cdot E^2} \cdot \frac{1}{\cos^2\theta_{MP}} \tag{4}$$

The coordinate ρz is replaced by the reduced coordinate: $x = \rho z / \rho z_m \quad \in [0, 1]$

As the transmission law has nearly the same form for primary energies between 5 and 30 keV, the dependence on E_0 is eliminated. The fraction of electrons scattered back from the thin layer $d\rho z$ into the semi-space is

$$dT = -T(\rho z) \cdot N \cdot \sigma_{\mathrm{TOT}} \cdot \frac{1}{\cos\theta_{MP}} \cdot \rho z_m \cdot dx,$$

$T(\rho z)$ fraction of transmitted electrons.

For ρz_m we applied Zeller and Ruste's [10] range, which is the most probable energy range based on a decceleration law of the Thomson-Whiddington type:

$$\rho z_m = \frac{E_0^2}{2\pi \cdot e^4 \cdot N \cdot Z \cdot \ln\left[\frac{1+\psi^2}{\psi^2}\right]}$$

where N is the number of atoms per unit of mass depth and
$\psi = 0.175$

With Eq. (2) the backscattered fraction is described by

$$dT = -T(x) \cdot \frac{1}{[1 - MPP(x)]} \cdot \frac{Z}{8 \cdot \ln\left[\frac{1+\psi^2}{\psi^2}\right]} \cdot \frac{1}{\cos^3\theta_{MP}} \cdot dx \tag{5}$$

We assumed that all electrons scattered in the direction of the surface do not contribute to the transmission anymore. The probability for a backscattered electron near the surface of being rescattered into the forward direction is rather small since the path to the surface is short and the Rutherford cross section is small for comparatively high energies. At a greater mass depth the error induced by neglecting the fraction of rescattered electrons is partly compensated by the fact that the range of the electrons is small.

The fraction of electrons contributing to the lower part of the energy distribution should be eliminated from $T(\rho z)$ at each mass depth ρz since the mean free path is too small for electrons to penetrate deeper into the sample. The assumption to represent the energy distribution by its mean energy neglects this electron-loss. This fact is considered by the function $f(W_P)$. Moreover $f(W_P)$ considers errors arising from the application of the Thomson-Whiddington law. The fraction of electrons lost at a mass depth $\rho z = x \cdot \rho z_m$ in a layer of thickness dx is equal to

$$dT = -T(x) \cdot f(W_P) \cdot dx$$

The function $f(W_P)$ depends on the atomic number and the relative energy loss $W_P = E(\rho z)/E_0$.

$$f(W_P) = \frac{(2.45102 + 0.04441 \cdot Z)^2}{\left[W_P^2 + \dfrac{1}{1 - W_P}\right]} \tag{6}$$

The final result for the transmission coefficient is obtained as follows

$$T(x) = \exp\left[-\int_0^X N \cdot \sigma_{\mathrm{TOT}}(x) \cdot \frac{1}{\cos\theta_{MP}(x)} \cdot \rho zm \cdot dx - \int_0^X f(W_P) \cdot dx\right] \tag{7}$$

Results

We found that our transmission law is in good agreement with Monte Carlo results. In general our curves have the tendency to predict higher values for the transmission coefficient (Figs. 2–4) than the simulations by Hénoc and Maurice [6]. Nevertheless the correct variation with the tilt angle is derived up to angles of about 60°. In Figs. 2, 4, 7 a comparison is performed with the data-fit of Neubert and Rogaschewski's [5] measurements of Al and Au between 20 and 60 keV. For aluminum the comparison shows fairly good agreement with the applied Monte Carlo simulations. We believe that the discrepancies for gold in Fig. 4 may be explained by the different reduced mass depths which were applied. Neubert and Rogaschewski introduced the Bethe range while we preferred Zeller and Ruste's [10] range.

For normal incidence a comparison of experimental data and theoretical calculations is presented in Figs. 5–7 for aluminium, copper and gold. Zeller and Ruste's transmission law is close to experimental findings. In spite of its simple form, Kanaya and Okayama's [7] expression proves to be a good description of the transmitted fraction at normal incidence. Even experimental data from Seidel [3] and Cosslett and Thomas [1] show discrepancies for Au (Fig. 7). The results deduced from the applied Monte Carlo program and our new expression are always close to the published data, considering the large experimental uncertainties.

Conclusion

An analytical expression for a transmission law for oblique angle of incidence is presented. Our transmission coefficients can be calculated for all atomic numbers. A comparison to the Monte Carlo program of Hénoc and Maurice shows that the correct tendency of the transmission characteristics is reproduced. The comparison of our expression with experimental data and several theoretical findings at normal incidence proves the reliability of our results.

Acknowledgement. We acknowledge the support given by the Austrian Fonds zur Förderung der wissenschaftlichen Forschung (Project N° P10415-PHY).

References

[1] V. E. Cosslett, R. N. Thomas, *Br. J. Appl. Phys.* **1964**, *15*, 883.
[2] J. R. Young, *J. Appl. Phys.* **1956**, *28*, 524.
[3] H. Seidel, *Doctoral Thesis*, University Münster, Germany, 1972, p. 565.
[4] W. Oppel, *Doctoral Thesis*, University of Köln, Germany, 1973.
[5] G. Neubert, S. Rogaschewski, *J. Phys. D: Appl. Phys.* **1984**, 17, 2439.
[6] J. Hénoc, F. Maurice, *Electron Probe Quantitation*, Plenum, New York, 1991, p. 105.
[7] K. Kanaya, S. Okayama, *J. Phys. D: Appl. Phys* **1971**, *5*, 43.
[8] A. Ya. Vyatskin, A. N. Kabanov, K. A. Makarov, V. V. Trunev, *Radiotek, i Elektron.* **1970**, *15*, 565.
[9] A. Ya. Vyatskin, A. N. Pilyankevich, V. V. Trunev, *Fiz. Tverd. Tela* **1964**, *6*, 1563.
[10] C. Zeller, J. Ruste, *Revue de Physique Appliquée* **1964**, *6*, 1563.
[11] M. Andrae, K. Röhrbacher, P. Klein, J. Wernisch, *Mikrochim. Acta* **1996**, [*Suppl.*] *13*, 225.

Mikrochim. Acta [Suppl.] 13, 225–232 (1996)

Depth Distribution Function for Oblique Angle of Incidence

Michael Andrae*, Kurt Röhrbacher, Peter Klein, and Johann Wernisch

Institute for Applied and Technical Physics, Technical University of Vienna, Wiedner Hauptstrasse 8–10, A-1040 Wien, Austria

Abstract. Based on the multiple-reflection model first suggested by V. Lantto a new description of electrons penetrating with an oblique angle of incidence into the sample is reported. In contrast to most of the authors, who are trying to adapt models for normal incidence to the experimental results for tilted samples, our model describes the altered shape of the depth distribution function as a consequence of changed physical quantities due to different geometry. These quantities used for the multi-reflection model are: angular- and energy-distributions of backscattered and transmitted electrons, the backscattering- and the transmission-coefficient. The theoretically derived depth distribution functions are compared with results from Monte Carlo simulations.

Key words: quantitative electron probe microanalysis, tilted specimen, depth distribution function.

Since the beginning of electron-probe micro-analysis (EPMA) launched by Castaing [1] in 1951 a lot of experimental and theoretical work has been done regarding the depth distribution function of electrons impinging vertically on the surface. In the last decades, however, the demand for accurate correction procedures also for tilted specimens has grown. Packwood et al. [2] published modifications of the basic parameters used for their gaussian-model [3] for tilted samples. This model is only valid up to a tilt angle $\gamma = 30°$. The surface ionisation and the initial slope of the depth distribution function are assumed to be independent of γ. In addition the simplified-PAP-model (XPP) of Pouchou and Pichoir [4] was adapted to analysis at oblique angle of incidence by modification of several parameters [5]. These authors stated that measurements with tilted specimens are important in order to increase the emerging intensity of strongly absorbed lines. Merlet published a new procedure for electron probe quantification at variable angles of incidence [6]. Since experimental studies of the alteration of the depth distribution function by tilting the specimen are rare, Monte Carlo simulations must be used for a comparison with our depth distribution functions. Monte Carlo programs based on a single scattering

* To whom correspondence should be addressed

model which are accurate for normal incidence can be expected to be reliable also for oblique incidence, since only the geometry is changed. For a comparison with our new model we applied the Monte Carlo program 'mcsing' published by Hénoc and Maurice [7]. As mentioned above, measurements with oblique incidence are useful for the analysis of strongly absorbed lines. By tilting the sample the average ionisation depth is smaller and thus, for a given take-off angle, low energy radiation is less absorbed on its way to the detector resulting in a larger peak in the spectrum. For the thickness determination of a thin layer on a substrate, measurements at non normal incidence provide a higher intensity of the radiation generated in the layer. There will be more possibilities to compare theoretical predictions with the experiment, if depth distribution functions are also valid for tilted specimens. This increase of information can be used to simultaneously determine with high accuracy, for instance, the concentration in a sample and the thickness of a thin film.

To extend existing models for normal incidence to oblique incidence does not simply mean that some parameters have to be changed. For normal incidence, depth distribution functions in general have the same shape for all atomic numbers, with information from measurements with the tracer-technique and Monte Carlo simulations leading to parabolic [8], gaussian [3] or exponential models [4]. Finding a general shape for $\Phi(\rho z)$ for oblique incidence is not as simple. Depth distribution functions obtained from Monte Carlo simulations for high tilt angles have no maximum value in the sample. In general depth distribution functions for $\gamma \neq 0$ are

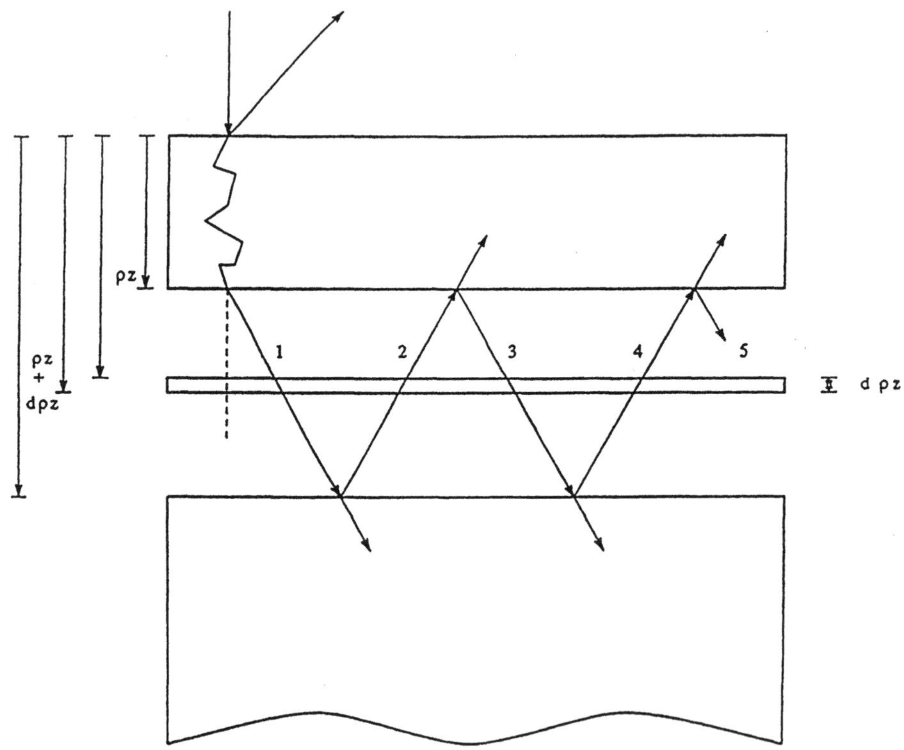

Fig. 1. Principle of the calculation of the depth distribution function. The intensity of the radiation produced in the infinitesimally thin layer at a mass depth ρz is calculated taking transmitted (*1*) and backscattered (*2–5*) electrons into account

quite different from those for normal incidence. That is why we used a multiple reflection model first presented by Cosslett and Thomas [9] which uses only physical quantities like angular- and energy-distributions as well as the transmission- and backscattering-coefficient. The shape of the obtained depth distributions results from the model applied. Therefore several physical quantities for tilted specimen had to be derived.

Theory

The model applied in this paper is based on the multiple reflection theory first published by Cosslett and Thomas [9] and later refined by Lantto [10] and August [11]. In spite of some simplifications, the calculated depth distribution functions $\Phi(\rho z)$ of August agree well with the experimental findings [12]. To determine the value of $\Phi(\rho z)$ at a certain mass depth ρz the sample is virtually cut along a plane parallel to the surface at the considered depth. The ionizations in an infinitesimally thin layer in the virtual gap (see Fig. 1) are caused by transmitted electrons (1. trajectory) and backscattered electrons (2. – 5. trajectory). The index of the trajectory describes how often an electron penetrates the infinitesimal layer. The intensity above the fifth trajectory is too low to be essential for the calculation of $\Phi(\rho z)$. The value of the depth distribution function at a mass depth ρz can be written as follows:

$$\Phi(\rho z) = \frac{\sum\limits_{k=1}^{\infty} \int_1^{U_0} \int_0^{\pi/2} N(U,\theta,\rho z)_k \frac{1}{\cos\theta} Q(U) d\theta\, dU}{\int_1^{U_0} \int_0^{\pi/2} \delta(\theta-\gamma)\delta(U-U_0) \frac{1}{\cos\theta} Q(U) d\theta\, dU} \approx \frac{\sum\limits_{k=1}^{5} \int_1^{U_0} \int_0^{\pi/2} N(U,\theta,\rho z)_k \frac{1}{\cos\theta} Q(U) d\theta\, dU}{Q(U_0)/\cos\gamma} \tag{1}$$

for $k = 1 \rightarrow N(U,\theta,\rho z)_k = \dfrac{d^2\tau}{d\theta\, dU}$,

for $k = 2 \ldots\ldots \infty \rightarrow N(U,\theta,\rho z)_k = \dfrac{d^2\eta}{d\theta\, dU}$,

with:

$\dfrac{d^2\tau}{d\theta\, dU}, \dfrac{d^2\eta}{d\theta\, dU}$	double differential angular and energy distribution of transmitted- and backscattered electrons,
$U = \dfrac{E}{E_c}$	overvoltage ratio,
E_c	critical energy of observed line,
U_0	overvoltage ratio with $E = E_0$ (initial energy),
$Q(U)$	ionization cross section,
γ	deviation from the normal incidence.

The denominator in (1) is proportional to the intensity produced in a self-supporting reference layer equivalent to the infinitesimal layer tilted by the same angle γ. The function $N(U,\theta,\rho z)_k$ has to be known for every mass depth and trajectory.

For transmitted electrons Monte Carlo simulations show that the angular dependence of energy distributions is small. Therefore $\frac{d^2\tau}{d\theta\, dU}$ in (1) is replaced by the total angular $\frac{d\tau}{d\theta}$ and energy $\frac{d\tau}{dU}$ distribution.

$$\int_1^{U_0}\int_0^{\pi/2}\frac{d^2\tau}{d\theta\, dU}\frac{1}{\cos\theta}Q(U)d\theta\, dU \approx \frac{1}{\tau}\int_1^{U_2}\frac{d\tau}{dU}Q(U)dU\cdot\int_0^{\pi/2}\frac{d\tau}{d\theta}\frac{1}{\cos\theta}d\theta$$

$$\frac{d^2\tau}{d\theta\, dU}\approx\frac{1}{\tau}\frac{d\tau}{d\theta}\frac{d\tau}{dU}\qquad\frac{d\tau}{d\theta}=\int_1^{U_0}\frac{d^2\tau}{d\theta\, dU}dU\qquad\frac{d\tau}{dU}=\int_0^{\pi/2}\frac{d^2\tau}{d\theta\, dU}d\theta\tag{2}$$

τ transmission coefficient [12].

The angular distribution is by definition a δ-function at the surface and reaches a saturation function at the diffusion depth:

$$\frac{d\tau}{d\theta}\propto\sin\theta\cdot\cos\theta\cdot\exp\left[\frac{-(\theta-\mu)^2}{\sigma^2}\right]\tag{3}$$

with:

$\sigma^2 = 5500\cdot[\tanh(a\cdot x)]^b$,

a, b functions depending on atomic number,

$x = \frac{\rho z}{\rho z_m}$ reduced mass depth,

ρz_m range.

The parameter μ can be calculated from the most probable angle θ_{MP} [13]:

$$\left(\frac{d\tau}{d\theta}\right)'_{\theta MP}=0\rightarrow\mu=\theta_{MP}-\left(\frac{1/2-\sin^2\theta_{MP}}{\sin\theta_{MP}\cdot\cos\theta_{MP}}\right)\cdot\sigma^2\tag{4}$$

A new expression for the total energy distribution of transmitted electrons was deduced from measurements of Reimer et al. [4], Cosslett and Thomas [15], Shimizu et al. [16] and Fitting [17].

$$\frac{d\tau}{dW}=\begin{cases}A\cdot W^B & 0<W<W_P\\ A\frac{W_S-W}{W_S-W_P}\cdot W_P^B & W_P<W<W_S\\ 0 & W_S<W<1\\ & W=\frac{U}{U_0}\end{cases}\tag{5}$$

The most probable energy W_P is given by:

$$W_P = 1-(0.81921-0.00598\cdot Z)\cdot X^{1.49396\cdot f(\gamma)}\tag{6}$$

$f(\gamma)=[\cos\gamma]^{a\cdot\gamma^2+b\cdot\gamma+c}$

a, b, c functions depending on atomic number,

γ tilt angle in degrees

W_S is the energy where the energy distribution is zero [18]:

$$W_S = 1-0.9\cdot(1-W_P)\exp[-0.0582\cdot Z]\tag{7}$$

The exponent B in Eq. (5) is the parameter which influences the shape of the left part of the curve and is given by:

$$B = \frac{\log\left(\frac{1}{2}\right)}{\log\left(\frac{W_S}{2 \cdot W_P} + \frac{1}{2} - \frac{\mathrm{FWHM}}{W_P}\right)} \tag{8}$$

$\mathrm{FWHM} = \beta_1 \cdot \tanh[\beta_2(1 - W_P)]$

FWHM full width at half maximum

β_1, β_2 functions depending on atomic number

The parameter A follows from the normalisation of $\frac{d\tau}{dU}$:

$$\int_1^{U_0} \frac{d\tau}{dU} \cdot dU = \tau \rightarrow A = \tau \cdot \left[W_P^B \cdot \frac{E_0}{E_c} \left\{ \frac{W_P}{B+1} + \frac{W_S - W_P}{2} \right\} \right]^{-1} \tag{9}$$

If electrons are backscattered (trajectories 2–5), the angular dependence of energy distributions, especially for light elements, cannot be neglected. For a certain angle of incidence α and initial energy E_i the integral in (1) for k > 1 is calculated as follows:

$$\int_1^{U_0} \int_0^{\pi/2} \left(\frac{d^2\eta}{d\theta\, dU}\right)_{\alpha,E_i} \frac{1}{\cos\theta} Q(U) d\theta\, dU = \int_0^{\pi/2} \left(\frac{d\eta}{d\theta}\right)_{\alpha,E_i} \frac{1}{\cos\theta} \int_1^{U_0} \left(\frac{d\eta}{dU}\right)_{\alpha,E_i,\theta} \cdot Q(U) \cdot dU \cdot d\theta \tag{10}$$

$\left(\frac{d\eta}{d\theta}\right)_{\alpha,E_i} = \int_1^{U_0} \left(\frac{d^2\eta}{d\theta\, dU}\right)_{\alpha,E_i} \cdot dU$ total angular distribution,

$\left(\frac{d\eta}{dU}\right)_{\alpha,E_i,\theta}$ energy distribution in direction θ, normalized with respect to unity.

The total angular distribution is taken from Klein et al. [19] and the angular dependent energy distributions from Andrae et al. [20]. Figure 2 shows the two

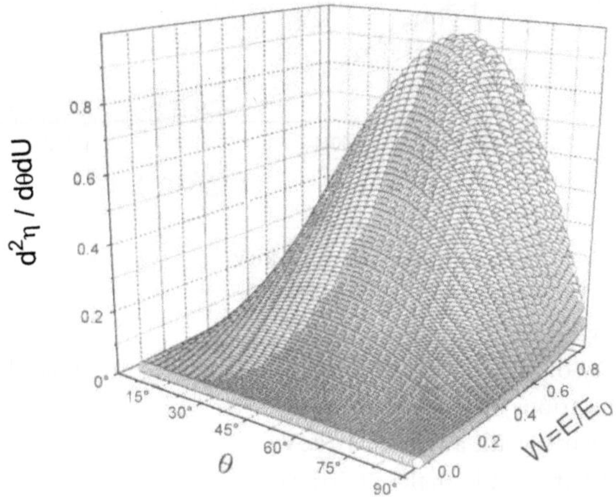

Fig. 2. Double differential angular and energy dependent backscattering coefficient of aluminium with an initial energy $E_i = 10$ keV and an angle of incidence $\alpha = 35°$

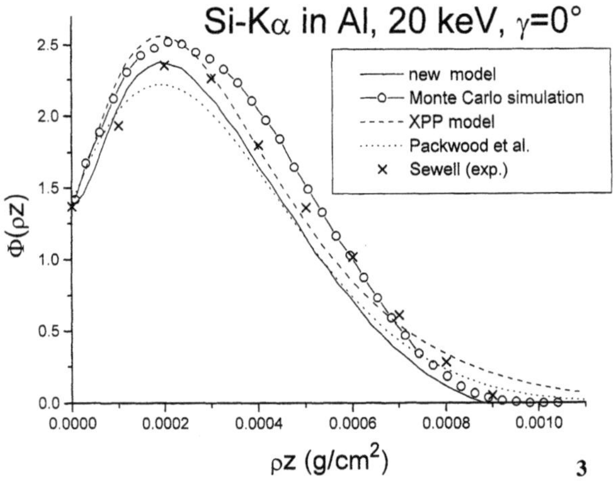

3

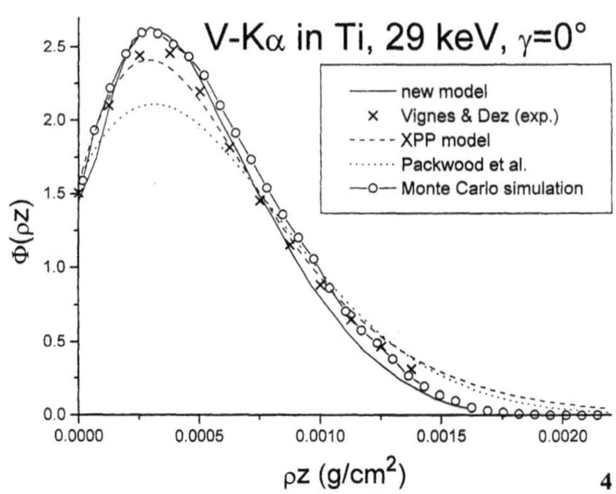

4

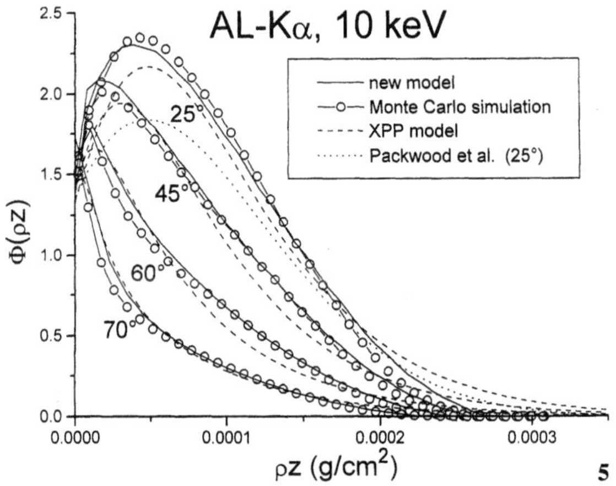

5

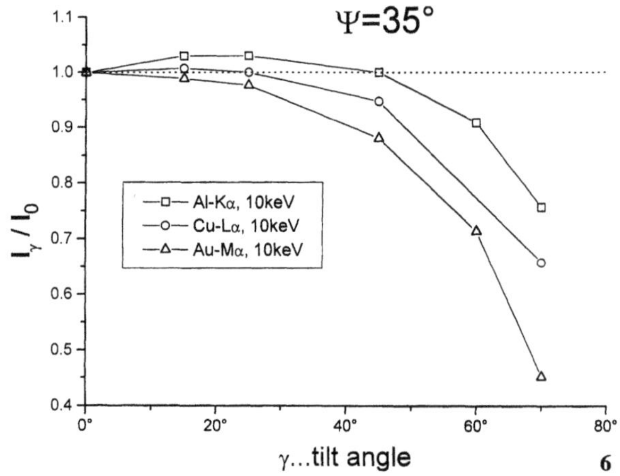

6

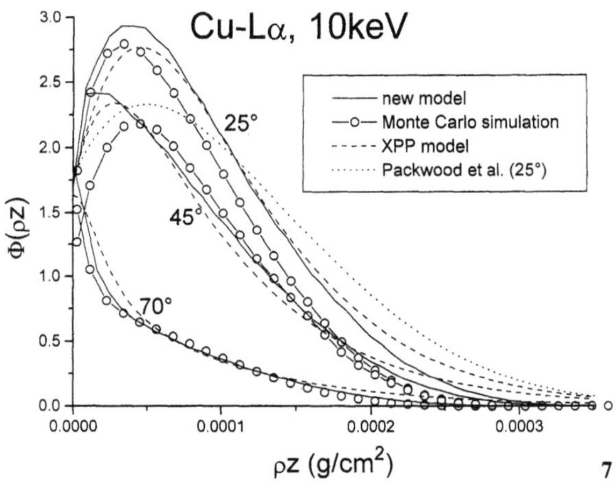

7

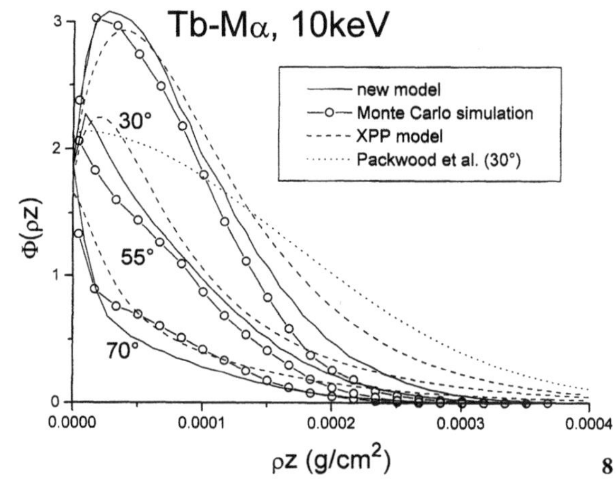

8

Fig. 3. Depth distribution function of Si-Kα in Al for normal incidence at 20 keV

Fig. 4. Depth distribution function of V-Kα in Ti for normal incidence at 29 keV

Fig. 5. Depth distribution functions of Al-Kα at various tilt angles at 10 keV

Fig. 6. Influence of the tilt angle γ on the intensity of radiation I_γ leaving the probe with an take-off angle $\Psi = 35°$ calculated with the new model for various atomic numbers

Fig. 7. Depth distribution functions of Cu-Lα at various tilt angles at 10 keV

Fig. 8. Depth distribution functions of Tb-Mα at various tilt angles at 10 keV dimensional angular- and energy-dependent plane $(d^2\eta/d\theta\,dU)_{\alpha,E_i}$ of backscattered electrons of aluminum with an initial energy $E_i = 10\,\mathrm{keV}$ and an angle of incidence $\alpha = 35°$. Integral (10) must be calculated for every energy and angle of the last trajectory; a correlation has to be performed. In a first approximation only the angular distribution of the last trajectory is used for the correlation. Since the dependence of the total angular distribution of backscattered electrons on the initial energy is weak, the error induced is small

$$\int_0^{\pi/2} \frac{d\tau}{d\alpha} \cdot \int_0^{\pi/2} \left(\frac{d\eta}{d\theta}\right)_{\alpha,E_i} \frac{1}{\cos\theta} \int_1^{U_0} \left(\frac{d\eta}{dU}\right)_{\alpha,E_i,\theta} \cdot Q(U) \cdot dU \cdot d\theta \cdot d\alpha \qquad (11)$$

By analogy higher trajectories are calculated from the angular distribution of the last trajectory. Stepping through the sample the whole depth distribution function is obtained.

Results

Since measurements of depth distribution functions of tilted specimens are rare, we used the Monte Carlo program published by Hénoc and Maurice [7] for comparison with our new model. The quality of Monte Carlo results for tilted specimens can be tested by comparing depth distribution functions obtained for normal incidence with experimentally derived functions and other models. Since only the geometry in the single scattering Monte Carlo program is changing for tilted specimens, we can expect results reliable for $\gamma = 0°$ to be reliable also for $\gamma \neq 0°$.

Figures 3 and 4 show a comparison of depth distribution functions calculated with our new model, the XPP-model of Pouchou and Pichoir [4], the gaussian model of Packwood et al. [3] and the already mentioned Monte Carlo program with experimental results from Sewell et al. [21] for a Si-tracer (Kα) in Al at 20 keV and Vignes and Dez [22] for a V-tracer (Kα) in Ti at 29 keV. In both cases all the models mentioned except that of Packwood et al. – which has the general tendency to predict too small maximum values – show comparable results. The alteration of the depth distribution function by tilting the sample is shown for Kα-radiation of aluminum in Fig. 5. The maximum of the function is shifted to the surface with increasing tilt angle until there is no maximum value in the sample. The primary intensity is proportional to the area of $\Phi(\rho z)$ divided by $\cos(\gamma)$. In Figs. 5, 7 and 8 the area under the depth distribution function is always smaller for a tilted sample. This can be understood from the definition of the depth distribution function, which refers to a self supporting layer tilted by the same angle γ (see Eq. (1)). It should not be concluded from Fig. 5 that the signal emerging from a tilted specimen is always smaller than at normal incidence. On the contrary, for small tilt angle ($\gamma < 15°$) especially for light elements the measured signal increases slightly due to the fact that the average ionisation depth decreases and absorption is weaker. For higher angles of incidence the count-rate is smaller than for electrons impinging normally as the backscattering coefficient increases. Figure 6 illustrates the influence of the count-rate on the tilt angle of several atomic number (13, 29, 79) with a take-off angle $\Psi = 35°$.

In Figs. 7 and 8 the reliability of the new model is tested for copper-Lα and terbium-Mα. A comparison with the model of Packwood et al. [2] for small angles γ ($<30°$) and of Pouchou and Pichoir [5] for tilted specimens is performed. The results of our new model are reliable up to $\gamma = 70°$, which is sufficient for analytical applications since the signal decreases markedly for higher tilt angles (see Fig. 6). For higher atomic numbers the maximum of the derived functions is deeper in the sample than our Monte Carlo program predicts. It should be mentioned that our model is based exclusively on basic parameters which were derived first and then implemented into our new model. We did not manipulate our depth distribution functions to obtain a better correspondence with Monte Carlo results.

Conclusion

A new $\Phi(\rho z)$ model for oblique angle of incidence is presented. Starting from a multiple reflection model depth distribution functions can be calculated from angular and energy distributions. A comparison with Monte Carlo simulations measurements and other models shows that our model is reliable up to a tilt angle of about 70°.

Acknowledgement. We acknowledge the support given by the Austrian "Fonds zur Förderung der wissenschaftlichen Forschung" (Project N° P10415-PHY).

References

[1] R. Castaing, *Thesis*, Univ. Paris, 1951.
[2] R. Packwood, C. Parker, V. Moore, Microbeam Analysis, (D. E. Newbury, ed.), p. 258.
[3] R. Packwood, G. Remond, J. D. Brown, *Proc. 11th ICXOM*, London, Ont., 1987, p. 274.
[4] J. L. Pouchou and M. A. Pichoir, Microbeam Analysis, (D. E. Newbury, ed.), 1988, p. 315.
[5] J. L. Pouchou, M. A. Pichoir, D. Boivin, *Proc. of the 12th Int. Con. on X-Ray Optics and Microanalysis*, Krakow, 1989, p. 52.
[6] C. Merlet, *Microbeam Analysis*, 1994, p. 213.
[7] J. Hénoc, F. Maurice, *Electron Probe Quantitation*, Plenum, New York, 1991, p. 105.
[8] J. L. Pouchou, M. A. Pichoir, *Electron Probe Quantitation*, Plenum, New York, 1991, p. 31.
[9] V. E. Coslett, R. N. Thomas, *Br. J. Appl. Phys.* **1965**, *16*, 779.
[10] V. Lantto, *J. Phys. D* **1979**, *12*, 1181.
[11] H.-J. August, J. Wernisch, *X-Ray Spectrom.* **1991**, *20*, 131.
[12] H.-J. August, J. Wernisch, *X-Ray Spectrom.* **1991**, *20*, 141.
[13] M. Andrae, P. Klein, K. Röhrbacher, J. Wernisch, *Mikrochim. Acta* **1996**, *[Suppl.] 13*, 217.
[14] L. Reimer, K. Brockmann, U. Rhein, *J. Phys. D* (*Appl. Phys*), **1978**, *11*, 2151.
[15] V. E. Cosslett, R. N. Thomas, *Br. J. Appl. Phys.* **1964**, *15*, 883.
[16] R. Shimizu, Y. Kataoka, T. Matsukawa, T. Ikuta, K. Murata, H. Hashimoto, *J. Phys. D.* (*Appl. Phys.*) **1975**, *8*, 820.
[17] H.-J. Fitting, *Phys. Stat. Sol.* **1974**, *A 26*, 525.
[18] H.-J. August, J. Wernisch, *Ultramicroscopy*, **1990**, *32*, 113.
[19] P. Klein, M. Andrae, K. Röhrbacher, J. Wernisch, *Scanning* **1996**, *18*(*6*), in press.
[20] M. Andrae, K. Röhrbacher, P. Klein, J. Wernisch, *Scanning* **1996**, *18*(*6*), in press.
[21] D. A. Sewell, G. Love, V. D. Scott, *J. Phys. D* **1985**, *18*, 1233.
[22] A. Vignes, G. Dez, *J. Phys. D* **1968**, *1*, 1309.

Mikrochim. Acta [Suppl.] 13, 233–240 (1996)

Simulation of EDS Spectra Using X-RES Software

Mimoun Aouine*, Claude Esnouf, and Thierry Epicier

Groupe d'Etudes de Métallurgie Physique et de Physique des Matériaux, Institut National des Sciences Appliquées-Bât. 502, 69621 Villeurbanne Cedex, France

Abstract. This work is related to the general framework of activities concerning microanalysis in electron microscopy. Especially, X-ray spectroscopy is a very powerful tool used by physicists and metallurgists. Its application field is commonly the study of new materials; since they are more and more complex, their development requires a detailed microstructural characterization. Our work brings a contribution to the methodology of analyzing X-ray spectra acquired in a transmission electron microscope. We have proposed a method which allows complex cases, frequently encountered during microscopic observations of real materials, to be evaluated. Such cases define a common class of problems, that is the very localized chemical analysis near an interface, or, more generally, chemical segregations. The problem has been approached by means of simulation of the X-ray emission, coupled with a mathematical deconvolution method of spreading effects of the electron beam. A specific software program has been developed on PC-type computers. The use of the Visual-Basic tool kit has allowed a friendly user interface to be written. The method has been applied to several cases of microanalysis of complex microstructures [1]; a test case will be presented in this contribution. All these applications show the interest of our approach, and point out the advantages offered by the software, which might thus be of a great help for microscopists.

Key words: EDS microanalysis, Monte Carlo simulation, high spatial resolution.

The X-ray spectroscopy applied on thin specimen offers the possibility to analyse chemically a very small volume of matter. Hence, this method, together with EELS and atomic probe techniques, is considered as one of the best microanalysis methods, in terms of high spatial resolution. However, the choice of experimental conditions and the interpretation of results require great care and rigour in the case of X-ray microanalysis. Owing to the smallness of the analysed volume, the X-ray emission remains weak, and the acquired spectrum can be disturbed by parasitic

* To whom correspondence should be addressed

effects, or by the background signal. Nevertheless, this technique approaches an absolute measurement of the composition of thin homogeneous foils when X-ray absorption and fluorescence can be neglected. In the case of non-homogeneous specimens (for example, when it is desired to determine the concentration profile of some element near an interface) several difficulties appear. On the one hand, the probe width, and its broadening in the sample, may result in the mixing of separate signals, produced from each side of the interface. On the other hand, the quantification software provided by constructors is not always well adapted to complex foil geometries.

Therefore, we have developed a methodology for the exploitation of EDS spectra, with a dedicated attention to the case of interfaces. Several programs have then been written, based on the Monte Carlo simulation of the X-ray emission, and taking into account the local physico-chemical conditions. Coupled with a numerical deconvolution, this methodology aids in the resolution of many problems met by users. Then, it offers possibilities for the acquisition of information at a scale finer than that normally fixed by the beam size.

Physical Processes: Simulation by the Monte Carlo Method

This work focuses on the spatial resolution aspects in the field of EDS microanalysis technique using an TEM/STEM machine. The limitation of the resolution is fixed by the size of the incident electron probe and its broadening in the sample, which depends on the energy of incident electrons, the thickness of the sample, and the nature and distribution of elements in the material.

In the case of an heterogeneous sample, the exact knowledge of the X-ray emission distribution passes through the modelling of the generation of the X-ray and the simulation of trajectories [2–3]. The modelling has to take into account electron exchanges across the interface, and has to adapt step by step the variations of the chemical composition and of the related absorption effects. In the present developments, we have only considered the elastic diffusion for the description of the electron trajectories. Indeed, the distribution of elastic events, as described by the Rutherford cross-section and based on a Coulombian potential between the electron and the positive charge of the atomic nucleus, involves high angular deflections (possibly up to 180°), compared to inelastic interactions, which produce small deflections (a few degrees). Consequently, the elastic interaction dominates the spatial distribution of electrons. Moreover, the description of inelastic events would require a different statistical approach; in practice, an inelastic interaction is produced for each elastic interaction event, as postulated within the Bethe's laws [4]: that is, trajectories will be calculated assuming continuous energy losses.

At last, it must be mentioned that the procedure described above treats the material as an amorphous agglomeration of atoms, which means that crystallographic aspects cannot be considered here; thus, to ensure the validity of the method, it is necessary to avoid the presence of any strong diffracted wave.

Figure 1 represents the flow chart of the complete simulation of an electron trajectory, as well as of the X-ray emission. In a first step, one has to define parameters related to the nature of the sample, such as the chemical composition and the thickness(es) of the foil. Then, the energy of the primary electron, the angle of

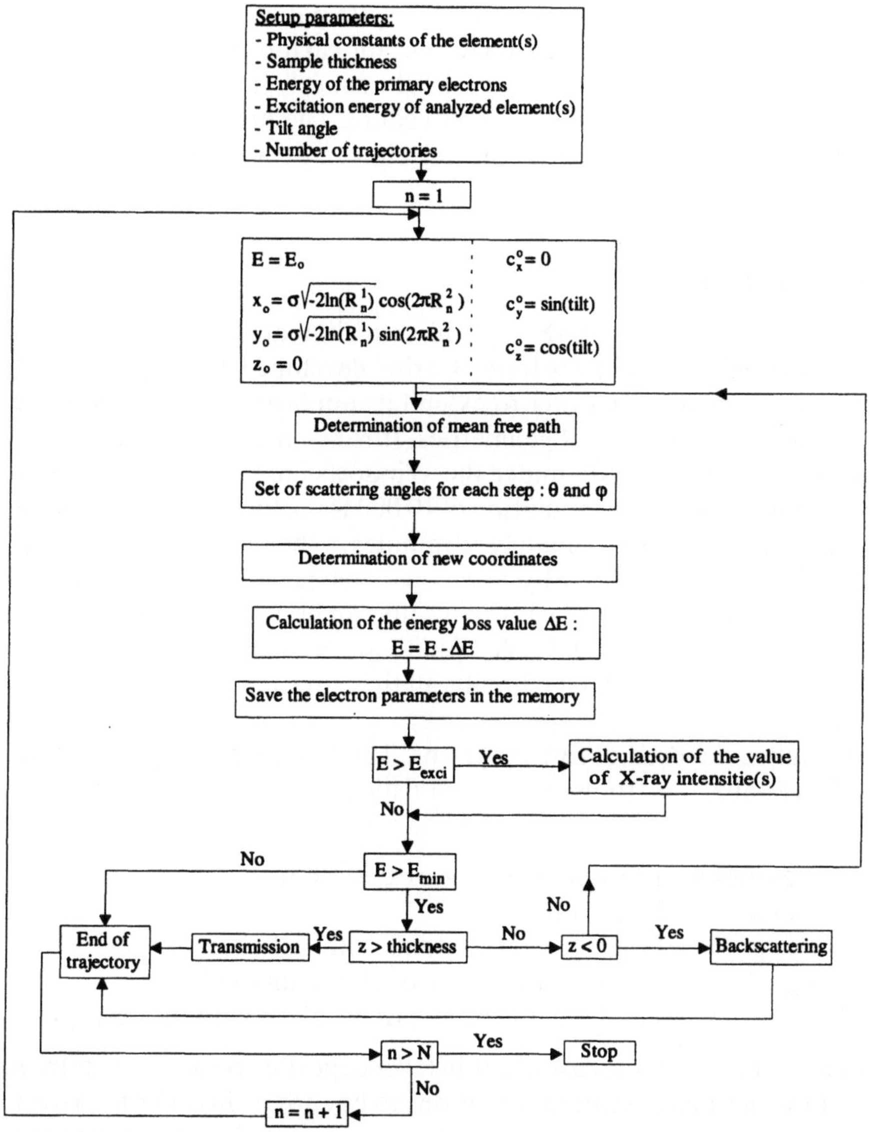

Fig. 1. Schematic flow chart of a simplified treatment of the electron trajectory

impact, the size of the probe* and the number of trajectories are fixed. A single electron of energy E_0 arrives at the sample surface in (x_0, y_0, z_0), along a direction defined by its director cosines. A random number R_n is generated for the determination of the length of the step s_1. Two other random numbers are used to fix the diffusion and azimuthal angles. After the first step, the new coordinates of the electron and its energy E_1 are stored. If the coordinate z_1 reached in depth is included in the interval $[0, t]$ (where t is the thickness of the sample), conditions that govern the X-ray emission are examined. If the electron energy is greater than one of the

* Here, the probe size is the mean square deviation σ; the electron distribution in the beam has been assumed to be of a gaussian from: $I(r) \propto \exp(-r^2/2\sigma^2)$

shell levels (all thresholds K and L of elements are systematically considered), one determines the X-ray emission rate of each element for each threshold. Values are saved and injected in a procedure that displays simultaneously the photon spectrum. The next step consists in repeating the same procedure. The processing of a single trajectory is ended when the electron exits the foil, either by transmission ($z_n > t$) or by backscattering ($z_n < 0$).

Presentation of the Software

The present software, named X-RES (X-ray emission simulation), is based on the SS_MC programs (single scattering Monte Carlo) developed by Joy [5], with numerous further developments in order to extend its application to real cases, i.e. complex microstructures and thin foil geometries. It runs on PC-type machines (a 486-DX computer is recommended), under the Windows environment (Microsoft Visual Basic software). Typically, a simulation of 10,000 trajectories needs an hour in the case of a bulk sample, and approximately 10 minutes in the case of a thin foil, with a 486-DX2 66 MHz machine. The physical constants have been preloaded in Windows formatted files (extension .INI). The results can be saved as .BMP files for the graphical displays, and text (.TXT) files (that can be imported by Microsoft EXCEL) for the various calculated data: finally, all these output files can be directly sent to a printer.

X-RES is actually composed of five programs. The first three programs correspond to simple visualization tools and to statistical parameter measurements such as:

- the electron trajectories distribution in a bulk or thin sample,
- the rate of the backscattered electrons,
- the X-ray intensity of K and/or L-levels for atoms within the target,
- the electron energy distribution transmitted through the thin foil,
- the value of the k-factors [6] and the functions $\varphi(\rho z)$ with or without absorption.

A fourth program processes the case of a thin layer deposited on a substrate (bulk or thin foil). It allows optimal operating conditions to be chosen, in order to provide the chemical analysis of the deposited layer; it also provides a help in the determination of its chemical composition, if the thickness is known, and, reciprocally, allows its thickness measurement if the composition is known.

The last and biggest program treats the case of complex interfaces, and is described in the next section.

Exploitation of X-Ray Spectra Acquired Near Interfaces

Procedure

The analysis of an interface between two compounds M ($A_xB_yC_zD_tE_u$ – up to five chemical elements can be included in any material that can be handled with X-RES –) and M' ($A'_{x'}B'_{y'}C'_{z'}D'_{t'}E'_{u}$) generally consists in the determination of the concentration gradient of one or several chemical element(s) $X(Y, Z, \ldots)$ ($X - Y, Z, \ldots -$ = any of the $A/B/C/D/E$ or $A'/B'/C'/D'/E'$ elements present) near and/or through the interface M/M' by the STEM/EDS technique. What makes the interpretation tricky is that the experimental X-ray profile partially extends into each compound M and M' across the interface, even if no interdiffusion occurs through the interface. This 'spreading' of the exact concentration profile is due two effects: the electron probe profile, and electron/matter interaction which causes a beam spreading into the foil.

Simulation Procedure

Generally, the above treatment consists in the exploitation of 'linescan' data. The principle of the Monte Carlo simulation method is to calculate a theoretical profile from a real profile, and to compare the result with the experimental acquisition through an iterative 'trial-and-error' procedure (note that the opposite way is not possible). When possible, this approach is made easier if data used for the simulation has previously been refined by a mathematical deconvolution of the experimental profile from the effects of the probe size and the transverse spreading of the beam in the sample* [7]. This further numerical step, developed in [1], is not presently included in X-RES.

This procedure has been applied to several experimental cases [1]. The methodology is presented in Fig. 2, and can be summarized as follows:

- (i) a 'seed' concentration profile $C_0(x)$ is deduced from the experimental data, corresponding to the 'linescan' acquisition of X-ray photons across the interface.
- (ii) when applicable, a mathematical deconvolution provides a refined profile $C_1(x)$, which, at this step, constitutes an approach to the final solution.
- (iii) $C_1(x)$ (or $C_0(x)$, deduced from $C_0(x)$ assuming an arbitrary compression along x), is injected in a first Monte Carlo simulation.
- (iv) the new result is the simulated profile $C(x)$ which is compared to the experimental profile $C_0(x)$. If the agreement is judged insufficient, the step (iii) and (iv) are iteratively repeated, taking into account refinements of the structural parameters, such as the densities and the local concentrations.

Further inputs into the Monte Carlo simulation are of course the values of the probe size and the thickness of the sample, which must then be measured or estimated by any suitable method.

The SiC/Al Interface: a Test Example

The SiC/Al interface is an interesting test case, because no interdiffusion of any element is expected. The studied material is a metallic matrix composite, made of an aluminum matrix reinforced by SiC whiskers exhibiting an hexagonal section. In this case, the electron beam can be easily oriented parallely to the interface plane from a diffraction/tilting work: the axis of the fibers is aligned along the **c**-axis of the

* The formula used is that given by P. Doig: $b(t) = 2.35(\sqrt{(2\sigma^2 + \beta t^3)/2} - \sigma)$. The Monte Carlo simulations fit well such an evolution and confirm the expression for β [7].

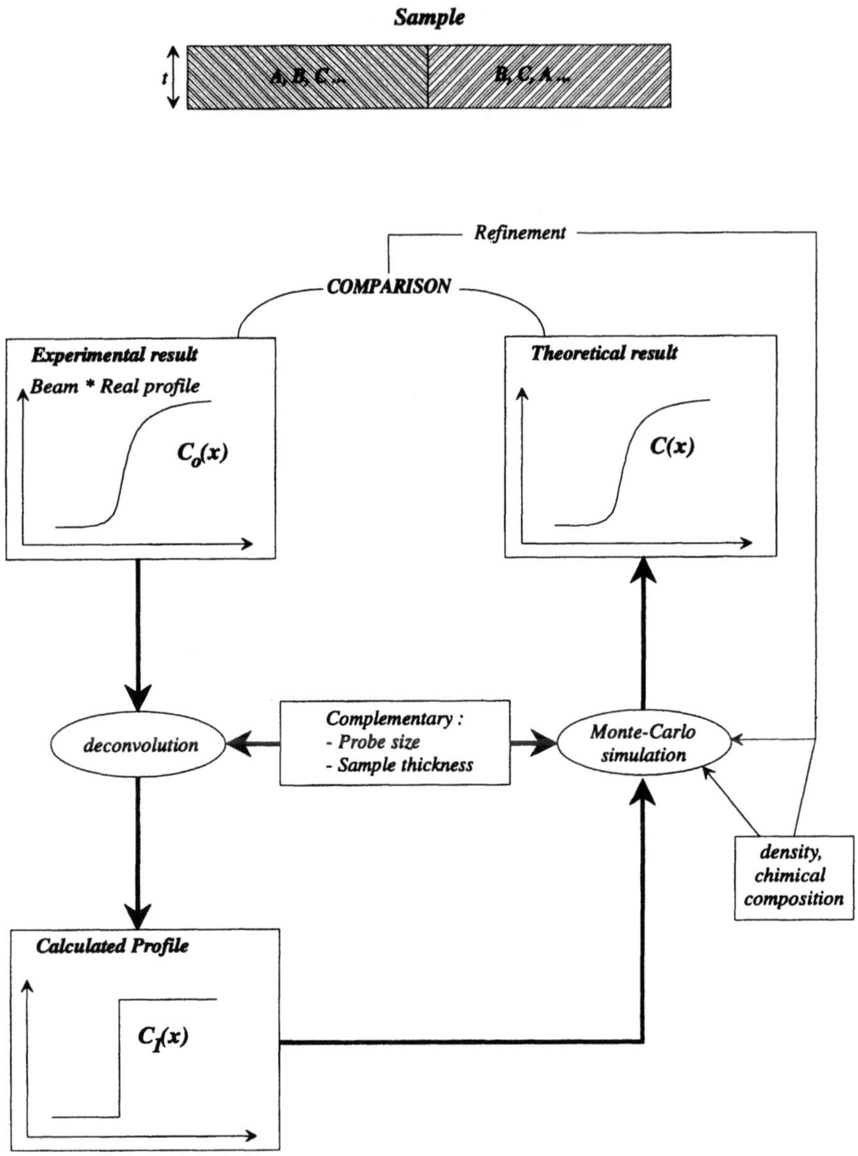

Fig. 2. Quantitative determination of a concentration profile in EDS analysis

hexagonal α-SiC lattice, and the whiskers have then to be oriented along the [0001] zone axis.

The chemical analysis near the interface has been made using the STEM mode in the following conditions: incident electron energy of 200 keV, and probe size of 22 nm (this value corresponds to the half-width of the beam spreading, i.e. 2.35 σ). The probe size was estimated experimentally: the video signal was recorded in the STEM mode, during a line-scan experiment across an ideally abrupt interface between an edge-on [100]-MgO cubic microcrystallite and the vacuum. The take-off angle value was 25 degrees, and the thickness of the sample was measured by electron energy loss spectroscopy: the ratio between the intensities under the plasmon peaks and the whole loss spectrum respectively, gives the thickness of the analyzed region

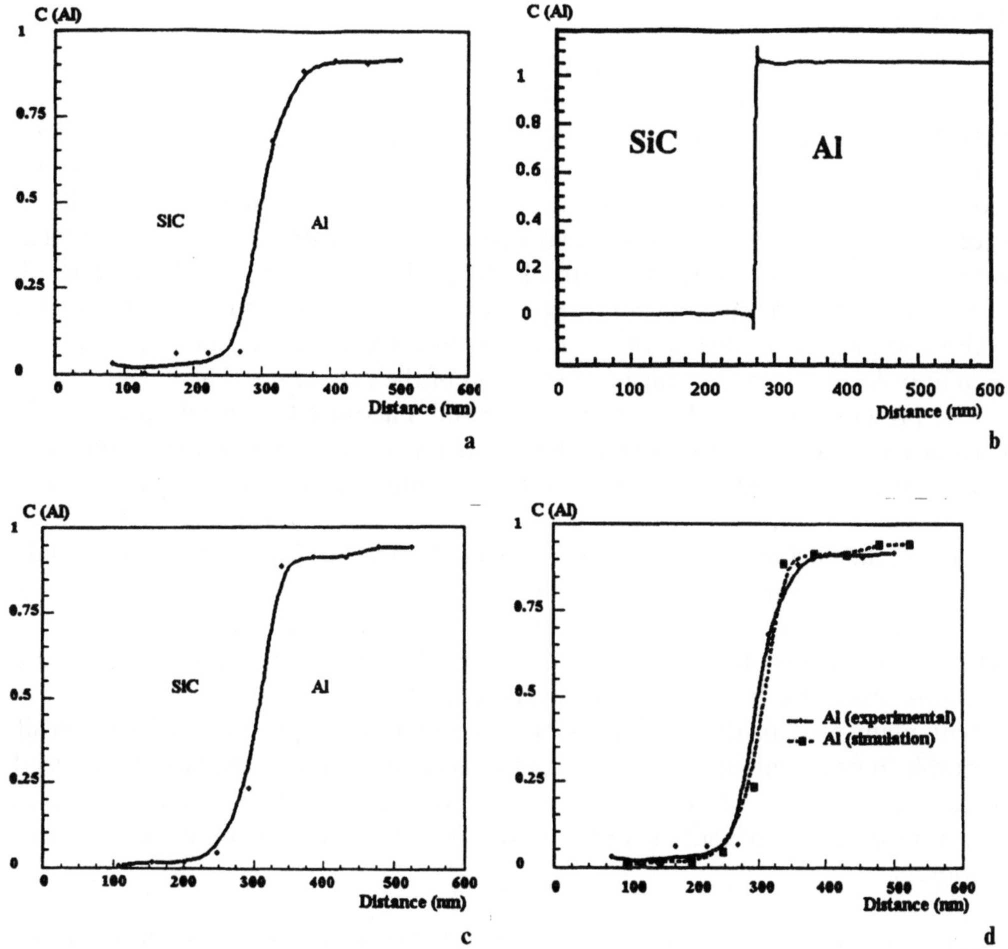

Fig. 3. Al concentration profile across an SiC/Al interface: **a** experimental profile, **b** deconvolution of the experimental profile, **c** theoretical profile, **d** comparison between experimental and theoretical profile. Simulation (energy: 200 keV, thickness: 400 nm, probe size $d_{1/2}$: 22 nm, take off angle: 25°, number of trajectories: 20000)

[8]. A value of 400 nm was deduced from measurements in the aluminum matrix near the interface.

The experimental profile of the element Al (Fig. 3a) exhibits an apparent gradient of concentration which spreads up to 35 nm into the SiC region. From this experimental profile, a deconvolution of the previously mentioned physical effects (Fig. 3b) has clearly confirmed the initial hypothesis, that it is a chemically abrupt interface.

In order to validate this result, the deconvoluted profile is injected into a Monte Carlo simulation step, that outputs the simulated profile shown in the Fig. 3c. From the superimposition of both experimental (Fig. 3a) and simulated (Fig. 3c) profiles (Fig. 3d), it can be concluded that the chemical gradient at this Al/SiC interface is actually abrupt.

Conclusion

The present work has described the principles and the main characteristics of the X-RES software, dedicated to the analysis of EDS experiments in a transmission electron microscope. This software allows various problems, met in classical EDS microanalysis, to be solved. In the case of interface analysis, the efficiency of the method can be improved by coupling X-RES with a preliminary deconvolution procedure. This mathematical approach also shows that using the beam broadening formula given by Doig leads to a simulated profile very close to the true result. Nevertheless, in general and complex cases, the deconvolution is not possible; X-RES can then be applied alone, and trial-and-error iterations have to be undertaken in order to approach the exact solution of the problem studied.

The present work has been applied to results acquired on a microscope with rather large probe sizes; although modern instruments deliver nanometric probes, it remains of the greatest importance to apply a numerical approach such as that proposed in X-RES, since spreading effects due to the beam interaction with the sample will then become predominant. Clearly, the need for high spatial microanalysis work is increasing in the course of TEM studies of advanced materials and multi-materials. This observation is reinforced by the fact that data acquisition and their processing become more and more use-friendly in the present computer programs developed by constructors of X-ray EDS devices.

Our main goal in developing X-RES was indeed to propose software easily adaptable to practical microanalysis situations encountered in the field of materials sciences, as it has been done in the field of high resolution TEM imaging, where a large number of use-friendly simulation programs are now available on personal computers.

Acknowledgements. The authors would like to thank D. C. Joy for the use of its SS_MC program.

References

[1] M. Aouine, *PhD Thesis* INSA de Lyon, 1994.
[2] D. B. Williams, A. D. Jr. Roming, *Ultramicroscopy* **1989**, *30*, p 38.
[3] J. M. Titchmarsh, *32ième Colloque Annuel SFME, Rouen-Mont St Aignan, 30 June–3 July*, 1992, pp. 3–4.
[4] H. Bethe, *Z. Physik* **1932**, *76*, 293.
[5] D. C. Joy, EUREM 88, York, England, 1988, pp. 23–32.
[6] G. Cliff, G. W. Lorimer, *J. Microscopy* **1975**, *103*(2), 203.
[7] P. Doig, D. Lonsdale, P. E. J. Flewitt, in: *Quantitative Microanalysis with High Spatial Resolution, London, March 25–27, 1981* (G. W. Lorimer, M. H. Jacobs, P. Doig eds.), The Metals Society, London, 1981, pp. 41–46.
[8] R. F. Egerton, *Electron Energy Loss Spectrometry in Electron Microscope*, Plenum, New York, 1986, 410 pp.

Mikrochim. Acta [Suppl.] 13, 241–249 (1996)

On the Use of the GeLα Line in Thin Film X-Ray Microanalysis of $Si_{1-x}Ge_x$/Si Heterostructures

Aldo Armigliato[1,*], Thomas Lewis[1,2], and Rodolfo Rosa[3]

[1] CNR-Istituto LAMEL, via P. Gobetti 101, I-40129 Bologna, Italy

[2] Institut für Angewandte und Technische Physik, Technische Universität Wien, Wiedner Hauptstrasse 8–10, A-1040 Wien, Austria

[3] Dipartimento di Scienze Statistiche, Università di Bologna, via Belle Arti 41, I-40126 Bologna, Italy

Abstract. The Monte Carlo simulation has been applied to the determination of composition and thickness of thin $Si_{1-x}Ge_x$ alloys in cross-sectioned, pseudomorphic $Si_{1-x}Ge_x$/Si heterostructures. The experimental intensity of the SiKα, GeLα and GeKα X-rays have been measured in an analytical electron microscope and ratioed. Two methods have been applied: the '2-tilt angle method', which is based on the convergence of the I(SiKα)/I(GeLα) ratios at two different tilt angles, and the '2-ratio method', which uses the I(SiKα)/I(GeKα) and the I(SiKα)/I(GeLα) ratios at a fixed angle. It is found that both methods give solutions, but not in the whole Ge concentration range investigated (from 5 up to 30 at.%). The best way is to apply the '2-tilt angle method', but using I(SiKα)/I(GeKα) ratios.

Key words: silicon-germanium alloys, composition determination, thin films, X-ray microanalysis, Monte Carlo method.

In thin film X-ray microanalysis, performed in the analytical electron microscope, methods based on the simultaneous determination of composition and thickness have proved to be more advantageous than the usual Cliff-Lorimer procedure [1]. In previous papers [2, 3] we have successfully exploited the Monte Carlo simulation to determine the Ge concentration and the local thickness in cross sections of $Si_{1-x}Ge_x$ heterostructures, which are presently employed in the fabrication of high frequency bipolar transistors. The method involves the measurement of the I(SiKα)/I(GeKα) intensity ratio at two different angles of incidence of the electron beam ('2-tilt angle' method). An approach, alternative to Monte Carlo, is the differential X-ray absorption method, proposed by Horita et al. [4], who performed

* To whom correspondence should be addressed

Table 1. Nominal layer thickness (nm) and Ge concentration (at.%) of the investigated samples

Sample	Growth	t(Si-Ge)	t(buffer)	C(Ge)
SIGE1	APCVD	130	25	5.5
SIGE2	MBE	500	100	10
SIGE3	MBE	500	100	15
SIGE4	MBE	52	—	21
SIGE5	RTCVD	47	—	33

the analysis of a ternary system at normal incidence, by including L and M lines in their measurements.

To assess the practicability of the GeLα line in our Si-Ge alloys (with Ge concentration up to 30 at.%) we have applied our method by using either I(SiKα)/I(GeLα) intensity ratio at 0° and 20°, or the I(SiKα)/I(GeKα) together with the I(SiKα)/I(GeLα) ratios at a fixed incidence angle ('2-ratio' method).

Experimental

The $Si_{1-x}Ge_x$/Si heterostructures were fabricated by either solid source MBE or CVD on (001) substrates. The nominal features of the specimens investigated are reported in Table 1. The details of the MBE growth have been previously described [3]. For the heterostructures grown by CVD, the substrates were loaded into a horizontal reactor, where a Si buffer was deposited by a gas mixture H_2/HCl/SiH_4; the Si-Ge layer was then grown using UHP quality silane and germane [5].

Transmission electron microscopy (TEM) cross sections were prepared by a standard procedure [6], involving mechanical lapping and ion beam milling to perforation. In Fig. 1 is reported a TEM micrograph of a typical $Si_{1-x}Ge_x$/Si heterostructure.

For the X-ray analysis, a Philips CM30 TEM, equipped with an EDAX PV9900 energy dispersive spectrometer (EDS) was employed. The accelerating voltage was 300 kV, the spot size at the specimen level 10 nm, obtained in the nanoprobe mode. A Gatan liquid-nitrogen cooled, double-tilt holder was used, to prevent contamination during the spectra acquisition.

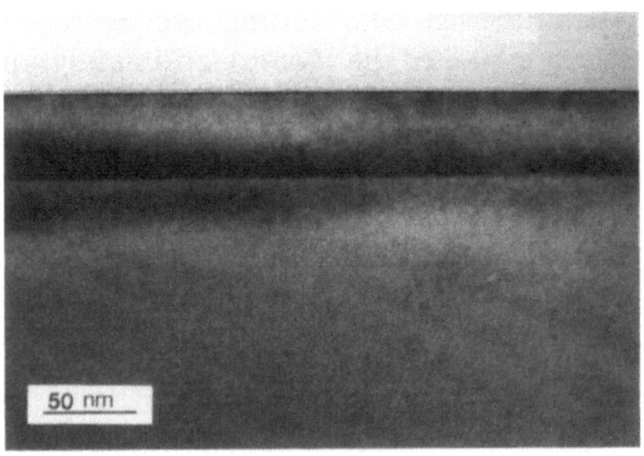

Fig. 1. Cross sectional TEM micrograph of a typical $Si_{1-x}Ge_x$/Si heterostructure (sample SIGE4). Here the thickness of the Si-Ge film is of only 42 nm, which is below the critical thickness for the corresponding Ge concentration (21 at.% nominal, see Table 1). In fact, no misfit dislocation is visible at the $Si_{1-x}Ge_x$/Si interface

Methods of X-ray Quantitative Analysis

Monte Carlo Method

The '2-tilt angle' method. This analytical method has been described elsewhere [7, 8]. It consists of the measurement of the intensity of the I(SiKα) and I(GeKα) (or I(GeLα)) X-ray intensities, as generated from the silicon and germanium atoms in an area of the Si-Ge film in the TEM cross sections, at two different tilt angles, i.e. 0° and 20°. The tilt axis is perpendicular to the Si-Ge/Si interface, whereas the plane containing the incident and the detected X-rays is always parallel to this interface. In this way the analyzed volume remains confined within the Si-Ge film.

From the net X-ray intensities, the ratios $R_m^K = I(\mathrm{SiK}\alpha)/I(\mathrm{GeK}\alpha)$ (m = measured), or $R_m^L = I(\mathrm{SiK}\alpha)/I(\mathrm{GeL}\alpha)$, are computed at 0° and 20° tilt angle. In order to determine simultaneously both the composition and the local sample thickness of the films, two computer programmes are employed. The first code (CARLONE) is a Monte Carlo simulation based on the single scattering approach and the continuous energy-loss approximation. The elastic cross section is calculated by the Wentzel potential, empirically corrected according to Arnal et al. [9]. The ionization cross section by Mott and Massey [10] is chosen. All the adopted formulae are relativistically corrected to account for an electron energy of 300 keV.

The fluorescence yield ω and the relative intensity factor a of Kα (or Lα) radiation to total K shell (or L shell) X-ray intensity have been taken from Bambynek et al. [11] and from Schreiber and Wims [12], respectively. For an L line (such as GeLα) the Coster-Kronig transitions [13] must be taken into account. In Table 2 are reported the values of ω and a employed in the present work.

The correction factor ε for the absorption of the emitted X-rays in the detector (which includes the beryllium window, the gold contact and the silicon dead layer) is rather critical, particularly when the GeLα line (E = 1.188 keV) is involved. From the measurement of the intensity ratio I(CuL)/I(CuK) in a thin copper film [14] a Be thickness of 11 μm is obtained, which is a reasonable figure. As to the mass absorption coefficients μ/ρ, to be used in the calculation of both ε and the X-ray absorption in the specimens, we have considered the tabulations by Veigele [15] and Heinrich [16] and assumed the values that yielded Ge concentrations in closest agreement with the results obtained by other techniques (Rutherford backscattering spectrometry (RBS) or double crystal X-ray diffractometry (DCXD)). The values of ε and μ/ρ(Si, Ge) are reported in Table 3.

CARLONE, finally, generates two sets of computed ratios $R_c^K = I(\mathrm{SiK}\alpha)/I(\mathrm{GeK}\alpha)$ (or $R_c^L = I(\mathrm{SiK}\alpha)/I(\mathrm{GeL}\alpha)$) for 0° and 20°, as a function of Ge concentration C and local mass thickness ρt.

Table 2. Fluorescence yield ω and relative intensity factor a values employed in this work

X-ray line	ω	a
GeLα	0.01	0.729
SiKα	0.047	0.966
GeKα	0.54	0.868

Table 3. Detector efficiency ε and mass absorption coefficients in the Si-Ge films (cm^2/g) employed in this work

X-ray line	ε	μ/ρ(Si)	μ/ρ(Ge)
GeLα	0.344	986	1262
SiKα	0.779	328	4717
GeKα	0.994	34.7	41.2

The computed sets R_c^K (R_c^L) are given as input to the code ROSIN to minimize the difference $|R_c^K - R_m^K|$ ($|R_c^L - R_m^L|$) at the two angles. As described in detail in ref. [8], two characteristic curves corresponding to the two tilt angles are generated in the (C, ρt) space, whose intersection yields the required values of the unknowns. The effect of the experimental errors, estimated around 2%, on the final result is evaluated by the procedure discussed in detail in ref. [8]. First a set of simulated measurements is generated from the one actually performed, in order to obtain 50 'measured' values of each R_m^K (R_m^L). Afterwards, such artificially generated data are subjected to the bootstrap procedure. In such a way two bunches of 100 characteristic curves are obtained. All the intersecting points, say $(C^*_{Ge})_b$ and $(\rho t^*)_b$ $(1 \leqslant b \leqslant 100)$, yields the bootstrap estimates of the variability of C_{Ge} and ρt, respectively, according to:

$$\sigma_\theta = \left[\sum_{b=1}^{B} (\theta_b^* - \theta^*)^2/(B-1) \right]^{1/2} , \qquad (1)$$

where $\theta^* = \sum_{b=1}^{B} \theta_b^*/B$. In our case θ stands for C_{Ge} or ρt, and $B = 100$. An example of how the bootstrap method works is reported in *Results and Discussion* (see Fig. 3).

The '2-ratio' method. This method differs from the above described '2-tilt angle' one in that the experiments are performed at a fixed angle (e.g. at normal incidence) and all the three intensity values I(GeLα), I(SiKα) and I(GeKα) are exploited. In this case the programme ROSIN is used to determine the C and ρt values that minimize the differences $|R_c^K - R_m^K|$ and $|R_c^L - R_m^L|$. Of course the measurements and the calculations can be repeated at a different angle, in order to check the validity of the results.

The DXA Method

The differential X-ray absorption method (DXA), described by Morris et al. [17] and Horita et al. [4] makes use of the difference in absorption between K and L (or L and M) lines emitted simultaneously from the same element. Following Horita [4] and using all the three X-ray intensities, e.g. I(GeLα), I(GeKα) and I(SiKα) at a fixed angle of incidence, we computed the expressions:

$$C_{Ge}/C_{Si} = K^K_{GeSi}(I(\mathrm{GeK}\alpha)/I(\mathrm{SiK}\alpha))\,\mathrm{F}(\rho t)^K , \qquad (2)$$
$$C_{Ge}/C_{Si} = K^L_{GeSi}(I(\mathrm{GeL}\alpha)/I(\mathrm{SiK}\alpha))\,\mathrm{F}(\rho t)^L ,$$
$$C_{Ge}/C_{Ge} = K^{LK}_{GeGe}(I(\mathrm{GeK}\alpha)/I(\mathrm{GeL}\alpha))\,\mathrm{F}(\rho t)^{LK} = 1 ,$$

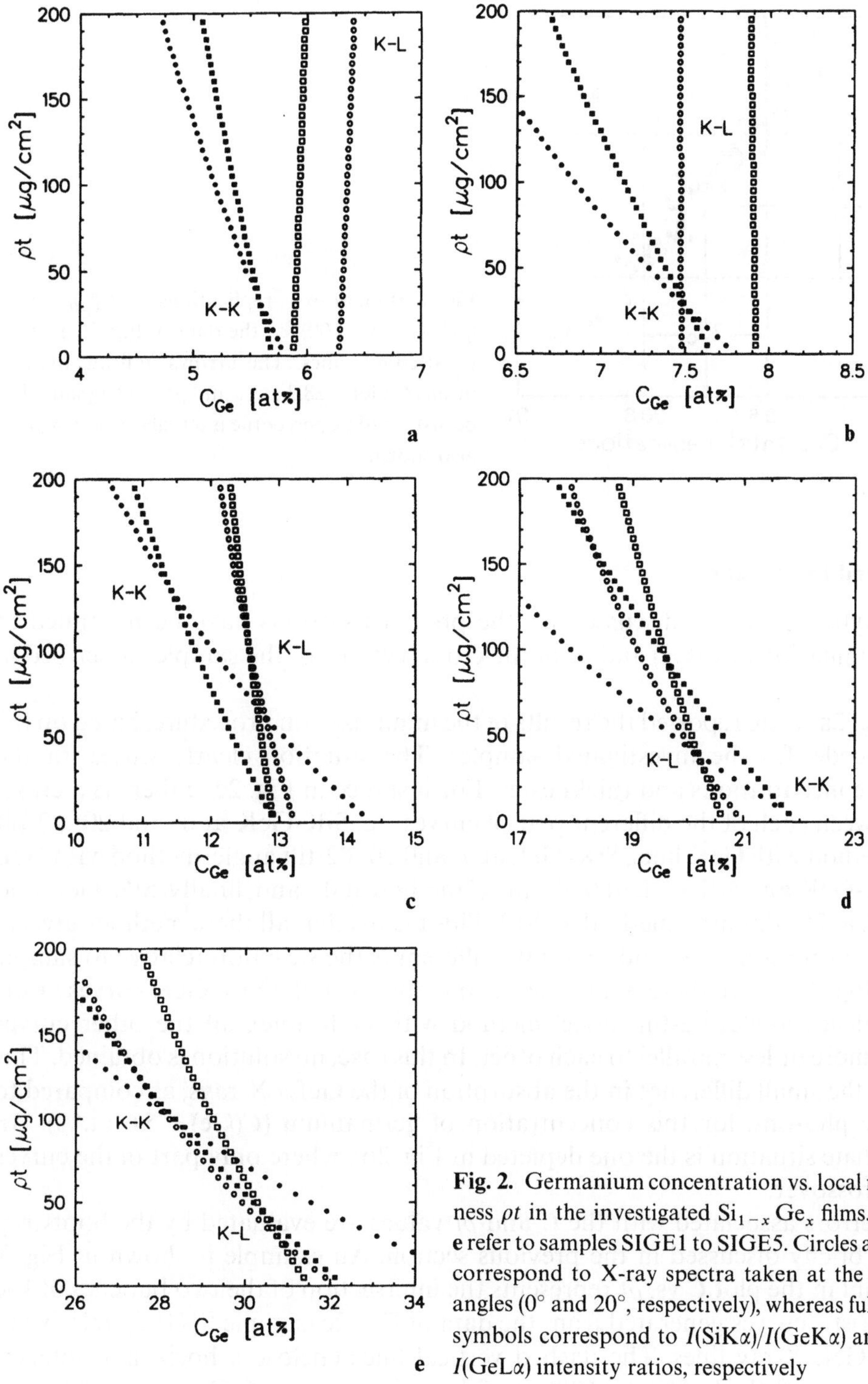

Fig. 2. Germanium concentration vs. local mass thickness ρt in the investigated $Si_{1-x}Ge_x$ films. Plots **a** to **e** refer to samples SIGE1 to SIGE5. Circles and squares correspond to X-ray spectra taken at the two tilting angles (0° and 20°, respectively), whereas full and open symbols correspond to $I(\mathrm{SiK}\alpha)/I(\mathrm{GeK}\alpha)$ and $I(\mathrm{SiK}\alpha)/I(\mathrm{GeL}\alpha)$ intensity ratios, respectively

where the terms K are the well known Cliff-Lorimer factors [1] and $F(\rho t)$ represents the absorption correction [18]. The second (or the third) expression in eq. (2) is used to determine the film thickness, wherefrom the Ge concentration is obtained according to the first one. However, due to the presence of exponentials in $F(\rho t)$, the complexity can be solved by employing an iterative procedure.

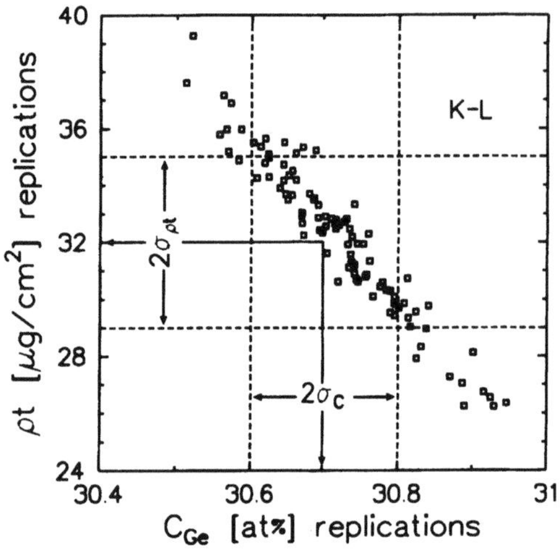

Fig. 3. Bootstrap replications $(C^*_{Ge})_b$ and $(\rho t^*)_b (1 \leqslant b \leqslant 100)$ for the data of Fig. 2e, relative to SiK-GeL lines. The arrows indicate the solutions $C(\text{Ge}) = 28.3$ at.% and $\rho t = 32\ \mu\text{g/cm}^2$. The bootstrap 68% percentile intervals (σ_C and $\sigma_{\rho t}$) are also shown

Results and Discussion

The analytical methods discussed in tthe previous sections have been applied to the determination of C(Ge) and ρt in the cross sections of the samples described in Table 1.

In Fig. 2a–e are reported the results of the minimization procedure, based on the ROSIN code, for the investigated samples. The situation clearly varies for the different concentrations and thicknesses. For instance in Fig. 2c–e there is a crossover between each of the different pair of curves, i.e. SiK-GeK at 0° and 20° ('2-tilt angle' method with GeK line), SiK-GeL at 0° and 20° ('2-tilt angle' method with GeL line), SiK-GeK and SiK-GeL at 0° ('2-ratio' method at 0°) and, finally, SiK-GeK and SiK-GeL at 20° ('2-ratio' method at 20°). This means that all these methods give, in this case, a solution for C and ρt. Quite different is the situation relative to sample SIGE1 (Fig. 2a). Here there is a crossover only between the SiK-GeK curves at the two incidence angles ('2-tilt angle' method with GeK line), all the other curves running more or less parallel to each other. In this case, no solution is obtained. This is due to the small difference in the absorption of the GeLα X-rays, as compared to the SiKα photons, for this concentration of germanium ($C(\text{Ge}) = 5.6$ at.%). An intermediate situation is the one depicted in Fig. 2b), where only part of the curves have a crossover.

The errors associated with the C and ρt values are evaluated by the bootstrap method, briefly discussed in the previous section. An example is shown in Fig. 3. Each point in the plot C vs. ρt represents the intersection of the two bunches of 100 characteristic curves generated from the data of Fig. 2e (sample SIGE5), relative to the SiK-GeL X-ray lines. The dashed vertical lines enclose a horizontal interval which represents the bootstrap 68% confidence interval for $C(\text{Ge}) = 28.3$ at.% (see arrow); similarly, the dashed horizontal lines enclose a horizontal interval which represents the bootstrap 68% confidence interval for $\rho t = 32\ \mu\text{g/cm}^2$ (see arrow).

All the Ge concentration and thickness values obtained by these different methods, together with the associated errors, as deduced from the bootstrap method, are reported in Table 4. As it can be seen, the concentration values are generally

Table 4. Germanium concentrations (at.%, first line) and local TEM sample thicknesses ($\mu g. cm^{-2}$, second line) as obtained by the different methods in the investigated Si-Ge alloys

Sample	Monte Carlo method '2-tilt angle' K-K	K-L	'2-ratio' 0°	20°
SIGE1	5.60 ± 0.02	no sol.	no sol.	no sol.
	28 ± 5			
SIGE2	7.49 ± 0.03	no sol.	7.5 ± 0.03	no sol.
	30 ± 4		32 ± 4	
SIGE3	11.46 ± 0.05	12.61 ± 0.06	12.73 ± 0.04	12.85 ± 0.02
	134 ± 4	98 ± 13	66 ± 2	10 ± 2
SIGE4	21.68 ± 0.07	20.15 ± 0.08	20.00 ± 0.06	19.53 ± 0.05
	8 ± 2	39 ± 5	43 ± 2	96 ± 3
SIGE5	28.3 ± 0.1	30.7 ± 0.1	28.03 ± 0.08	29.74 ± 0.05
	100 ± 2	32 ± 3	105 ± 2	60 ± 2

Sample	DXA method 0°	20°	RBS or DCXD
SIGE1	no. sol.	no sol.	5.5 ± 0.3
SIGE2	7.5	no sol.	7.7 ± 0.5
	31		
SIGE3	12.9	12.9	12.7 ± 0.5
	45	4	
SIGE4	20.1	19.7	21 ± 0.5
	41	87	
SIGE5	28.1	30.9	33 ± 0.5
	103	22	

DXA = differential X-ray absorption method; RBS = Rutherford backscattering spectrometry; DCXD = double crystal X-ray diffraction.

determined with a very high precision (better than 1%) and also the accuracy is quite good: in fact they agree with the ones found by the Rutherford backscattering spectrometry (RBS) or the double crystal X-ray diffractometry (DCXD) within less than 10%. Also the difference between the C values obtained with the two Monte Carlo-based methods and the DXA one is of the order of 10% or better. Instead, the thickness values are generally in a large mutual disagreement, not only among the different methods, but also within them (compare e.g. the ρt values obtained by the '2 tilt angle method' assuming either the GeKα or the GeLα line in the intensity ratios). This is probably due to the larger uncertainties in the physical parameters involved in the quantitation. Luckily, however, the local sample thickness in a TEM cross section is a parameter of no practical significance, the relevant thickness being that of the original Si-Ge film in the heterostructure, which is immediately deduced from the width of the corresponding stripe in the image (see Fig. 1).

In any case, it results from the data in Table 4 that only the '2-tilt angle' method, based on the SiK-GeK intensity ratio yields a solution in all the composition range considered. All the other methods, which include the GeLα line, are sometimes unable to give a result. This occurs in a range of Ge concentrations from about 5 to 8 at.%. As mentioned above, this is due to the small difference in the absorption of the GeLα X-rays, with respect to the SiKα photons, in this type of samples. The same difficulty is of course encountered by the DXA method (see Table 4), which also exploits the GeLα line.

Conclusions

It has been found that (i) for Ge concentrations of 5 to 8 at.% the '2-tilt angle' method, applied to the GeL line, does not work, and also the '2-ratio' one yields poor results, and (ii) outside this range both methods give Ge concentrations in fair agreement with the ones obtained by the GeKα-based 2-tilt angle procedure, but very different thickness values. The main reason for these discrepancies lies (i) in he sample absorption of the GeLα X-rays, which is much closer to the one of the SiKα line than the GeKα photons, and (ii) in the poorly known values of the physical parameters relative to L-lines. Therefore our experiments have demonstrated that for Si-Ge alloys the most accurate analyses are obtained by the '2-tilt angle' method involving K lines.

Acknowledgements. The authors are indebted to S. S. Iyer (IBM T. J. Watson Research Institute, Yorktown Heights, USA) and to B. Dietrich (Institut für Halbleiterphysik, Frankfurt/Oder, Germany) for supplying the heterostructures. The skilful technical assistance of F. Corticelli, A. Garulli and D. Govoni is also gratefully acknowledged.

References

[1] G. Cliff, G. W. Lorimer, *J. Microscopy* **1975**, *103*, 203.
[2] A. Armigliato, M. Servidori, F. Cembali, R. Fabbri, R. Rosa, F. Corticelli, D. Govoni, A. V. Drigo, M. Mazzer, F. Romanato, S. Frabboni, R. Balboni, S. S. Iyer, A. Guerrieri, *Microsc. Microanal. Microstruct.* **1992**, *3*, 363.
[3] A. Armigliato, D. Govoni, R. Balboni, S. Frabboni, M. Berti, F. Romanato, A. V. Drigo, *Mikrochim. Acta* **1994,** *114/115*, 175.
[4] Z. Horita, K. Ichitani, T. Sano, M. Nemoto, *Phil. Mag.* **1989**, *A59*, 939.
[5] Th. Morgenstern, I. Babanskaya, G. Morgenstern, K. Schmalz, P. Gaworzewski, P. Zaumseil, D. Krüger, K. Tittelbach-Helmrich, H. Kühne, *Solid State Phenomena* **1993**, *32–33*, 409.
[6] A. Garulli, A. Armigliato, M. Vanzi, *J. Microsc. Spectrosc. Electron.* **1985**, *10*, 135.
[7] R. Rosa, A. Armigliato, *X-ray Spectrom.* **1989**, *18*, 19.
[8] A. Armigliato, R. Rosa, *Ultramicroscopy* **1990**, *32*, 127.
[9] F. Arnal, J. L. Balladore, G. Soum, P. Verdier, *Ultramicroscopy* **1977**, *2*, 305.
[10] N. F. Mott, H. S. W. Massey, *The Theory of Atomic Collisions, 3rd Ed.*, Oxford University Press, Oxford, 1965.
[11] W. Bambynek, B. Craseman, R. W. Fink, H. U. Freund, H. Mark, C. D. Swift, R. E. Price, P. V. Rao, *Rev. Mod. Phys.* **1972**, *44*, 716.
[12] T. P. Schreiber, A. M. Wims, *X-ray Spectrom.* **1982**, *11*, 42.
[13] D. Coster, R. de L. Kronig, *Physica* **1935**, *2*, 13.

[14] A. O. Sandborg, *EDAX Editor* **1984**, *12*, 2.
[15] W. M. J. Veigele, *Atomic Data* **1973**, *5*, 51.
[16] K. F. J. Heinrich, in: *Scanning Electron Microscopy and X-ray Microanalysis* (J. I. Goldstein, D. E. Newbury, P. Echlin, D. C. Joy, C. Fiori, E. Lifshin, eds.), Plenum, New York, 1992, Ch. 14.
[17] P. L. Morris, M. D. Ball, P. J. Statham, In: *Electron Microscopy and Analysis 1979, Inst. Phys. Conf. Ser* **1980**, *52*, 413.
[18] J. I. Goldstein, J. L. Costley, G. W. Lorimer, S. J. B. Reed, In: *Scanning Electron Microscopy 1977, Vol. I* (O. Johari ed.), IITRI, Chicago, 1977, p. 315.

Mikrochim. Acta [Suppl.] 13, 251–257 (1996)

Computer Simulations of the X-Ray Intensity Distribution from Submicron Particles Embedded in a Matrix

Alexander Berner* and Igor Levin

Department of Materials Engineering, Technion – Israel Institute of Technology, Haifa 32000, Israel

Abstract. Computer simulations of the X-ray intensity distribution from submicron particles embedded in a matrix were performed. The results of simulations confirmed the possibility of approximation of the real X-ray distribution by a finite distribution with the same integral and mean value. It was shown that the mean particle size can be calculated from the coefficient of variation of the measured mass concentration with a reasonable degree of accuracy. In addition, the possibility to determine the coefficient of variation of the particle size from the skewness of the concentration distribution was shown.

Key words: EPMA, X-ray intensity distribution, submicron particles, computer simulations.

Micro-particles in a matrix can significantly change the properties of a material, and thus the characterization of such particles is an important objective of microstructural analysis. The present work is devoted to the determination of the mean parameters of the particle population (mean size, coefficient of variation of a particle volume, number of particles per unit volume, etc.) by electron probe microanalysis (EPMA). The conventional method to solve this problem is to analyze the images obtained from either electrons or characteristic X-rays from the particles [1–4]. However, a problem arises for the case of a high number density of submicron particles per unit volume [5].

Previously, the method for the fast determination of the mean size of submicron particles randomly distributed in a matrix was developed and tested experimentally [5]. In this method a variance of the characteristic X-ray intensity distribution measured from the particles was analyzed and related to the mean size of the particles. However, the method was based on a number of assumptions, a part of which could not be verified experimentally. To validate these assumptions, a computer program for the simulation of the distribution of the characteristic X-ray intensity from the submicron particles embedded in a matrix was developed.

* To whom correspondence should be addressed

Theory and Formulation of a Problem

Consider some volume Ω of a heterogeneous material composed of a matrix and randomly distributed micro-particles of one and the same composition. Assume that the volume Ω is much larger than the volume of a single particle. Then the mass concentration in the volume Ω of an element present only in the particles is given as

$$C = \frac{C_0\rho_0}{\rho} \cdot \frac{\sum_{i=1}^{n} \omega_i}{\Omega} = \frac{C_0\rho_0}{\rho} \cdot \frac{n\omega}{\Omega}, \tag{1}$$

where C_0 is the concentration of the measured element in the particle, ρ_0 is the density of the particle, ω_i is the volume of a single particle or of a part of a particle in the volume Ω, n is the number of the particles in the volume Ω, ρ is the density of the volume Ω and ω is the average volume of the particles in the single volume Ω.

The mass concentration in the volume Ω is a function of two random variables n and ω. Consider only the case when the volume fraction of the particles is much less than unity. Then the random variables n and ω can be regarded as independent. In reference [5] it was shown that under this condition the coefficient of variation of the density is much less than the coefficients of variation of n and ω, and therefore ρ can be considered as a constant. So the mass concentration in the volume Ω is a multiplication of two independent random variables. For this reason all the statistical moments of a random variable C are functions of the statistical moments of n and ω.

The distribution of n in the volume Ω can be modeled by a binomial distribution [5]. The volume Ω can be subdivided into two sub-populations, one of which is the particles (type 1) and the other a matrix (type 2). From this volume a sample Ω/ω is selected by a process of random trials. The probability q of a type 1 outcome is constant and equal to the volume fraction of the particles in the sample. Then the statistical moments of n can be expressed in terms of N and V, where N is the average number of the particles in the volume Ω and V is the mean particle volume in Ω, where an average is taken over multiple volumes Ω.

It was shown previously [5], that neglecting the particle size distribution, the mean particle size d can be determined from Eq. (2):

$$d = \sqrt[3]{V} = \sqrt[3]{\frac{\bar{C}\Omega}{K - \bar{C}}\, \sigma^2_{r,c}, \mathrm{n}} \tag{2}$$

where $K = C_0\rho_0/\rho$, $\bar{C}$ is the mean concentration of element of interest in the specimen and $\sigma_{r,c}$ is coefficient of variation of the concentration in volume Ω

It should be noted that Eq. (2) was derived without any additional assumptions or simplifications. Eq. (2) reflects the fact that for the sample composed of microparticles randomly distributed in a matrix the variance of a concentration distribution includes a part connected to the particle size and under certain conditions of sampling this part of the variance can be dominant. So the mean particle size can be strictly determined from Eq. (2).

However, an application of the indicated approach in EPMA requires that several serious assumptions should be made. First, one that concerns the concept of the excited volume. Eq. (2) is based on an analysis of concentration distribution for some finite volume Ω. However, the volume excited by electrons does not have sharply defined boundaries. Rather it can be characterized by the three-dimensional distribution of X-ray production. In reference [5] it was proposed to approximate this real X-ray distribution by a finite distribution with the same integral and the maximum possible number of coincident statistical moments with the real distribution.

Another important point is a substitution of the statistical moments of the X-ray intensity distribution for the statistical moments of the concentration distribution. In Eq. (2) $\sigma_{r,c}$ stands for the coefficient of variation for the concentration distribution. However, only the coefficient of variation for the distribution of the characteristic X-ray intensity $\sigma_{r,i}$ can be measured. In reference [5] it was supposed that $\sigma_{r,i} = \sigma_{r,c}$ for the case where C is much less than unity and the excited volume is much larger than a volume of a single particle.

The method for the determination of the mean particle size based on these assumptions was developed and verified experimentally [5]. Results of determinations were in a good agreement with results obtained by some direct methods. However the assumptions providing the basis of the proposed method could not be proved experimentally.

The main goal of the present work is a verification of the indicated assumptions by the computer simulations of the X-ray intensity distribution from particles embedded in a matrix. In addition, a possibility to use the statistical moments of high orders for determination of the coefficient of variation of particle size was considered.

Computer Simulation

Distribution of X-Rays

The main purpose of computer simulations is to compare the statistical moments for the real distribution of the characteristic X-ray intensity from particles randomly distributed in a matrix with the statistical moments for the distribution of concentration in some equivalent finite excited volume. Generally, formulae describing the X-ray distribution for a heterogeneous specimen composed of microparticles randomly distributed in a matrix are lacking at present. However, for the volume fraction of particles which is much less than unity and the mean volume of a single particle which is much less than the excited volume, a lateral distribution of electrons and X-rays in a heterogeneous specimen can be replaced by that for the uniform homogeneous matter. A lateral distribution was taken as a product of X-ray distributions in depth $\Phi(z)$ and in transverse plane $\Phi(r, \varphi)$, where r, φ and z are cylindrical coordinates.

The real $\Phi(z)$ was chosen as two straight lines within the framework of a quadrilateral model [6]. A radial Gaussian function was used for the X-ray distribution in the transverse plane [7]:

$$\Phi(r) = (r/\sigma_r^2) \exp(-r^2/2\sigma_r^2), \tag{3}$$

where σ_r is the standard deviation of the distribution. Then statistical moments of the distributions $\Phi(z)$ and $\Phi(r)$ can be calculated analytically.

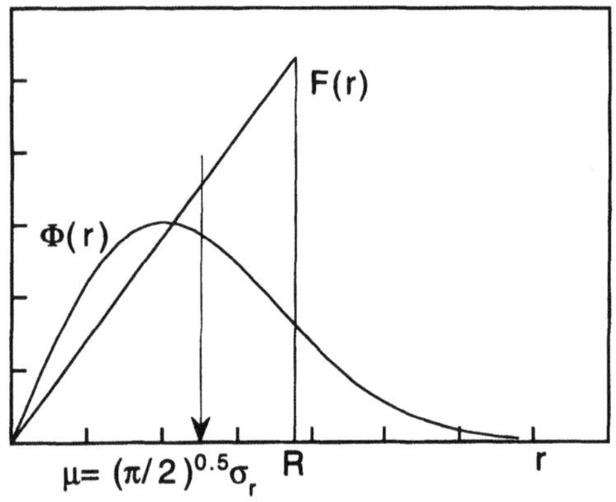

Fig. 1. A schematic representation of the real X-ray distribution in the transverse plane $\Phi(r)$ and of the equivalent distribution $F(r)$

Excited Volume

In order to determine volume Ω, the real excited volume was approximated by a finite uniform distribution with the same integral. The finite distribution was taken as the product of uniform X-ray distributions in depth $F(z)$ and in transverse plane $F(r)$. The real excited volume was modeled by a cylinder with a radius R and a height H. The parameters R and H were calculated to obtain the same mean value as for the real distribution in the appropriate direction. The equivalent distribution $F(r)$ was found as

$$F(r) = \frac{16r}{9\pi\sigma_r^2}, \qquad r \in [0, R], \qquad R = \frac{3}{2}\sqrt{\frac{\pi}{2}}\sigma_r. \tag{4}$$

In this case

$$\int_0^\infty \Phi(r)dr = \int_0^R F(r)dr = 1 \tag{5}$$

and

$$\int_0^\infty r\Phi(r)dr = \int_0^R rF(r)dr = \sqrt{\frac{\pi}{2}}\sigma_r. \tag{6}$$

A schematic representation of functions $\Phi(r)$ and $F(r)$ is presented in Fig. 1. The function $F(z)$ was expressed as

$$F(z) = 1/H, \qquad z \in [0, H], \tag{7}$$

where the parameter H was calculated from the depth distribution $\Phi(z)$ in a similar way.

Number of Particles in Ω

The number of particles in the selected volume Ω was obtained according to the binomial distribution:

$$P(n) = C_N^n q^n (1-q)^{N-n}, \tag{8}$$

where $P(n)$ is the probability of n successes in a fixed number of outcomes N, C_N^n is the number of combinations of N items, n at a time and q is the probability of success at one trial. In our case $N = \Omega/V$, where V is the mean particle volume and q is equal to the volume fraction of particles in the sample.

To calculate a value of a random variable n, pseudorandom numbers x rectangularly distributed on the interval (0, 1) were used. The sequence of pseudorandom numbers was obtained by the multiplicative congruential method [8]. The period of the series was equal to 2^{31}. The value of n was determined from the formula:

$$\sum_{i=0}^{n-1} P(i) \leqslant x \leqslant \sum_{i=0}^{n} P(i), \tag{9}$$

Shape, Size and Orientation of Particles

It was shown previously [5] that the statistical moments of a distribution of a mass concentration in the volume Ω depend only slightly on the shape and orientation of particles. For this reason cubic particles were used for simulations and one of the cubic faces was parallel to the sample surface. Then the integrals of all the X-ray distributions considered can be calculated analytically.

Gaussian and log-normal distributions were used to determine a single particle volume ω_i in accordance with formula:

$$\int_0^{\omega_i} \varphi(\omega)d\omega = x(\omega_i), \tag{10}$$

where $\varphi(\omega)$ is the selected distribution for the single particle volume, x is the rectangularly distributed pseudorandom number. Coordinates r and z of the particle center were obtained from the uniform distributions over the equivalent cylindrical excited volume.

Simulation of an X-Ray Intensity and k-Ratio

Integration of the real X-ray distribution was performed taking into account X-ray absorption in a specimen. The take-off angle of the X-ray radiation was taken to be equal to 90°. After integration, the data were converted to the X-ray intensities and k-ratios. For this purpose two additional parameters were used: the mean value of counts, corresponding to a background measured from a pure matrix (BG), and the peak-to-background ratio for the particle (PB). Then the mean number of the X-ray quanta was obtained from the Eq. (11):

$$I = \mathrm{BG}\cdot(1 + y\cdot \mathrm{PB}), \tag{11}$$

where y is the integral of the X-ray distribution. The number of X-ray quanta was assumed to be governed by Poisson statistics. Bulk uniform material of the particle composition was used as a standard.

Algorithm of Simulations

1. Input of initial data.

   The following input parameters were used:

- volume fraction of particles (0.05 and 0.01);
- coefficient of variation of a single particle volume (0, 30, 75, 150 and 300% relative);
- distribution for the single particle volume (Gaussian or log-normal);
- ratios of the standard deviations of functions $\Phi(r)$ and $\Phi(z)$ to the mean particle size ranged from 0.5 to 10;
- absorption parameters for a matrix and for a particle;
- mean background for a pure matrix (1 and 10);
- peak-to-background ratio for the standard (20 and 100);
- total number of particles (10000).

2. Calculation of parameters R and H for the equivalent finite uniform distributions $F(r)$ and $F(z)$ and calculation of the equivalent excited volume Ω.
3. Calculation of a number of particles in the volume Ω.
4. Determination of particles coordinates (r, z) relative to the probe position and calculation of the particle sizes (ω_i).
5. Integration using the real and the equivalent X-ray distributions.
6. Determination of the simulated k-ratios for the given excited volume.
7. Repetition of steps 3 through 6.
8. Calculation of the statistical moments of the simulated distributions for the concentration and for the k-ratio.

Results and Conclusions

The simulations clearly show that values of coefficient of variation σ_r and skewness Sk for the concentration distribution in the equivalent excited volume Ω are very

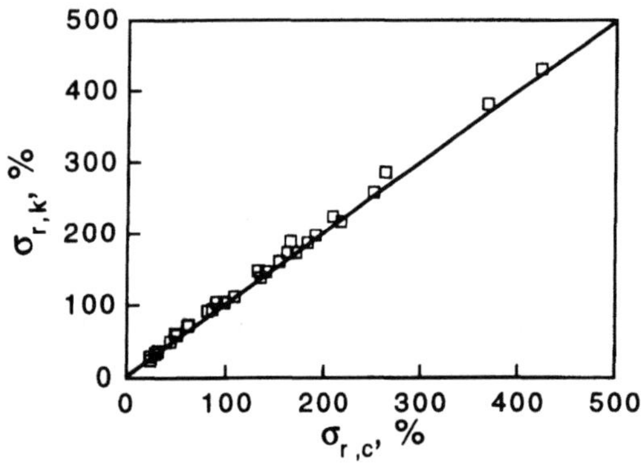

Fig. 2

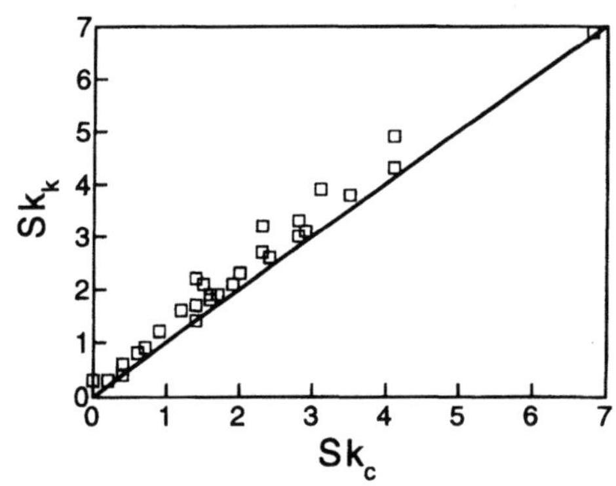

Fig. 3

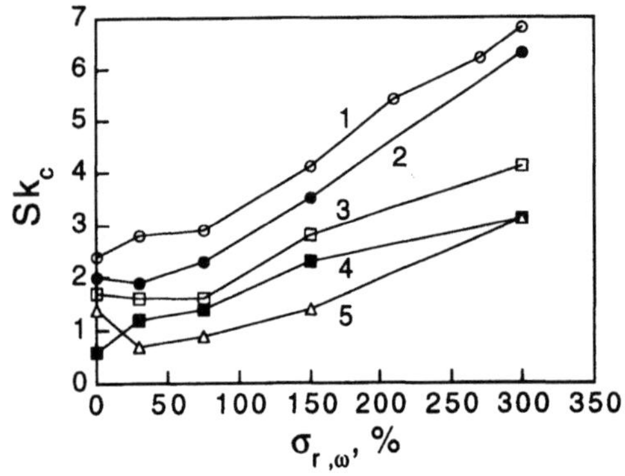

Fig. 4

Fig. 2. Correlation between the coefficients of variation for the concentration distribution $\sigma_{r,c}$ and for the k-ratio distribution $\sigma_{r,k}$

Fig. 3. Correlation between the skewnesses of the concentration distribution Sk_c and the k-ratio distribution Sk_k

Fig. 4. Dependences of the skewness on the coefficient of variation of the single particle volume, Ω/V: *1* 2.8, *2* 4.9, *3* 11.1, *4* 22.2, *5* 56.0

Table 1. Mean particle size $\langle d \rangle$ as determined from Eq. (2) for different values of $\sigma_{r,\omega}$, %: 1–0, 2–30, 3–75, 4–150, 5–300. The volume fraction of the particles is 0.05 and the nominal particle size d_0 is 1

Ω/N	1	2	3	4	5
347	1.0	1.1	1.3	1.6	2.7
56	1.0	1.0	1.1	1.5	2.2
22.2	1.0	1.0	1.0	1.4	1.8
11.1	0.9	0.9	0.9	1.2	1.7
4.9	0.8	0.8	0.9	1.1	1.6
2.8	0.8	0.8	0.8	1.0	1.4

close to the corresponding parameters for the k-ratio distribution for all the experimental conditions considered (Figs. 2, 3). While the skewness systematically deviates from the ideal value, the deviation is always much smaller than the error in skewness determination. Based on these results, it was concluded that the error introduced into the calculation of $\sigma_{r,c}$ and Sk_c by substitution of the finite distribution for the X-ray intensity for the real distribution can be neglected for all the experimental conditions.

The results presented in Table 1 demonstrate that the mean particle size can be calculated from Eq. (2) with a reasonable accuracy (less than 40%), even for the values of $\sigma_{r,\omega}$ as large as 300%, while for the values of $\sigma_{r,\omega}$ less than 100% the error is within 10%. These conclusions are in good agreement with results obtained previously [5].

Computer simulations showed that skewness of the concentration distribution Sk_c depends on the coefficient of variation of the single particle volume $\sigma_{r,\omega}$ (Fig. 4). This dependence is approximately linear for the small values of Ω/V and apparently can be used for the determination of $\sigma_{r,\omega}$. Research is currently being undertaken to develop the measurement procedure.

Thus both the mean particle size and its variance can be calculated from the coefficient of variation and the skewness of the measured concentration of the element contained in the particle.

Acknowledgements. This work was supported by the Israel Ministry of Science and Technology.

References

[1] Digiscan/FDC, Operators Manual, *EDS Microanalysis Software*, LINK Systems, High Wycombe, Bucks., England, 1988.
[2] P. Van Espen, *Mikrochim. Acta* **1994**, *114/115*, 129.
[3] A. I. Berner, O. P. Sideleva, *Zavod. Lab.* **1984**, *50*, 27.
[4] M. P. Jones, *J. Phys. (Paris) Colloq.* **1984**, *C2*, 211.
[5] A. Berner, I. Levin, L. Klinger, D. G. Brandon, *X-Ray Spectrom.* **1995**, *24*, 13.
[6] D. Sewell, G. Love, V. Scott. *J. Phys. D.* **1985**, *18*, 1245.
[7] A. I. Berner, O. P. Sideleva, B. N. Yurchenko, *High-Purity Subst.* **1990**, *4*, 151.
[8] G. Dahlquist, A. Bjorck, *Numerical Methods*, Prentice-Hall, Inc. Englewood Ceiffs, New-Jersey, 1974.

Mikrochim. Acta [Suppl.] 13, 259–269 (1996)

Determination of Rare Earth Elements in Biological and Mineral Apatite by EPMA and LAMP-ICP-MS

Simon Chenery[1,*], Terry Williams[2], Tracey A. Elliott[3], Peter L. Forey[3], and Lars Werdelin[4]

[1] Analytical Geochemistry Group, British Geological Survey, Keyworth, Nottingham NG12 5GG, U.K.

[2] Department of Mineralogy, The Natural History Museum, Cromwell Road, London SW7 5BD, U.K.

[3] Department of Palaeontology, The Natural History Museum, Cromwell Road, London SW7 5BD, U.K.

[4] Department of Palaeozoology, Swedish Museum of Natural History, Box 50007, Stockholm, Sweden

Abstract. Procedures are outlined, and a comparison made for the determination of rare earth elements (REE) in apatite by two separate microanalytical techniques: electron probe microanalysis (EPMA), and laser ablation microprobe-inductively coupled plasma-mass spectrometry (LAMP- ICP-MS). Well-characterised, compositionally-homogeneous mineral apatite grains were used as a reference material, bulk REE determinations having previously been made by instrumental neutron activation analysis (INAA). Good agreement was found between the two microanalytical techniques, and with the data by INAA. Detection limits for LAMP-ICP-MS are comparable with INAA (typically 0.5–5 μg/g in the apatite matrix), and are generally 1 to 2 orders of magnitude better than EPMA.

The techniques were applied to the microanalysis of fossilised bone recovered from contrasting terrestrial depositional environments at Olduvai Gorge (Tanzania), and to modern and fossil fish teeth from marine and freshwater environments. The data indicate that the microdistribution of REE in these biological apatites is not homogeneous, and that the relative abundances of these elements can provide information on the palaeoenvironment during the fossilisation processes.

Key words: electron microprobe, laser ablation microprobe, REE, fossil bone.

---

* To whom correspondence should be addressed

Mineral apatite, essentially $Ca_5(PO_4)_3(OH, F, Cl)$, is a relatively common accessory mineral which occurs in many rock types which have crystallised from magmatic or hydrothermal fluids. Studies of the distribution of REE in mineral apatite provide information on the fractionation of such fluids within the earth's crust or within a magma chamber.

Biological apatite in living organisms, such as bone and teeth, has a much less ordered structure, although is still essentially a hydroxyapatite $Ca_5(PO_4)_3.OH$, but contains also much organic material such as collagen and proteins. Biological apatite preserved in the fossil record has undergone variable loss of the organic material, some degree of recrystallisation and exchange of inorganic components with the depositional or fossilisation environment. Several studies of REE in fossilised bone and teeth samples have been undertaken from a range of environments [1–3] and have provided information on the burial history of the bone or teeth [4] or the palaeoenvironment [5].

However, the abundances of total REE in both natural and mineral apatite samples generally do not exceed 1–2 wt.%, which poses a challenge for analysis by

Table 1. Elements analysed and interferences observed in analysis of apatite by EPMA

Element	Line	Crystal	Interfering element	% Interference
La	$L\alpha$	PET 002	no interferences observed	
Ce	$L\alpha$	PET 002	Ba	29.6
			Sr	0.22
Pr	$L\beta$	LIF 200	no interferences observed	
Nd	$L\alpha$	LIF 200	Ce	0.40
Sm	$L\alpha$	LIF 200	Ce	1.65
Eu	$L\alpha$	LIF 200	Pr	14.9
			Nd	1.27
			Mn	0.51
Gd	$L\alpha$	LIF 200	Ce	6.95
			La	0.15
Tb	$L\alpha$	LIF 200	Sm	0.75
			Fe	0.25
			Ce	0.08
Dy	$L\alpha$	LIF 200	Mn	31.4
			Eu	0.81
Ho	$L\beta$	LIF 200	Dy	0.46
Er	$L\alpha$	LIF 200	Tb	7.00
			Fe	0.35
Tm	$L\alpha$	LIF 200	Sm	8.69
			Dy	1.09
Yb	$L\alpha$	LIF 200	Tb	0.44
			Dy	0.42
			Sm	0.28
Lu	$L\alpha$	LIF 200	Dy	4.48

EPMA. Usually microanalysis of apatite is undertaken using techniques such as ion probe [6] or PIXE [7]. Because such techniques are not widely accessible, we outline procedures for the more readily available technique of EPMA, and the relatively new, but potentially powerful technique of LAMP-ICP-MS.

Although EMPA is a mature microanalytical technique, the analysis of low concentrations of the REE in natural samples is not routine because of the presence of many mutually-interfering peaks, and problems of assigning background positions in a matrix comprising all of the REE. In contrast, LAMP-ICP-MS is a technique potentially capable of analysing many elements at trace concentrations with a spatial resolution of 5–10 μm. Analysis of the REE in natural samples by LAMP-ICP-MS poses quite different problems from those of EPMA, and these are outlined here.

Table 1 details the elements, lines and crystals used in the procedure, together with empirically determined values for the mutually interfering REE, and values for several other elements often associated with apatite. The interference values given in Table 1 should be used only as a guideline, as absolute values will depend primarily on the resolution of individual analysing crystals, and will differ for different microprobes.

The selection of background positions are critical for accurate analysis, particularly at low concentration levels, and Table 2 details those positions which are interference-free for apatities and other rare earth-bearing minerals. (It is important however, to check whether any unsuspected lower-order lines interfere with the selected background positions when analysing minerals of unusual major element compositions.)

Detection limits for the REE depend on several factors: the spectral interferences and overlap corrections necessary; the operating conditions, such as accelerating voltage and specimen current used; the counting times for peak and background positions, and the ability of the mineral itself to stand up to high probe currents without specimen damage occurring. Detection limits for apatite have been initially calculated on the 6σ "limit of determination" [14]. These values have been empirically determined by calculating the background values at the peak positions from interpolation of the overall background spectra. Under these operating conditions,

Table 2. Background positions used for REE in analysis of apatite by EPMA

Wavelength	Crystal	Sin theta ($\times 10^5$)	Theta
2.68931	PET 002	30935	18.02
2.60531	PET 002	29775	17.32
2.70473	LIF 200	67170	42.20
2.60728	LIF 200	64750	40.35
2.24086	LIF 200	55650	33.81
2.08180	LIF 200	51700	31.13
1.82812	LIF 200	45400	27.00
1.55028	LIF 200	38500	22.64

Table 3. LAMP-ICP-MS operating conditions

Laser	Spectron SL803 Nd: YAG
Wavelength	266 nm (quadrupled from 1064 nm)
Maximum energy	70 mJ (energy changed with size of crater desired)
Mode	fully Q-switched, TEM00
Pulse Length	10 ns
Laser repetition rate	10 Hz
Microscope	Leitz Aristomet
Laser focusing objective	25X or 36X
Spectrometer	VG PlasmaQuad 2+
Forward power	1350 W
Injector gas flow rate	1 L min^{-1}
Data acquisition mode	peak jumping
Points per peak	3
Dwell time per peak	10 ms
Total integration time for all	30 s
isotopes	–

Table 4. Interferences observed in the analysis of apatite by LAMP-ICP-MS

Interferent ion	Interfered ion
$^{135}Ba^{16}O^+$	$^{151}Eu^+$
$^{142}Ce^{16}O^1H^+$	$^{159}Tb^+$
$^{143}Nd^{16}O^+$	$^{159}Tb^+$
$^{144}Nd^{16}O^+$	$^{160}Gd^+$
$^{160}Dy^+$	$^{160}Gd^+$
$^{146}Nd^{16}O^+$	$^{162}Dy^+$

detection limits as low as 0.015 wt.% are achievable for some elements, although more realistic values in natural apatite containing several REE typically range from 0.03–0.07 wt.% (Table 4). Where large interference corrections are necessary, for example in the measurement of Eu and Gd using Lα lines in minerals enriched in the lower atomic number REE, detection limits (and standard errors) are much larger and will depend on the concentration of the interfering element.

Experimental

LAMP-ICP-MS Instrumentation

The laser ablation microprobe (LAMP) consists of a Spectron Nd: YAG laser operating in the far-UV (266 nm), linked to a high quality Leitz microscope and is reported in UK Patent 9106337.0, Serial 2254444, 1992. The advantage of using a UV laser rather than a conventional Nd: YAG laser operating in the IR is better absorption of the radiation by geologically and environmentally interesting phases such as calcite and silicate minerals low in iron and magnesium. A VG PlasmaQuad 2+ spectrometer, with a modified sample introduction system (described below) was used in peak jumping mode to achieve the best detection limits.

In this paper, we describe analytical procedures for the two techniques in the analysis of well-characterised mineral apatite grains, and illustrate their application in the analysis of fossilised biological apatite.

Methodology

EPMA. Although EPMA is a well-established technique for the microanalysis of many minerals [8], quantitative analyses of the REE in minerals is not routine and poses specific analytical problems. Several published procedures refer in general to analysis of REE in minerals [9–11], and specifically to the overlap and interference problems [12–14]. The preferred method is to use a wavelength-dispersive (WD) rather than energy-dispersive (ED) microprobe, because of spectral interference problems (outlined below) and because detection limits by WD are 1 to 2 orders of magnitude better than by ED [14]. Some of the additional problems of energy-dispersive analysis of rare-earth minerals have also been discussed [15].

An analytical procedure has been developed for the analysis of REE in mineral and biological apatites based on a Cameca SX50 four spectrometer wavelength-dispersive microprobe located at the Natural History Museum, London. The operating conditions used in this study were: 25 kV accelerating voltage; 150 nA current measured on a Faraday cup; 10 μm spot size; count times of 100 seconds on both peak and backgrounds. Standards used for the REE were a series of single REE-Ca-Al-silicate glasses prepared by Peter Hill (University of Edinburgh), where concentrations of each REE vary from 11 to 22 wt.%. A well-characterised natural fluorapatite was used as the standard for Ca, P and F; natural celestine for Sr; synthetic barium fluoride for Ba, and pure metals for Mn and Fe. Matrix corrections were made using a PAP $\phi\rho z$ model.

Two major problems arise in the analysis of REE in minerals by EPMA: 1) the presence of interfering peaks, both from the REE themselves, and from other commonly-occurring elements such as Mn, Fe, Sr and Ba, 2) the choice of background positions in a crowded region of the energy spectra. These problems are illustrated in WD spectra using a PET 002 diffracting crystal for REE-rich and REE-poor regions of bone apatite samples (Fig. 1). The presence of detectable concentrations of Ba in the sample not only produces a significant interference at the Ce Lα wavelength, but also decreases available background positions.

LAMP-ICP-MS. The ICP-MS is capable of operating in either scanning mode or peak jumping mode. In scanning mode the whole mass spectrum from Li (at 7 amu) to U (at 238 amu) can be covered. However, because the laser ablation produces a small discrete signal for only a few seconds, much time is wasted acquiring redundant data in the scanning mode. Therefore, peak jumping to only those isotopes of interest provides a better signal/background ratio and consequently improvements in detection limits and reproducibility.

During ICP-MS analysis the test material under investigation is mounted in a Perspex cell placed under the microscope. The sample can be illuminated in either transmitted or reflected light. After choosing the area of interest and placing this under the cross-wires, the laser is fired through the microscope, in this study for 30 seconds. The ablated material is transported to the ICP-MS instrument by a continuous flow of argon through the cell and connecting tubing, which connects into the back of the ICP torch.

In the ICP torch the sample is vaporised and ionised by a plasma. The ionised material passes into the vacuum region of the mass spectrometer where a quadrupole selects ions of an appropriate mass to change ratio. This quadrupole can be rapidly scanned or peak jumped from masses as light as 7 (lithium) up to at mass 238 (uranium). An electron mutiplier then detects the ions and an associated multi-channel analyser counts them.

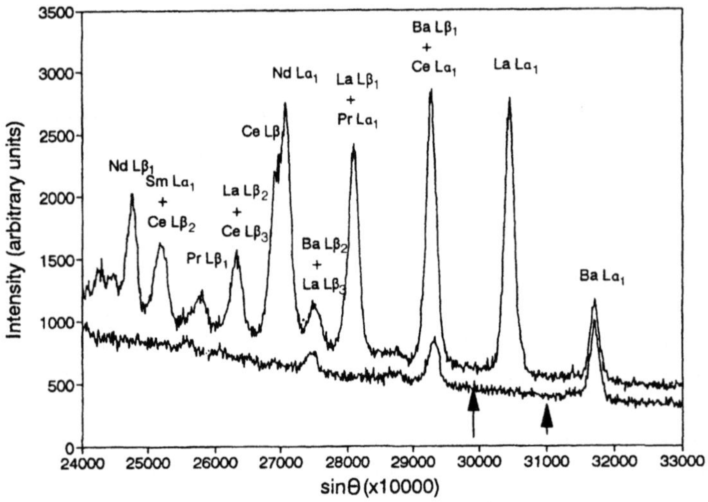

Fig. 1. Wavelength-dispersive EPMA spectra for REE-rich and REE-poor fossil bone apatite samples. Arrows indicate background positions selected. Operating conditions: 20 kV, 100 nA, PET 002 analysing crystal

To overcome difficulties in optimising the instrument in this configuration, particularly when measuring transient signals, a dual gas flow sample introduction system has been adopted in which flow from the ablation cell is merged with the carrier gas from a conventional nebulizer/spray chamber. These two gas flows are combined concentrically in a small glass mixer placed between the nebuliser and the torch. The advantage of this configuration is that it permits simultaneous introduction of laser ablated solids and nebulized aqueous solutions, which enter an identical bulk plasma environment. As a consequence, the ICP mass spectrometer can be tuned and directly calibrated using aqueous standards just as if solution nebulisation was being performed. This procedure solves the problem of finding suitable solid calibration standards, which may not be homogeneous on a scale of a few micrometres. This 'dual gas flow' technique of calibration has previously been described by one of the authors [16–18] and a similar system described by Moenke-Blankenburg et al. [19]. LAMP-ICP-MS operating details are given in Table 3.

It is necessary to use an internal standard to compensate for the poor precision in the analytical signal caused by variations in the mass of the material ablated. This is normally a minor isotope of a major elemental component, the concentration of which is known from stoichiometry or determined by an independent analytical technique [17, 18]. In this study the Ca concentration has been assumed from the stoichiometry of apatite or determined by EPMA.

Some small interferences are found in LAMP-ICP-MS, these are normally caused by oxide or hydride species but occasionally by hydroxides. Interferences corrected for are given in Table 4.

Laser ablation sampling of the test materials and detection limits. One of the fundamental properties of LAMP-ICP-MS is that, for a given set of operating parameters, the detection limits of the system are directly proportional to the mass of material sampled. Therefore for a particular application a compromise may have to be made between the desired spatial resolution and detection limits. Material for the bones and teeth was sampled in two fashions, either as discrete spots or as rasters. The advantage of sopt analysis is the high spatial resolution, crater sizes may be varied between 40 μm diameter and greater than 50 μm deep to smaller than 1 μm diameter and deep. However, this may not give enough material to obtain the very best detection limits needed, or if the material is heterogeneous, may give a biased sample. In this situation the beam may be manually rastered and a significantly larger mass of material removed. The detection limits and repeatability are also improved by limiting the number of isotopes determined in any one acquistion to 12 to 20 and peak jumping just between those elements [20].

For an ablated mass equivalent to a crater 20 μm in diameter and 10 μm deep, the following list gives theoretically calculated detection limits, in ppm, for a wide variety of elements: Mg: 10; Al: 40, Ca: 140, Fe: 15, Cr: 10, Mn: 5, Co: 1, Cu: 5, Zn: 2, Sr: 0.6, Ba: 0.4, Pb: 0.6, REE: 0.2–2, Th: 0.2, U: 0.4.

Results and Discussion

Analysis of 'Homogeneous' Separated Apatite Grains by EPMA, LAMP-ICP-MS and INAA

A number of individual apatite grains have been analysed by LAMP-ICP-MS during 4 separate analytical sessions to monitor accuracy and reproducibility. Table 5 summarises the results and compares them with previous INAA data. The reproducibility is proportional to the concentration with RSDs of 19–51% for REEs > 10 μg/g, RSDs of 23–50% for 1–10 μg/g and 68–88% for < 1 μg/g. Also some of the variance for Gd is due to improved corrections for light REE oxide interference i.e. $^{144}Nd^{16}O^+$ on $^{160}Gd^+$.

A comparison with INAA and EPMA data shows good agreement between all three techniques for Nd and Sm; reasonable agreement with La by INAA and EPMA, Tb and Yb by INAA and not such good agreement for Ce and Eu. The small difference for Yb between LAMP-ICP-MS and INAA suggests that a small correction for the $^{143}Nd\ ^{31}P^+$ ion is required. As yet no interferences have been found to account of the apparently high Ce and Eu values.

Table 5. Comparative data (μg/g) by LAMP-ICP-MS and EPMA for mineral apatite

	LAMP-ICP-MS			EPMA	INAA	
	(mean)	(std. dev.)	(n)	(mean)	(std. dev.) (n = 24)	(mean) (n = 2)
La	1270	240	(31)	1160	170	1101
Ce	2880	856	(31)	1920	400	1831
Pr	233	59	(31)	< 500[a]	–	n.a.
Nd	595	126	(31)	670	190	604
Sm	65	17	(31)	< 300	–	69
Eu	19.0	5.9	(31)	< 700	–	14.4
Gd	38.1	15.4	(25)	< 500	–	< 40
Tb	3.3	1.5	(25)	< 500	–	3.6
Dy	15.2	7.8	(15)	< 300	–	n.a.
Ho	1.9	0.5	(15)	< 300	–	n.a.
Er	5.6	2.8	(15)	< 400	–	n.a.
Tm	0.9	0.6	(25)	< 400	–	< 1
Yb	4.2	1.8	(25)	< 300	–	3.05
Lu	1.0	0.9	(25)	< 400	–	< 0.7

Major element concentrations by EPMA: CaO = 55.6%; P_2O_5 = 39.4%; BaO = 0.21%; SiO_2 = 0.19%, F = 3.57%.

[a] Detection limits by EPMA are the 6σ "limit of determination" based on 100 s count time, 25 kV and 150 nA probe current, and take into account potential overlaps from a typical natural apatite matrix.

It is proposed, in future, that other REE-rich mineral phosphates will be investigated to try and identify the potential causes of systematic differences between analytical methods and to provide reference materials for phosphate mineral analysis.

Applications

REE in fossil bone samples. The distribution of REE across the cortices of fossil bone apatite from Olduvai Gorge, Tanzania (1.8-1.2 mA years) were analysed by LAMP-ICP-MS. Previous work on separated sections of the cortices by INAA have shown these elements to be heterogeneously distributed [1], but in a systematic manner, such that a concentration profile exists with the outer portion of the fossil bone apatite enriched in REE relative to the centre of the cortex.

Although the concentration levels of REE at the extreme outer portions are sufficiently high to allow analysis by EPMA, the central regions require a more sensitive microanalytical technique. Ion probe microanalysis is one analytical option (although at high cost); LAMP-ICP-MS is a second, lower cost option, and was used in this study.

Fossil bone samples of an equid from the oxic "Red Beds" (No. OLD.1 – a fluvial environment), and a reptile from anoxic "Green Beds" (No, OLD-8.10 – a lacustrine environment) were analysed across their cortices. The distribution of the REE within the cortices generally showed the anticipated enrichment at outer and inner portions, i.e. closest to the bone/sediment interface, and depletion in the central regions. However, the spatial distribution of Ce differs significantly from the other REE, it having a much flatter profile across the bone cortex in the sample from the oxic depositional environment. This is best seen in conventional chondrite-normalised plots [where the concentrations of individual REE are normalised (i.e. divided by) the concentrations observed in chondritic meteorities] where large positive Ce anomalies occur in the central regions of the bone cortex (Figs. 2a, 2b).

Ce can exist in two oxidation states (3 + and 4 +) in natural surface environments – the other REE being stable only in the 3 + state. To interpret the anomalous distribution of Ce it is suggested that in the oxic depositional environment of the "Red Beds", Ce occurred predominantly as Ce^{4+}, (which result in the large positive Ce anomalies observed, Fig. 2a). In the anoxic lacustrine environment, Ce occurred predominantly as Ce^{3+} giving much smaller Ce anomalies (Fig. 2b). Ce in these contrasting states was then incorporated into the bone apatite in relatively differing amounts during the fossilisation process, thus the magnitude of the Ce anomaly reflects the redox potential of the depositional environment.

REE in fossil fish teeth samples. Several studies have also been undertaken of the REE distribution in fossilised fish remains (i.e. teeth and bone) [6, 22, 23] which have attempted to derive palaeosalinity or other palaeoenvironmental information. In this study, samples of fossil (and Recent) freshwater and marine fish teeth were analysed for the REE by LAMP-ICP-MS. In the Recent samples, all the REE were below detection limits, however, variable concentrations of these elements were observed in the fossil samples.

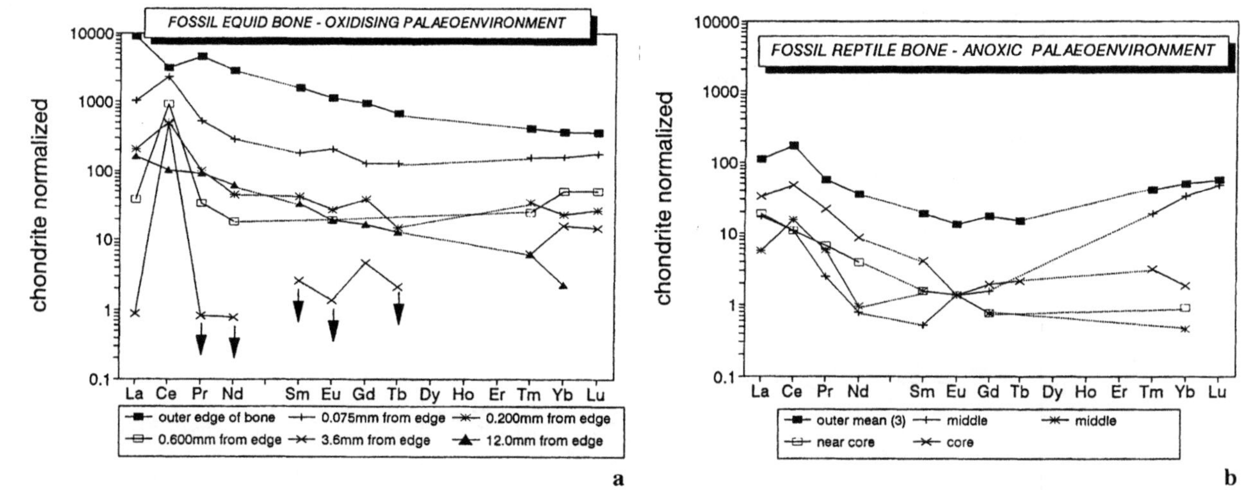

Fig. 2. REE data for fossilised bone samples from contrasting depositional environments at Olduvai Gorge, Tanzania. **a** Chondrite-normalised plots for sections acoss the bone cortex of sample OLD.1. **b** Chondrite-normalsied plots for sections acoss the bone cortex of sample OLD-8.10

Fossil samples analysed: Pliocene ray tooth (*Aetobatus* sp. BM No. P63630) from Lee Creek Mine, Yorktown Formation – open marine environment; a Pleistocene Great White Shark tooth (*Carcharodon carcharias* sp. P63629) from Lee Creek Mine, James City Formation – shallow marine bay environment; a Pleistocene pike tooth (*Esox lucius* West Runton, Norfolk) – freshwater river environment. The data shown as chondrite-normalised REE plots in Fig. 3 illustrate the large variations in absolute abundances between marine samples, and also subtle differences in the chondrite-normalised patterns. Large differences in absolute abundances may be a function of the relative ages of the samples, the older Pliocene ray tooth being enriched in REE relative to the younger Pleistocene shark tooth. However, uptake of trace elements into fossil bone and teeth is now considered to be a relatively rapid process taking place early in the diagenetic history of the sample. A more likely interpretation for the observed variation is that ray teeth have a significantly more porous structure than pike or shark teeth and is thus more susceptible to REE uptake and incorporation during the fossilisation process. It was not possible in this case to correlate the REE chondrite-normalised patterns with salinity type, and the observed patterns probably reflect some degree of diagenetic overprint which has destroyed any original signal from the burial environment.

Advantages and Limitations of EPMA and LAMP-ICP-MS Techniques

This study has allowed a comparison to be made of the advantages and limitations of EPMA and LAMP-ICP-MS for the determination of REE in mineral and biological apatites.

EPMA: Quantitative with good matrix correction procedures, detection limits typically > 200 µg/g allowing measurement of only lightest REE in studied apatite,

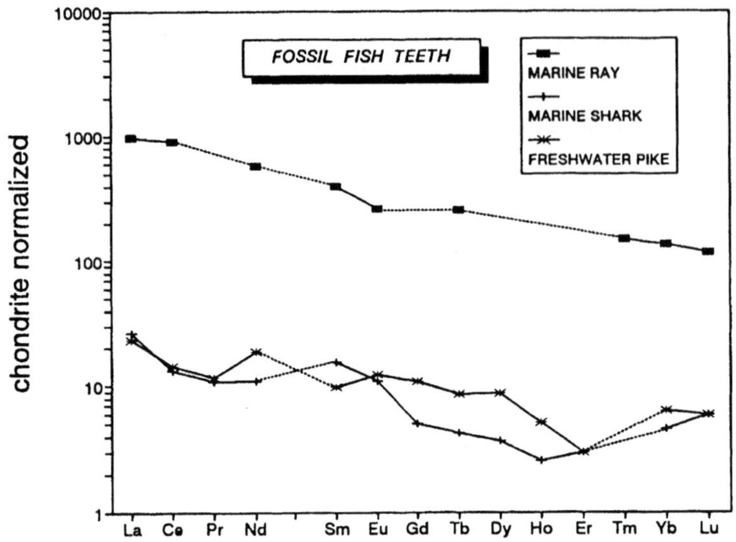

Fig. 3. Chondrite-normalised plots for fossil fish teeth. Samples analysed: Pliocene ray tooth (*Aetobatus* sp. BM No. P63630) from Lee Creek Mine, Yorktown Formation – open marine environment; a Pleistocene Great White Shark tooth (*Carcharodon carcharias* sp. P63629) from Lee Creek Mine, James City Formation – shallow marine bay environment; a Pleistocene pike tooth (*Esox lucius*, No. 40307–8, West Runton, Norfolk) – freshwater river environment

elemental mapping possible, isotopic information not possible, flat, polished, coated surface needed.

LAMP-ICP-MS: Requires an internal standard for quantitative analysis. Matrix effects (fractionation) have been observed by others [21], detection limits $< 2\,\mu g/g$ for $20\,\mu m$ diameter crater and measurement of all REEs possible in studied mineral apatite, no elemental mapping possible, isotopic information possible, no special sample preparation required.

Conclusions

Microanalytical techniques on biological apatite can provide important information on aspects of the depositional environment, and can assist in reconstruction of the palaeoenvironment.

Analysis for the REE in biological apatite by EPMA can be achieved if the total concentrations of these elements are at the 1 wt.% level. Advances in instrument stability and reliability have enabled unattended overnight operation to become routine, and advances made in computational power have been essential in assisting the collation and interpretation of the large amounts of data generated. It is anticipated that future trends in EPMA will build on these improvements, and with higher kV equipment probable, this technique may enter the field of "trace" microanalysis, where improvements in detection limits to $20–50\,\mu g/g$ are foreseen. LAMP-ICP-MS is already capable of achieving these levels of detection, and in the analysis of REE in mineral or biological apatite has many advantages over that of EPMA.

Techniques such as ion probe and PIXE, although providing significant improvements in detection limits for the REE compared with EPMA, are relatively inaccessible, and are unlikely to be used for routine microanalysis. The low detection limits (typically < 1 μg/g) already obtainable by LAMP-ICP-MS for modest crater sizes (20 μm) combined with good selectivity which minimises interference corrections thus provides a real alternative to these techniques for applications such as the determination of REE in biogenic apatites. The new generation of high sensitivity ICP-mass spectrometers promises to reduce detection limits by an order of magnitude for the same crater size, or to provide similar detection limits on a crater size of half the diameter and depth. In addition, the relatively lower cost of LAMP-ICP-MS should ensure that this technique will become increasingly available in the near future, and will mature to become complementary to EPMA.

Acknowledgements. This paper is published with the permission of the Director, British Geological Survey (Natural Environment Research Council). The authors thank Jennifer Cook (BGS) for advice and support of this research. This work has been part-funded by a grant from Swedish Natural Science Research Council (No. 06407) to L.W.

References

[1] P. Henderson, C. A. Marlow, T. I. Molleson, C. T. Williams, *Nature* **1983**, *306*, 358.
[2] J. Wright, H, Schrader, W. T. Holser, *Geochim. Cosmochim. Acta* **1987**, *51*, 631.
[3] K. Toyoda, M. Tokonami, *Nature* **1990**, *345*, 607.
[4] C. T. Williams, P. J. Potts, *Archaeometry* **1988**, *30*, 237.
[5] C. T. Williams, in: *Trace Elements in Environmental History* (G. Grupe, B. Herrman, eds.), Springer, Berlin Heidelberg New York Tokyo, 1988, p. 27.
[6] P. Grandjean, F. Albarede, *Geochim. Cosmochim. Acta* **1989**, *53*, 3179.
[7] T. A. Elliott, G. W. Grimes, *Nucl. Instr. Meth. Phys. Res.* **1993**, *B77*, 537.
[8] S. J. B. Reed, *Electron Microprobe Analysis, 2nd Ed.*, Cambridge University Press, 1993, 326pp.
[9] R. Åmli, W. L. Griffin, *Am. Mineral.* **1975**, *60*, 599.
[10] R. A. Exley, *Earth Planet. Sci. Lett.* **1980**, *48*, 97.
[11] J. B. Dawson, J. V . Smith, I. M. Steele, *Chem. Geol.* **1994**, *17*, 285.
[12] P. L. Roeder, *Can. Mineral.* **1985**, *23*, 263.
[13] P. L. Roeder, D. MacArthur, X.-P. Ma, G. R. Palmer, A. N. Mariano, *Am. Mineral.* **1987**, *72*, 801.
[14] P. J. Potts, *A Handbook of Silicate Analysis*, Blackie, Glasgow, 1987, 622pp.
[15] D. G. W. Smith, S. J. B. Reed, in: *Proc. Instit. Phys. Conf. Series 61: Electron Microscopy and Analysis Group Conference 1981* (M. J. Gorinje, ed.), University of Cambridge, 1982, 159.
[16] S. Chenery, J. Cook, *J. Anal. Atom. Spectr.* **1993**, *8*, 299.
[17] X. Querol, S. Chenery, in: *European Coal Geology* (M. K. G. Whateley, D. A. Spears, eds.), Geol. Soc. Spec. Publ. 82, 1995, p147.
[18] S. Chenery, J. M. Cook, M. Styles, M., E. M. Cameron, *Chem. Geol.* **1995**, *124*, 55.
[19] L. Moenke-Blankenburg, T. Schumann, D. Günther, H.-M. Kuss, M. Paul, *J. Anal. At. Spectrom.* **1992**, *7*, 251.
[20] L. Moenke-Blankenburg, D. Gunther, *Chem. Geol.* **1992**, *95*, 85.
[21] S. Chenery, J. M. Cook, F. Poitrasson, in: *European Winter Conf. on Plasma Spectroscopy*, 1995 (Abstract).
[22] P. Grandjean, H. Cappetta, A. Michard, F. Albarede, *Earth Planet. Sci. Lett.* **1987**, *84*, 181.
[23] H. F. Shaw, G. J. Wasserburg, *Geochim. Cosmochim. Acta* **1985**, *49*, 503.

Mikrochim. Acta [Suppl.] 13, 271–277 (1996)

Quantitative Analysis of the Compound Layer of Plasma Nitrided Pure Iron

Jan D'Haen*, Marc D'Olieslaeger, Luc De Schepper, and Lambert M. Stals

Institute for Materials Research, Materials Physics Divison, Limburgs Universitair Centrum, Universitaire Campus, Wetenschapspark 1, B-3590 Diepenbeek, Belgium

Abstract. Pure iron is plasma nitrided in a low pressure triode ion plating equipment. During this process a compound layer, containing different Fe–N phases depending on the plasma process parameters used, is formed at the surface.

In this paper, the nitrogen concentration of the different phases is determined by measuring the backscatter yield of the different layers in a cross-sectional sample. Under appropriate instrumental conditions, the backscatter yields of elements and compounds can be determined quantitatively in a scanning electron microscope by measuring the specimen current. A simple relation between the backscatter coefficient of a binary target and those of its constituents is used for a quantitative analysis of the binary Fe–N phases.

The strength of the specimen current measurement technique for this kind of application lies in the high spatial resolution which can be obtained using a small spotsize of the primary electron beam. This high resolution is required because the different Fe–N layers are very thin. Because the limitations of photon counting statistics are not involved, meaningful measurements can be performed very rapidly. As compared to X-ray analysis techniques, this is an advantage because in this way carbon contamination during the experiment can be minimised.

Key words: Fe-N, nitrogen concentration, plasma nitriding specimen current, electron backscatter yield.

Several techniques, based upon the interaction of an electron beam with a specimen, are available to investigate the chemical composition of a material. When an electron beam hits the specimen several signals like X-rays, photons and electrons escape from the sample. Mostly X-rays are used to characterize the chemical composition of the specimen (X-ray spectrometry: EDS and WDS). If only a general idea of the sample composition is necessary and if a lower accuracy is allowed (>0.1 wt%), an EDS analysis in a conventional SEM can be used. If however a quantitative analysis of light elements is desired very often a WDS analysis

* To whom correspondence should be addressed

(EPMA) is used. The major disadvantages of a WDS analysis are the very high focussing requirements, the time required to perform the analysis, the formation of a carbon contamination layer during the analysis and the need for standards in quantification [1]. Especially soft X-rays like N–K_α and O–K_α lines are absorbed markedly in a carbon layer [2]. Also the very high probe current, especially required for light element analysis (up to 500 nA), is a large disadvantage since it reduces the lateral resolution considerably (probe size up to 0.5 μm).

Some of these disadvantages can be avoided if not X-rays but electrons are used in the chemical composition analysis. Especially backscattered electrons are very sensitive to the atomic number and less to the surface roughness and the crystal orientation. In absence of the surface roughness and of a difference in crystal orientation, the backscatter yield is found to increase monotonically with increasing atomic number. This allows an immediate semi-quantitative analysis of the specimen composition when the backscatter yield can be determined. According to Castaing [3], the backscatter yield η for a binary alloy, like the Fe–N binary system, can be written as (c_A: concentration of element A):

$$\eta_{\mathrm{alloy}} = c_A\eta_A + (1 - c_A)\eta_B. \tag{1}$$

Backscattered electrons can be detected directly with BSE detectors, or indirectly by measuring the specimen current [4]. In general, there exist two different methods to measure the specimen current: the grid method and the composite target method [5]. According to Heinrich [6] the target current technique is a valuable tool for the characterization of small areas of binary specimens if the substituent elements differ considerably in atomic number. As for WDS analysis, extreme flatness of the samples is required. In contrast with WDS, meaningful measurements can be performed very rapidly since the limitations of photon counting statistics are not involved [6] and the reduced probe size allows the analysis of a layered structure with a higher resolution. Comparing both techniques, the specimen current technique seems very adequate to determine the nitrogen concentration in the thin (several microns) compound layer of plasma nitrided pure iron.

Experimental

In this paper the specimen current technique based upon the composite target method was used. As mentioned by Colby [5] the composite specimen method of Heinrich allows a more complete understanding of the different electron currents in the vicinity of the sample and it circumvents the errors inherent to the grid method. The composite target is extended from two [6] to three parts, which can be biased independently. Also, the currents through the different parts can be measured independently (Fig. 1). Figure 1a shows the experimental setup in top-view. The corresponding side-view is given in Fig. 1b. The specimen is mounted on position 1 and has dimensions of 7 × 1 mm^2 only. This small size has the beneficial effect that very few electrons, backscattered or generated at the walls of the SEM, also called peripherical current [3], will hit the sample. The specimen is biased at + 50 V in order to avoid the escape of secondary electrons. In the middle of the target, at position 2, is an area which is biased at − 300 V. Its function is to avoid extraction of secondary electrons from the sample. The small distance between this area and the two areas of the composite target causes high electrical fields (1000–3000 V/cm) to be compared with the collection field of the Everhart-Thornley (ET) detector (100 V/cm) [7]. The outer part of the target, at position 3, is very large as compared to the sample and is biased at + 50 V. Due to the large size most of the electrons, backscattered or generated elsewhere in the

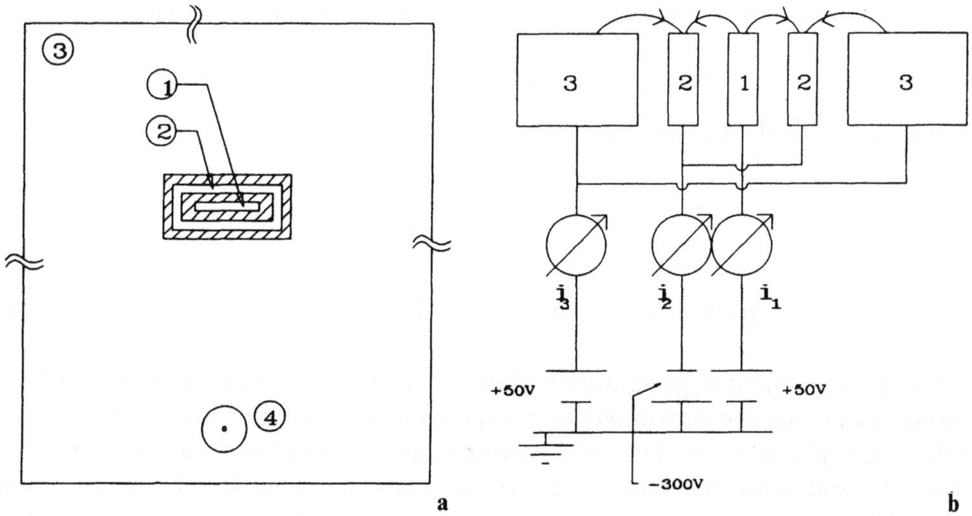

Fig. 1. Schematic drawing of the extended composite target: top-view (**a**) and side-view (**b**)

SEM, will be collected. In order to determine the beam current, a Faraday cup has been placed at position 4. One restriction is however that the beam stability cannot be measured by the Faraday cup during the measurement of the specimen current on a sample. By applying the mentioned voltages to the different parts of the composite target it is possible to avoid the emission of secondary electrons from the specimen and to reduce the influence of the peripherical currents on the specimen current.

Pure iron samples were plasma nitrided in a nitrogen-argon atmosphere using a low pressure Balzers triode ion plating equipment [8]. After the nitriding process the samples were polished in cross-section after having been coated with a Cu–Ni layer to protect the edges.

The specimen current measurements are carried out in a Philips SEM 535-M using an accelerating voltage of 20 kV and a spotsize of 200 nm diameter as a compromise between a small interaction volume and a high current intensity. The measurement of the primary electron beam current with the Faraday cup revealed a slightly decreasing intensity as a function of time. The specimen current signal is recorded during a linescan over the nitrided sample in cross-section. The specimen is scanned several times to obtain an idea of the drift in the specimen current due to a possible change in the primary electron beam current.

Results and Discussion

If the specimen current can be measured properly it is possible to determine the backscatter yield of the specimen directly. To avoid charging of the sample the netto flux of the electron beams has to be zero according to Thevenin's theorem. The specimen current is equal to [5]:

$$i_{SC} = i_B - i_{BS} - i_{SE} + i_{BS}^{(2)} + i_{SE}^{(2)} + i_{SE}^{(3)} \qquad (2)$$

where i_B is the incident primary electron beam current, i_{BS} the backscattered electron beam current, i_{SE} the secondary electron beam current, $i_{BS}^{(2)}$ and $i_{SE}^{(2)}$ the backscattered respectively secondary electron beam current generated by the backscattered electrons and $i_{SE}^{(3)}$ the SE current caused by the SE impinging upon the walls. Using the notations δ, δ_W, η and η_W for respectively the secondary yields, the backscatter yields of the specimen and the walls (index W), the following relations can be written:

$$i_{BS} = \eta i_B, \quad i_{SE} = \delta i_B, \quad i_{BS}^{(2)} = \eta_W i_{BS}, \quad i_{SE}^{(2)} = \delta_W i_{BS}, \quad i_{SE}^{(3)} = \delta_W i_{SE} \tag{3}$$

The backscatter yield can then be expressed as:

$$\eta = 1 - \delta - \frac{i_{SC}}{i_B} + \frac{i_{BS}^{(2)} + i_{SE}^{(2)} + i_{SE}^{(3)}}{i_B}$$

or

$$\eta*(1 - \eta_W + \delta_W) = 1 - \delta*(1 - \delta_W) - \frac{i_{SC}}{i_B} \tag{4}$$

With some experimental precautions these relations can further be simplified. Secondary electrons are prevented to escape from the sample due to the bias of +50 V, which yields $\delta = 0$. This bias voltage causes the attraction of secondary electrons generated by the backscattered electrons at the walls of the specimen chamber. However, the contribution of these peripherical currents (δ_W and η_W) is minimized to less than 1% of the total specimen current due to the small specimen size, the high electrical field caused by the area biased at −300 V and the large size of the outer part of the target. The backscatter yield can thus be approximated by:

$$\eta = 1 - \frac{i_{SC}}{i_B} \tag{5}$$

The thus obtained experimental values correspond well with the backscatter yields for iron, nickel and copper tabulated in literature. The experimental setup for the determination of backscatter yields is similar to the configuration used by Bishop [9]. In this way the need of a grid, to screen the influence of the walls, can be avoided.

The extended composite target method was used to determine the nitrogen concentration profile in the plasma nitrided layers using the backscatter yields for nitrogen and iron. The backscatter yield for pure iron can be determined from the inner part of a nitrided specimen. Due to the lack of an appropriate standard for the determination of the backscatter yield of nitrogen, a gas nitrided pure iron specimen has been used. This result is checked with the pure iron specimen which has been plasma nitrided for 9 hours. From optical microscopy it is known that both specimens have a dual phase compound layer. The outer part of the compound layer is the hexagonal ε-$Fe_{2-3}N$ phase in which a concentration gradient is possible. The inner part of the compound layer is the cubic γ'-Fe_4N phase which has a very narrow (±0.1 wt%) concentration range around 5.8 wt% N.

Positioning the primary electron beam on the γ'-Fe_4N phase, knowing the backscatter yield for pure iron and using the relation for the backscatter yield for a binary alloy, the backscatter yield for nitrogen can be determined. This value is used to determine the nitrogen concentration profile in the ε-$Fe_{2-3}N$ phase. The nitrogen concentration can be calculated by:

$$\text{wt\%N} = 100*\frac{\eta_{Fe} - \eta}{\eta_{Fe} - \eta_N} \tag{6}$$

In the previous paragraph it has been tacitly assumed that the specimen current is only a function of the chemical composition of the specimen. The specimen current is

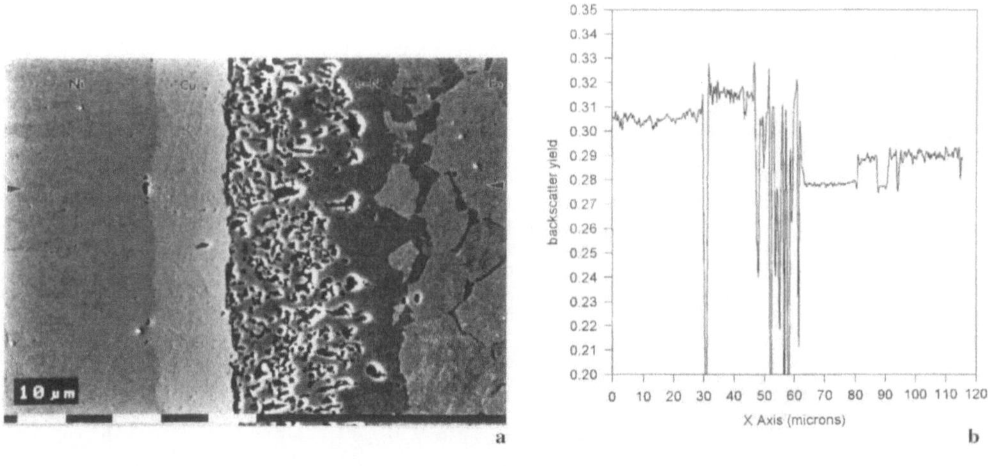

Fig. 2. Cross-section of a gas nitrided pure iron sample: (B)SE image (**a**) and backscatter yield curve (**b**)

however very sensitive to the surface roughness, which makes a careful polishing of the samples very important. Another problem, which cannot be avoided, is the dependence of the backscatter yield upon the crystallographic orientation, a phenomenon used in electron channeling contrast imaging. To get an idea of the spread upon the backscatter yield due to the above mentioned problems, the specimen current has been measured on the pure iron specimen at 32 positions with random grain orientations. At 20 kV and a 200 nm spotsize the iron backscatter yield is determined as 0.289 ± 0.003.

Figure 2a shows the SE image of the gasnitrided specimen which has been used as standard. Although the specimen is biased at +50 V, is it possible to form an image using the ET-detector as BSE detector. The differences in intensity on such a BSE image correspond then to differences in atomic contrast. The specimen current signal reveals an inverted contrast as compared to the SE image [3]. To reveal the same contrast the specimen current signal is often inverted. The same effect can be obtained by observing the backscatter yield curve instead of the specimen current signal. Figure 2b shows the backscatter yield curve, calculated with relation [5], corresponding with Fig. 2a. The curve is a horizontal intersection of the image and is marked by the arrows and runs through the pore between the Ni and Cu layer. Both figures show the same atomic number dependence. Both curves show from the left to the right the Ni, Cu, Fe–N and Fe layers. The Fe–N layer is a dual phase with a thick porous ε-$Fe_{2-3}N$ layer and a thin γ'-Fe_4N layer. The interface between the compound layer and the pure iron substrate is not straight due to a pronounced grain boundary diffusion process. The γ'-Fe_4N phases along the grain boundaries, also called fingers, are clearly visible in the BSE image and the backscatter yield curve due to their lower intensity. The γ'-Fe_4N layer and the nitride fingers are used further to determine the backscatter yield for nitrogen. From Fig. 2b the backscatter yield for Ni, Cu and Fe can be determined directly ($\eta_{Ni} = 0.306$, $\eta_{Cu} = 0.316$, $\eta_{Fe} = 0.288$). The backscatter yield for γ'-Fe_4N is 0.277 ± 0.001. Assuming that the γ'-Fe_4N phase contains 5.8 wt% N the backscatter yield value for nitrogen can be calculated with relation (1) to be 0.098 ± 0.017, assuming no spread upon the iron

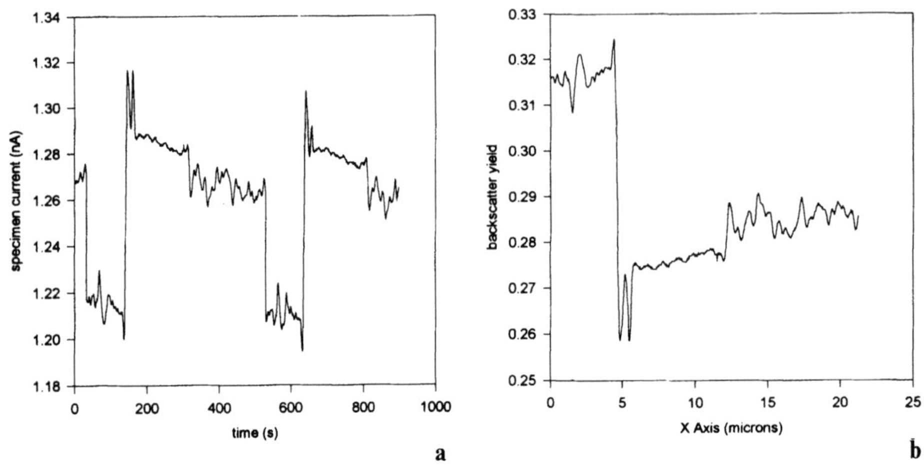

Fig. 3. Cross-section of a plasma nitrided pure iron sample: specimen current (**a**) and backscatter yield curves (**b**)

backscatter yield ($\eta_{Fe} = 0.288$). A small uncertainty in the experimental backscatter yield value causes thus a larger uncertainty upon the η_N value. This uncertainty is a serious problem in the absolute determination of the nitrogen concentration in a nitride layer. The total curve, in Fig. 2b, shows ripples because it is nearly impossible to polish the heterogeneous and rather soft parts in the specimens perfectly. The compound layer shows in the γ'-Fe_4N phase almost no ripples in the backscatter yield curve due to its higher hardness. Figure 3 shows the specimen current Fig. (3a) and the corresponding backscatter yield curve Fig. (3b) for the plasma nitrided pure iron specimen. Figure 3a shows a continuous decrease of the intensity due to a decreasing primary electron beam intensity. In Fig. 3b a correction for the decrease has been carried out assuming a linear decrease as a function of time. The backscatter yield curve reveals at the left the copper layer and at the right the iron substrate. Again the ripples in the curve are present due to a minor surface roughness. In the middle of the curve an increasing part is visible which corresponds with the Fe–N compound layer. This curve shows almost no ripples due to the higher hardness. In contrast with the gas nitrided standard the plasma nitrided specimen contains no pores in the compound layer. The almost linearly increasing backscatter yield of the compound layer towards the substrate corresponds to a linearly decreasing nitrogen concentration. Using relation (6) it would be possible to calculate the nitrogen concentration as a function of the position in the compound layer. At the moment however it is not possible to calculate the nitrogen concentration exactly because the primary electron beam current is not stable enough to calculate the nitrogen backscatter yield with a high enough accuracy.

Conclusions

The extended composite method, with the appropriate choice of the bias voltages, presented in this paper is very suited to determine the backscatter yield of pure elements. If the primary beam current could be stabilised with a high precision or could be measured during the experiment simultaneously with the specimen current, it would be possible to determine the concentration of the two elements in a binary alloy with a high accuracy.

The specimen current measurement can be a very good and extremely fast alternative, in comparison with WDS, for the measurement of the nitrogen concentration in nitrided pure iron if the prerequisite for the primary beam is fulfilled.

The nitrogen backscatter yield has to be confirmed in future work by measuring other binary alloys with nitrogen as one of the elements.

Acknowledgements. The authors wish to thank M. A. J. Somers (T. U. Delft, The Netherlands) for coating the nitrided specimens with a Cu–Ni layer, and D. Polus for technical assistance in the development of the extended composite target.

Thanks are also due to the Belgian Science Supporting Institute "Interuniversitair Instituut voor Kernwetenschappen" for financial support under contract no. IIKW-4.00004.91, no. IIKW-4.00002.91 and no. IIKW-4.00001.96. This text presents research results of the Belgian Programme on Interuniversity Attraction Poles initiated by the Belgian State Prime Minister's Office (Brussels), Science Policy Programming, Scientific responsibility is assumed by the authors.

References

[1] E. Heikinheimo, *Proc. EMAS'94, Electron Probe Microanalysis of Materials Today – Practical Aspects, 15–17 June*, 1994, *Kirkkonummi, Finland, Part VIII.*

[2] P. Karduck, *Proc. EMAS'94, Electron Probe Microanalysis of Materials Today – Practical Aspects, 15–17 June 1994, Kirkkonummi, Finland, Part V.*

[3] R. Castaing, in: *Advances in Electronics and Electron Physics, Vol. XIII* (L. L. Marton, C. Marton, ed.), Academic Press, New York, 1960, p. 317.

[4] J. I. Goldstein, D. E. Newbury, P. Echlin, D. C. Joy, C. E. Fiori, E. Lifshin (eds.), *Scanning Electron Microscopy and X-ray Microanalysis*, Plenum, New York, 1981.

[5] J. Colby, *Electron Probe Microanalysis* (A. J. Tousimis, L. Marton, eds.), Academic Press, New York, p. 177.

[6] K. Heinrich, *X-ray Optics and Microanalysis, 4th Intl. Cong. on X-Ray Optics and Microanalysis* (R. Castaing, P. Deschamps, J. Philibert, eds.), Hermann, Paris, 1966, p. 159.

[7] D. E. Newbury, D. C. Joy, P. Echlin, C. E. Fiori, J. I. Goldstein (eds.), *Advanced Scanning Electron Microscopy and X-ray Microanalysis*, Plenum, New York, 1986.

[8] J. D'Haen, C. Quaeyhaegens, G. Knuyt, M. D'Olieslaeger, L. M. Stals, *Surf. Coat. Techn.* **1995**, *74–75*, 405.

[9] H. E. Bishop, in: *X-ray Optics and Microanalysis, 4th Intl. Cong. on X-Ray Optics and Microanalysis* (R. Castaing, P. Deschamps, J. Philibert, eds.), Hermann, Paris, 1966, p. 153.

Mikrochim. Acta [Suppl.] 13, 279–287 (1996)

Correction of the Edge Effect in Auger Electron Microscopy

Luděk Frank* and František Matějka

Institute of Scientific Instruments, Czech Academy of Sciences, Královopolská 147, 612 64 Brno, Czech Republic

Abstract. The edge effect, which manifests itself especially in an edge over-brightening in images formed by shallow escape depth signals, is addressed. Attempts had been made to utilize the ratio technique which was developed for suppression of other spurious contrasts in Auger micrographs; it consists in relating the Auger signal to a suitable energy filtered background signal. A systematic comparison of a quantified edge effect in both the Auger images and the energy selected background images is reported here. Surface steps on Si and W surfaces and a W terrace edge on a Si substrate were used at 3 to 20 keV primary electron energy, both low and high energy Auger electrons of Si and W were registered together with a background of up to 3 keV. Results have shown that the edge effect is less pronounced with the background so that any ratioing leads to an undercorrection. The best suppression (nearly acceptable for low energies, both primary and Auger ones) is obtained when using the background between 50 and 200 eV as the reference signal.

Key words: edge effect, Auger electron microscopy, ratioing techniques.

Any surface chemical heterogeneity of a specimen under observation in a microprobe or an analytical SEM is usually connected with some surface relief, i.e. the boundaries between the surface areas of different compositions are often formed by some surface steps or terrace jumps. The chemical image contrast, produced by Auger electrons, is then combined with the topographical contrast which is to be suppressed in order to get the net information. If the step is sharp and deep enough, its image is accompanied by a set of phenomena forming the so called edge effect.

The nub of the phenomenon is the difference between the great depth of penetration of exciting primary electrons and a shallow escape depth of energy-characterized analytical signals mediated mainly by Auger electrons. The exciting particles are entering the signal-giving subsurface layer not only after the initial impact but also after being backscattered from the depth. These diffused electrons "illuminate" the surface from the inside and their sources are omnidirectional and

* To whom correspondence should be addressed

not distant from the surface, in contradiction to the nearly parallel primary illumination. Thus, the backscattered illumination is quasihomogeneous over the real surface while the primary illumination is homogeneous over the surface projection into the observation plane. For a real non-smooth surface, any inclined facet represents, owing to the backscattered illumination, some additional signal source, projected into the plane of observation, which has an inclination dependent intensity. The contrasts actually observed are then influenced by angular acceptance of a detector.

In conventional scanning electron microscopy (SEM), the "edge effect" is the term most often used for signal enhancement caused by diffusively scattered electrons passing through an increased area, i.e. edge sidewalls [1]. It is considered as a useful, and often dominating, part of the topographic contrast but sometimes attempts are made to avoid it by using a lower primary energy [2].

The edge effect in Auger electron images, i.e. the basic edge signal enhancement due to some emission from the sidewall which cuts the interaction volume, was reported by Shimizu et al. [3]. They demonstrated a larger manifestation of the effect for low energy Auger electrons. The image of a heterogeneous specimen, i.e. a chemical edge, was simulated by Tuppen and Davies [4]. They recognized several possible phenomena near the terrace edge and pointed out a wide spread of signal variations to a distance a few times longer than the overlayer thickness, and some shift of the variations towards the overlayer. Umbach et al. [5] considered the primary energy dependence of the phenomenon and found it increasing with increasing energy between 10 and 30 keV for a 0.6 μm Au overlayer on Si. ElGomati et al. [6] presented a detailed analysis of sub-effects contributing to the edge effect for general excitation and take-off geometry and compared successfully their Monte-Carlo simulations with the experimental data taken with 1.4 μm Al and 640 nm Au overlayers on Si. They have pointed out that the usual ratioing techniques suppressing the topographical contrast are not successful here; some parts of this contrast effect are compensated but other artefacts are introduced. ElGomati et al. [7] proposed recently a novel method, based on Monte Carlo calculations, of detecting and correcting the edge effects in scanning Auger microscopy. Their method relies on using energy backscattered electrons which have an energy of 75% of the primary energy and hence are relatively surface sensitive. These electrons are claimed to have a behaviour similar to that of Auger electrons, and the ratioing with respect to this signal is to be usable for the correction of the edge artefacts.

The authors of this paper proposed recently a method of quantification of the edge effect consisting in spatial integration, along the line-scans taken across an edge, of the signal curve offsets from its trivial rectangular behaviour [8]. The main finding was that even for steps significantly higher than the penetration depth, the edge effect intensity continues to vary with the step height and it is much less dependent on the penetration depth, i.e. the primary energy.

In this work which is the extension of the earlier paper [8], systematic comparisons were made between the edge effect in Auger electron micrographs and in images taken in the energy filtered background across the whole energy range available, both for steps at homogeneous surfaces and a "heterogeneous" edge of a terrace of different composition. To quantify the comparison, the previously published method [8] was used. The purpose was to identify an energy of the background at

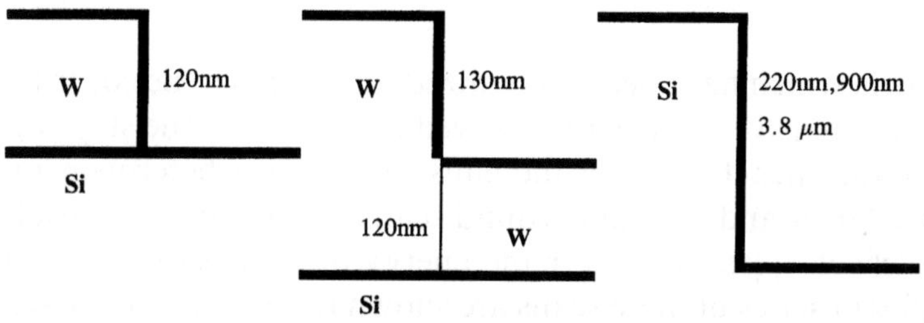

Fig. 1. The specimen geometry

which the edge effect intensity is nearest to that in an Auger image so that the ratioing is the most successful in suppressing the effect.

Experimental

The effect was examined at surface steps on Si and W substrates and at the W overlayer edge on a Si substrate (see Fig. 1). All specimens were prepared on standard silicon substrates used in the semiconductor industry. The geometrical shapes of individual testing structures were created by using electron lithography either with a positive (PMMA) or a negative (PCMS on the chlorostyren basis) resist. The resist exposition was made in the Tesla BS 600 e-beam vector scan lithograph.

Flat depletions etched into the Si substrate were prepared by using two methods: Shallow structures, up to 200 nm, were made by plasma etching in RF plasma in the mixture $CF_4 + O_2$ (15%) in a planar reactor through the resist mask. The deep relief structures, 1 to 4 μm, were anisotropically etched into the Si [110] surface in the aqueous solution of KOH (46 mass per cent) at 80 °C. The surrounding Si surface was protected by a layer of thermally grown SiO_2 250 nm thick which was finally removed in HF. The [110] orientation of the Si substrate was chosen in order to obtain edges perpendicular to the original surface. For the same reasons, the edges of the rectangular openings in the oxide mask were oriented perpendicularly to crystallographic directions [*111*] and [1*11*] for this etching. The metal (tungsten) relief was prepared by plasma etching of a layer deposited onto a silicon substrate by vacuum evaporation (about 250 nm). The etching was performed in the same plasma as before in a barrel reactor. Etching conditions (gas flow, pressure, power) were selected so that a minimum etching rate for silicon is achieved after the metal layer is passed through. The thicknesses of the layers and the step heights were measured by using the Taylor and Hobson Stylus instrument Talystep.

The experiments were made in the Perkin-Elmer 595 Auger microprobe. The electron impact was 30° from the normal but in the plane of the step sidewall. Owing to a limited point resolution of the device, a low beam current had to be used, namely 2 to 3 nA, which gives a resolution within 0.6 to 0.25 μm for a 3 to 20 keV primary energy. Five line-scans across the edge were always recorded with some intermediate lateral shifts and afterwards mutually fitted and summed, in order to eliminate undesired topography details. Micrographs of the structures and an example of the rough data and their accumulation are shown in [8].

The following structures were available: steps on the Si surface with heights of 0.12 μm, 0.9 μm and 3.8 μm, steps on the W surface 0.13 μm high and sharp edges of a 0.12 μm high W terrace on the Si surface. The Auger line-scans across the edges were recorded by using of Si LMM 89 eV, Si KLL 1617 eV, W NOO 164 eV and W MNN 1729 eV peaks (nominal energy positions were taken from a table in [9]). The experiments were carried out at 3, 10 and 20 keV of the primary energy. In addition to the Auger scans (published and discussed in [8]), line-scans in the energy filtered background were registered within the whole energy range available with PHI 595, i.e. 0 to 3000 eV. In all pictures, the specimens are oriented so that the upper plane above the step or the terrace is to the left from an edge.

Results and Discussion

From all the subphenomena which have been recognized (see [6, 8]) to be contributors to the edge effect, two intensive ones were observed only, namely the step edge enhancement and the mutual penetration of the emissions near a heterogeneous terrace edge, both for the Auger and the background scans. Important new results presented here are edge effect appearances for the energy filtered background at various energies. Examples of series of these scans are shown in Fig. 2 for the Si step edge (Figs. 2 a–c) and for the W/Si terrace edge (Fig. 2d). The step allocated enhancements shown in these pictures are typical for all examined cases.

The measure of the step edge enhancement is high for true secondaries but it still slightly increases with the increasing energy and reaches its flat maximum somewhere between 40 and 200 eV. Further, the enhancement continuously decreases. At lower primary energies and low detection energies, the position of the "enhancement peak" is shifted a bit (roughly by 0.2 μm for 3 keV primary energy) toward the upper step plane, with respect to its "rightmost" (toward the substrate) position. The exact edge position is naturally not known and relative movements within the line-scan sets can be observed only. For higher detection energies, the peak moves away from the step up to the rightmost position at about 50 eV and then continuously returns. For higher primary energies, the movements are smaller because the relatively small part of the whole background energy range can be examined. The low energy shift toward the upper plane direction is not obvious but the continuous shift at higher detection energies is present. Explanation of these observed phenomena is not straightforward: the peak position for secondary electrons (10 eV) should fit the position for the backscattered electrons having the energy at which the secondary electron yield is maximum, i.e. around 300 eV for Si. This agrees with the observations but otherwise, the peak movement away from the edge for low loss backscattered electrons contradicts the basic realizations about the sidewall crossing the interaction volume where the electron energy decreases with the distance from the impact point (so that the low loss electrons can be found near the impact point and their peak should be rightmost shifted). To explain the observations will require a further study including Monte Carlo simulations of the experiment.

For the heterogeneous W/Si edge, a slope of the signal falling down from a tungsten level to a silicon level is the only interesting observable feature. The variations observed at various background detection energies can be understood as the opposite to the previous case: The signal step is the lowest and "leftmost" placed at some tens of eV and it grows toward both sides (more to the true secondaries) and moves away from the terrace. These shifts are not longer than 0.1 μm (for the 0.12 μm terrace height).

The energy filtered background scans were compared with the Auger scans; they are also shown using thick lines in Fig. 2. A general result of the comparison is that the step edge enhancement is always smaller in line-scans taken in the energy filtered background than in Auger electron scans so that a full correction of the effect by ratioing is impossible. The relatively best result of correction is obviously obtained for the maximum effect in the background, i.e. somewhere in the range 50 to 200 eV. Promises for the correction of the signal penetration at heterogeneous edges cannot be judged without quantification.

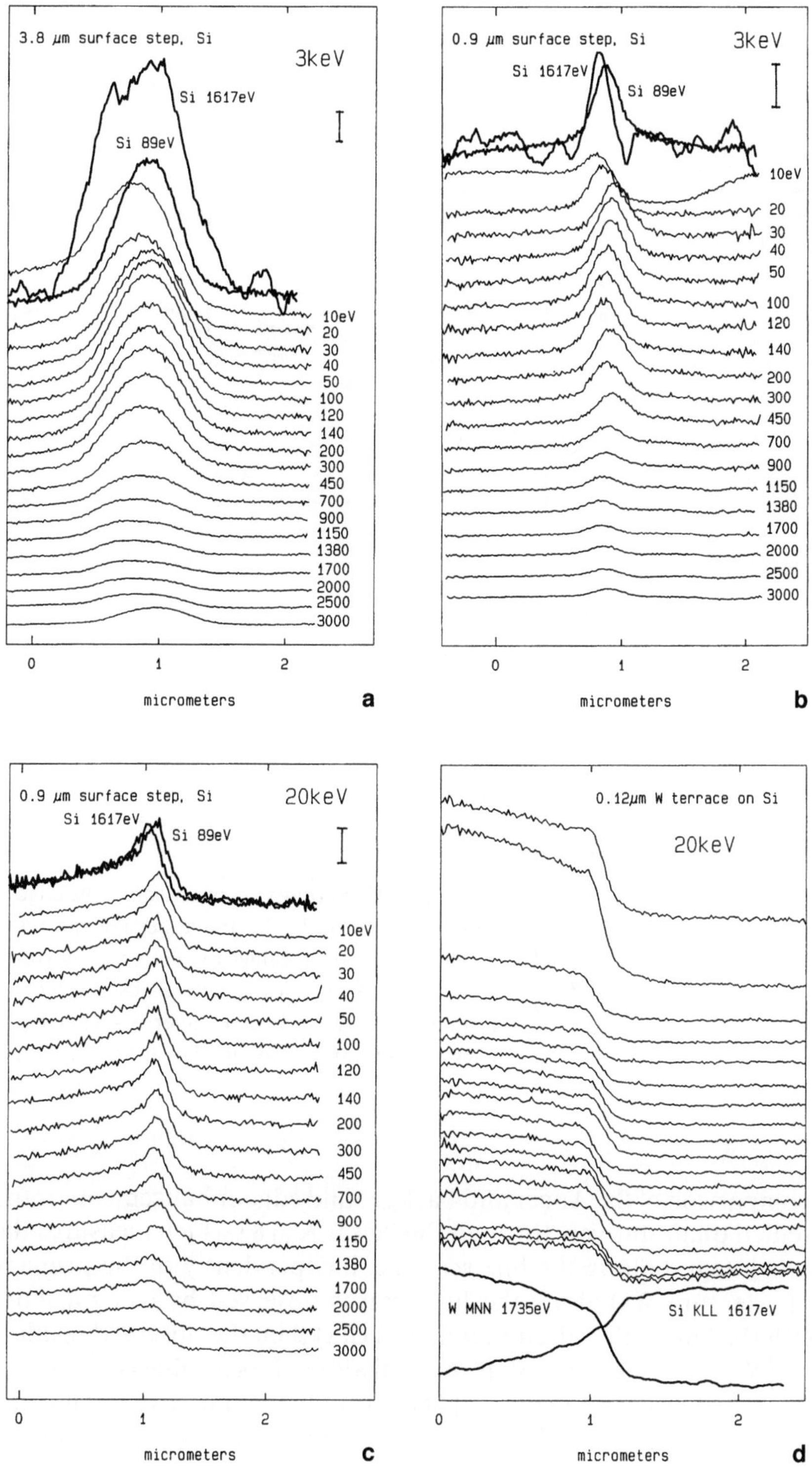

Fig. 2. Appearance of the edge effect on various structures, in the Auger electron emission and the energy filtered background. Thick line Auger scans are labelled by the element and the energy, whereas thin line background scans by the energy only. **d** Background energies are the same as in (**a**)–(**c**). **a**–**c** All curves were scaled to a same constant level behind the edge and then vertically shifted in order to avoid the overlaps. Thus, the vertical scale is that of a relative or percentage signal change and the vertical marker in the top right corner corresponds to the 50% change. In **d** no scaling was made, only vertical shifts – the edge slope is an important feature here

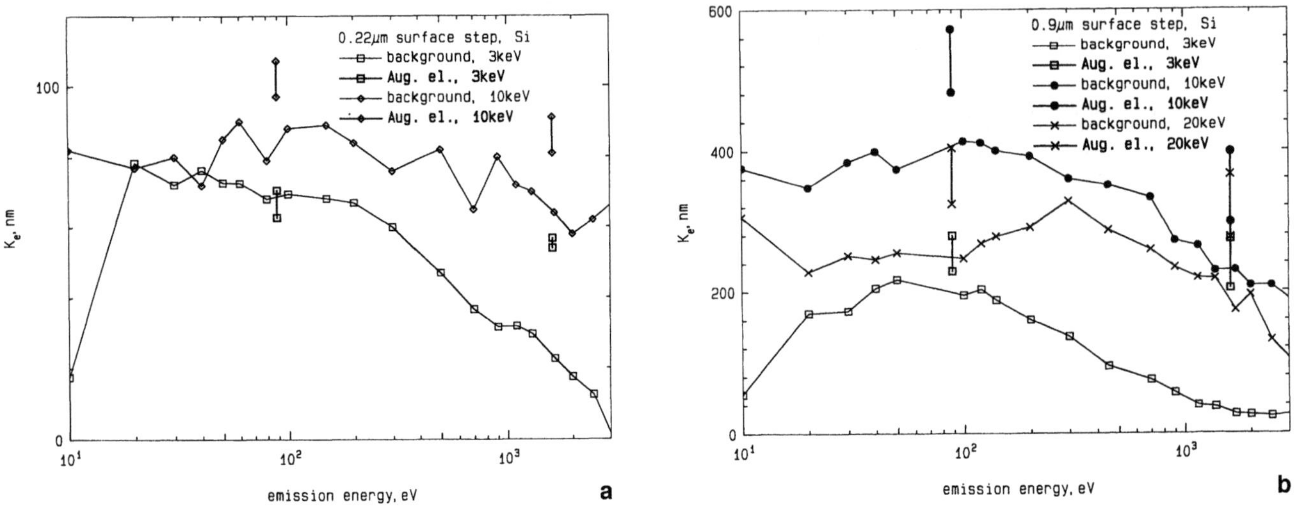

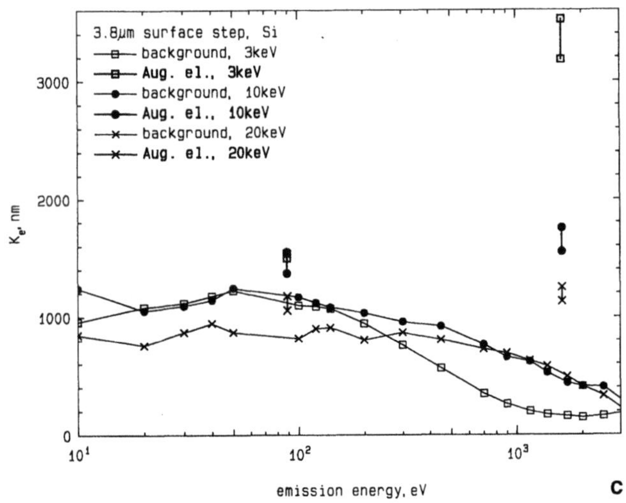

Fig. 3 a–c. Quantitative comparison of the edge enhancements on the Si surface steps of various heights. The Auger values are shown at corresponding peak energies as vertical abscissas corresponding to the measurement errors while the background values are simply plotted against the energy.

In Fig. 3, the comparison of the Auger and background edge enhancements at the step edges is presented quantitatively through the factor K_e [8] which measures the enhancement by its integral along the line-scan, i.e. independently of the primary beam spotsize. (If a feature $S(x)$ along the line-scan is observed by using a finite spot-size beam with the lateral distribution function $C(x)$ then the observed result is smeared to $S'(x) = \int S(x')C(x - x'')dx'$ but its integral along the scan does not change because $\int S'(x)dx = \int S(x') \int C(x - x')dxdx' = \int S(x')dx'$.) One can easily see that the edge enhancement in the background line-scans is everywhere lower than that of Auger scans.

Examples of the practical ratioing are shown using sets of ratioed scans in Fig. 4. Their appearance naturally confirms what has already been discussed about the enhancement heights but the mentioned horizontal shifts also play a role. In Fig. 4a, the best correction is achieved at 50 eV of the reference signal energy-this is once more shown quantitatively in Fig. 5a. In Fig. 4b, none of the corrected scans is acceptable, partially because of the mutual horizontal shifts of the ratioed features

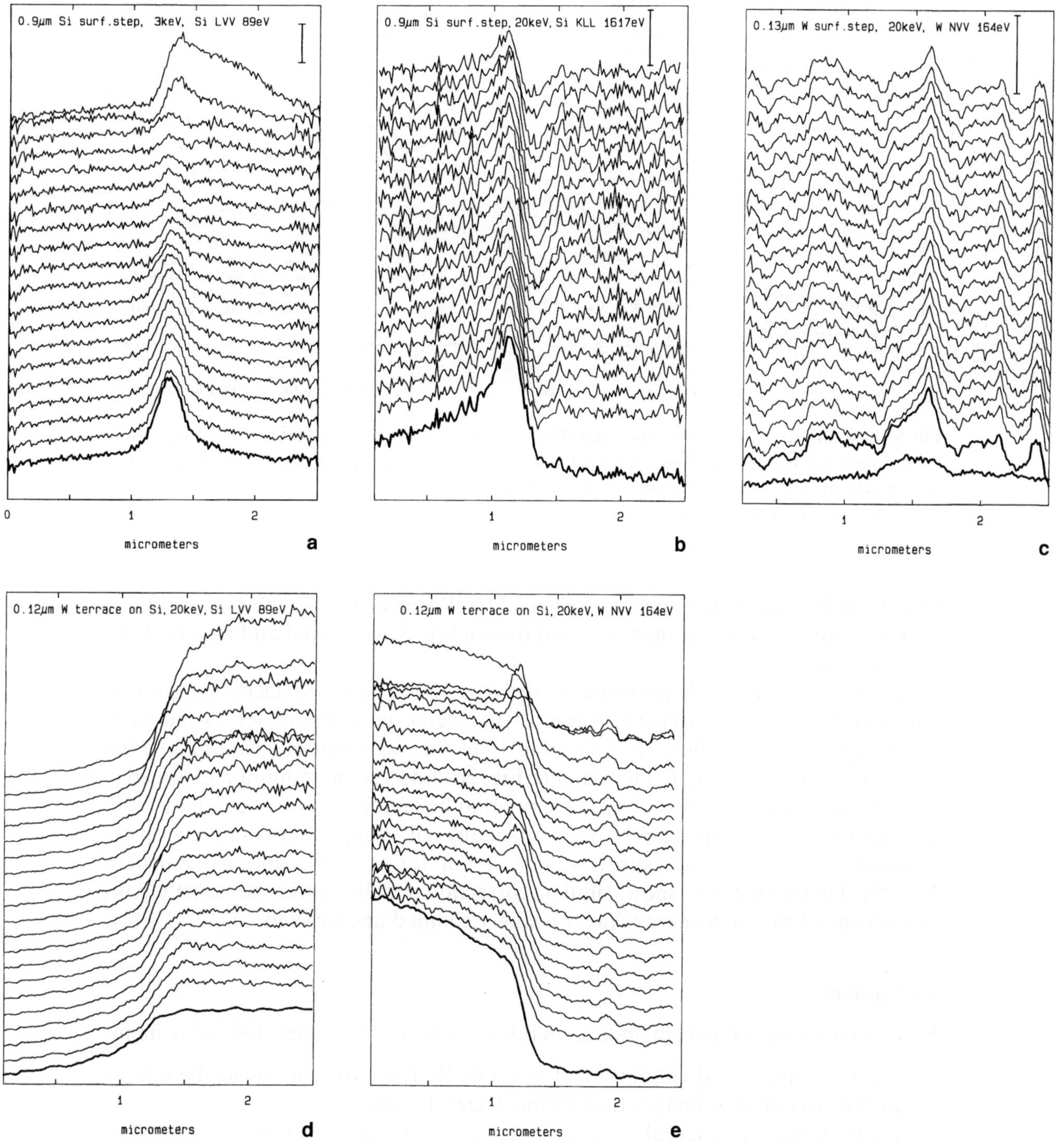

Fig. 4. Sets of the ratioed linescans for which the background emission filtered at various energies is used as the reference, together with the original Auger scan (thick line; the second bottom thick line in (**c**) is the background scan at 200 eV, i.e. near behind the peak). In each picture, the same vertical scale is used for all curves. **a**–**c** The enhancement measure is shown equally as in Fig. 1, as regards the vertical scale. **d** and **e** No vertical scale is shown because it is unimportant, only the edge slope

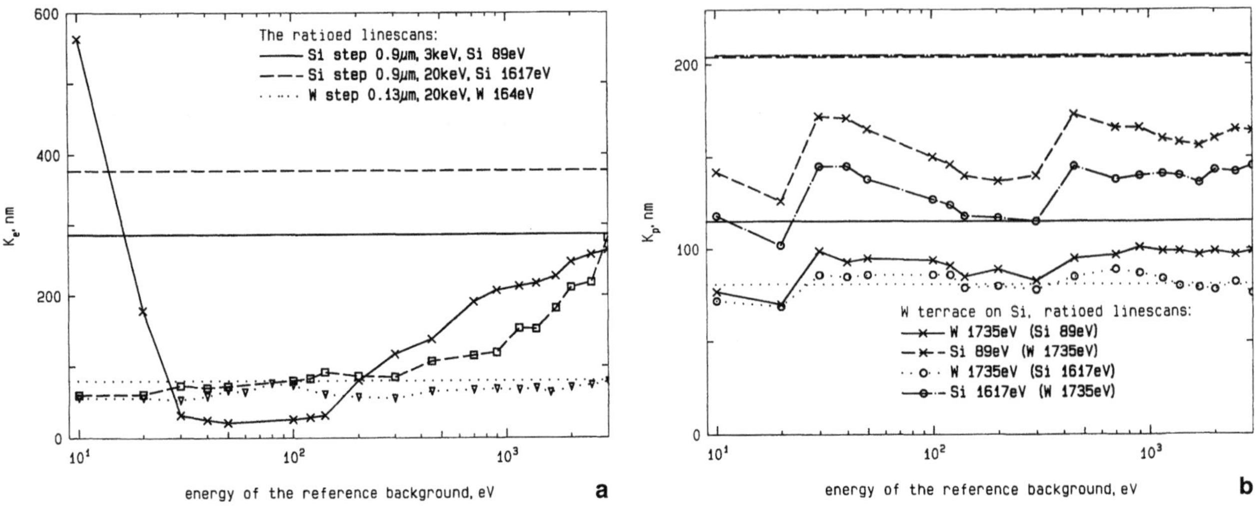

Fig. 5. Quantitative expression of the edge effect in ratioed scans plotted against the energy of the filtered reference background and compared with the original values for noncorrected Auger scans, shown as horizontal lines of the same type. **a** Examples of the homogeneous surface steps shown in Fig. 3; **b** the heterogeneous terrace edge, all four combinations of the Auger scans recorded in the experiment

which produce new wave shaped artefacts. The W step in Fig. 4c exhibits negligible variations only; this experiment suffered from a low Auger signal and insufficiently smooth surfaces.

The terrace edges on heterogeneous surfaces should be made steeper by ratioing which is in this case expected to suppress the signal penetration across the edge. From Figs. 4d and 4e, the scans ratioed to 100 eV background or so seem to be the steepest. The quantitative comparison in Fig. 5b shows some minimum of the edge effect between 100 and 300 eV and some peculiar behaviour at lowest energies. Note that the signal penetration can be observed at any pair composed of one Auger peak from each of the terrace and substrate side, so that four combinations are shown in Fig. 5b. Of them, three show similar correction results while the fourth pair, composed of both high energy Auger peaks, remained uncorrected.

Conclusions

When summarizing the main findings of this work we can formulate following items:

a) The step edge signal enhancement is generally less pronounced in the energy filtered background images than in the Auger images.

b) The background signal enhancement is largest at energies just behind the range of the true secondaries, i.e. somewhere between 50 and 200 or 300 eV.

c) The ratioing of Auger scans with respect to the energy filtered background is the most successful in suppressing the edge effect on homogeneous surface steps at low energies. See the suppression of the K_e factor indicated by three curves in Fig. 5a: to 7% of the original value at 50 eV, or to 16% at 10 eV (both for Si surface) or to 66% at 30 eV for W surface).

d) The suppression of the signal penetration at the heterogeneous terrace edges is also best at low energies; it falls to within 56 to 72% for the first three curves in Fig. 5b between 200 and 300 eV.

In practice, it seems to be worth trying to suppress the edge effect by ratioing, first of all, at the homogeneous surface steps by using the reference signal around 50 to 200 eV, and one can probably expect more acceptable results for low atomic number specimens. Such a procedure can be very simply introduced into the operation software of Auger microprobes.

Acknowledgements. The PHI 595 experiments were made at the Institute of Physics, Technical University Clausthal, Germany; cordial thanks are due to Professor E. Bauer and Dr. A. Pavlovska. Assistance of Dr. J. Dupák, ISI Brno, in specimen preparation is acknowledged.

References

[1] L. Reimer, *Scanning Electron Microscopy*, Springer, Berlin Heidelberg New York Tokyo, 1985, p. 231.
[2] D. Pejchl, I. Müllerová, L. Frank, *Czech. J. Phys.* **1993**, *43*, 983.
[3] R. Shimizu, T. E. Everhart, N. C. MacDonald, C. T. Hovland, *Appl. Phys. Lett.* **1978**, *33*, 549.
[4] C. G. Tuppen, G. J. Davies, *Surf. Interf. Anal.* **1985**, *7*, 235.
[5] A. Umbach, A. Hoyer, W. H. Brünger, *Surf. Interf. Anal.* **1989**, *14*, 401.
[6] M. M. ElGomati, M. Prutton, B. Lamb, C. G. Tuppen, *Surf. Interf. Anal.* **1988**, *11*, 251.
[7] M. M. ElGomati, A. M. D. Assad, T. A. ElBakush, M. Prutton, *Proc. Multinat. Congr. El. Microsc.*, Ital. Soc. El. Microsc., Parma, 1993, p. 109.
[8] L. Frank, F. Matějka, *J. Microscopy*, **1995**, *180*, 158.
[9] T. Sekine, Y. Nagasawa, M. Kudoh, Y. Sakai, A. S. Parkes, J. D. Geller, A. Mogami, K. Hirata, *Handbook of Auger Electron Spectroscopy*, JEOL, Tokyo, 1982, p. 188.

Mikrochim. Acta [Suppl.] 13, 289–298 (1996)

Low Energy Imaging of Nonconductive Surfaces in SEM

Luděk Frank, Martin Zadražil*, and Ilona Müllerová

Institute of Scientific Instruments, Czech Academy of Sciences Královopolská 147, 612 64 Brno, Czech Republic

Abstract. If any not absolutely smooth and homogeneous nonconductive surface is observed in SEM, a locally variable charging appears which produces local electrical fields that destroy the image. The surface charge depends on the total electron yield σ, i.e. the number of outgoing electrons divided by the number of the incoming ones, on the primary current, resistance between the field of view and the earth, surrounding gas pressure and time. A charge balance is established between primary and emitted electron currents, the specimen current taking charges off the surface, and the charged particle bombardment from the atmosphere, which can be neglected in high vacuum. σ changes with the landing energy of primary electrons: near to zero energy, $\sigma < 1$, and the surfaces charge negatively. For nearly all materials, between two characteristic energy values E^I and E^{II}, σ exceeds unity and the charge is positive, while above E^{II}, $\sigma < 1$ again and a negative charge appears. E^I can be found usually between 80 and 300 eV and E^{II} amounts to one or a few keV and depends on the electron impact angle. A novel method is proposed consisting in approaching one of the critical landing energies E^I or E^{II} where $\sigma \to 1$ and the surface charge formation is suppressed to a level corresponding to the intensity of spontaneous discharging. The retarding field principle of low energy SEM realization, which minimizes the point resolution changes with the landing energy, makes both E^I and E^{II} equally well accessible. In order to find the critical energy without getting the specimen intensively charged, we recommend to measure, quickly enough from the point of view of the charging process duration, the time dependence of the signal in a single pixel by stepwise approaching the landing energy of the slowest change, i.e. the acceptably approximated critical energy. A model of the image signal vs. time development is presented, together with a description of algorithms and software for the critical energy measurement. The preliminary results presented confirm our basic ideas regarding the method.

Key words: low-energy SEM, nonconductive specimens, insulators, slowelectrons, noncharging SEM.

---

* To whom correspondence should be addressed

When a solid specimen is bombarded by charged particles from its environment, its charge balance is influenced by the incoming species. In any experiment of this kind and particularly in electron microscopy, the incoming particles cause emission of other charged particles and they are also partly reflected or backscattered. In the electron microscope, the specimen is bombarded by an intensive electron beam exciting the useful signal and, in the dependence on the surrounding gas pressure, by electrons and ions from the atmosphere. With the exception of the booming environmental electron microscopy where the gas atmosphere surrounding the specimen has a pressure of hundreds of Pa or even more and where the charged particles impact from the atmosphere is significant or even decisive, in all other kinds of electron microscopy performed in vacuum conditions ranging from 10^{-2} Pa for old and cheap instruments up to 10^{-8} Pa in true UHV devices for surface studies, the charge coming in from the environment can be neglected. Attempts to compensate the charge by an auxiliary charge source, such as an auxiliary electron gun at another suitable energy or an ion gun, were successful in principle [1] and this method represents the only alternative to that described here.

If the specimen is conductive and held at the microscope ground potential, the excess charge is led this way, producing the so-called specimen current utilizable as an imaging signal. If the specimen is not conductive enough, the charge remains localized within the beam interaction volume, i.e. in SEM near the surface, and its field exhibits a spurious influence on the image geometry. This phenomenon is well known to any electron microscopist and the usual way to avoid the problems is to cover the specimen with a thin conductive layer; this is the standard specimen preparation step, e.g. in the biomedical practice. The usually used metal coverage a few tens of nm thick can obviously hide some specimen details, and processes which occur during the coverage formation, including absorption of the kinetic energy of the metal species, can alter the properties of sensitive specimens. The coverage can also be destructive from the point of view of an interoperational check in the semiconductor technology. There hardly exists any optimum coverage thickness between too thick layers obscuring the specimen observation and too thin films not securing the conductivity. For these reasons, the problem of charging continuously attracts high attention.

Welter and McKee [2] imaged nonconductors in a field emission gun SEM at a high frame frequency, and they claim establishment of a charge balance when the scan rate is sufficiently faster than the discharge rate given by the specimen resistivity ρ and permittivity ε. Similar conclusions are formulated in [3] where also the frame time T is recommended to be much shorter than $\rho\varepsilon$. The dependence of the equilibrium voltage on the angle of incidence is pointed out, being caused by the tilt angle dependence of the secondary electron yield; some mean value is used in [3] for imaging ferroelectric insulating crystals. A useful simple "condenser" model is presented in this work which helps to explain the basic dependences. Many useful data regarding the critical energies for metals are published in [4]. The measurement was based on the detection of the specimen current in the dependence on the landing energy and showed E^{I} to be between 100 and 240 eV and E^{II} between 1.6 and 8.3 keV for some metals between Fe and Au, while lighter metals have $\sigma_{\mathrm{m}} < 1$. For Ni, E^{II} was found between 1.6 and 4.6 keV for the angle of incidence between 0° and 60° while E^{I} exhibits fluctuation changes only between 0.17 and 0.2 keV. The crystallographic orientation changes between light and dark grains in polycrystalline metals amount to even more than 1 keV (e.g. 2.30 and 3.44 keV for Cu). The contamination layers were found to significantly decrease E^{II} to one half or even less (from 5.81 to 2.9 keV for Ag or from 2.74 to 0.8 keV for Cu). The corresponding changes in E^{I} are observed to be a bit smaller, of both directions and without clear correlation to the main contaminant (oxygen vs. carbon). A lot of data concerning the height and position of the secondary electron maximum for elements and also for some insulating compounds are given in [5]. Physical processes connected with the charging are discussed in [6] together with the main solutions to reduce the charging.

Principle of Noncharging Electron Microscopy

As already mentioned above, the total electron yield from a solid, i.e. the average number of emitted electrons per one impacting electron, varies significantly with the energy of the impacting electrons. Figure 1 shows a typical dependence of this kind, provided the maximum yield σ_m exceeds unity. This is not always the case: from many data published in [4], one can find that for most elements, $\sigma_m > 1$, but with some exceptions for some light elements and elements at the beginnings of the periodic table rows. Nevertheless, for typical insulating compounds [5], the maxima σ_m exceed unity much more expressively. It follows also from the practical electron microscopical experience regarding attempts to image insulators that a curve like that shown in Fig. 1 can really be obtained; this can be deduced from observations of the charging phenomena described below.

Suppose now that $\sigma_m > 1$ for a specimen under observation in the electron microscope. For any energy of the impacting electrons, except for the two values E^{I} and E^{II} indicated in Fig. 1 where $\sigma = 1$, the outgoing charge differs from the incoming one. The difference is dissipated in the specimen and creates a local

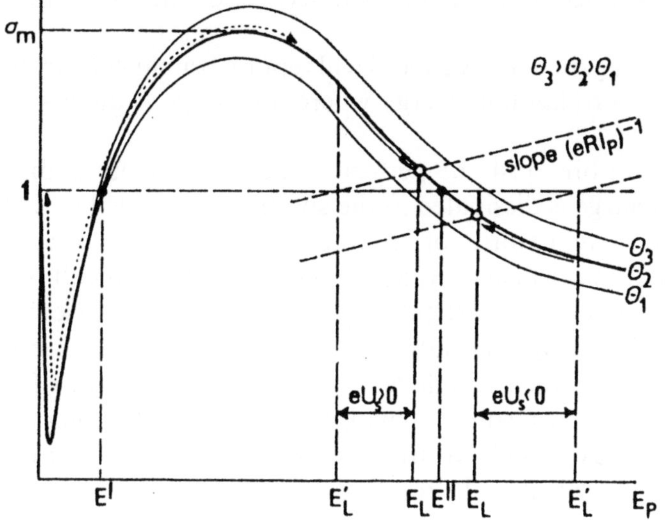

Fig. 1. Schematic representation of the dependence of the total electron yield on the electron impact energy, with indications of the movements along the curve due to the charging

electrostatic field which forces the charge to flow to nonilluminated parts outside the field of view and toward the specimen depth and consequently to ground. Nevertheless, except for very good conductors, at a high electron illumination density some local charging can always be observed in the electron microscope. For less conductive and insulating materials, the charge remains localized for a time relatively long compared with the important experiment times, e.g. the pixel frequency in SEM, and its field destroys the image through influencing the trajectories of both the illumination and the signal electrons. The sign of the rest charge depends on the position of the $\sigma(E)$ with respect to the unity level at the given energy.

Above E^{II}, the negative charge is formed while the number of outgoing electrons is lower than that of the incoming electrons. The field of this charge, if it remains localized at the surface, decelerates the additionally incoming electrons so that their effective landing energy lowers. Consequently, the emission caused by slower electrons increases as indicated by the curve in Fig. 1 and this development continues up to the balance at E^{II} where the surface charge stops to change. The actually reached energy, i.e. the surface potential, is influenced by the specimen conductivity or the resistance R of the viewfield with respect to earth, as shown in Fig. 1. The original electron energy E_L' changes into the final landing energy E_L.

If the original impact energy is between E^I and E^{II}, then $\sigma > 1$ and a positive charge is formed at a less conductive surface. The field of this charge attracts both the primary and the emitted electrons so that the primary electrons impact with a higher effective energy. Because the decisive portion of the emitted electrons is true secondaries with energies mostly in units of eV, they are easily recaptured by the surface field and the charge balance is restored by a surface potential of few volts only. The situation changes when the specimen forms the cathode in the cathode lens in a scanning low energy electron microscope (SLEEM) [7–9]. Then a field of a few of kV/mm above the surface combines with the field of the surface charge. Near the margins of the viewfield, the effective electrostatic field can reach high angles with respect to the optical axis but near the axis both contributions are simply summed so that a landing energy near E^{II} is again established, now from the opposite direction. One can expect some effective "lowering" in the $\sigma(E)$ curve in the dependence on the viewfield dimensions and the surface (lateral) conductivity, i.e. of the radial field component.

Below E^I, the surface charges again negatively and will repulse primary electrons up to mirror conditions with a zero landing energy where no charge is dissipated again.

Noncharging microscopy can obviously be performed at the electron landing energies E^I and E^{II} where no charge is dissipated in the specimen, provided $\sigma_m > 1$ which can be expected for a decisive majority of insulators. Two obstacles exist to materialize this: measurement of the critical energies E^I and E^{II} is very difficult (see [10]), and the critical energies are too low, particularly E^I, from the point of view of the achievement of an acceptable image resolution in SEM. The second issue is appropriately solved by introducing the cathode lens assembly into the SEM [9] which enables one to achieve a similar resolution (within a factor 3 to 4) along the whole landing energy scale from, say, 15 keV to 10 eV. The proposed method for the critical energy measurement is described in the following section.

Measurement of the Critical Energy

The methods of measurement of critical energies are summarized in [10]. A critical energy for a conductive specimen can easily be adjusted by measuring the specimen current; at a critical energy it is zero. Otherwise, a computation or Monte Carlo simulation of the total electron yield [11] is generally available which can bring at least some estimation of E^{II}. (It has been pointed out on the basis of computations [11] that the mere 3 nm layers of oxide or of carbon shift E^{II} for silicon by 100 eV in opposite directions.) Methods of measurement suitable for insulators [10] are complicated and inconvenient in practice; the only reasonable method consists simply in series of trials when the landing energy is varied and a change in the image appearance is observed after a quick shift of the viewfield or a quick change in the magnification.

As mentioned above, one can expect dependences of the critical energies on several specimen and configuration parameters, i.e. the electron impact angle, crystallographic orientation, and surface contamination. The time development of the charging depends also on the primary current and its density. The impact angle dependence which is the most important factor limiting the convenience of the method in imaging real nonsmooth specimens is restricted to E^{II} only because it is caused by the tilt agle dependence of the secondary electron emission. For E^{I}, the electron penetration depth and the escape depth of secondaries are already comparable or the latter is larger, so that the tilt angle influence disappears.

The description of the charging processes in the previous section induces a method of measurement which is an improvement on the "trial" method mentioned above: any type of the charging, dependent on the value of the primary electron energy with respect to the actual values of E^{I} and E^{II}, is connected with a shift along the $\sigma(E)$ curve, i.e. with some change in the emitted signal. The disadvantage of the method of trials is that large areas of the specimen are charged before a suitable energy can be determined. But a quick sensing of the initial signal change by using a computer controlled measurement from a pixel can produce the necessary information in a very short time and from a small specimen area.

Let us now quantitatively estimate the phenomena occurring in connection with the charging: An insulating specimen of the thickness d and permittivity $\varepsilon = \varepsilon_r \varepsilon_0$ is put onto the holder held at the potential $-U_0$. The holder is positioned at a distance of $L \gg d$ from the grounded anode. The specimen surface potential becomes $-U'_0 = -U_0(1 - d/L\varepsilon_r)$ and the landing electron energy $E'_L = E_p - eU'_0 - eU_S$ where U_S is the surface potential due to charging. In the steady state, the incoming charge given by the difference between primary and emitted currents equals to the current flowing away through the specimen resistance. This is expressed by the equation $[\sigma(E_L) - 1]I_P = (E_L - E'_L)/eR$, where E_L is the final landing energy, I_P is the primary current and $R = \varrho d/\pi a^2$ is the leakage resistance between the irradiated surface and the conductive specimen holder (ϱ is the resistivity and a is the viewfield radius). Using the Taylor expansion around $\sigma(E^{II})$ we get for E_L the relation

$$\frac{E_L - E'_L}{E_L - E^{II}} \simeq eRI_P \left[\frac{d\sigma}{dE}\right]_{E=E^{II}} \tag{1}$$

This equation determines the final surface potential due to charging,

$U_S = (E_L - E'_L)/e$. Obviously, $E_L \rightarrow E^{II}$ for good insulators. Another existing equilibrium landing energy is only the zero energy for the mirror conditions.

Within the transition period, the emitted signal changes from $I(0) = I_P\sigma(E'_L)$ to $I(\infty) = I_P\sigma(E_L)$. For purposes of the critical energy detection, the initial development of the emitted current dI/dt is important. The charge accumulation is simply given by the relation $dQ/dt = \{\sigma[E(t)] - 1\}I_P$. Let us suppose that, at higher magnifications, the charged area (the viewfield) is significantly smaller than the insulator thickness. The condenser capacitance can be neglected with respect to the absolute capacitance of the charged area (related to ground) as regards the surface potential formation, so that the potential of a charged area is $U_S = Q/(2\varepsilon_0 a)$ and $dQ/dt = (2\varepsilon_0 a/e)(dE/dt)$. The time behaviour of the electron landing energy $E(t)$ is then given by

$$t = \frac{2a\varepsilon_0}{eI_P}\int_{E'_L}^{E}\frac{dE}{\sigma(E) - 1} \tag{2}$$

In a near neighbourhood of a critical energy E_C, the function $\sigma(E)$ can be expanded as $\sigma(E) \simeq 1 + (d\sigma/dE_C)(E - E_C)$. After substitution into (2) we get the exponential function $E(t) \sim E'_L + \text{const} \times \exp(t/\tau)$ and the same is true for the signal function $I(t)$. The time constant τ is then

$$\tau \simeq \pm\frac{2\varepsilon_0 a}{eI_P}\left[\frac{d\sigma}{dE}\right]^{-1}_{E=Ec} \tag{3}$$

In the scanning mode, we have to suppose that the charge decay during the frame time T is negligible, $T \ll \rho\varepsilon$. The constant τ has a sign depending on the sign of the integral in (2), so that it is positive around E^I and negative in the vicinity of E^{II}. Figure 2 shows schematically the signal development for various primary electron energies.

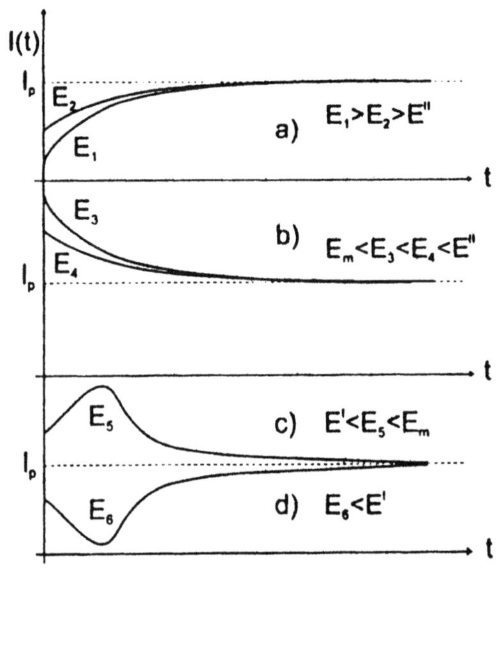

Fig. 2. Scheme of the time development of the SLEEM image signal from a nonconductive specimen in the various electron energy intervals

Estimating $d\sigma/dE$ at $E = E^{II}$ to $-5 \times 10^{-4}\,\mathrm{eV}^{-1}$ [10] and having, for example, $I_P = 1\,\mathrm{nA}$, $\varepsilon_r = 10$ and $a = 10\,\mu\mathrm{m}$, we get $\tau = 3\,\mathrm{msec}$ and for $d = 100\,\mu\mathrm{m}$ and $\varrho \geqslant 10^{10}\,\Omega\mathrm{m}$, E_L is very close to E^{II}.

Mainly owing to its angular dependence, E^{II} varies across a non-smooth specimen and local charges keep arising even though the average E^{II} is adjusted. The neighbourhood of E^I does not suffer from this effect but the mode is unstable, as indicated in Figs. 1 and 2c, d: any initial difference from E^I is amplified during the beginning stage. No data about the behaviour of $\sigma(E)$ near E^I seem to be available; from the published plots it follows that a substantially larger slope can be expected, so that $\tau^I < \tau^{II}$.

Experimental

Low energy SEM can be performed in two ways: either a low energy gun can be incorporated into the microscope or the electrons can be, after being focused to the final spot at the nominal high energy, decelerated within the cathode lens to a low landing energy – this brings a well suppressed lowering of the point resolution with the decreasing energy so that instead of the several orders of magnitude a factor of about 3 or 4 is between the resolutions at energies differing by four orders of magnitude [9]. In this case, the specimen forms the cathode of the cathode lens to which a potential difference similar to the SEM accelerating voltage is applied.

We adapted several SEM's to SLEEM (scanning low energy electron microscopy) and verified the resolution–energy dependence [12]. The dedicated high-vacuum SLEEM intended for the development of the presented method was based on the Tesla BS 343 Mini-SEM, equipped with a thick wall behanite specimen chamber and a prototype of a fully digitally controlled SEM electronic console, both manufactured by Tescan Brno. The configuration of the specimen chamber is schematically shown in Fig. 3. The ultimate vacuum achieved sofar in the specimen chamber is 5×10^{-5} Pa and the operating vacuum is 1 to 2×10^{-4} Pa. The nominal spot size of the original microscope was 16 nm at 15 keV only (for TV scan rate) and is planned to be improved by replacing the column. The resolution at 10 eV is estimated to be slightly above 50 nm but has not been reliably tested yet.

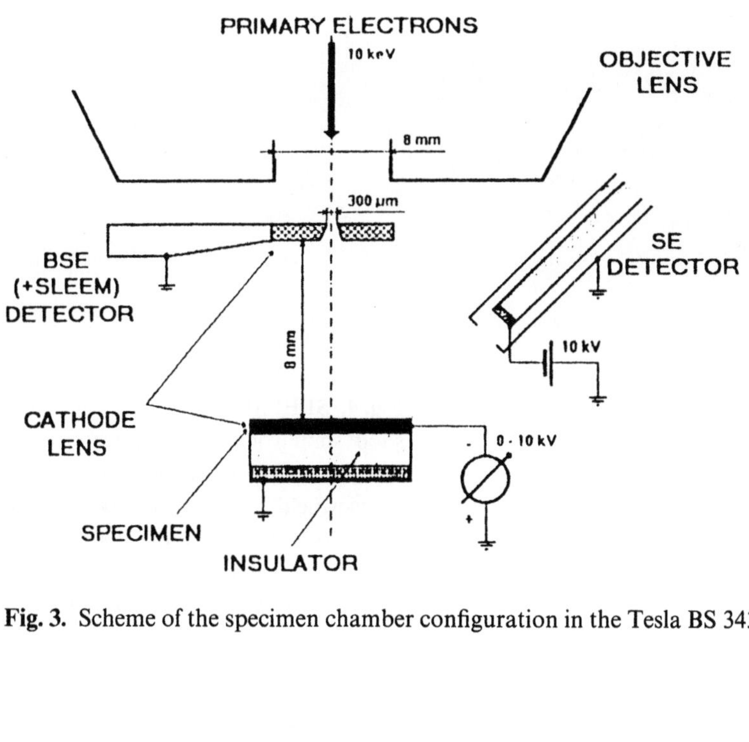

Fig. 3. Scheme of the specimen chamber configuration in the Tesla BS 343 SEM adopted to SLEEM

As shown in Fig. 3, a coaxial bored scintillator based backscattered electron detector is used as the SLEEM detector (see [9]). The emitted electrons are accelerated within the cathode lens and collimated to the axis so that a high detection efficiency is achieved, although afterwards a great portion of the signal is lost owing to the narrow edge connection to the lightpipe. Still the signal is larger at low energies than in the original SEM mode for both secondary and backscattered electron imaging.

In the frame of the project of noncharging microscopy, the SEM control computer is being complemented with a specially designed software a part of which has already been put into operation in cooperation with Tescan Brno. It consists of an interaction selection of a special scanning mode (scanning in a strip at the top side of the viewfield only) and of a selection of a large number of the measurement points (pixels) within the whole scanning field. In each point, as many as 5000 measurements of the detector signal can be performed in regular adjustable time intervals; the shortest one is now 16 μs long and it will be shortened. In addition to the discrete points, points along a line can be defined. Measurement in each point is stored as a signal vs. time curve, together with the measurement parameters. "One shoot snap" mode is available from the whole viewfield; after the snap the scanning of the strip is restored. In the near future, the software will be complemented with procedures for processing the signal vs. time curves and for controlling an electronically adjustable high voltage supply for specimen biasing.

The first experiments were performed on teflon specimens partially covered by a 20 nm Au layer. Experiments using other technical insulators are in progress.

Preliminary Results

Figure 4 shows the micrograph of the partially gold-layer-covered teflon specimen: the repeatedly scanned area in the top (approx. one fourth of the picture height) is charged up but below this area the left part of the uncoated teflon surface at 1500 eV

Fig. 4. SLEEM micrograph (the "first shoot snap") of a teflon surface, covered with 20 nm Au on the right hand side of the viewfield, taken at the 1500 eV landing energy, the field width 300 μm. The uppermost strip was repeatedly scanned during the critical energy measurement

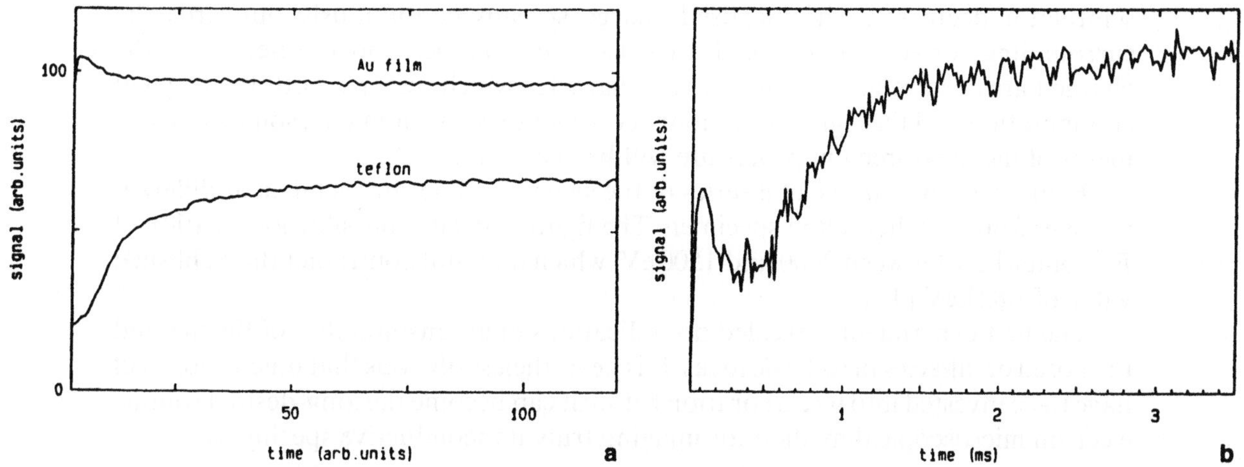

Fig. 5. a The typical behaviour of the signal from the uncoated and coated parts of the teflon specimen; **b** an irregular signal development (discussed in the text)

landing energy is not destroyed by any surface charge and exhibits significant contrast with respect to the Au coated right part.

The signal vs. time measurement is illustrated by its typical appearance in Fig. 5a. Although some traces of a charging effect are visible even on the Au covered surface, they are negligible with respect to changes in the signal from the uncoated surface. Figure 5b illustrates the main issue connected with the present version of the procedure: the charging development here is more complicated than expected, namely the current increase, faster than expected, is followed by a signal drop and only then the expected signal behaviour follows. This can be explained by the radial electrostatic field formed by the highly localized charge in the illuminated pixel: this field combines with the axial cathode lens field and the resulting field divergent from the pixel exhibits a less efficient collimation of the emitted electrons toward the optical axis than the pure axial field, so that the amount of electrons striking the

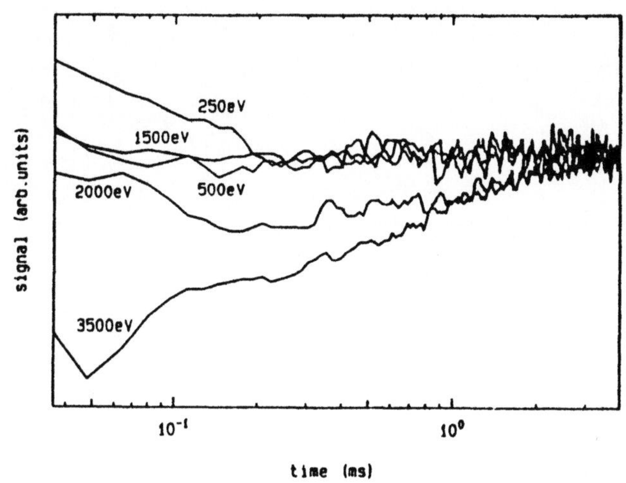

Fig. 6. Set of the signal vs. time measurements for various landing energies, taken with another teflon specimen

scintillator decreases. The localized charge spreads continuously outwards the surrounding surface, the charged area diameter enlarges and the field near the charged area centre gets again its axial character which restores both the original configuration and the signal behaviour. To avoid these phenomena, some improvements of the measurement procedure will be necessary.

Figure 6 is an example of a series of the signal vs. time curves taken at different energies from another teflon specimen. The figure indicates possible localization of E^{II} somewhere between 2000 and 1500 eV, which does not contradict the published value of 1.82 keV [11].

The first experiments revealed no indications of an unsuitability of the method proposed or fallacies in its basic ideas. It is nevertheless obvious that much effort will have to be invested into its elaboration before it can become the long-desired routine electron microscopical method for imaging truly nonconductive specimens.

References

[1] P. Morin, M. Pitaval, E. Vicario, *J. Phys. Electron. Sci. Instrum.* **1976**, *9*, 1017.
[2] L. M. Welter, A. N. McKee, *Scanning Electron. Microsc.* **1972**, *I*, 161.
[3] R. Le Bihan, *Proc. 7th Int. Vac. Congr.*, Vienna, 1977, p. 2351.
[4] H. E. Bauer, H. Seiler, *Scanning. Electron. Microsc.* **1984**, *III*, 1081.
[5] H. Seiler, *J. Appl. Phys.* **1983**, *54*, R1.
[6] J. Cazaux, P. Lehuede, *J. Electron. Spectrosc. Rel. Phenom.* **1992**, *59*, 49.
[7] R. S. Paden, W. C. Nixon, *J. Phys. Electron. Sci. Instrum.* **1968**, *1*, 1073.
[8] I. Müllerová, M. Lenc, *Mikrochim. Acta [Suppl.]* **1992**, *12*, 173.
[9] I. Müllerová, L. Frank, *Scanning* **1993**, *15*, 193.
[10] L. Reimer, U. Golla, R. Böngeler, M. Kässens, B. Schindler, R. Senkel, *Optik* **1992**, *92*, 14.
[11] D. C. Joy, *Scanning* **1989**, *11*, 1.
[12] I. Müllerová, L. Frank, *Mikrochim. Acta* **1994**, *114/115*, 389.

Mikrochim. Acta [Suppl.] 13, 299–305 (1996)

Investigation of the Bonding Mechanism of Glass Ceramic Layers on Metal Alloys

Martina Frigge* and Günter Völksch

Otto-Schott-Institut, Friedrich-Schiller-Universität Jena, Fraunhoferstr.6, D-07743 Jena, Federal Republic of Germany

Abstract. A compound material from a metal alloy coated with a bioactive glass ceramic has high mechanical strength and shearing resistance due to the metal substrate and can be used as implant material for bone substitution. The thin glass ceramic layer allows a direct intergrowth between bone and implant without connective tissue. For practical application of this compound material it is important to know the bonding mechanism between the metal and the glass ceramic layer.

Metal alloys were coated with bioactive glass ceramic by sintering in air at 700 °C. Two different techniques were used to study the interface: Scanning electron microscopy at cross-sections prepared by ion beam slope cutting and transmission electron microscopy at thin cross-sectional specimens. Both methods were combined with energy dispersive X-ray spectroscopy. During sintering in air the metal surface is oxidized and an interlayer is formed which consists of oxidic components of the metal and the glass ceramic. This interlayer results in good adhesion of the glass ceramic on the metal alloy.

Key words: glass ceramic coating, bone substitution, interface.

Special properties are necessary to make a material useful for bone substitution [1]. First of all, it has to show biocompatibility, i.e. the material should be non-toxic and should not release toxic ions upon reaction with the human body. The material has to be resistant against corrosion caused by the human body fluid. Depending on the position in the body, it has to show particular mechanical strength and toughness.

Materials are characterized as bioactive, if they show good intergrowth with the bone without forming connective tissue [2]. Such materials form strong bonds with the bone. Therefore, the implant can stay in the body for a very long period of time. Unfortunately, most materials which show bioactive behaviour, such as certain glasses, ceramics and glass ceramics have poor mechanical properties, so they can only be used for small implants which are not mechanically loaded. For larger

* To whom correspondence should be addressed

implants, biocompatible metals and alloys are used which have the required mechanical properties.

The best way to get an implant that meets the aforementioned demands is to coat a biocompatible metal with a layer of a bioactive material. The essential prerequisite about this is a good adhesion between metal and coating. The coating has to be long term stable to avoid contact of the bone with the metal to prevent the formation of connective tissue. In order to attain the best mechanical properties, the coating has – depending on the formed reaction layer with the bone – to be as thin as possible, so that the mechanical properties of the layer are negligible. A certain porosity of the coating is of advantage to allow the bone to grow into the bioactive layer.

There are lots of different procedures to coat a metal with a layer of glass, glass ceramic or ceramic. In most cases a heat treatment is necessary during the coating process. To avoid thermal stresses in the material, which decreases the adhesion, the thermal expansion coefficients of metal alloy and layer should be similar.

For further improvements of the implants it is necessary to investigate the interface reaction between the materials. Therefore, this reaction zone was studied using two techniques: scanning electron microscopy (SEM) at cross-sections which were prepared with the ion beam slope cutting method [8] and transmission electron microscopy (TEM) at thin cross-sections.

The distribution of the elements across the interface was observed using energy dispersive X-ray analysis (EDX).

Experimental

A cobalt-chromium alloy, widely used as bioinert implant [1], was coated with a bioactive SiO_2-free phosphate glass ceramic [3, 4]. This combination was selected since both materials have similar thermal expansion coefficients. Mechanical testing proved sufficient mechanical properties of the joint for the planned application [5]. The compositions of both materials and their thermal expansion coefficients are shown in Table 1. After heat treatment for controlled crystallization, the *bulk* glass ceramic, for which in vitro and in vivo tests proved bioactive properties [4], contains the main crystalline phases $Ca_{10}(PO_4)_6\,(OH,F)_2$ (apatite), $AlPO_4$ and $NaZr_2(PO_4)_3$. The thin glass ceramic *layer* (about 200 μm) on the metal substrate contains the same crystalline phases after sintering and subsequent heat treatment.

Metal substrate and glass ceramic were joined as follows: the glass powder was mixed with isopropanol and brought onto the sandblasted metal surface. The sintering of the coating shows best performance at 700 °C in air and subsequent heat treatment for crystallization at 550 °C. For SEM investigations, which were combined with EDX, a smooth cross-section of the interface is essential. Mechanical grinding and polishing was not successful, due to the vastly different mechanical properties

Table 1

	Composition in wt%	Thermal expansion coefficient
Glass ceramic	23.2 P, 11.9 Ca, 11.3 Na, 5.0 Al, 3.4 Zr, 43.4 O, 1.8 F [6]	$14–18 \cdot 10^{-6}\,K^{-1}$ [7] (20–300 °C)
Metal alloy	48.8 Co, 19.6 Cr, 17.9 W, 10.2 Ni, 1.9 Fe, 1.6 Mn	$14.7 \cdot 10^{-6}\,K^{-1}$ (100–300 °C)

of the joined materials. Ion beam slope cutting [8, 9] proved to be the best method to get smooth sections of subsurfaces: a parallel ion beam strikes on the surface which is partly covered by a screen and removes material from the exposed area and a slope is formed which contains the interface. A very small incidence angle of 10° was used to stretch the interfacial area.

The slope area was studied using a scanning electron microscope (Zeiss DSM 940A) which is equipped with an EDX spectrometer (Oxford Instruments, eXL 10). Elemental mappings from the main constituents of alloy and glass ceramic were acquired at the interfacial area.

For TEM investigations, cross-sections were prepared as described by Strecker et al. [10]. To avoid destruction of the interface during ion milling, this process was carried out using a small incidence angle of about 10°. Since the different components of the specimen are eroded at different rates, the specimen was oscillated about 20° around perpendicular incidence. By this arrangement the glass ceramic is shadowed by the metal and preferential sputtering of the layer is avoided. In this case glass ceramic coated polished metal surfaces were studied using a Hitachi H-8100 transmission electron microscope (200 kV) equipped with an EDX spectrometer (Oxford Instruments, ISIS). EDX spectra were collected in the glass ceramic as a function of the distance from the interface.

Results and Discussion

In Fig. 1 a SEM-micrograph (backscattered electrons – BSE) of the slope is shown. In the lower part the Co-Cr alloy can be seen, the upper part corresponds to the glass ceramic. Due to sandblasting before coating the interface between these materials is not a straight line. The surface of the ion beam cut slope is smooth enough for EDX analysis in the transition region between metal and glass ceramic. In contrast to the majority of the mechanically prepared specimens, no gap is visible. The stripes perpendicular to the interface are caused by the ion beam.

The interface is well-defined and clearly visible due to the high contrast caused by differences in the average atomic number of metal alloy and glass ceramic.

Figure 2 depicts elemental mappings of oxygen, sodium, aluminium, phosphorus, calcium, chromium and cobalt and tungsten. It should be mentioned that the $P_{K\alpha}$ peak overlaps with the $Zr_{L\alpha}$ and it is only possible to detect the sum of intensities of $P_{K\alpha}$ and $Zr_{L\alpha}$ line. In this case the elemental mapping was indicated as $P_{K\alpha}$ since the P content in the glass ceramic is much higher than the Zr content. The

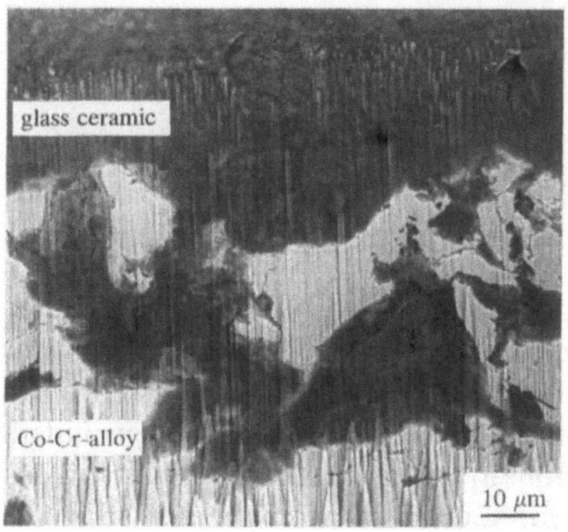

Fig. 1. Scanning electron micrograph (backscattered electrons) of the slope

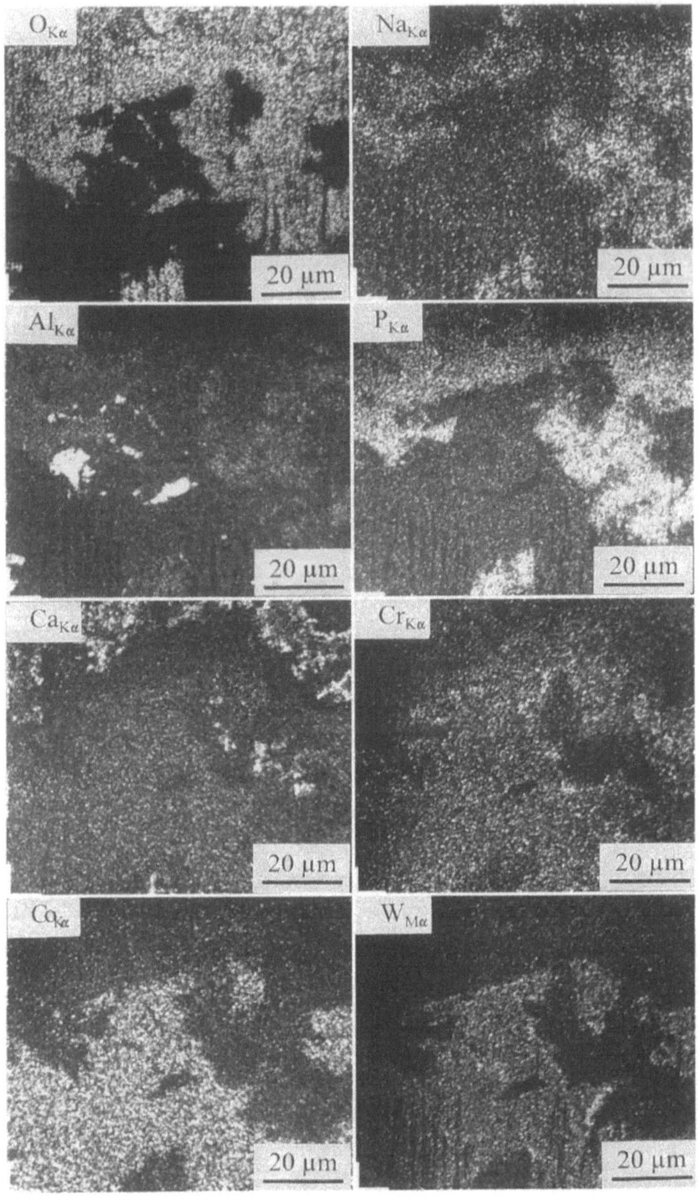

Fig. 2. EDX mapping at the slope using the K_α lines of O, Na, Al, P, Ca, Co and Cr and the M_α line of W (excitation energy 15 keV)

elemental mappings do not appear as sharp as the BSE image because of larger excitation volumes for X-ray emission. The excitation energy was 15 keV, i.e. the excitation area has a diameter of about 1–5 μm and is dependent on both the energy of the radiation and the matrix material.

The elemental mappings of Co and W appear sharp at the interface. In contrast the $Cr_{K\alpha}$ intensity shows a gradual change across the interface. The components of the glass ceramic have the following distribution at the interface: phosphorus is detectable in larger amounts in the glass ceramic and is homogeneously distributed. The same can be stated for oxygen. Some regions at the metal surface exhibit high Al and O contents. They were identified as corundum grains which remained in the

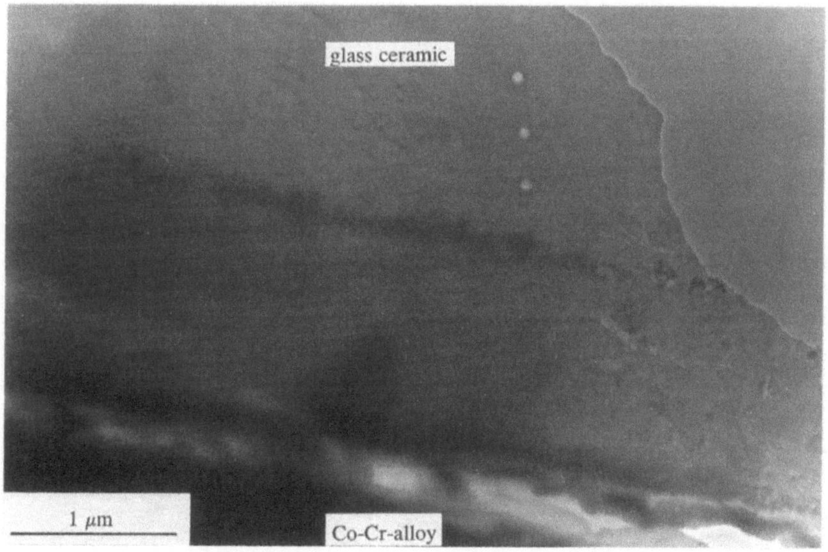

Fig. 3. Transmission electron micrograph of the analysed area of the interface

metal surface after sandblasting. The intensity differences of aluminium in the metal and glass ceramic are small. Ca is not homogeneously distributed in the glass ceramic because the particular crystalline phases have different Ca contents. It is remarkable that the Ca content near the interface is lower than in regions further away from the boundary. An interfacial region is formed which consists mainly of O, Cr and P (or Zr). This region has a width of 8–15 μm in the micrograph. Taking into account the gentle slope of this particular specimen, this layer has a thickness of about 1.5–2.5 μm.

The advantage of TEM investigation is the better spatial resolution of EDX due to a smaller excitation volume which corresponds practically to a cylinder with the diameter of the electron beam, in this case about 60 nm. In contrast to the SEM measurements, the $Zr_{K\alpha}$ peak ($E_0 = 15.77$ keV) is detectable in the TEM because of the higher excitation energy of 200 kV. A TEM micrograph of the interface is depicted in Fig. 3. The dark part corresponds to the metal which is not thin enough to become transparent for electrons. A gap is visible in some parts of the interface due to the preferred removal of material in this area in the course of TEM sample preparation. The light circles mark some areas from which spectra were taken.

Figure 4 contains a series of spectra collected at different distances from the interface in the glass ceramic. In order to get spectra that can be compared, they were normalised with respect to the background.

The series of spectra shows the following changes in intensities of the elements as a function of the distance from the interface: the Co content does not change significantly. The intensity of $Cr_{K\alpha}$ decreases with increasing distance from the interface into the glass phase. The Ca content near the interface is significantly lower than in the bulk material. The relative intensities of Na, Al and P do not show remarkable changes. It is worth mentioning, that the intensity of $Zr_{K\alpha}$ decreases with larger distances from the interface. The results for the components of the metal are in agreement with those obtained from SEM investigations. The reason for the

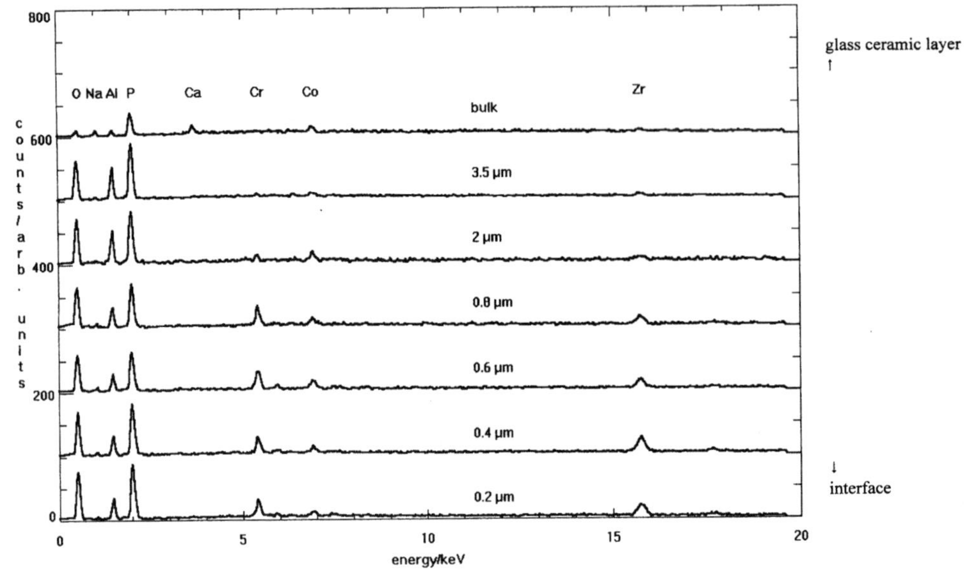

Fig. 4. EDX spectra of the TEM sample in the area of the glass ceramic layer as a function of the distance from the interface

detection of small amounts of Co everywhere in the sample is presumably to be found in the ion beam thinning. Some material is removed by the ion beam and deposited at the surface of the specimen.

From these results, the process of sintering can be explained as follows: at higher temperatures the metal surface is oxidized. The oxidic components react with the glassy coating and form an interlayer. Sintering tests in vacuum led to bad adhesion of the layers [6] and prove the essential role of oxygen. The enrichment of Cr as constituent of the alloy in the interlayer can be explained by the high oxidation potential of this element. Due to its high oxidation potential, Cr is preferably oxidized and known to diffuse into glass at higher temperatures [11, 12].

Conclusions

SEM and TEM investigations displayed an interlayer between the bulk metal and the glass ceramic layer without any gap between these two materials. This study shows that the metal surface is oxidized during sintering in air and that the oxidic components react with the glassy coating and form an interlayer. The oxidic layer is rich in Cr, P and Zr and guarantees a good adhesion between metal and coating. This result is in accordance with the mechanical measurements [5]. Since the interlayer does not have sharp borders, it allows a gradual change between the metallic bonding in the alloy and the structure of the glass ceramic coating.

From the point of view of material science, the material is well suitable as bone implant. Further efforts are necessary to investigate the formation mechanism of the interlayer with respect to the influence of the ion diffusion from the alloy into the glass on the crystal phase formation near the interface.

Acknowledgements. The authors wish to thank Prof. W. Hauffe and A. Grahl (TU Dresden) for preparing the sample by the ion beam slope cutting method and Dr. T. Höche for his help in preparing the TEM sample and carrying out the TEM investigations.

References

[1] M. Semlitsch, H. G. Willert, *Medizinisch-Orthopädische Technik* **1976**, *96*, 86.
[2] M. Ungethüm, U. Fink, *Z. Orthop.* **1988**, *126*, 697.
[3] W. Vogel, W. Höland, *Angew. Chem.* **1987**, *99*, 541.
[4] P. Wange, J. Vogel, L. Horn, W. Höland, W. Vogel, *Silicates Industriels* **1990**, *55*, 231.
[5] P. Wange, M. Frigge, G. Völksch, C. Jana, W. Götz, *Advances in Science and Technology 12: Materials in Clinical Applications* (P. Vincienzi, ed.), Faenza, 1995.
[6] M. Frigge, A. Grahl, S. Däbritz, W. Hauffe, G. Völksch, *Beitr. Elektronenmikroskop. Direktabb. Oberfl.* **1994**, *27*, 131.
[7] P. Wange, G. Carl, K. Naumann, J. Vogel, W. Vogel, W. Götz, W. Höland, *Silicates Industriels* **1991**, *56*, 21.
[8] W. Hauffe, *Beitr. Elektronenmikroskop. Direktabb. Oberfl.* **1990**, *23*, 305.
[9] M. Frigge, A. Grahl, S. Däbritz, W. Hauffe, G. Völksch, *Glastech. Ber. Glass Sci. Technol.* **1994**, *67C*, 396.
[10] A. Strecker, U. Salzberger, J. Mayer, *Prakt. Metallogr.* **1993**, *30*, 482.
[11] F. Hong, D. Holland, *J. Non-Cryst. Solids* **1989**, *112*, 357.
[12] A. J. Sturgeon, D. Holland, G. Partridge, C. A. Elyard, *Glass Technol.* **1987**, *27*, 102.

Mikrochim. Acta [Suppl.] 13, 307–316 (1996)

Monte Carlo Method for Quantitative Analysis of Bulk and Layered Samples

Ondrej Gedeon[1,*], Václav Hulínský[1], Karel Jurek[2], and Marta Šimečková[2]

[1] Department of Glass and Ceramics, Institute of Chemical Technology, Technická 5, 166 28 Prague, Czech Republic

[2] Institute of Physics, Academy of Sciences of Czech Republic, Cukrovarnická 10, 162 00 Prague, Czech Republic

Abstract. The present article describes a Monte Carlo method for routine quantitative microanalyses of bulk and stratified samples. The model is based only on elemental processes without experimental adjustments. Our approach is tested on data sets of bulk borides and carbides as well as on prepared layers of Au on Pt and Au on Si, where k-ratios were measured. Results point out that the proposed model is able to give results comparable to those obtained by empirical or semi-empirical approaches.

Key words: EPMA, Monte Carlo, correction procedure, layered samples.

There are two distinct trends in quantitative electron probe analysis: the empirical and semi-empirical correction methods used in practice and Monte Carlo methods mainly used for the determination of general trends of such quantities as backscattering coefficients, surface ionization and electron range. But it was not clear so far if the theory based on first principles is able to predict the same or comparable quantitative results as empirical methods. Several authors expressed their belief [19] that the Monte Carlo method can not achieve such excellent results and can serve only as a good qualitative indicator. We decided to clarify this question and to establish a correction programme fully based on first principles. The first attempts were based on the Rutherford and "Bethe" cross-sections [8, 9]. Results were very encouraging especially because they were obtained without any adjustment to experimental data. It was also found [8] that there is no influence of deep atom shell ionization or secondary electrons on the distribution functions calculated by this model. In this paper, we propose to improve the model by introducing the Mott cross-section for elastic scattering and the Rao-Sahib and Wittry energy-loss approximation for low electron energy. To test the accuracy of the model we first calculated k-ratios for bulk borides and carbides samples. To extend and test the model on layered samples we prepared layers of gold on platinum and silicon

* To whom correspondence should be addressed

substrates, then we have measured k-ratios of these samples that were compared to results of model calculations.

Experimental

Monte Carlo Simulation

The various Monte Carlo algorithms are well known and therefore our MC model will be described briefly. We assume a bulk target with a flat surface exposed to the normal incident electron beam. The actual path Λ between two scattering events is for bulk samples given by

$$\Lambda = \Lambda_0 \ln R, \tag{1}$$

where Λ_0 is the mean free path and R is a random number. For a layered sample the actual path is determined as described by Murata et al. [17]. After traversing a distance Λ the electron in question will be scattered either elastically or inelastically. At each scattering centre a random number is used to determine the type of the scattering and in the case of elastic scattering another random number predicts the scattering element. The elastic scattering is described by the Mott cross-section according to values tabulated by Riley et al. [23]. For energies not available in the tables we have carried out a linear interpolation between two adjacent energies or an extrapolation for energies below 1 keV.

While using Mott cross-section for elastic scattering is straightforward and is generally accepted as the best choice, the situation in the case of inelastic events is much more complicated. Many processes, such as volume and surface plasmon excitations, deep energy shell ionizations, influence of secondary electrons, interaction of primary electron with the valence band, contribute to it but many are poorly known quantitatively. Moreover, introducing many processes drastically increases computing time. To find a simple and acceptable formula describing the inelastic scattering we assumed that the primary electron can lose only a distinct amount of energy ΔE which is the same for each incident electron and independent of the element. On the other hand, the form of the cross-section is restricted by the requirement that it must imply the same stopping power as the Bethe law. To conform to these assumptions the inelastic differential cross-section σ^{in} is

$$\sigma^{in}(E, E') = \frac{B(E)}{\Delta E}\,\delta(E - E' - \Delta E), \tag{2}$$

with E, E', ΔE the energy of the electron before and after scattering, and energy loss, respectively. The quantity $B(E)$ is given by the Bethe energy-loss approximation [4], which for low energies ($E < 6.338J$) is replaced by the formula according to Rao-Sahib and Wittry [20]. The mean ionization potential J given by Berger and Seltzer [3] was used.

The parameter ΔE in Eq. (2), the energy loss per inelastic scattering, does not follow from the model but is considered here as a free parameter. Nevertheless, a variation of ΔE over a wide range does not affect the distribution functions significantly [8]. If we assume small ΔE, i.e. the inelastic cross section is much larger than the elastic one, the calculated energy loss between two elastic events is close to that obtained from Bethe's law. When assuming larger ΔE (the inelastic cross section is smaller than the elastic one) a few elastic collisions occur between two inelastic events. The electron in question loses a higher portion of energy per inelastic event but its path between two inelastic events is longer. However, in the case of elastic scattering, the influence of the energy of the primary electron on its angular distribution is rather weak and therefore a not too extreme increase of ΔE does not change the distribution function significantly. Of course, the lower the primary electron energy, the smaller value of ΔE should be chosen.

The elastic and inelastic cross-sections fully determine the electron transport in the sample. To calculate the X-ray production we have to complete the description of elemental processes by an ionization cross section. Among a few tested: Mott-Massey [16], Worthington-Tomlin [26], Pouchou-

Pichoir [19], Hutchins [12], we have found that the Gryzinski [11] cross-section gives the best agreement between calculated distribution functions $\varphi(\rho z)$ and experimental ones [10].

In the case of stratified samples with high atomic number we have taken into account the continuum fluorescence. To incorporate the phenomenon we used the cross-section by Rossouw and Whelan [24] for photon generation. Edge jumps were calculated according to Poehn et al. [18].

On the basis of the above mentioned elemental processes we have developed a correction programme for quantitative analysis. For each analyzed element α and for each iteration step the photon production Q_α is calculated by means of Monte Carlo procedure. The photon production is then normalized to unit concentration. Simulating the photon production in a standard then yields the k-ratio k_α

$$k_\alpha = \frac{c_\alpha Q_\alpha(c)}{c_\alpha^{st} Q_\alpha^{st}(c^{st})} \tag{3}$$

where the superscript "*st*" denotes quantities related to the standard. The values for standards are calculated in advance and can serve as a library. The iteration procedure comes from Eq. (3)

$$c_\alpha^{(n)} = k_\alpha c_\alpha^{st} \frac{Q_\alpha^{st}(c^{st})}{Q_\alpha(c^{(n-1)})} \tag{4}$$

where n indicates the number of iteration.

Taking the inelastic cross-section in the form of delta function with strictly given energy loss has another advantage. The primary electron loses its energy independently of the type of scattering atom and therefore the ionization probability per ionization event can be calculated for a particular type of atom on the basis of its ionization cross-sections and the mass absorption coefficient. It implies that the assessment of trace element takes the same amount of computer time as any other element which is especially advantageous for a routine use of the model.

Preparation of Layered Samples

Thin gold layers were deposited on Si and Pt substrates by magnetron sputtering. Silicon substrates were prepared by mechanical polishing; the platinum substrates were obtained by magnetron sputtering on polished silicon. A set of thin films of increasing thickness was prepared by progressive masking. Their thicknesses were individually measured by a stylus method. However, the precision of the individual measurement was not high enough so that the sputtering rate was calculated by dividing the total thickness by discharge current and sputtering period while sputtering conditions were kept constant (especially argon pressure of 2×10^{-2} Pa). The sputtering rates were found to be constant within $\pm 2\%$, except for the very thin films. The thicknesses of all films were calculated assuming proportional dependence of the thickness on time and glow discharge current. Mass thickness was calculated using the same density as for bulk metal.

Intensities of generated X-rays were measured with an JEOL JXA-733 electron probe X-ray microanalyser at 10, 20 and 30 keV, respectively. Polished specimens of pure elements were used as standards to calculate k-ratios of Lα and Mα lines (ratio of an intensity from the layered sample to an intensity of the bulk sample). Dependence of k-ratios versus thickness should give a smooth curve so that random experimental errors could be revealed and more reliable results were obtained compared to large but inconsistent collection of analysis of not well defined specimens.

Results and Discussion

In this section we summarize the experimental and computed results for bulk and stratified samples separately. It is worth to stress that all calculated results are obtained without any experimental adjustment and are based only on the above discussed elemental processes.

Bulk Samples

The programme was tested on Bastin and Heiligers borides [1] and carbides' data sets [2] using their mass absorption coefficients apart from a few cases where the results were inadequate: coefficients for Ta(Mα)/Ta (Mα line of tantalum in tantalum) and W(Mα)/W were taken from [5] and those for Ti(Kα)/C, V(Kα)/C, Ta(Mα)/C, W(Mα)/C were taken as proposed in [15]. Borides of nickel and cobalt were excluded from the set for reasons discussed by Pouchou and Pichoir [19]. For all samples, the calculated k-ratio k(calc) was compared with the measured k-ratio k(exp). The number of simulated electron trajectories was in all cases 10 000 to keep the root-mean-square (RMS) coming from statistics below 0.3%. The results for all elements considered are summarized in Fig. 1. The RMS for this set is 1.99% and the mean value is 0.998, implying that the calculated results are fully comparable with the best ones obtained by empirical or semi-empirical approaches [15, 19].

In the case of the analyses of boron and carbon we have obtained worse results. RMS and mean value are 5.90% and 1.041 respectively for boron; and 4.00% and 0.993 for carbon. The main reason for this discrepancy lies, as we believe, in the uncertainty of absorption coefficients because the electron transport is described by the same equations for carbon and boron as for companion elements. This is, for example, confirmed by the results of carbon boride where the RMS and mean value for boron are 1.17% and 1.007, respectively. The large improvement of the results could also be achieved by adjusting the absorption coefficients so

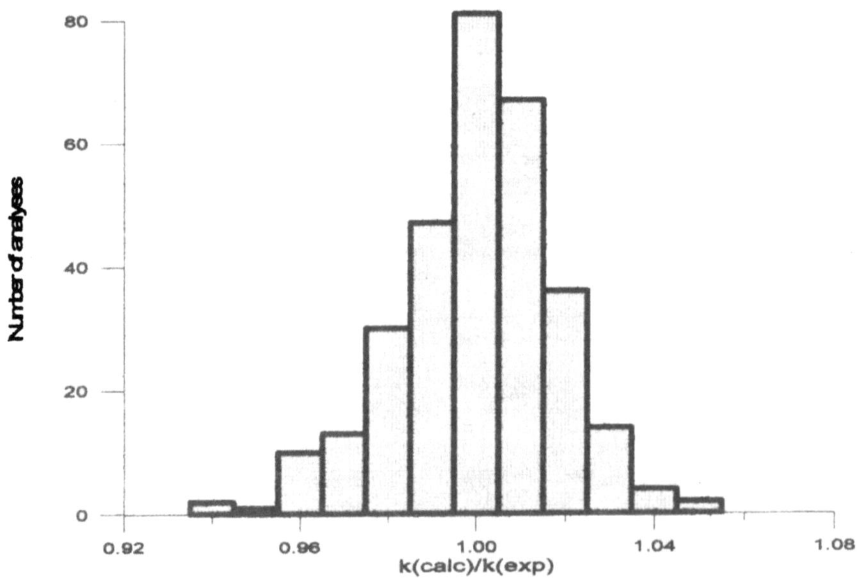

Fig. 1. Error histogram for elements present in the borides' and carbides' data sets (Experimental values are taken from [1, 2])

that the calculated values were as close as possible to the experimental ones. However, we do not think it is the best way to determine these coefficients. The mean value of boron (borides set) also shows a systematic shift in the k-ratios, again most probably due to incorrect absorption coefficients. On the other hand it should be kept in mind that in the case of large absorption coefficients even a small discrepancy in the calculated $\phi(\rho z)$ curve can cause a great deviation of k-ratio from the experimental results and therefore one can expect higher RMS-values. We can conclude that the present model, purely based on elemental processes, is able to predict results fully comparable to other approaches. Due to its nature, it is very universal, reproducible, easily updated to new achievements and we assume, that the maximum error will not exceed the maximum error obtained in the presented systems.

Layered Samples

Figures 2 to 5 show the results for the Mα and Lα lines in the Au/Pt system measured at various voltages. In this case there is a very small difference between the atom number of the film and the substrate. This means that the curves are very close to the integral distribution functions and reflect the adequacy of their modeling in the systems with high atom number. Comparing the calculated values with experimental, it can be deduced that the relative difference is generally below 20%, except for very small thicknesses and high voltage (the largest relative deviations is in the case of Au(M_α)/Pt – 30 keV – 30 nm and equals 49%). However, in this case, experimental points deviate from the expected shape of the line due to insufficient precision of the thin film thickness determination. The PAP (programme STRATA) gives slightly better results overall, likely because it is also based on the experimental points. It is

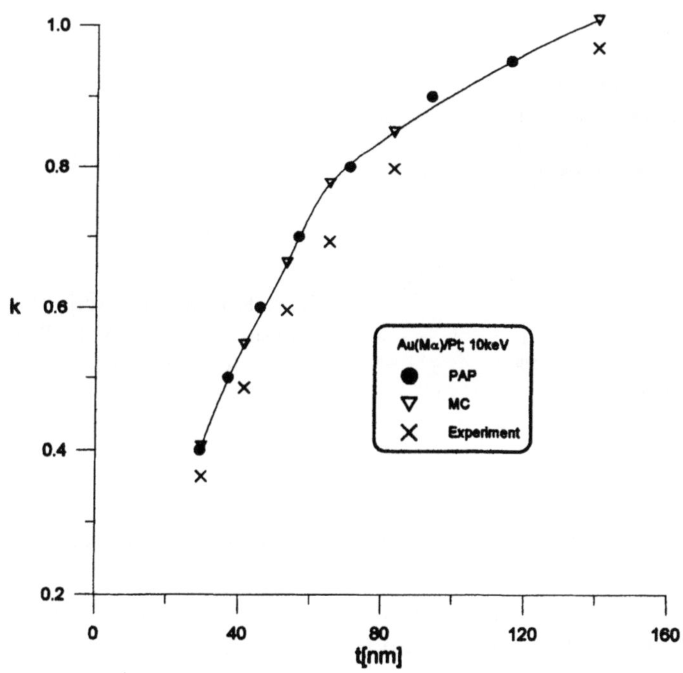

Fig. 2. K-ratio of the Au(Mα) line on Pt substrate; 10 keV; PAP denotes values obtained by programme STRATA

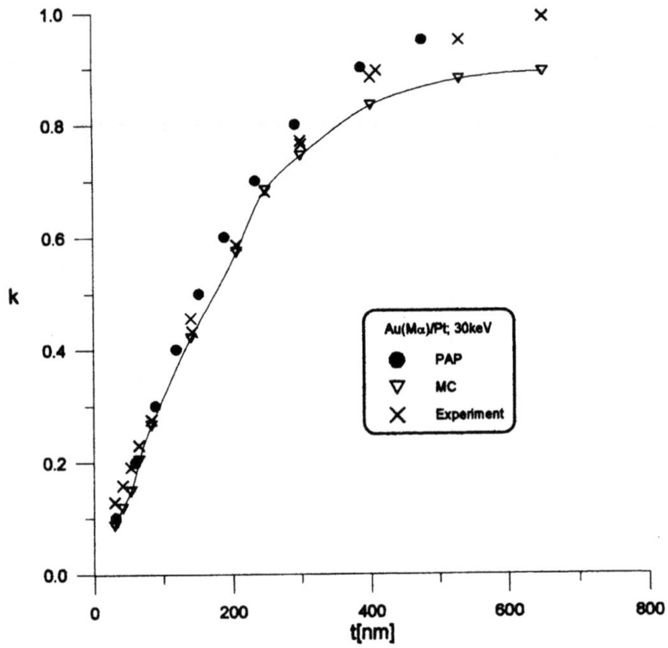

Fig. 3. K-ratio of the Au(Mα) line on Pt substrate; 30 keV; PAP denotes values obtained by programme STRATA

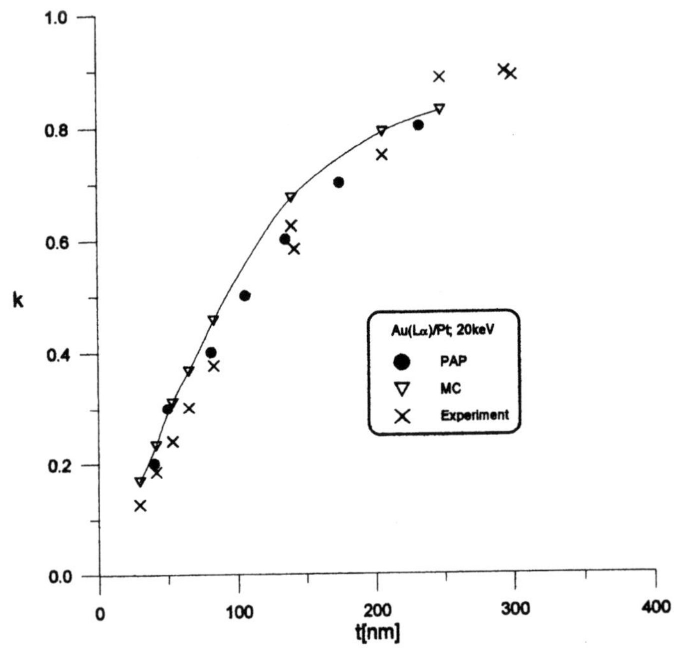

Fig. 4. K-ratio of the Au(Lα) line on Pt substrate; 20 keV; PAP denotes values obtained by programme STRATA

not surprising especially in the case of Lα lines where continuum fluorescence must be taken into account.

Figures 6 to 8 show the results for the more interesting system Au/Si, where the difference in atom numbers between the Au film and the Si substrate is large. Again

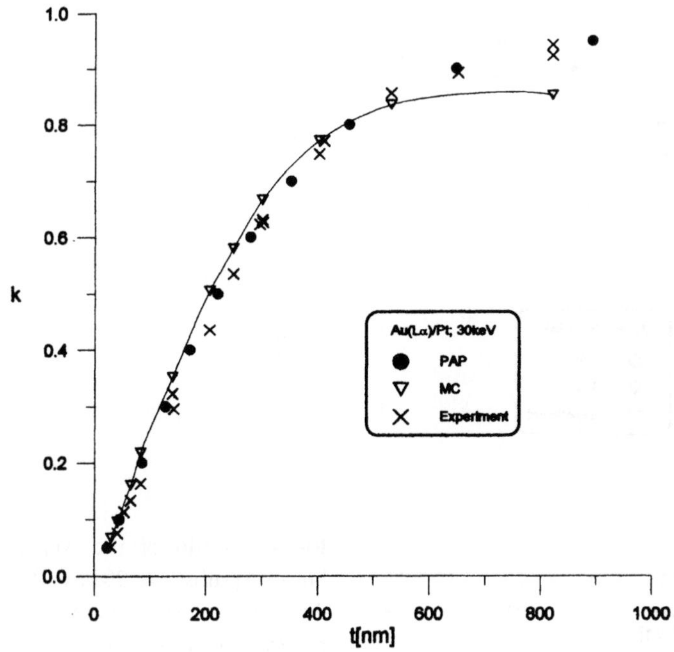

Fig. 5. K-ratio of the Au(Lα) line on Pt substrate; 30 keV; PAP denotes values obtained by programme STRATA

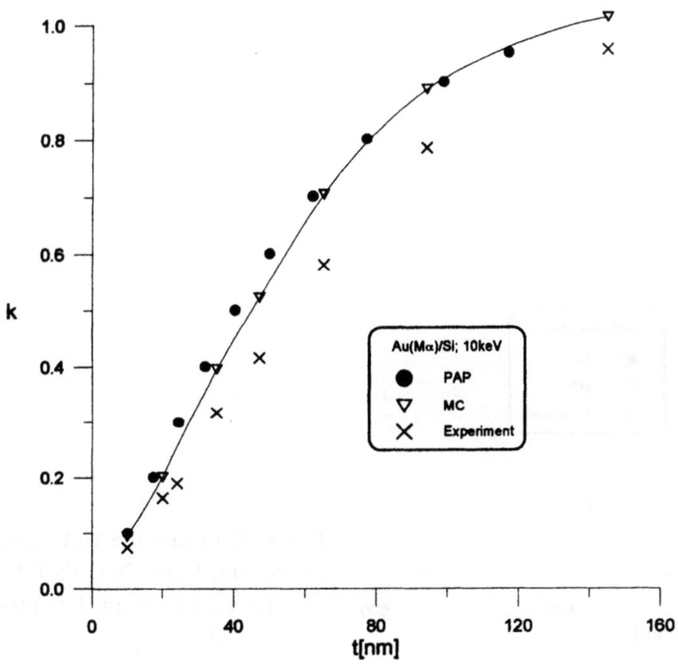

Fig. 6. K-ratio of the Au(Mα) line on Si substrate; 10 keV; PAP denotes values obtained by programme STRATA

we can repeat all reasons for deviations from the experimental results as in the former case. The deviation of PAP from the experiment increases while the differences of the model from experiment are approximately the same as in the former case. This indicates the universality of the model and its independence on the samples.

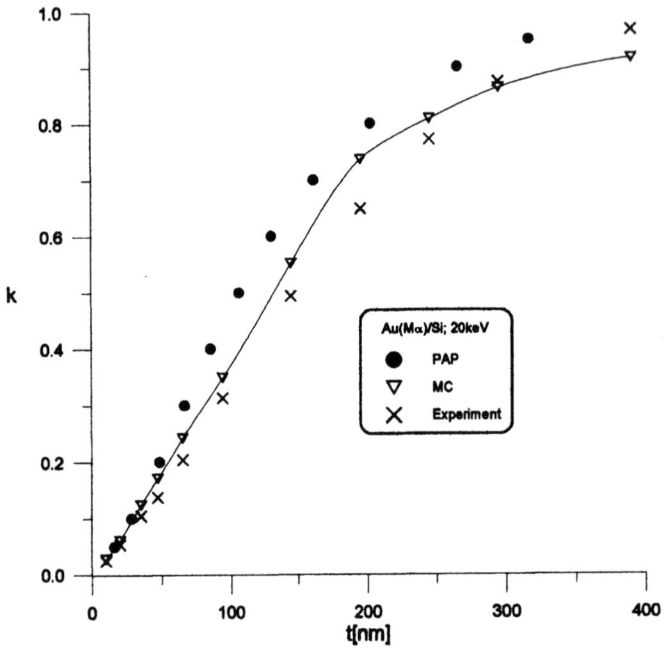

Fig. 7. K-ratio of the Au(Mα) line on Si substrate; 20 keV; PAP denotes values obtained by programme STRATA

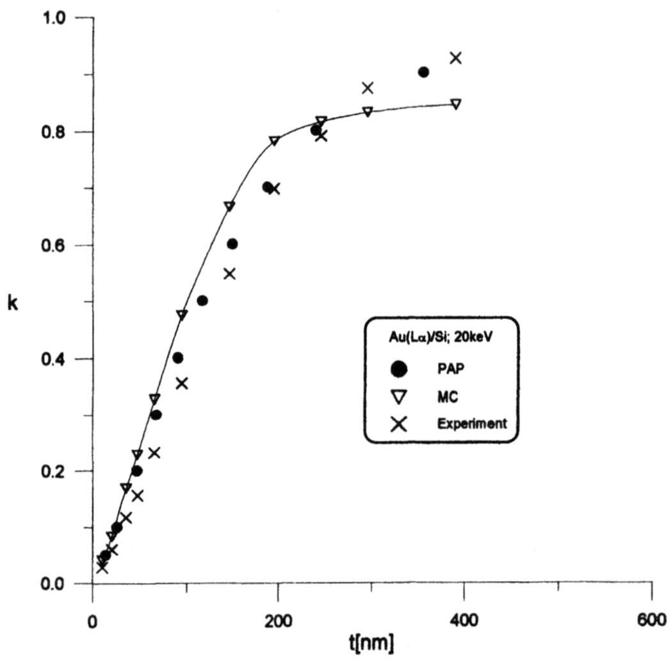

Fig. 8. K-ratio of the Au(Lα) line on Si substrate; 20 keV; PAP denotes values obtained by programme STRATA

The maximal deviation is about 25%, independent of the type of film and substrate and dominantly comes from the modeling of the elementary processes. In contradiction to bulk samples, using bulk standards does not compensate for systematic errors coming from the description of the elemental processes.

Conclusion

Although the model looks very simple, it is able to produce acceptable results. It is supposed that the zig-zag trajectory is determined by the elastic scattering, and the energy deceleration by the inelastic-cross section in the form of a delta function. The angular part of inelastic events are neglected for the sake of simplicity. This fact implies the limitation of the present model which lies on the lower side of energy of the primary electron. Except for the part of energy where the angular deviation coming from the inelastic events is not negligible the simulation of electron transport is as adequate as the precision of the two elemental processes. The energy loss ΔE must be chosen so as not to be too high. The weakest point of the present model is the ionization cross section which should be replaced by the tabulated values in the next step. All three processes have a complementary influence on the result (in some cases continuum fluorescence must be included) and therefore it is very difficult to solve the particular deviations from the experiment and it is in some extent only a speculation. The samples on which the model was tested were chosen so that they include the limitating cases in current microprobe analyses. It is quite expected that the model will work for more complex systems, e.g. systems with elements present both in the film and in the substrate or systems with a larger numbers of layers, with comparable precision as was achieved in the chosen samples.

Acknowledgements. This work was financially supported by Grant Agency of Czech Republic through the grant N. 104/93/2150.

References

[1] G. J. Bastin, H. J. M. Heiligers, *Report Eindhoven Univ. of Techn. Netherlands* 1986.
[2] G. F. Bastin, H. J. M. Heiligers, *Report Eindhoven Univ. of Techn. Netherlands* 1984.
[3] M. J. Berger, S. M. Seltzer, *Nat. Res. Council. Publ.* **1964**, *1133*, 252.
[4] H. Bethe, *Ann. Phys.* **1930**, *5*, 325.
[5] L. S. Birks, *Electron Probe Microanalysis*, Wiley-Interscience, New York, 1971.
[6] J. D. Brown, L. Parobek, *6th ICXOM* (G. Shinoda, K. Ohra, T. Inochawa, eds.), University of Tokyo Press, Tokyo, 1972, p. 163.
[7] L. Parobek, J. D. Brown, *X-Ray Spectrom.* **1978**, *7*, 26.
[8] O. Gedeon, V. Hulinsky, *9th Czechoslovak Spectroscopic Conf.*, Ceské Budejovice, 1992.
[9] O. Gedeon, V. Hulinsky, *Mikrochim. Acta* **1994**, *114*/115, 305.
[10] O. Gedeon, *Doctoral Thesis*, Prague, 1994.
[11] M. Gryzinski, *Phys. Rev.* **1956**, *A 14*, 531.
[12] G. A. Hutchins, *Characterisation of Solid Surfaces*, Plenum, New York, 1974.
[13] P. Karduck, W. Rehbach, in: *Electron Probe Quantitation* (K. F. J. Heinrich, D. E. Newbury, eds.), Plenum, New York, 1991.
[14] P. Karduck, N. Amman, *19th Ann. Meeting of MSC*, 1991.
[15] C. Merlet, *Mikrochim. Acta* **1994**, *114/115*, 363.
[16] N. F. Mott, H. S. W. Massey, *The Theory of Atomic Collision*, Oxford University, London, 1949.
[17] K. Murata, H. Kawata, K. Nagami, *Scanning Microscopy* [*Suppl.*] **1978**, 1.
[18] C. Poehn, J. Wernisch, W. Hanke, *X-Ray Spectrometry* **1985**, *14* (3), 120.
[19] J. L. Pouchou, F. Pichoir, in: *Electron Probe Quantitation* (K. F. J. Heinrich, D. E. Newbury, eds.), Plenum, New York, 1991, p. 31.

[20] T. S. Rao Sahib, D. W. Wittry, *Appl. Phys.* **1974**, *45*, 5060.
[21] W. Rehbach, *Doctoral Thesis*, Aachen, 1988.
[22] L. Reimer, E. R. Krefting, in: *Use of Monte Carlo Calculations in Electron-Probe Microanalysis and Scanning Electron Microscopy* (K. F. J. Heinrich, D. E. Newbury, H. Yakowitz, eds.), NBS Spec. Publ. 460, 1976.
[23] M. Riley, C. J. Mac Callum, F. Biggs, *At. Data Nucl. Data Tabl.* **1975**, *15*, 443.
[24] C. J. Rossouw, M. J. Whelan, *J. Phys. D. Appl. Phys.* **1979**, *12*, 797.
[25] D. A. Sewell, G. Love, V. D. Scott, *J. Phys. D. Appl. Phys.* **1985**, *18*, 1233.
[26] C. R. Worthington, S. G. Tomlin, *Proc. Phys. Soc.* **1956**, *A 69*, 401.

Mikrochim. Acta [Suppl.] 13, 317–324 (1996)

SIMS Linescan Profiling of Chemically Bevelled Semiconductors: A Method of Overcoming Ion Beam Induced Segregation in Depth Profiling

Ching-Ming Hsu* and David S. McPhail

Department of Materials, Imperial College of Science, Technology and Medicine, Prince Consort Rd., London SW7 2BP, U.K.

Abstract. Copper implanted p-type silicon samples were investigated using secondary ion mass spectrometry (SIMS) depth profiling with positive oxygen primaries incident at a variety of angles. Beam induced segregation leads to considerable distortion of the profile if the beam was incident at angles between normal and 30°. These results were consistent with SIMS work on thin films of copper on p-type silicon. The copper implanted material was chemically bevelled. SIMS linescan profiles of the chemically bevelled samples yielded accurate copper profiles free of beam induced segregation. The pile-up of the Cu at the SiO_2/Si interface during oxygen depth profiling at normal incidence was observed by chemically bevelling a SIMS crater that had been synthesised with an oxygen primary beam. The location of the copper atoms with respect to the altered layer (i.e. the internal profile) was then found by linescanning down the bevel plane.

Key words: copper, segregation, SIMS, bevelling, linescanning.

Ion beam induced impurity segregation occurs during SIMS depth profiling of certain impurities in silicon, for example copper, silver and palladium, when using reactive primary ions such as oxygen and nitrogen. Gross distortions of the true distribution can then arise. Two main arguments have been used to explain impurity segregation behaviour in silicon under reactive primary beam bombardment. They are (1) the thermodynamic model [1, 2] which suggests that the tendency to segregate depends on the relative heats of formation of the compounds synthesised during reactive sputter profiling. The heat of formation of SiO_2, for instance, is more negative than that of CuO or CuO_2, and as a consequence Cu atoms will move out of the altered layer of SiO_2 and will migrate towards the Si substrate; (2) the field-induced migration model [3, 4] which suggests that the displacement of impurity ions is associated with the build up of an electrical field across the synthesised SiO_2

---

* To whom correspondence should be addressed

layer. In the case of Cu in Si under positive oxygen ion bombardment, a positive charge will build up on the SiO_2 surface due to enhanced electron emission on the sputtered surface leading to a large field across the altered layer. Positive copper ions will then be driven to the oxide/silicon interface. The extent of the impurity segregation has been found to be dependent upon primary beam energy and angle of beam incidence [5, 6], and the type of impurity atoms [7]. The pile-up of Cu at the SiO_2/Si interface during oxygen depth profiling has been reported by Boudewijn [8] and Kilner [6] using Cs^+ internal depth profiling on the O_2^+ sputtered crater. Other work has been conducted with thin films of Cu deposited upon a Si substrate [5].

In this study, a B doped p-type Si substrate implanted with ^{63}Cu was investigated using SIMS depth profiling. Oxygen primary beams at two beam energies and a series of angles of incidence were used to measure the copper distribution. The results were compared with those obtained by Homma [5] (Cu thin film deposited p-type Si) and Kilner [6] (Cu implanted n-type Si). We also synthesised shallow bevels in the implanted material using a chemical bevelling technique we have developed in our laboratory. The bevelled samples were subsequently linescanned in the SIMS machine and the results compared to the depth profiles. This 'bevel and linescan' approach was originally developed for high depth resolution studies of quantum wells and delta doped layers [9, 10] and depth resolutions of 2 nm have been achieved with aluminium deltas in gallium arsenide. Here, however we were primarily interested in the use of a 'bevel and linescan' approach to eliminate artefacts associated with beam induced segregation.

Experimental

Samples were prepared by the implantation of ^{63}Cu (200 keV/10^{15} cm^{-2}) into a B doped p-type (100) Si wafer (resistivity 3 Ω cm). 4 mm × 8 mm rectangles were scribed out for the chemical bevelling process. The apparatus used for the chemical bevelling is shown in Fig. 1. Samples were lowered into the etching

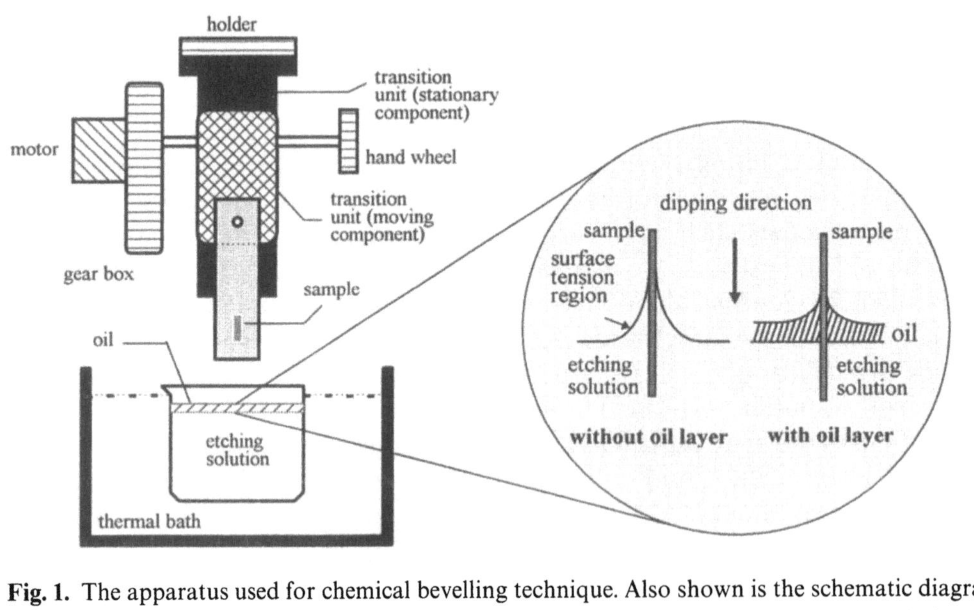

Fig. 1. The apparatus used for chemical bevelling technique. Also shown is the schematic diagram of the sample/solution interface both with and without an oil layer on top of the etching solution

solution at a rate of 250 μm/min to a depth of 5 mm and were then swiftly removed, rinsed in the water and then cleaned with acetone in an ultrasonic bath. A layer of vegetable oil above the etchant was used to prevent etching of the exposed silicon material by fumes and to reduce the surface tension between the sample and the etchant, thus yielding an abrupt bevel-surface interface. The etchant used was HF-HNO_3-CH_3COOH with various concentration ratios of 1:3:6, 1:3:7, 1:3:8, leading to etch rates in the Si of 0.70, 0.50 and 0.31 μm/min respectively. The bevel magnifications produced were approximately 3.6×10^2, 5.0×10^2 and 8.0×10^2. The 1:3:8 etchant has been used for two dimensional dopant profiling in silicon [11, 12] and for selective etching of B doped Si to investigate the implant induced secondary defects [13].

SIMS analyses of the Cu implanted samples were performed in an Atomika 6500 instrument. Depth profiles of the implanted material were conducted with an $^{16}O_2^+$ primary beam (10 keV and 15 keV) rastered over a 200 μm × (200 μm/cos θ) area at angles of incidence (θ) ranging from 0° and 50° to the surface normal. Linescan profiling on bevelled samples was carried out using a 15 keV $^{16}O_2^+$ primary beam incident either at 0° or at 30° to the surface normal; the beam width during imaging and linescanning was 7 μm. Linescan lengths of 300 μm, 400 μm and 500 μm respectively were used for the bevels produced with the 1:3:6, 1:3:7 and 1:3:8 solutions. Ion images of the bevels were employed to obtain a 3-D profile.

One SIMS depth profile was conducted with the beam incident normally using a 2 mm × 2 mm raster. The profile was terminated before the peak of the copper implant had been reached (at a depth of ~ 100 nm) and part of the crater was then chemically bevelled using 1:3:10 HF-HNO_3-CH_3COOH solution to a depth of 280 nm. Thus the altered layer at the bottom of the crater was projected as a wide surface stripe on the bevel plane. The bevelled samples was then ion imaged and linescan profiled with a 10 keV oxygen primary beam to locate the copper atoms with respect to the beam induced altered SiO_2 layer (i.e. to determine the copper internal profile).

Results and Discussion

Figure 2 illustrates the depth profiles of ^{63}Cu implants in B doped p-type Si obtained at a variety of angles of incidence of the primary ion beam. Also shown in the inset of the graph is the decay length on the trailing edge of the implant, i.e. the depth in nm over which the signal decays by a factor e, as a function of the angle of incidence of the primary beam for beam energies of 10 keV and 15 keV. It can be seen that the decay length drops rapidly from 10° to 30° and is essentially constant thereafter. It would appear that beam induced segregation distorts the implant shape for angles up to 30° and that for higher angles of incidence the SIMS profile is essentially correct. The maximum intensity at the peak of the copper implant was obtained at an angle of incidence of 30°. The relationship between decay length and angle is similar to that observed from SIMS depth profiles of thin films of copper on Si [5]. However, the decay lengths of the Cu implant at angles greater than 30° are approximately 2.5 times those obtained from the thin films. Clearly the values derived from the thin film are associated with beam induced mixing whereas the values derived from the implant are characteristic of the implant distribution.

Figure 3 shows the result of linescanning a bevelled sample. (1:3:7 HF-HNO_3-CH_3COOH etching solution was used in this case). The linescan data has been depth calibrated using the bevel shape, which was measured using a Talystep surface profilometer. The details of the linescan profile conversion process have been discussed elsewhere [9]. The linescan profile is compared with the profile from the TRIM simulation and the SIMS depth profile obtained at 42° with a 15 keV beam. In

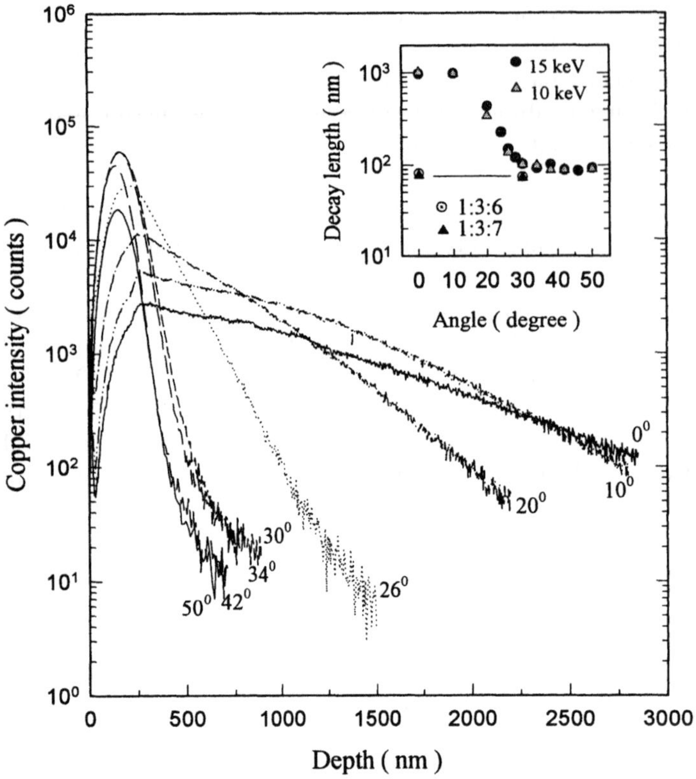

Fig. 2. Depth profiles of a ^{63}Cu implant in B doped p-type Si at a variety of angles of beam incidence to the sample normal. The inset shows the decay length as a function of the angle of incidence of the beam both for beam energies of 10 keV and 15 keV. Decay length observed using the bevel and linescan approach are also included in this inset

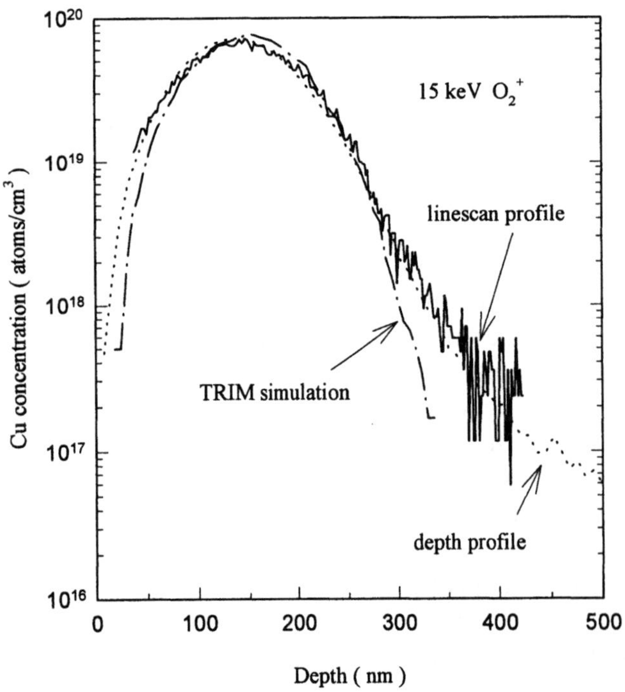

Fig. 3. The converted and depth calibrated linescan profile from the bevel (chemically bevelled 1-3-7 HIF-HNO_3-CH_3COOH), compared with the profile from a TRIM simulation and a SIMS depth profile analysed at 42°

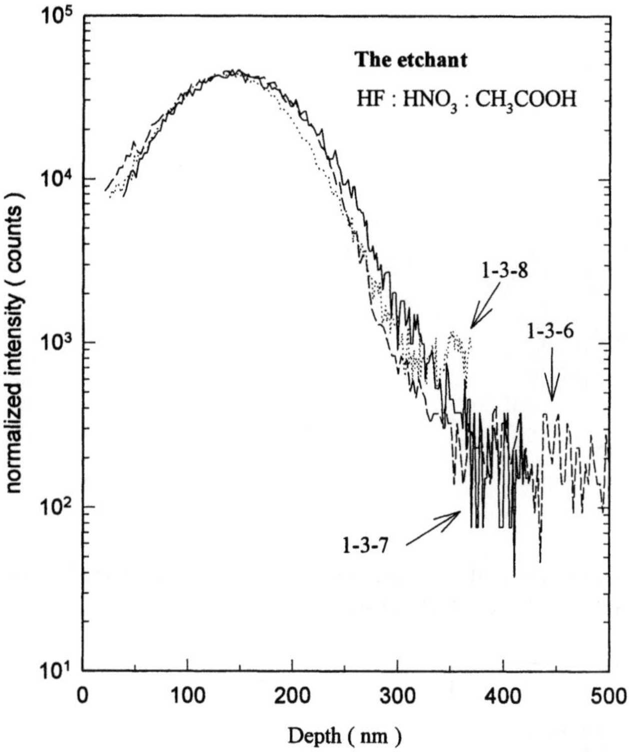

Fig. 4. The converted and depth calibrated linescan profiles from three bevels chemically bevelled with 1-3-6, 1-3-7 and 1-3-8 HF-HNO_3-CH_3COOH solution. The bevel magnifications were respectively, 3.6×10^2, 5.0×10^2 and 8.0×10^2

TRIM simulation, a number of energetic ions are in sequence implanted into a target and causes a series of cascade of residual atoms. The implanted ions loss energies and finally sit in the target at a depth from the surface. The ions can be counted as a function of depth so a distribution profile of the ion is obtained. It can be seen from the figure that all three profiles are very similar in shape indicating the accuracy of the bevel and linescan approach. Furthermore the shape of the linescan profiles does not seem to depend upon the bevel magnification, as shown in Fig. 4. The other main advantage of the bevel and linescan approach is the speed of acquisition of data. A linescan can be completed in a few minutes whereas the depth profiles can take an hour or more. The main disadvantage of the bevel and linescan approach compared to depth profiling is the relatively poor sensitivity due to the low analytical volume per data point. Typically the intensity of the Cu signal at the peak of the implant in a depth profile was greater that in linescan profiling by two orders of magnitude.

The impurity intensity can be increased by increasing the beam current, thus increasing the analytical volume. Care must be taken, however, since profile distortions associated with a lateral shift in the image will occur if too much material is removed. We also found that the signal intensity was greater if the imaging beam was incident at 30°. The Cu profiles obtained at normal incidence and 30° were identical in all other respects. The measured decay lengths for these linescan profiles, which are included in the inset to Fig. 2, are the same as those measured by depth profiling above 30°.

Figure 5 shows a $^{63}Cu^+$ image (400 μm × 400 μm) of a bevel synthesised in the copper implanted silicon with 1:3:7 HF-HNO_3-CH_3COOH. The bevel was 4 mm long and 0.9 μm deep. The uniformity of the image clearly indicates that the chemical

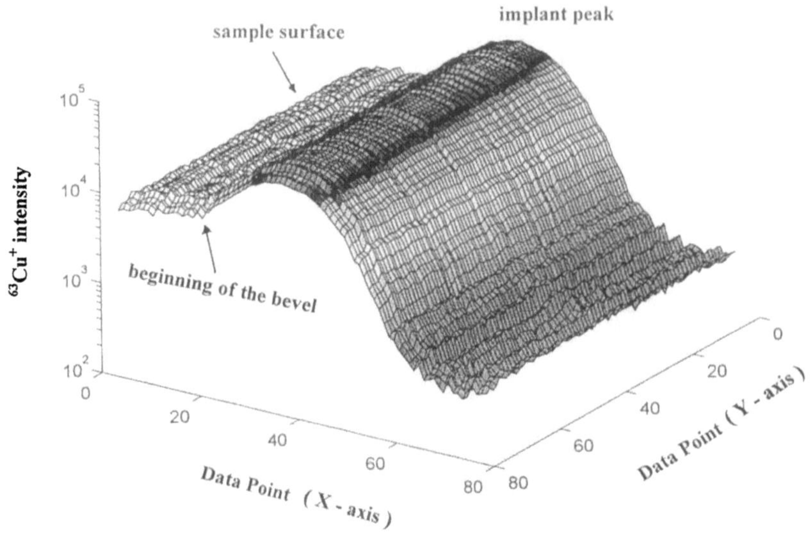

Fig. 5. A 3-D Cu implant profile data processed from the image data using Matlab. The imaged sample was chemically bevelled with 1:3:7 HF-HNO_3-CH_3COOH in this case

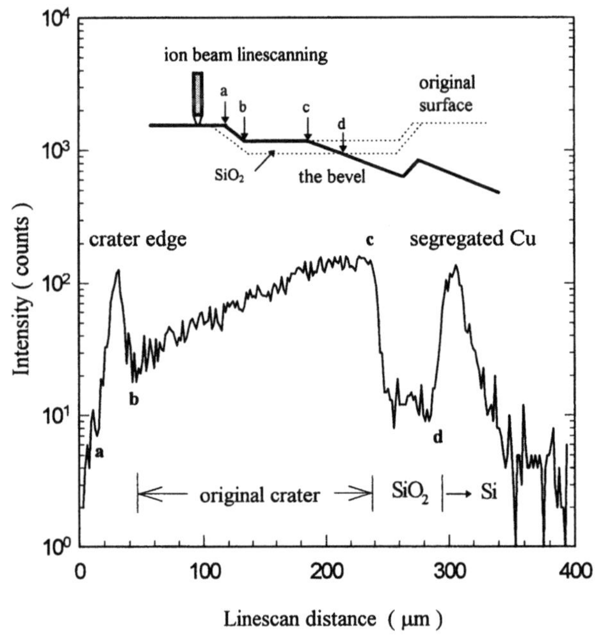

Fig. 6. The linescan profile of the chemically bevelled crater. The crater was produced by depth profiling to the peak of the copper implant using a 15 keV O_2^+ primary beam rastered over a 2 mm square. Part of the crater was then chemically bevelled to reveal the altered layer as a wide surface stripe (*a–d* discussed in the text)

bevelling process has been well controlled with no lateral variations in etch rate on the bevel plane.

Copper segregation in SIMS depth profiles is thought to be associated with the low propensity of the copper atoms to sit in the altered SiO_x layer and it is of interest therefore to stop a SIMS depth profile and measure the internal copper profile through the altered layer i.e. the distribution of copper atoms through the SiO_x and into the silicon. The schematic diagram of the ion beam linescanning on the synthesised crater (2 mm × 2 mm) after the chemical bevelling process is shown in the inset of Fig. 6. The linescan started on the original surface, went down the side-wall of the crater, along the crater base and then down the bevel. As the beam was scanned across the side-wall of the crater $a \rightarrow b$ a copper peak was observed. This peak is associated, we believe with the Cu atoms driven out of the oxide to the oxide/silicon interface at point *a*. The beam was now scanned across the base of the crater (which is oxidised) from $b \rightarrow c$ and the copper signal increased with depth reflecting the implant shape. The increasing Cu intensity from point *b* down to point *c* is associated with unevenness in the crater base. At point *c* the imaging beam meets the bevelled section. The region from $c \rightarrow d$ represents the altered layer itself, projected onto the bevel surface. The altered layer was projected on the bevel ($M = 5 \times 10^2$) as a surface approximately 40 microns wide, suggesting a true thickness of 80 nm. This thickness was significantly larger than that reported elsewhere [14], and the discrepancy is still not fully understood. The concentration of copper in this region inside the oxide is low and a large copper peak was observed at point *d*. The copper distribution was consistent with rejection of copper atoms from the oxide and a build up at the interface between the altered layer and the underlying substrate (point *d*). Thereafter the secondary ion intensities followed the shape of the trailing edge of the Cu implant profile. This profile was entirely consistent with the view that copper is driven out of the oxide layer, as it forms, and had a good agreement with the internal profile obtained by depth profiling an altered layer with a Cs^+ beam [6].

Conclusions

SIMS depth profiling of copper implants in p-type silicon using an oxygen primary ion beam leads to serious profile distortions, when the beam is incident at angles between 0° and 25°. The bevel and linescan approach has been shown to yield accurate copper depth profiles consistent both with TRIM predictions and data from SIMS depth profiles obtained at angles of incidence above 30°. The bevel and linescan approach is also rapid, thus representing a saving in machine time, but does suffer from a lower sensitivity than SIMS depth profiling. The internal profile of copper atoms in a SIMS crater (produced with oxygen primaries) has been investigated by the bevel and linescan approach. Copper is driven out of the oxide altered layer during the build up period, with Cu atoms near the surface driven onto the surface; these atoms appear as a pre-peak in the linescan. Cu atoms below the depth of the implanted oxygen primaries are driven downwards to the interface between the SiO_x and the Si and appear as a Cu peak at that interface during the linescan.

Acknowledgement. We wish to thank Dr. T. Tate, Department of Electrical Engineering, Imperial College, London, for the implantation of Cu in Si substrates.

References

[1]. Y. Homma, K. Wittmaack, *Appl. Phys. A*, **1990**, *50*, 417.
[2] S. M. Hues, P. Williams, *Nucl. Intrum. Meth. Phys. Res. B*, **1986**, *15*, 206.
[3] K. Wittmaack, Y. Homma, *Appl. Phys. Lett.* **1991**, *58(19)*, 2138.
[4] C. J. Vriezema, K. T. F. Janssen, P. R. Boudewijn, *Appl. Phys. Lett.* **1989**, *54(20)*, 1981.
[5] Y. Homma, T. Maruo, *Surf. Interf. Anal.* **1989**, *14*, 725.
[6] J. A. Kilner, D. S. McPhail, S. D. Littlewood, *Nucl. Instr. Meth. Phys. Res. B* **1992**, *64*, 632.

[7] K. Wittmaack, *Nucl. Instr. Meth. Phys. Res. B* **1987**, *19/20*, 484.
[8] P. R. Boudewijn, C. J. Vriezema, in: *SIMS VI* (A. Benninghoven, A. M. Huber, H. W. Werner, eds.) Wiley, Chichester, 1988, pp 499–502.
[9] C. M. Hsu, D. S. McPhail, *Nucl. Instr. Meth. Phys. Res. B* **1995**, *101*, *August(I)*, 427.
[10] C. M. Hsu, V. K. M. Sharma, M. J. Ashwin, D. S. McPhail, *Surf. Interf. Anal.* **1995**, *23*, 665.
[11] W. Vandervorst, T. Clarysse, P. De Wolf, L. Hellemans, J. Snauwaert, V. Privitera, V. Raineri, *Nucl. Instr. Meth. Phys. Res. B* **1995**, *96*, 123.
[12] L. Gong, S. Petersen, L. Frey, H. Ryseel, *Nucl. Instr. Meth. Phys. Res. B* **1995**, *96*, 133.
[13] C. Spinella, V. Raineri, M. Saggio, V. Privitera, S. U. Campisano, *Nucl. Instr. Meth. Phys. Res. B* **1995**, *96*, 139.
[14] J. A. Kilner, G. P. Beyer, R. J. Chater, *Nucl. Instr. Meth. Phys. Res. B* **1994**, *84*, 176.

Mikrochim. Acta [Suppl.] 13, 325–332 (1996)

Experimental Verification of Theoretical Models Simulating the Temperature Increase in EPMA of Glass

Vaclav Hulinsky[1,*], Karel Jurek[2], and Ondrej Gedeon[1]

[1] Department of Glass and Ceramics, Institute of Chemical Technology, Technicka 5, 166 28 Prague 6, Czech Republic

[2] Institute of Physics, Czech Academy of Sciences, Cukrovarnicka 10, 162 53 Prague 6, Czech Republic

Abstract. The temperature field in glass exposed to electron beams of different diameters and deposited energy was calculated. A cylindrical shape of the excited volume was chosen for the heat source, having a diameter equal to the diameter of the beam and a height equal to the maximum range of electrons calculated with the Kanaya and Okayama formula. The grid method for the numerical calculation of the partial differential equation for the heat transport was used. The calculated temperature field was verified by means of specially prepared glass-ceramic samples that crystallize over a suitable temperature range. Scanning electron images of the cross-sections of the exposed areas have given satisfactory results, in good agreement with calculated temperatures. It has been found that at normal conditions of EPMA when the absorbed current does not exceed 10 nA the maximum temperature in glass remains far under the transformation interval of the glass and cannot be the main reason for the rapid escape of alkali ions from the excited volume.

Key words: temperature field, grid method, glass-ceramics, electron bombardment.

Electron microprobe analysis is one of the oldest methods using a focused electron beam having an energy typically from 1 to 30 kV and a diameter from 5 nm to some micrometers. For more than thirty years it is known that alkali oxide silica glasses are unstable under the electron beam, since their composition in the excited volume changes [1–5]. The aim of this contribution is to calculate the temperature increase in a sample irradiated by an electron beam and experimentally verify this calculation. This can possibly help to separate the effect of increased temperature and the effect of electric charging in the excited volume. These two effects have been treated simultaneously in the course of studying alkali escape from the excited volume in the

---

* To whom correspondence should be addressed

published literature [6–9] so it was not possible to make unambiguous conclusions. The interactions of charged particles, such as electrons, can change the structure and chemical composition of the sample studied. About 40–80% of the power delivered into the excited volume [10] by the electron probe is transformed into heat. Heat is generated in the excited pear-shaped volume below the surface and its quantity is determined both by irradiation parameters such as the accelerating voltage, the primary current, the beam diameter and by a material parameter "f" defined as the fraction of the primary electron energy absorbed in the sample. The quantity (1 − f) represents back scattered or secondary electrons. The temperature increase can be calculated by the equation for thermal conduction representing the mathematical expression of the energy conservation law. For steady state and the electron probe at rest

$$-\operatorname{div}(\lambda \text{ grad } T) = V \tag{1}$$

where λ is the thermal conductivity, T is the absolute temperature, V is the power density absorbed in the sample.

The partial differential equation has, in the general case, a numerical solution; an analytical one can be obtained only for special boundary conditions, namely symmetry and simple form of the heat source. The temperature rise in the center of the probe has been calculated for the first time by Castaing [11] in the case of a hemispherical homogeneous source in a semi-infinitive homogeneous sample with its outer surface kept at room temperature. The solution of such a simple spherically-symmetric problem gives the formula

$$\Delta T = \frac{3}{4\pi} \frac{UI}{\lambda r_0} \tag{2}$$

ΔT is the maximum temperature rise in the center of the probe, r_0 is the probe radius, λ is the thermal conductivity of the sample, U is the accelerating voltage and I is the probe current. Primary electrons deposit a uniform amount of power into a hemisphere whose radius is taken as being equal to that of the probe. Castaing suggested that this radius is actually greater so the calculated temperature is overestimated. In spite of the fact that the electron density in the excited volume under the probe is distinctly greater compared to that on the anode of a common X-ray tube, Castaing stated: "It might therefore be feared a priori that the sample might become seriously overheated at the analyzed point. This is, in fact, not the case since the very small diameter of the bombarded region permits very strong cooling by thermal conduction. "Reimer [10], Dudek [12] and Zaveta et al. [13] have derived similar formulae for the maximal temperature increase of a plane specimen surface assuming that the power $P = fUI$ is homogeneously dissipated inside a hemisphere of radius R/2 where R is the electron range in given material. The outer edge of the semi-infinite sample is at room temperature. Radiation from the surface to vacuum is neglected. The solution of the equation of thermal conduction in this case gives the following expression for the maximal temperature increase in the center on the surface of the irradiated area

$$\Delta T = \frac{3}{2\pi} \frac{fUI}{\lambda R} \tag{3}$$

where symbols are the same as in the previous case.

The problem of temperature increase was treated by many authors for different geometries of the sample and for different initial and boundary conditions that are necessary for solving the partial differential equation of thermal conduction and for different materials and irradiation parameters [6, 14–19]. It is not the aim of this work to give a complete survey about these works but we hope that the most important are included here.

Experimental

Numerical Solution for Glass Heated by an Electron Beam

The differential partial Eq. (1) can be solved numerically by the finite difference method also called grid method. In our case a square grid was selected. During the calculation process the derivations of div (λ grad T) are supplied with the differences in this grid. The following boundary conditions were chosen

$$\frac{\delta T}{\delta z}(z,r)_{z=0}=0; \qquad \frac{\delta T}{\delta r}(z,r)_{r=0}=0;$$

$$T(z,r)_{z=Z}=T_0; \qquad T(z,r)_{r=R}=T_0;$$

The first condition reflects the assumption of zero heat transfer from the surface; the second one in the first row is implied by the symmetry of the problem and the last two conditions determine the boundary where the temperature falls to the room temperature.

T_0 is room temperature, Z and R were established to be at least ten times greater than the coordinates of the heat source. We consider steady state and heat conduction only i.e. $\frac{\delta(\rho c_v T)}{\delta \tau}=0$,

where τ is the time, ρ is the density and v is the specific heat at constant volume. The calculated area is symmetrical to the z axis, the sample is semi-infinite, the power fUI is fully absorbed in the sample and is converted into heat only, the thermal conductivity coefficient for the glass under study depends on the temperature [20] as follows:

$$\lambda = 1.3953 - 2.8591\,10^{-3}T + 5.2335\,10^{-6}\,T^2, \tag{4}$$

where λ[W m^{-1} K^{-1}] is the thermal conductivity, T [K] is the temperature.

The conducting layer on the surface plays no role in heat transport and the electron beam is at rest. A homogeneous cylindrical source was used in our calculations. The diameter of the cylindrical source is given by the diameter of the beam and its depth is determined as D_{max} where D_{max} was calculated after Kanaya and Okayama [21].

Experimental Verification of the Calculated Temperature Field

It is evident that direct temperature measurement in the excited volume is possible only at very high currents of the electron beam, as has been shown by Zaminer et al. [22]. Moreover, it is not possible to study the temperature field created inside and outside the excited volume. Our idea was to use a material that can change its structure with the temperature; this change must be instant, irreversible and observable with sufficient resolution.

The material exhibiting structural changes in the temperature range 560–1000 °C may be glass-ceramics. From a large series of glass-ceramics materials we have experimentally chosen the glass-ceramic with composition 57.5% SiO_2, 10% Al_2O_3, 10% MgO, 5% CaO, 7.5% Na_2O, 10% CaF_2.

This glass was melted from the high purity batch at 1470 °C in a Pt crucible. The homogeneous glass melt was poured out on a metal plate to prevent crystallization. The plate of glass was tempered at 550 °C for 16 h. The glass was transparent containing no bubbles or cords and had a weak blue colour.

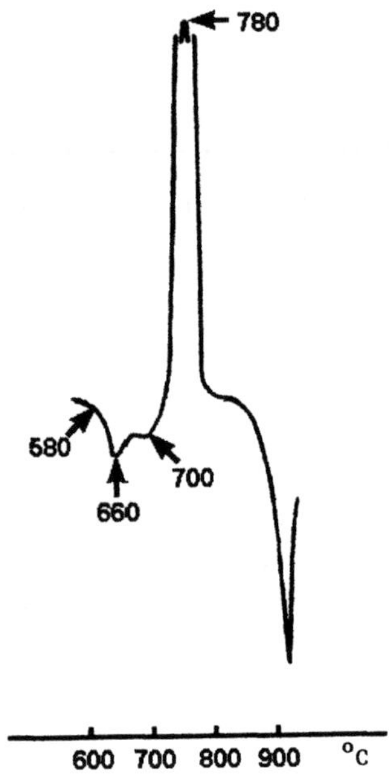

Fig. 1. The DTA curve of the glass-ceramics used for experiment

The gradient crystallization experiment, microscopic investigation, X-ray diffraction and DTA measurements were established with this glass.

Gradient crystallization has shown the following results: below 690 °C the glass remains transparent, in the temperature range 690–1050 °C the glass crystallizes, above 1050 °C deformation occurs. At 1150 °C the glass starts melting.

Microscopy has shown that at 750 °C uniform spherolites, composed of diopside crystals in their center surrounded with pyroxene crystals in the outer zone, are created. The character of the spherolites is conserved up to 850 °C. At 900 °C spherolite is composed of individual rods of pyroxene. Above 900 °C the plagioclase appears creating individual bars and dendrites. In the temperature range 1050–1100 °C the melting of crystals begins and in the glass melt only the biggest diopside crystals remain.

The DTA curve of this glass is shown in Fig. 1 indicates an extended transformation minimum in the range 560–700 °C resulting from the superposition of two minima typical for glass exhibiting phase separation. At 700 °C begins the sharp increase of the curve creating a peak at 780 °C that is connected with the large amount of heat released in the course of crystallization of the glass. At about 950 °C the endothermic minimum appears connected with the melting of plagioclase and later of pyroxene.

The kinetics of crystallization may be determined from density measurements. The results of these measurements are shown in Fig. 2. It is worthwhile to bear in mind that the size of crystals depends also on the quantity of mass available in the volume having a sufficiently high temperature for the crystallization.

Prisms of polished glass-ceramics were exposed to an electron beam at different absorbed currents, accelerating voltage and diameter of the beam. A steady beam has been used for exposure that took 5–10 min; in other experiments the beam was moving at a rate of 2 μm/min. The exposed glass-ceramics was carefully polished on the side perpendicular to the surface where the beam was impinging on the sample. The whole operation was optically controlled. The polished cross-sections in the center of the excited volume were etched with a solution of 2% hydrofluoric acid in water for 10 s. Dry samples were coated with a thin layer of gold and observed in a scanning electron microscope SEM.

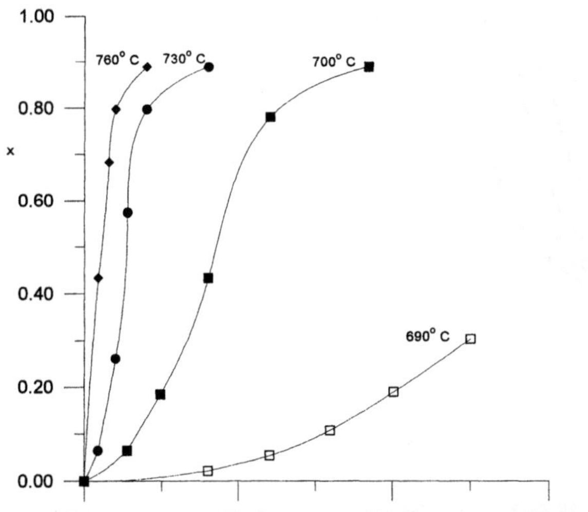

Fig. 2. The kinetics of crystallization of the glass-ceramics. X is the crystallization ratio against time in minutes

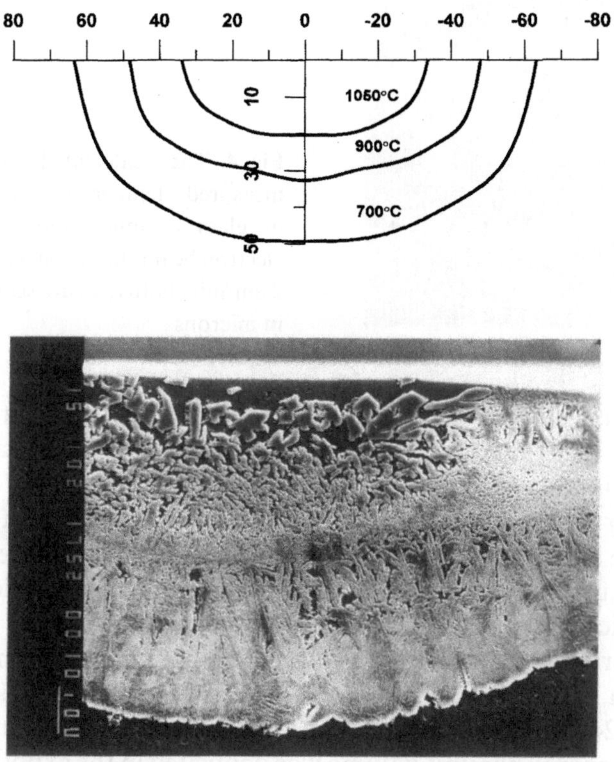

Fig. 3. The calculated and measured temperature field in glass-ceramics; stationary electron beam, both axes are scaled in microns

Results and Discussion

Figure 3 represents the calculated temperature field together with the SEM picture at the same scale for the glass-ceramics exposed to a stationary beam of 200 μm diameter, accelerating voltage 50 kV and absorbed current 7560 nA. Coming from the top edge of the SEM picture one can see the following areas; the first one is the dark area in which the glass-ceramics was heated above the melting point of diopside crystals. This area continuously grades into the larger area containing big diopside

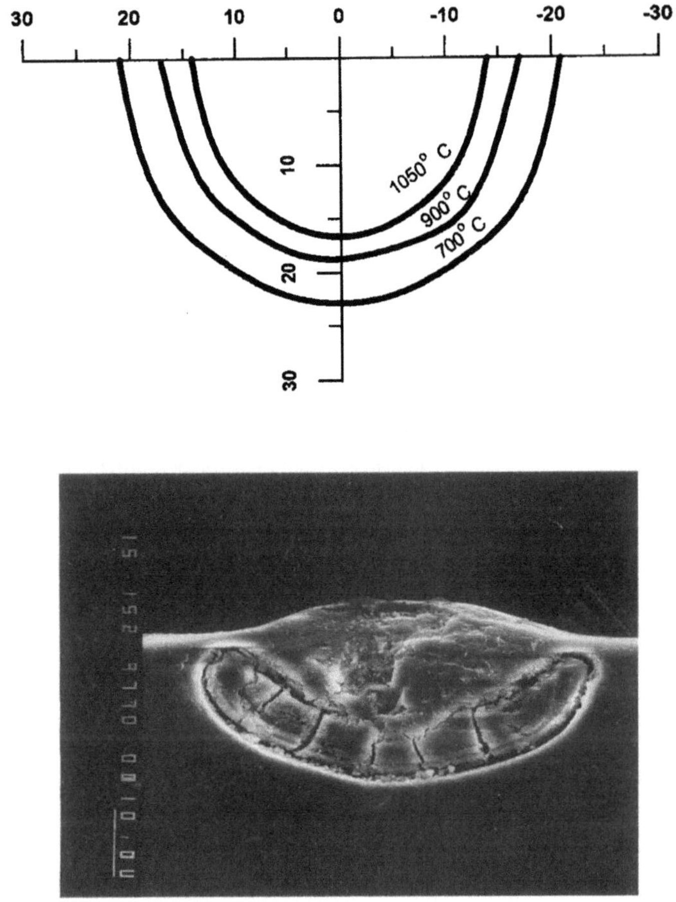

Fig. 4. The calculated and measured temperature field in glass ceramics under an electron beam shifting at a rate 2 μm/min; both axes are scaled in microns

crystals that become denser and smaller with increasing depth. The third area contains individual rods of pyroxene that convert to spherolites near the lowest part of the crystallized volume. This area is separated from the unchanged glass by a sharp boundary. The upper figure represents the numerical solution of the temperature field. In spite of the fact that the model is rather simplified it is apparent that the calculated isothermal curves and the temperature regions of crystallization and melting in the glass ceramics, described above, are in good agreement.

Figure 4 represents glass-ceramics exposed to a beam of 26 μm in diameter that was moving on the surface at a speed of 2 μm/min, accelerating voltage was 50 kV and the absorbed current was 3240 nA. Compared to Fig. 3, a smaller diameter of the beam was used, so that a higher energy per unit volume was deposited in the exposed area. For this reason the picture looks slightly different from the previous one. The first zone near the top level contains entirely melted crystals. Because of very high temperature a bulge of melted glass is formed. The second zone, consisting of diopside crystals, is extremely narrow and therefore invisible at this magnification. The third zone, containing very small spherolites, looks like a broken roll. Agreement with the calculation, bearing in mind the simple model, is still satisfactory. In all cases, probably due to the lateral spread of the electric field which caused the migration of alkali ions, as shown in [23], both the crystallization temperature and

Table 1. Maximum temperature calculated in the excited volume for glass-ceramics at different irradiation parameters of the electron beam

Diameter [μ]	Current [nA]	Acc. voltage [V]	Temperature [°C]
50	68	50	30
50	1000	50	327
100	58	50	23
100	126	50	37
100	234	50	69
100	1000	50	222
200	1000	50	140

the crystallization rate can be influenced by these alkali migration effects; they may change the composition of the glass-ceramics substantially over a period of time comparable that to needed for crystallization of the glass-ceramics.

Numerical simulation has the advantage that it is possible to calculate not only the maximum temperature, as was done in previous works, but to map the temperature field that may be created over sufficiently greater dimensions than the excited volume itself. This method has enabled to calculate the problem not only for the focused beam and the spherical or hemispherical heat source. For the defocused beams and various irradiation parameters, calculated maximum temperatures for glass-ceramics are shown in Table 1. We can conclude that our calculations of the temperature field, created in the glass-ceramics under the electron beam, were reliably verified by means of characteristic temperatures of this system derived from DTA measurements and microscopic observations of the morphology of crystals that developed in individual samples of glass-ceramics exposed to different temperatures.

References

[1] J. L. Lineweaver, *J. Appl. Phys.* **1963**, *34*, 1786.
[2] A. K. Varshneya, A. R. Cooper, M. Cable, *J. Appl. Phys.* **1966**, *37*, 2199.
[3] M. P. Borom, R. E. Hanneman, *J. Appl. Phys.* **1967**, *38*, 2406.
[4] L. F. Vassamilet, V. E. Caldwell, *J. Appl. Phys.* **1969**, *40*, 1673.
[5] P. J. Godhew, J. E. C. Gulley, *Glass Technol.* **1974**, *15*, 123.
[6] G. S. Almasi, J. Blair, R. E. Ogilvie, *J. Appl. Phys.* **1965**, *36*, 1848.
[7] R. Hayami, H. Wakabayashi, *J. Cer. Soc. Jap.* **1970**, *78*, 138.
[8] H. J. Dudek, *Glasstechn. Berichte* **1968**, *41*, 10.
[9] F. Autefage, *Bull. Mineral.* **1980**, *103*, 623.
[10] L. Reimer, *Scanning Electron Microscopy*, Springer, Berlin Heidelberg New York Tokyo, 1985.
[11] R. Castaing, *Electron Probe Analysis, Advances Elect. Elect.* **1960**, *13*, 317.
[12] H. J. Dudek, *Z. Angew. Physik* **1971**, *18*, 331.
[13] K. Zaveta, K. Jurek, V. Kambersky, P. Kral, *J. Magnetism Mag. Mat.* **1990**, *86*, 254.
[14] B. Gale, K. F. Hale, *Br. J. Appl. Phys.* **1961**, *12*, 115.
[15] L. G. Pittaway, *Br. J. Appl. Phys.* **1964**, *15*, 967.
[16] J. Ling, *Br. J. Appl. Phys.* **1967**, *18*, 991.

[17] J. Ziller, W. Uhlmann, *Proc. 4 Tagung Mikrosonde*, Dresden, 1978.
[18] W. Uhlmann, U. Cobet, *Proc. 3. Tagung Mikrosonde*, Berlin, 1975.
[19] V. Hulinsky, J. Matousek, J. Damaskova, *Proc. 7. Tagung Mikrosonde*, Dresden 1988.
[20] A. Blazek, J. Ederova, J. Endrys, J. Kada, *Silikaty* **1981**, *XXV*, 359.
[21] K. Kanaya, S. Okayama, *J. Phys. D. Appl. Phys.* **1963**, *14*, 617.
[22] Ch. Zaminer, B. Bottcher, *Mikrochimica Acta [Suppl.]* **1974**, *V*, 57.
[23] K. Jurek, V. Hulinsky, O. Gedeon, *Mikrochim. Acta [Suppl.]* **1996**, *13*, 339.

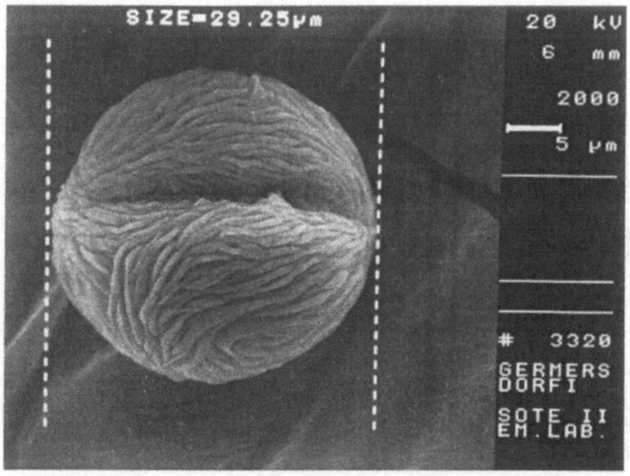

Fig. 1. Spherical cherry pollen. Diameter is 30 μm. Magn.: 2000x

The present study is an approach to the analytical characterisation of the principal elements constituting the mineral components of pollens from different fruit species. This type of investigation has very few matches in literature, despite the importance that some mineral elements have in the germination of the pollen grain. Investigations and knowledge about organic constituents of pollens (such as proteins, free amino acids, sugars, lipids and starch) are more extensive, connected with their importance in causing allergic reactions and their biological function [1]. The different kinds of organic materials, which constitute the majority of the pollens' mass as a matrix, are composed of various very light elements (H, C, N, O). These light elements are not measured by EDS. Mineral elements are measured in such an organic matrix, where quantitation with a conventional ZAF correction procedure is not applicable [7, 8].

Pollen grains are spherical or oval particles, in our samples their size ranges from 19 to 31 μm in diameter, and from 34 to 68 μm in length. The micrograph shows a medium size spherical pollen in Fig. 1. The water content of pollens is fairly low compared to other biological material, dry mass is usually about 95% of their wet mass.

Experimental

Pollen containing anthers were collected from ten different variety of fruit species such as apple (Malus domestica Borkh.), pear (Pyrus communis L.), cherry (Cerasus avium L.), sour cherry (Cerasus vulgaris Mill.), European plum (Prunus domestica L.), Japanese plum (Prunus salicina Lindl.), almond (Amygdalus communis L.), apricot (Armeniaca vulgaris Lam.), quince (Cydonia oblonga Mill.) and peach (Persica vulgaris Mill.). Collected anthers were kept at room temperature and allowed to dry for 24 hours prior to preparation. Pollen grains were shaken from the air-dried anthers onto aluminium specimen stubs mounted with double-faced conductive adhesive tape. No carbon coating was applied because of the high mass absorption coefficient of carbon for some of the interesting element X-rays [2], and because negligible charging was experienced. X-ray analysis was carried out in a Zeiss 940 DSM scanning electron microscope operated at 20 keV accelerating voltage equipped with a Si(Li) detector and a Link AN10/55S energy dispersive analyser system. The detector-specimen distance was 24 mm

Mikrochim. Acta [Suppl.] 13, 333–338 (1996)

Quantitation of Mineral Elements of Different Fruit Pollen Grains

Mária J. Bucsek[1,*], József Nyéki[2], Zoltán Szabó[2], and Anna Kádár[1]

[1] 2nd Central Electron Microscope L., 2nd Department of Pathology, Semmelweis University of Medicine, Üllõi út 93, H-1091 Budapest, Hungary

[2] Department of Horticulture, Georgikon Faculty of Agronomy, Pannon University of Agricultural Sciences, Ujmajori u.2. H-8361, Keszthely, Hungary

Abstract. In our investigations we accomplished quantitative measurements using energy dispersive X-ray analysis on mineral elements of pollen grain samples belonging to different fruit species such as apple (Malus domestica Borkh.), pear (Pyrus communis L.), cherry (Cerasus avium L.), sour cherry (Cerasus vulgaris Mill.), European plum (Prunus domestica L.), Japanese plum (Prunus salicina Lindl.), almond (Amygdalus communis L.), apricot (Armeniaca vulgaris Lam.), quince (Cydonia oblonga Mill.) and peach (Persica vulgaris Mill.). Pollen grains are particles but, as a result of their size ranging in our samples from 19.2 to 31.2 μm in diameter, they are considered as bulk samples. Link's ZAF-P/B program was used for the quantitative evaluation of data. The main constituents were Mg, P, S, Cl, K, Ca, while others such as Na, Al, Si and Cu, whose intensity in the spectra did not exceed the standard deviation, were excluded from further quantitative evaluation. Differences in the elemental concentrations of different fruit pollens were significant ($p > 5\%$).

Key words: energy dispersive X-ray analysis, ZAF/PB quantitation, mineral elements, pollen.

The distinctive characteristics of pollen grains are examined by light (LM) and scanning electron microscopy (SEM). SEM equipped with an energy dispersive detector (EDS) allows us to simultaneously study not only morphological features but also to determine and quantify their different mineral components. There are only few data about the mineral composition of pollen grains in literature. Moreover, these deal mainly with qualitative determination. It is difficult to collect sufficient quantity of pollen for conventional chemical determination and analysis of all constituents. However, quantitative X-ray microanalysis gives us the unique opportunity to measure all constituents on one single pollen grain.

* To whom correspondence should be addressed

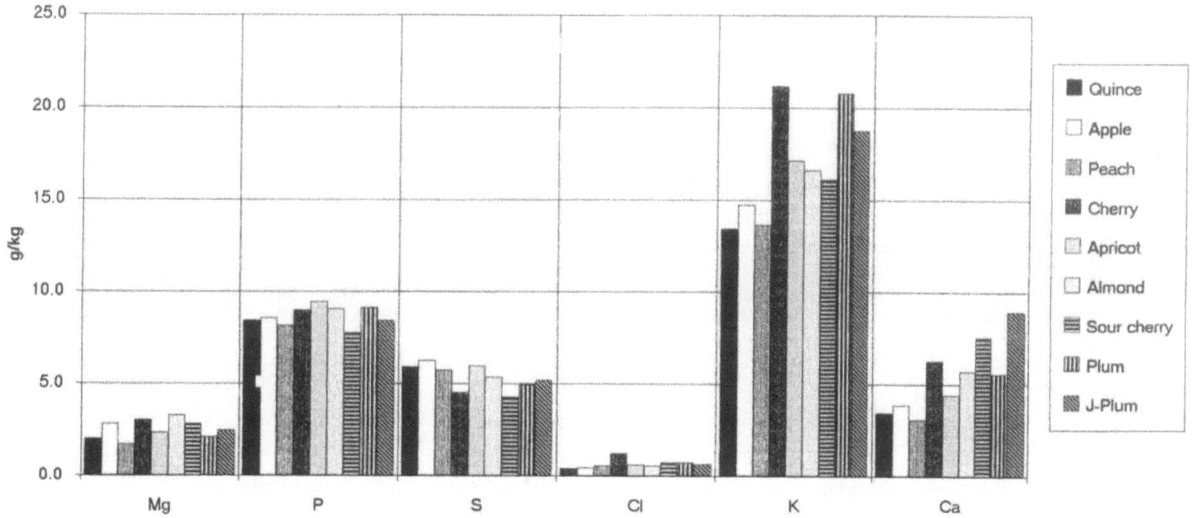

Fig. 2. Elemental concentrations in different species of pollen grains

with a take-off angle of 35°. A rastered beam magnification of 10000 × (corresponding to an 9.2 × 9.2 μm area) was used. The counting time was 60 s. Charging was checked for at the high energy end of the X-ray spectra background (Fig. 4). Each spectrum was collected from a different pollen grain. The number of varieties measured ranged between 2 and 9 for the ten fruit species. Within each variety 10–15 measurements were performed so the number of measurements per species ranged between 20 and 104. Energy calibration was made for each group of spectra of one variety with reference to a cobalt standard.

The X-ray intensities in the specimens were compared by means of the ZAF/PB program, which contains correction procedures developed for electron beam analysis of particles, rough specimens and bulk biological samples. The principle behind the ZAF/PB quantitation method is to use the peak-to-local background ratio as a measure of elemental concentration which is less dependent on changes in local geometry. As the surface of pollen grains shows considerable micro-roughness, this method gives more appropriate results, than the use of net peak intensities only [9]. Peak-to-local background ratios for each element, derived from the spectrum analysis, are used to calculate elemental concentrations with a modified ZAF routine [3] by the following procedure. Escape peaks are removed and the spectrum subjected to a digital filter to produce a series of profiles which are independent of background. These profiles are modified to match the current spectrometer gain and resolution and then subjected to the same filter. Peak areas are then determined from a least squares fit of the filtered profiles to the filtered spectrum. Fit index is then calculated over the filtered region only and not over the whole spectrum. The complete profiles of peak plus background are used in this process. The background subtracted profiles are then scaled to match the unknown spectrum and stripped away from it and the background under the peaks is determined. The continuum is calculated from the first estimate of concentration and the true background fitted to it. The ratio Z^2/A, absorption, and fluorescence corrections are calculated in an iterative process using ZAF type calculations based on a series of standard factors previously stored on disc. Such information includes details of sample and spectrometer geometry, pure element peak profiles prepared from standard spectra and the previously calculated C, H, O and N matrix composition from the known organic constituents of the pollen grains. When convergence has been achieved, the peak areas, peak to background ratios and corrected concentrations are printed out [10].

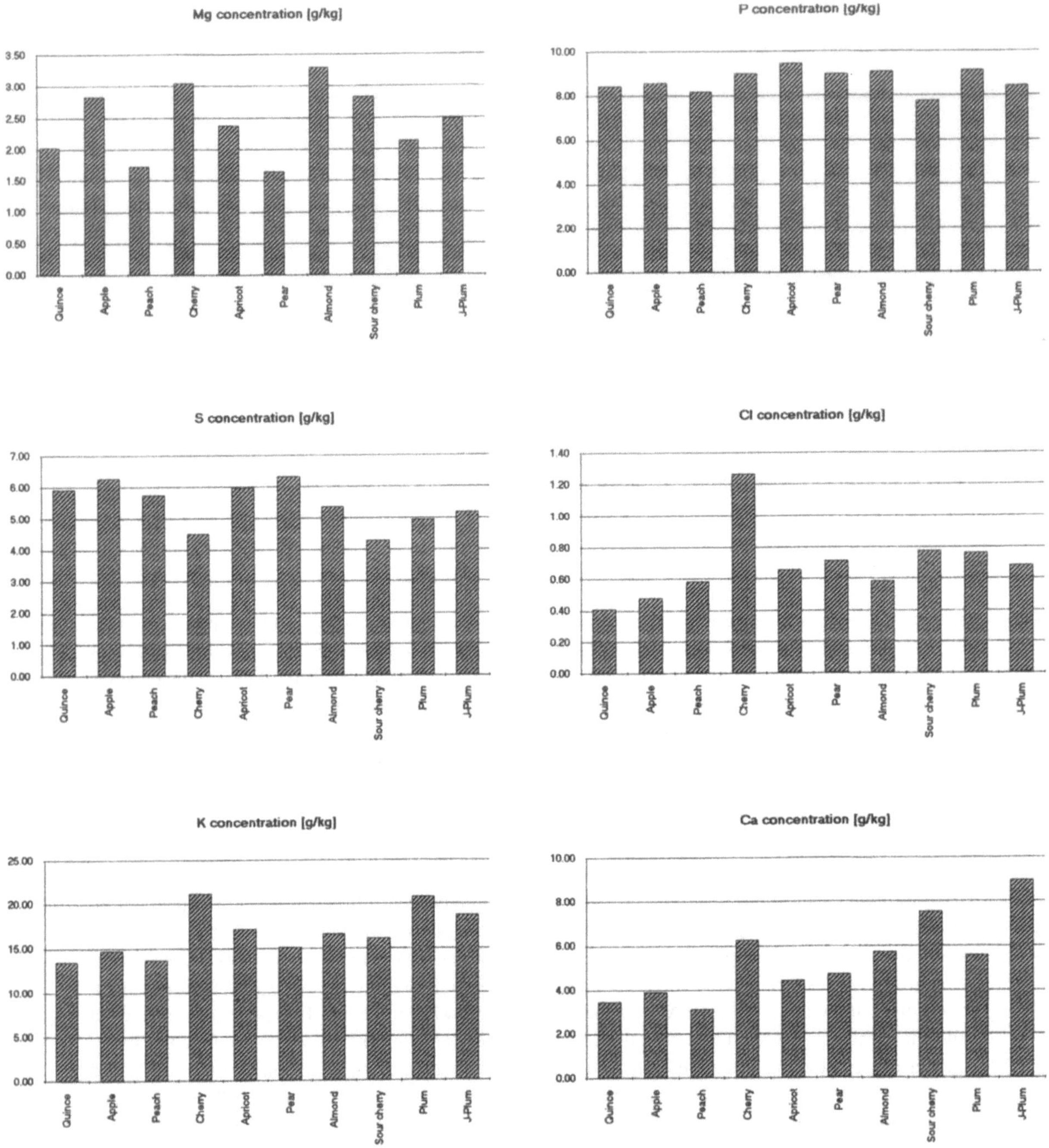

Fig. 3. Main elemental concentrations in different fruit pollen grains

Results and Discussion

In the spectra of all pollen samples characteristic peaks of Na, Mg, Al, Si, P, S, Cl, K, Ca, and Cu were detectable with different peak intensities. In Fig. 5 one of the spectra can be seen. The Na, Al, Si and Cu peak intensities were so small that they did not

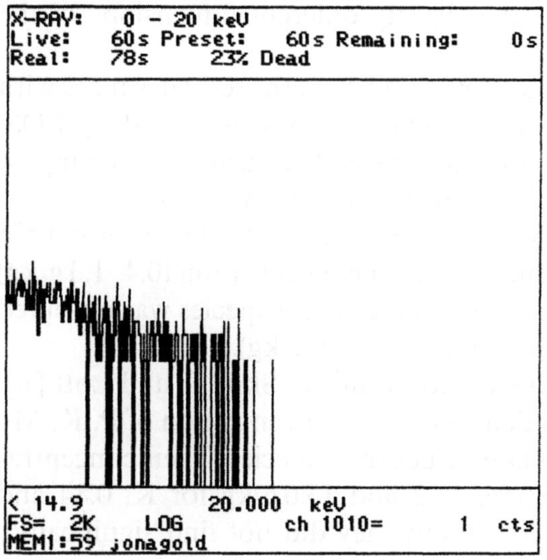

Fig. 4. End of an X-ray spectrum background representing electrostatic charging

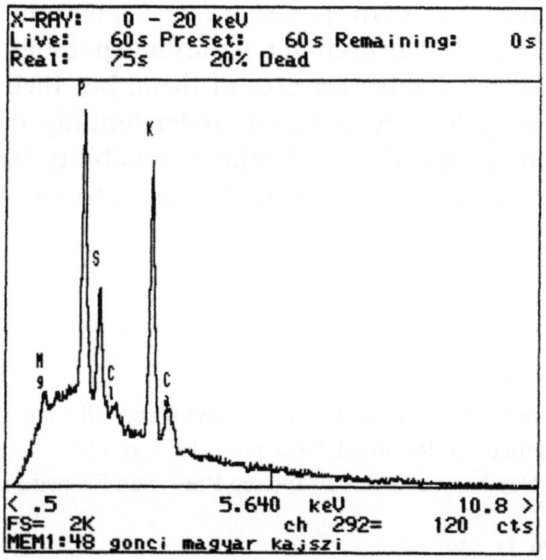

Fig. 5. Pollen spectrum of an apricot variety

exceed the 2σ (standard deviation) of the background. Therefore these elements were excluded from the quantitation procedure.

For the quantitative electron probe analysis of particles their size range has to be considered. If the dimensions of a particle approach the interaction volume, escaping electrons cause a lower intensity in the particle compared to a bulk sample. The diameter of the estimated analytical volume for Ca at 20 keV in bulk biological tissue, (where $\rho = 1\ g/cm^3$) is near to 20 μm [4]. In our investigations the length of pollens was larger than 20 μm, only some of the smallest diameters were very close to this value. When larger particles, which have dimensions in excess of 20 μm are analysed at 20 keV, the analytical volume is the same as in a bulk sample [5]. For this reason and for uniformity of treatment, all pollen grains in our sample were considered as bulk samples disregarding the slightly different dimensions of the smallest ones from the given value.

The evaluated elements constituted together 3.5–4.8% of the total dry mass with the matrix of 95.2–96.5% consisting of organic material, not measured by EDS. Distribution of the elemental concentrations among pollen grains belonging to different fruit species can be seen on Fig. 2. K has the highest concentrations (13.5–21.4 g/kg), followed by P (7.8–9.5 g/kg), S (4.3–6.3 g/kg), Mg (1.6–3.3 g/kg), Ca (3.1–8.9 g/kg), and finally Cl having the smallest mean concentration (0.4–1.3 g/kg) (Fig. 3). The smallest concentration range among the different species was observed for P (7.8–9.5 g/kg), the most deviating one for Ca (3.1–8.9 g/kg).

Our measurements are in good agreement with results of Skilnyk and Lott [6]. They used neutron activation analysis to determine the concentration of P, K, Mg and Ca in pollen samples of two closely related Cucurbita species. Their concentration values were 1.28 and 1.78 g/kg for Mg, 13.2 and 13.0 g/kg for K, 0.84 and 1.17 g/kg for Ca and 5.32 and 7.10 g/kg for P, but they did not find significantly different concentrations for the two closely related species.

Through analysis of variance (ANOVA), a statistically significant difference was observed in the elemental concentrations of pollens belonging to these different fruit species ($p > 5\%$). It means, that the mineral content of the pollen grain is representative for the species. Taking this into consideration it seems that not only morphological characteristics of the pollen surface are distinctive features of them, but their mineral composition too. This studies may lead to a better understanding of pollination and fertilization. Moreover, our results offer an additional possibility for identifying them in different fields, for example in horticultural science, forensic medicine and allergology.

References

[1] R. Lunden, *Svensk Kemisk Tidskrift* **1954**, *66*, 201.

[2] A. T. Marshall, in: *Principles and Techniques of Scanning Electron Microscopy. Biological Applications, Vol. 4* (M. A. Hayat, ed.), Van Nostrand Reinhold, New York, 1975, p. 127.

[3] P. J. Statham, in: *Microbeam Analysis* (D. E. Newbury, ed.), San Francisco Press, San Francisco, 1979, p. 247.

[4] N. C. Barbi, *Scanning Electron Microsc.* **1979**, *II*, 659.

[5] J. A. Small, K. F. J. Heinrich, D. E. Newbury, R. L. Myklebust, *Scanning Electron Microsc.* **1979**, *II*, 801.

[6] H. R. Skilnyk, J. N. A. Lott, *Can. J. Bot.* **1992**, *70*, 491.

[7] A. Boekestein, A. M. Stadhouders, A. L. H. Stols, G. M. Roomans, *Scanning Electron Microsc.* **1983**, *II*, 725.

[8] G. M. Roomans, *Scanning Microscopy* **1990**, *4*, 1055.

[9] A. Boekestein, *Mikrochim. Acta* **1992**, [*Suppl. 12*], 205.

[10] P. Echlin, E. Taylor, *J. Microsc.* **1986**, *141*, 329.

Mikrochim. Acta [Suppl.] 13, 339–347 (1996)

Electron Beam Induced Migration of Alkaline Ions in Silica Glass

Karel Jurek[1,*], Václav Hulínský[2], and Ondrej Gedeon[2]

[1] Institute of Physics AVCR, Cukrovarnická 10, 162 53 Prague 6, Czech Republic
[2] Institute of Chemical Technology, Technická 5, 166 28 Prague 6, Czech Republic

Abstract. Sodium and potassium silica glasses were irradiated with a 50 keV electron beam and the migration of alkalis was studied on cross-sectioned specimens. Alkaline ions moved from the grounded (i.e. electrically connected with the ground) surface deeper into the sample even at low temperatures due to the electric field and at about half the electron range colloidal particles with a diameter in the range from 10 to 200 nm were created. Near the surface alkali depleted layers with modified structure were formed.

Key words: alkaline ion migration, silica glass, electron beam, X-ray microanalysis.

During irradiation of silica glass with an electron beam, alkaline elements migrate from the excited volume as can be observed by the decrease of the characteristic X-radiation with irradiation time. This phenomenon causes problems during X-ray microanalysis and has been described by various authors, e.g. Lineweaver [1]. Till now, no consistent model explaining the origin, ways and directions of migration, where and in which state the alkali atoms are deposited was proposed even though a rather comprehensive study was made by Yamamoto et al. [2]. During our studies of temperature increase due to electron bombardment, structural changes were observed which could not be explained by thermal diffusion. These changes were systematically studied using a broadened electron beam in order to eliminate boundary effects. In this way the problem could be simplified to a one-dimensional one, i.e. only changes in depth are important. In this case the excited volume has approximately the shape of a flat cylinder with the ratio of its height to its diameter of approx. 1:10.

Experimental

First special pre-nucleated glass S1, then float glass and finally glasses Na17 and K15 were studied. Their composition is presented in Table 1. Samples were prepared in the shape of rectangular prisms

* To whom correspondence should be addressed

Table 1. Composition of investigated glasses (weight percentage)

	S1	Float	Na17	K12
SiO_2	56.3	73.0	65.0	79.5
CaO	11.9	9.8	17.5	8.5
Na_2O	7.3	13.9	17.5	–
K_2O	–	0.06	–	12.0
MgO	9.8	1.5	–	–
Al_2O_3	9.8	1.6	–	–
F	4.8	–	–	–

with two perpendicular and polished sides. Both this sides were coated either with a ca 30 nm thick carbon film for X-ray microanalysis or an 15 nm thick gold film for scanning electron microscopy. The surface of the glass was exposed to a 50 keV electron beam of 200, 100 and 50 μm diameter and currents in the range of 60 nA to 3 μA for time intervals from 10 s to 30 min. Cross-sections through the centers of exposed areas were made by accurate grinding and polishing. Concentration profiles were measured using a KEVEX ED spectrometer at 10 keV (soda glass) or 15 keV (potassium glass) and a beam current of 2 nA on a JEOL JXA-733 electron microprobe. To observe structural changes, the specimens were etched in 1% HF for 20 s and coated with a Au film. The glass plates, 7 × 7 × 1 mm in size, were irradiated using step scanning in such a way that the entire surface (7 × 7 mm^2) became exposed. The glass Na17 was exposed by a beam creating an absorbed current of 1 μA during 3 min at each step. For the glass K15 the absorbed current was 3 μA and the exposure time 3 min at each step.

Results

From the series of differently exposed glasses those exposures were selected for which no melting or softening of the glass could occur. Calculations according to [3] showed that the maximum temperature was 327 °C for an electron beam of 50 μm diameter and 1 μA. Under such conditions the thermal diffusion is sufficiently small and can be neglected. Scanning electron microscopy of the cross-sections of glass S1 revealed a compact, several micrometers thick, layer at the surface followed by a strongly damaged layer. The boundary between both areas is quite sharp (Fig. 1). The thickness of the surface layer increases with the absorbed current. For an electron beam of 62 nA and 100 μm diameter the surface layer is nearly invisible but the "porous" layer below can be distinguished (Fig. 2). The float and Na17 glasses exhibited similar behaviour, but the growth of the surface layer was slower (Fig. 3). The time dependence of the surface layer thickness is shown in Fig. 4. All three glasses exhibit the same tendency, the maximum thickness is reached after 15 or 20 min. It is interesting to compare e.g. the two curves for Na17 glass which were obtained at the same current density but different beam diameters (100 and 200 μm).

X-ray microanalysis revealed depletion or even absence of Na in the surface layer. Line profile of Na in the Na17 glass after 15 min exposure by a 200 μm and 1 μA beam (Fig. 5) revealed significant enrichment of Na under the surface layer, confirming migration of Na atoms in the depth of the specimen. If this migration is driven by an electric field, the Na atoms must be ionised, however it is not clear in

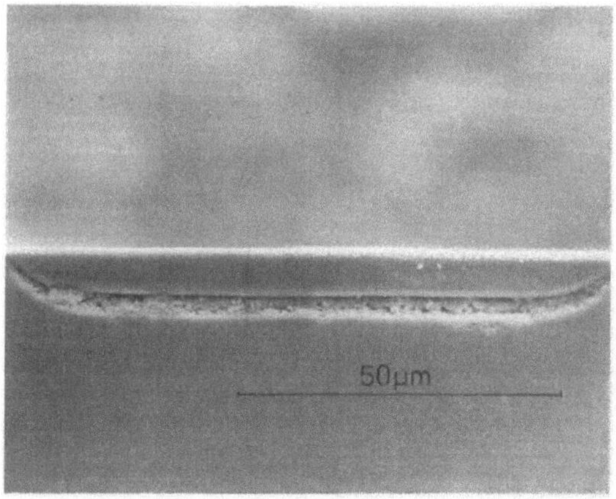

Fig. 1

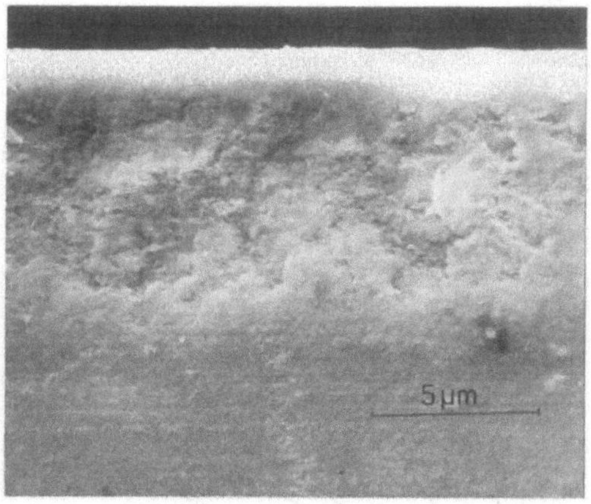

Fig. 2

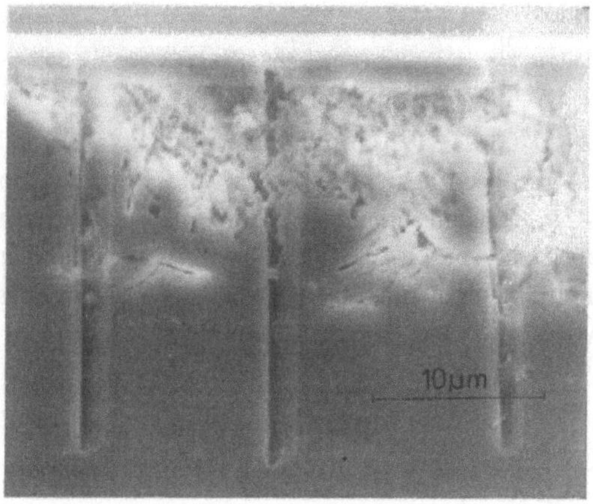

Fig. 3

Fig. 1. Glass S1, beam diameter 100 µm, absorbed current 234 nA, exposure 5 min
Fig. 2. Glass S1, beam diameter 100 µm, absorbed current 58 nA, exposure 5 min
Fig. 3. Glass float, beam diameter 100 µm, absorbed current 126 nA, exposure 10 min

which state they remain in the region of increased concentration. To reveal this, a glass plate of 7 × 7 mm was exposed over the entire area which caused yellowish brown colouring. The optical transmittance spectrum in the visible range was measured. The spectrum, similar to that in [4, 5] was characteristic for scattering on colloidal metallic particles whose size range was estimated to 10–200 nm. X-ray diffraction under low impact angle showed some small peaks which could not be identified but they indicate the presence of a crystalline phase in the surface layer and thus the possibility of a phase transition which could cause the sharp boundary between surface layer and the deeper layer.

The glass K12 also exhibits migration of potassium ions during electron bombardment and the creation of a potassium depleted surface layer. Due to the lower migration velocity it was possible to study its growth in more detail. In Fig. 6 it is possible to see traces evidently created during K-ions migration caused by bombardment with an electron beam of 200 µm diameter and 1 µA current after 2, 3, 5 and 10 min, respectively. Using an 100 µm beam and the same current, migration was much faster as can be seen in Fig. 7. After 1 min exposure a rather compact surface

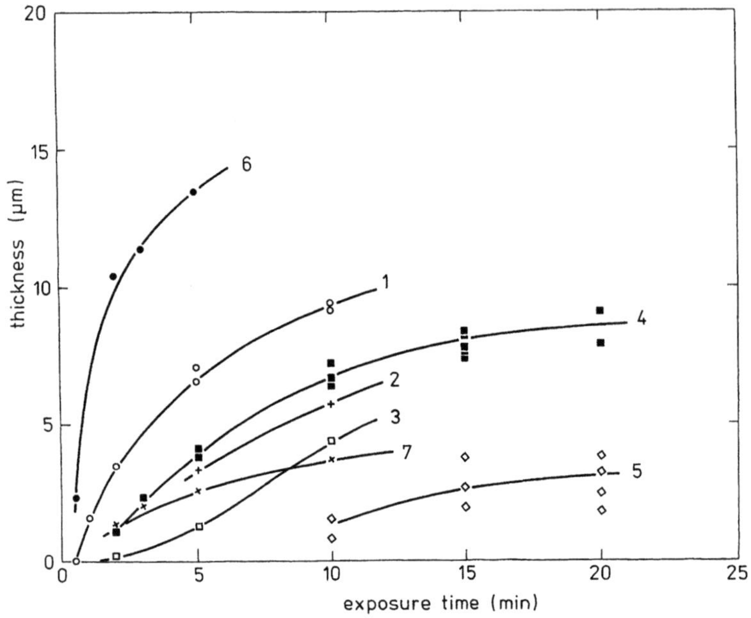

Fig. 4. Time dependence of the surface layers growth. *1* Glass S1, beam diameter 100 µm, absorbed current 270 nA. *2* Glass S1, beam diameter 100 µm, absorbed current 110 nA. *3* Glass float, beam diameter 100 µm, absorbed current 250 nA. *4* Glass Na17, beam diameter 200 µm, absorbed current 1 µA. *4* Glass Na17, beam diameter 200 µm, absorbed current 1 µA. *5* Glass Na17, beam diameter 100 µm, absorbed current 254 nA. *6* Glass K12, beam diameter 50 µm, absorbed current 1 µA. *7* Glass K12, beam diameter 100 µm, absorbed current 1 µA

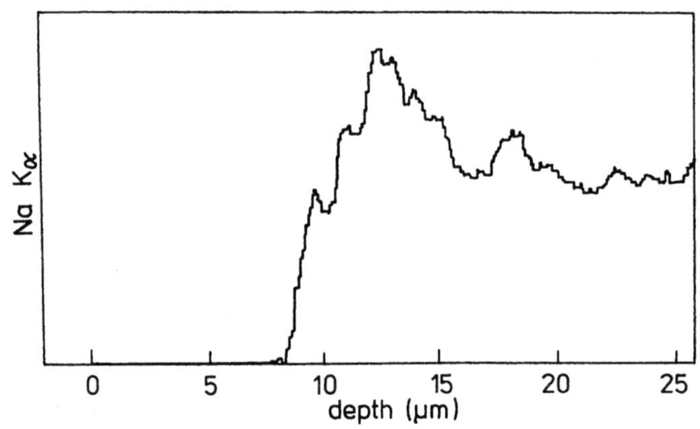

Fig. 5. Glass Na17, beam diameter 200 µm, absorbed current 1 µA, exposure 15 min., line profile of Na

layer about 6 µm thick is revealed and under it "bubbled" glass. However, the boundary between the two regions is not yet sharp. The corresponding concentration profile of K (Fig. 7b) shows a local maximum at a depth of about 1 µm and another maximum at about 6 to 7 µm. After 2 min exposure the surface layer becomes compact and sharply separated from the bottom (Fig. 7c). The K profile

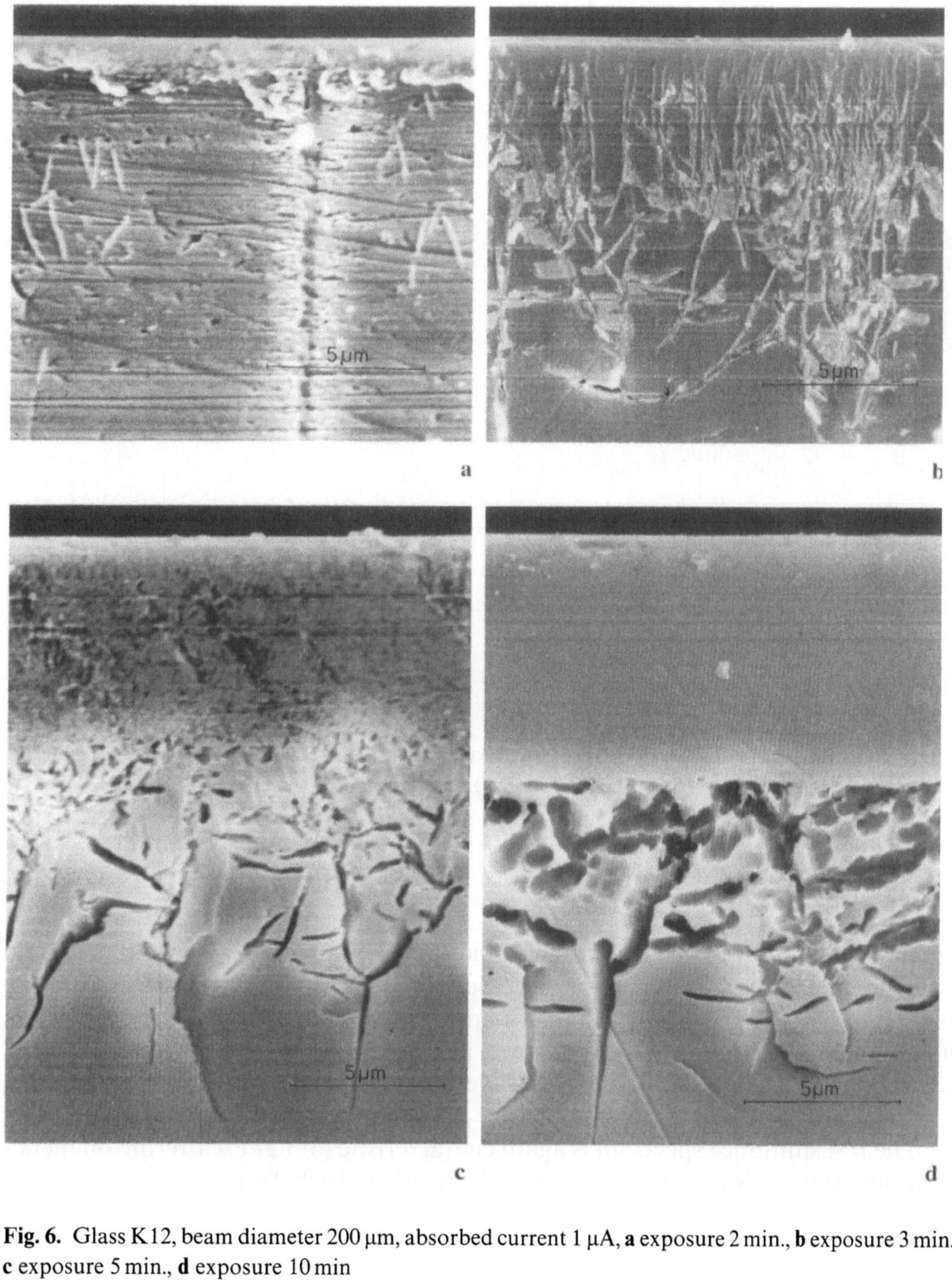

Fig. 6. Glass K12, beam diameter 200 µm, absorbed current 1 µA, **a** exposure 2 min., **b** exposure 3 min., **c** exposure 5 min., **d** exposure 10 min

(Fig. 7d) shows that the concentration of K is nearly zero up to a depth of approx. 1.3 µm, then about 10% to 15% of the original bulk concentration up to a depth of 7 µm, then the concentration increases steeply and peaks at 9.5 µm depth. Using the same beam diameter, but with an absorbed current of 253 nA significant changes in the glass occured after 5 min exposure. Here the formation of the surface layer is

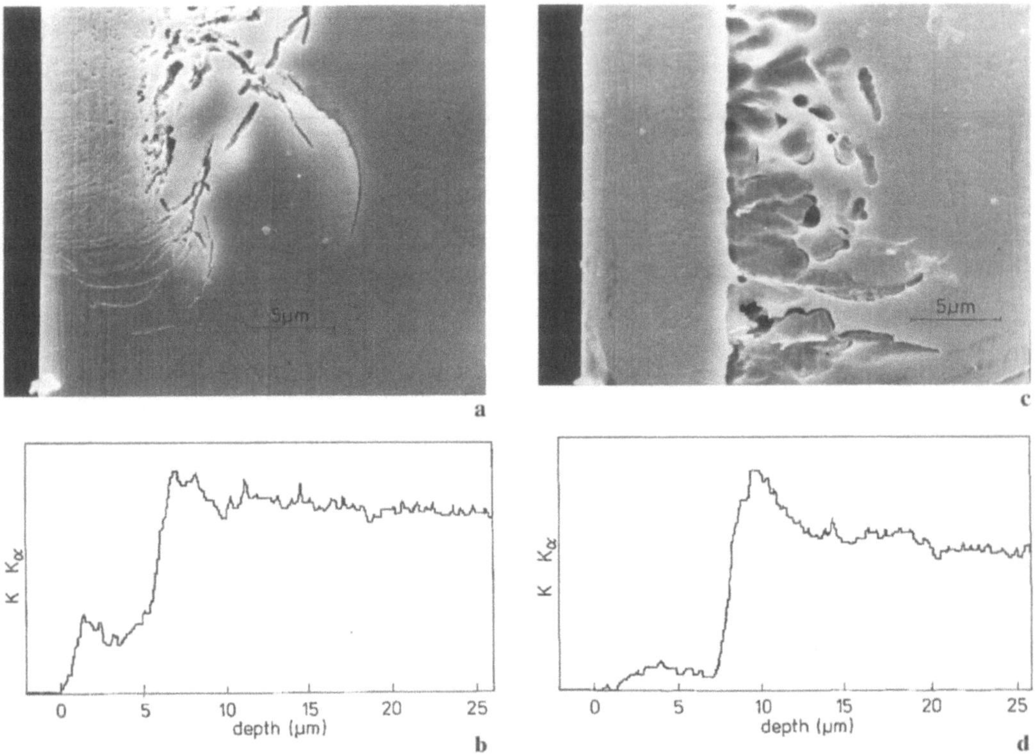

Fig. 7. Glass K12, beam diameter 100 μm, absorbed current 1 μA, **a** exposure 1 min., **b** corresponding line profile of K, **c** exposure 2 min., **d** corresponding line profile of K

observed starting with the formation of individual channels (Fig. 8a). After 10 min exposure the density of channels is increased (Fig. 8b) and the deteriorated area can be seen. At 50 μm and 68 nA the traces occur after 10 min exposure (Fig. 9). After 20 min the sharp boundary starts to grow at the bottom of the surface layer which, however, remains porous. The growth time of the surface layers for the K12 glass is also presented in Fig. 4.

The transmittance spectrum is again characteristic for light scattering on metallic colloidal particles with a size ranging from 10 nm to 200 nm.

Discussion

All results confirm that alkaline ions migrate from the grounded (connected with the ground electrically) surface to the depth of the specimen due to the electric field created by the injection of high energy electrons which is in agreement with Yamamoto et al. [2]. These ions then condensate at some depth into neutral colloidal particles, as was deduced from optical transmittance spectra. The presence of Na^+ ions stated by Pederson [6] can be explained by surface reactions induced by

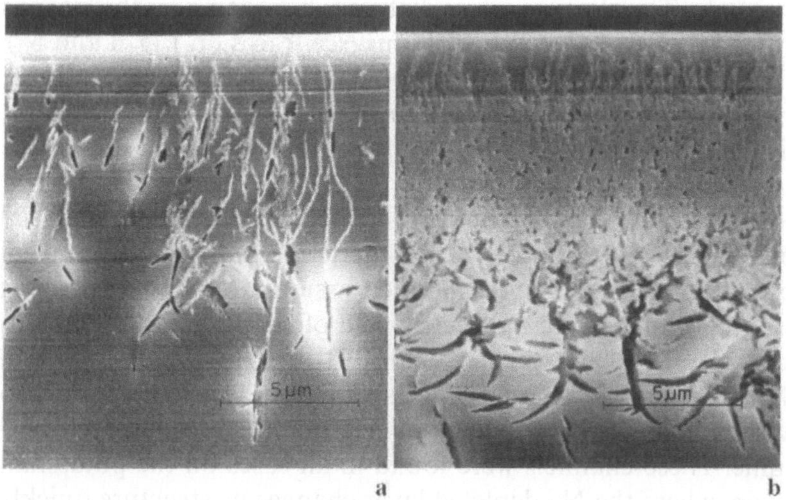

Fig. 8. Glass K12, beam diameter 100 µm, absorbed current 253 nA, **a** exposure 5 min., **b** exposure 10 min

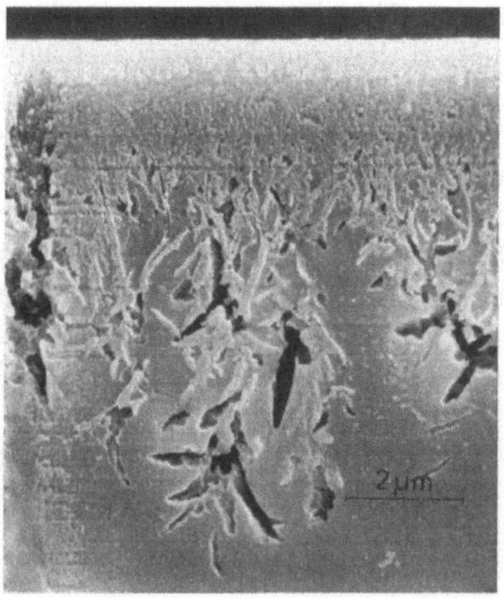

Fig. 9. Glass K12, beam diameter 50 µm, absorbed current 68 nA, exposure 10 min

the XPS method used. This opinion is supported by results presented by Ishiyama [4] and Whitkop [7].

Concentration profiles show that the region of maximum Na or K concentration lies between about half and two thirds of electron range which was in this experiment approx. 20 µm as calculated according to Kanaya and Okayama [8]. However, the precise analysis is not possible because this area, uncovered by grinding and polishing reacts quickly with air moisture which causes severe corrosion of the glass so that only "pores" remain. The electron microprobe analysis damages the glass as well even when perfomed under moderate conditions (see traces on Fig. 3 and 6a). The very beginning of the migration was observed on the K12 glass where channels, revealed by quicker etching, probably facilitate movement of alkali ions. Relevant concentration profiles show faster migration at some depth while at the surface the concentration remains temporary higher (Fig. 7b).

Cazaux [9] tried to estimate the electric field created in a dielectric material by an homogeneously charged cylinder. This model suggests migration of positive ions in the depth, but under oversimplified assumptions. The transport of Na^+ ions in glass under influence of an electric field of 50 kV/cm at a temperature of 80 °C was also described by Regnier et al. [10]. At the anode a several micrometer thick Na- free layer was created. The effect of the electric field can also be seen in Fig. 7a where channels in the shape of electric lines are observed outside the excited volume. These channels were found in all cases on the potassium glass. On the soda glass, where the Na depleted layer changes its structure quickly and becomes compact, only the expansion out of the excited volume can be seen.

The origin of free ions and the creation of channels remain to be explained. Free ions can be created by electron impact, but it is more difficult to explain the channels. Unfortunately, X-ray microanalysis does not have the sufficient resolution to explain, why the channels appear white. It might be caused by a change of the local charge density which would cause the change in secondary electron emission. The length of the channels is too long for normal diffusion. The photomicrographs show that their formation is random at the beginning and their number increases with time. The structure of the remaining glass breaks down and becomes more compact so that even crystallisation can occur. At low beam currents and large diameters about 25% to 30% of the alkalis remain in the surface layer even after long exposure times. In the less stable Na17 glass, Na is nearly completely depleted from the surface layer. In the K12 glass, at a beam diameter of 50 μm and a current 1 μA, a potassium free layer is created after 30 s, while at a current 68 nA this layer is not formed even after 30 min. In the first case, the increased temperature, estimated to be 327 °C, could contribute to a faster migration of K ions and the breakdown of the original structure.

Neither electron beam broadening below the surface of the specimen nor significant reduction of the primary electron range due to the electric field, as proposed in [11], were observed.

Conclusion

Present experiments proved that alkaline ions in glass migrate even at low temperature towards the depth of the specimen whose surface is coated by a conductive layer and grounded when it is irradiated by a high energy electron beam. In the present work 50 keV electrons were used to enlarge the excited volume and thus to facilitate observation and analysis. The alkaline ions migrate to a certain depth due to the electric field where they condensate into colloidal particles of various sizes which cause brown colouring of the glass. Depletion of the surface layer may cause its structure to change.

Acknowledgements. The authors are indebted to Dr. K. Polák for optical transmission spectra measurements, to Mr. J. Fryštacký for precise sample preparation (both Institute of Physics of Academy of Sciences of Czech Republic) and to Dr. R. Kužel (Faculty of Mathematics and Physics of Charles University) for X-ray diffraction measurement.

References

[1] J. L. Lineweaver *J. Appl. Phys.* **1963**, *34*, 1786.
[2] H. Yamamoto, S. Isobe *Proc. Tenth Inter. Congr. Glass*, Kyoto, 1974, p. 1–9.
[3] V. Hulínský, O. Gedeon, K. Jurek, *Mikrochim. Acta* [*Suppl.*], **1996**, *13*, 325.
[4] M. Ishiyama, J. Chiba, Y. Suzuki *Proc. Tenth Inter. Congr. Glass*, Kyoto, 1974, p. 5–1.
[5] R. G. Gossink, T. P. A. Lommen, *Appl. Phys. Lett.* **1979**, *24*, 444.
[6] L. R. Pederson, *Surf. Sci.* **1982**, *119*, L307.
[7] P. G. Whitkop, *Surf. Sci.* **1981**, *110*, 261.
[8] K. Kanaya, S. Okayama, *J. Phys. D. Appl. Phys.* **1963**, *14*, 617.
[9] J. Cazaux, *J. Appl. Phys.* **1986**, *59*, 1418.
[10] P. Regnier, Y. Serruys, A. Zemskoff, *Phys. Chem. Glass.* **1986**, *27*, 185.
[11] J. Graham, C. R. M. Butt, R. B. W. Vigers, *X-ray Spectrom.* **1984**, *13*, 126.

Mikrochim. Acta [Suppl.] 13, 349–353 (1996)

Application of the Boltzmann Transport Equation in the Thickness Determination of Thin Films

Grigoris Kaltsas, Nikos Glezos*, Evangelos Valamontes, and Androula G. Nassiopoulos

Institute of Microelectronics, NCSR "Demokritos", P. O. Box 60228, Aghia Paraskevi, Attikis 15310, Greece

Abstract. The diffusion approximation of the Boltzmann transport equation for an electron beam incident on a sample, is used to derive an analytical expression for the depth distribution of X-ray production $\Phi(\rho z)$. This solution is used as a starting point for a numerical evaluation of $\Phi(\rho z)$ in the case of a multicomponent sample. Results are applied to the prediction of the signal ratio involved in coating thickness measurements. Good agreement is found with experimental results and an exponential decay law is established.

Key words: Boltzmann equation, thin film thickness, x-ray depth distribution.

The determination of thickness of coatings is of importance in microelectronics and non-destructive methods are preferred for obvious reasons. The two main methods used for this purpose are slow energy electron loss spectroscopy (SEELS) and electron probe microanalysis (EPMA). Using EMPA the X-ray signal from an element contained in the thin film I_f is compared to the signal from a known bulk standard of the same composition I_b. The curves I_f/I_b vs. e-beam energy for different thin film thicknesses are used as a reference for the specification of an unknown thickness [1, 2]. The specific form of these curves and consequently the energy range at which measurements should be done can be predicted by Monte Carlo (MC) calculations [3, 4]. In this article an analytical approach is proposed which is much faster than MC and more evident. This approach is based on an analytical solution of the Boltzmann transport equation developed by one of the authors [5, 6] for electron beam lithography (EBL). Furthermore, it is shown that a part of the experimentally obtained signal ratio curve follows a well defined exponential law.

Description of the Model

The Boltzmann transport equation describing an electron beam incident vertically on a sample can be reduced to a diffusion equation with simple boundary conditions

* To whom correspondence should be addressed

[5, 6]:

$$\frac{\partial f}{\partial E} = \rho^2 \frac{\lambda(E)}{3} \left(\frac{dE}{ds}\right)^{-1} \frac{\partial^2 f}{\partial(\rho z)^2} \tag{1}$$

where $f(z, E)$ is the electron distribution as a function of depth z in the sample and electron energy E, ρ is the material density, dE/ds is the Bethe energy loss rate of the electron along its path [7]:

$$\frac{dE}{ds} = 78500 \frac{Z\rho}{AE} \ln\left(\frac{1.166E}{J}\right) \tag{2}$$

where J is the mean ionization potential and λ is the energy dependent diffusion length [5, 6]:

$$\lambda(E) = \frac{0.51AE^2}{Z^2\rho}\left(\ln\frac{1+\beta}{\beta} - \frac{1}{1+\beta}\right)^{-1} \tag{3}$$

where Z is the atomic number, A the atomic weight and β the screening parameter for electron scattering on the target atoms of the sample:

$$\beta = 0.00545\ Z^{2/3}/E \tag{4}$$

In (2), (3) and (4) E is in keV and λ in µm. Equation (1) is directly solvable for a point beam incident vertically on the sample with appropriate boundary conditions [5, 6]. The result is:

$$f(\rho z, E) = \frac{1}{\sqrt{\pi}\sigma(E)} e^{-(\rho z)^2/4\rho^2\sigma^2(E)} \left[1 - \frac{\sqrt{\pi}\sigma(E)}{\Delta} e^{\zeta^2} \operatorname{erfc}(\zeta)\right] \tag{5}$$

with:

$$\zeta = \frac{(\rho z)}{2\sigma(E)\rho} + \frac{\sigma(E)}{\Delta} \tag{6}$$

and $\Delta = 2/3\ \lambda(E_0)$. The gaussian parameter $\sigma(E)$ is given by:

$$\sigma^2(E) = \int_E^{E_0} \frac{\lambda(E')}{3} \left(\frac{dE'}{ds}\right)^{-1} dE' \tag{7}$$

Expression (5) describes the electron distribution in a homogeneous material as a function of the electron energy and depth. Using (5) one may derive the depth distribution function of X-ray production $\Phi(\rho z)$:

$$\Phi(\rho z) = \text{constant x} \int_{Ec}^{E_0} f(\rho z, E)\, G(E, E_c) dE \tag{8}$$

where $G(E, E_c)$ is the Gryzinski ionisation cross-section and E_c the binding energy. The constant depends upon the fluorescence yield and the nature of the material and it is irrelevant if signal ratios are calculated. In Eq. (8) the direction of electron movement is not taken into consideration. One may include this feature in the same way as in EBL by calculating the lateral distribution of the electrons [6].

In order to show the relation of this model to that proposed by Packwood and Brown [8, 9] we observe that the z-dependence of $\Phi(\rho z)$ is similar to $f(\rho z, E)$. The maximum value of $\sigma(E)$ is obtained for $E = E_c$. Substituting (2) and (3) in (7) one finds:

$$\sigma^2(E_c) = 2.17 \times 10^{-6} \frac{A^2}{Z^3 \rho^2} \int_{E_c}^{E_0} \left(\ln \frac{1+\beta}{\beta} - \frac{1}{1+\beta} \right)^{-1} \frac{E^3 dE}{\ln(1.166E/J)} \tag{9}$$

This expression scales with an exponent <4 for E_0 and -3 for Z. Packwood and Brown specified corresponding values 3.5 for E and -2.32 for Z using a random walk approach and fitting to data.

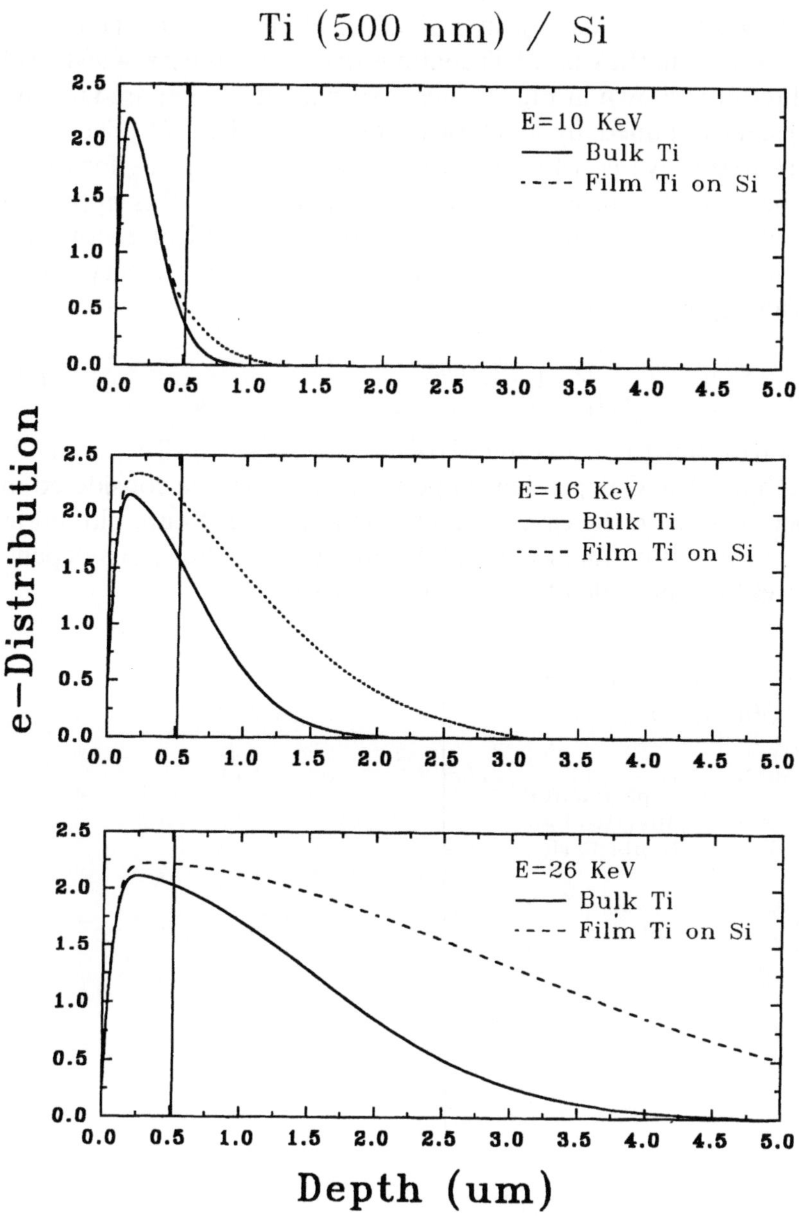

Fig. 1. Electron density vs. depth for different e-beam energies

Next, we consider the case of a multi-layered sample. When E is close to the initial e-beam energy E_0, Eq. (5) is a valid solution for the first layer since electrons have not yet penetrated the second layer. For the rest of the energy range Eq. (5) is used as an initial condition for the numerical evaluation of the diffusion Eq. (1) [5, 6]. Results of such a calculation are shown in Fig. 1 in the case of Ti(500 nm) film over bulk Si for three different e-beam energies 10,16, and 26 keV. The quantity $\int f(\rho z,E)dE$ is plotted vs. depth.

Application to Coating Thickness Measurement

In order to calculate the thickness of a coating the X-ray intensity I_f of an element contained in the coating is compared to the intensity I_b of a bulk material of the same composition. This was done in the case of Ti coatings over Si for the kα (4.964 keV) line of Ti. A typical result is shown in Fig. 2 where experiment is compared to MC calculations and results obtained by numerical solution of Eq. (1). The most common procedure is to specify the critical value of the electron energy E_{cr} for which the film thickness becomes the production range [10]. However this approach requires several measurements in the neighbourhood of the critical energy. By inspecting the experimental data it was found that the ratio I_f/I_b for energy values close to E_{cr} follows an exponential law:

$$\frac{I_f}{I_b} = 1 - e^{-C/E^{\alpha}} \tag{10}$$

where C depends upon the film thickness and α is of the order of 2.0 in all experiments examined by the authors. The C dependence upon thickness will be considered in a forthcoming paper. Results in the case of several Ti films over Si are plotted in Fig. 3 with $\alpha = 2.7$. The points of intersection of the lines with the ln(E) axis depend upon the film thickness and provided the calibration curve.

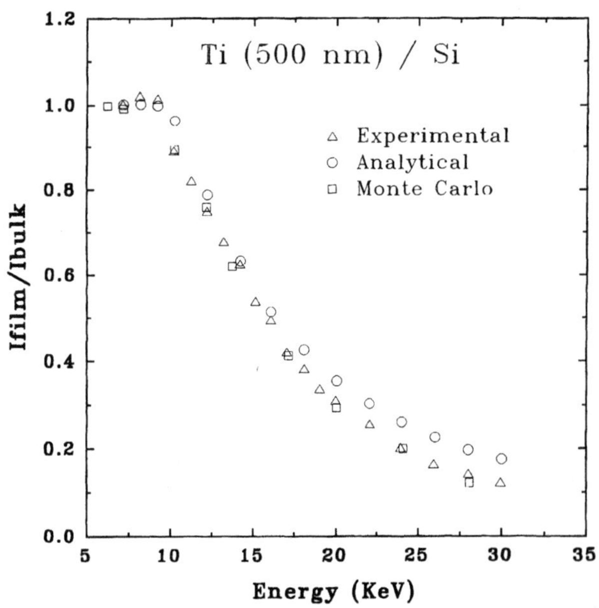

Fig. 2. Signal ratio vs. e-beam energy

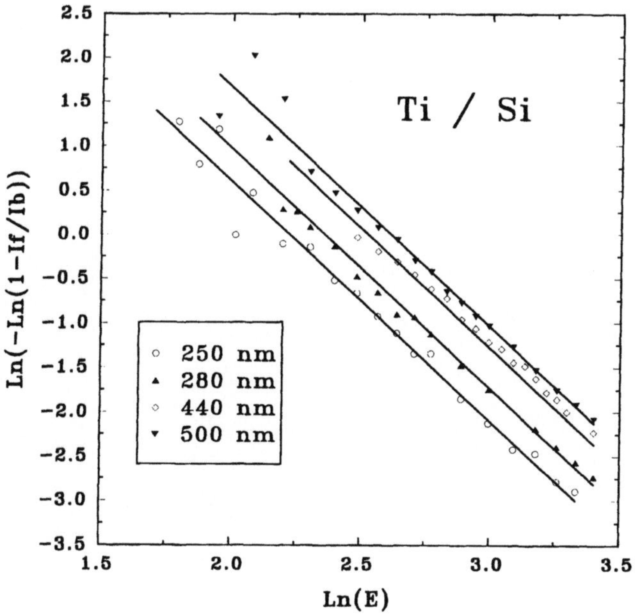

Fig. 3. Signal ratio vs. energy for various Ti films over Si substrate

Conclusions

The analytical model proposed in this article is based upon a fundamental transport equation for electrons incident vertically on a sample. The resulting $\Phi(\rho z)$ is of the same shape as that obtained previously by a combination of probability arguments and semi-empirical methods. This approach is much faster for X-ray signal evaluation than MC calculations as shown previously in EBL.

The experimental method evaluation of thickness of coatings becomes more precise with the observation that for energies larger than a critical value an exponential law is followed. The region for measurements may be easily predicted by the analytical approach discussed. The energy dependence of the signal ratio in Eq. (10) is consistent with the form of the distribution given by Eq. (5) and it will be examined in more detail in the future.

References

[1] W. E. Sweeney, Jr. R. E. Seebold, L. S. Birks, *J. Appl. Phys.* **1960**, *31(1)*, 1061.
[2] H. E. Bishop, D. M. Poole, *J. Appl. Phys. D6* **1973**, *1142*.
[3] J. L. Pouchou, F. Pichoir, *La Recherche Aerospaciale* **1973**, *5*, 349.
[4] A. G. Fizgerald, A. D. Gillies, H. L. L. Watton, *Surf. Interf. Anal.* **1960**, *16*, 163.
[5] N. Glezos, I. Raptis, D. Tsoukalas, M. Hatzakis, *J. Vac. Sci. Technol. B* **1992**, *10*(7), 2606.
[6] I. Raptis, N. Glezos, M. Hatzakis, *J. Vac. Sci. Technol. B* **1993**, *11*(*6*), 2774.
[7] H. Bethe, *Ann. Phys.* (*Leipzig*) **1930**, *5*, 325.
[8] R. H. Packwood, J. D. Brown, *X-ray Spectrom.* **1981**, *10*, 138.
[9] J. D. Brown, R. H. Packwood, *X-ray Spectrom.* **1982**, *14*, 187.
[10] W. Reuter, J. D. Kuptsis, A. Lurio, D. F. Kyser, *J. Appl. Phys.* **1978**, *11*, 2633.

Mikrochim. Acta [Suppl.] 13, 355–361 (1996)

Characterisation of the Shape of Microparticles via Fractal and Fourier Analyses of Scanning Electron Microscope Images

Volodymyr V. Kindratenko, Pierre J. M. Van Espen*, Boris A. Treiger, and René E. Van Grieken

Micro- and Trace Analysis Center (MiTAC), Department of Chemistry, University of Antwerpen (UIA), B-2610 Antwerpen, Belgium

Abstract. Two different approaches based on Fourier and fractal analyses, are applied for particle shape characterisation. The advantages and limitations of both approaches as well as their fields of application are discussed. Examples of the application of these approaches for particles of various types are given.

Key words: shape, fractal dimension, Fourier analysis, particle characterisation, image analysis.

Computer controlled scanning electron microscopy (CC SEM) in combination with energy-dispersive X-ray microanalysis (EDX) has proven to be a powerful tool for analysing microscopic particles. The method provides both compositional (via EDX spectra) and morphological (via SEM images) characteristics of particles. In practice, the chemical composition is used as the main source of information, e.g. to classify the particles according to their origin. However, there are situations where the shape of the particles is of primary importance. For example, the shape and size of so-called 'T-grain' silver halide crystals determine the performance of the photographic emulsion [1]. Therefore, the ability to classify the crystals based on their shape is of interest for evaluating photographic materials. As another example, the shape of some atmospheric microparticles is indicative for their origin, e.g. two types of aerosol particles, spherical fly ash and irregularly shaped soil dust particles, originate from different processes, but have similar compositions [2, 3]. Again, only the shape of such microparticles allows us to differentiate between them and to make source apportionment.

The problem is how to reveal and represent the shape in a form suitable for particle characterisation. At the current stage of shape analysis one can distinguish

* To whom correspondence should be addressed

a large number of characterisation methods, the most commonly applied being 1) shape factors based [4], 2) chain coding ([5], p. 205) 3) Fourier analysis ([5], p. 206), and 4) fractal analysis ([5], p. 162). Here we examine the applicability of fractal analysis for describing the shape of individual aerosol particles and Fourier analysis for describing the shape of silver halide crystals.

Methods

Preprocessing of SEM Images

SEM images consist of pixels where the intensity (colour) of each pixel is proportional to the number of the back-scattered electrons, emitted from the corresponding point at the surface of a material. Some of them represent the corresponding points of the microparticles (foreground pixels), the others represent background. In the present study we used images with 256 grey levels. For the extraction of the shape of particles only foreground pixels are of interest. In order to separate foreground from background pixels, the thresholding technique [6] was applied. A binary image is obtained with only two intensity values. In such images the shape information is contained in the particle edge. The border following algorithm [7] was used for finding and storing a connected list of coordinates of the particle edge. Thus, the shape (silhouette edge) is defined as the list of the connected coordinates of the particle edge.

Description of Particle Shape Using Fourier Coefficients

The particle's silhouette edge can be described in polar coordinates as a function of radius R vs. the sweep angle θ from the center of gravity of the particle [8]. This function can be expanded in a Fourier series:

$$R(\theta) = R_0 + \sum_{n=1}^{\infty} (A_n \cos(n\theta) + B_n \sin(n\theta))$$

where R_0 is the radius of a circle having the same area as the profile, generated by the particle's silhouette edge; θ is the polar angle measured from an arbitrary reference line; A_n and B_n are the Fourier series coefficients. Each term of this series is called a harmonic. Beddow [9] proposed to use the harmonic amplitude $A_n^2 + B_n^2$ that indicates the contribution of the n-th component in the overall shape where n is the harmonic order. The more complicated the shape is, the more harmonics should be taken into account for the description of the shape. Only the first few are responsible for the 'general' shape and they can be used for the purpose of identification [8]. E.g., for a particle with a pronounced hexagonal shape, only the first six harmonics are important (Fig. 1B); a particle with a pronounced triangular shape can be reconstructed quite well by first six coefficients too (Fig. 1C).

Mostly convex shapes can be described correctly by this method. In the other case it is possible to obtain different radii for the same angle. Some extensions of this techniques for the case of particles with a concave shape can be found in the literature [10].

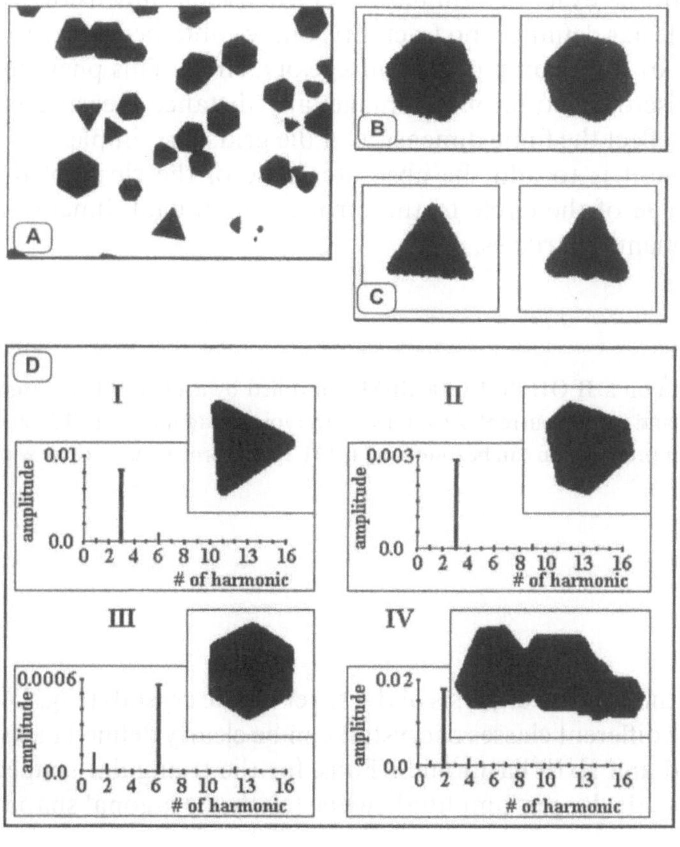

Fig. 1. **A** 'T-grain' silver halide crystals (SEM backscattered electron image). **B** The hexagonally shaped silver halide crystal (left) and restored hexagon by using the first 6 Fourier coefficients (right). **C** The triangular silver halide crystal (left) and restored triangle by using the first 6 Fourier coefficients (right). **D** Amplitude of the Fourier coefficients for different crystal shape

Description of Particle Shape Using Fractal Dimension

Fractal dimension is a concept to describe the measure of the 'irregularity' of irregular or self-similar shapes. Self-similarity means that any part of the shape looks the same as any other part and they all look the same as the whole structure. Real particles are not fractal from a mathematical point of view because self-similarity cannot be observed over the entire range of sizes – only over some limited intervals. Also, a particle can have several intervals where one can observe the self-similarity. Different intervals correspond to so-called 'textural' and 'structural' fractal dimensions [11].

The concept of fractal dimension can be applied to particle characterisation in the following way. The perimeter P of the contour of a particle depends on the length L of the 'yardstick' with which we make the measurements: $P \sim L^{\alpha}$. The fractal dimension of the contour of the particle in this case is $1-\alpha$. Thus, the calculation of the fractal dimension is based on the estimation of the length of the silhouette edge at different resolutions and the fractal dimension is obtained from a graph in which the logarithm of the length estimate is plotted vs. the logarithm of the resolution (Richardson plot) [12]. A 'Hybrid' method [13] for fractal dimension calculation is used in this work. Measurements were performed with a 'yardstick' from 0.02 to 0.2 of the maximal distance between points of the silhouette edge of the microparticles (maximum Feret diameter).

There are difficulties with the exact determination of the fractal dimension for computerised images. A circle has definitely no fractality, but we obtained a textural fractal dimension equal to 1.01 for a computerised image of a circle. This phenomenon is explained by the discrete nature of the image: any distance measure on a regularly spaced grid will reflect the finite dimension of the grid. The simplest way to take this error into account is to add the absolute value of the slope of the Richardson plot for the image of the circle to the error in the fractal dimension determination of the experimental particles.

Experimental

The measurements were performed on a JEOL JSM-6300 SEM controlled by a PGT IMIX system using an electron energy of 20 keV and a beam current of ca. 1 nA and a typical magnification of 5000 x. Details about the samples and their preparation can be found in [1, 14]. Image processing is done with software written by the authors.

Results and Discussion

Silver Halide Crystals

An example of the results of the Fourier analysis of the silver halide crystals (Fig. 1A) is given in Fig. 1D. The three different classes of crystals can be clearly defined based on analysis of only the third and sixth harmonics. Thus, for the triangular shaped crystals the 3rd harmonic has the largest amplitude while for the hexagonal shaped

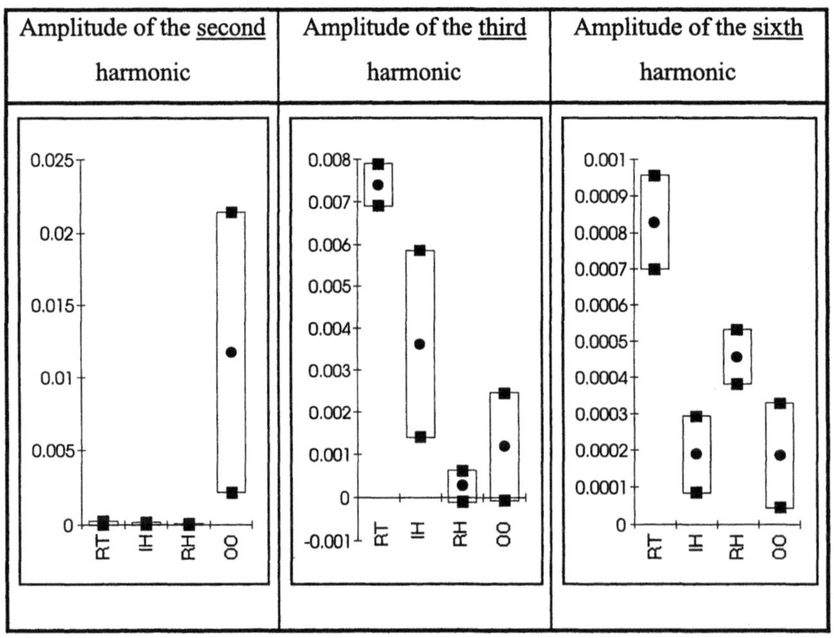

Fig. 2. Average values of the amplitude and standard deviation of the second, third and sixth harmonics for the four different types of silver halide crystals. *RT* triangles, *IH* irregular hexagons, *RH* hexagons, *OO* overlapping crystals

crystals the 6th harmonic has the largest amplitude. Overlapping crystals can be revealed by the second harmonic. Small crystals usually have a large amplitude of the first harmonic, but they can be eliminated from the analysis using geometrical parameters such as the radius of the equivalent circle.

The analysis results for the silver halide crystals are summarised in Fig. 2. Nine crystals of triangular shape, 13 crystals of irregular hexagonal shape, 13 crystals of perfect hexagonal shape and 10 sets of overlapping crystals were analysed; the average values and standard deviations of the second, the third and the sixth coefficients are shown.

Aerosol Particles

A typical example of a spherical fly ash particle, its EDX spectrum and the results of the fractal analysis are presented in Fig. 3A. A backscattered electron image of a soil

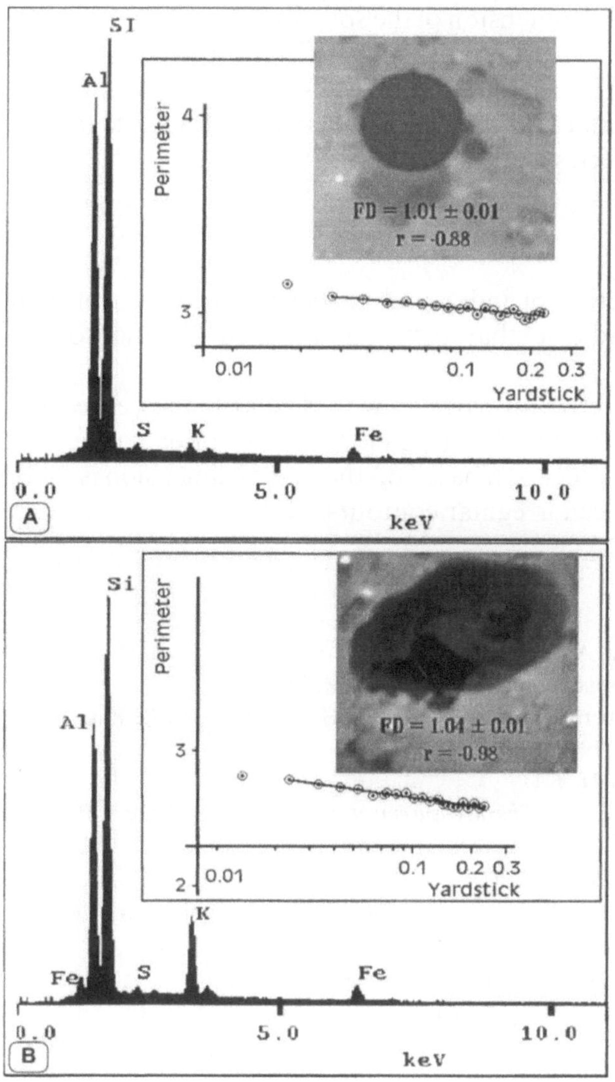

Fig. 3. A Spherically shaped fly ash particle (SEM backscattered electron image, EDX spectrum and results of the fractal analysis). **B** Soil dust particle (SEM backscattered electron image, EDX spectrum and results of the fractal analysis). **r** the coefficient of correlation for the given regression straight line, *FD* fractal dimension. The yardstick is expressed in units of the maximum Feret diameter

Table 1. Fractal dimension and standard deviation for different type of individual aerosol particles

Type of aerosol particle	Number of particles studied	Textural fractal dimension
Spherically shaped fly ash	35	1.01 ± 0.01
Soil dust	37	1.04 ± 0.01

dust particle, its EDX spectrum and the results of fractal analysis are shown in Fig. 3B. Comparison of Figs. 3A and 3B leads to the conclusion that it is impossible to judge whether a particle belongs to the fly ash class or the soil dust class solely on the base of the composition. At the same time, fractal analysis reveals that spherically shaped fly ash particles possess no fractality, whereas the 'textural' fractality of soil dust particle is evident.

The results of the analysis of 72 individual aerosol particles are summarised in Table 1. The mean value of the fractal dimension of the soil dust particles is 1.04 with a standard deviation of 1.01. This value is statistically different from the average value of 1.01 ± 0.01 obtained for 35 fly ash particles. The absence of fractality for the spherically shaped fly ash particles and the presence of only textural fractality for soil dust particles holds for the all particles analysed.

Conclusion

Two algorithms for shape description of individual particles were tested on real samples. The results obtained indicate that they can be used for the correct classification of the particles.

Fourier description is very attractive for particles with regular shapes, especially for those with axes of symmetry. Normally, the first 6 to 8 Fourier coefficients reproduce the shape quite well. Classification based on the fractal dimension is more suitable for particles with pronounced irregular contours.

References

[1] I. Geuens, *Ph.D. Thesis*, University of Antwerp, 1993.

[2] A. R. Ramsden, M. Shibaoka, *Atmospheric Environment* **1982**, *16*, 2191.

[3] G. L. Fisher, B. A. Prentice, D. Silberman, J. M. Ondov, A. H. Blermann, R. C. Ragaini, A. R.McFariand, *Environ. Sci. Technol.* **1978**, *12*, 447.

[4] A. I. Medalia, *Powder Technol.* **1970/71**, *4*, 117.

[5] J. C. Russ, *Computer-Assisted Microscopy: The Measurement and Analysis of Images*, Plenum, New York, 1990, pp. 205, 206, 162.

[6] A. D. Brink, *Pattern Recognit. Lett.* **1989**, *9*, 335.

[7] A. Rosenfeld, A. C. Kak, *Digital Picture Processing, 2nd Ed., Vol. 2*, Academic Press, London, 1982, p. 219.

[8] B. Raeymaekers, P. Van Espen, F. Adams, *Mikrochim. Acta* **1984**, *II*, 437.

[9] J. K. Beddow, G. C. Philip, A. F. Vetter, M. D. Nasta, *Powder Technol.* **1977**, *18*, 19.

[10] J. K. Beddow (ed), *Particle Characterization in Technology, Vol. 2, Morphological Analysis*, CRC, Boca Raton, Florida, 1984, pp. 3–14.

[11] B. Kaye, *A Random Walk Through Fractal Dimensions*, VCH, New York, 1989.
[12] B. B. Mandelbrot, *Science* **1967**, *156*, 633.
[13] N. N. Clark, *Powder Technol.* **1986**, *46*, 45.
[14] V. V. Kindratenko, P. J. Van Espen, B. A. Treiger, R. E. Van Grieken, *Environ. Sci. Technol.* **1994**, *28*, 2194.

Mikrochim. Acta [Suppl.] 13, 363–376 (1996)

Calculation of the Surface Ionisation Using Analytical Models of Electron Backscattering

Peter Klein, Michael Andrae, Kurt Röhrbacher, and Johann Wernisch*

Institute for Applied and Technical Physics, Technical University of Vienna, Wiedner Hauptstrasse 8-10, A-1040 Wien, Austria

Abstract. The value of the surface ionisation for oblique incidence is calculated by testing various new approaches. A model based on a multiple reflection theory developed at the Institute for Applied and Technical Physics at the Technical University of Vienna is extended to oblique incidence by the use of a new fit for the angular distribution of backscattered electrons (according to data published by Kanter) under non-normal incidence. Furthermore two analytically derived models (proposed by Werner and Niedrig) are tested for their applicability to calculate the surface ionisation by numerical integration of the double differential backscattering coefficient versus angle and energy. The combined model of Werner is extended to oblique incidence to make a direct comparison possible. The results are compared with expressions published by Pouchou, by Cazaux and by Merlet; reasons for discrepancies are discussed; commonly used simplifications are reviewed critically.

Key words: surface ionisation, angular distribution, backscattered electrons, oblique incidence, analytical models.

In EPMA (electron probe micro analysis) correction programmes the surface ionisation plays a very important role; either as a fit parameter for the depth distribution function e.g. in Pouchou and Pichoir's model [9] or as a point of reference in experimental measurements of the depth distribution function using the sandwich technique (tracer method) proposed by Castaing [17].

The surface ionisation $\Phi(0)$ is defined as the ratio of the primary ionisations in a surface layer of infinitesimal mass thickness to the primary ionisations of a self-supporting layer of identical mass thickness and composition under equal conditions of electron bombardment. This definition is equivalent to that of the Auger backscattering factor $R + 1$ as pointed out by Cazaux [5].

* To whom correspondence should be addressed

This implies that due to electron backscattering from the sample a value greater than 1 is achieved: the self-supporting film suffers from a lack of backscattered electrons which would cause additional ionisations.

August [1] gives an excellent review and comparison of existing expressions and measurements of the surface ionisation and also provides a theoretically 'exact' formula to calculate this quantity for normal incidence:

$$\Phi(0) = 1 + \frac{\int_0^{\pi/2} \int_1^{U_0} \frac{d^2\eta}{dU \cdot d\Omega} \cdot Q(U) \cdot dU \cdot \frac{1}{\cos(\theta)} \cdot d\Omega}{Q(U_0)} \tag{1}$$

$\frac{d^2\eta}{dU \cdot d\Omega}$	normalized energy and angular distribution of backscattered electrons; $d\Omega = 2\pi \sin\theta \, d\theta$ because of rotational symmetry in case of normal incidence
U	overvoltage ratio E/E_c (E: electron energy, E_c: critical energy)
$Q(U)$	ionisation cross section
$1/\cos(\theta)$	path elongation factor

For the ionisation cross section Pouchou and Pichoir's expression will be used [7].

The path elongation factor should quantify the geometrical fact that electrons traversing the infinitesimally thin layer under a certain angle may cause more ionisations due to their longer path in the layer.

Eq. (1) is the mathematically and physically exact expression of the calculation of the surface ionisation; the only problem is finding an expression for the double differential backscattering coefficient to be integrated analytically.

To make life easy many authors e.g. August [1], Cazaux [5], Merlet [11], Tirira Saa [15] make the approximation (implicitly or explicitly) to separate the angular and energy distribution of the backscattered electrons:

$$\frac{d^2\eta}{dU \cdot d\Omega} \rightarrow \frac{d\eta}{d\theta}\frac{d\eta}{dU} \tag{2}$$

This approximation on the one hand implies that there is only one energy distribution for all electrons emerging from the sample under different angles or on the other hand that all electrons emerging from the sample with different energies have the same angular distribution. It is one aim of this paper to review these assumptions critically.

Extension of Oblique Incidence

The extension of Eq. (1) to oblique incidence (angle of incidence α, measured from the surface-normal) results in (without any simplifications):

$$\Phi_\alpha(0) = 1 + \frac{\int_0^{2\pi} \int_0^{\pi/2} \int_1^{U_0} \frac{d^2\eta}{dU \cdot d\Omega} \cdot Q(U) \cdot dU \cdot \frac{1}{\cos(\theta)} \cdot d\Omega}{\frac{Q(U_0)}{\cos(\alpha)}} \tag{3}$$

with the additional φ-integration in the numerator because of the loss of rotational symmetry due to non-normal incidence and a factor $1/\cos(\alpha)$ in the denominator

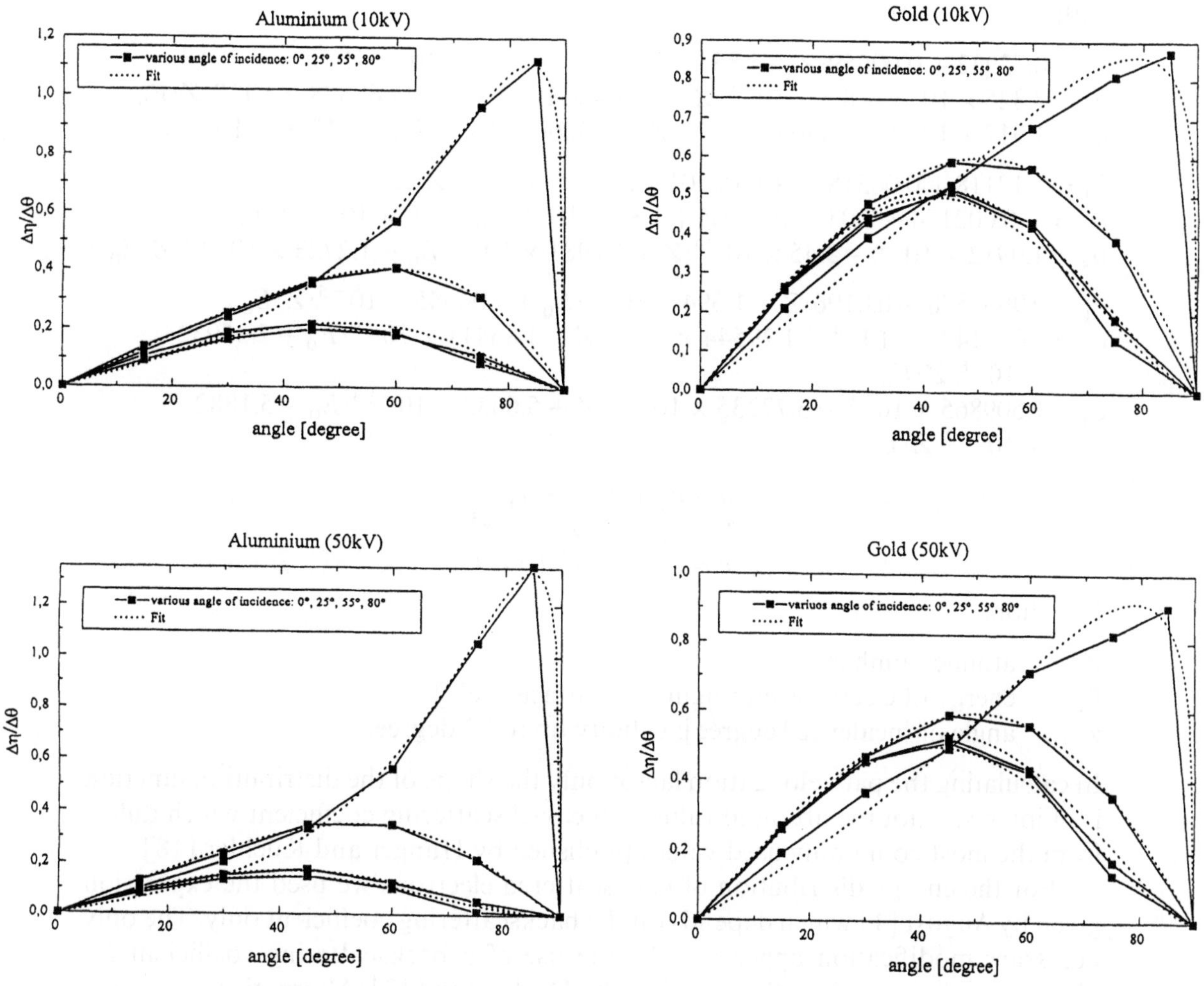

Fig. 1. Comparison of the new fitting function and measurements by Kanter [10] for aluminum (10 and 50 kV) and gold (10 and 50 kV) for 0°, 25°, 55° and 80° (angle of incidence)

which represents the path-elongation in the self-supporting layer and therefore an increase of ionisations.

Following August's treatment in separating the angular- and energy- distributions of backscattered electrons, an expression for the angular distribution of backscattered electrons for non-normal incidence is required to calculate the path elongation factor; since no expressions representing the tendencies found by measurements (eg. Kanter [10]) were in existence, we made a fit for the angular distribution depending on the atomic number, the acceleration energy of the impinging electrons and the angle of incidence (measured to surface normal). The function found to describe the curves of Kanter best is given by (Fig. 1):

$$\frac{d\eta}{d\theta} = \mathrm{a}\cdot\sin(\theta)\cdot\cos^{\mathrm{b}}(\theta)\cdot\exp(-\mathrm{c}\cdot\cos(\theta)) \qquad (4)$$

with

$$a_1 = 0.72932 + 0.00659 \cdot Z - 5.833 \times 10^{-4} \cdot E_0 + 8.33 \times 10^{-5} \cdot Z \cdot E_0$$
$$a_2 = 7.325 \times 10^{-3} - 8.5 \times 10^{-5} \cdot Z - 2.944 \times 10^{-4} \cdot E_0 + 4.662878 \times 10^{-6} \cdot Z \cdot E_0$$
$$a_3 = 1.917 \times 10^{-4} - 2.4405 \times 10^{-6} \cdot Z + 6.3066 \times 10^{-6} \cdot E_0 - 8.1279 \times 10^{-8} \cdot Z \cdot E_0$$

$$b_1 = 1.12118 + 0.00318 \cdot Z + 0.02201 \cdot E_0 - 2.5 \times 10^{-4} \cdot Z \cdot E_0$$
$$b_2 = -0.002174 + 8.03 \times 10^{-5} \cdot Z + 1.87 \times 10^{-4} \cdot E_0 - 3.0 \times 10^{-6} \cdot Z \cdot E_0$$
$$b_3 = 1.9912 \times 10^{-6} - 3.45 \times 10^{-9} \cdot Z + 1.7981 \times 10^{-8} \cdot E_0 - 1.2723 \times 10^{-10} \cdot Z \cdot E_0$$

$$c_1 = 0.9965578 - 0.0106 \cdot Z + 1.594 \times 10^{-3} \cdot E_0 + 4.04545 \times 10^{-5} \cdot Z \cdot E_0$$
$$c_2 = 1.0241477 \times 10^{-4} - 1.29544 \times 10^{-6} \cdot Z - 1.86411 \times 10^{-6} \cdot E_0 + 2.35932 \times 10^{-8} \cdot Z \cdot E_0$$
$$c_3 = 1.509865 \times 10^{-8} - 3.77235 \times 10^{-11} \cdot Z + 5.6632 \times 10^{-10} \cdot E_0 - 5.1882 \times 10^{-12} \cdot Z \cdot E_0$$

$$a = a_1 + a_2 \cdot \alpha + a_3 \cdot \alpha^2$$
$$b = b_1 - b_2 \cdot \alpha - b_3 \cdot \alpha^3$$
$$c = c_1 + c_2 \cdot \alpha^2 + c_3 \cdot \alpha^4$$

Notation:

Z atomic number
E_0 energy of electrons impinging on surface [keV]
α angle of incidence [degree]; validity up to 87 degree.

In calculating the path elongation factor only the shape of the distribution function is of interest – not the absolute value of the backscattering coefficient which differs from the most commonly used values published by Hunger and Küchler [18].

For the energy distribution of backscattered electrons we used the expression given by August [1] which depends on the backscattering coefficient only. The only necessary modification appears to be the use of a backscattering coefficient for oblique incidence such as the one given by Darlington [12]. Alternatively, expressions derived by Czyzewski and Szymanski [14] or Merlet [21, 22] were programmed and gave no significant deviations; these expressions depend directly on the angle of incidence α, which, however, causes no remarkable effect on the position of the maximum of the energy distribution.

The results obtained by the application of all mentioned simplifications and assumptions are shown in Fig. 2, Fig. 3a and Fig. 3b compared with values obtained by calculations by Pouchou et al. [8], Cazaux [5] and Merlet [21, 22] on the one hand, by measurements and Monte-Carlo simulations by Sewell [6] (for Au and Al only), by experiment by Benhayoune [21] and by Monte-Carlo simulations using a programme published by Hénoc [20] on the other hand. The experimental values were provided with error bars of 5% considering the principal uncertainties connected with the tracer technique.

A generally satisfactory performance can be observed although towards higher angles of incidence and increasing atomic number certain deviations can be noticed. The reason seems to be the path-elongation factor: Cazaux and Merlet make only very crude approximations about the shape of the angular distribution of the backscattered electrons and therefore neglect the tendencies found by Kanter: for low atomic numbers a nearly optical reflection-law may be observed; for high atomic

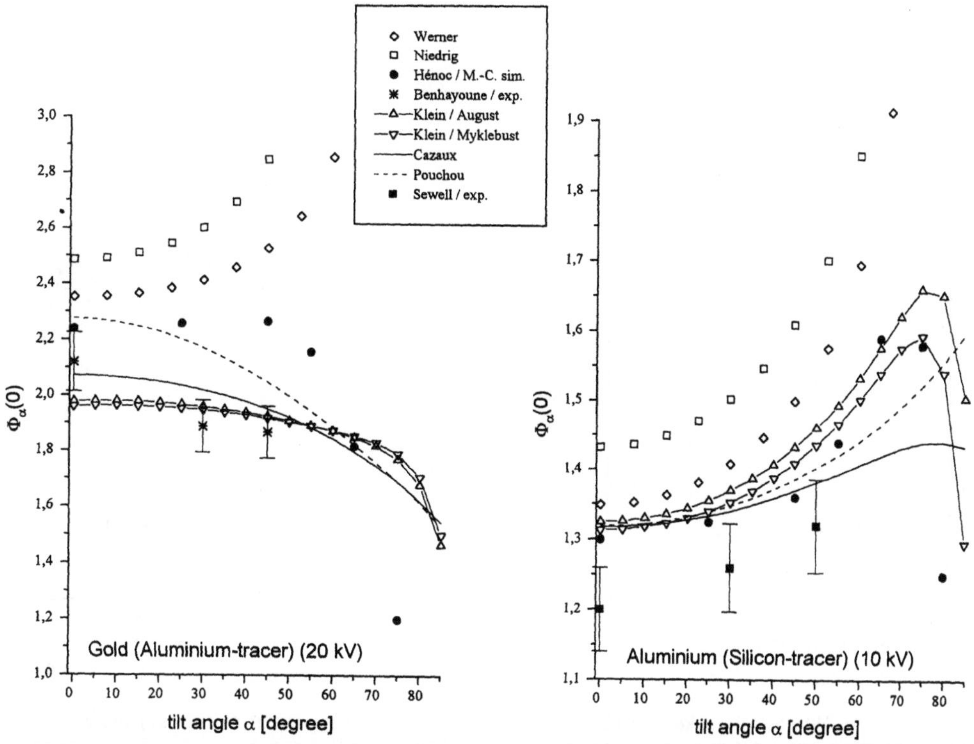

Fig. 2. Comparison of the angular dependence of the surface ionisation value obtained by Sewell (experimental, [6]), Sewell (Monte-Carlo sim., [6]), Benhayoune (experimental, [21]), Hénoc (Monte-Carlo-sim., [20]), Cazaux (theor., [5]), Pouchou (theor., [8]), evaluating analytical expressions derived by Werner (analyt., [4]) and Niedrig (analyt., [3]) and the new expression calculated by modification of a multiple reflection model (August (theor., [1]) using energy distributions of backscattered electrons proposed by August [1] and Czyzewski [14] gold (20 keV, tracer aluminum with $E_c = 1.559$ keV) and aluminum (10 keV, tracer silicon with $E_c = 1.838$ keV). [The analytical expressions are evaluated for pure elements !]

numbers (e.g. gold) diffuse electrons play a more and more dominant role and therefore embed the distinct peak found for e.g. aluminum. Figure 8 shows that a comparison of the path-elongation factors reveals deviations of up to approx. 20% (high atomic numbers), but additionally, deviations resulting from Cazaux's assumption concerning the energy spectra of backscattered electrons are observed: Cazaux considers the distribution to be a δ-function situated at a mean energy which is obviously not realistic. This way lower values for the fraction of the ionizing electrons and for the weighted ionisation cross section are calculated.

The expression by Pouchou et al. [8] is a fit to data obtained by Monte-Carlo simulations; Pouchou admits, however, that '···it is not sure that the values are perfectly reliable.' (p. 55) Especially the missing maximum for lower atomic numbers in his data-fit seems to be caused by unrealistic values provided by his multiple scattering Monte-Carlo code for grazing incidence.

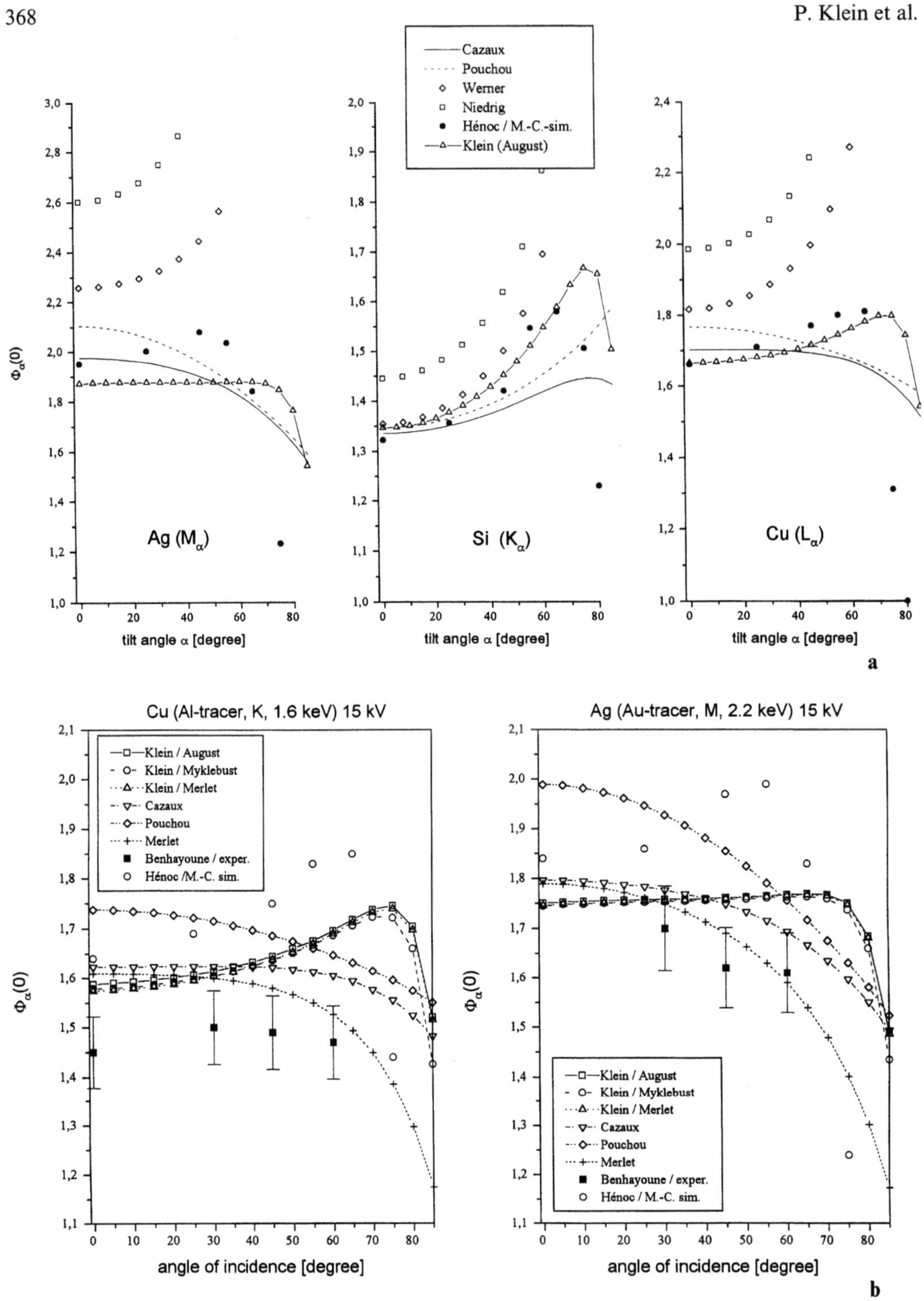

Fig. 3. **a** Comparison of the angular dependence of the surface ionisation value; theoretically derived expressions of Cazaux [5] and Pouchou [8], evaluated analytically derived expressions by Werner [4] and Niedrig [3], Monte-Carlo simulations after Hénoc [20] and the new approach. ($E_0 = 10\,\text{kV}$, pure elements without any tracers). **b** Comparison of the angular dependence of the surface ionisation value; theoretically derived expressions by Cazaux [5], Pouchou [8] and Merlet [21, 22], the new approach using different energy distributions for backscattered electrons (August [1], Czyzewski [14], Merlet [21, 22]) and experimental values by Benhayoune [21]. Copper (15 kV, aluminum-tracer with $E_c = 1.559$) and silver (15 kV, gold-tracer with $E_c = 2.119$)

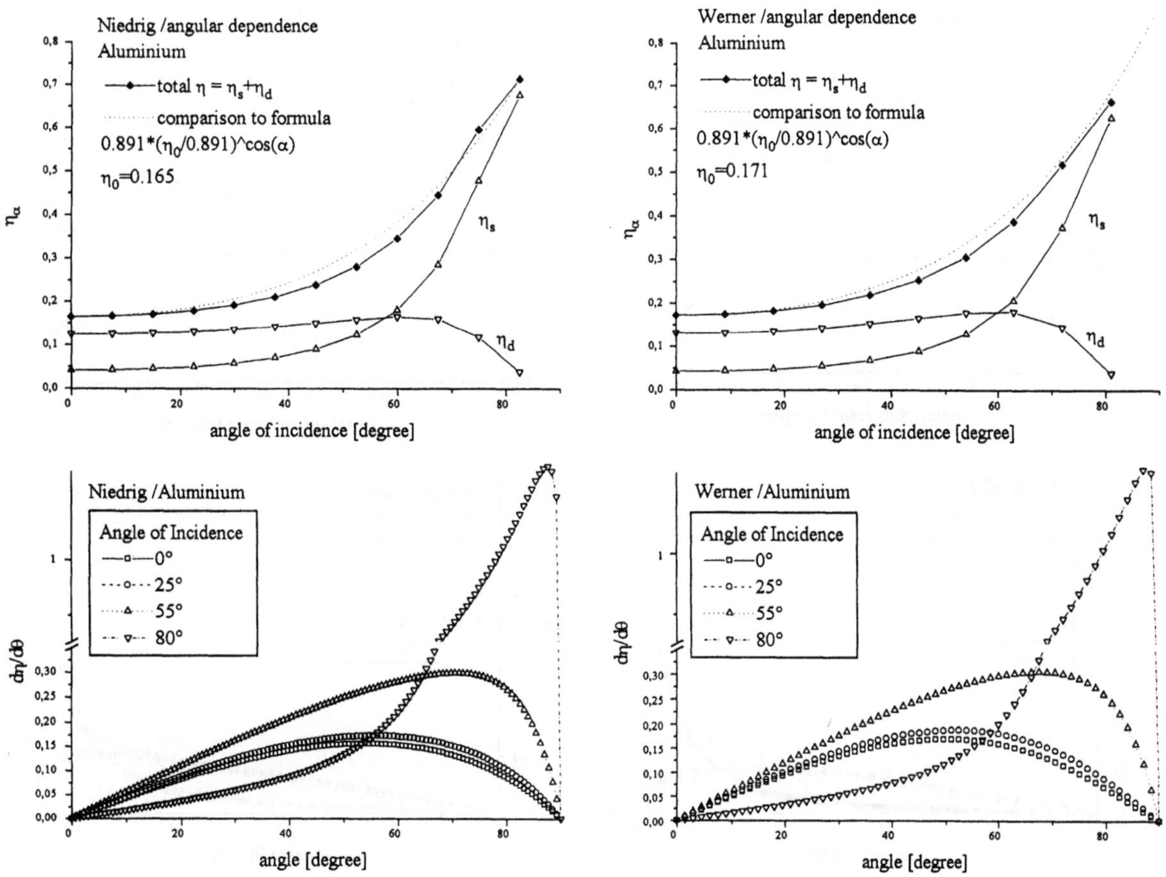

Fig. 4. Angular distribution and angular dependence of the total backscattering coefficient for aluminum (compared to [18]) evaluating analytical expressions by Niedrig [2] and Werner [4] (modified for oblique incidence by the author) for 10 kV (left) and 50 kV (right)

More General Approaches

Niedrig presented [3] analytically derived models of electron backscattering which – as a result – give expressions for this function

$$\frac{d^2\eta}{dU \cdot d\Omega}$$

which may be used to calculate the surface ionisation.

The models with the most realistic performance are two combined models; one was proposed by Niedrig himself and the other one by Werner [4]. 'Combined' should mean regarding two scattering concepts – single Rutherford scattering and diffusion – in the models.

Although both models cannot describe exactly the experimental findings especially concerning angular distributions of backscattered electrons, rather good results may be achieved with regard to determining the surface ionisation value for normal incidence.

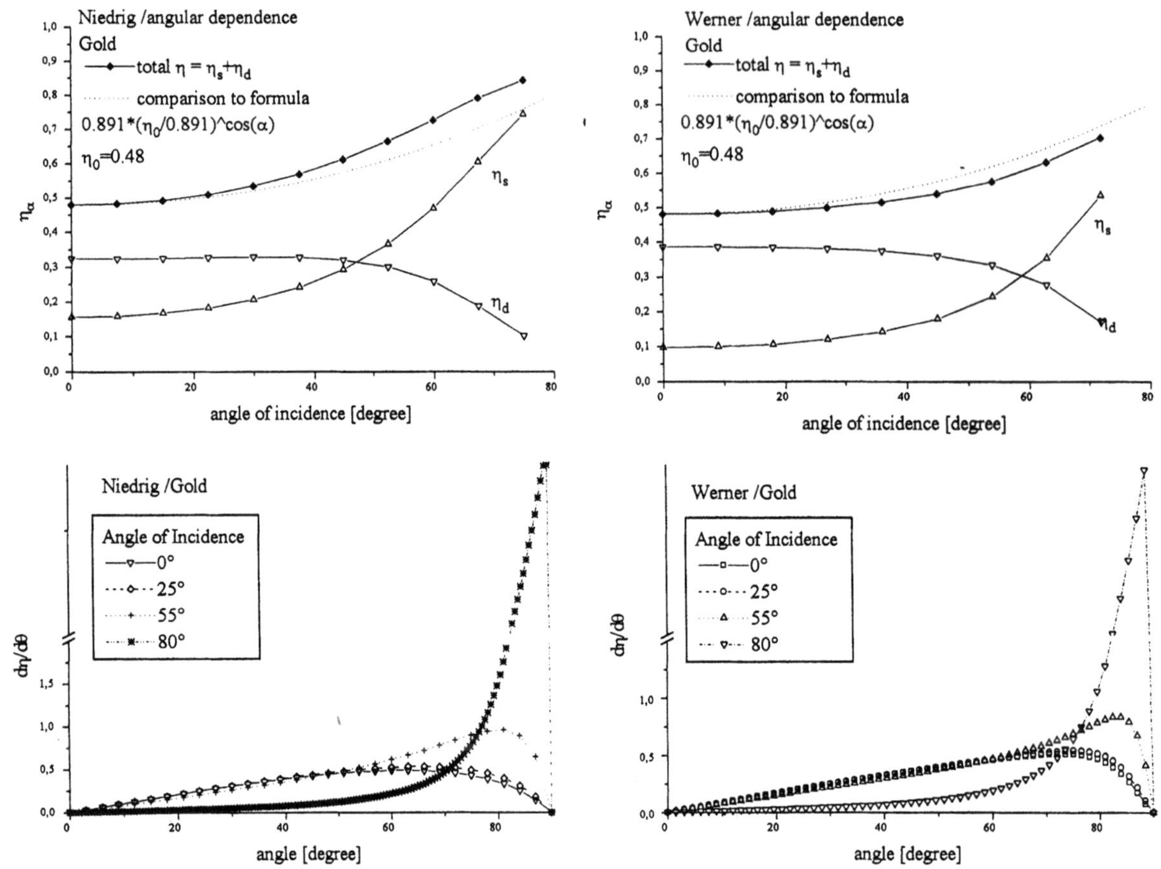

Fig. 5. Angular distribution and angular dependence of the total backscattering coefficient for gold (compared to [18]) evaluating analytical expressions by Niedrig [2] and Werner [4] (modified for oblique incidence by the author) for 10 kV (left) and 50 kV (right)

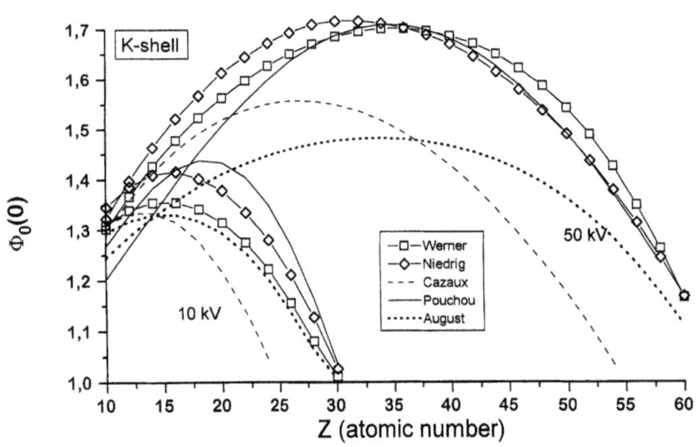

Fig. 6. Comparison of atomic number dependence (equivalent to the overvoltage ratio in respect to the K-shell edge energy) of the different approaches by Werner [4], Niedrig [2], Cazaux [5], Pouchou [8] and the new expression for normal incidence

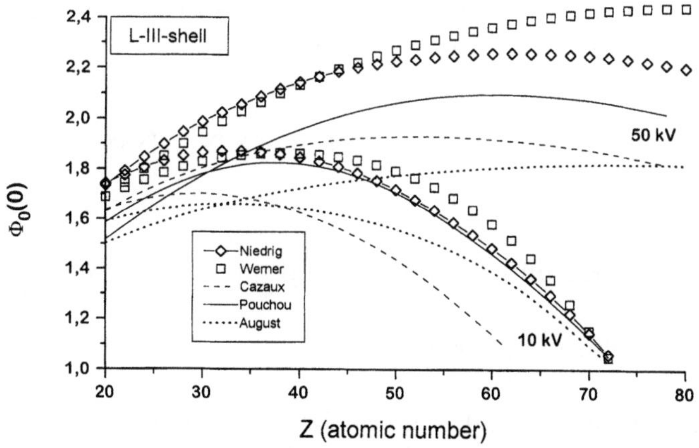

Fig. 7. Comparison of atomic number dependence (equivalent to the overvoltage ratio in respect to the L-III-shell edge energy) of the different approaches by Werner [4], Niedrig [2], Cazaux [5], Pouchou [8] and the new expression for normal incidence

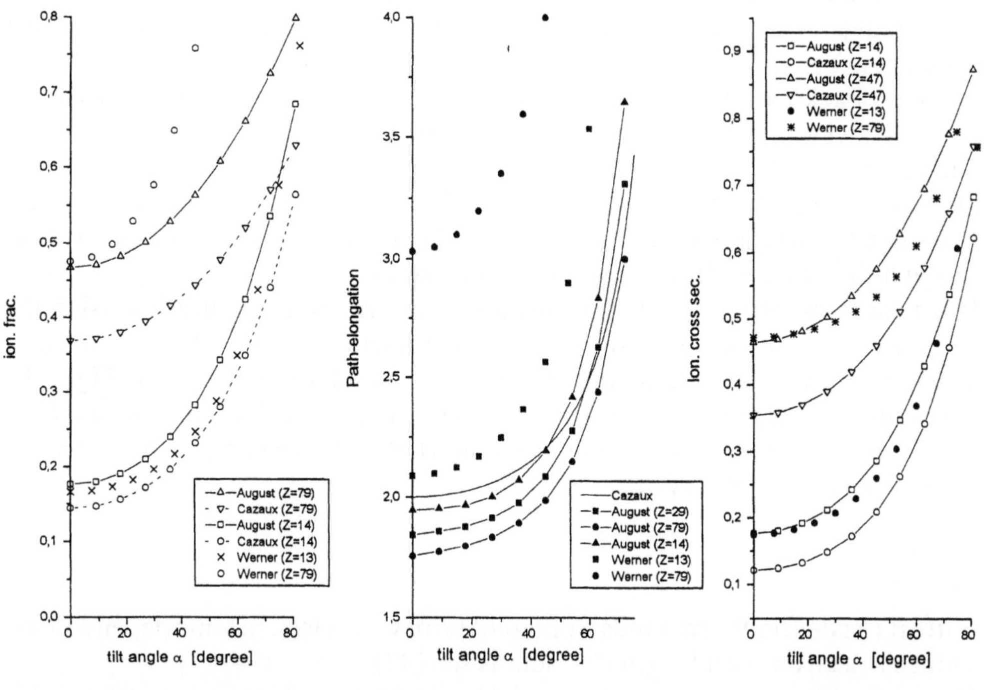

Fig. 8. Comparison of the single elements of the stepwise integration/evaluation of the surface ionisation integral: the path-elongation factor; the fraction of the electrons that have enough energy to ionize; the weighted ionisation cross section

Niedrig's Model

Niedrig combines the single Rutherford scattering with a diffusion-term of a most simple (and similar) mathematical structure to make analytical integration possible. For details of the derivation see [2]; the interesting expression is:

$$\frac{d^2\eta_s^N(\theta,\varphi,\varepsilon,\alpha)}{d\varepsilon\cdot d\Omega} = \frac{4a_s\cdot\varepsilon}{2\pi}\cdot\left[\frac{\cos(\alpha)+\varepsilon^2\cdot\cos(\theta)}{\cos(\theta)+\cos(\alpha)}\right]^{a^N-1}\cdot\frac{\cos(\theta)}{(\cos(\theta)+\cos(\alpha))\cdot f^2(\theta,\varphi,\alpha)} \qquad (5a)$$

$$f(\theta,\varphi,\alpha) = 1 - \sin(\theta)\cos(\theta)\sin(\alpha) + \cos(\theta)\cos(\alpha)$$

$$\frac{d^2\eta_d^N(\theta,\varphi,\varepsilon,\alpha)}{d\varepsilon\cdot d\Omega} = \frac{2a_d^N\cdot\varepsilon}{2\pi}\cdot\left[\frac{\cos(\alpha)+\varepsilon^2\cdot\cos(\theta)}{\cos(\theta)+\cos(\alpha)}\right]^{a^N-1}\cdot\frac{\cos(\theta)}{(\cos(\theta)+\cos(\alpha))} \tag{5b}$$

ε	E/E_0
a_s, a^N, a_d^N	constants, depending on atomic number (index N ⋯ Niedrig's model)
α	angle of incidence
$\frac{d^2\eta_s^N(\theta,\varphi,\varepsilon,\alpha)}{d\varepsilon\cdot d\Omega}$	single scattering contribution after Niedrig
$\frac{d^2\eta_d^N(\theta,\varphi,\varepsilon,\alpha)}{d\varepsilon\cdot d\Omega}$	contribution by diffusion after Niedrig

The integration of this expression over θ and φ gives the energy distribution of backscattered electrons. A comparison with experimentally found distributions reveals a rather unrealistic shape (no maximum); the angular distributions (by integration over ε) also show a quite unsatisfactory behaviour in comparison with measurements by Kanter [10] or Seidel [13].

The total backscattering coefficient, however, as well as the angular dependence of the total backscattering coefficient show a rather satisfying agreement with the formula derived by Neubert and Rogaschewski [16] (see Figs. 4 and 5).

Despite all these disadvantages this expression is analytically derived from basic relations like the Rutherford cross-section or the Thomson-Whiddington law and was not manipulated to fit the experimental findings.

It is even more surprising that a calculation of the surface ionisation using this expression gives comparable results for normal incidence. The surface ionisation integral may only be evaluated numerically. The results are shown in Fig. 6 for K-shell critical energies and Fig. 7 for L-III-shell critical excitation energies. Especially for low overvoltage-ratios a remarkable agreement with Pouchou's expression [8] can be observed.

Werner's Model

In contrast to Niedrig Werner uses a different term with a logarithmic dependence of the reduced electron path length (for details see [4]).

Originally, Werner developed his model for normal incidence; it was modified by Rogaschewski [19] for oblique incidence to calculate energy-spectra. We present our adaption for oblique incidence to obtain angular distributions regarding Werner's approach for the calculation of the parameters a^W, a_d^W using a different system of coordinates and notation (in comparison to [19]): (quotations from Werner [4] p. 418):

$$\frac{s_0}{\lambda_s} = \int_{\pi/2}^{\pi} \frac{d\sigma}{d\Omega}\cdot 2\pi\cdot\sin\vartheta d\vartheta = 4\pi Q_\pi \rightarrow \frac{4\pi Q_\pi}{\cos^2(\alpha)} \text{ because of}$$

$$\int_0^{2\pi}\int_{\pi/2}^{\pi} \frac{4\cdot\sin(\theta)}{(1-\sin(\theta)\sin(\alpha)\cos(\varphi)+\cos(\theta)\cos(\alpha))^2}\cdot d\theta\cdot d\varphi = \frac{4\pi}{\cos^2(\alpha)} \tag{6}$$

$$\frac{s_0}{\lambda_D} = \int_0^{\pi}(1-\cos\vartheta)(1-F)^2\frac{d\sigma}{d\Omega}\cdot 2\pi\cdot\sin\vartheta d\vartheta = 8\pi Q_\pi \rightarrow 8\pi Q_\pi$$

(the contribution of diffuse electrons shows no dependence on the angle of incidence) to be continued on p. 419 (Eq. 6a) {Eq. 6b remains unchanged}:

$$\chi_S = \int_0^t \frac{S_0}{\lambda_S} \cdot dt' \rightarrow \frac{a_s}{\cos^2(\alpha)} \cdot Q_\pi$$

$$a_{S\alpha} = a_{S0}/\cos^2(\alpha) \quad a_{D\alpha} = a_{D0} \quad a^W_{d\alpha} = 8/5 \cdot a_{D\alpha} \quad \text{and} \quad a^W_\alpha = a^W_{d\alpha} + a_{S\alpha}.$$

Following Werner's calculation without changes leads to equation 13a (p. 421) which has to be multiplied (according to the already made assumptions concerning the parameters a_S) by the factor $\cos^2\alpha$.

A last modification has to be applied to formula 22 on p. 425: due to the different conditions with respect to non-normal incidence and therefore a different geometry the relation between the path-length before and after the scattering event with regard to energy loss gives:

$$t = (1 - k^2)\frac{\cos(\theta)}{\cos(\theta) - 1} \rightarrow t = (1 - k^2)\frac{\cos(\theta)}{\cos(\theta) + \cos(\alpha)}$$

The formulas of interest including all modifications concerning oblique incidence:

$$\frac{d^2\eta^W_s(\theta,\varphi,\varepsilon,\alpha)}{d\varepsilon \cdot d\Omega} = \frac{4a_s \cdot \varepsilon}{2\pi} \cdot (1 + a^W_d \tau_m) e^{-(a^W - 1)\tau_m} \frac{\cos(\theta) \cdot \cos^2(\alpha)}{(\cos(\alpha) + \cos(\theta)) \cdot f^2(\theta,\varphi,\alpha)} \tag{7a}$$

$$\frac{d^2\eta^W_d(\theta,\varphi,\varepsilon,\alpha)}{d\varepsilon \cdot d\Omega} = \frac{a^{w^2}_d \cdot \varepsilon}{2\pi} \cdot \tau_m e^{-(a^W - 1)\tau_m} \frac{\cos(\theta)}{(\cos(\alpha) + \cos(\theta))} \tag{7b}$$

$$\tau_m = -\ln\left(1 - (1 - \varepsilon^2)\frac{\cos(\theta)}{\cos(\theta) + \cos(\alpha)}\right) \quad f(\theta,\varphi,\alpha) = 1 - \sin(\theta)\cos(\varphi)\sin(\alpha) + \cos(\theta)\cos(\alpha)$$

ε	E/E_0
a_S, a^W, a^W_d	constants, depending on atomic number and α (index W ··· Werner's model)
α	angle of incidence
$\frac{d^2\eta^W_s(\theta,\varphi,\varepsilon,\alpha)}{d\varepsilon \cdot d\Omega}$	single scattering contribution after Werner
$\frac{d^2\eta^W_d(\theta,\varphi,\varepsilon,\alpha)}{d\varepsilon \cdot d\Omega}$	contribution by diffusion after Werner

Figures 4 and 5 show the angular distributions under oblique incidence for gold and aluminum assuming the above mentioned modifications and the angular dependence of the total backscattering coefficient, compared to a formula proposed by Neubert and Rogaschewski [16].

In contrast to Niedrig's model this approach shows a much better performance concerning energy distributions of backscattered electrons: a more or less pronounced maximum may be observed, which stands comparisons to experimental values [3].

The comparison of angular distributions for backscattered electrons with experimental data shows the limitations of this model: strongly peaked distribution functions – especially for high atomic numbers over 50° angle of incidence (measured from the surface normal). Concerning the evaluation of the surface ionisation comparable values may be achieved only for normal incidence, which may be seen in Fig. 6 (K-shell) and Fig. 7 (L-III-shell).

For rather high angles of incidence the calculated values exceed a realistic range; a possible reason can be found in the wrong shape of the angular distribution and therefore an unrealistic path elongation by overestimation of angles near to the surface (comparable already to grazing incidence). Measurements by Kanter show a distinct peak for aluminum with a decreasing tendency towards high atomic numbers whereas Werner's and also Niedrig's distribution functions show the opposite: an already overestimated peak for aluminum with an increasing tendency for high atomic numbers.

Concerning the simplifications mentioned above a direct comparison is now possible: on the one hand the direct evaluation of the integral (formula 1) and on the other hand fractional calculation by separately dealing with angular and energy distributions:

$$\Phi_\alpha(0) \xrightarrow{???} \frac{\int_0^{\pi/2} \frac{d\eta}{d\theta} \cdot \frac{1}{\cos(\theta)} \cdot d\theta}{\int_0^{\pi/2} \frac{d\eta}{d\theta} \cdot d\theta} * \frac{\int_1^{U_0} \frac{d\eta}{dU} \cdot Q(U) \cdot dU}{\frac{Q(U_0)}{\cos(\alpha)}} \tag{8}$$

The comparison shows that – within the same method ! – the separate calculation gives values which are only slightly less than the 'exactly' calculated ones. It has to be pointed out – however – that a nearly identical result was achieved by the same basic formula just differently integrated (Fig. 9).

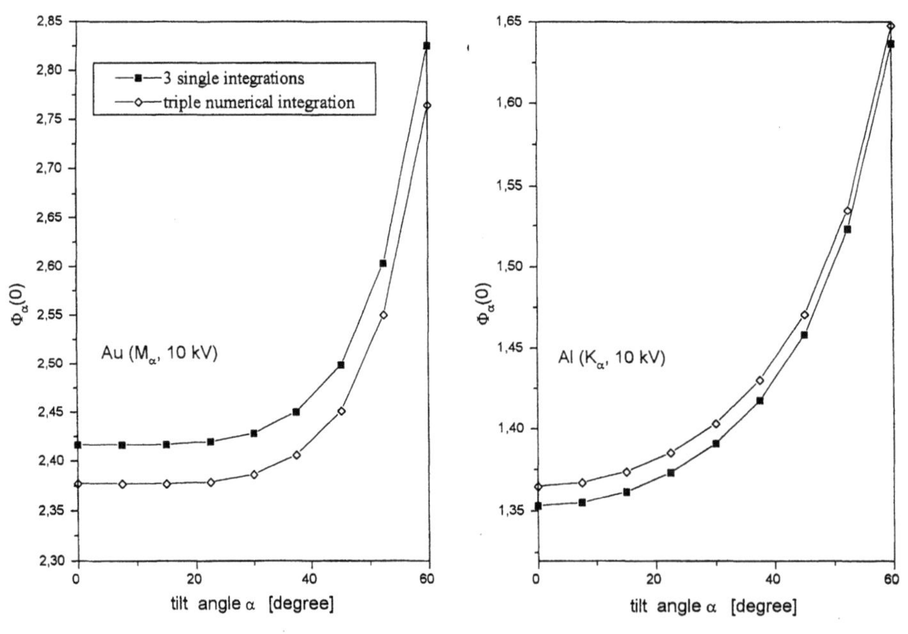

Fig. 9. Comparison of the different evaluation methods; one triple numerical integration of formula (7a) and (7b) versus three single integrations after formula (8)

These observations should nevertheless be a hint that simplified models for the calculation of the surface ionisation give lower values due to simplifications (and combinations) concerning the separation of the angular and energy distribution of backscattered electrons.

Conclusion

During the last few years tilting of the specimens has become a commonly used technique in EPMA for various purposes. Therefore a modification of the correction-programmes in existence is necessary to meet the new geometry. A modification of the calculation of the surface ionisation derived from a multiple scattering model proposed by August is presented, which is compared to already adjusted expressions; rather satisfying results are achieved; furthermore two analytically derived expressions (Niedrig, Werner) are used to calculate the value of the depth distribution function at the surface of the sample; due to limitations in these models more (normal incidence) or less (oblique incidence) agreement can be observed.

Generally, it has to be stated that the expressions in existence differ quite remarkably and only very little reliable experimental values have been published so far. The new approach, however, should be the most reliable one due to the implementation of a realistic expression for the path elongation factor based on experimental values for angular distribution of backscattered electrons.

Acknowledgement. We acknowledge the support given by the Austrian 'Fonds zur Förderung der wissenschaftlichen Forschung' (Project P10415-PHY).

References

[1] H.-J. August, J. Wernisch, *Scanning* **1990**, *12*, 14.
[2] H. Niedrig, *J. Appl. Phys.* **1982**, *53(4)*, R15.
[3] H. Niedrig, in: *Proceedings of Conf. on Electr.-Beam Interactions with Solids for Microsc., Microanal. and Microlith.* (D. Kyser, D. E. Newbury, H. Niedrig, R. Shimizu, ed.), SEM Inc., AMF O'Hare, IL 60666 U.S.A., Asilomar, 1992, pp. 51–68.
[4] U. Werner, H. Bethge, J. Heydenreich, *Ultramicroscopy* **1982**, *8*, 417.
[5] J. Cazaux, *Microsc. Microanal. Microstruct.* **1992**, *3*, 271.
[6] D. A. Sewell, G. Love, V. D. Scott, *J. Phys. D. Appl. Phys.* **1987**, *20*, 1567.
[7] J. L. Pouchou, F. Pichoir, in: *11th ICXOM, London Ontario, 1986* (J. D. Brown, R. H. Packwood, eds.), London, Ont., 1987, p. 249.
[8] J. L. Pouchou, F. Pichoir, D. Boivin, in: *Proceedings of the 12th Intern. Cong. on X-ray Optics and Microanalysis*, (S. Jasienska, L. J. Maksymovicz, eds.), *Vol. 1*, 1989, pp. 52–59.
[9] J. L. Pouchou, F. Pichoir, F. Girard, *J. Micosc. Spectrosc. Electron.* **1980**, *5*, 425.
[10] H. Kanter, *Ann. Physik* **1957**, *20*, 144.
[11] C. Merlet, *X-Ray Spectrometry* **1992**, *21*, 229.
[12] E. H. Darlington, *J. Phys. D* **1975**, *8*, 85.
[13] H. Seidel, *Thesis*, University of Münster, 1972.
[14] Z. Czyzewski, H. Szymanski, *Proc. 10th Int. Cong. Electron Microscopy, Vol. 1*, Offizon Paul Hartung, Hamburg, 1982, p. 61.

[15] J. H. Tirira Saa, J. A. Riveros, *X-Ray Spectrometry* **1987**, *16*, 27.
[16] G. Neubert, S. Rogaschewski, *J. Phys. D. Appl. Phys.* **1984**, *17*, 2439.
[17] R. Castaing, J. Descamps, *J. Phys. Radium* **1955**, *16*, 304.
[18] H. J. Hunger, L. Küchler, *Phys. Stat. Sol.* (*a*) **1979**, *56*, K45.
[19] S. Rogaschewski, *Phys. Stat. Sol.* (*a*) **1983**, *79*, 149.
[20] J. Hénoc, F. Maurice, *Electron Probe Quantitation*, Plenum, New York, 1991, pp. 105–143.
[21] H. Benhayoune, O. Jbara, C. Merlet, J. Cazaux, *Inst. Phys. Conf. Ser.* **1992**, *130*, 391.
[22] C. Merlet, *X-Ray Spectrometry* **1992**, *21*, 229.

Mikrochim. Acta [Suppl.] 13, 377–389 (1996)

Thickness Determination of Thin Insulating Layers

Peter Klein, Kurt Röhrbacher, Michael Andrae, and Johann Wernisch*

Institute for Applied and Technical Physics, Technical University of Vienna, Wiedner Hauptstrasse 8–10, A-1040 Wien, Austria

Abstract. Thin insulating films (Si_3N_4 on titanium-substrate) with various thicknesses (40–360 nm) are investigated with EPMA (electron probe micro analysis). A carbon- and, alternatively, gold-coating is applied to avoid charging phenomena and beam deflection.

The alteration of the shape of the depth distribution function due to the insulating property is considered by applying an electrostatic model (proposed by Cazaux) of a charged dielectric cylinder. A Monte-Carlo simulation is performed simulating the charging of the specimen by assuming a precharge and modification of the mean free path in the insulator. Experimentally derived k-ratios for Si_3N_4-films are verified quantitatively.

The applicability of the results to other insulating materials – especially with regard to different preparation-techniques – is discussed.

Key words: Si_3N_4, insulators, thin films, thickness determination, Monte-Carlo simulations.

Insulating materials have been a source of problems in electron probe microanalysis. Because of the charge build-up inside the non-conducting specimen several effects can be observed at the surface:

- beam deflection,
- charge build-up and deceleration of impinging electrons,
- even dielectric breakdown and destruction of the specimen.

Thin insulating layers are a common element in the architecture of integrated circuits and modern computer chips. For destruction-free analysis a microprobe can be applied. Commonly used EPMA-correction procedures for measured k-ratios (sample/standard) deal only with conducting samples, and do not consider any charging of the specimen.

Besides the problems concerning image disturbances, quantitative analysis is strongly affected: electrons, which form the probe current, are not conducted away

* To whom correspondence should be addressed

from the point of impact but remain there and build up a more or less strong field; the electric field originating from the first incoming electrons modifies the trajectories of the subsequently impinging electrons; the overall penetration behaviour is extremely disturbed. As a consequence, the depth distribution function of characteristic X-ray production $\Phi(\rho z)$ (the number of ionisations in a certain mass-depth normalised to a self-supporting infinitesimally thin layer under identical experimental conditions) is deformed. The basis of all thin-film correction procedures, however, is the exact knowledge of the shape of the depth distribution function.

Many papers have been written about observation conditions of insulating specimens [1, 2], methods to improve the quality of images [3] and theory of charging mechanisms [4–12], which cover a wide field of effects and corresponding explanations.

Ishibashi et al. [13] and Reimer et al. [14] performed a theoretical derivation of a threshold condition as a function of the angle of incidence (surface tilt) and the acceleration voltage concerning the observability of an insulator's surface due to negative charge build-up in a low-voltage SEM (scanning electron microscope).

Most of the publications dealing with the analysis of insulating specimens do not pay attention to possible distortions of the depth distribution function $\Phi(\rho z)$. Generally, it is claimed that by providing a conductive coating no further problems should occur.

Willich and Obertop [15] used an empirical correction of a gold coating thickness to obtain – by linear extrapolation – correct *k*-ratios for a-Si:C:N:O:H films; for evaluating the well-known PAP matrix correction model was used without any modifications.

Pouchou [16] mentioned a 'coating-option' in his Quantex + /XPP program that corrects absorption effects caused by conducting films of known thickness. The effect of a carbon coating of 47 nm on the absorption of O-Kα-radiation is up to 35% (4 kV acceleration energy). The general shape of the $\Phi(\rho z)$-function, however, is assumed to stay unchanged.

Colby [17] performed an analysis of several different types of insulating thin films including Si_3N_4; he claimed that in the case of a conductive substrate no coating is necessary especially when low accelerating potentials are used. Any charge build-up should result in a self-regulating system due to an increase in secondary electron emission. The films used for the analysis were prepared either pyrolytically by reaction of silane and ammonia or by DC plasma discharge from $SiBr_4$.

Calibration curves for the analysis of geological insulating samples have been published by Albee and Ray [18] and Bence and Albee [19].

Krause et al. [20] reported a new method for measuring the thickness of thin-film insulators (polymers) by determining experimentally a third (beyond the LVSEM range) cross-over energy and evaluating this quantity by a calibration curve (cross-over energy vs. film thickness). This cross-over energy should be indicated by no visible surface charging (resp. sign-reversal of the observed charge); the fields caused by trapped electrons acting inside the sample and their influence on X-ray production are not estimated.

Packwood et al. [21] just used the expression by Packwood and Brown [22] and a Gaussian concentration distribution for silicon to evaluate SiO_2-layers without applying any conductive coating. To simulate film-substrate combinations, a linear weighting of $\Phi(\rho z)$-curves was established, which might cover up distortions due to charging. Other authors (e.g. [23–27]) just mentioned the application of a conducting layer without considering implications with the exception of absorption effects.

Only few authors refer to the *possibility* of errors caused by the use of non appropriate shapes of $\Phi(\rho z)$. Brown [28], for example, expected – in case of an insulating specimen – the $\Phi(\rho z)$-curve to 'move closer to the surface' (p. 1418) but claimed that conventional quantification provides satisfying results.

Bastin and Heijligers [29] published an excellent review of phenomena and problems occuring in connection with non-conductive specimens in EPMA; they report measurements of Si_3N_4-specimens produced: a) by chemical vapor deposition, exhibiting the lowest conductivity, b) by arc-melting under nitrogen pressure, and finally, c) a hot isostatically pressed Si_3N_4 specimen, with gradually increasing conductivity. The difficulty of specifying the conductivity of a specimen – especially the influence of small fractions of impurities or beam-induced lattice defects – is pointed out and the possible distortion in the $\Phi(\rho z)$-curves is offered as an explanation for a rise of up to 15% above the expected k-ratios for various insulating specimens.

Le Gressus et al. [30] tried to combine theoretical models of the electric field caused by a trapped charge distribution with applied analysis of insulating specimens. They also assumed a 'skewing of the $\Phi(\rho z)$-curve towards the surface, leading to a reduction of the absorption coefficients' (p. 163) confirming the presumptions by Brown.

Fialin [31] pointed out the problems involved in the usage of an unchanged depth distribution function and performed an empirical correction of parameters used in the Philibert-Tixier ZAF-model; the obtained correction was tested on measurements of synthetic glasses and minerals. The measured k-ratios were reported to be generally higher than predicted by usual ZAF correction methods.

Some theoretical work has been done [7, 9, 10, 30, 32, 33, 34] to describe the characteristics of the internal field and potential caused by trapped electrons. The field and potential created by a uniformily charged cylinder characterised by its dielectric constant, ε_r is calculated for various specimen configurations. The presented explanatory model, however, cannot give any applicable quantifications on the forces acting inside the sample.

Experimental

Two series of specimens were available: Si_3N_4-layers (in steps of 40 nm) sputtered on a titanium-substrate (HF-magnetron-sputtering) and 6 Si_3N_4-films prepared by plasma enhanced chemical vapor deposition (PECVD) on titanium. The accuracy of the given layer-thicknesses was checked by a Dektak profilometer. To avoid surface charging effects in the SEM and to restore the short wavelength cut-off [29], different conductive coatings were applied: 25 nm carbon- and alternatively 4 nm gold-films. As a standard pure silicon -coated and uncoated- and pure titanium was used. The EDS-spectra of the sputtered layers revealed a less than 5 weight-%-fraction of argon (the carrier gas argon is used for the sputtering process).

The order of magnitude of the resistivity of the mentioned specimens is reported to be $> 10^9\,\Omega$cm (PECVD) and $> 10^7\,\Omega$cm (HF-magnetron sputtering) respectively. The order of magnitude of the relative dielectric constant ε_r is about less than 7 depending on the process parameters. The density of the PECVD-prepared specimens is between 2.2 g/cm^3 and 2.35 g/cm^3 (a mean value of 2.3 g/cm^3

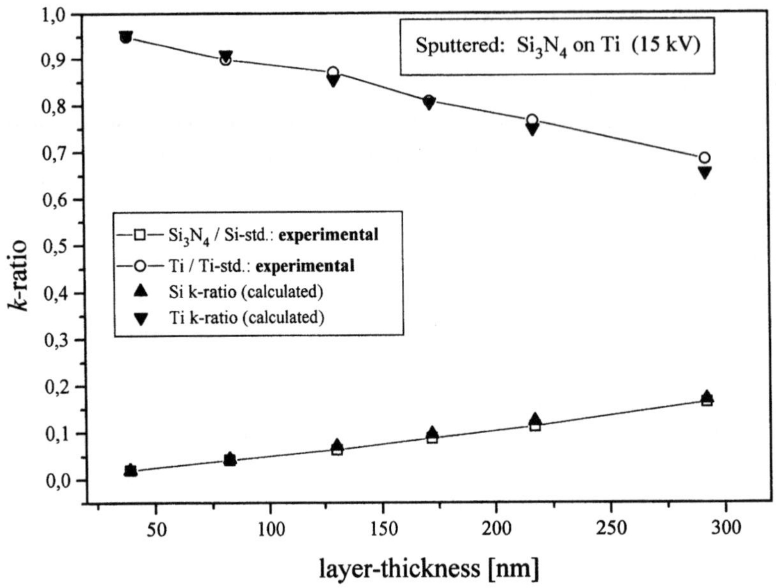

Fig. 1. Comparison of the measured and conventionally calculated (thin-film program [35, 36]) k-ratios for Si_3N_4-layers on titanium prepared by *sputtering*. ($E_0 = 15\,\text{keV}$)

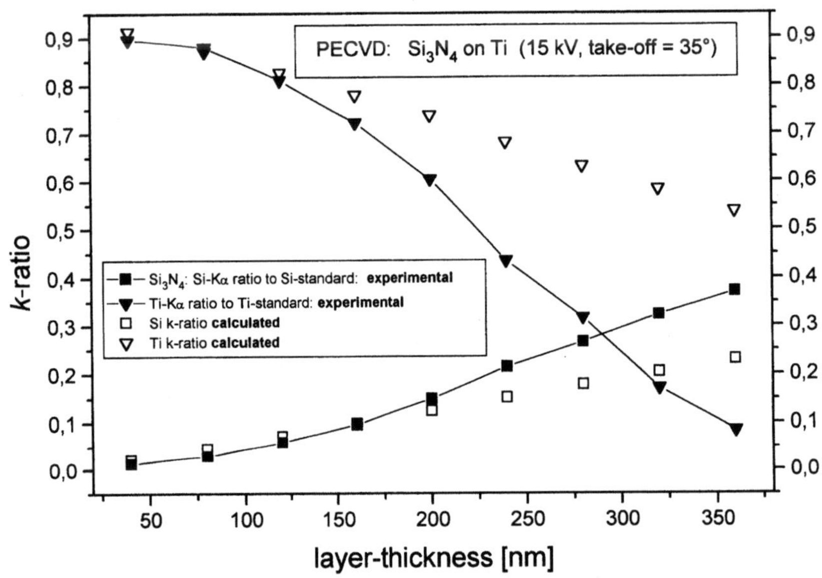

Fig. 2. Comparison of the measured and conventionally calculated (thin-film program [35, 36]) k-ratios for Si_3N_4-layers on titanium prepared by *PECVD*. ($E_0 = 15\,\text{keV}$)

was taken for the calculations) whereas the sputtered layers exhibit a slightly higher density of $2.4\,\text{g/cm}^3$.

Conventional evaluation of the measured k-ratios for the *sputtered* layers – either of silicon or of titanium – with an established thin-film-analysis-programme ([35], [36]) provides excellent results within usual error-rates (Fig. 1).

The specimens prepared by PECVD show a significant difference especially for higher thicknesses: starting at about 160 nm, the measured k-ratios show a dramatic increase in comparison with the

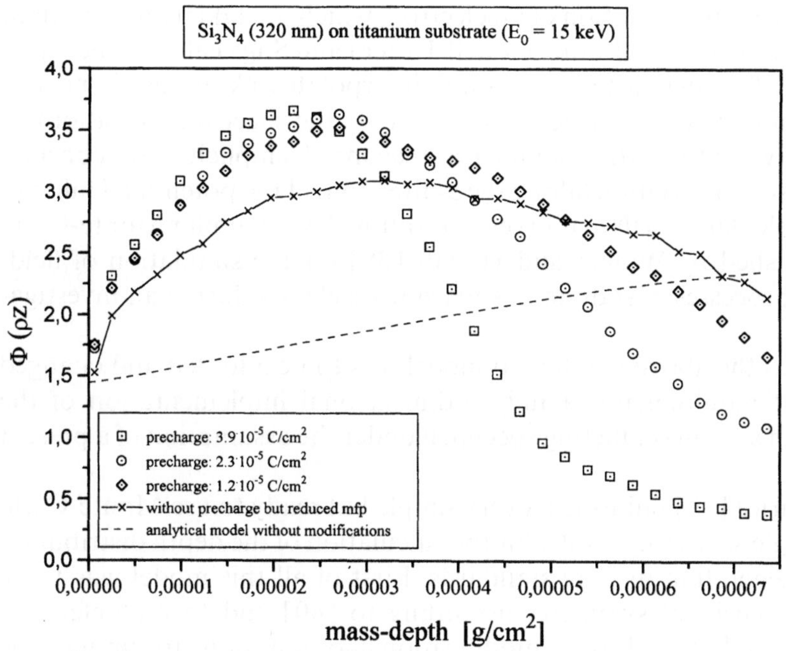

Fig. 3. Depth distribution function for a 320 nm Si_3N_4-layer on titanium substrate considering various precharge-doses in comparison to the analytical model by August and Wernisch [35]. Acceleration voltage: 15 kV

sputtered layers and with calculated values. At 15 kV accelerating voltage the effect is impressive; for 10 kV the effect is less pronounced but still observable. The *k*-ratios for the substrate material titanium show an equivalent decrease until the titanium signal nearly vanishes (Fig. 2).

The observed discrepancy may be qualitatively explained by the build-up of a strong internal field, which should cause a distortion of the depth-distribution function giving values up to 50% higher for the counting rate of the detected element.

The electrical conductivity of insulating layers strongly depends on the microstructure and therefore on the preparation process: while PECVD guarantees a perfect amorphous structure of thin films due to the low energy of the depositing atoms, sputtered layers show a polycrystalline structure, which obviously results in a significant increase of the conductivity especially for currents of the order $6 \cdot 10^{-11}$ A as were used for the measurements. Therefore the sputtered, conductive Si_3N_4-films, which are of nearly identical chemical composition as the PECVD-prepared ones, allows the direct comparison of one and the same material in an obviously conductive and an insulating appearance.

Since no existing correction procedure can explain these measurements, Monte-Carlo-simulations are the only way to gain information about the shape of the depth distribution function inside an irridated insulating thin film.

Monte-Carlo Simulation of Insulating Layers on Conducting Substrate

Kotera and Suga [37] published an interesting model for a Monte-Carlo approach taking internal fields and influence on the electron path into account. They assume a 'frozen-in charge' or precharge to form a constant internal field. This charge distribution formed by the first implanted electrons causes – via an electric field and a potential – influences on the motion of an electron, which are taken into acccount in every trajectory step (single scattering model). Kotera and Suga calculate field and potential for 40*40*20 virtual lattice points and interpolate (taking the six nearest lattice points neighbouring the lattice point next to the electron's position) 3-dimensionally in between the lattice points. For reasons of simplicity they consider only the one-dimensional variation (along the z-direction) of the potential. Unfortunately they do not calculate any $\Phi(\rho z)$ curves or k-ratios. A very similar Monte-Carlo approach was published by Walker and Howitt [38] for the simulation of field-induced migration processes of sodium in soda silicate glasses during an investigation in a SEM.

For our purposes the above mentioned model was to be adopted and changed due to the different requirements: a fully 3-dimensional implementation of the electrostatic forces in a non-conducting specimen under the assumption of a present precharge.

Hénoc and Maurice [39] published a very simple but ready to use Monte-Carlo code adopting a single scattering model for the calculation of the depth distribution function of X-ray generation in pure standards. First of all this model was to be extended to a 3-dimensional symmetry according to [40] and to multielement-specimens according to [41]. A further modification had to be done to simulate the film-substrate-interface especially with regard to the different pathlengths. A comparison of the performance of the modified Monte-Carlo programme with an approved analytical thin film model [35] shows good results.

Precharge

The simulation of the insulating property of a thin film is obtained by a frozen-in charge as proposed by Kotera and Suga [37] or Walker and Howitt [38] and a modification of the mean free path. Therefore the coordinates of the final position – assuming that an electron is trapped when it has come to rest in the sample [38] – reached by the first thousand electrons within the insulating layer are stored in a temporary file. After 'charging' the specimen in the above mentioned way, the actual simulation is performed. The electric field and potential created by the present frozen-in charge distribution is shown in Fig. 4; due to the boundary conditions the potential vanishes at the metal-insulator interface. To maintain the correct field- and potential-distribution, image charges have to be assumed; the theoretical background depending on different boundary conditions is given in [30] and [32].

The motion of an electron (mass m, charge q, z-direction) in a uniform external electric field (E_z) can be described one-dimensionally (z-component) as follows:

$$m\ddot{z} = q \cdot E_z$$

$$z = \frac{1}{2} \cdot \frac{q}{m} \cdot E_z \cdot t^2 + \{v_{0z} \cdot t + z_0\} \tag{1}$$

The unknown time t (defined as the time between two scattering events) can be

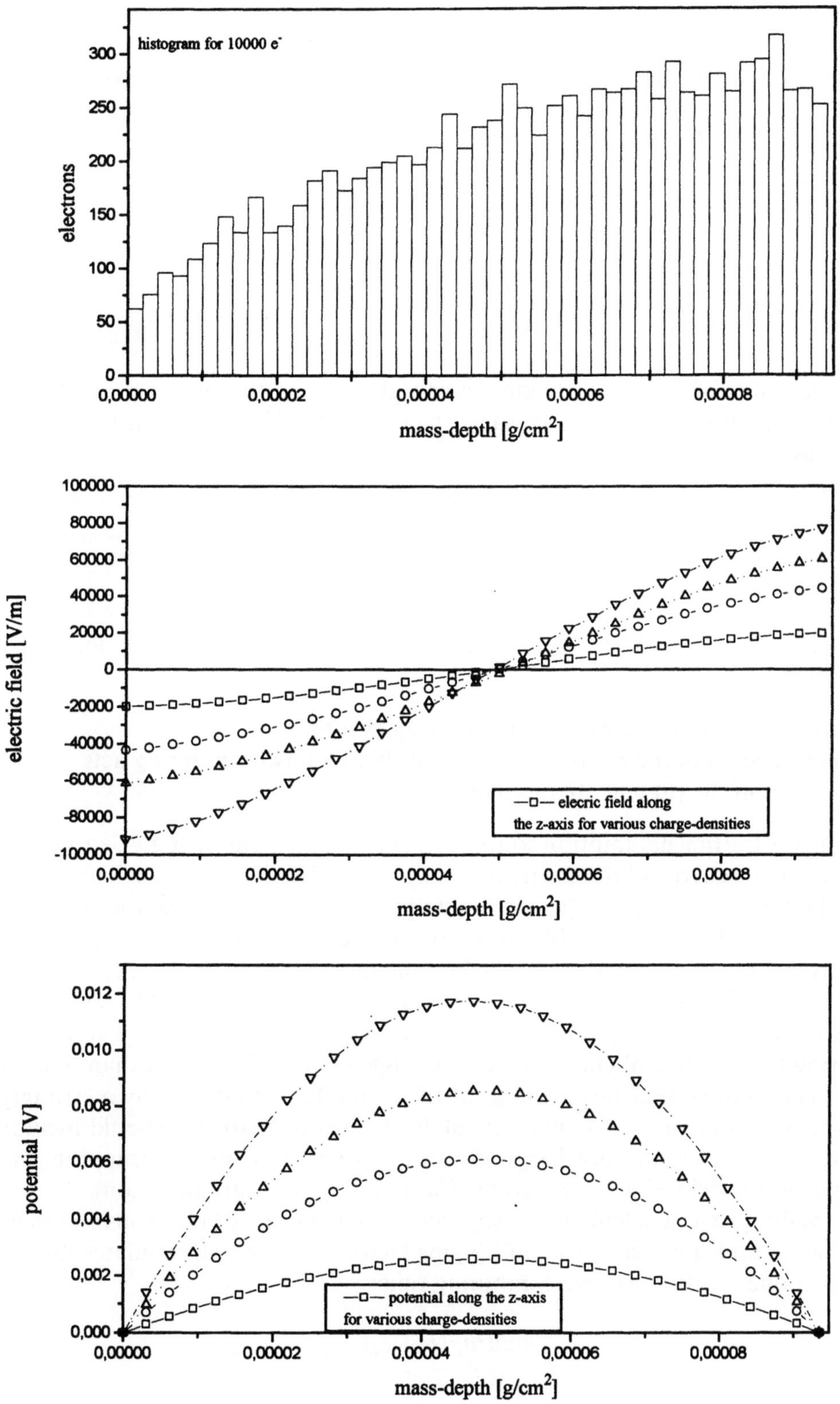

Fig. 4. Results of the Monte Carlo simulation: the potential and field created by the trapped electrons in the insulating Si_3N_4-layer (covered by a conductive surface layer; on a conductive substrate) and a histogram showing the distribution of trapped electrons, which are assumed to create the stationary internal field

calculated from the actual free path λ and the kinetic energy of the electron:

$$v \cdot t = \text{free path } \lambda \text{ between two scattering events}$$
$$E_{kin} = \frac{mv^2}{2} \rightarrow v = \sqrt{\frac{2E_{kin}}{m}}$$
$$\Rightarrow t = \frac{\lambda}{\sqrt{\frac{2E_{kin}}{m}}} \tag{2}$$

Within the range of 5–20 keV of kinetic energy (E_{kin}) of the observed electrons, a relativistic treatment is not necessary: $(v/c)^2 \leqslant 0.04$. Therefore Eq. (1) can be rewritten as

$$z = \frac{1}{2} \cdot \frac{q}{m} \cdot E_z \cdot \frac{\lambda^2}{\frac{2E_{kin}}{m}} = \frac{q}{4} \cdot \frac{E_z}{E_{kin}} \cdot \lambda^2 \tag{3}$$

Similar equations are derived for x and y coordinates. Eq. (3) is valid under following assumptions:

– the field E_i is constant between two scattering events,
– the kinetic energy of the electrons is constant between two scattering acts,
– no dynamics in the present charge-pattern.

The validity of the first assumption concerning the internal field strongly depends on the actual kinetic energy of the electron, and on the free path, which is calculated by the use of random numbers. At the maximum position of the potential the variation of the electric field is not negligible, even for relatively short paths (see Fig. 4); this induces an error, which should, however, be of no major influence on the calculation of the depth distribution function. The second assumption with regard to the energy of the electron – neglecting inelastic scattering for this purpose – is justified by considering the amount of energy loss at a single step, which is less than 5% on average. The third supposition – the basis of the model – is admittedly a strongly simplifying view of charge dynamics in solids. A correct treatment should include a time-dependence of the precharge electron distribution due to trapping *and* detrapping in the field of other electrons (Cazaux private communication).

The resulting field is calculated by superposition of all stored electrons and their image charges and enhanced by a multiplying factor (*mult*), which simulates several electrons at one position to save computing time:

$$\vec{E}(\vec{r}) = \sum_i \frac{mult \cdot q_i \cdot (\vec{r} - \vec{r}_i)}{|(\vec{r} - \vec{r}_i)|^3} = \sum_i \vec{E}_i \tag{4}$$

The influence of the internal field on the kinetic energy of an electron is regarded in the same way as mentioned by Kotera and Suga [37]:

$$E_2 = E_1 - \Delta E - (U_2 - U_1) \cdot q \tag{5}$$

where U_1 denotes the potential at the beginning of the step of interest, U_2 at the end of it and ΔE the energy loss due to inelastic scattering in between. E_1 and E_2 are the kinetic energy of the electron [keV] before and after having travelled the pathlength. The potential U is again calculated using the superposition of point charges and their corresponding images. To reach consistent *SI*-units the electric field and potential has to be multiplied by the factor

$$k = \frac{1}{4\pi\varepsilon_0\varepsilon_r} = \frac{1}{4\pi \cdot 8.854 \times 10^{-12} \cdot \varepsilon_r \dfrac{C}{Vm}} \tag{6}$$

Including Eqs. (1) to (6) in the Monte-Carlo program means a summation over all stored electrons and their image charges at each scattering event which is of enormous effect on the computing time even on modern workstations. In order to perform a simulation to obtain a reliable depth-distribution a minimum of at least $1 \cdot 10^4$ electron trajectories is required to reach acceptable statistics on random numbers.

Mean Elastic Free Path

The mean elastic free path Λ_{el} is usually calculated by (see for example [39])

$$\Lambda_{\mathrm{el}}^{-1} = \sigma_{\mathrm{tot}}^{\mathrm{el}} = \sum_i \rho N_A c_i \sigma_i^{\mathrm{el}}/A_i \quad \text{with} \quad \sigma_i^{\mathrm{el}} = \left(\frac{Z_i e^2}{2E}\right)^2 \cdot \frac{\pi}{\alpha_i \cdot (1+\alpha_i)} \tag{7}$$

Λ_{el}	mean elastic free path	ρ	density [g/cm^3]
$\sigma_{\mathrm{tot}}^{\mathrm{el}}$	total elastic scattering cross section	N_A	Avogadro's number
c_i	concentration [weight-%]	Z_i	atomic number of element i
α_i	screening parameter $\alpha_i = (h^2/8\mathrm{mEa}_0^2) \cdot Z_i^{2/3}$	E	energy [eV]
		A_i	atomic weight of element i

The mean free path λ used in the Monte-Carlo simulation is derived by $\lambda = -\Lambda_{el} \cdot \ln(\mathrm{R})$ with a random number $R \in [0, 1]$.

Classical textbooks of solid state physics (e.g. [42]) give a derivation for the mean free path of conduction electrons in *conductive* materials:

$$\textit{mean free path} = v \cdot \tau \text{ with relaxation time } \tau = \frac{m}{\rho \cdot n \cdot e^2} \tag{8}$$

ρ resistivity
n conduction electron density

under certain assumptions (see [42] pp. 8–9). In the case of an insulating specimen, the mean free path would be – according to Eq. (8) – indefinite: ρ tends to infinity whereas n drops to zero. The resistivity depends on various parameters beyond experimental control such as impurities, lattice defects, crystallographic structure, size of crystals, preparation conditions, storage conditions – just to mention a few. This means the resistivity of each specimen (especially of highly insulating materials with $\rho \geqslant 10^9\ \Omega\mathrm{cm}$) can differ by some orders of magnitude.

The density of conduction electrons for insulating materials is theoretically zero (for an ideal insulator); in EPMA one can assume that electrons forming the probe current behave as hot electrons in the conduction band [33].

The mean free path of electrons in amorphous insulating specimens is hardly an investigated quantity; most classical textbooks on solid state physics disregard the different scattering probability. Kittel [43] just points out that in amorphous insulators no band-structure, Bloch-waves or any periodic structure can be assumed. Furthermore he states that in the case of amorphous semiconductors a dramatic increase in scattering events occurs and that the mean free path decreases to the order of dimension of the microstructure (some tenth of a nm).

Results

Monte-Carlo simulations of insulating Si_3N_4-layers on titanium were performed considering various precharges and mean free electron paths. The results of the simulations can be seen in Figs. 3 and 4. The cited assumptions concerning the maximum position of the depth-distribution function ('moving closer to the surface') are confirmed. The quantification of the observed effects still remains problematic. The calculated k-ratios

$$k = \frac{\int_0^{\rho z_F} \Phi_F(\rho z)\cdot\exp(-\chi\cdot\rho z)\cdot d\rho z}{\int_0^{\infty} \Phi_{St}(\rho z)\cdot\exp(-\chi\cdot\rho z)\cdot d\rho z} \tag{9}$$

using the Monte-Carlo results for $\Phi_F(\rho z)$ depend strongly on the assumed precharge and on the mean free path. Figure 3 shows the depth-distribution function for a 320 nm Si_3N_4 layer under several conditions. The internal field and potential is given in Fig. 4; the theoretical predictions by Le Gressus et al. [30] and Cazaux [32] are confirmed qualitatively.

The first impinging electrons are assumed to form the precharge due to trapping by defects (an insulator without defects would not be charged [32]) until a stationary regime is reached; taking the current of $6\cdot 10^{-11}$ A we get a flow of $\approx 3\cdot 10^8$ electrons per second (reduced by a fraction of 20% as backscattered electrons). Assuming a precharge of $1.2\cdot 10^{-5}\ \frac{C}{cm^2}$ (once adjusted and kept constant for all other specimens!) and (empirically) a mean free path 25% in comparison to conductive samples the measured k-ratio can be calculated.

Figure 5a and 5b show a comparison between the calculations carried out and the experimental findings. The Monte-Carlo simulated k-ratios for silicon (Fig. 5b) are too low although they show an improvement in comparison to the analytical model [36]. Figure 5a shows a comparison of the titanium k-ratios; in this case a generally differing tendency can be observed. One reason for this inconsistency is the generally worse performance of the Monte-Carlo programme for the simulation of covered substrates. Another reason is the incapability of the Monte-Carlo model to describe the transition from insulating layers (> 120 nm) to sufficiently conducting thin films. This deficiency is much more pronounced for the titanium k-ratios as for the silicon values (note the vertical scale!).

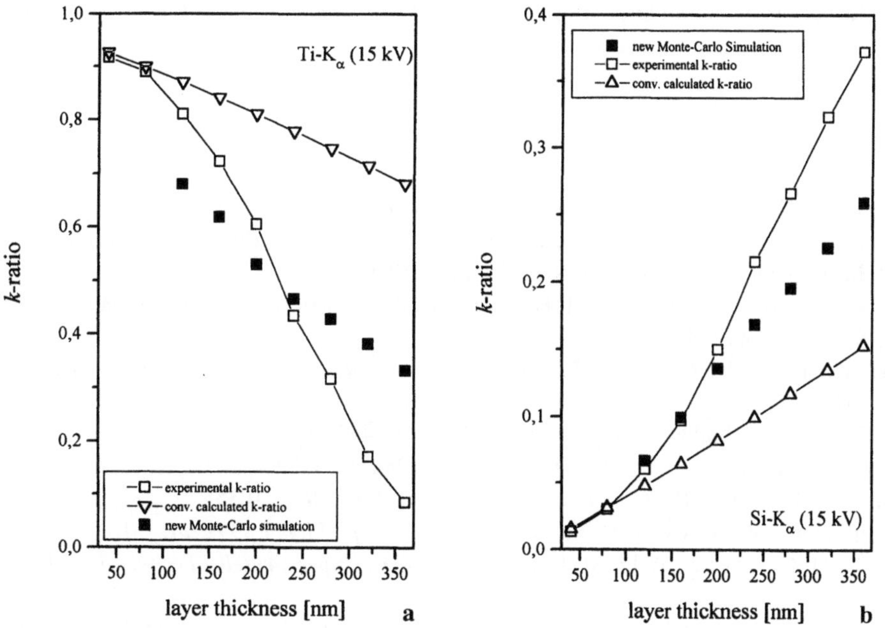

Fig. 5. Comparison of the k-ratios derived by experiment, Monte Carlo simulation and the analytical model [35, 36] for titanium (**a**) and silicon (**b**). Acceleration voltage: 15 kV. Precharge: $1.2 \cdot 10^{-5} \cdot C/\mathrm{cm}^2$

The present model does not claim to describe electron dynamics in insulating specimens; the qualitative and quantitative findings, however, should mobilize even more research efforts to develop realistic explanatory models concerning the influence of an internal charge build-up on the depth-distribution function in order to guarantee a correct evaluation and a valid interpretation of experimental data.

Conclusion

The alteration of the shape of the depth-distribution function in insulating specimens and the errors, which may occur, are commonly neglected. A conductive coating is used to restore conductivity of the surface of the investigated specimen indicated by no visible charging effects; internal fields caused by trapped electrons are usually neglected. Measured k-ratios, however, indicate a major influence of the non-conductive property on the results achieved. By the use of Monte-Carlo simulations the changes concerning the depth-distribution function are explained. Qualitative assumptions from literature can be conformed; quantitative predictions, however, depend strongly on the actual physical properties (conductivity, dielectric constant, density, way of preparation) of the investigated specimen. A valid simulation model, which could explain the effects quantitatively, has not been found so far.

Acknowledgement. We acknowledge the support given by the Austrian "Fonds zur Förderung der wissenschaftlichen Forschung" (Project Nr. P10415-PHY). Additionally we would like to thank

Univ. -Prof. Dr. R. Chabicovsky and Dipl. -Ing. H. Schalko (*Institut für Allgemeine Elektrotechnik und Elektronik*) for the preparation of the specimens.

References

[1] T. J. Shaffner, J. W. S. Hearle, *Scanning Electr. Microsc.* **1976**, *Part I*, 61.
[2] T. Ichinokawa, M. Iiyama, A. Onoguchi, T. Kobayashi, *Jap. J. Appl. Phys.* **1974**, *13(8)*, 1272.
[3] P. Morin, M. Pitaval, E. Vicario, *J. Phys. E. Sci. Instr.* **1976**, *9*, 1017.
[4] S. Matsuoka, H. Sunaga, R. Tanaka, M. Hagiwara, K. Araki, *Transactions on Nuclear Science, Vol. NS-23(5)*, 1976, pp. 1447–1452.
[5] R. Tanaka, H. Sunaga, N. Tamura, *IEEE Transactions on Nuclear Science, Vol. NS-26(4)*, 1979, pp. 4670–4675.
[6] B. Gross, J. Dow, S. V. Nablo, *J. Appl. Phys.* **1973**, *44(6)*, 2459.
[7] D. M. Taylor, A. A. Al-Jassar, *J. Phys. D: Appl. Phys.* **1984**, *17*, 1493.
[8] G. H. Curtis, R. P. Ferrier, *Br. J. Appl. Phys.* (*J. Phys. D*) **1969**, *2(2)*, 1035.
[9] V. I. Arkhipov, A. I. Rudenko, G. M. Sessler, *J. Phys. D: Appl. Phys 1* **1991**, *24*, 731.
[10] H.-J. Fitting, H. Glaefke, W. Wild, M. Franke, W. Müller, *Experimentelle Technik der Physik*, **1979**, *27/1*, 13.
[11] J. P. Vigouroux, J. P. Duraud, A. Le Moel, C. Le Gressus, D. L. Griscom, *J. Appl. Phys.* **1985**, *57(12)*, 5139.
[12] D. A. Berkley, *J. Appl. Phys.* **1979**, *50(5)*, 3447.
[13] Y. Ishibashi, T. Kodama, H. Oiwa, Y. Uchikawa, *Scanning* **1992**, *14*, 219.
[14] L. Reimer, U. Golla, R. Böngeler, M. Kässens, B. Schindler, R. Senkel, *Optik 92*, **1992**, *1*, 14.
[15] P. Willich, D. Obertop, *Mikrochim. Acta* **1989**, *II*, 233.
[16] J.-L. Pouchou, *Mikrochim. Acta* **1994**, *114/115*, 33.
[17] J. W. Colby, *Adv. X-Ray Analysis* **1968**, *11*, 287.
[18] A. L. Albee, L. Ray, *Anal. Chem.* **1970**, *12*, 1408.
[19] A. E. Bence, A. L. Albee, *J. Geology* **1968**, *76*, 382.
[20] S. J. Krause, J. Mohr, G. H. Bernstein, D. K. Ferry, D. C. Joy, *Microbeam Analysis-1989* **1989**, 459.
[21] R. H. Packwood, G. Remond, P. H. Holloway, *Surface Interface Anal.* **1988**, *11*, 127.
[22] R. H. Packwood, J. D. Brown, *X-Ray Spectrometry*, **1981**, *10*, 138.
[23] E. Kohlhaas, F. Scheiding, *Arch. Eisenhüttenwes.* **1970**, *41/1*, 97.
[24] W. Weisweiler, *Arch. Eisenhüttenwes.* **1974**, *45/5*, 287.
[25] W. Weisweiler, *Arch. Eisenhüttenwes* **1978**, *49/12*, 555.
[26] J. Graham, C. R. M. Butt, R. B. W. Vigers, *X-Ray Spectrometry* **1984**, *13(3)*, 26.
[27] J. M. Dijkstra, G. F. Bastin, H. J. M. Heijligers, N. Ammann, P. Karduck, *Beitr. Elektronenmikrosk. Direktabb. Oberfl.* **1992**, *25*, 141.
[28] J. D. Brown, *Spectrochim. Acta* **1983**, *38B 11/12*, 1411.
[29] G. F. Bastin, H. J. M. Heijligers, in: *Electron Probe Quantitation*, Plenum, New York, 1991, pp. 163–175.
[30] C. Le Gressus, F. Valin, M. Gautier, J. P. Duraud, J. Cazaux, H. Okuzumi, *Proc. of the 1st European Workshop in Modern Developments and Application in Microbeam Analysis*, EMAS and Univ. of Antwerp, 1989, pp. 151–180.
[31] M. Fialin, *X-Ray Spectrometry* **1988**, *17*, 103.
[32] J. Cazaux, *J. Appl. Phys.* **1986**, *59(5)*, 1418.
[33] J. Cazaux, C. Le Gressus, *Scanning Microscopy* **1991**, *5*, 17.
[34] J. Cazaux, *J. Microsc. Spectrosc. Electron* **1986**, *11*, 293.
[35] H. J. August, J. Wernisch, *X-Ray Spectrometry* **1991**, *20*, 131.
[36] H. J. August, J. Wernisch, *X-Ray Spectrometry* **1991**, *20*, 141.

[37] M. Kotera, H. Suga, *J. Appl. Phys.* **1988**, *63*(*2*), 261.
[38] T. M. Walker, D. G. Howitt, *Scanning* **1989**, *11*, 5.
[39] J. Henoc, F. Maurice, *Electron Probe Quantitation*, Plenum, New York, 1991, pp. 105–143.
[40] D. F. Kyser, K. Murata, *IBM J. Res. Develop.* **1974**, *18*(*4*), 352.
[41] R. L. Myklebust, D. E. Newbury, H. Yakowitz, *NBS Speical Publication 460*, *Proc. of a Workshop on the Use of Monte Carlo Calculations in EPMA and SEM*, 1976, pp. 105–128.
[42] N. W. Ashcroft, N. D. Mermin, *Solid State Physics*, Saunders, 1976.
[43] C. Kittel, *Introduction to Solid State Physics*, *6th Ed.*, Wiley New York, 1989.
[44] J. L. Pouchou, F. Pichoir, *Electron Probe Quantitation*, Plenum, New York, 1991, pp. 31–75.

Mikrochim. Acta [Suppl.] 13, 391–398 (1996)

High Energy and Angular Resolution Dynamic Secondary Ion Mass Spectrometry

Alexandr A. Kosyachkov

Department of Metal Surface Physics, Institute of Metal Physics, Ukrainian National Academy of Sciences, 36 Vernadskogo, Kiev 252680, Ukraine

Abstract. Energy and angular spectra of secondary ions ejected from solid surfaces under a few-keV ion beam bombardment are discussed with regard to application for studying the physical processes involved in sputtering and secondary ion emission.

Key words: ion sputtering, secondary ion emission, atomic collisions, surface analysis.

Study of energy and angular regularities of secondary ion emission (SIE) caused by ion beam bombardment of solid surfaces favors both a development of the surface diagnostics by secondary ion mass spectrometry (SIMS) and a better understanding of sputtering [1–4] and ionization [5, 6] mechanisms. The yield of energy-angular- and mass-separated secondary ions is small. However, an appropriate sensitivity to the surface and adsorbate states can be achieved by low fluence static SIMS [6]. Dynamic SIMS allows higher energy and angular resolutions which are preferable when studying the physical processes involved in sputtering and SIE phenomena: penetration of bombarding ions into solids, atomic collision cascades, ejection of surface atoms, their ionization and movement in the near-surface space.

Such investigations of the ionization process have been discussed in detail in several reviews [5] and broadly recognized since the dependence of the ionization probability of sputtered atoms on their velocity is a basic assumption of various ionization models. Some peculiarities of the other processes are briefly described in the present work on the basis of our original results as well as of those previously published, but in not completely available Ukrainian and Russian journals.

Instruments

A few ultra-high vacuum spectrometers with energy-, angular- and mass-resolutions have been specially designed to provide a free choice of the directions of both incident and ejected ions by independent rotations of the bombarded sample and analyzing system [7–9]. Figure 1 schematically represents this system [9], appropriate for analysis of charged particles, ions and electrons, either ejected or scattered by solid surfaces. In the case of SIMS, mass separated ions with energies 1–10 keV

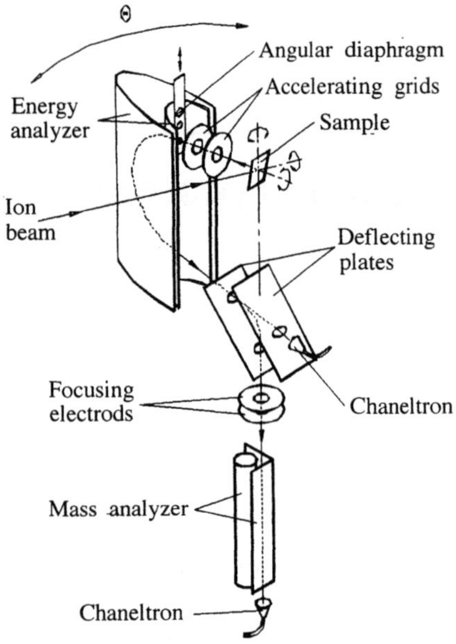

Fig. 1. Schematic representation of an analyzing system with energy-, angular-and mass-resolutions

bombard a sample which can be rotated about three axes. The secondary ions ejected in the direction of the rotatable analyzing system move in a field-free space, undergo an acceleration between the spherical grids and penetrate into the energy analyzer through an appropriate angular diaphragm. Being angular- and energy-analyzed, the ions pass the deflecting plates to the upper channeltron or, after deflection and additional focusing, to the mass separator. The system rotation axis coincides with the mass separator and the latter is fixed relative to the vacuum chamber. In other instruments [7, 8], the energy and mass analyzers are mechanically coupled and rotated together. The energy-angular- and mass-resolutions of the system shown in Fig. 1 are high, 0.05–1.0 eV, 1–6° and 1 a.m.u. respectively.

Applications

Ion Channeling in Single Crystals

Penetration of bombarding ions into single crystals strongly depends on the beam orientation relative to the transparent directions of the crystal lattice. Existing theories cannot properly describe the channeling process for the range of practically important incident energies, below 10 keV [3]. Being sensitive to the primary energy deposited near the surface, secondary ion and sputtering yields are suitable for experimental investigation of a few-keV ion penetration [2–4], in particular, of individual contributions of axial and planar channels to the ion channeling process [10, 11]. Figure 2 illustrates the influence of the Ar^+ bombarding direction on the SIE and sputtering yields from Nb(111) surfaces. The minima in the azimuthal dependencies correspond to deeper penetration of the primary ions into the crystal. Certain channels or their combinations, marked in the insets, result in different decreases of the ejection yields. Comparative measurements of the yield decrease at different bombarding directions allow estimates of the individual contributions and show that the channeling along the $\{110\}$ planes dominate over axial $\langle 100 \rangle$ and $\langle 111 \rangle$ channeling in bcc Nb crystals [11]. Marlowe simulations [10] indicate that

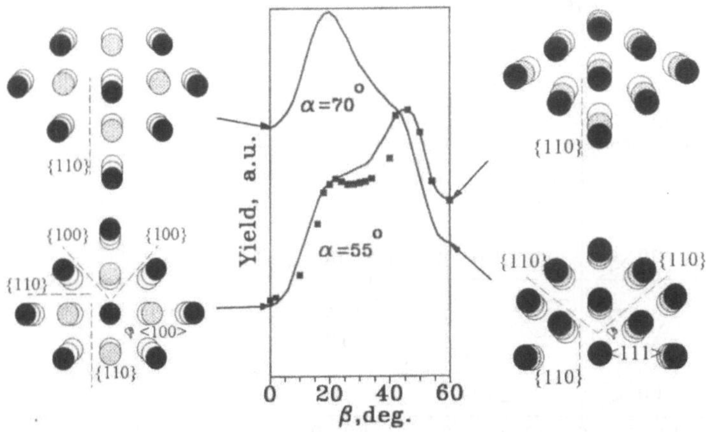

Fig. 2. Measured SIE (solid lines) and calculated for sputtering (squares) yields along the normal to the Nb(111) surface versus azimuth (β) and polar (α) angles of 5 keV-Ar^+ ion bombardment. Top views of the surface are shown in the insets together with the Miller indices of axial (symbols) and planar (dotted lines) chennels

penetration of Ar^+ ions into fcc Cu crystals also occurs along the most transparent $\langle 110 \rangle$ and $\langle 100 \rangle$ axial channels as well as due to $\{111\}$ planar channeling.

Another indication of high sensitivity to ion penetration depth is observed as drastic variations of secondary ion yield as a function of orientation of both polar and axial channels relative to the crystal surface, for example, more than a factor of 2 for the $\{110\}$ plane [11]. Moreover, the influence of the channel slope on the penetration depth appears to be comparable with that of the channel transparency.

Anisotropic Collision Cascades

Energy and angular distributions of sputtered particles are strongly influenced by atomic collision cascades and can be used to study collision processes [2–4]. The dependence of the angular distribution on the ejection energy is of considerable interest for such investigations because it indicates partial preservation of the incident momentum by the recoil flux [12] and, therefore, is suitable for revealing anisotropic contribution to sputtering and SIE.

Figure 3 illustrates the anistropy caused by variations of the surface structure, composition and bombarding conditions. Angular distributions depend weakly on the emission energy at normal bombardment of amorphous silicon (Fig. 3a) whose atoms are mainly sputtered in 3–8 cascade collisions [13]. They become asymmetric at $\alpha = 60°$ and the maximum shifts to larger polar angles with increasing ejection energy as shown in Fig. 3b. The energy dependent shift indicates considerable preservation of the incident momentum in the near surface cascades which become shorter in this case, 2–5 collisions [13]. Similar energy dependent shifts are observed for polycrystalline metal surfaces and also attributed to the incident momentum preservation in insufficiently developed collision cascades [15, 16]. This anisotropy is particularly noticeable for multicomponent materials. Considerable mass

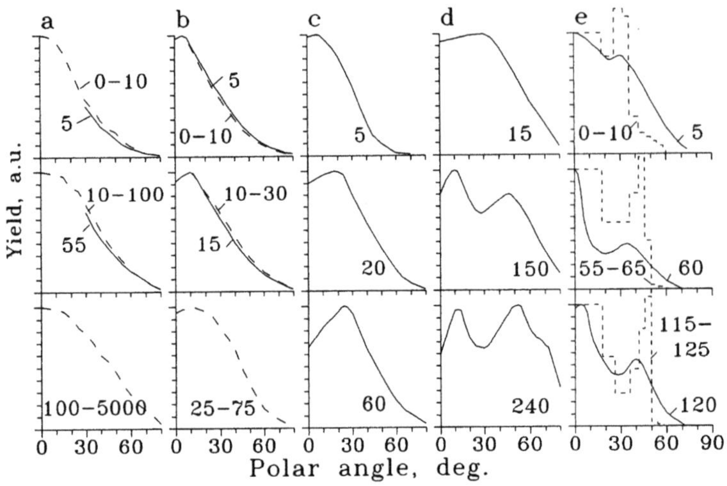

Fig. 3. Angular distributions measured for SIE (solid lines) and calculated for sputtering (dotted lines) at different ejection energies (marked in eV) under 5–6 keV Ar^+ ion bombardment at normal (**a**) and 60° (**b**–**e**) incident angles: Si^+ from amorphous silicon (**a** and **b**), Ba^+ from $YBa_2Cu_3O_{7-x}$ (**c**), Nb^+ from Nb(110) (**d**) and Al^+ from Al(100) (**e**)

differences of components, for example, of polycrystalline Y-Ba-Cu-O ceramics provide rapid dissipation of the projectile energy in the surface, with an extremely anisotropic near surface flux with short collision chains, 2 collisions before sputtering [13], and, as a consequence, both mass and energy dependent angular distributions partially shown in Fig. 3c.

The recoil flux in single crystals has similar features with that in random surfaces which can also be revealed in the energy-angular investigations. Figure 3d and e illustrate shifts of the background together with the maxima superimposed on the background of preferential emission along low index crystal axes (the Wehner spots [17]) under off-normal bombardment of the fcc Al(100) [14] and bcc Nb(110) and single crystal surfaces. Again, short collision chains dominate under these bombarding conditions [11] which favors the Wehner spot formation in a few scattering events by the surface ejection mechanisms [2, 4]. Similar energy dependent shifts were observed at off-normal bombardment of different grains of Mo [18] and Nb single crystals by Ar^+ ions.

The Surface Binding Energy

Following Thompson [19], a maximum in the energy distribution of sputtered atoms occurs at the energy $E_m = E_b/2$, where E_b is the surface binding energy. There were a number of successful attempts to evaluate the binding energies from energy distributions of neutrals sputtered from clean and oxidized suurfaces as discussed in ref. [20]. A significant correlation is also observed in SIMS experiments between E_m and the heat of atomization of oxides [21] and carbides [22].

Surface chemical reactions change both the surface collision cascades and binding forces. However, a qualitative correlation seems to exist between different

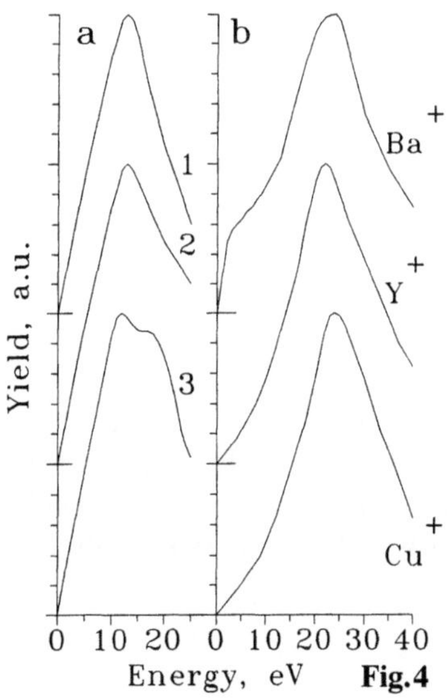

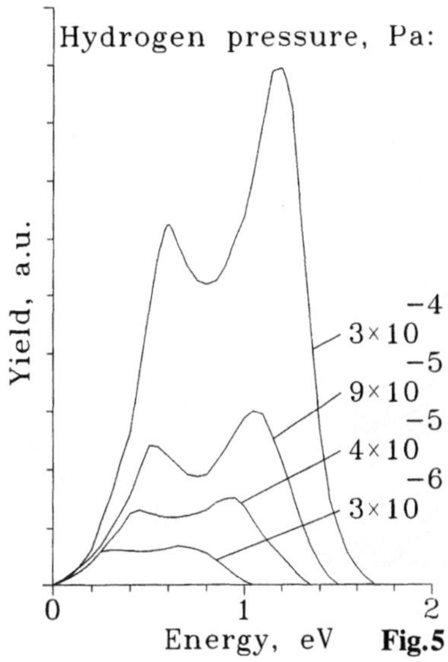

Fig. 4. Energy distributions of secondary ions at 5 keV Ar^+ ion bombardment of different surfaces: (**a**) Mo^+ from clean (*1*), oxygen exposed at 4×10^{-4} Pa (*2*) and oxidized (*3*) Mo(100); (**b**) different ions from $YBa_2Cu_3O_{7-x}$

Fig. 5. Energy distributions of Ar^+ secondary ions from Mo(100) surface bombarded by 6 keV Ar^+ ions at different hydrogen pressures

energy states of atoms in oxygen deficient and saturated Me—O bonds in oxidized surfaces and two-peak energy distributions of SIE from these surfaces. Figure 4 illustrates formation of such a distribution for the Mo(100) surface. The energy distribution of the Mo^+ secondary ions becomes broader under oxygen adsorption as compared with the clean metal surface. After oxidation of the surface, depth profiling of the oxide/substrate interface reveals a two-peak energy structure. The high energy peak can be attributed to SIE from saturated Mo—O bonds. Similar, but not so well, energy resolved flat distributions were observed for neutral Mo atoms sputtered from oxygen covered polycrystalline molybdenum surfaces [23].

The energy spectra of secondary ions ejected from bulk oxides of Y-Ba-Cu-O are shown in Fig. 5. Two-peaks are inherent to the Ba component of the ceramic only [24]. The peak energy positions correlate reasonably with the surface binding energies for Ba atoms in clean and oxidized barium surfaces, 1.8 eV and 22 eV respectively, which were calculated from similar energy spectra of neutrals [25]. Relatively high values of the binding energies for BaO and other oxides are supported by thermodynamic calculations as discussed in [20].

Electrostatic Interactions of Secondary Ions

The electrostatic potentials of both the surface and primary ion beam influence the initial energies and trajectories of secondary ions. These effects are theoretically treated [26] and shown to be important for high fluency ion bombardment and low

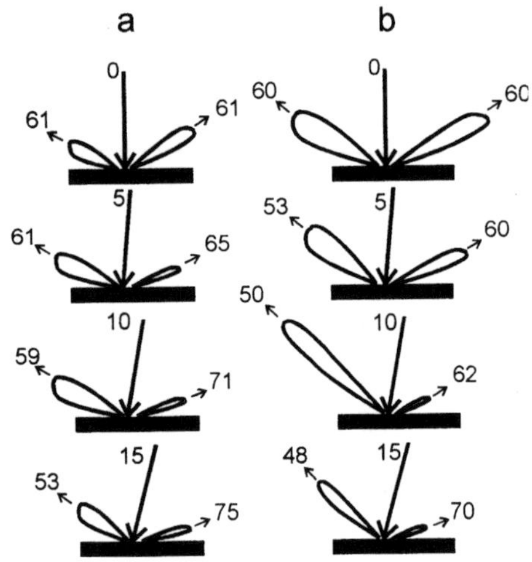

Fig. 6. Polar angle distributions of Ar^+ secondary ions at 2 keV (**a**) and 6 keV (**b**) Ar^+ ion bombardment of a Ni(100) surface. Bombarding and emission polar angles (deg.) are marked. The primary ion current is 2 μA

velocities of outgoing secondary ions. An appropriate image-potential correction to low energy emission is used to improve energy distributions of secondary ions ejected from clean metal surfaces [27]. Final energy-angular characteristics of SIE are also discussed with regard to the surface work function [28].

The work function variations just of the ion bombarded surface can be measured as shifts of secondary ion energy spectra caused by the contact potential changes between the surface and energy analyzer [29]. The specific energy distributions of secondary ions of argon, which are implanted under the Ar^+ ion beam bombardment and simultaneously sputtered from the surface [30], are of particular interest for this technique. Figure 5 illustrates the shifts of the energy distribution of the Ar^+ secondary ions during adsorption of hydrogen at the Mo(100) surface. It is seen, that the peaks are narrow and well defined on the energy scale providing determination of the work function variations with an accuracy of 0.05 eV [31].

Low energy Ar^+ secondary ions are exposed to strong electrostatic repulsion caused by the space charge of the bombarding beam. It gives an opportunity for experimental investigations of the ion repulsion [32]. Figure 6 illustrate how much the trajectories of the Ar^+ secondary ions are influenced by the Ar^+ primary ion beam. The latter splits the angular distributions of the ejected Ar^+ ions into two narrow petals which angular positions are effected by the incident beam direction. Simple correction formulas are proposed [33] in order to account for both the primary ion beam and surface image potentals.

Conclusions

High energy-and angular-resolved SIMS is proven to be useful for studying the near surface processes caused by ion bombardment of solids. Angular regularities of SIE reveal individual contributions of axial and planar channels to the ion channeling process as well as an anisotropy of the near surface collision cascades. Energy distributions of SIE provide information on the ionization mechanisms and global surface characteristics: the atomic energy states and work function. Energy and angular spectra of low energy emission are well suited for studying electrostatic effects in the vicinity of ion bombarded surfaces.

References

[1] P. Sigmund, in: *Sputtering by Particle Bombardment I* (R. Behrisch, ed.), Springer, Berlin 1981, p. 9.

[2] M. T. Robinson, in: *Sputtering by Particle Bombardment I* (R. Behrisch, ed.), Springer, Berlin Heidelberg New York, 1981, p. 73.

[3] H. E. Roosendaal, in: *Sputtering by Particle Bombardment III* (R. Behrisch, ed.), Springer, Berlin Heidelberg New York, 1991, p. 219.

[4] W. O. Hofer, in: *Sputtering by Particle Bombardment III*, (R. Behrisch, K. Wittmaack, eds.), Springer, Berlin Heidelberg New York Tokyo, 1991, p. 15.

[5] M. L. Yu, p. 91; M. L. Yu, D. Lang, *Nucl. Instr. Meth. Phys. Res.* **1986**, *B14*, 403; G. Slodzian, *Phys. Scripta* **1983**, *T6*, 54; G. Blaise, A. Nourtier, *Surf. Sci.* **1979**, *90*, 495; P. Williams, *Surf. Sci.* **1979**, 588; K. Wittmaack, in: *Inelastic Ion-Surface Collisions* (N. H. Tolk, J. C. Tully, W. Heiland, C. W. White, eds.) Academic Press, New York, 1977, p. 153; V. T. Cherepin, *Microzond Analysis*, Naukova Dumka, Kiev, 1992, 343 pp.

[6] N. Winograd, *Progr. Solid State Chem.* **1982**, *13*, 285.

[7] R. A. Gibbs, N. Winograd, *Rev. Sci. Instrum.* **1981**, *52*, 1148.

[8] H.-J. Lane, J. W. Amg, B. Winger, R. G. Cooks, in: *Abstr. 11th Int. Mass Spectrometry Conf.* Bordeaux, 1988, p. 117.

[9] V. T. Cherepin, A. A. Kosyachkov, I. N. Dubinsky, V. E. Isyanov, *Prib. Tekh. Exper.* (*Russia*) **1986**, *1*, 155; *Sci. Instrum.* (*Poland*) **1986**, *1*, 1.

[10] M. Hou, W. Eckstein, *Nucl. Instr. Meth. Phys. Res.* **1986**, *B13*, 324.

[11] A. A. Kosyachkov, V. T. Cherepin, S. M. Chichkan, Poverhnost (Russia) **1991**, *12*, 105; S. J. Tsanev, D. S. Karpuzov, A. A. Kosyachkov, *Nucl. Instrum. Meth. Phys. Res.* **1992**, *B71*, 381.

[12] L. B. Sanders, *Thesis*, Univ. Leiden, 1968.

[13] A. A. Kosyachkov, *Nucl. Instr. Meth. Phys. Res.* **1992**, *B72*, 40; *Phys. Metals* **1992**, *11*, 190.

[14] I. R. Chakarov, D. S. Karpuzov, A. A. Kosyachkov, V. T. Cherepin, *Proc. Conf. on Ion Implantation and Ion Beam Equipment* (D. S. Karpuzov, I. V. Katardjiev, S. S. Todorov, eds.) World Scientific, Singapore, 1990, p. 509; A. A. Kosyachkov, V. T. Cherepin, *Poverhnost* (*Russia*) **1991**, *2*, 153.

[15] Y. Matsuda, Y. Yamamura, C. Honda, S. Matsubaguchi, M. Maeda, K. Muraoka, M. Akazaki, *Jpn. J. Appl. Phys.* **1986**, *25*, 1256.

[16] J. Dembowski, H. Oechsner, Y. Yamamura, M. Urbassek, *Nucl. Instr. Meth. Phys. Res.* **1987**, *B18*, 464.

[17] G. K. Wehner, *J. Appl. Phys.* **1955**, *26*, 1055.

[18] I. R. Chakarov, V. T. Cherepin, D. S. Karpuzov, A. A. Kosyachkov, *Nucl. Instr. Meth. Phys. Res.* **1989**, *B39*, 81.

[19] M. W. Thompson, *Philos. Mag.* **1968**, *18*, 377.

[20] R. Kelly, *Nucl. Instr. Meth. Phys. Res.* **1987**, *B18*, 388.

[21] Y. Okayima, *J. Appl. Phys.* **1984**, *55*, 230.

[22] V. T. Cherepin, A. A. Kosyachkov, M. A. Vasiljev, *Phys. Stat. Solid.* **1978**, *A50*, K113.

[23] H. Gnaser, M. Saidoh, J. Von Seggern, W. O. Hooer, *Nucl. Instr. Meth. Phys. Res.* **1986**, *B15*, 169.

[24] A. A. Kosyachkov, V. T. Cherepin, S. M. Chichkan, G. A. Klimenko, *Superconductivity Physics, Chemistry, Technique* (*Russia*) **1990**, *3*, 1964; A. A. Kosyachkov, *Nucl. Instr. Meth. Phys. Res.* **1992**, *B65*, 546.

[25] D. Grischkowsky, M. L. Yu, A. C. Balant, *Surf. Sci.* **1983**, *127*, 315.

[26] P. Bryce, J. C. Kelly, *J. Phys. C. Solid State Phys.* **1972**, *5*, 1604.

[27] M. J. Vasile, *Surfac Sci.* **1982**, *115*, L141.
[28] K. Wittmaack, *Phys. Scripta* **1983**, *T6*, 71.
[29] G. Blais, G. Slodzian, *Surf. Sci.* **1973**, *40*, 708.
[30] A. A. Kosyachkov, V. T. Cherepin, *Solid State Commun.* **1989**, *69*, 659.
[31] A. A. Kosyachkov, *The 8-th Internat. Confer. on Secondary Ion Mass Spectrometry, SIMS-VIII* (Intern. Congress Centre RAI, ed.) Amsterdam, The Netherlands, 1991, p. 81; *Poverhnost* (*Russia*) **1994**, *4*, 25.
[32] A. A. Kosyachkov, *Poverhnost* (*Russia*) **1989**, *9*, 128.
[33] M. L. Adamovich, A. A. Kosyachkov, V. T. Cherepin, *Poverhnost* (*Russia*) **1992**, 7, 27.

Mikrochim. Acta [Suppl.] 13, 399–408 (1996)

EPMA and Mass Spectrometry of Soil and Grass Containing Radioactivity from the Nuclear Accident at Chernobyl

Odile Lefèvre, Maria Betti, Lothar Koch, and Clive T. Walker*

The European Commission, Joint Research Centre, Institute for Transuranium Elements, P.O. Box 2340, D-76125 Karlsruhe, Federal Republic of Germany

Abstract. Electron microprobe analysis and mass spectrometry results for top soil and grass contaminated by nuclear fallout from the Chernobyl accident are presented. The material, which was collected in July 1990, was obtained from the International Atomic Energy Agency in Vienna. The soil sample was taken from the Novozybkov district of Bryansk in Russia about 150 km north-east of Chernobyl. The grass sample was collected in the Polesskoe district of Kiev in the Ukraine about 100 km south of the reactor. Trace element and isotope analysis of the bulk samples was performed by glow discharge mass spectrometry. Secondary ion mass spectrometry was to used to establish the locations of certain radionuclides. The mineral phases in the soil sample were identified by electron probe microanalysis. It is confirmed that ^{137}Cs was mainly responsible for the radioactivity in both samples. The grass, however, contained 2.5 times more ^{137}Cs than the soil. Most of this caesium was evidently incorporated in the plant tissue. Radioactive ^{129}I was detected in alkali feldspar and other silicates in the soil sample. It is assumed that this iodine had been incorporated in the crystal lattice of the minerals by an ion exchange mechanism. In addition, particles containing ^{238}U and ^{235}U, the two principal uranium isotopes in UO_2 nuclear fuel were found in both the soil sample and the grass sample. Far more of these hot particles were present in the soil sample, however. The discovery of fuel particles has significance, because they may act as long-term reservoirs of radionuclides, slowly releasing their fission products over many years to the invironment.

Key words: Chernobyl, soil, grass, electron probe microanalysis, mass spectrometry.

When reactor No 4 at Chernobyl exploded at 1.23 a.m. (local time) on April 26 1986 about eight tons of radioactive material escaped into the atmosphere. This included 15–20 MCi of iodine-131 and 1–2 MCi of cesium-137, corresponding to 20–25% of

* To whom correspondence should be addressed

the reactor's inventory of these two radionuclides at the time of the accident. The radioactivity was spread by the wind over Byelorussia and the Ukraine, and traces of it reached as far west as France and Italy. Of the material ejected into the atmosphere 0.3–0.5% was deposted on site, 1.5–2.0% was deposited within 20 km and the remainder was dispersed beyond 20 km [1]. The deposition pattern in the Chernobyl region was dominated by gravitational settling of coarser material released during the first five days following the accident. As a consequence of the accident, soil and ground water within 30 km radius of the Chernobyl plant was severely contaminated (radiation dose levels exceeded 100 μSvh^{-1}) and 135,000 people had to evacuated from an area of 800 km^2 around the plant.

This paper presents microprobe analysis and mass spectrometry results from top soil and grass contaminated by nuclear fallout from the Chernobyl accident. The material, which was collected in July 1990, was obtained from the International Atomic Energy Agency (IAEA) in Vienna. The soil sample (Ref. IAEA-375) was taken from the Novozybkov district of Bryansk in Russia about 150 km north-east of Chernobyl. The concentration of activity in the soil was 6.5×10^3 Bq kg^{-1}. The grass (Ref. IAEA-373) was collected in the Polesskoe district of Kiev in the Ukraine about 100 km south of the reactor. Its radioactivity was 1.5×10^4 Bq kg^{-1}. These levels of radioactivity are at least seven times higher than normal. For example, activity values in the range of 400 to 850 Bq kg^{-1} are reported for shale (a group of sedimentary rock) and values of around 150 Bq kg^{-1} are quoted for quartz sand and limestone.

Cesium-137 is mainly responsible for the radioactivity of both samples, although the material is said to contain one or more hot-particles of fallout origin. These may be defined as particles with high specific activity of about 3×10^8 Bq kg^{-1} and consisting of non-volatile radioactive substances that are only slightly soluble in water [2]. Two basic types of hot-particles have been identified: particles containing a mixture of fission products and particles with just a few fission product elements such as Ru, Ce and Sr [1].

Experimental

Sample Preparation

Two different methods of preparation were employed. For glow discharge mass spectrometry (GDMS) a 500 mg sample of material was compacted into a disc 13 mm in diameter and 2 mm thick using an automatic hydraulic press (3630X, SPEX, USA). A load of eight tons was applied for five minutes. For secondary ion mass spectrometry (SIMS) and electron probe microanalysis (EPMA) about the same weight of material was taken. This was first dried at 50 °C for 24 h and then mixed with epoxy resin. When the mass had hardened, after a further 24 h, it was introduced into an epoxy mount. The material was then ground and polished using standard techniques. To avoid charging effects the samples were coated with a conducting film of carbon about 20 nm thick.

Glow Discharge Mass Spectrometry

Trace element analysis of the bulk material was carried out using a VG 9000 glow discharge mass spectrometer. The machine which is installed in a glove box [3] consists of a d.c. glow-discharge ion source coupled to a double-focusing mass spectrometer of reverse (Nier-Johnson) geometry. This provides high transmission ($>75\%$) and sensitivity whilst operating at a high mass resolution of 5000 with 10% valley definition. A discharge current of 1.2 mA and a voltage of 1.3 kV was used. Ion detection is accomplished by a dual detection system consisting of a Faraday cup for the measurement of large ion currents (typically $>10^{-13}$ A) and a transverse mounted Daly detector [4] for the measurement of small ion currents.

Glow discharge mass spectrometry was carried out in two steps. First, a full mass spectrum was acquired from each sample at a mass resolution greater than 5000. At such a high mass resolution, the analytes of interest could be separated from the molecular isobaric interferences caused by the complex matrices of the samples. From the mass spectrum the elements of interest were selected. Isotopes were analysed at a mass resolution of 1000 in order to obtain flat top peaks, which are required for accurate isotope ratio measurements. For each isotope 30 scans were made consisting of 50 points measured with dwell times of 160 ms (Faraday cup) and 100 ms (Daly detector). Quantification was made using relative sensitivity factors (RSFs) obtained from other matrices. Prior to analysis, the samples were sputtered for 15 min to remove any surface contamination. A conducting binder was not employed. To ensure that the samples were conductive a secondary cathode made of tantalum was used [5–7]. A stable d.c. discharge was obtained after 3 minutes.

Secondary Ion Mass Spectrometry

Secondary ion mass spectrometry was used to identify the location of the radionuclides cesium-137, iodine-129 and uranium-238 in the soil and grass amples.

A Cameca IMS 4f secondary ion mass spectrometer was used. This instrument has a double-focusing mass spectrometer which allows rapid switching between masses. It also has micro-focusing ion sources (Cs^+ and duoplasmatron with O_2^+, O^-, Ar^+) that can be used either in the ion microscope or ion microprobe mode.

For the investigation of the soil and grass samples the instrument was used in the microscope mode. A O^- primary ion beam was employed at an energy of 6 keV. Positive secondary ions at beam currents of 20–40 nA (typical spot size $\leqslant 25\,\mu m$) were collected from a circular area 60 μm in diameter. The instrument was operated at a mass resolution of 600 with an energy window of 50 eV. Ion images showing the distribution of all the major constituents and the locations of cesium-137, iodine-129 and uranium-238 in areas 150 and 200 μm square were produced using a standard acquisition time of 300 s. In the case of the grass sample the potassium-39 signal was used to image the individual blades and sheaths of grass.

Electron Probe Microanalysis

Electron probe microanalysis was used to identify the different mineral phases in the soil sample. A Cameca MS46 electron microprobe analyser fitted with a Tracor TN 5500 energy dispersive analyser and installed in glove-box to make possible the analysis of alpha-active materials (substances containing plutonium, americium or curium) was employed for this purpose.

A combination of energy dispersive and wavelength dispersive analysis was employed. Energy dispersive analysis was used to obtain a comprehensive picture of the sample composition. X-ray spectra were acquired from the different mineral phases at an electron acceleration potential of 15 keV and a beam current of 1 nA. The acquisition time was set at 200 s. For quantification the PRZ option in the standardless analysis (SQ) program was used and the constituent elements were in put as stoichiometric oxides.

Wavelength dispersive analysis was used to produce X-ray maps showing the distribution of the different mineral phases in an area 0.4 × 0.3 mm square. Regions of interest in the sample were selected using the electron absorbed (specimen current) image, although owing to the similarity in atomic number the individual mineral phases could not be distinguished. The X-ray maps were produced at an

Table 1. Glow discharge mass spectrometry results for the soil sample (figures are in ng/g unless otherwise indicated)

Element/radionuclide	Mass concentration
Si	40.0%
K	20.0%
Al	12.0%
Na	0.7%
Mg	0.5%
Ca	0.4%
Fe	0.4%
Ti	0.3%
Zr	[a]
Ba	0.1%
O	22.0%
C	3.0%
Sr-90	0.001
Ru-106	[a]
I-129	0.0002
Cs-137	0.002
Th-232	4950
U-234	1.0
U-235	30
U-238	2000
Pu-239	0.001
Pu-240	0.001

[a] not detected.

electron acceleration potential of 15 keV and a beam current of 50 nA. The acquisition time was varied in accordance with the signal intensity.

Results

Soil

The results of GDMS are shown in Table 1. It is seen that Si, K and Al were the principal cations in the soil. Most of the radionuclides measured were present in amounts smaller than 1 ppb. Only ^{232}Th, ^{238}U and ^{235}U were found in significant amounts (at the ppm level). The concentrations of U and Th measured were within the range at which these elements occur in nature. For example, 0.03–68 ppm Th and 0.14–45 ppm U are found in common metamorphic rocks and common sedimentary rocks contain up to 80 ppm U [8]. The fact that the concentration of ^{232}Th in the sample was much higher than the concentration of ^{238}U (4950 ng/g compared with 2000 ng/g) indicates that most of that measured was indeed natural and did not

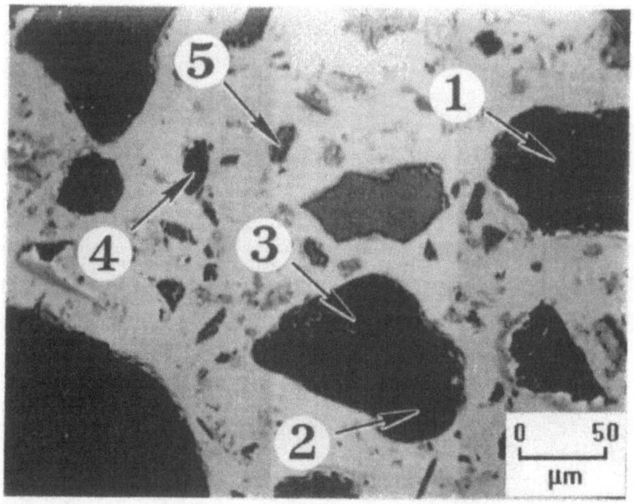

Phase no.	Name	Chemical formula
1	quartz	SiO_2
2	ba-bearing potassium feldspar	$(Ba)KAlSi_3O_8$[a]
3	potassium feldspar	$KAlSi_3O_8$
4	garnet (type almandine)	$Fe_3Al_2Si_3O_{12}$
5	plagioclase	$(NaAlSi_3O_8, CaAl_2Si_2O_8)$
	anatase, brookite or rutile[b]	TiO_2

[a] Barium 1.2 wt%.
[b] Small inclusions only.

Fig. 1. Minerals in the soil sample as identified from their chemical compositions measured by electron probe microanalysis. Upper: Specimen current image

result from the Chernobyl accident. In contrast, it would appear that a large part of the uranium measured came from nuclear fuel. Not only is the measured $^{235}U/^{238}U$ ratio of 0.015 twice as high as that found in nature, but fuel enriched to 1.8% was irradiated in the Chernobyl reactor. Further, the measured $^{235}U/^{238}U$ ratio of 0.015 indicates that the fuel from which the fallout derived had been irradiated to a low burn-up of 4–5 GWd/tU. At the time of the accident the reactor contained nuclear fuel with a burn-up in the range zero to 27 GWd/tU.

The detection of carbon is attributed to organic matter in the soil. The EPMA results are given in Fig. 1. Six different minerals were identified of which quartz and feldspar appeared to be the most abundant. The results of the SIMS investigation are presented in Fig. 2. The figure contains ion images for ^{29}Si, ^{129}I and ^{238}U. It can be seen that iodine is present at the same location as silicon. Moreover, a small particle containing uranium is also revealed. Cesium-137 was not detected in the regions examined by SIMS. The intensity of the ion peak at mass 137 matched the abundance of that isotope in natural barium.

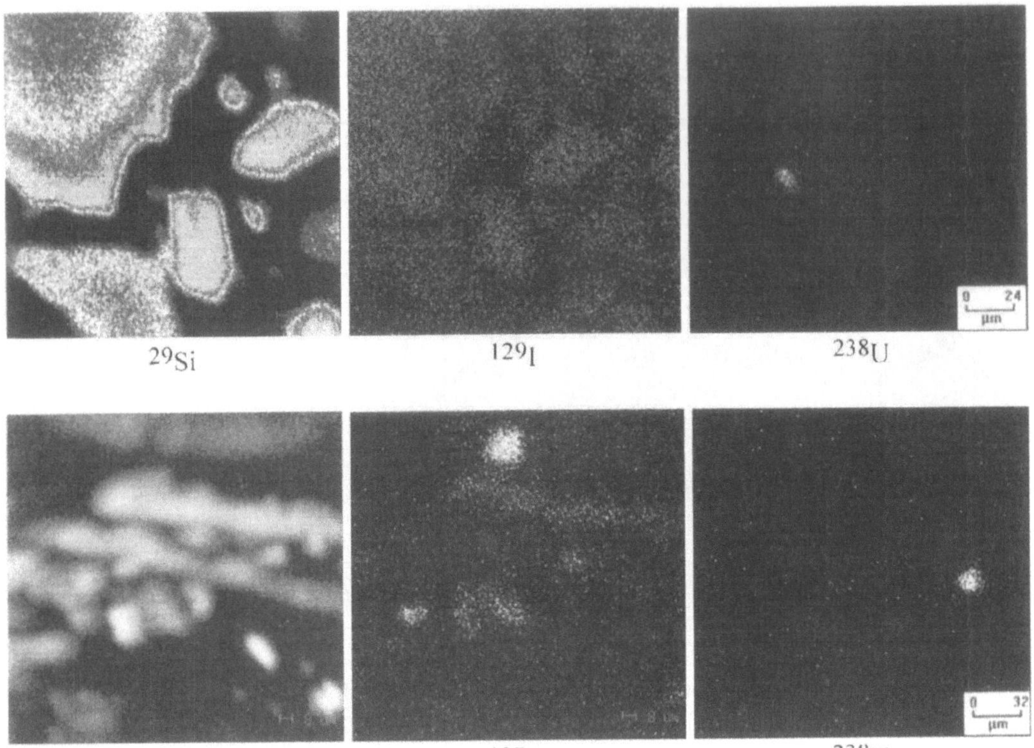

Fig. 2. Ion images from the SIMS investigations showing iodine and uranium in the soil sample (upper) and caesium and uranium in the grass sample (lower)

Grass

Glow discharge mass spectrometry results for selected radionuclides are reported in Table 2. Relatively high concentrations of ^{232}Th, ^{235}U and ^{238}U were measured (50, 5 and 330 ng/g respecively). The measured $^{235}U/^{238}U$ ratio is again ca. 0.015 suggesting that most of the uranium derived from the Chernobyl acident. The level of ^{90}Sr was around 0.001 ng/g, and the concentrations of both ^{137}Cs and ^{239}Pu did not exceed 0.005 ng/g. Iodine-129 was not detected. Ion images from the SIMS investigation are shown in Fig. 2. It is seen that ^{137}Cs was present in the individual blades and sheathes of grass that constituted the sample. It is also evident that two particles of extraneous matter are present in the area studied. One particle contains ^{137}Cs; the other, which is noticeably smaller, contains ^{238}U.

Discussion

Novozybkov lies in a region of loamy and sandy soils which are used principally for the cultivation of potatoes. GDMS revealed that, in addition to ^{137}Cs, the soil contained 0.0002 ng/g ^{129}I (see Table 1). The SIMS results (Fig. 2) indicate that this iodine was incorporated in the silicon containing minerals in the soil. Presumably,

Table 2. Glow discharge mass spectrometry results for the grass sample

Radionuclide	Concentration, ng/g
Sr-90	0.001
Ru-106	[a]
I-129	[a]
Cs-137	0.005
Th-232	50
U-235	5
U-238	330
Pu-239	0.005
Pu-240	0.001

[a] not detected.

the incorporation of iodine in the mineral phases occurred by an ion exchange mechanism. This is a characteristic property of clay. Since X-ray diffraction was not employed, it is not possible to state whether the silicon containing phases analysed were clays. Certainly, the feldspars identified by EPMA (Fig. 1) along with other silicates are clay forming minerals. On the other hand, clays have a characteristic structure. In clays the constituent minerals occur as booklike particles, aggregates of flakes or bundles of lathes and needles that are of the order of microns in diameter. No evidence of such structures was found in the soil sample.

The grass from Polesskoe contained about 2.5 times more ^{137}Cs than the soil from Novozybkov (cf., concentrations in Tables 1 and 2). This is consistent with what is known about the behaviour of cesium. That it is absorbed by plants and that it is removed from the top soil at a rate that is dependent on the soil characteristics. Cesium may be removed from top soil by wind and water erosion, by the action of burrowing organisms, such as worms, by leaching and also by plant uptake. Bishop noted [9] that earth worms could move up to 10 kg·m^{-2} of soil a year in established pastures (corresponding to a soil depth of 14 mm). From the studies of Frissel and Pennders (see e.g., ref. [10]) it appears that as a result of leaching, cesium migrates to deeper soil strata at a speed of 0.3 to 0.5 cm per year. The distribution coefficient, K_d, which is defined as the ratio of the concentration of the radionuclide in the soil phase to that in the solution phase, is acknowledged to be one of the best indicators of leaching behaviour in soils in temperate regions. Cesium is characterised by a high K_d indicating that it is leached relatively slowly from soil. K_d values for agricultural soils published by the IAEA [11] show that cesium is removed ten times faster from sandy and organic soils than from loamy and clay soils.

It is probable that most of the cesium measured in the grass sample was incorporated in the plant tissue. The absence of ^{129}I in the sample would suggest that rainfall had washed most of the water soluble contaminants, including cesium, from the grass surfaces. Though the relatively high concentrations of ^{238}U and ^{232}Th in the sample and the detection of a particle containing cesium would suggest that some surface contamination was still present when the grass was collected. The two routes for the incorporation of radionuclides by plants are foliar absorption and root uptake. According to Watkins [12] foliar absorption may occur by way of the stomata or across the cuticular. Clearly, the longer a contaminant remains on the plant surface the more likely it is to be absorbed. Thus rainfall is an important factor in determining the degree to which radionuclides are absorbed by plant surfaces. The rate of uptake of Cs by the plant root system is dependent not only on the depth to which the roots penetrate, but also on the soil type, the soil pH and the fraction of organic matter in the soil. The uptake of Cs from sandy soil is about five times higher than from clay soil, whereas the uptake from loam soil is about twice as high as from clay soil [13]. A representative soil-to-plant transfer factor ($Bq \cdot kg^{-1}$ dry crop per $Bq \cdot kg^{-1}$ dry soil) for Cs and grass is 0.21 [14]. This compares with 0.26 for alfalfa fodder, 0.03 for barley grain and 0.018 for wheat grain.

Particles of nuclear fuel were an important component of the Chernobyl fallout in Russia and the Ukraine. Particles containing ^{238}U, the principal uranium isotope in UO_2 fuel, were seen in both the soil sample and the grass sample. From the GDMS results it would appear that these particles contained about 1.5 wt% ^{235}U. Moreover, the GDMS results also indicate that these hot-particles were far more abundant in the soil sample (2000 ng/g ^{238}U was measured in the soil compared with 300 ng/g in the grass). It is considered that surface contamination was the source of the fuel particles in the grass sample. Pickup from soil taken with the grass sample can be ruled out. The sample contained just 0.01 wt% Si (GDMS result) which establishes that the amount of soil present was negligible. The finding of such hot-particles in the samples has significance because they may act as long-term reservoirs of radionuclides which under specific conditions are released to the local surroundings.

The activity of the radionuclides in the soil and grass samples derived from their concentrations measured by GDMS are compared with the values reported by the IAEA in Tables 3 and 4. It is seen that good agreement exists between the activities

Table 3. Activity of the radionuclides measured by GDMS in the soil sample

Radionuclide	Radioactivity, $Bq \cdot Kg^{-1}$	
	From GDMS concentrations	Reported by the IAEA[a]
Sr-90	5000	99.6
I-129	0.0013	0.0017
Cs-137	6400	4925
Th-232	20	20.7
U-234	23	25.0
U-235	2.1	[b]
U-238	23	22.6
Pu-239	2	[b]
Pu-240	8	[b]

[a] After correction for three years of isotopic decay.
[b] Not reported.

Table 4. Activity of the radionuclides measured by GDMS in the grass sample

Radionuclide	Radioactivity, $Bq \cdot Kg^{-1}$	
	From GDMS concentrations	Reported by the IAEA[a]
Sr-90	5000	1311.9
Cs-137	16000	12349
Th-232	0.2	[b]
U-235	0.36	[b]
U-238	3.8	[b]
Pu-239	12	[b]
Pu-240	8	[b]

[a] After correction for three years of isotopic decay.
[b] Not reported.

given by the IAEA and those derived from the GDMS measurements. The GDMS data confirm that ^{137}Cs was mainly responsible for the radioactivity of both samples, and that the radioactivity caused by ^{137}Cs was approximately a factor of 2.5 higher in the grass. The activities reported by the IAEA were corrected for isotopic decay in the time elapsed between the certification date (31 December 1991) and when GDMS was carried out (February 1995). For most of the isotopes the activity estimated from the GDMS concentration represents an upper limit derived from the limit of detection for the technique. The activity was calculated using the mass equivalent to an activity of 1 Bq, which was determined from the half-life of the isotope.

Conclusion

Even though the soil and grass samples were collected roughly 150 and 100 km from the site of the accident, the level of radioactive contamination was such that it could easily be detected by GDMS and SIMS. Cesium-137 was mainly responsible for the radioactivity of both samples, but the grass contained about 2.5 times more ^{137}Cs than the soil. Most of this cesium was evidently incorporated in the plant tissue. The soil sample contained radioactive ^{129}I. This isotope was detected in alkali feldspar and other silicates in the soil sample. It is assumed that the iodine had been incorporated in the crystal lattice of the minerals by an ion exchange process. Particles containing ^{238}U and ^{235}U, the two principal uranium isotopes in UO_2 fuel, were also found in both the soil sample and the grass sample. Far more of these hot-particles were present in the soil, however. The discovery of fuel particles has significance, because they may act as long-term reservoirs of radionuclides, slowly releasing their fission products over many years to the local surroundings.

Acknowledgements. The authors wish to thank J. J. Le Goux and M. Schuhmacher of Cameca, France, for their assistance with the SIMS investigations, and D. Ackermand of the Christian-Albrechts-University to Kiel, Germany, for his help in identifying the minerals in the soil sample.

References

[1] L. J. Appleby and S. P. Luttrell, in: *Radioecology after Chernobyl* (F. Warner, R. M. Harrison, eds.), Wiley, Chichester, 1990, p. 49.

[2] O. V. Ruimantsev, L. M. Kitrov, in: *Proc. Seminar on Comparative Assessment of the Environmental Impact of Radionuclides Released During Three Major Nuclear Accidents: Kyshtym. Windscale and Chernbyl, Report EUR 13574, vol. 1*, Commission of the European Communities, 1990, p. 157.

[3] M. Betti, G. Rasmussen, T. Hiernaut, L. Koch, D. M. P. Milton, R. C. Hutton, *J. Anal. At. Spectrom.* **1994**, *9*, 385.

[4] N. R. Daly, *Rev. Sci. Instrum.* **1960**, *31*, 264.

[5] D. P. M. Milton, R. C. Hutton, *Spectrochim. Acta* **1993**, *48*, 39.

[6] W. Schelles, S. de Gent, V. Muller, R. van Grieken, *Appl. Spectr.* **1995**, *49*, 939.

[7] M. Betti, G. Rasmussen, L. Koch, in: *Proc. Winter Conf. on Plasma Spectrometry*, San Diego, USA, 1994, p. 329.

[8] K. H. Wedepobl (ed.), *Handbook of Geochemistry*, Springer, Berlin Heidelberg New York, 1969–78, Vol. 2/5.

[9] G. P. Bishop, *Review of Biosphere Information: Biotic Transport of Radionuclides as a Result of Mass Movement of Soil by Burrowing Animals*, UK Nirex Ltd, Harwell, Report NSS/R194, 1989.

[10] M. J. Frissel, R. M. J. Pennders, in: *Ecological Aspects of Radionuclide Release* (P. J. Coughrey, J. N. B. Bell, T. M. Roberts, eds.), Blackwell, Oxford, 1983.

[11] *Handbook of Parameter Values for the Prediction of Radionuclide Transfer in the Terrestrial and Freshwater Environments*, IAEA, Vienna, 1991.

[12] B. M. Watkins, *Interception, Retention, Absorption and Translocation of Radiounuclides by Vegetation Following Application in Irrigation Waters*, UK Nirex Ltd. Harwell, Report NSS/R221, 1990.

[13] R. Kirchman, J. N. B. Bell, P. J. Coughtrey, M. Frissel, T. E. Hakonson, W. C. Hanson, D. Horrill, B. J. Howard, L. J. Lane, C. Myttenaere, W. L. Robinson, C. Ronneau, G. Shaw, W. R. Schell, J. Van den Hock, A. Konoplyov, N. Zezina in: *Radioecology after Chernobyl* (F. Warner, R. M. Harrison, eds.), Wiley, Chichester, 1990, p. 118.

[14] M. J. Frissel, K. E. van Bergeijik, *Mean Transfer Values Derived by Simple Regression Analyses*, RIVM-LSO. Bilthoven, Netherlands, Report IV, IUR Working Group on Soil-to Plant Transfer, 1989.

Mikrochim. Acta [Suppl.] 13, 409–417 (1996)

Application of a New Monte Carlo Simulation Algorithm to Electron Probe Microanalysis

X. Llovet[1], A. Riveros[2], and F. Salvat[3,]*

[1] Serveis Científico-Tècnics, Universitat de Barcelona, Lluís Solé i Sabarís, 1-3, E-08028 Barcelona, Spain

[2] Facultad de Matemática, Astronomía y Física, Universidad Nacional de Córdoba, Medina Allende y Haya de la Torre, 5000 Córdoba, Argentina

[3] Facultat de Física (ECM), Universitat de Barcelona, Diagonal 647, E-08028 Barcelona, Spain

Abstract. A new general purpose code for Monte Carlo simulation of electron-photon showers is applied to a variety of electron transport problems involving electron energies in the keV range. The underlying scattering model combines elastic scattering cross-sections calculated by the partial wave method with inelastic cross-sections obtained from Liljequist's generalized oscillator strength model. The reliability of the trajectory generation algorithm has been verified through a comparison of simulation results with measured backscattering coefficients and spatial dose distributions. The number of inner-shell ionizations in a trajectory segment is evaluated by means of various approximate total ionization cross-sections which are commonly used in the literature. Simulated depth distributions of ionization, in the form of $\Phi(\rho z)$-functions, are compared with available data from tracer experiments. Calculated ionization distributions for thin films on substrates are presented.

Key words: electron probe microanalysis, Monte Carlo simulation, inner-shell ionization, electron transport.

The spatial distribution of ionizations caused by the electrons impinging on the specimen is the primary quantity for quantitative electron probe microanalysis (EPMA). For homogeneous samples and normal incidence, this distribution is conventionally expressed as the $\Phi(\rho z)$-function (see e.g. ref [1]), which gives the density of ionizations as a function of the mass depth ρz. This function can be either measured experimentally or calculated using Monte Carlo simulation methods (comprehensive reviews have been given by Karduck and Rehbach [2] and Mackenzie [3]). A variety of simple analytical approximate expressions for $\Phi(\rho z)$ have been proposed [3], which make it possible to estimate the elemental concentrations from the measured X-ray intensities by means of relatively simple, analytical algorithms.

* To whom correspondence should be addressed

The generalization of this procedure to heterogeneous specimens (e.g. thin layers on substrates or multilayered structures) or special geometries (e.g. oblique incidence) is difficult, mainly because of the lack of knowledge about the spatial distribution of ionizations. In this context, Monte Carlo simulation is a valuable tool, not only to determine the required ionization distributions but also to check the reliability of approximate formulations.

The most accurate Monte Carlo codes currently in use (see e.g. ref. [4]) perform "detailed" simulation, i.e. all interaction events experienced by an electron are described in chronological succession. In the energy range of interest (say 1–50 keV) the dominant interactions are elastic scattering and inelastic collisions. Bremsstrahlung emission, although responsible for the spectral background [5], has a negligible stopping effect at these energies and can be disregarded when simulating the electron tracks. In order to ensure the reliability of the simulation results, there is a tendency to use the most accurate differential cross-sections (DCS) available. As these are normally given in numerical form, the simulation must be performed on the basis of systematic interpolation on extensive data bases. This complicates the code considerably and, moreover, means that it can only be applied to the materials for which cross-sections tabulated with the required detail are available.

In this paper we present results obtained with a general purpose Monte Carlo simulation package based on an alternative approach. The code, named PENELOPE (PEN-etration and Energy LOss of Positrons and Electrons), was developed at the University of Barcelona to simulate electron-photon showers in arbitrary materials and for a wide range of energies ($\sim$1 keV to $\sim$100 MeV). A detailed description of the adopted DCSs and the simulation algorithm has been given elsewhere (see Baró et al. [6] and references therein). PENELOPE is coded as a FORTRAN subroutine package, which generates particle histories according to built-in differential cross-sections. The user must only provide a main program that follows the evolution of the tracks and keeps score of relevant quantities.

The aim of the present work is to provide convincing evidence of the fact that, in many cases of interest in electron microscopy and microanalysis, PENELOPE yields results that are sufficiently accurate for practical purposes. A brief description of the code is given. Benchmark comparisons of simulation results and experimental data are presented.

The Simulation Code PENELOPE

In various aspects, the structure of PENELOPE is similar to that of the "condensed" simulation codes used in high energy physics and clinical dosimetry [7]. Although the code transports electrons, positrons and photons, only the electron transport part has been used in the present calculations. Secondary electrons generated by the primary particle and its descendants are followed down to the adopted absorption energy.

Electron (and positron) histories are simulated by means of a mixed procedure. Interactions of the electrons with the medium are classified as hard events, in which the angle deflection θ and the energy loss W are greater than certain cutoffs θ_S and W_S, and soft interactions (with $\theta < \theta_S$ or $W < W_S$). Hard events are simulated in detail; the effect of the usually many soft interactions that occur between a pair of consecutive hard events is described as a single "artificial" event in which the particle is deflected and loses energy according to a multiple scattering theory. Between each pair of consecutive (hard or artificial) events the particle "flies" freely with a well-defined energy. The angle cutoff θ_S is automatically adjusted by the program in such a way that the average deflection in each hard elastic event is nearly independent of the electron (or position) energy, the energy loss cutoff W_S is directly set by the user. The practical advantage of mixed simulation is that the calculation of soft events is largely simplified. As the generation of these events takes up a considerable fraction of the computer time, mixed simulation is normally much faster than detailed simulation.

The atomic DCSs adopted in PENELOPE are simple analytical functions of the scattering angle and the energy loss. These DCSs have a physically reasonable shape and allow random sampling of the scattering angle and energy loss in individual interactions in a completely analytical way, so that sampling errors that could originate from numerical interpolation are completely avoided. Although these analytical DCS differ in detail from the actual ones, they lead to essentially the same simulation results, provided only that the scattering is multiple (i.e. the average number of events along an electron track must be of the order of 20 or larger). The validity of the whole simulation scheme rests on the fact that multiple scattering distributions are determined by the values of a few integrals of the DCSs; the effect of finer details is washed out after repeated interactions. This statement can be "experimentally' verified through Monte Carlo simulation; a formal proof can also be given on purely theoretical grounds [8]. Thus, the simulation code can be largely simplified by using analytical approximate DCSs leading to the correct values of these relevant integrals. The price paid for this simplicity is that simulation results are reliable only under multiple scattering conditions.

The DCS for single elastic scattering is expressed as a combination of the Wentzel (screened Rutherford) DCS and a fixed-angle scattering process. The analytical DCS contains three parameters that are determined in such a way that the mean free path between collisions and the mean and the variance of the angular deflection in each elastic collision are identical with the values obtained with the realistic numerical DCS. The required information reduces to the mean free path λ_{el} and the first λ_1 and second λ_2 transport mean free paths. These quantities are continuous functions of the energy, that have been calculated (and stored in a database) for the elements $1 \leqslant Z \leqslant 92$ and energies from 100 eV to 100 keV using the Dirac partial wave method with the free-atom Dirac-Hartree-Fock-Slater field corrected for exchange effects [9].

The DCS for inelastic collisions [10] is based on a simple generalized oscillator strength model proposed by Liljequist [11]. In this model the ionization of each atomic electron shell is described by a single oscillator with strength f_i equal to the number of electrons in the shell, and resonance energy $W_i = aU_i$, where U_i is the ionization energy of the shell and a is an adjustable parameter. Excitations of valence electrons (with binding energy less than 30 eV) are accounted for by a single oscillator with oscillator strength f_{val} and resoanance energy W_{val}. The value of a is determined in such a way that the mean excitation energy I recommended by Berger and Seltzer [12] is exactly reproduced. Excitations on the Bethe ridge are described by means of the Møller DCS (electrons) and the Bhabha DCS (positrons). Density effect corrections are also included. The model permits the evaluation of the inelastic mean free path and the stopping power analytically. The calculated stopping powers normally agree with those recommended by the ICRU [12] to within 0.05% for energies above 10 keV. Secondary electron emission in electron and positron hard events is also simulated according to the Liljequist oscillator model.

Simulation Results and Discussion

In the simulations reported here, the number of generated primary tracks is of the order of 100,000 for medium and high atomic number elements and 200,000 for light elements. The statistical uncertainties of the results are normally less than 1%.

The reliability of the trajectories generated by PENELOPE has been validated by benchmark comparisons with experimental transport data. Two relevant examples are given here.

Backscattered Electrons

In Fig. 1, simulated backscattering fractions for mono-energetic electron beams impinging normally on homogeneous elemental samples of different atomic number are compared with values measured by different groups [13–17]. It is worth pointing out that the backscattered fraction is the result of a delicate competition between elastic and inelastic scattering. Moreover, electrons are backscattered after travelling different path lengths inside the sample. Those with large paths experience multiple collisions and are thus well described by the theory. However, a fraction of

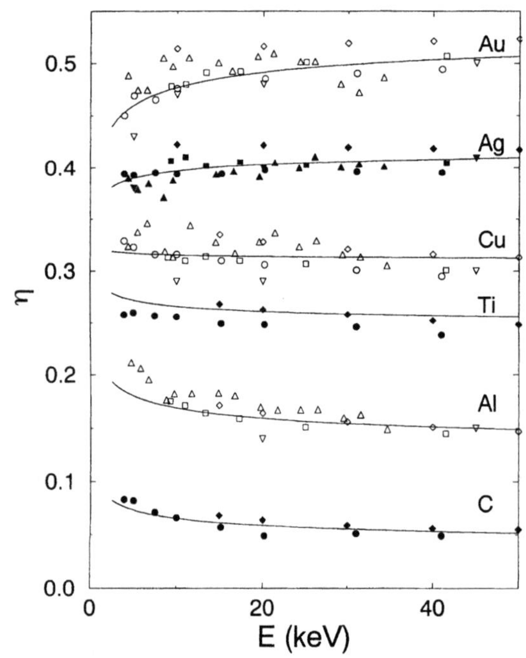

Fig. 1. Backscattering coefficient as a function of incident energy. Experimental data are from Hunger [13] (circles), Dresher et al. [14] (squares), Neubert and Rogaschewski [15] (diamonds), Massoumi et al. [16] (triangles) and Coslett and Thomas [17] (inverted triangles). Solid lines represent Monte Carlo results

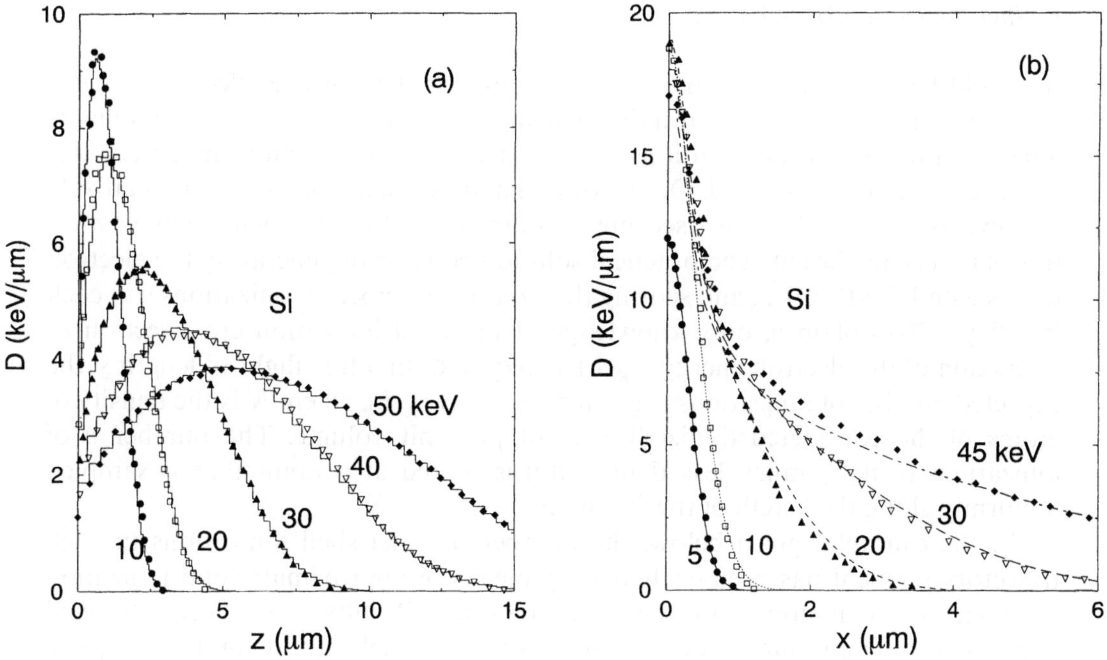

Fig. 2. Depth-dose distributions (**a**) and lateral dose distributions (**b**) in silicon for electron beams of the indicated energies at normal incidence. Experimental data are from Werner et al. [18]. Histograms (**a**) and curves (**b**) are Monte Carlo results

electrons leave the sample after travelling comparatively short path lengths and experience only single or plural scattering. Under these circumstances the validity of the approximate elastic cross-sections and the simulation scheme is open to question. Therefore, the backscattering coefficient represents a stringent global test of the simulation algorithm. The spread of experimental data from different authors in Fig. 1 indicates that these are affected by considerable instrumental uncertainties. Simulation results are seen to agree reasonably well with the "average" of the experiments.

Dose Distributions

As a second example, simulated depth- and lateral-dose distributions for electrons with different energies in silicon (normal incidence) are compared with experimental data of Werner et al. [18] in Fig. 2. These distributions were measured from the current induced by the electron beam in a p-n junction. The simulated depth-dose distributions practically coincide with the experimental results. In order to account for the finite width of the junction and the electron diffusion length, simulated lateral-dose distributions have been convoluted with a Gaussian profile (with a fixed width, independent of the beam energy). The resulting curves agree satisfactorily with the measured data, the differences are probably due to spatial inhomogeneities of the junction. Thus, it may be concluded that PENELOPE gives a realistic description of electron trajectories and spatial dose distributions.

K-Shell Ionization Distributions

It should be noted that PENELOPE accounts for the average effect of inelastic collisions on the primary electron through an *approximate* DCS. To obtain reliable inner-shell ionization distributions, a more realistic cross-section model for this process should be considered. On the other hand, K-shell ionizations are relatively rare events, so that the direct scoring of ionizations along the generated electron tracks is very inefficient. The practical solution consists in generating the electron tracks with PENELOPE, and scoring the *expected* number of ionizations n in each free flight. To obtain n, only knowledge of the total ionization cross-section as a function of the electron energy, $\sigma_K(E)$, is required. In a free flight of length s, the expected number of ionizations is given by $n = s \cdot N\sigma_K(E)$, where N is the density of atoms of the considered (ionized) element per unit volume. The number n of ionizations (usually much less than unity) is scored at a point that is sampled uniformly along the length of the free flight.

In the examples given below, the number of inner-shell ionizations in each trajectory segment has been evaluated by means of approximate total ionization cross-sections commonly used in the literature. Results have been obtained with three different simple analytical formulae, namely, those of Powell [19], Worthington and Tomlin [20] and Gryzinsky [21]. We have also used the K-shell ionization cross-section proposed by Mayol and Salvat [22], which is based on the Born approximation and includes exchange and low energy (Coulomb) corrections. This cross-section is expected to be more accurate than the analytical formulae, especially near the ionization threshold.

Simulated depth distributions of ionization for copper and magnesium tracers in an aluminium matrix, obtained with the aforesaid total ionization cross-sections, are compared with experimental data [23, 24] in Fig. 3. The plotted $\Phi(\rho z)$ functions are defined as the ratio between the density of ionizations per unit depth in the semi-infinite specimen and that of an isolated film of negligible thickness. With this normalization, the value of the $\Phi(\rho z)$ function at the surface ($z = 0$), which coincides with the so-called backscattering factor, is independent of the adopted cross section. The interesting feature here is the clear influence of the adopted ionization cross section on the shape of the $\Phi(\rho z)$ function. Owing to the scarcity and limited accuracy of experimental data available, a definite conclusion about the relative accuracy of the various cross sections cannot be drawn. The cross section of Mayol and Salvat [22] seems to yield results slightly closer to the experiment, as could be expected from its physical content, and is the one used in the simulations that follow.

Simulated depth distributions of K-shell ionization of Zn atoms in a copper matrix for 20 kV and 30 kV electron beams are compared with experimental data of Brown and Parobek [25] in Fig. 4b. Apart from a slight difference in the positions of the maxima, the agreement between simulation and experiment is satisfactory. Figure 4a sets out $\Phi(\rho z)$ functions for K-shell ionization of carbon atoms embedded in a silver matrix for different beam energies. The simulation reveals that the value of $\Phi(0)$ is practically independent of the electron energy. This differs from the measured data [26]. Very likely, the variation of the measured $\Phi(0)$ with energy can be attributed to the finite thickness of the tracer layer used in the experiments.

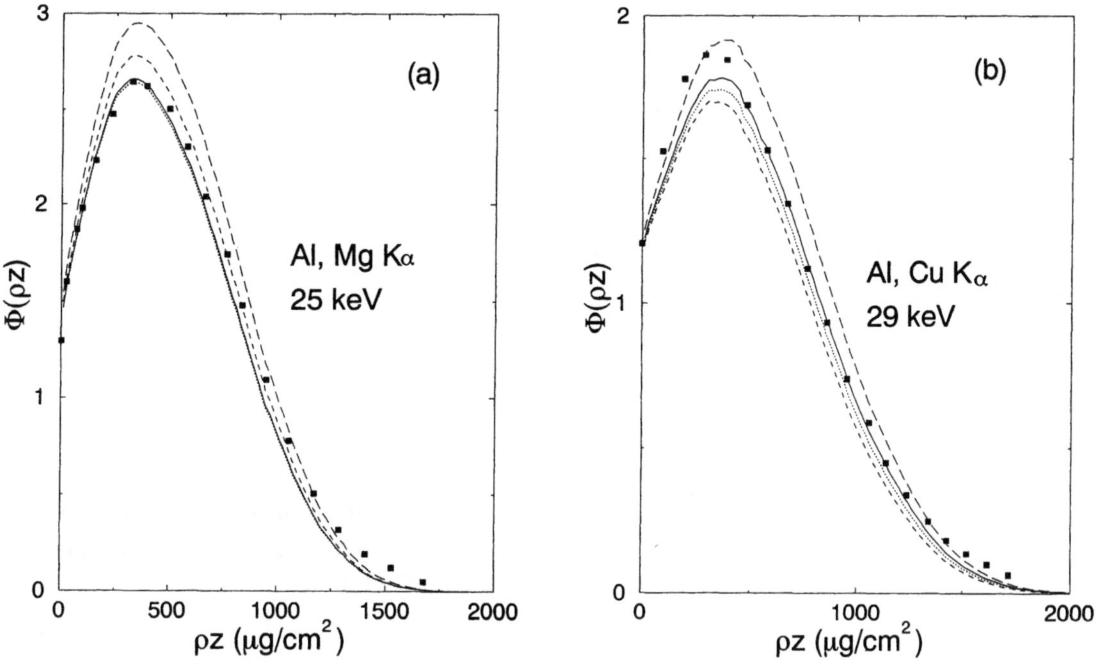

Fig. 3. Depth distributions of K-shell ionizations of Mg (**a**) and Cu (**b**) in an Al matrix for 25 and 29 kV electrons, respectively. Experimental data are from refs. [23] and [24]. Lines represent Monte Carlo results using different total ionization cross sections: Mayol-Salvat (solid), Powell (dotted), Gryzinsky (dashed) and Worthington-Tomlin (long dashed)

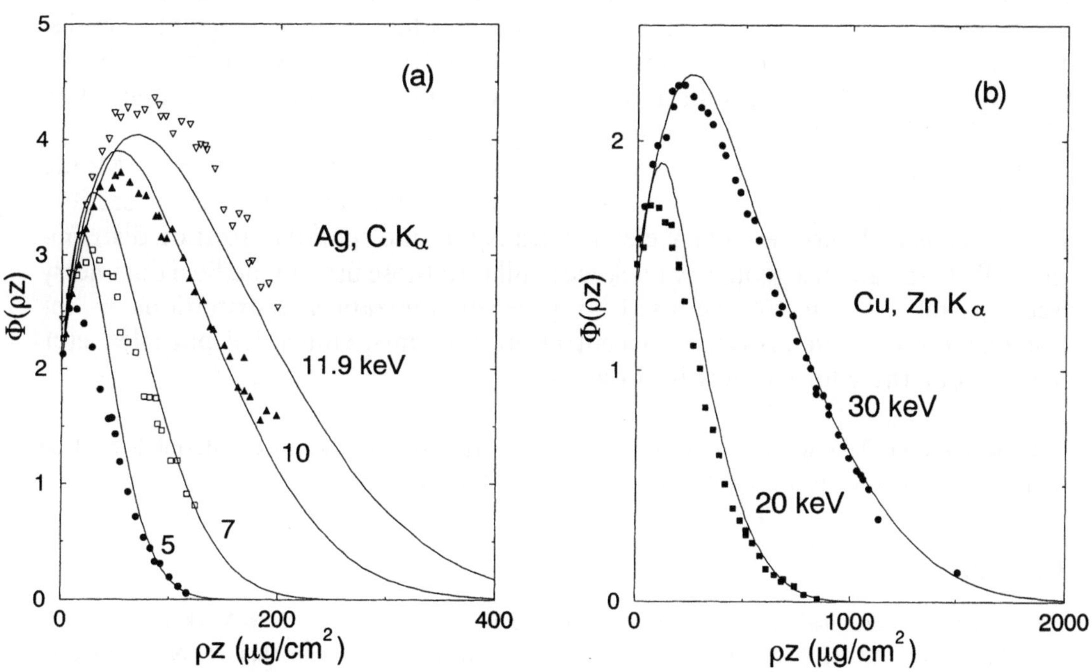

Fig. 4. Depth distribution of K-shell ionizations for a C tracer in an Ag matrix (**a**) and Zn tracer in Cu (**b**). Experimental data in part (**a**) are from Rehbach [26]; those in part (**b**) are from Brown and Parobek [25]. Solid lines represent results from Monte Carlo calculations

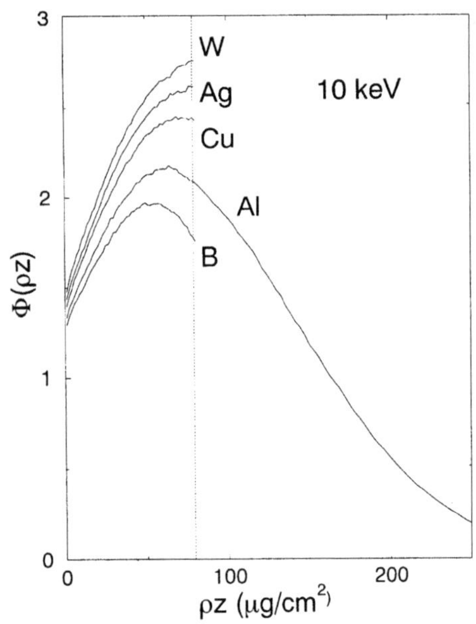

Fig. 5. Simulated depth distribution of ionizations for 10 keV electrons in a 300 nm Al film on semi-infinite B, Al, Cu, Ag and W substrates

Simulated $\Phi(\rho z)$ functions for a 300 nm thick aluminium layer on substrates of different atomic numbers are displayed in Fig. 5. There is a marked influence of the substrate composition on the film ionization distribution. The average number of ionizations per electron increases with the atomic number of the substrate and, at the same time, the point of maximum ionization gradually shifts to larger depths. Evidently, these facts are direct consequences of the stronger backscattering in heavy elements (see Fig. 1). Our results agree closely with those obtained by Karduck et al. [27] using a completely different Monte Carlo code.

In conclusion, we have shown that PENELOPE, a general-purpose Monte Carlo code, when combined with a sufficiently accurate ionization cross section, gives a realistic description of kV electron transport and spatial ionization distributions. With the aid of a geometry package similar to those used in medical dosimetry (see e.g. ref. [7]), it can be used to generate ionization distributions in inhomogeneous specimens (with concentration gradients, embedded particles, etc). Work along these lines is now in progress.

Acknowledgement. This work has been partially supported by the Comisión Interministerial de Ciencia y Tecnología (Spain) under contract PB92-0801-C02-01.

References

[1] L. Reimer, *Scanning Electron Microscopy*, Springer, Berlin Heidelberg New York Tokyo, 1985.

[2] P. Karduck, W. Rehbach, in: *Electron Probe Quantitation* (K. F. J. Heinrich, D. E. Newbury, eds.) Plenum, New York, 1991.

[3] A. P. Mackenzie, *Rep. Prog. Phys.* **1993**, *56*, 557.

[4] L. Reimer, D. Stelter, *Scanning* **1986**, *8*, 265.

[5] Z.-J. Ding, R. Shimizu, K. Obori, *J. Appl. Phys.* **1994**, *76*, 7180.

[6] J. Baró, J. Sempau, J. M. Fernández-Varea, F. Salvat, *Nucl. Instr. Meth. B* **1995**, *100*, 31.
[7] T. M. Jenkins, W. R. Nelson, A. Rindi (eds.), *Monte Carlo Transport of Electrons and Photons*, Plenum, New York, 1988.
[8] J. M. Fernández-Varea, R. Mayol, J. Baró, F. Salvat, *Nucl. Instr. Meth. B* **1993**, *73*, 447.
[9] F. Salvat, R. Mayol, *Comp. Phys. Comm.* **1993**, *74*, 358.
[10] F. Salvat, J. M. Fernández-Verea, *Nucl. Instr. Meth. B* **1993**, *73*, 447.
[11] D. Liljequist, *J. Phys. D Appl. Phys.* **1983**, *16*, 1567.
[12] M. J. Berger, S. M. Seltzer, *National Bureau of Standards Report NBSIR 82-2550*, Washington, 1982 (also available as *ICRU Report 37*, 1984).
[13] H. Hunger, L. Küchler, *Phys. Status Sol.* **1979**, *A56*, K45.
[14] H. Drescher, L. Reimer, H. Seidel, *Z. Angew Phys.* **1970**, *29*, 331.
[15] G. Neubert, S. Rogaschewski, *J. Phys. D Appl. Phys.* **1984**, *17*, 2439.
[16] G. Massoumi, N. Hozhabri, W. Lennard, P. Schultz, *Phys. Rev. B* **1991**, *44*, 3486.
[17] V. Cosslett, R. Thomas, *Br. J. Appl. Phys.* **1965**, *16*, 779.
[18] U. Werner, F. Koch, G. Oelgart, *J. Phys. D Appl. Phys.* **1988**, *21*, 116.
[19] C. J. Powell, in: *Electron Impact Ionization* (T. D. Märk, G. H. Dunn, eds.), Springer, Wien New York, 1985.
[20] M. Gryzinsky, *Phys. Rev.* **1965**, *A138*, 336.
[21] C. R. Worthington, S. G. Tomlin, *Phys. Rev.* **1956**, *43*, 839.
[22] R. Mayol and F. Salvat, *J. Phys. B At. Mol. Opt. Phys.* **1990**, *23*, 2117.
[23] R. Castaing, J. Descamps, *J. Phys. Radium* **1955**, *16*, 304.
[24] R. Castaing, J. Henoc, in: *4th ICXOM* (R. Castaing, P. Deschamps, J. Philibert, eds.), Hermann, Paris, 1966.
[25] J. D. Brown, L. Parobek, in: *6th ICXOM* (G. Shinoda, K. Ohra, T. Ichinokawa, eds.), *University of Tokio Press, Tokio, 1972.*
[26] W. Rehbach, *Ph.D. Thesis*, Aachen, 1988.
[27] P. Karduck, N. Ammann, W. P. Rehbach, *Microbeam Analysis* (J. R. Michael, P. Ingram, eds.), San Francisco Press, San Francisco, 1990.

Mikrochim. Acta [Suppl.] 13, 419–424 (1996)

Topography Development on Single Crystal MgO Under Ion Beam Bombardment

N. J. Montgomery* and D. S. McPhail

SIMS Group, Materials Department, Imperial College of Science, Technology and Medicine, Prince Consort Road, London SW7 2BP, England, U.K.

Abstract. Magnesium oxide (100) single crystals were analysed by secondary ion mass spectrometry (SIMS) using a xenon ion beam incident over a wide range of angles. Optical and atomic force microscopy (AFM) were used to assess the ion beam induced surface topography that developed in the SIMS craters. Severe topography developed over a range of angles between 30° and 45° and increased with increasing total ion dose. AFM was used to show that this topography first develops as etch pits which subsequently coalesce into a ripple like structure. The MgO sputter yield as a function of primary ion angle of incidence was calculated and showed a large amount of variability at angles where topography developed. The sputter yield at different positions on the same crystal was found to vary by up to 50% when the beam was incident at 40°. The sputter yield was found to correlate to crater bottom surface roughness and the crater depth.

Key words: SIMS, MgO, ripple topography, combined techniques (AFM, SIMS).

Secondary ion mass spectrometry is a surface/near surface analysis technique with an optimal depth resolution of a few nanometers due to primary ion beam mixing [1]. This depth resolution can be severely degraded by a number of factors, one of the more important being the development of topography in the crater bottom during analysis. Previous depth resolution studies have mainly been carried out on semiconducting materials [2]; there have been relatively few studies on ceramics [3].

Single crystal (100) MgO is increasingly used as a substrate material when growing $YBa_2Cu_3O_7$ based superconducting thin films because of its good lattice match. These films can be analysed using SIMS [4] in order to assess the quality of the films, in particular the amount of inter-diffusion of the different elements across the YBCO/MgO interface. Any development of surface topography will degrade the quality of the information obtained using SIMS by either broadening the interfaces or causing uncertainty in the depth calibration process.

* To whom correspondence should be addressed

In this paper a systematic study of the behaviour of (100) MgO under 15 keV xenon ion beam bombardment was carried out in order to define optimal conditions for SIMS analysis with respect to angle of incidence. At angles where topography does develop an explanation for the variability in sputter yield obtained under identical operating conditions is given. A possible theory for the development of the etch pits into a continuous ripple structure is also discussed.

Experimental

All experiments were carried out using an Atomika 6500 Ion Microprobe instrument. The primary ion beam conditions were a 15 keV xenon ion beam run at 50 nA current with a beam width of about 80 μm. The (100) single crystal MgO samples were commercially grown with a surface finish of under 2 nm root mean square (R.M.S.) (measured via AFM). Analyses were carried out under a vacuum lower than $1.3 \cdot 10^{-6}$ Pa and using electron beam flooding to allow for correct charge compensation across the analysed area. Craters were 200 μm × 200 cos θ μm, where θ is the angle of incidence with respect to the normal, and analyses were run for an hour unless otherwise specified. The angle of incidence was correct to within less than 0.5°.

AFM and crater depth measurements (using a Talystep) were carried out ex-situ shortly after removing the samples from the SIMS machine. Yield measurements were made by dividing the total number of atoms removed (based on a unit cell size of $(4.212\,\text{Å})^3$) by the total number of xenon atoms received (calculated using the primary ion beam current).

Results and Discussion

Figure 1 shows the amount of surface roughness present after analysis when the primary ion beam angle of incidence θ (measured to the surface normal) was changed

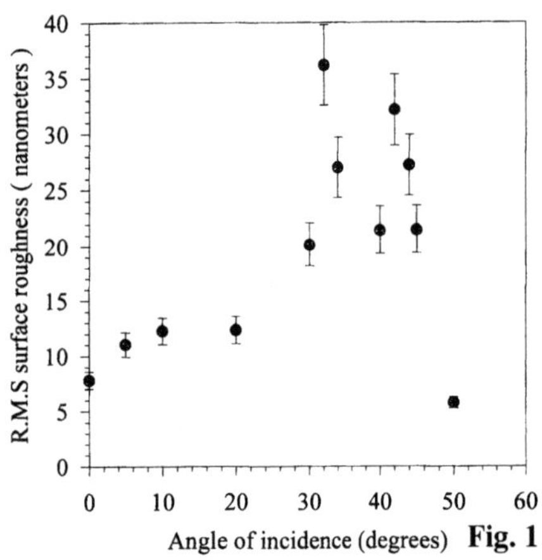

Fig. 1

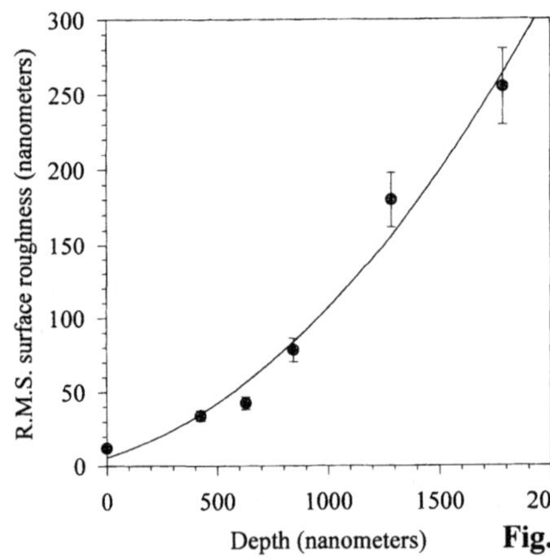

Fig. 2

Fig. 1. The variation of the crater bottom surface roughness of MgO for increasing angles of incidence θ (normal incidence = 0°) of the primary ion beam. The same total primary ion dose was used for all these analyses. The vertical error bars represent the variability observed when the experiment was repeated several times

Fig. 2. The evolution of the crater bottom surface topography in MgO for increasing total ion dose (or crater depth) with the beam incident at 38°. A second order least squares fit gives a correlation coefficient $R^2 = 0.9803$

from normal incidence up to 50°. The ion beam dose was 1.125×10^{15} Xe atoms for these analyses. In all cases the R.M.S. surface roughness in the crater bottom has increased from its original surface value of 2 nm. For measurements carried out below 20° this can be explained by the surface damage caused by the primary ion beam when it hits the sample surface. The energy the ion possesses is dispersed in the near surface region displacing atoms from their lattice positions, thus creating small scale random topography; this is especially true for heavy primary ions like Xe. At high angles of incidence (around 50° and above) the same phenomenon occurs, though the size of the topography has decreased because the penetration depth of the ions is shallower. In the intermediary region between 30° and 45°, the surface roughness is bigger and shows greater scatter when the experiment is repeated. AFM images of the surface indicate the presence of numerous pits across the crater bottom. The evolution of this topography with increasing ion beam dose was subsequently studied in more depth at 38° incidence and is shown in Fig. 2. Here the total amount of time the MgO was exposed to the ion beam was varied between 1/2 hour and 2 hours. The evolution of this surface topography displays a square dependence with respect to depth indicating an inherent worsening of the surface features observed with time. The use of AFM to visualise the way in which this topography evolved confirmed that the surface morphology tends towards a ripple like structure which increases in amplitude as the total ion dose increases (see Fig. 3).

The sputter yield of MgO shows a large variation when the angle of incidence is varied as seen in Fig. 4. There seems to be one region between 0° and 30° where values obtained were reproducible. This is also true at approximately 42° and above, though in this case the yield rises more sharply with increasing angle of incidence. However at the intermediate angles between 32° and 40° each yield has a large amount of scatter. For example, seven experiments were carried out at an angle of incidence of 40°. Under identical conditions of irradiation, i.e. beam energy, current, time and angle of incidence, the crater depths varied from around 700 nm to 1100 nm leading to a range of sputter yields between 0.95 and 1.55 for MgO/Xe. These intermediate angles are the same as the range of angles shown in Fig. 1 where there is a significant amount of surface roughness developing in the crater bottom. The data obtained from one particular angle (in this case 40°) was analysed in more detail to extract any trends in the results. First the sputter yield was plotted as a function of the R.M.S. crater bottom surface roughness (Fig. 5). Here a distinct trend was observed where high surface roughness at the end of the measurement was associated with a low MgO sputter yield. A correlation can also be made between sputter yield and the final crater depth (Fig. 6) where the shallower craters have a lower sputter yield. It must be emphasised that all these analyses were carried out under identical conditions with the same total ion dose (within experimental error). Clearly the rate of evolution of topography varies from place to place on the same crystal. If topography is severe then the final crater depth is shallower leading to a reduced MgO sputter yield. Whereas if the yield is high, the crater is much deeper and shows less final surface roughness. These variations cannot be explained by changes in the SIMS analysis conditions (e.g. the beam current was stable during each analysis session to better than 5%). It seems as though the final result depends on how quickly the surface roughness develops for a particular run. This surface roughness

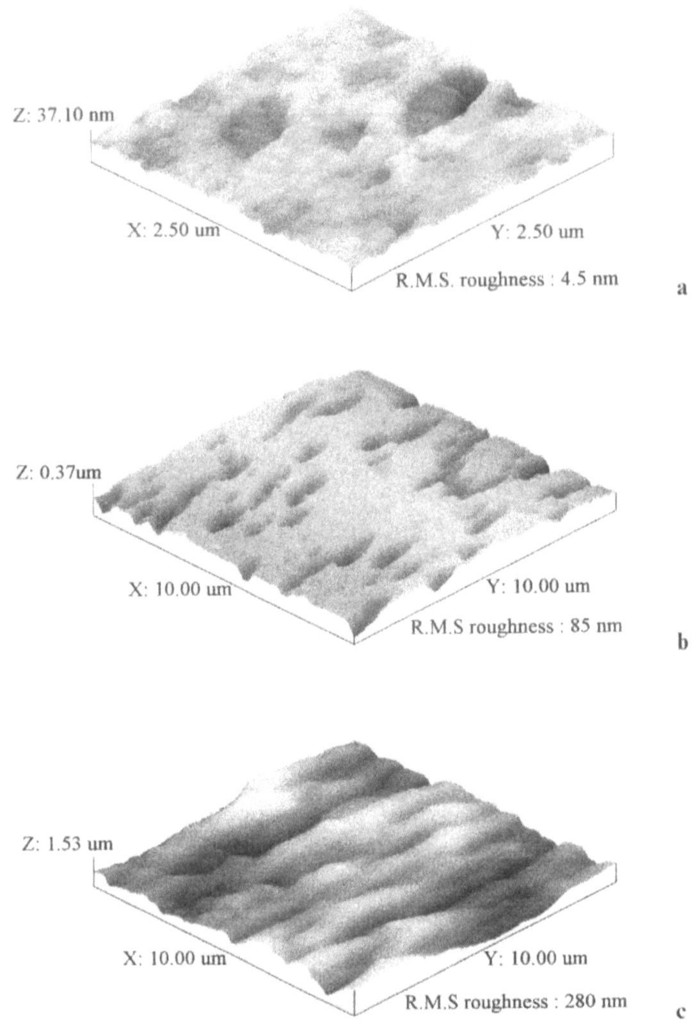

Fig. 3. A series of AFM images showing the evolution of surface topography with time from a series of etch pits to a ripple like structure. **a** $\frac{1}{2}$ hour, **b** 1 hour, **c** 2 hours. Obtained at 38° incidence

develops in a random manner and as yet the process which actually initiates the development of topography is unknown. This could be due to the small amount of topography created because of the ion beam bombardment or because of surface irregularities. At angles between 30° and 45° this roughness is prone to develop into a random array of small etch pits which are perpendicular to the primary ion beam direction. These features grow because of the yield differences that occur due to the etch pits presenting a range of angles to the primary ion beam. Prolonged exposure to the ion beam leads to these etch pits coalescing into a ripple like structure which will continue to grow as the ion dose received is increased. It seems probable that the sputter yield will decrease as the analysis proceeds in cases where significant topography is present.

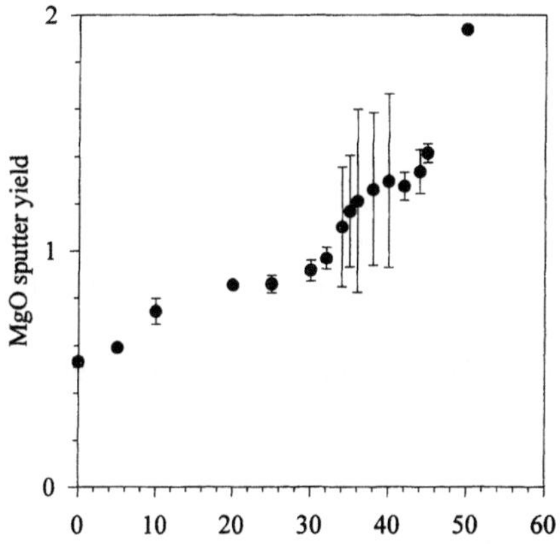

Fig. 4

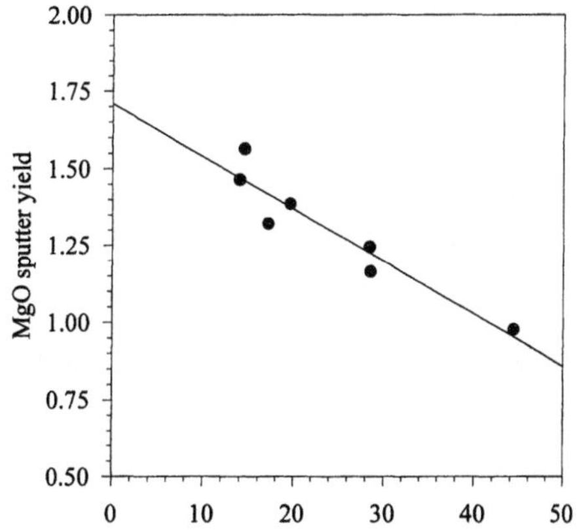

Fig. 5

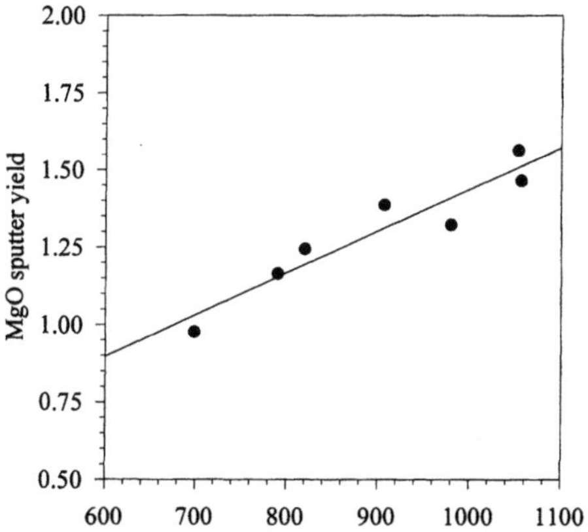

Fig. 6

Fig. 4. The variation of the average MgO sputter yield as a function of angle of incidence. Error bars represent the amount of scatter in the results for each particular angle. A variation of 50% is observed at 40°

Fig. 5. The correlation between the MgO sputter yield and crater bottom topography for a number of analyses conducted under identical conditions at an angle of incidence of 40°

Fig. 6. The correlation between the MgO sputter yield and crater depth for a number of analyses conducted under identical conditions at an angle of incidence of 40°

Conclusions

When analysing single crystal (100) MgO with SIMS it has been found that crater bottom topography develops between angles of incidence from 30° up to 45°. These angles must be avoided in order to obtain high depth resolution SIMS results. It is noted that positive primary ion beams are often incident at these angles on Cameca type instruments. The large scatter in the sputter yield at these angles of incidence is associated with the development of topography which worsens with increasing total ion dose. The actual initiating factor is as yet unknown but may be the small scale random topography caused by the ion beam bombardment sputtering process and the rapid change in sputter yield with angle which occurs around these angles.

Acknowledgements. This work has been funded by EPSRC.

References

[1] S. Hofmann, *Appl. Surf. Sci.* **1993**, *70/71*, 9.
[2] H. W. Werner, *Surf. Inter. Anal.* **1982**, *4*(*1*), 35.
[3] A. M. C. Kilner, J. A. Kilner, J. C. Elliott, G. Cressey, S. D. Littlewood, *Surf. Inter. Anal.* **1991**, *17*, 158.
[4] Y. Li, J. A. Kilner, T. J. Tate, M. J. Lee, Y. Li, R. E. Somekh, P. Przyslupski, *J. Appl. Phys.* **1995**, *78*, in press.

Mikrochim. Acta [Suppl.] 13, 425–433 (1996)

Determination of SPM TIP Shape Using Polystyrene Latex Balls

Péter Nagy[2,*], Géza I. Márk[1], and Erzsébet Balázs[1]

[1] MTA KFKI, Research Institute for Materials Science, H-1525 Budapest, P.O. Box 49, Hungary
[2] MTA KKKI, Research Institute for Chemical Science, H-1525 Budapest, P.O. Box 17, Hungary

Abstract. In scanning probe microscopy (SPM) imaging of the structure of the tip apex at a mesoscopic scale is extremely important. The present paper offers a simple way for estimating the shape of SPM tips. A standard sample consisting of a close packed monolayer of small polystyrene latex balls is used to characterise the tip shape. The image formation process is simulated by a matrix "convolution" method and the dependence of the SPM image on the tip shape is studied by calculating images of the test sample. Real pictures taken on the same test sample were used to calculate the shape of the tip utilising the known shape and size of the latex balls.

Key words: SPM, STM, AFM, image formation, tip.

Since their inception during the last decade, the efficiency of scanning probe microscopy (SPM) has inevitably been compared with traditional electron microscopy. Contrary to electron microscopy, very little or even no sample preparation is needed in the case of SPM investigations. This is an obvious advantage for biological samples when tedious preparation procedures may destroy the actual structure of the sample. The less complicated preparation on non life-science samples makes it convenient to apply SPM for other fields, too. The ability to investigate structures in their native environment, as in fluids or simply in air, is also very attractive. Furthermore, different types of the SPM family are capable of resolutions down to the atomic level for both conductive or non-conductive samples. To achieve the same resolution in electron microscopy makes it necessary to apply a very keen preparation technique, and to carry out more sophisticated experiments on expensive equipment. A further common advantage of all SPM techniques is both the high field of view and the high depth resolution. Nevertheless, the investigation of rough structures with SPM can produce distorted lateral features on steeply sloped samples although both the vertical and horizontal resolutions are excellent. Image distortions in SPM pictures are, in part, due to an inherent fault of the imaging process: while an electron beam can be extremely focused, an SPM probe always has

* To whom correspondence should be addressed

finite dimensions. In this report we show how the dimensions of the SPM tip can produce distortions in imaging and a numerical procedure will be proposed for producing a more accurate representation of the sample.

SPM images are computer reconstructions from digital data acquired during grid scanning of a probe tip across the sample. The sample topography is obtained by either keeping a specific tip-sample interaction at a constant level, or probing the strength of the interaction when the tip travels at constant height. Ideally, a single atom at the very end of the tip is responsible for the image formation. In real cases however, the interacting probe end is not a single atom [1]. Therefore, depending on the sample topography the interacting area moves around the tip. In many cases, e.g. when the periodicity of a flat sample is investigated, step heights or adsorbate coverage on flat samples are studied, the tip does not represent any problem because the measurements are unaffected by the non-ideal interactions between the structured tip and the sample. But in the case of rough samples, various distortions are obtained. An obvious case is when a cylindrical object, e.g. a tobacco mosaic virus or an M-13 bacteriofag [2], on a flat surface appears much wider than its height. Furthermore, in all cases where a feature on the sample has a steeper slope than that of the tip surface, the imaging process, changes the roles of the probe and the sample, the feature on the sample images the tip. This process is the so called "self imaging" and may be caused both by protrusions [3–8] or by holes [9].

Many authors have demonstrated the practical importance of tip shape reconstruction on specially prepared samples. Using mathematical morphology considerations it can be proven that there exist an upper and lower limit for the possible surface variations of a sample [10]. Assuming a model for the tip shape, the distance between these limits can be minimised and depending on the reality of the tip model an optimal sample surface can be fitted and a stray band may be calculated.

Similarly to electron microscopes, the SPM is able to zoom in over a very wide range of magnification, but the SPM uses a solid probe for acquiring the image. This probe should be short and "sharp on atomic scale" in order to obtain images of high magnification, but long and considerably rigid for large area scans. The rigidity of the tip in image formation is crucial. Elastic deformation of the tip during the scan leads to images with more enhanced distortions. The images taken in trace and in retrace motion become slightly shifted in the direction of motion of the tip, the amount of the shift depends on the strength of friction between sample and tip. This effect is especially pronounced in AFM images [11], where the friction or lateral force is always present, and the tip cantilever construction is inherently elastic in transverse direction, too.

Our attempt was to deal with the image distortion caused by the finite size of the tip. In a first step we constructed the mathematical model of the shape of the tip and that of the sample, and then applying these models we constructed the image of the model sample surface mimicking the scanning process. The next step was to reconstruct the shape of the real tip from an image of a real sample which corresponded to our model.

A program for calculating and visualising the model for the tip and for the sample as well as the tip reconstruction procedure was set up using the PV-WAVE programming environment [13] and the SXM-SHELL developed for evaluation of SPM images [14].

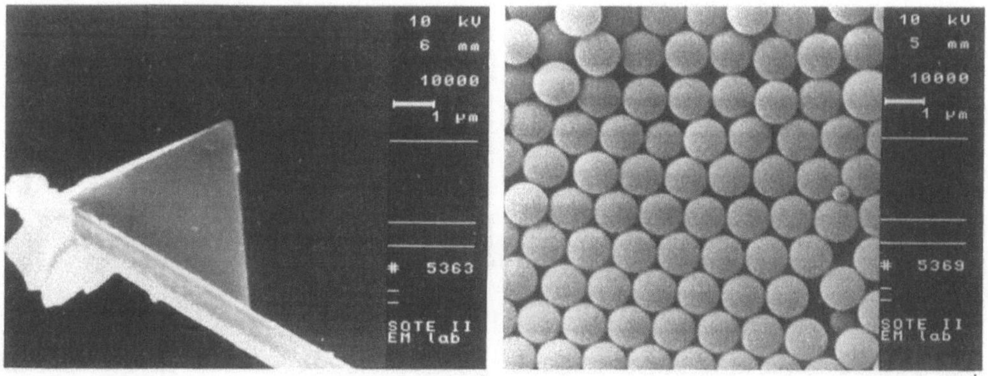

a b

Fig. 1. **a** SEM image of AFM tip, **b** SEM image of sample

Experimental

For the experimental work a commercial SPM instrument (both AFM and STM, Nanoscope III, Digital Instruments) was used. The AFM was equipped with a 125 µm scanner (J type). The images were recorded in the constant force mode using commercially available AFM tips made of silicon nitride. The tips have a pyramidal shape of 5 µm edge length with a slope of approximately 60°. Figure 1a displays the scanning electron microscopic image of the cantilever tip. The substrate is fixed in the tip holder in such a way, that the base of the tip declines 10° to the horizontal plane. A further small declination can be caused by the tripod fixation of the AFM head in both X and Y directions which was not considered in our calculations.

Latex balls were used as sample material. Close packed monolayers of these balls give a sample surface of very different slopes in an ordered way. So the detection of the tip shape is relatively easy, hence a possibility for its reconstruction is furnished. The latex balls are components of commercial paints: the glass transition temperature can be lowered to a temperature below the room temperature. In this case the suspension dries forming a film layer. However, at temperatures below the glass transition temperature the balls preserve their shape [12].

The latex balls are commercially available as water suspension for immunolabel purposes [15]. The original water suspension contains 2.5% solid latex consisting of 1.053 µm diameter latex balls. The standard deviation in diameter is 0.001 µm. For AFM sample preparation the original water suspension was diluted in ratio 1:3 with double distilled water and a small amount of an aqueous solution of sodium hexametaphosphate (calgon) was added to lower the wetting angle on the glass. Glass cover sheets were cleaned in chromosulphuric acid and purified with distilled water. The suspension was dropped on a freshly cleaned glass plate, and dried under constant vibration. This process resulted in a nearly perfect close packed layer of latex balls. Fig. 1b displays a scanning electron microscopic image of this sample.

Imaging Model

Image formation in SPM is a "convolution" process between tip and sample. On atomic scale images the image is formed by "one" atom at the very tip. In this case the image quality is a real proof of tip quality, the deviation from it causes a clearly detectable decrease in image quality. On a mesoscopic scale and, especially, on rough samples, image formation is not a "one atom process" but a geometric "convolution" of sample and tip shape. In this "convolution" the sample and the tip are equivalent,

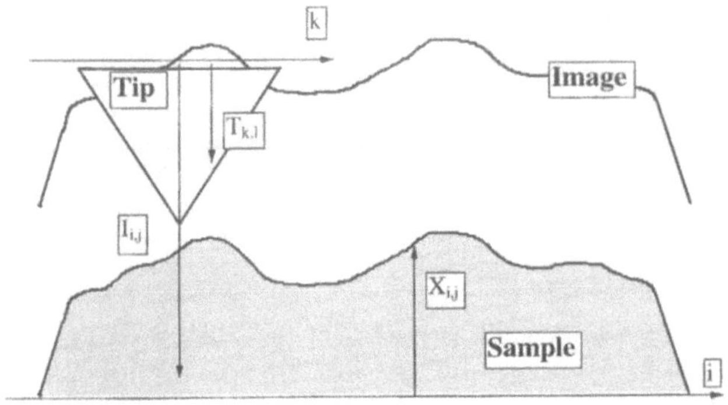

Fig. 2. Image formation in an SPM device

it is upon us to consider the sample or the tip as the imaging probe. The only advantage the tip has is its sharpness. If the sample has very sharp features, we obtain clear images of the tip [7].

Figure 2 shows the tip during acquisition of one line scan. Let us represent the sample with the matrix X_{ij}, the elements of which are the sample heights over a grid lying in the x-y plane. Let the grid spacing be equal to the pixel spacing of the image. The tip is represented by a matrix $T_{k,l}$ containing the tip height values over the same grid. During image acquisition the tip is gliding over the sample. In our model the middle of the tip matrix is repositioned over the grid points of the sample. All positions correspond to an image pixel and in all positions the image height (I) is the maximum of the sums of heights $X + T$ in coinciding grid points of tip and sample, or:

$$I_{i,j} = \max_{k,l}(X_{i+k,j+l} + T_{k,l})$$

This is a very straightforward procedure for computer simulation. We constructed a model of our sample consisting of two-dimensional close packed balls of 1 µm diameter. The tip represented an AFM tip with 10° declination and a pyramidal form of 5 µm square base and 5 µm height. Sample and tip are shown in Fig. 3, the

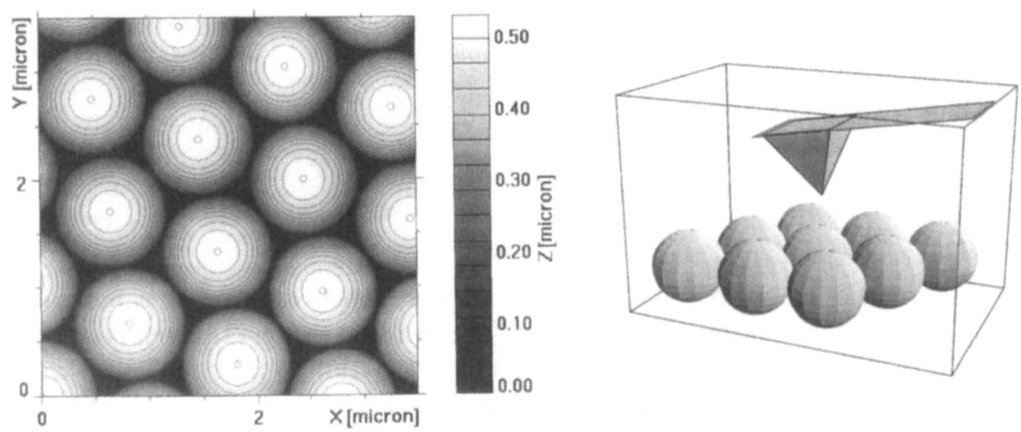

Fig. 3. The calculated model of the sample (left) and 3D view of tip and sample (right)

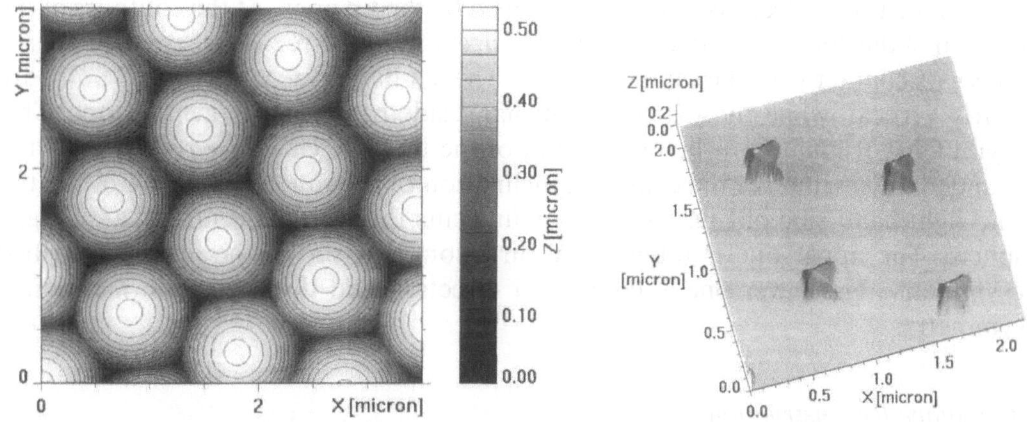

Fig. 4. The model image (left) and the reconstructed tip apices using this image in 3D view (right)

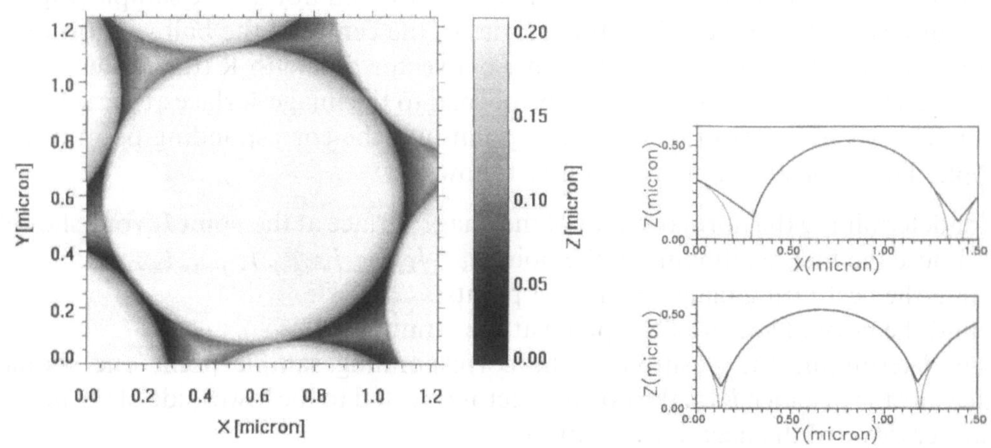

Fig. 5. The difference between model sample and its image, "pots" (left) and line cuts of a sphere of the model image (right)

simulated image is on the left hand side of Fig. 4. The simulated image differs considerably from the model of the sample. On the edge areas of the spheres, where they represent steep slopes, the side walls of the tip glide on a certain point of the sample. This causes nearly flat areas with the slope of the tip sides. An additional asymmetry is introduced by the fact, that the tip is mounted in 10° declination. This causes a flat surface on the right hand side of every ball. This asymmetry can be seen on the left hand side of Fig. 5, showing the line cuts through a ball in X, and in Y directions, respectively.

In everyday practice, the question arises how much an image is distorted. The distortion can be calculated by subtracting the model sample from the model image. The surface obtained this way is shown in the left hand side of Fig. 5. It is known that while taking the scan, in certain areas not the tip apex but its side walls get in contact with the sample. The regular arrangement in the image is very similar to pots in bottom-view. The bottom of the "pots" are flat, indicating that a large part of the image is distortion free because it was imaged by the tip apex. At the contours of the bottom of the "pots" the edge is rounded off, showing the region where the tip apex looses its contact with the sample surface. The left hand side walls of the "pots" are nearly vertical, while those at the right hand side have a more gentle slope. This asymmetry originates in the declination of the tip. The changing thickness of the rounded area witnesses the changing inclination of the touching surface of the tip. The right hand side of Fig. 5 shows the line cuts through the centre of an imaged sphere. The upper one is taken in X direction, and shows the above described asymmetry. The lower one is taken in Y direction and it has the just mentioned symmetry.

Tip Shape Reconstruction

To reconstruct the tip and sample, as a first step we need a procedure for determining the tip surface on the basis of an image. This is easy, if we invert the image formation process in our mind. Let the latex ball image the tip and not tip the sample. Figure 6 demonstrates this process. The trajectories of the centre of the ball will form the image surface. At every point of the sample a vector of length R (the radius of the balls) is constructed. This vector must be normal to the image surface at the selected point. This vector transforms the image point into the corresponding point of the sample. The reconstruction process is as follows:

- For determining the normal vector of the image surface at the point $I_{k,j}$ one should
  - choose the first neighbours of the point $I_{k,j}$: $I_{k+1,j}$, $I_{k-1,j}$, $I_{k,j+1}$, $I_{k,j-1}$,
  - find the best fitting plane for these 4 points,
  - find the normal vector of the plane at the point $I_{k,j}$.
- For determining the position of the corresponding sample point one should measure the distance R on the normal vector directed to the lower side of the image surface (toward the negative Z values).
- The resulting point is considered as one point of the tip surface.
- One should repeat this calculation for all image pixels, resulting in a set of sample points.

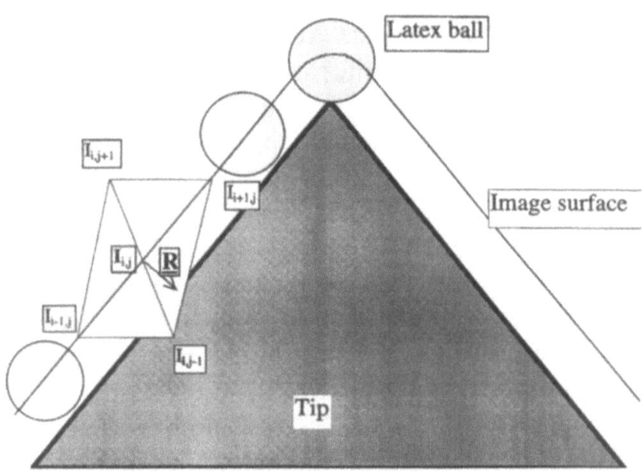

Fig. 6. Tip reconstruction

These points do not lie on an equidistant grid, hence the surface points in the original equidistant X-Y grid will be obtained by linear interpolation.

Because the tip is imaged by all the spheres, this procedure leads to a regular arrangement of the transformed tip apices with the same symmetry as the latex ball monolayer has.

In order to show the efficiency of the suggested procedure in a first step the simulated image was reconstructed. On the right hand side of Fig. 4 the reconstructed tip is shown at an enlarged scale. The three bumps in the image are the reconstruction of the tip on 3 of the balls. The bumps are identical, because the image is perfectly periodic by construction. The 100–200 nm size of the bumps corresponds to that part of the tip, which takes part in the SPM image formation process. The edges of the tip are well reconstructed, but only a much smaller part of the side walls is shown because the areas not touching the spheres are not reconstructed.

Results and Discussion

Two different AFM tips were used for taking our images. Figure 1a shows an SEM image of a nearly perfect tip. It has a pyramidal form as described above, with an apex radius of about 80 nm (measured in an SEM image after demounting the tip). The left hand side of Fig. 7 shows the image taken by this tip. The shape of the balls is less regular than in the simulated image, caused by the sampling noise of the image. The image was taken at 256*256 pixels resolution. On the other hand the asymmetry in the shape of the spheres is evident. Similarly to the model-picture the more gentle slopes found at the right hand sides of the spheres are due to the declined tip. On the right hand side of Fig. 7 the bumps representing the reconstructed tip are shown on an enlarged scale. The bumps have different features on their surfaces due to the instrumental noise. The size of the bumps is about 100 nm larger than that of the simulated tip, indicating the finite size of the real tip apex. The lack of edges of the pyramids and the noise bands between the neighbouring bumps can be explained by the poor resolution of the image.

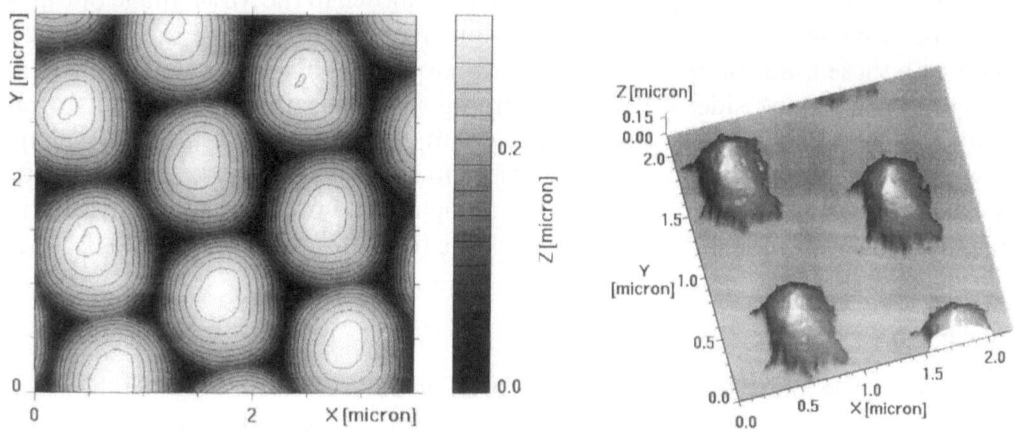

Fig. 7. Image taken with the "good" tip (left) and the reconstructed tip apex using this image (right)

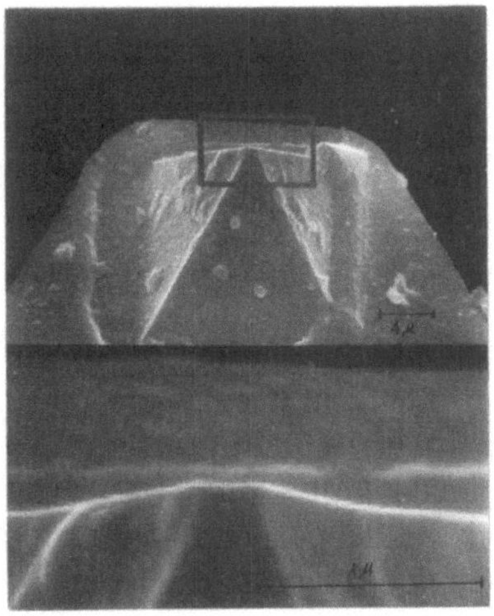

Fig. 8. SEM image of the "bad" tip

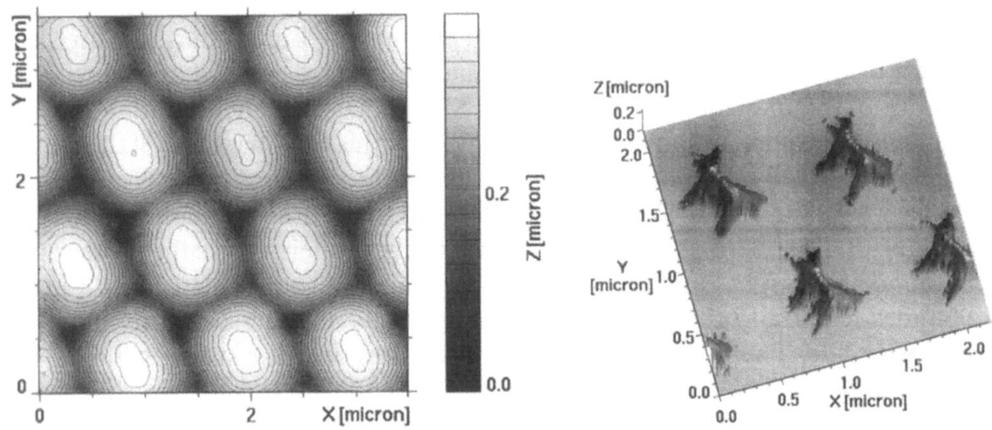

Fig. 9. Image taken with the "bad" tip (left) and the reconstructed tip apex using this image (right)

The second tip is an irregular one. Its shape is much more like a house roof than a pyramid: it has no perfect apex, but an edge as is shown in the SEM image of Fig. 8. This type of wrong tips occurs relatively frequently in our batch of tips. In images taken with these kinds of tips, the image distortion is not always obvious, as can be observed on left hand side of Fig. 9. Assuming that the sample was not distorted during the measurement, the tip reconstruction procedure leads to the tip shape shown on the right hand side of Fig. 9. This is in surprisingly good agreement with the SEM picture of Fig. 8. One can easily recognise the presence of the pyramidal edges as well as some decrease in noise level due to the higher resolution (512*512).

Conclusion

We modelled the image formation process of SPM devices. As an example the image of a monolayer of close packed spheres was simulated. Based on this simulation a procedure was set up to reconstruct SPM tip shape using a standard sample consisting of polystyrene latex balls. The method reveals the form of the tip apex on a 100–500 nm scale, hence it fills the gap between atomic scale investigations and usual SEM imaging. Our attempt is to continue this work and set up a procedure for the reconstruction of the sample, too.

Acknowledgement. The work was supported by the OTKA grant Nos T4324, and T017626.

References

[1] L. P. Biro, G. I. Mark, E. Balazs, in: *NATO-ASI School on Nanophase Materials* (G. C. Hadjipanayis, R. W. Siegel, eds.), Kluver, 1994, pp. 205–208.
[2] T. Thundat, X. Y. Zheng, S. I. Sharp, D. P. Allison, R. J. Warmack, D. C. Joy, T. L. Ferrell, *Scanning Microscopy International, vol. 6*, 1992, pp. 903–910.
[3] A. Bottino, G. Capannelli, A. Grosso, O. Monticelli, O. Cavallerri, R. Rolandi, R. Soria, *J. Membrane Sci.* **1994**, *95*, 289.
[4] Z. Wang, C. Bai, C. Dai, G. Huang, P. Zhang, Y. Zhang, W. Chu, L. Qiar, Y. Wang, *J. Vac. Sci. Technol. B* **1994**, *12*, 2456.
[5] M. A. George, A. Burger, W. E. Collins, J. L. Davidson, A. V. Barnes, N. H. Tolk, *J. Appl. Phys.* **1994**, *76*, 4099.
[6] K. L. Westra, D. J. Thomson, *J. Vac. Sci. Technol. B* **1994**, *12*, 3176.
[7] F. Atamny, A. Baiker, *Surface Sci.* **1995**, *323*, L314.
[8] Y. Umehara, Y. Ogiso, K. Chihara, K. Mukusa, P. E. Russel, *Rev. Sci. Instr.* **1995**, *66*, 269.
[9] T. O. Glasbey, G. N. Batts, M. C. Davies, D. E. Jackson, *Surface Sci.* **1994**, *318*, L1219.
[10] J. S. Villarrubia, *Surface Sci.* **1994**, *321*, 287.
[11] J. O. Tegenfeldt, L. Montelius, *Appl. Phys. Lett.* **1995**, *66*, 1068.
[12] V. V. Granier, *Tappi J.* **1994**, *77*, 220.
[13] PW-WAVE Advantage, Visual Numerics Inc., 1993, Suite 400, 9990 Richmond Avenue, Houston, Texas 77042, USA.
[14] SXM-SHELL U. Basel, Inst. Physik, Kondensierte Materie, Switzerland.
[15] Polysciences Inc., Polybead Polystyrene Microspheres.

Mikrochim. Acta [Suppl.] 13, 435–442 (1996)

Combined Characterization of Nanostructures by AEM and STM

Vladimir P. Oleshko[1,*,**], Renaat H. Gijbels[2], Wim A. Jacob[3], and Michael V. Alfimov[1]

[1] N.N. Semenov Institute of Chemical Physics, Russian Academy of Sciences, 117421 Moscow, Russia

[2] Department of Chemistry and [3] Department of Medicine, University of Antwerp (UIA), B-2610 Antwerpen, Belgium

Abstract. Analytical electron microscopy methods (electron energy-loss spectroscopy (EELS), zero-loss electron spectroscopic imaging (ZLESI) and energy-dispersive X-ray spectroscopy (EDX) in conjunction with scanning tunneling microscopy (STM) have proven useful for the combined characterization of the morphology, topography, crystallinity, electronic structure and elemental composition of small Ag and Ag_2S particles, silver dendritic crystals and nanocrystalline films. Nanotopography of deposited Ag films has been detected by STM with a lateral resolution up to 0.2 nm. EELS spectra of submicron and nano-sized particles contained low intensity bands, which were assigned to the Ag $M_{4,5}$-edge at about 367 eV, $SL_{2,3}$-edge at 165 eV and OK-edge at 532 eV. The energy-loss near-edge structure of the delayed Ag $M_{4,5}$-edge concerned with transitions from Ag $3d^{3/2}$ and $3d^{5/2}$ electronic states above Fermi level has been evaluated.

Key words: Ag and Ag_2S small particles, AEM (EELS/ZLESI,EDX) STM.

Imaging of nanostructures, e.g., small particles, nanophases and nanocrystalline films has been performed with various types of microscopes due to its significance in fundamental problems (quantization and nonlinear optical effects) and numerous technical applications (catalysis, microelectronics, etc.). Analytical electron microscopy (AEM) can combine imaging in the conventional transmission (CTEM), scanning transmission (STEM) or scanning (SEM) modes with a variety of signals, contrast effects and analytical facilities. For small particles, a number of publications has been devoted to electron-loss spectroscopic (EELS) studies of plasmons between 5 and 25 eV energy-loss, which imply collective oscillations of the quasi-free electron gas [1, 2]. Elemental analysis can be performed by recording the core edges (EELS)

* Present address: Department of Chemistry, University of Antwerp (U.I.A.), Universiteitsplein 1, B-2610 Antwerpen-Wilrijk, Belgium

** To whom correspondence should be addressed

and characteristic X-rays (energy-dispersive X-ray spectroscopy (EDX)). Moreover, the detailed profile of the energy-loss spectrum at the core threshold, energy-loss near-edge structure (ELNES), is very sensitive to the local density of vacant electron states with given quantum symmetry. The great advantage of EELS and EDX spectroscopies is that this type of analyses can be made on an object down to the submicrometer and even subnanometer sizes [3, 4].

With the recently developed scanning tunneling microscopy (STM) and atom force microscopy it became possible, additionally to CTEM/STEM, to study both periodic and aperiodic structures in the 3D-real space with lateral resolution up to 0.1–0.2 nm and vertical resolution up to 0.01–0.005 nm [5, 6]. This is much better than even for dedicated SEMs equipped with field-emission guns. In principle, STM measurements can be performed under ambient conditions down to atomic resolution without further sample preparation. Therefore it would be promising to combine the versatile AEM facilities and scanning probe techniques in order to characterize many-particle-systems at the nanostructural level. In this study EELS, EDX, zero-loss electron spectroscopic imaging and selected-area diffraction (ZLESI/ZLSSAED), and STM were applied for the combined structural and analytical characterization of a series of ultradisperse species (Ag and Ag_2S nanoparticles, submicrometer-sized Ag dendritic crystals and nanocrystalline Ag films) which can show pronounced size effects and transformations as demonstrated by unique photocatalytic activities [7] and surface enhanced Raman scattering [8].

Experimental

Island and continuous films were prepared by thermal evaporation of Ag powder at 10^{-3} Pa onto carbon supports and freshly cleaved mica. Ag colloids were obtained by reduction of a $AgNO_3$ solution in the presence of 0.1% polyvinyl alcohol. Self-supported dendritic crystals were prepared *in situ* by dark reduction of a saturated $AgNO_3$ solution on a carbon film. Nano-sized Ag_2S particles were obtained by fast mixing of $Na_2S{\cdot}9H_2O$ and $AgNO_3$ solutions in the presence of 1% gelatine. Particles were deposited on microscopic grids covered with carbon or holey Formvar films.

Ag films were imaged in air at 1 nA tunneling current and 10–100 mV bias voltage with a home-made STM operating in the constant-current mode [9]. The quality of the tungsten tip was tested by obtaining atomically resolved images on cleaved highly oriented pyrolytic graphite. A ZEISS EM 902 electron microscope with an integrated energy filter was used for EELS/ZLESI measurements. Smoothing was applied to remove noise from the spectrum profiles by the use of Savitsky-Golay filtering [10].

Results and Discussion

Figure 1 presents ZLESI and STM images of a series of Ag and Ag_2S species which demonstrate a variety of morphologies of ultradisperse matter governed by the chosen preparation techniques. Selected examples, i.e. non-aggregated thermally deposited Ag nuclei at an initial stage of formation of an island film (a), randomly aggregated Ag colloids (b), Ag dendritic crystals (c), Ag nanocrystalline films (d, e) and non-aggregated Ag_2S colloidal nanoparticles (f), as one can suppose, may be useful to evaluate the potentialities of the applied techniques in regard to the characterization of nanostructures at different stages of the evolution from the ultradisperse state via random aggregation and crystallization to the bulk state.

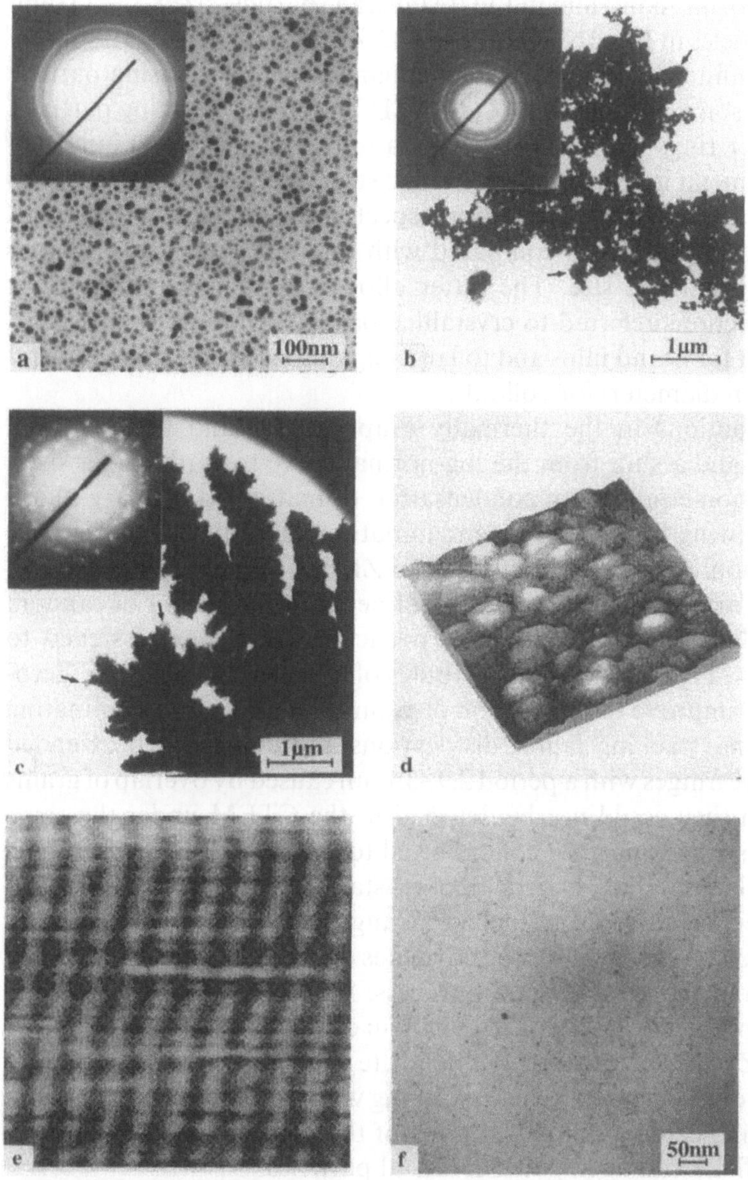

Fig. 1. Electron and STM images of Ag and Ag_2S species. Left insets in the micrographs a–c present ZLSSAED patterns. **a** ZLESI, Ag island film (nucleation and initialization of island structure). **b** ZLESI, randomly aggregated Ag colloidal particles stabilized with 0.1% PVA. Arrows show globules of the protective polymer. **c** ZLESI, self-supported Ag dendritic crystals grown on carbon support. Arrows show Ag nuclei 20–50 nm in size formed in the vicinity of an edge of the bulk dendrite. **d** STM, 3D-view of a Ag island film (coalescence of islands and channel formation), $521 \times 512\ nm^2$. **e** STM, top view of the Ag film, $2.1 \times 2.1\ nm^2$. **f** ZLESI, non-aggregated Ag_2S nanoparticles stabilized with 1% gelatine

Zero-loss micrographs of an island film in Fig. 1a (average particle size 8.8 ± 1.0 nm) and of colloidal Ag particles in Fig. 1b (average size 42.7 ± 2.2 nm) revealed a slightly higher contrast than unfiltered ones. The effect enhanced with increasing particle size. Bend contours on some particles and ZLSSAED patterns containing partially discrete Debye-Scherrer rings (see insets in Fig. 1a and c, respectively) indicated a crystallization of the metal in a face centered cubic structure. Diffraction patterns revealed no significant internal differences with respect to the positions of the rings. However, an improved resolution as compared with unfiltered ones was demonstrated with increasing particle size. The latter allowed to detect more clearly numerous discrete reflections referred to crystalline particles of more than several nanometers in diameter for island films and to larger aggregated crystallites (several hundreds nanometers in diameter) for colloids.

Particle size distributions in the thermally evaporated island films showed a bimodal appearance and a shift from the log-normal curve towards larger sizes. This was related to a non-equilibrium condensation of matter on primary nuclei tentatively with multitwins [11]. Subsequent initiation of Ostwald coalescence resulted in the formation of island structures. The ZLESI micrograph in Fig. 1b shows the topological arrangement of a statistical network consisting of random aggregates and irregular chains of coagulated particles, which were assigned to quasi-fractal structures [17]. Arrows show globules of a protective polymer. Zero-loss filtering enabled to improve the resolution of grain boundaries and dominating crystalline defects (twins, stacking faults, dislocations and complicated extended defect structures). Moiré fringes with a period 2.9–3.1 nm caused by overlap of grains were observed, though they could not be detected in the CTEM under the same conditions. The size distribution in the colloid fitted to the log-normal curve.

Ag dendritic crystals exhibited 3D-growth of crystalline matter preferentially in lateral directions. Their anisotropic branched arrangement (see Fig. 1c) was influenced mainly by diffusion of silver ions to active sites on the substrate in the course of the fast reduction and the crystallization process. Random branches probably grew from numerous sphere-like nuclei of 20–50 nm in diameter formed at the initial nucleation centres. A ZLSSAED pattern of a dendrite shown in the inset in Fig. 1c demonstrates double point reflections due to twinning with a mirror reflection about the $\{111\}$ planes; this indicated a higher ordering of the matter as compared with polycrystalline island films and aggregated colloidal particles.

The large scale STM topographic images of Ag films on mica, exhibiting (111) surfaces, revealed characteristic atomic flat protrusions of 10–30 nm in diameter and of 3–10 nm in height (see Fig. 1d). Some elongated regions of 100–200 nm in length and of 10–30 nm in width were randomly observed as was shown by profile analysis. In line with electron microscopic observations of earlier stages of the film formation, this points out that islands grow vertically and laterally. Consequently they become elongated and join to form a network structure by coalescence of secondary nuclei.

An assignment of small scale images (see Fig. 1e) was more complicated. It is known that high-resolution STM imaging of silver surfaces in air is still rather difficult [12] because of a number of factors, i.e. chemisorption of air components, surface diffusion effects, etc. Better results on the STM visualization of the metal films and particles with atomic resolution are achieved in ultra-high vacuum [13], though topographic images demonstrating atomic resolution were recorded in air

[5, 14, 15]. Therefore STM experiments were done in dry atmosphere with freshly prepared Ag films. One can suppose that high-resolution STM images will reveal a distorted lattice from a semiconducting oxidized surface layer of about monolayer thickness. This is in agreement with data of EELS analyses of the island films and colloids. Non stable I-V curves indicated the presence of contaminants and surface instability mainly due to oxidation processes and adsorption of water vapour [16]. Another reason for the increased noise of tunnel current and image distortions may be a surface diffusion of Ag atoms additionally induced by interaction with the tip.

So, in addition to the evaluation of morphologies, inner crystalline structure and size distributions by energy-filtering TEM, STM has provided substantial information on the nanotopography of thermally deposited Ag films which can be used for a detailed quantitative description of 3D-nucleation and growth stages of the films. STM measurements can be made down to atomic resolution even under ambient conditions, however the effect of the surface contamination should be taken into account.

A local coverage of the clusters in the island film of about $1.7\,10^3\,\mu m^{-2}$ already allowed the detection of plasmon excitations and of the low intensity Ag M ionization edges in line with data of EDX/STEM spot analysis. In order to avoid interference in the low-loss region of silver and carbon support, self-supported Ag dendrites were studied. Recorded EELS spectra (see Figs. 2a–c) contained a series of plasmon losses and inner-shell elemental bands attributed to silver (Ag $M_{4,5}$-edge at about 367 eV), oxygen (O K-edge at 532 eV) and carbon (C K-edge at 284 eV, not shown). In the case of self-supported dendritic crystals, thin film structures formed by numerous sphere-like nuclei of several nanometers in size in the vicinity of the bulk crystal edge and/or between neighbouring dendritic branches (see Fig. 1c) have been found the most suitable for EELS measurements. Figure 2a shows in the low loss region a main maximum at 26.6 eV, besides the broader losses at 5 eV, 10.5 eV, 20 eV and 33 eV, respectively. The broadened maximum at about 5 eV (± 2 eV) is probably due to an overlap of the interband transition of Ag expected at 3.9 eV, the volume plasmon loss at 3.7 eV and the surface plasmon loss at 3.6 eV [18]. Hence, the multiple losses between 3 and 40 eV in low-loss spectra were referred to oscillation of the imaginary part of the energy-loss function Im $[-1/\varepsilon(E)]$ caused by interband electron transitions of underlying 4d electron shells of low binding energy mixed with the conduction band [19]. Weak bands between 50 and 80 eV obviously can be attributed to the minor delayed Ag $N_{2,3}$-edge at about 60 eV. Unfortunately, the bulk and surface plasmons in the nanoparticles deposited on thin films were masked by the carbon $\pi + \sigma$ plasmon in the support. A shift of the plasmon maximum with decreasing Ag particle size to lower energy-losses (Ag_2S particles exhibited similar behaviour) was referred to the contribution of surface plasmons (see Fig. 2b). Results obtained point out that different shape of the low-loss spectra in Figs. 2a and b is affected by the influence of the carbon support, particle packing density and morphology. The net intensity of the Ag $M_{4,5}$-edge increased with the size of the analyzed particles (see Fig. 2c). The presence of oxygen obviously indicated a partial oxidation which occurred more intensively on unprotected particles. The smoothing allowed to reveal a weak Ag M_3-edge at 572 eV and Ag M_2-edge at 602 eV.

The delayed Ag $M_{4,5}$-edge arises due to a centrifugal potential barrier making it difficult for $3d^{3/2}$ and $3d^{5/2}$ electrons with non-zero initial angular momentum

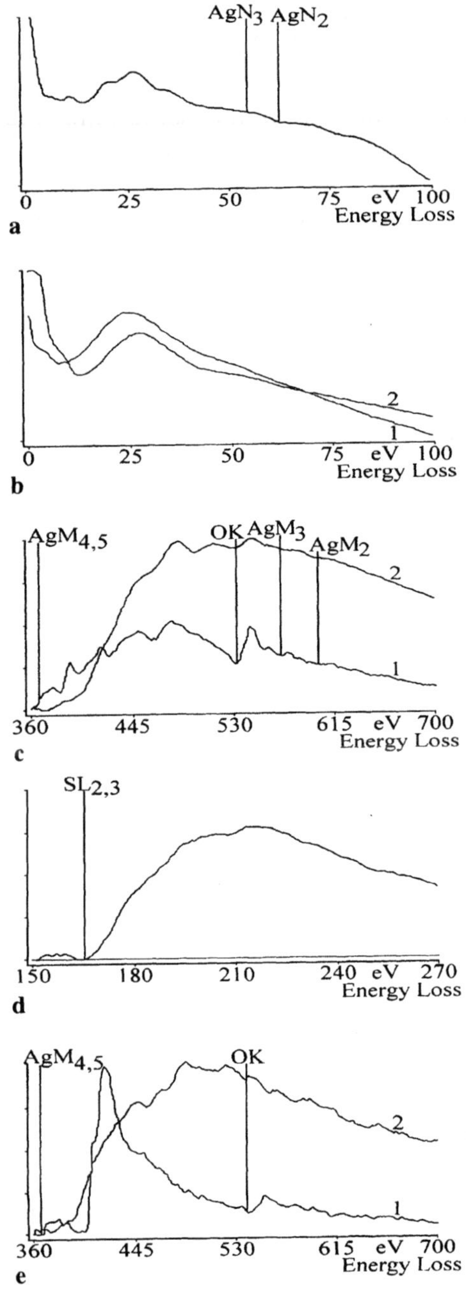

Fig. 2. EELS spectra of Ag and Ag_2S species. The net Ag $M_{4,5}$-edge and the S $L_{2,3}$-edge were obtained by background subtraction from initial spectra following smoothing. **a** Thin film area formed by nuclei in the vicinity of a wedged edge of a self-supported Ag dendrite. **b** Low-loss region; Ag nanoparticles in the island film (curve 1) and aggregated Ag colloid (curve 2). **c** Net Ag $M_{4,5}$-edge, Ag nanoparticles in the island film (curve 1) and aggregated Ag colloid (curve 2). **d** Net S $L_{2,3}$-edge, Ag_2S nanoparticles. **e** Net Ag $M_{4,5}$-edge, Ag_2S nanoparticles (curve 1) and aggregated Ag_2S colloid (curve 2)

($1 = 2$) to make a transition to a continuum state of low kinetic energy close to the edge threshold at 366.7 eV [19]. The edges were usually delayed by 28–33 eV. Calculations of the Ag $M_{4,5}$-edge profile according to the Hartree-Slater method [20] show that atomic effects appear to dominate the EELS spectrum as compared with solid state factors. At low continuum energies and momentum transfer the edge intensities are small because only the 3d-4p transition contributes at the threshold. The intensity then rises steeply after the energy exceeds a certain value to enable the continuum wave functions to overlap with the 3d orbital [19]. Optically forbidden transitions 3d-4s give rise to a maximum at nonzero momentum transfer near threshold, but these are swamped by the 3d-4f transitions as energy increased. However, estimations based on atomic models [20] assume the target to consist of isolated atoms and cannot predict band structure and bonding effects in actual solids. So, the Ag $M_{4,5}$-edge ELNES exhibited modulations caused by spin-orbital splitting of $3d^{3/2}$- and $3d^{5/2}$-states. Besides, a shift of the plasmon losses and a decrease of the Ag $M_{4,5}$-edge intensity with decreasing particle size might be explained by a narrowing and lowering of the 3d-band with respect to the Fermi level.

A ZLESI image of Ag_2S clusters with an average size of 3.9 ± 0.6 nm deposited on a holey Formvar film is shown in Fig. 1f. Even after filtering, clusters were seen as low contrast sphere-like particles without visible crystalline structure. Their size distribution demonstrated a better fit with a normal curve rather than with a log-normal one. In EELS spectra of Ag_2S particles the S $L_{2,3}$-edges were detected at 162–165 eV (see Fig. 2d). The S $L_{2,2}$-edges excited by $2p^{3/2}$ and $2p^{1/2}$ transitions exhibit no deviations from the rounded profile of the delayed by more than 10–20 eV; this results from the centrifugal barrier preventing overlap between the initial and final 2p-wave functions, particularly for final states with a large angular-momentum quantum number.

The Ag $M_{4,5}$-edges for Ag_2S specimens (see Fig. 2e) were delayed by 18–36 eV. Two relatively sharp peaks (ELNES) at 403 eV and at about 425 eV (Ag M_4-edge) appeared in the spectrum of Ag_2S nanoclusters (curve 1) due to the splitting of states induced by 3d-4f ($\Delta 1 = +1$) transitions to unoccupied states just above the edge threshold. Unusual Ag $M_{4,5}$-edge profiles as compared to aggregated Ag_2S and Ag colloids and thermally deposited Ag nanoparticles, in principle, might be caused by several factors such as differences in nearest-neighbour coordination of silver ions and their valence state and bonding. In the particular case of Ag_2S nanoparticles, this might be a proposed size-dependent transformation of the Ag_2S nanoclusters from the monoclinic modification to the superionic cubic one with disordered Ag^+ sublattice. Perhaps, an enhanced concentration of mobile Ag^+ cations could constitute electron traps. Then interstital Ag^+ ions could combine with the traps to form a silver excess. Furthermore, it is known [21], that non-stoichiometric cubic $Ag_{2+\delta}S$ ($\delta \leqslant 0.002$) is characterized by high electronic conductivity about $1.3 \cdot 10^3\,\Omega^{-1}\,cm^{-1}$ that is 10^6 times higher than in monoclinic Ag_2S. Therefore the probable transformations could result in an increase of the unoccupied density of states and the density of the $3d^{3/2}$ and $3d^{5/2}$ states in Ag^+ ions. This hypothesis seems to be attractive but, it requires further elaboration. In X-ray spectra of Ag_2S clusters acquired in STEM and CTEM modes the S Kα line at 2.31 keV was recorded clearly. Because of the lower coverage of the clusters (about $16.5\,10^2\,\mu m^{-2}$) and smaller sizes than of those for Ag nanoclusters, larger areas of analyses up to 30–100 nm in diameter and broader electron probes were necessary to acquire reasonable intensities of the Ag L-series (Ag $L\alpha_1$ at 2.98 keV and Ag $L\beta_1$ at 3.15 keV).

Conclusions

As was shown on selected examples of Ag and Ag_2S small particles, Ag nanocrystalline films and dendritic crystals, nanostructures formed in the course of transfer from the ultradisperse state via random aggregation and recrystallization to the bulk state can be successfully characterized by combined application of AEM techniques (EELS, ZLESI/ZLSSAED, EDX) and STM. This allows the following detailed description of morphologies and surface nanotopographies, elemental compositions, crystalline and electronic structures, sequence of the particle aggregation, crystallization and thin film growth (nucleation and formation of island structure; coalescence and recrystallization of islands and aggregates, channel and continuous film formation).

Acknowledgements. The authors are thankful to Ir. R. De Keizer (Agfa-Gevaert N. V., Belgium) who provided some samples, to Prof. V. Panov and Dr. I. Yaminsky (Moscow State University, Russia) for large scale STM images of Ag films and helpful comments. One of the authors (V.P.O.) was supported by the Belgian Federal Services for Scientific, Technical and Cultural Affairs of the Prime Minister's Office through IUAP-III, Convention-49 and by the Russian Foundation for Basic Researches, grant No 95-03-08252.

References

[1] D. Ugarte, C. Colliex *Z. Phys. D* **1989**, *12*, 333.
[2] S. Holst, W. Legler, *Z. Phys. D* **1993**, *25*, 261.
[3] J. I. Goldstein, D. B. Williams, *Microbeam Analysis* **1992** *1*, 29.
[4] N. D. Browning, S. J. Pennycook, *Microbeam Analysis* **1993**, *1*, 19.
[5] B. Sattler, *Z. Phys. D* **1991**, *19*, 287.
[6] S. E. Roark, K. I. Rowden, *Chem. Phys. Let.* **1993**, *212*, 50.
[7] V. Oleshko, R. Gijbels, W. Jacob, M. Alfimov, *Microbeam Analysis* **1995**, *4*, 1.
[8] J. A. Creighton, *Spectroscopy of Surface Vol. 16*, (R. J. H. Clark, R. E. Hester, eds.), Wiley, New York, 1988, pp. 38–43.
[9] Y. N. Moiseev, V. I. Panov, S. V. Savinov, I. V. Yaminsky, *J. Vac. Sci. Technol.* **1994**, *B12*, 1690.
[10] A. Savitsky, M. Golay, *Analy. Chem.* **1964**, 36.
[11] B. D. Hall, M. Flueli, R. Monot, J.-P. Borel, *Z. Phys. D* **1989**, *12*, 97.
[12] V. P. Oleshko, V. Kindratenko, R. Gijbels, P. Van Espen, W. Jacob, *Mikrochim. Acta [Suppl.]* **1996**, *13*, 443.
[13] J. P. Rabe, *Angew. Chem. Int. Ed. Engl.* **1989**, *28*, 117.
[14] M. Poengen, J. F. Wolf, J. Frohn, M. Giesen, H. Ibach, *Surf. Sci.* **1992**, *274*, 430.
[15] M. Dietterle, T. Will, D. M. Kolb, *Surf. Sci.* **1995**, *327*, L495.
[16] M. O. Delcourt, F. Yala, E. Merlen, S. Remita, N. Keghouche, E. Delariue, *Scanning* **1994**, *8*, 449.
[17] T. David, J. P. Dufour, G. Chabrier, *Thin Solid Films* **1995**, *256*, 268.
[18] H. Seiler, U. Haas, K.-H. Kortje, B. Ocker, *Microsc. Microanal. Microstruct.* **1991**, *2*, 191.
[19] R. F. Egerton, *Electron Energy-Loss Spectroscopy in the Electron Microscope*, Plenum, New York, 1986, pp. 205–208.
[20] R. D. Leapman, P. Rez, D. F. Mayers, *J. Chem. Phys.* **1980**, *72*, 1232.
[21] M. K. Van Doorselaer, *J. Phot. Sci.* **1987**, *35*, 42.

Mikrochim. Acta [Suppl.] 13, 443–451 (1996)

Study of Quasi-Fractal Many-Particle-Systems and Percolation Networks by Zero-Loss Spectroscopic Imaging, Electron Energy-Loss Spectroscopy and Digital Image Analysis

Vladimir P. Oleshko[1,*,**], Volodymyr V. Kindratenko[2], Renaat H. Gijbels[2], Pierre J. M. Van Espen[2], and Wim A. Jacob[3]

[1] N.N. Semenov Institute of Chemical Physics, Russian Academy of Sciences, 117421 Moscow, Russia

[2] Department of Chemistry and [3] Department of Medicine, University of Antwerp (U.I.A.), B-2610 Wilrijk-Antwerpen, Belgium

Abstract. Submicron colloidal Ag particles and nano-sized filaments forming a statistical percolation network during "*in situ*" development of double structure tabular microcrystals of AgBr(I) emulsions have been studied by electron energy-loss spectroscopy and zero-loss electron spectroscopic imaging (EELS/ZLESI). Image analysis has shown that random quasi-fractal clusters were formed in the colloid. ZLESI has been applied to characterise the morphology and defect structure of aggregated particles and filaments. Their energy-loss spectra revealed plasmon excitations and interband 4d electron transitions between 4–32 eV energy-loss. To study the cluster structure and its relation to the physical properties, fractal analysis including estimations of cluster fractal dimensions and of density autocorrelation functions has been performed. Mechanisms of fractal aggregation based on known models of diffusion limited aggregation, cluster-cluster aggregation and percolation are discussed.

Key words: small particles, EELS/ZLESI, image analysis, fractal clusters, percolation networks.

Nowadays coagulated fractal and quasi-fractal structures have been an area of active research [1–3]. They can result from a variety of different processes: film deposition on a solid surface, adsorption of molecules by porous matter, solidification of colloids, crystallization of ultradisperse powders, various biophysical processes, etc. As a rule, fractal properties at objects having porous, ramified, rough

* Present address: Department of Chemistry, University of Antwerp (U.I.A.), Universiteitsplein 1, B-2610 Antwerpen-Wilrijk, Belgium

** To whom correspondence should be addressed

or sparse and tenuous structures are usually associated with nonequilibrium growth [4]. The unique physical and chemical properties, different from those of both gases and condensed media make fractal clusters of great fundamental interest. Moreover, fractal properties can be used for the characterisation of many-particle-systems and microanalysis of surfaces and various disperse substances as well. Volume and surface plasmon excitations, in particular, are determined both by properties of the individual particles and by collective effects due to interaction among the particles densely packed in aggregates. To understand how fractal geometry affects electronic properties of aggregated matter, electron-optical and optical methods can be applied [3, 5].

Here we present some data on formation of quasi-fractal clusters in aggregated Ag colloids and of percolation filament networks during development of composite tabular microcrystals of Ag(Br, I) emulsions studied by electron energy-loss spectroscopy (EELS), zero-loss electron spectroscopic imaging (ZLESI) and image analysis.

Experimental

A suspension of colloidal Ag particles was obtained by reduction of a diluted aqueous solution of $AgNO_3$ in the presence of 0.1% polyvinyl alcohol (PVA). In another experiments statistical networks of Ag filaments were produced by a 2-minute "*in situ*" treatment, with the Kodak D-19 developer, of double structure tabular microcrystals; the latter had a size of 7–10 μm and consisted of AgBr cores and Ag(Br, I) shells containing 5–6 mol. % of AgI. Samples for observations were prepared by deposition of particles and developed crystals on object grids covered with thin carbon films. A ZEISS EM 902 electron microscope with an integrated prism-mirror-prism energy filter operated at 80 kV accelerating voltage was used for EELS/ZLESI measurements. EELS spectra were acquired using a photomultiplier serial detector with a collection half-angle of 11 mrad. The exit energy-selecting slit in the image mode was 10 eV. The analyses were performed with an energy resolution of 1.5–2.5 eV on the full-width at half-maximum of the zero-loss peak. EDX spot analysis of the specimens was performed in the STEM mode at 80 kV accelerating voltage using a JEOL 1200EX electron microscope equipped with an EM-ADID10 scanning observation device and a TRACOR NORTHERN TN5500 energy dispersive X-ray analyzer.

Digital images of particles were acquired via a SIT TV camera by a KONTRON IBAS 2000 image analysis system. Quantitative image processing was done to perform fractal analysis of particle aggregates as well as to estimate the parameters of the size distribution of the particles. Selected fields in the latter case contained 500 distinguishable particles.

In order to separate particles from their background, a technique described in [6] has been applied. The method is based on the idea of a similarity of the original image and obtained from its binary ones. The similarity is measured in terms of the correlation, so from all possible binary images obtained from the original one using different global threshold values only the image having highest correlation between them and the original one is used as the binary image. The sensitivity of the results to the accuracy of the background subtraction technique was tested and described in an earlier publication [7]. A border following algorithm [8] has been used for the contour extraction. Estimation of fractal dimensions, $\boldsymbol{D}$, has been performed using the following techniques: (1) the "hybrid" method searching the relation, $P \sim S^{1-D}$, between the perimeter of the cluster's projection on an image plane, $\boldsymbol{P}$, and a "yardstick" size, $\boldsymbol{S}$, used for the measurement of the perimeter [9]; (2) the box counting method based on the relation, $N \sim S^{-D}$, between the number of boxes, $\boldsymbol{N}$, covering the cluster projection contour on the image plane and their side, $\boldsymbol{S}$ [10]; (3) the covering set method using the relation, $N \sim S^{-D}$, between the number of occupied sites, $\boldsymbol{N}$, covering the cluster projection and their side, $\boldsymbol{S}$ [10]; (4) the Forrest-

Witten technique [11] based on the relation, $N \sim S^D$, between the number of primary particles, N, within a square of side, S, plotted on the projection plane of the cluster. The density autocorrelation analysis was done to evaluate the relation, $C(r) \sim r^{D-d}$, between the average density, $C(r) = 0.5\, prNS \langle p(r + r')p(r') \rangle$, of occupied points in the projection plane of the cluster at a distance r. Here N is the total number of occupied sites, $p(r')$ is the density at a reference point vector r'; $p(r + r')$ is the density at vector $r + r'$ equal 1 (occupied point) or 0 (free point); d is dimension of space ($d = 2$). Testing of some techniques on a Sierpinsky carpet with the fractal dimension 1.89 gave the next results: 1.80 ± 0.19 (the covering set method); 1.89 ± 0.13 (the Forrest-Witten technique); 1.91 ± 0.11 (the pair correlation function analysis). Analysis of the triadic Koch curve with theoretical fractal dimension 1.26 gives the following results: 1) 1.25 ± 0.07 (the "hybrid" method); 2) 1.26 ± 0.15 (the box counting method).

Results and Discussions

Ag Colloids

TEM micrographs of colloidal Ag particles are shown in Fig. 1. A zero-loss filtered image (b) demonstrated a higher contrast and better resolution than a corresponding unfiltered one (a) because the blurring and the chromatic aberration caused by inelastically scattered electrons in the specimen is avoided. This allowed to observe more clearly grain boundaries and crystalline defects, e.g. twins and stacking faults. Figure 1c shows a low magnification view of a cluster formed by colloidal particles. The inset presents a discrete ring zero-loss spectroscopic selected-area electron diffraction pattern of aggregated particles pointing to a face centered cubic polycrystalline structure. Zero-loss filtering in this case removed the inelastically scattered electrons, which contribute to the background, and increased the contrast of the diffraction patterns. The particle size distribution could satisfactorily be fitted to a log-normal curve with a mean size of 42.7 ± 2.2 nm.

Usually the particles in such collids are prevented from aggregation by repulsive forces between double layers at the particle surfaces. These forces overcome the Van der Waals attraction. Addition of ions to the colloidal system affects the space charge of the particle surfaces and induces aggregation. Initially this leads to coagulation aggregates. Neighbouring particles in the aggregates are separated by thin ion shells at their surface. By aging of the colloidal system the neighbouring particles coalesce partially and build a common grain boundary. The pictures in Fig. 1 demonstrate the topological arrangement of a statistical network consisting of random aggregates and irregular chains of coagulated particles. The aggregate shown in Fig. 1c apparently consists of three parts formed at earlier stages by cluster-particle aggregation. The actual aggregation cannot adequately be described by a model of cluster-particle (CPA) or cluster-cluster (CCA) aggregation only. Real processes may involve these two mechanisms occurring simultaneously or sequentially and/or more complicated aggregation mechanisms. The average density of the aggregate, estimated by the autocorrelation density technique, decreased with increasing radius of gyration in accordance with the CCA model which preferentially contributes to the final stage.

An essential feature of the studied colloids is a power law relationship $M \sim R^{D_\beta}$ between the mass, M, (or the number of particles) and the radius of gyration, R, of the aggregate (see Fig. 2a). Here D_β is the mass fractal dimension characterising a given

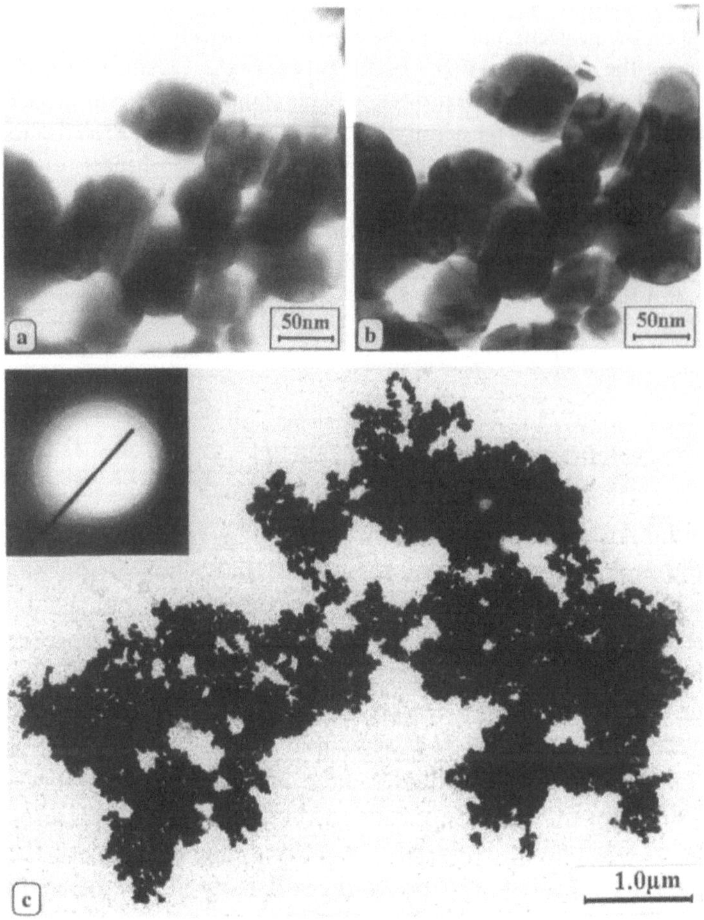

Fig. 1. TEM images of colloidal Ag particles stabilized with a protective polymer: **a** CTEM image; **b** ZLESI image; **c** ZLESI image of a fractal cluster formed by the particles. Left inset shows a zero-loss spectroscopic selected-area electron diffraction pattern

population of aggregates. It is significant that at the early stages of aggregation clusters forming by the CPA mechanism contain a small number (100–300) of specks and therefore cannot be considered as fractals in the strict sense of the word [12]. In this case the parameter D_β describes the distribution of the substance in the system. However, the high values of the correlation coefficient of linear regression (0.98) indicate that even in this range the fractal approach is most appropriate for describing disordered aggregated structures. Analysis of a population of 39 clusters gave the value $D_\beta = 1.81 \pm 0.11$. This result can be compared with the value 1.81 ± 0.05 [13] obtained for computer simulated diffusion limited aggregation (DLA) growth of fractal clusters along random linear trajectories. The model developed for describing the 3D cluster aggregation was shown to be in agreement with results obtained by the Forrest-Witten technique [11]. Since we consider preferentially 2D aggregated structures, the cluster mass determined from the projection will be approximately equivalent to the total mass of the object in three

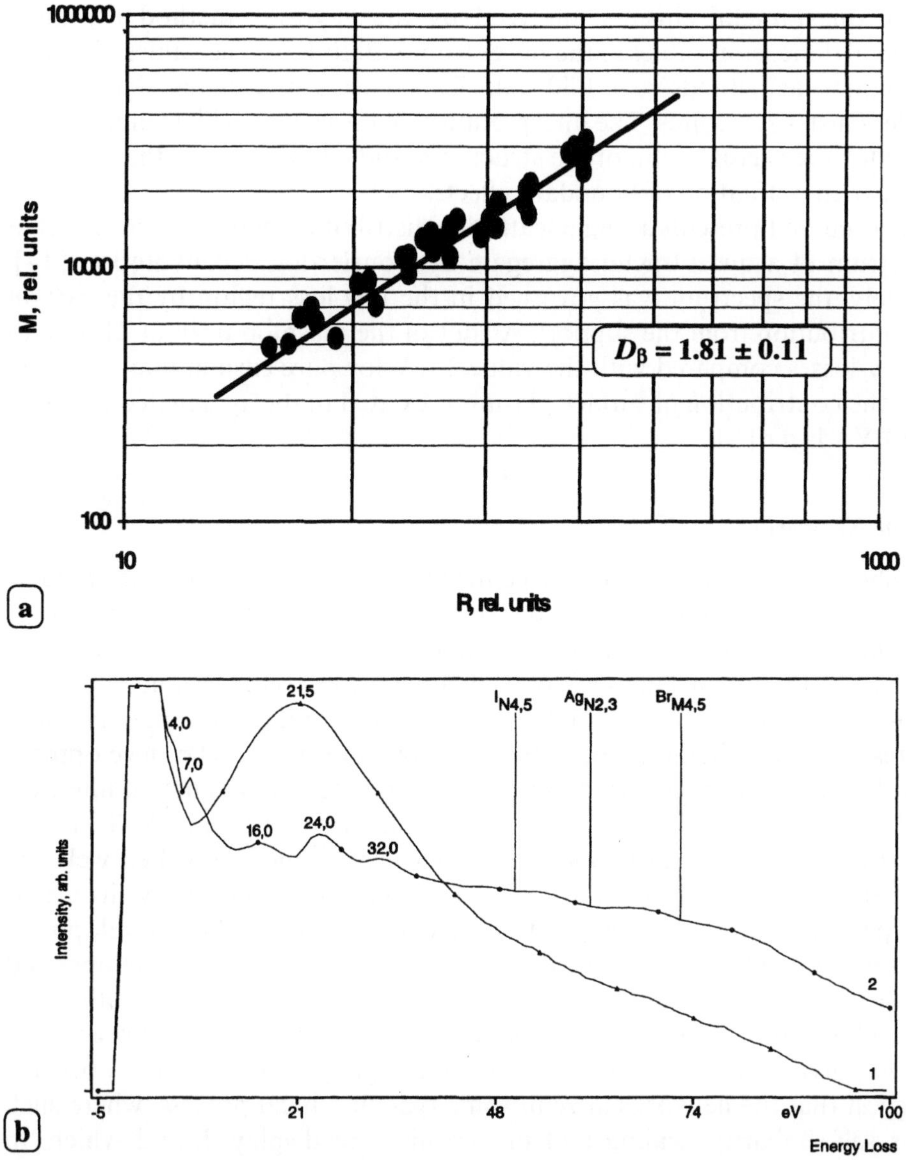

Fig. 2. **a** Mass dependence, M, of the clusters vs. their size, R; **b** single scattering distributions derived from initial EELS spectra of colloidal Ag particles (curve 1) and filaments (curve 2)

dimensions. Also, the radius of gyration, obtained from the projection, will be essentially the same value determined from the 3D structure. Thus the projection demonstrates the same behaviour as the corresponding 3D cluster. This may explain why our data are in good agreement with results for the simulated 3D cluster aggregation. The values of the fractal dimension obtained by the Forest-Witten technique, which characterise the internal automodeling of separate aggregates (known as the box fractal dimension, D_α, ($D_\alpha = D_\beta$ for self-similar fractals), were in the range 1.86–1.91. Some differences between D_β and D_α were referred to a disturbance of the inner cluster automodeling.

Most of quasi-fractal clusters analyzed by the "hybrid" method exhibited contour fractal dimension values in the range from 1.20 to 1.35. The values obtained by the box counting methods were 1.30–1.45. This reflects an irregular shape of the outer border cluster structures. The interpretation of these data, which, in principle, can be used for the classification of the structures studied, requires further analysis including also an evaluation of boundary effects.

Figure 2b, curve 1 presents a single scattering distribution derived from an initial EELS spectrum of a quasi-fractal aggregate by Fourier-log deconvolution [14]. Unfortunately, the spectrum was governed in the low-loss region by the carbon plasmon contribution from the support. A shift of the plasmon peak at 21.5 eV to lower energy-losses compared with the value 25 eV for pure carbon film probably was due to the contribution of surface plasmons excited in the aggregated particles and in the PVA layer [5].

Ag Filament Networks

Figure 3a shows a ZLESI image of the central part of a double structure tabular microcrystal after a 2-minute "*in situ*" treatment with the Kodak developer D-19. Development of crystals was usually accompanied with an intensive etching and resulted in the formation of a network of Ag filaments in the core region [15]. Here again zero-loss filtering allowed to improve the contrast and resolution of filament fine features, i.e. grain boundaries, twins and stacking faults. The development process in the shell regions occurred more slowly and etching was not so intensive because of the lower solubility of $AgBr_{1-x}I_x$ as compared with pure AgBr. In the central part of the crystal, Ag filaments and particles formed branched web-like statistical networks. Generation of disordered aggregated structures by silver filaments and particles is obviously related to the stochastic nature of the development process in specific conditions, i.e., the local excess of the active volume developer and the high overpotential of the anode step of the reaction, the occurrence of both direct chemical and solution physical development mechanisms, and the presence of a protective polymer layer on the crystal surfaces [16]. Moreover, the obtained results suggest that the networks arise in some type of critical process, where such features as self-similarity, scaling and universality are displayed, and which, in principle, may be suitably described by percolation [17]. The power-law behaviour of the silver mass, $\boldsymbol{N}$, (the number of occupied sites) vs. the square size, $\boldsymbol{S}$, shown in Fig. 3b points out that the networks belong to random fractals. The fractal dimension, $\boldsymbol{D}_\alpha$, determined from the slope of the curve was 1.95 ± 0.11. Numerical computations of the network growth within a simplified invasion percolation model [18] shown in Fig. 3 describe, respectively, an initial stage (c) and a final stage (d) of the aggregation process with a hexagonal lattice. At the final stage a simulated extended network obviously exhibits the morphology of filaments similar to that on the ZLESI image in Fig. 3a. The value of $\boldsymbol{D}_\alpha$ for the simulated network shown in Fig. 3d was 1.94 ± 0.11 in satisfactory agreement with the experimental value (see Fig. 3b). It was also found that in both cases the autocorrelation function, $\boldsymbol{C}(\boldsymbol{r})$, demonstrates a similar behaviour. Based on the obtained results, one can conclude that this model can be applied to describe the filament network growth in the course of the photographic development process.

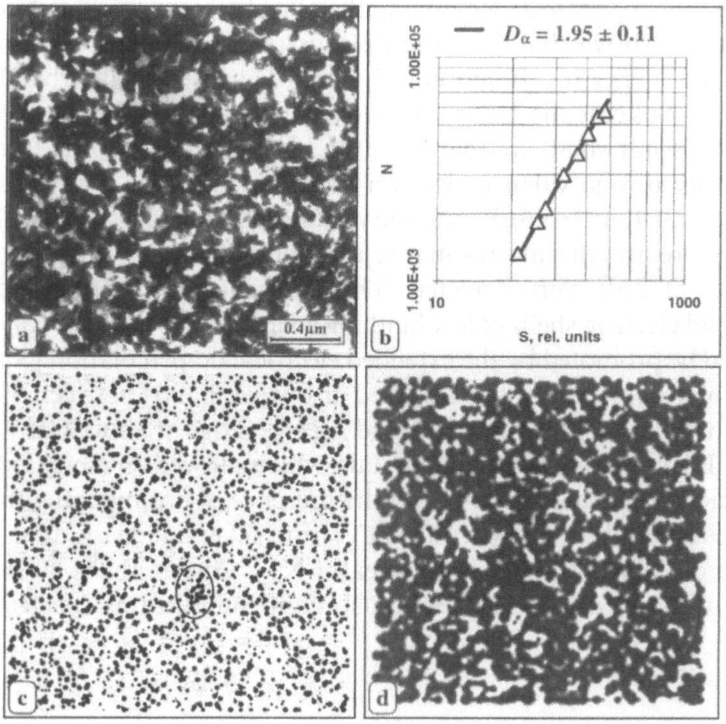

Fig. 3. **a** ZLESI micrograph of a network of Ag filaments formed in the course of a 2-minute development of a double structure Ag Br(I) tabular microcrystal; **b** the number of occupied sites, N, vs. the square size, S, plotted on the image of the network; **c** computer simulation of growth of a percolation network (initial stage, a primary cluster is marked by the ellipse, probability of the occupation of a lattice site, $p = 0.2$); **d** computer simulation (final stage, formation of the branched network, $p = 0.7$). The line in graph (**b**) corresponds to the experimental image (**a**), points obtained by the treatment of simulated image (**d**) are shown by triangles

Numerous chemically produced contact areas within numerous filaments allowed to reduce the total electrical resistivity of the system. Above the percolation threshold they formed extended conducting paths within the network of randomly connected filaments. The correlation length of such paths increased with increasing filling factor and could promote the direct electrical *dc* conductivity [3]. On the other hand, it should be noted that the growth of the developed silver phase involves a series of complicated cooperative transformations of preferentially ionic bonding in silver halide to metal bonds within the networks of nano-sized silver filaments and particles. This leads to an overlapping of molecular orbitals of primary silver clusters and to the formation, in the growing filaments, of an extended band structure involving, in particular, a narrow 4d band separated from the conduction band and a number of surface and defect states. Drastic changes of the electronic structure, obviously affected in this case by the high dispersity of the matter, could be characterised by EELS.

The enlarged low-loss part of the spectrum of filaments formed in the crystal core (the recorded area was 0.8 μm in diameter) contained a series of energy-loss maxima between 4 and 32 eV (see Fig. 2b, curve 2). Removing the plural scattering contribution from the initial spectrum shows that these multiple losses are due to oscillation of the imaginary part of the energy-loss function $\mathrm{Im}[-1/\varepsilon(E)]$ caused by interband electron transitions [15]. In this case the energy of the transition from the 4d band to the Fermi level ($\geqslant 3.9$ eV) is not small compared to the plasma energy of the free electron gas with a density of one valence electron per atom of silver (9 eV). As a result, the surface loss ($\varepsilon = -1, h\nu = 3.6$ eV) and the volume loss ($\varepsilon = 0, h\nu = 3.8$ eV) are close together. Therefore, the plasma resonance at about 4.0 eV as well as the "free-electron" resonance at 7 eV were detected. This was referred to collective excitations of underlying 4d electron shells of low bindiing energy mixed with the conduction band which could be promoted by the extended electrical dc-percolation [3].

Energy-loss spectra of developed crystals revealed also weak bands between 50 and 80 eV, which were attributed to the minor delayed Ag $N_{2,3}$-edge at about 60 eV. This was confirmed by EDX spot analysis of the crystals, which showed an increased intensity of the Ag L-series (Ag Lα at 2.98 keV) and the presence of bromide (the Br Lα at 1.48 keV) and traces of iodide (the I Lα at 3.94 keV) from silver halide residuals. In accordance with data of the EDX analysis, the weak I $N_{4,5}$-edge at about 50 eV and the Br $M_{4,5}$-edge at about 72 eV [15] could also contribute in this spectral region.

Conclusions

This study shows that EELS/ZLESI and digital image analysis are useful for the characterisation of coagulated Ag colloids and of chemically generated statistical networks of Ag filaments formed during "*in situ*" development of composite Ag(Br, I) tabular microcrystals. Data presented point out that aggregated Ag particles of 42.7 ± 2.2 nm in size deposited onto an inert carbon support form random quasi-fractal clusters. The mass fractal dimension of the clusters was estimated as 1.81 ± 0.11 in line with data on simulation of the fractal cluster growth according to the DLA model.

Box fractal dimension of a random network of nano-sized Ag filaments formed in the course of development of double structure tabular microcrystals of Ag(Br, I) emulsions was determined as 1.95 ± 0.11 in satisfactory agreement with numerical computations of the network growth within the invasion percolation model. Energy-loss spectra of the percolation networks revealed plasmon excitations and interband electron transitions in the range from 4 up to 32 eV energy-loss. This was attributed to collective excitations of underlying 4d shells of low binding energy which probably were promoted by the extended electrical dc-percolation.

Acknowledgements. One of authors (1) was supported by the Belgian Federal Services for Scientific, Technical and Cultural Affairs (DWTC/SSTC) of the Prime Minister's Office through IUAP-III, Convention-49 and by the Russian Foundation for Basic Researches, grant No 95-03-08252.

References

[1] J.-P. Chevalier, C. Colliex, M. Tence, *J. Microsc. Spectrosc. Electron.* **1985**, *10*, 417.
[2] M. Tence, J.-P. Chevalier, R. Julien, *J. Phys.* **1986**, *47*, 1989.

[3] D. Schonauer, M. Quinten, U. Kreibig, *Z. Phys. D* **1989**, *12*, 527.
[4] L. M. Sandler, *Nature* **1986**, *322*, 789.
[5] H. Seiler, U. Haas, B. Ocker, K.-H. Kortje, *Faraday Discuss.* **1991**, *92*, 121.
[6] A. D. Brink, *Pattern Recognit. Lett.* **1989**, *9*, 335.
[7] V. V. Kindratenko, P. J. Van Espen, B. A. Treiger, R. E. Van Grieken, *Environ. Sci. Technol.* **1994**, *28*, 2194.
[8] A. Rosenfeld, A. C. Kak, *Digital Picture Processing, Vol. 2*, Academic Press, London, 1982, p. 219.
[9] N. N. Clark, *Powder Technol.* **1986**, *46*, 45.
[10] T. Cleary, R. Samson, J. W. Gentry, *Aerosol Sci. Technol* **1990**, *12*, 518.
[11] S. R. Forrest, T. A. Witten, *J. Phys. A Math. Gen.* **1979**, *12*, L109.
[12] V. P. Oleshko, M. V. Alfimov, X.-F. Cui, S.-E. Wang, *J. Phot. Sci.* **1994**, *42*, 110.
[13] P. Meakin, *J. Col. Interf. Sci.* **1984**, *102*, 491, 505.
[14] R. F. Egerton, Z. L. Wang, *Ultramicroscopy* **1990**, *32*, 137.
[15] V. Oleshko, R. Gijbels, W. Jacob, F. Lakiere, A. Van Daele, E. Silaev, L. Kaplun, *Microsc. Microanalys. Microstruct.* **1995**, *6*, 1.
[16] V. P. Oleshko, M. V. Alfimov, *J. Imag. Sci. Techn.* **1994**, *38*, 162.
[17] T. Nakayama, K. Yakudo, L. Orbach, *Rev. Mod. Phys.* **1994**, *66*, 381.
[18] D. Wilkinson, M. Barsony, *J. Phys.* **1984**, *A17*, L129.

Mikrochim. Acta [Suppl.] 13, 453–462 (1996)

Calculation of Bremsstrahlung Spectra for Multilayer Samples

Arnold Pfeiffer[1], Christian Schiebl[1], and Johann Wernisch[2,*]

[1] Digital Equipment Österreich AG, Campus-based Engineering Center, Favoritenstr. 7, A-1040 Wien, Austria

[2] Institute for Applied and Technical Physics, Technical University of Vienna, Wiedner Hauptstrasse 8-10, A-1040 Wien, Austria

Abstract. Investigations and new developments of a simple multiple electron scattering model by August are the basis for further calculations of characteristic and continuous depth distribution functions. The validity of the predicted depth distributions has been demonstrated by comparisons with other models. Therefore, application of the simple electron scattering model in correction programs for the determination of elemental concentrations and film thicknesses is justified. Within the scope of these correction calculations, accurate knowledge of the excited breamsstrahlung spectrum is essential for quantitative analysis, especially in the field of EDS and peak-to-background measurements. Using the description of the continuous depth distribution, a calculation method for breamsstrahlung spectra applicable to any bulk or multilayer sample is presented.

Key words: bremsstrahlung spectra, continuous radiation, depth distribution function, thin films, background correction.

The knowledge of the bremsstrahlung spectrum is necessary for the quantitative analysis of the elemental concentration or the mass thicknesses of a sample. In electron micro probe analysis its accurate description is used for background subtraction methods, thin film analysis, in the so called "peak-to-background" method, and for the determination of the limit of detection. Due to the importance of the bremsstrahlung spectrum, a great number of descriptions have been published so far (e.g. [7, 9, 10, 11, 15, 21, 22, 24]), but the resulting intensity distributions of the generated bremsstrahlung, which have been presented up till now, are only valid for homogeneous bulk samples.

In the case of heterogeneous structures such as multilayer samples, an accurate description of the generation of continuous radiation at a certain mass depth and the

* To whom correspondence should be addressed

absorption in the sample on the way to the detector is indispensable for the calculation of the whole bremsstrahlung spectrum. This requires knowledge of the depth distribution function for continuous radiation $\Phi_v(\rho z)$. Based on a simple electron scattering model, August developed an analytical model for the calculation of the continuous depth distribution function [3, 4]. Applying a refined version of this model it is possible to calculate the intensity distribution of the bremsstrahlung spectra for any multilayer multielement structure [17].

Depth Distribution for Continuous Radiation

The calculation of the depth distribution function for continuous radiation is based on the physical principles of a simple electron scattering model [2, 8, 17]. This semiempirical model quantifies the mean number of electron trajectories passing through an infinitesimally thin layer $\Delta\rho z$ at a certain mass depth ρz by calculating the physical quantities of the mean energy and angular distributions and the transmission and backscattering coefficients of the electrons depending on the acceleration energy of the primary electrons, the tilt angle, the composition and the structure of the sample. This number of electron trajectories is related to the number of characteristic and continuous X-rays generated at the mass depth ρz.

The decisive physical quantity, whether characteristic or continuous radiation is taken into consideration, is the inserted cross-section. The differences in shape of the calculated depth distribution functions and the mean depth of generation of the radiation are caused exclusively by the corresponding cross-sections [4]. According to Powell [20], most of the published ionization cross-sections reach their maximum value at overvoltage ratios $U_c = E/E_c$ (E is the electron energy and E_c the edge energy of the ionized shell) between 2.7 and 4.0. Contrary to this, the bremsstrahlung cross section is monotonously increasing with decreasing overvoltage ratio $U_v = E/E_v$ (E is the electron energy and E_v the energy of the generated continuous radiation). These circumstances result in a mean generation depth of the continuous radiation, which is located deeper in the sample as compared to the characteristic radiation [3, 4]. This fact is proven by experimentally determined peak-to-background measurements of Birks et al. [6] for varying take-off angles.

The ionization cross-section employed for the calculation of the characteristic depth distribution can be taken from Pouchou [19], which leads to acceptable results for depth distributions of characteristic radiation for overvoltage ratios between $1.0 \leqslant U_c \leqslant 25.0$. The cross-section for bremsstrahlung generation is taken from Kramers [15], although it has been shown in various publications (see e.g. [24]) that this cross-section is inadequate for low radiation energies, very low and high atomic numbers. Furthermore, the anisotropy of the bremsstrahlung is not taken into account. But the application of the Kramers cross-section leads to analytical solutions of integrals appearing in the model of August [3], and therefore to time-saving calculations, which is of great importance for the application in iterative procedures and correction calculations.

Figures 1 and 2 show a comparison of the characteristic and continuous depth distribution functions for Si and W at various acceleration energies. For clarity the continuous depth distribution functions are normalized to the area of the characteristic distributions. The continuous distributions are calculated for a radiation energy

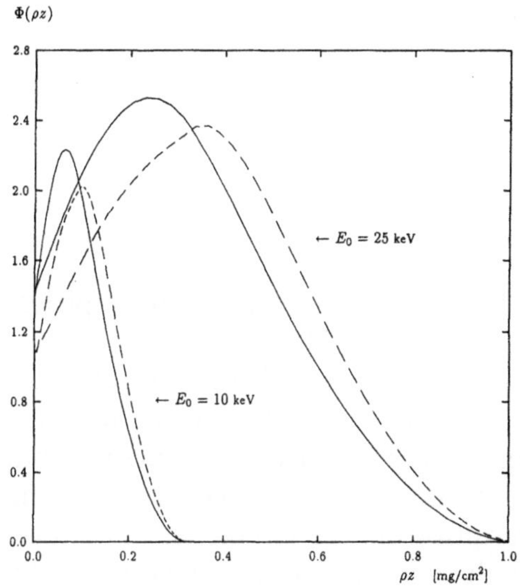

Fig. 1. Comparison of continuous depth distribution functions in Si at $E_0 = 10\,\text{keV}$ and $E_0 = 25\,\text{keV}$ for $E_v = 1.739\,\text{keV}$ (dashed liines ---) with characteristic depth distribution functions for Si K_α (solid lines —)

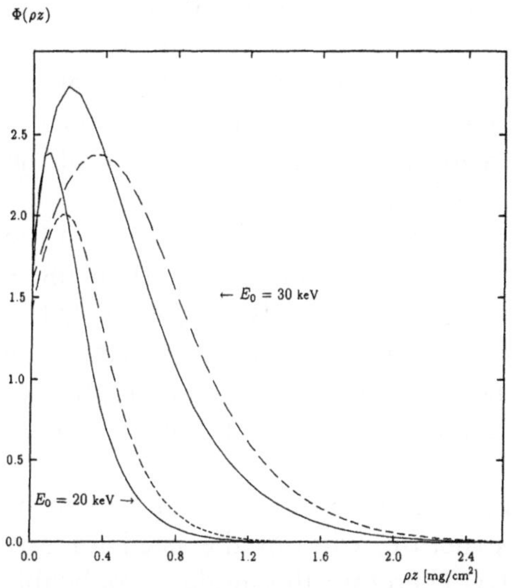

Fig. 2. Comparison of continuous depth distribution functions in W at $E_0 = 20\,\text{keV}$ and $E_0 = 30\,\text{keV}$ for $E_v = 8.398\,\text{keV}$ (dashed lines ---) with characteristic depth distribution functions for W L_α (solid lines —)

corresponding to the energy of the characteristic depth distributions. These figures emphasize the expected fact, that the maximum of the generation of continuous radiation lies at a deeper mass depth than the one for characteristic radiation.

Calculation of Bremsstrahlung Spectra

In the case of the descriptions of bremsstrahlung spectra, most of the published expressions are based on the original Kramers distribution [15], which represents the number of the generated continous radiation photons $n(E_v)$ of a bulk sample by the Eq. (1):

$$n(E_v) = k \cdot Z \cdot \frac{E_0 - E_v}{E_v} \tag{1}$$

E_v is the energy of the continuous radiation, E_0 the acceleration energy, k the Kramers constant and Z the atomic number of the bulk sample. In principle, the descriptions based on the Kramers distributions are more or less modifications and improvements of the atomic number and energy dependencies the kind of:

$$n(E_v) = k' \cdot Z^a \frac{(E_0 - E_v)^b}{E_v^c} \tag{2}$$

Examples of possible parametrizations for k', a, b and c, resulting from empirical and theoretical treatments can be found in [9, 10, 21, 22, 23].

For the calculation of the bremsstrahlung spectrum measured at the detector, the absorption of the continuous radiation in the sample on the way to the detector needs to be taken into account. These absorption corrections are deducible from models for characteristic radiation. Reed [22] and Statham [24] represent the absorption correction by using a modified description of the characteristic radiation absorption coefficient from Philibert [18], by reducing the Lenard's constant and by replacing E_c, the edge energy of the characteristic radiation, with E_v, the energy of the continuous radiation. The absorption correction of Anderson and Wittry [1] has been adopted by Rao-Sahib and Wittry [21]. A very simple method has been proposed by Ebel et al. [9], who introduced a mean absorption depth d_{eff} of the emitted radiation. d_{eff} is assumed to be constant for all radiation energies. Nevertheless, the results of this model in combination with a new parameterization of the Kramers distribution are in excellent agreement with measurements.

The calculations of depth distributions for continuous radiation show that the differences of the characteristic depth distributions are much greater than those assumed by Reed and Statham. Therefore, the use of characteristic depth distribution models for the description of the absorption of continuous radiations seem to be inadequate. Moreover, it is questionable whether the modification and adoption of these models for the calculation of the generation of the bremsstrahlung in layered structures would be possible and significant.

The depth distribution for continuous radiation according to August can be calculated for any bulk or layered sample structure, without the need for any corrective factors regarding the layered geometry. Therefore, this model is applicable for the calculation of the X-ray generation and intensity of continuous radiation by primary electrons in analogy to the so called $\Phi(\rho z)$-models, taking the atomic number and the absorption corrections in homogeneous and heterogeneous samples into account.

The investigated layered sample consists of t layers. The mass thickness of a specific layer i deposited at the mass depth ρd_i is d_i. The layer consists of m_i elements. The weight fraction of the element k in the layer i is $c_i(k)$.

A bremsstrahlung spectrum of a multilayer multielement sample taking the absorption in the sample into account is calculated with Eq. (3):

$$n(E_v) = const \cdot \sum_{i=1}^{t} Q_v(E_v, E_0) f_i(E_v) \cdot \int_0^{d_i} \Phi_i(\rho z) \exp(-\chi_i \rho z) d\rho z \tag{3}$$

E_v is the energy of the continuous radiation, E_0 the acceleration energy, Q_v the cross-section of bremsstrahlung generation, f_i the absorption of the radiation in the layers above i, $\Phi_i(\rho z)$ the depth distribution for the continuous radiation in this layer and $\exp(-\chi_i \rho z)$ the absorption of the radiation in the layer i. The constant *const* adjusts the count rate to the required units.

The absorption f_i of the radiation in the layers above i (for $i > 1$) is calculated with:

$$f_i = \exp\left(-\sum_{j=1}^{i-1} d_j \chi_j\right) \tag{4}$$

with

$$\chi_i = \frac{1}{\sin\psi}\left(\frac{\mu}{\rho}\right)_i.$$

ψ is the take-off angle to the detector and $(\mu/\rho)_i$ is the mass absorption coefficient for the continuous radiation of energy E_v in the corresponding layer. In the case of $i = 1$, which is the surface layer, f_1 is equal to unity.

The cross-section of the bremsstrahlung generation is taken from Kramers with a modified atomic number dependence:

$$Q_v(E_v, E_0) = k \cdot \frac{\bar{Z}_i^n}{\bar{A}_i E_v E_0} \tag{5}$$

In this equation $\bar{Z}_i$ and $\bar{A}_i$ are the mean atomic number and mean atomic weight the of the layer i. The exponent n depends on the mean atomic number of the layer i and varies from 2.18 to 2.23. k is the Kramers constant [15].

Results and Discussion

Bremsstrahlung Spectra of Bulk Samples

To prove the reliability of the continuous depth distribution function and the $\Phi(\rho z)$-model for the bremsstrahlung spectra, calculations for pure bulk samples are made. These calculations are compared with the bremsstrahlung distributions of Ebel et al. [9], which represent best-fits of measured spectra in the form of:

$$n(E_v) = const \cdot i \Delta E \Delta t Z^{1+n} \cdot \left(\frac{E_0}{E_v} - 1\right)^x \cdot \exp(-\chi \rho d_{\text{eff}}) \tag{6}$$

The fit parameters are *const*, x, d_{eff} and E_0. The number of photons is expressed by a reduced number of continuous radiation n_{red}, which is given by the measured count rate of bremsstrahlung photons divided by the detector efficiency of the measurement setup, the atomic number Z of the pure standard sample, the beam current i in [pA], the measurement time of data accumulation Δt and the energy interval ΔE in [eV]. The units of n_{red} are $10^6 \times [pA \cdot s \cdot eV]^{-1}$. The mass absorption coefficients are taken from McMasters et al. [16]. The correspondence of the reported fits to the experimental data is excellent. Ebel et al. reports for his procedure standard deviations σ_{st} from the measurements ranging from 3.3% to at maximum 7.4%.

To obtain a general view of the reliability of the atomic number and energy dependency, bremsstrahlung spectra for the bulk elements of C, Al, Cu, Sb and Au

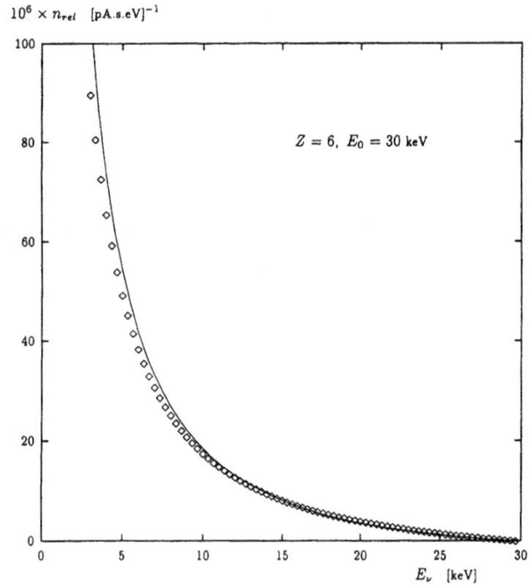

Fig. 3

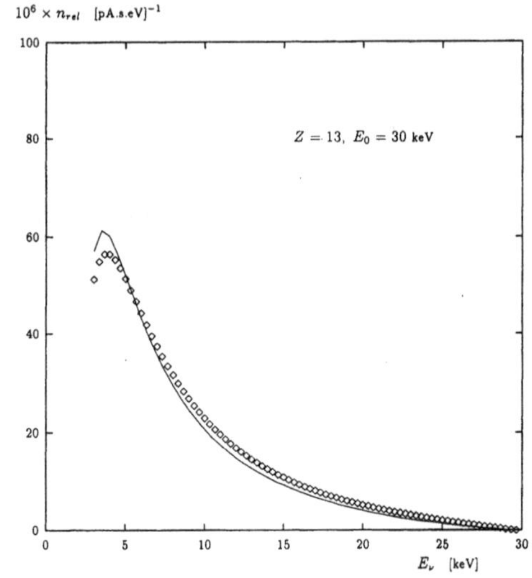

Fig. 4

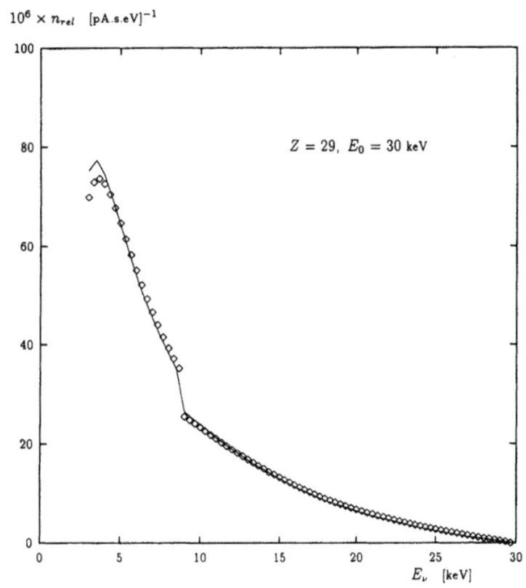

Fig. 5

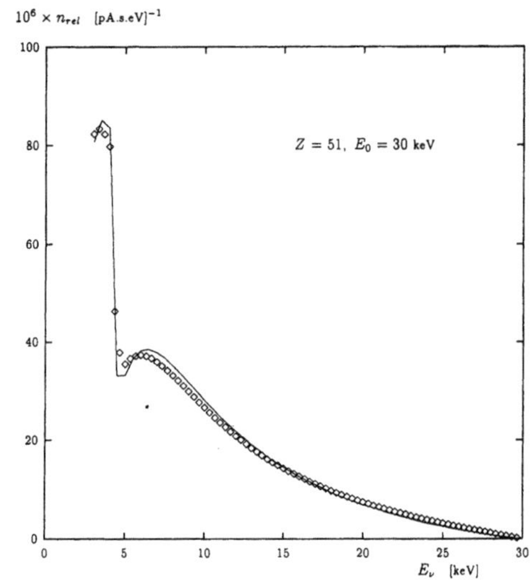

Fig. 6

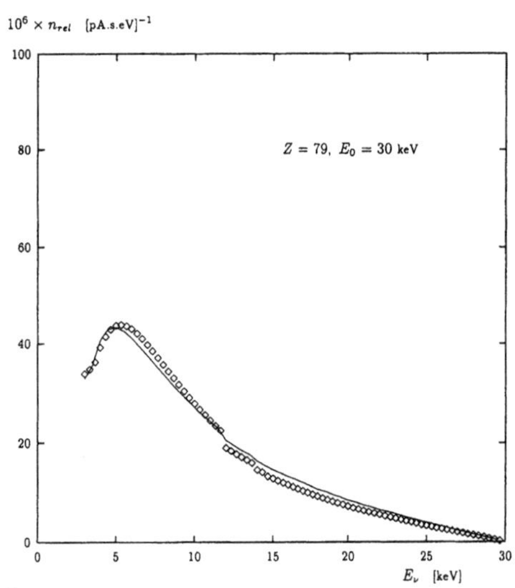

Fig. 7

Figs 3–7. Comparison of bremsstrahlung spectra of C, Al, Cu, Sb and Au for $E_0 = 30$ keV and radiation energies E_ν ranging from 3 to 30 keV with an take off angle being 30°: Best-fit calculations according to Ebel et al. [9] (solid line —) and calculation of the new model (◇◇◇) applying a depth distribution of continuous radiation are calculated for a radiation energy ranging from 3 to 30 keV at an acceleration energy of 30 keV and for a take off angle 30°. The required mass absorption coefficients are taken from Henke [12, 13]. The comparisons to the fitted measurement data by Ebel et al., which are presented in the Figs. 3–7 result in a remarkably good correspondence over the mentioned energy range. With the exception of the Au spectra, the new model computes slightly lower count rates for the lower energy regions. By way of contrast, the count rates for higher energies are a little bit higher. The Au spectra generally lie above the results of Ebel et al. for the whole energy range. For lower radiation energies, for which the Kramers cross-section is a rather poor approximation, prounced deviations measurements should be expected.

Bremsstrahlung Spectra of Layered Samples

In an analogous way Figs. 8–10 show the application of the $\Phi(\rho z)$-model for continuous radiation for the calculation of bremsstrahlung spectra for different layer samples. The count rate n_ν of the bremsstrahlung is expressed in units of $10^6 \times [pA \cdot s \cdot eV]^{-1} \cdot n_\nu$ of the pure element samples shown in the figures is equivalent to $n_{rel} \times Z$ of the corresponding spectra in Figs. 3–7.

Figure 8 shows bremsstrahlung spectra for Cu films deposited on Au. The density of Cu is assumed to be 8.951 g/cm^3. For samples with a Cu film of a thickness below 500 nm the maximum value of the bremsstrahlung spectrum (the maximum value for the Cu/Au sample with a 500 nm Cu film lies at $E_\nu = 4.8$ keV) exceeds the one of the pure Au bulk. The reason is that due to the additional copper film, the maximum of the continuous depth distribution of Au is reached and, therefore, the generation of continuous radiation of Au at low energies is increased. Furthermore, the mass absorption of the low energy radiation in the Cu film is less than in Au. In the case of higher radiation energies, the bremsstrahlung spectra of the film/substrate samples are below the spectrum of the Au bulk. For these radiation energies, less continuous radiation is generated in the Au substrate, because the

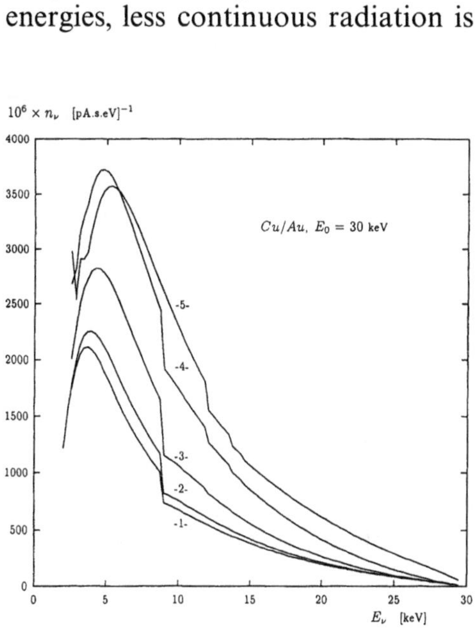

Fig. 8. Bremsstrahlung spectra of Cu deposited on Au for $E_0 = 30$ keV and radiation energies E_ν ranging from 2.6 to 30 keV: pure Cu (*-1-*), Cu film of 1500 nm (*-2-*), Cu film of 1000 nm (*-3-*) Cu film of 500 nm (*-4-*) and pure Au (*-5-*)

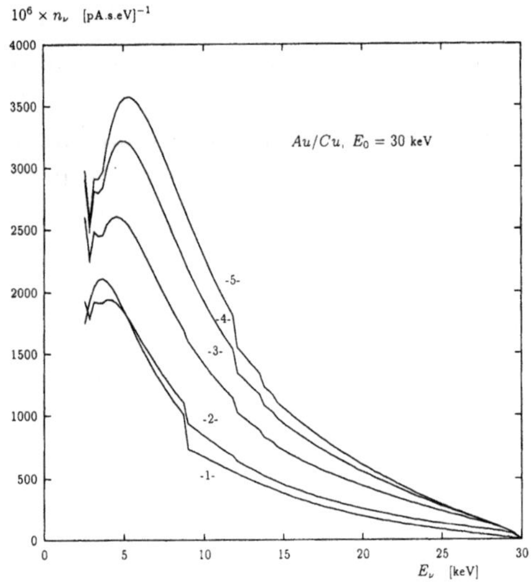

Fig. 9. Bremsstrahlung spectra of Au deposited on Cu for $E_0 = 30\,\text{keV}$ and radiation energies E_ν ranging from 2.6 to 30 keV: pure Cu (*-1-*), Au film of 100 nm (*-2-*), Au film of 300 nm (*-3-*), Au film of 500 nm (*-4-*) and pure Au (*-5-*)

number of high energy electrons which are able to penetrate the Cu film decreases with increasing film thickness. Due to the lower atomic number of Cu the bremsstrahlung intensity of the Cu film is clearly weaker than the one of a corresponding Au layer. Additionally, the mass absorption in Cu is increased for radiation energies, higher than the energy of Cu K edge at 8.979 keV. Summing up the parts of the continuous radiation for higher energies generated in the Cu film and the Au substrate results in a bremsstrahlung spectrum of lower intensity compared to the Au bulk. The spectra for the Cu film thicknesses of 1000 nm and 1500 nm are more and more identical with the bremsstrahlung spectra of the pure Cu bulk, but the count rate intensity of the film/substrate spectrum is higher. This is a result of the increased number of backscattered electrons from the Au substrate, which additionally generates continuous radiation in the Cu film layer.

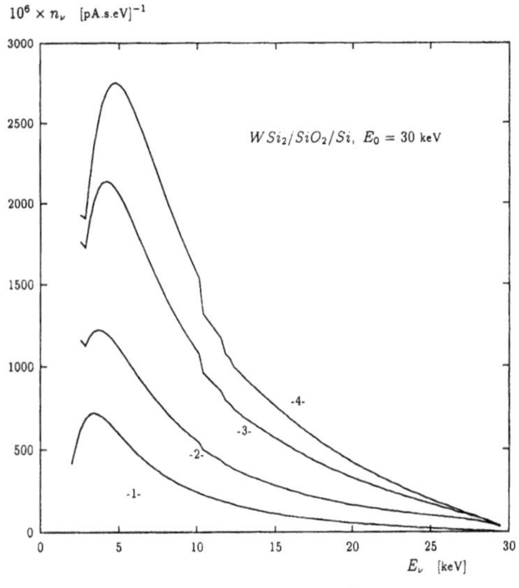

Fig. 10. Bremsstrahlung spectra of WSi_2 deposited on SiO_2 with a thickness of 100 nm deposited on Si for $E_0 = 30\,\text{keV}$ and radiation energies E_ν ranging from 2 to 30 keV: pure SiO_2 (*-1-*), WSi_2 film of 200 nm (*-2-*), WSi_2 film of 500 nm (*-3-*) and pure W (*-4-*)

An analogous presentation is Fig. 9 for the Au/Cu film/substrate sample. In this case the Au film thickness is varied. Due to the atomic number dependence of the bremsstrahlung cross-section the number of photons generated in the Au film steadily increases with increasing film· thickness. The bremsstrahlung spectrum changes from a Cu spectrum to a Au spectrum with increasing film thickness.

In principle, there are no restrictions on the model for the elemental composition of the layers and the substrate. Therefore it is possible to calculate bremsstrahlung spectra for samples which contain one and the same element in different layers. An example for such a case is the $WSi_2/SiO_2/Si$ sample, presented in Fig. 10.

Conclusion

The application of a continuous depth distribution function, which is calculated according to the electron scattering model of August, for the evaluation and description of bremsstrahlung spectra for pure bulk and multilayer-multielement samples is presented. The comparisons of pure bulk spectra calculated with the new model with results of Ebel et al. [9] demonstrate in principle the validity of the model for continuous radiation energies from 3 to 30 keV. For lower energies, additional measurements would be necessary to prove the correctness of the model. Especially the atomic number dependency and the absolute value *const* in the bremsstrahlung description should be investigated. Representative calculation results of bremsstrahlung spectra for film/substrate samples and multilayer samples are presented.

Due to the computation procedure of the model it can be easily adapted for background correction procedures and for additional peak-to-background calculations.

References

[1] C. A. Anderson, D. B. Wittry, *J. Phys. D* **1968**, *1*, 529.
[2] H. J. August, J. Wernisch, *X-Ray Spectrum.* **1991**, *20*, 131.
[3] H. J. August, J. Wernisch, *Scanning* **1991**, *13*, 207.
[4] H. J. August, *Mikrochim. Acta* [*Suppl.*] **1992**, *12*, 139.
[5] H. J. August, *Mikrochim. Acta* [*Suppl.*] **1992**, *12*, 147.
[6] L. S. Birks et al., *J. Appl. Phys.* **1965**, *36*, 699.
[7] M. G. Brunetto, J. A. Riveros, *X-Ray Spectrom.* **1984**, *13*, 60.
[8] V. E. Cosslett, R. N. Thomas, *Br. J. Appl. Phys.* **1965**, *16*, 779.
[9] H. Ebel et al., *X-Ray Spectrum.* **1989**, *18*, 89.
[10] H. Ebel et al., *Advances in X-Ray Analysis* **1992**, *35*, 721.
[11] C. E. Fiori et al., *Anal. Chem.* **1976**, **48**, 172.
[12] B. L. Henke, R. I. Elgin, *Adv. X-Ray Anal.* **1970**, *13*, 639.
[13] B. L. Henke, E. S. Ebisu, *Adv. X-Ray Anal.* **1974**, *17*, 197.
[14] G. A. Hutchins, in: *Characterization of Solid Surfaces*, Plenum, New York, 1974.
[15] H. A. Kramers, *Philos. Mag.* **1923**, *46*, 836.
[16] W. H. McMaster et al., *Compilation of X-Ray Cross-Sections*, UCRL-50174, Sect. II, Rev. 1., Lawrence Radiation Laboratory, University of California, Livermore, CA, 1969.
[17] A. Pfeiffer, *Doctorial Thesis*, Technische Universität Wien, 1995.

[18] J. Philibert, *3rd ICXOM*, Academic Press, New York, 1963. p. 379.
[19] J. L. Pouchou, *Proc. of 3rd EMAS 1993*, 1993, p. 67.
[20] C. J. Powell, *Rev. Mod. Phys.* **1976**, *48*, 33.
[21] T. S. Rao-Sahib, D. B. Wittry, *J. Appl. Phys.* **1974**, *45*, 5060.
[22] S. J. B. Reed, *X-Ray Spectrom.* **1975**, **4**, 14.
[23] D. G. W. Smith, C. M. Gold, *Microbeam Analysis*, San Francisco Press, San Franciso, 1979.
[24] P. J. Statham, *X-Ray Spectrom.* **1976**, *5*, 154.

Mikrochim. Acta [Suppl.] 13, 463–471 (1996)

Thickness Measurement of Thin Films by EPMA – Influence of $\phi(0)$, MAC's and Substrate

Peter Pölt* and Bernd Stöckl

Research Institute of Electron Microscopy, Technical University of Graz, Steyrergasse 17, A-8010 Graz, Austria

Abstract. Extremely thin films down to a thickness of 20 Å have been produced by sputtering and evaporation of various elements onto different substrates. The dependence of thickness measurements performed by EPMA on $\phi(0)$, the mass absorption coefficients for self-absorption, and the volume charge induced in the case of insulating substrates has been investigated. The use of different expressions for $\phi(0)$ led to thickness differences of about 10% maximum. For 20 Å films the results vary in extreme cases with nearly the same percentage as the mass absorption coefficient for self-absorption of the X-ray line used for the measurement. The reduced absorption in the film causes a change of the peak shape. No influence of volume charges was detected in case of insulating substrates within measurement accuracy. Generally an as high overvoltage ratio as possible is preferable for the measurements.

Key words: thin films, EPMA, $\phi(0)$, mass absorption coefficient, specimen charging.

In the last decade the determination of the thickness and the composition of thin films and multilayers on substrates as well as of unsupported thin films by EPMA has become commonplace [1–6]. This is due to the development of realistic models for the depth distribution function $\phi(\rho z)$, which represents the X-ray intensity per unit mass thickness relative to that produced in an isolated layer. It is usual in the analysis of bulk specimens to use standards as similar to the unknown as possible, for example an electrically nonconductive standard for a nonconductive unknown specimen. Thus a compensation for eventual shortcomings of the theoretical model underlying the calculation routines is possible. This is generally impossible in the analysis of thin films. It seems even less possible, the thinner the film. This lack of compensation between standard and unknown specimen may provide a lot of information on the accuracy and the importance of some of the basic expressions and parameters of the theoretical models and also the maximum error caused by them.

* To whom correspondence should be addressed

Besides, all these models are strictly speaking only valid for electrically conducting specimens, since it is nearly impossible to take into account effects as charge trapping, breakdowns etc., phenomena that are influenced by the intrinsic properties of the materials and not only by their chemical composition. Extremely thin conducting films on different types of substrates – conductors as well as insulators – may provide a method to estimate the influence of volume charges on the result of film thickness measurements and the applicability of the customary calculation routines for nonconductive specimens.

Preliminary results regarding all these points will be presented in the next sections. Preliminary, because a much greater variety of film-substrate combinations than analysed so far would be necessary to get an interpretation free of contradictions.

Experimental

Thin films of nickel, copper and gold on Si-wafers, NaCl-crystals and slide glasses have been fabricated by both sputtering with a Penning sputter coater [7] and evaporation. Simultaneously, the thickness of the films had been measured by a quartz thickness measurement device. According to literature an accuracy of ± 0.01 nm can be achieved [8], but for the device at disposal for the measurements an accuracy of around 0.1 nm will represent a more realistic value. The Si-wafers have been polished to remove surface contaminations and the oxide layer and were then immediately coated with the film. NaCl-crystals have been cleaved to get a clean surface. To get unsupported films we dissolved NaCl-crystals with the film in distilled water. The films were then placed on a copper grid like those used in TEM.

Measurements have been performed with a WDX-2A spectrometer from Microspec attached to a Philips SEM 505. Pure element standards have been used for the determination of the k-ratios. For calculation of the film thickness the programs PROZA, TFA and MLA, based on the work of Bastin et al. [1, 9, 22], all distributed by TECHAN B. V. and a self-written program based on the work by Hunger [5] have been used.

Results and Discussion

In film thickness measurements the ratio between the X-ray intensity of a film element and the intensity of the same element of a bulk standard can be used:

$$k = \frac{\int_0^{\varrho z} c_f \phi_f(\varrho z) \exp(-\chi_f \varrho z) d\varrho z}{\int_0^{\infty} c_s \phi_s(\varrho z) \exp(-\chi_s \varrho z) d\varrho z} \tag{1}$$

with c_f and c_s the concentrations of the element in respectively film and standard, ϱz the mass thickness of the film and $\chi = (\mu/\rho)\operatorname{cosec}(\Theta)$ (Θ = take-off angle).

In the case of an ultrathin monoelement film ($\rho z \to 0$) and a pure element standard this results approximately in

$$k = \frac{\phi_f(0)\varrho z}{\int_0^{\infty} \phi_s(\varrho z) \exp(-\chi_s \varrho z) d\varrho z} \tag{2}$$

Hence the result for the film thickness depends strongly on the accuracy of the surface ionization $\phi(0)$ and the mass absorption coefficient μ. As film elements Ni, Cu and Au had been chosen. A 20 Å layer of Ni was already sufficient to prevent surface charging in case of non-conductive specimens. This has been proven by measuring

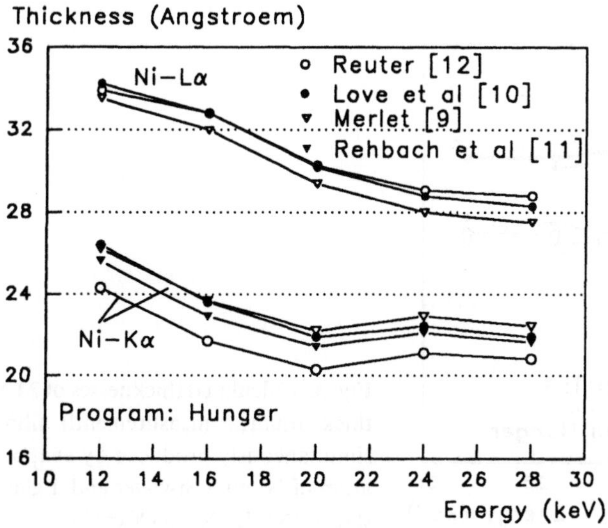

Fig. 1. Dependence of the calculated thickness of a 24 Å thick (quartz measurement!) Ni-film on a Si-wafer on $\phi(0)$ (μ(Ni–L_α/Ni): 3300 cm^2/g [TFA default value])

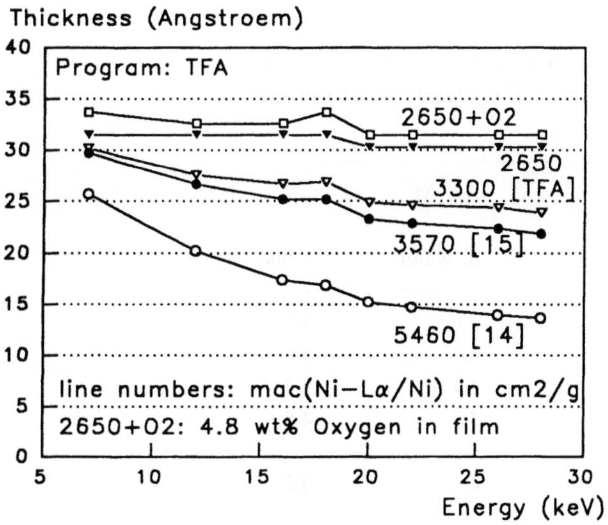

Fig. 2. Dependence of the calculated thickness of a 21 Å thick (quartz measurement!) Ni-film on a Si-wafer on the mass absorption coefficient for Ni–L_α in Ni. The mean thickness calculated with the Ni–K_α intensities: 17.8 Å

the high energy cut-off point of the continuous spectrum using an energy dispersive spectrometer. In case of 20 Å layers of Cu and Au on insulators an additional layer of 100 Å carbon was necessary. Thus in the latter cases the thickness of the conductive double layer system was much greater than that of the actual metal film of interest. Care had been taken to avoid too great differences in the mean atomic number of the substrates chosen. Both the backscatter coefficient and the penetration depth of the electrons correspond in the case of a 20 Å film on a substrate to that of the substrate. Thus for equal films on different substrates differences in the measured film thickness should be attributable to differences in the electrical conductivity of the substrates only and not to inaccuracies in the calculation of a substrate parameter as for example the backscatter coefficient.

As Figs. 1 and 3 and Tables 1 and 2 show, EPMA film thickness results – with the exception of measurements with the Ni–L_α (and also, though not demonstrated here,

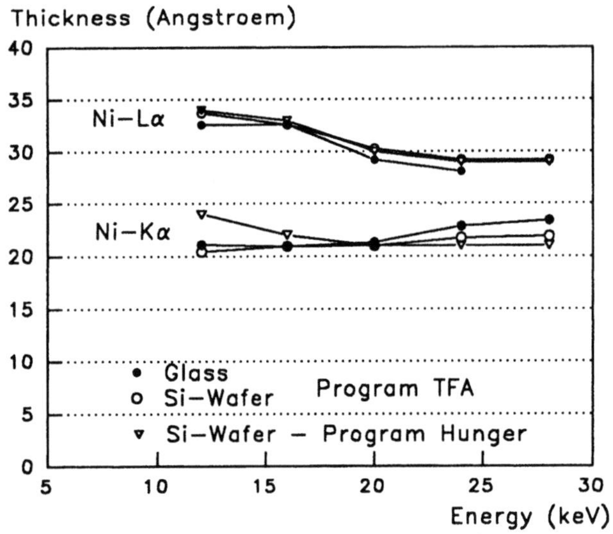

Fig. 3. Calculated thicknesses of 24 Å thick (quartz measurement!) films simultaneously produced by evaporation of Ni on a Si-wafer and a glass slide (μ(Ni–L_α/Ni): 3300 cm^2/g)

Table 1. Mass thickness ($\mu g/cm^2$) of Au-films simultaneously evaporated on a Si-wafer and a glass slide calculated by TFA (quartz measurement: 4.6 $\mu g/cm^2$)

Accel. voltage (kV)	Au–L		Au–M	
	Si-wafer	Glass	Si-wafer	Glass
8	—	—	4.6	4.7
12	—	—	4.5	4.6
16	3.9	3.5	4.4	4.5
20	3.9	3.9	4.3	4.5
24	4.2	4.3	—	—
28	3.9	3.9	4.1	4.2

Table 2. X-ray intensity ratios (Ni–K_α) and the mass thickness T ($\mu g/cm^2$) of a film on a NaCl-crystal before and after substrate removal calculated by TFA (quartz measurement result: 8.98 $\mu g/cm^2$)

Accel. voltage (kV)	NaCl-substrate		Unsupported	
	k-ratio	T	*k*-ratio	T
12	0.0561	7.72	0.0487	7.74
	—	—	0.0504	8.01
20	0.0140	8.04	0.0106	7.86
	0.0144	8.27	0.0110	8.18
28	0.0068	8.16	0.0050	8.13

the Cu–L_α) lines – are in good agreement with the quartz measurement results. Generally the thickness measured with the quartz was – independent of the film – maximally around 10%–20% higher than that obtained by EPMA. This may be due to an error of the quartz measurements, since no special calibration had been performed for the quartz measurement device. In air, partial oxidation of the films occured and probably of the substrates as well. Generally this has been neglected, because for very thin films the ionisation cross-section within the film is constant, absorption is negligible and the backscatter factor is determined by the substrate. Therefore, the film can be regarded approximately as a stack of films, each film consisting of one element. The contributions of the single film elements to the total film thickness are therefore summing up. This assumption enables a better comparison with the quartz measurements which relate to the film thickness before oxidation. Except for very low overvoltage ratios the programs TFA and Hunger give completely the same results (Fig. 3). For convenience in the figures, the mass thicknesses have been converted to linear thicknesses. Bulk densities have been used.

In Fig. 1 a decrease of the calculated thickness with energy is observed. In the case of the K_α-line this is typical for the program of Hunger. Results with the program TFA show the opposite behaviour (see Fig. 3). Different expressions for the ionisation cross section could be a possible cause for that. But since the expression used in TFA is not known to the authors this assumption could not be checked.

For a discussion of the rather strange behaviour of the results obtained with the Ni–L_α line see below. In the following the influence of $\phi(0)$, absorption and volume charge will be discussed in detail.

Ultrathin Films and $\phi(0)$

A great variety of expressions for $\phi(\rho z)$, the depth distribution function of X-ray production, can be found in the literature [10–13]. For the analysis of bulk specimens the special type of this equation is not so important as long as the integral of $\phi(\rho z)$ gives the correct value for the emitted intensity. In addition, division by the integral for a standard of similar composition will compensate for some possible shortcomings of a certain $\phi(\rho z)$-model. No such compensation is possible if the specimen is a thin film. On the contrary Eq. (2) shows that in this case $\phi(0)$ in the numerator is divided by the full integral of $\phi(\rho z)$. Thus $\phi(0)$ directly enters the film thickness with all the consequences.

$\phi(0)$ is generally regarded as dependent on the overvoltage ratio U_0 ($U_0 = E_0/E_c$) and the backscatter coefficient η. Two examples are given below:

$$\phi(0) = 1 + 2.8 * \left(1 - 0.9 * \frac{1}{U_0}\right) * \eta \quad \text{(see Reuter [13])}$$

$$\phi(0) = 1 + \frac{\eta}{1+\eta} * (I(U_0) + G(U_0) * \ln(1+\eta)) \quad \text{(see Love et al. [11])}$$

Despite the great difference in the complexity of these expressions and despite the fact that in Reuter's expression $\phi(0)$ does not go to 1 for $U_0 \rightarrow 1$, all these expressions only lead more or less to a parallel shift of the calculated thickness results as a function of the primary electron energy, as is demonstrated in Fig. 1 for a 24 Å Ni-film on a silicon wafer. In all cases the same expression for η has been used, the one recommended by Hunger [5]. Differences of around 10% maximum can be found for the results. For high overvoltage ratios the difference decreases. This is not surprising, since $\phi(0)$ is mainly regarded as proportional to $1/U_0$. A replacement of the $\phi(0)$ of Reuter, used by Hunger, by that of Love et al. would shift the EPMA-results nearer to those obtained by the quartz measurements.

Influence of Mass Absorption Coefficient, Peak Shape Changes

The same reflections as on $\phi(0)$ are valid for the mass absorption coefficient for self-absorption of the film material. Even for soft X-rays with very high mass absorption coefficients the absorption in a 20 Å layer is low compared to that in a bulk standard, because it is completely emitted from the surface. Thus an erroneous value of the mass absorption coefficient will fully propagate, without any compensation, into the result. Because of the incompletely filled 3d-shells, especially some of the mass absorption coefficients for self-absorption of the transition elements of the 4th period and some of the lanthanides are rather poorly known. Additionally, the mass absorption coefficients may also change according to chemical bond. Literature values ranging between 3300 and 5460 can be found for Ni–L_α in tables and publications [14–17]. As Fig. 2 demonstrates for a 21 Å Ni-layer on a silicon wafer, the results for the calculated thicknesses vary with nearly the same percentage as the mass absorption coefficient for Ni–L_α in Ni.

In Ni and Cu the L_3 absorption edge lies very close to the diagram L_α-line. Thus a high energy satellite peak could have its maximum or at least part of its emission band slightly higher than the L_3 absorption edge. High energy satellites are produced from transitions in multiply ionised atoms. The existence of such nondiagram lines is well-known for both the K-lines and the principal L- and M-lines [19]. According to White [20] more multiply ionized X-ray states are possible than single vacancy states. Doubly ionized states are in any case present as a result of the Auger effect. A high energy satellite peak in the case of Cu–L_α is also stated by Remond et al. [21].

The position of the L_3 absorption edge between diagram and nondiagram line results in strongly different mass absorption coefficients for self absorption for these lines. Additionally, these lines are not resolved even by the use of a wavelength dispersive spectrometer. Thus in a pure element bulk specimen, because of the extremely high absorption of the X-rays of this satellite, the main L_α-peak remains nearly undistorted by this additional line. The situation is completely different in the case of a thin layer. Because of the reduced absorption, the satellite will appear in nearly full height, overlap with the main peak and cause a severe change in the peak shape and even a peak shift. For the same reason the L_α to L_β peak intensity ratio for a pure element specimen differs from that of an oxide of the same element, although effects of chemical bonding may contribute.

Thus a high energy satellite seems to be responsible for the change in peak shape observed between a 20 Å Cu film and a pure standard (see Fig. 4). For Ni only a small peak broadening was observed, but instead a strong peak shift.

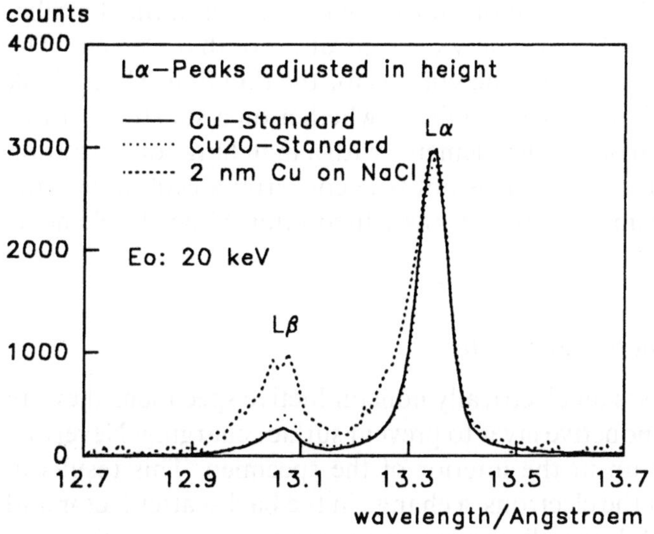

Fig. 4. Comparison of wavelength-scans on a Cu-Standard, a Cu_2O-Standard and a 20 Å thick Cu-film on a NaCl-crystal. A similar scan on a carbon coated NaCl-crystal has been used for background subtraction in case of the thin film

Partial oxidation of thin films in air is unavoidable. To secure that neither partial oxidation of the film nor the influence of the substrate was responsible for the observed effects, a wavelength scan of the L_α-peak was performed for a 100 Å Ni film on a NaCl-crystal. A slightly smaller peak shift than for the 20 Å layers was observed. The reason may be the already stronger effect of absorption for the nondiagram line. Then the crystal was dissolved and the measurement repeated for the unsupported film. A diffraction pattern of the film produced by TEM corresponded to that of pure Ni. Finally, the film was oxidized to full stoichiometry (900 °C, 1′30″, air) and the measurement repeated again. No substantial peak shift and change in peak shape was observed. Also measurements on a NiO standard displayed a much smaller peak shift than that observed for the film. To exclude an influence of the spectrometer scans on a pure Ni standard were performed at very high and extremely low count rates. Lkiewise no influence was observed.

Because of the different mass absorption coefficients for self-absorption for the diagram L_α-line and its high energy satellite neither simple peak to background measurements nor peak-area integrations will give correct results. Instead peak deconvolution would be necessary. All the results in Figs. 1–3 have been obtained by performing peak to background measurements with the L_α-lines. This is the reason for the decrease of the calculated thickness with energy. The choice of a too high mass absorption coefficient results in the calculation of a too high standard intensity, whereas because of the negligible absorption this has no effect on the film intensity. Thus a too small k-ratio and a too small thickness is calculated. This trend increases with increasing electron energy causing the decrease of the calculated thickness with increasing electron energy in Figs. 1–3. The steepness of the decrease depends on the mass absorption coefficient chosen. Because of the neglect of peak deconvolution the high contribution of the satellite to the film intensity leads to a too high film thickness. This effect is rather independent of electron energy. This is why the decrease of the measured film thickness starts at a too high value at low primary electron energy.

A peak change of the same kind would of course not only occur in the thin film, but also in a bulk compound with a very low concentration of that element. The arguments with respect to absorption are the same as for the thin film. Thus a bulk standard could be defined as similar to a thin film – whether on a substrate or free supporting – if it is a dissolution of the film elements with a matching concentration in a low absorption matrix. Of course this is not true concerning $\phi(0)$. In fact the program TFA treats very thin films on substrates as a dissolution of the film elements in the substrate [22].

Conductive Thin Films on Insulating Substrates

To perform quantitative analyses on electrically nonconductive specimens they are generally coated with a thin conductive layer to prevent surface charging. Nevertheless, a volume charge is built up in the interior of the specimen. This results in a reduced penetration depth of the electrons, a change in the backscatter factor and the amount of absorption [18]. For bulk specimens an increase in the backscatter coefficient and thus a decrease in X-ray generation could be partially compensated by a reduction of the absorption because of a shift of X-ray generation nearer to the surface. But for a thin film on a substrate the main parameter for the change in the film signal induced by the substrate is the backscatter factor. A compensation by change in the absorption does not take place for the film signal. Therefore, thickness measurements of thin conductive layers on substrates of different electrical conductivity should yield different results, if there is a substantial change of the backscatter coefficient induced by a volume charge. Also removal of the substrate and subsequent determination of the thickness of the unsupported film should lead to a different result. For thin films of Ni, Cu and Au fabricated on Si-wafers, NaCl-crystals and glasses as substrates, no definite substrate dependence was found (see Fig. 3 and Table 1). The film thickness results before and after substrate removal were also in close agreement, as is shown in Table 2 (when in this table more than one value is presented for a certain primary electron voltage, these values are measurements made at different times to get a feeling for the reproducibility and accuracy of the results). Because of the low count rates from these ultrathin films and thus bad counting statistics a charge induced thickness deviation smaller than 5–10% could be detected only with great effort.

Conclusion

Measurements of the thickness of extremely thin films on substrates by EPMA can be performed with high accuracy. It is advisable to use an as high overvoltage as possible. This reduces the dependence of the results not only on the special expression of $\phi(0)$ but quite generally on the kind of program used. Difficulties were encountered in connection with measurements with the L_α-lines of the transition elements of the 4th period. Generally for L_α-lines a low primary electron energy is preferable because of the accordingly lower absorption. Also the sensitivity of the w.d. spectrometer is, especially in the case of thin films, higher at low energies. But for the transition elements of the 4th period, high overvoltage ratios may lead to acceptable results even if no peak deconvolution is performed.

Acknowledgement. The authors would like to thank Mr. A. Brunegger for the careful preparation of the films.

References

[1] G. F. Bastin, J. M. Dijkstra, H. J. M. Heijligers, D. Klepper, *Microbeam Anal.* **1993**, *2*, 29.
[2] J. M. Dijkstra, G. F. Basin, H. J. M. Heijligers, D. Klepper, *Mikrochim. Acta* **1994**, *114/115*, 277.
[3] P. Willich, D. Obertop, *Surface Interface Anal.* **1988**, *13*, 20.
[4] J. L. Pouchou, F. Pichoir, *Scanning* **1990**, *12*, 212.
[5] H. J. Hunger, *Scanning* **1988**, *10*, 65.
[6] R. A. Waldo, *Microbeam Anal.* [*suppl.*] **1993**, *2*, S23.
[7] E. Jacopic, A. Brunegger, R. Essl, G. Windisch, *Proc. 9th ICEM, Toronto, Vol. 1* (J. M. Sturgess, ed.), Imperial, Toronto, 1978, p. 150.
[8] R. Flood, *Scanning Electron Microsc.* **1980**, *I*, 183.
[9] G. F. Bastin, H. J. M. Heijligers, J. M. Dijkstra, *Proc. XIIth ICEM, Seattle, Vol. 2* (L. D. Peachey, D. B. Williams, eds.), San Francisco Press, San Francisco, 1990, p. 216.
[10] C. Merlet, *Mikrochim. Acta* **1994**, *114/115*, 363.
[11] G. Love, M. G. Cox, V. D. Scott, *J. Phys. D. Appl. Phys.* **1978**, *11*, 23.
[12] W. Rehbach, P. Karduck, *Microbeam Anal.* (D.E. Newbury, ed.), San Francisco Press, San Francisco, 1988, p. 285.
[13] W. Reuter, *Proc. 6th ICXOM, Osaka* (G. Shinoda, K. Kohra, T. Ichinokawa, eds.), University of Tokyo Press, Tokyo, 1972, p. 121.
[14] G. Zschornack, *Atomdaten für die Röntgenspektralanalyse*, Springer, Berlin Heidelberg New York Tokyo, 1989, T7.
[15] B. L. Henke, E. M. Gullison, J. C. Davis, in: *Atomic and Nuclear Data Tables, Vol. 54, Nr. 2*, (A. Li-Scholz, ed.), Academic Press, New York, 1993, p. 240.
[16] J. L. Pochou, F. M. A. Pichoir, *Microbeam Anal.* (D. E. Newbury, ed.), San Francisco Press, San Francisco, 1988, p. 319.
[17] M. Watanabe, Z. Horita, M. Nemoto, *Proc. 13th ICEM, Vol. 1*, (B. Jouffrey, C. Colliex, eds.), Paris, Les Editions de Physique, Les Ulis, 1994, p. 605.
[18] M. Kotera, H. Suga, *Proc. 13th ICEM, Vol. 1*, (B. Jouffrey, C. Colliex, eds.), Paris, Les Editions de Physique, Les Ulis, 1994, p. 763.
[19] S. J. B. Reed, *Electron Microprobe Analysis*, Cambridge University Press, Cambridge, MA, 1993, p. 301.
[20] E. W. White, *Microprobe Analysis* (C. A. Andersen, ed.), Wiley Chichester, 1973, p. 352.
[21] G. Remond, C. Gilles, M. Fialin, O. Rouer, R. Marinenko, R. Myklebust, D. Newbury, *EMAS'95 Book of Abstracts*, European Microbeam Analysis Society, 1995, p. 133.
[22] G. F. Basin, H. J. M. Heijligers, *Atb. Metallurgie* **1990**, *3*, 72.

References

Mikrochim. Acta [Suppl.] 13, 473–477 (1996)

A Simple Procedure to Check the Spectral Response of an EDX Detector

Mathias Procop

Bundesanstalt für Materialforschung und -prüfung, Unter den Eichen 87, D-12205 Berlin, Federal Republic of Germany

Abstract. Recent advances in semiconductor detection systems have extended the accessible energy range down to 100 eV and allow the identification of light elements in energy dispersive X-ray spectra. A quantitative analysis on the basis of the K-line intensities of these elements, or of the low energy line intensities of heavier elements, is getting now strongly dependent on the spectral response of the detector in the region below 1 keV. The spectral response is not time constant but changes due to the growth of contamination layers on the detector window and/or the detector crystal. The periodic record of the spectrum of a freshly cleaved Si surface provides a simple check of the response function. As an example, spectra are given for a detector with ice contamination and for the same detector after removal of this layer. The sensitivity of the check can be increased dividing the spectrum by the detector window transmission. In a similar way, spectra can be corrected regarding the transmission of contamination layers before quantification as shown for boron nitride.

Key words: X-ray spectrometry, EOX, detector.

Recent developments in semiconductor X-ray detectors have remarkably increased detector's performance after a long period of stagnation. One of the important improvements is the extension of the accessible energy range down to energies below 1 keV for most modern systems equipped with ultra thin windows. This enables the detection of light elements. Moreover, the lateral resolution of an EDX mapping can be improved using low primary energies still exiting the low energy lines of the heavier elements.

The identification of lines and corresponding elements in the energy range below 1 keV in most cases causes no difficulties using modern detector systems. However, the quantification and identification of compounds represents often a serious problem. This may be attributed to variations in detector efficiency (counts/photon/channel) due to the formation of a contamination layer on the detector window (e.g. oil), or on the detector crystal itself (e.g. ice), changing the detector response at low energies. The problem is demonstrated for a BN spectrum in Fig. 1. The B/N intensity ratio is 0.12 with the detector in the as-received state.

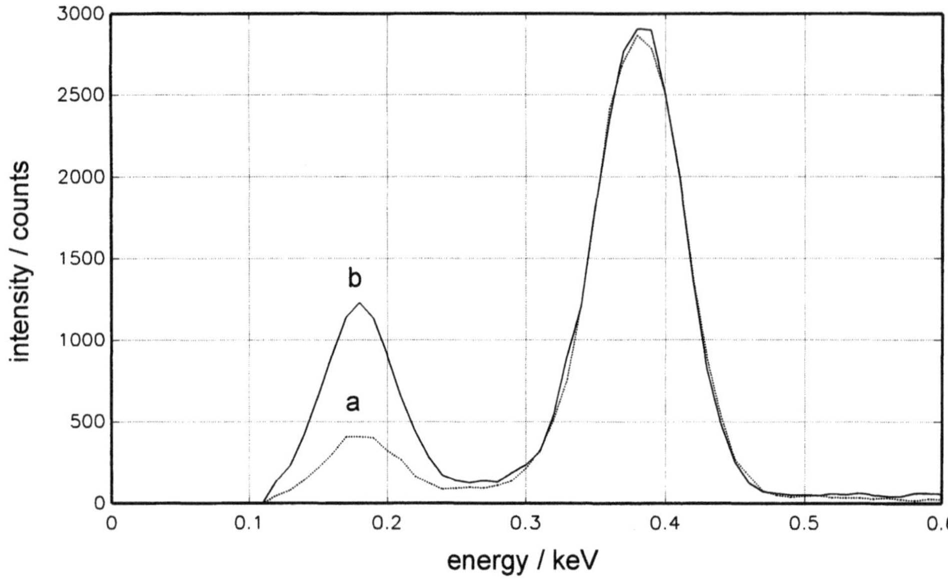

Fig. 1. 5 keV BN spectra recorded with the same detector before (*a*) and after heating and pumping out the system (*b*)

A standardless analysis gives a composition close to a nonexistent compound BN_2. The intensity ratio increased to 0.34 after heating and pumping out the system. A standardless analysis allows the correct identification as BN.

Difficulties of this kind may be avoided if a simple and quick check of the detector response in the low energy range would be available. It is shown that recording a silicon spectrum is a helpful tool.

Experimental

All measurements were performed using a commercial EDX-system connected to a SEM with a cold field emitter. The EDX software allows a standardless quantification on the basis of Schamber's method [1] using supplied reference spectra.

Results and Discussion

Contamination layers of oil or ice cause characteristic absorption edges in the X-ray spectrum below 1 keV. A material without characteristic lines in this energy range should be used to monitor these edges. Such a material is silicon. The tail of the Si-K line extends, depending on the detector type, only up to about 1.4 keV. Below this, each structure in the Bremsstrahlung spectrum of Si is caused by the detector itself. Silicon has the additional advantage to be easily available as single crystal wafer. Freshly cleaved (111) surfaces are smooth and without impurities within the EDX detection limit. An example is given in Fig. 2, spectrum a. The spectrum shows a strong O-K absorption edge, much stronger than the C-K edge from the polymer window. An ice layer covers the detector crystal. The spectrum reflects directly the thickness of contamination layers on the detector crystal or window. The intensity

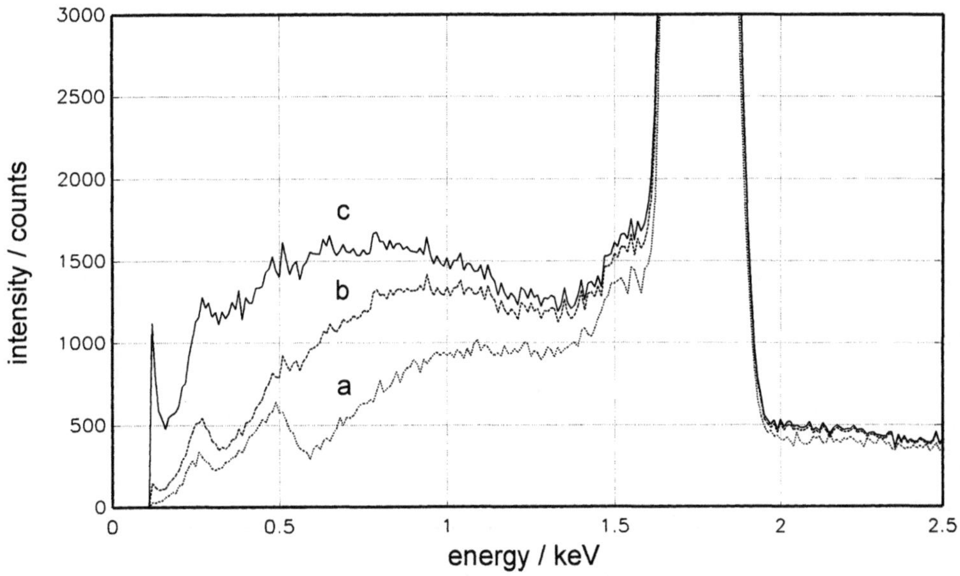

Fig. 2. 10 keV spectrum from a freshly cleaved Si(111) surface with the detector in the as-received state (*a*), after heating and pumping out the system (*b*), and after division of spectrum b by the window transmission (*c*)

jump at the edge and the corresponding absorption coefficients allow to estimate the thickness to be about 0.5 µm. This is an advantage over the often recommended measurement of the K/L intensity ratio of copper or nickel (see e.g. [2]). The K/L ratio allows to monitor only changes of the detector response. There is no value for a contamination free detector and it is impossible to test the performance of a new detector. Moreover, no conclusions concerning the kind of contamination and its reason can be drawn.

In most cases, the inspection of the low energy Si spectrum will be a sufficient check. Spectrum b in Fig. 2 was obtained after pumping the detector system down to 10^{-3} Pa at elevated temperature. The spectral response changed dramatically. The sensitivity is up to a factor of three higher. It is possible to further increase the sensitivity of the check and to find out whether the absorption edges still seen in spectrum b are from contamination layers or from the window material. In this case information about the window material is necessary.

The used detector system is equipped with a MOXTEK AP 1.3 polymer window. Fig. 3 shows the normalised transmission curve of this window type. The transmission was calculated on the basis of layer thicknesses given by the manufacturer and absorption coefficients for C, N, O, and Al [3]. The calculated transmission agrees well with the transmission given in the data sheet [4]. The transmission convoluted with the detector resolution (FWHM about 75 eV in this energy range) is also shown in Fig. 3. A channel by channel division of a Si X-ray spectrum by the convoluted window transmission should result in a structureless background below 1 keV in the absence of contamination on either the window or the crystal. Spectrum c in Fig. 2 is the result of such a correction of spectrum b for the window transmission. The absence of absorption edges indicates a contamination free

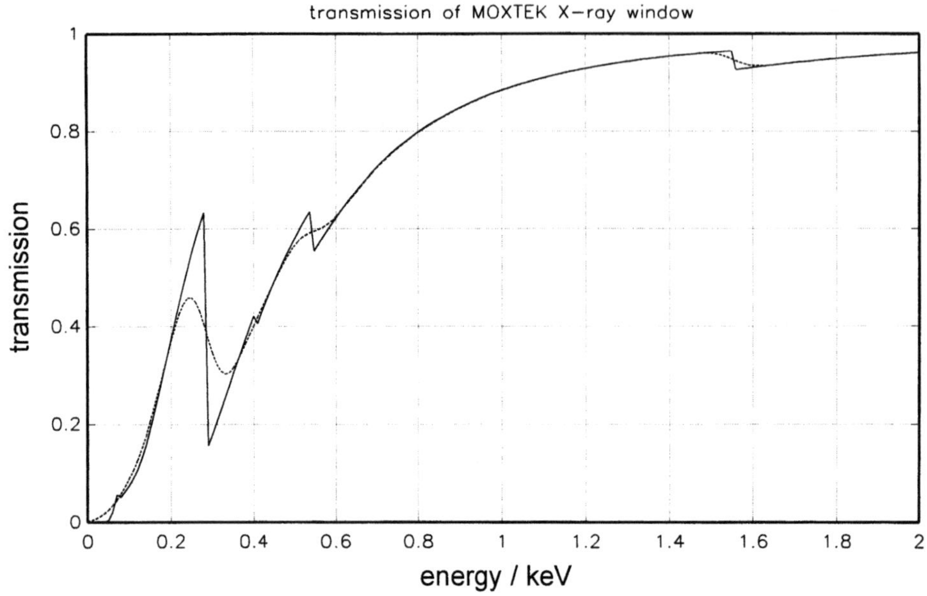

Fig. 3. Transmission of the MOXTEK AP 1.3 window calculated from the given layer thicknesses and convoluted with a detector resolution of 75 eV

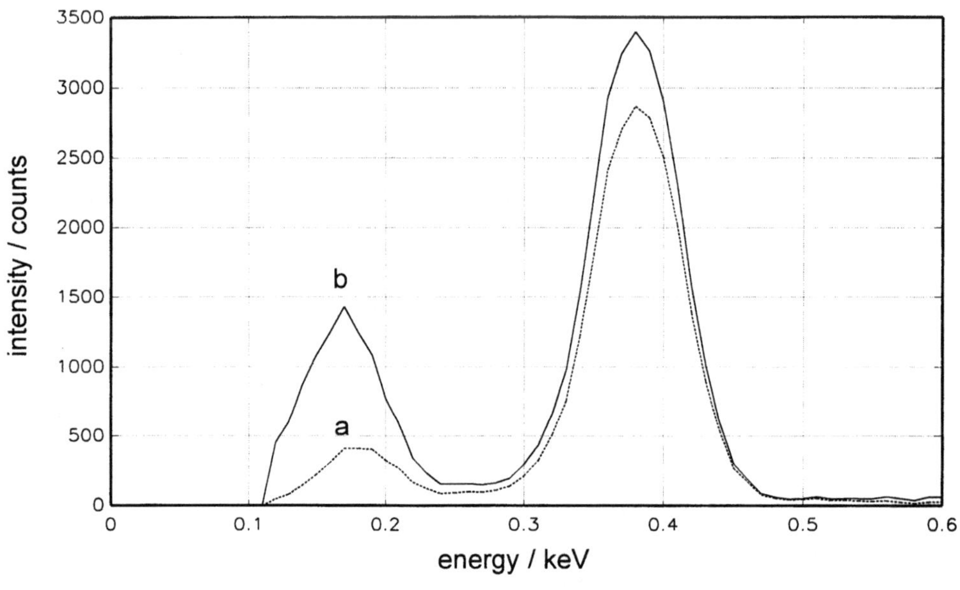

Fig. 4. 5 keV BN spectrum from a detector contaminated by an ice layer (*a*) and after division of the spectrum by the transmission of the contamination layer (*b*)

detector system. The remaining weak structure at the C-K edge can be explained by the kind of approximation of the absorption coefficients neglecting EXAFS [5] and by a slight deviation of the window thickness from the data sheet.

In a similar way spectrum a in Fig. 2 can be divided by the convoluted transmission of an ice layer. Fitting to spectrum b results in an ice layer thickness of

0.5 μm. This procedure is more exact than the estimation given above. The division of the BN spectrum recorded with the ice covered detector (Fig. 1, spectrum a) by the transmission of the ice layer gives the spectrum shown in Fig. 4. It is very similar to spectrum b in Fig. 1. Standardless quantification results in the true composition. This example demonstrates the possibility to correct the spectrum before quantification. This may be applied in case of only weak contamination not justifying such a rigorous procedure as heating and pumping out. Some commercial software package offer the possibility to include ice and oil contamination layer thicknesses in the quantification procedure. Layer thicknesses have to be determined before from the K/L intensity ratio in the copper spectrum with the problems mentioned above. The correction of the low energy part of the spectrum by a "contamination transmission curve" before quantification is an alternative way. It should be preferred in case of light element analysis. The periodic record of the spectrum of a freshly cleaved Si wafer enables a steady update of the contamination transmission curve.

Conclusions

The periodic record of the spectrum of a freshly cleaved Si(111) surface is recommended to check the possible growth of contamination layers of the window or crystal of an EDX detector. The thickness of a contamination layer can be estimated from the height of absorption edges in the low energy part of the Si spectrum. In case of known window transmission curve the sensitivity of the control increases when dividing the Si X-ray spectrum by the window transmission. A contamination free detector has to show the Si Bremsstrahlung spetrum without overlapping absorption edges. In case of known window transmission the transmission of the contamination layer can be calculated too. This allows a quantitative analysis on the basis of a spectrum after correction for the absorption in the contamination layer.

References

[1] J. J. McCarthy, F. H. Schamber, *NBS Special Publ. 604*, Gaithersburg, 1981, p. 273.

[2] S. J. B. Reed, *Electron Microprobe Analysis, 2nd Ed.*, University Press, Cambridge, 1993, pp. 110, 121.

[3] B. L. Henke, E. M. Gullikson, J. C. Davis, *Atomic Data* **1993**, *54(2)*, 181.

[4] Data Sheet of MOXTEK Inc., Orem, Utah 84057.

[5] See ref. [3], p. 202.

Mikrochim. Acta [Suppl.] 13, 479–483 (1996)

Virtual WDS

S. J. B. Reed* and A. Buckley

Department of Earth Sciences, University of Cambridge, Downing St., Cambridge CB2 3EQ, England, U.K.

Abstract. 'Virtual WDS' entails the synthesis of wavelength-dispersive spectra on a computer screen using stored experimental pure-element spectra. The principal object is to facilitate the selection of optimal positions for background measurements, in order to minimise the effect of any interfering neighbouring lines. This approach is much more effective than referring to wavelength tables and obviates the need to use 'real' instrument time for spectrometer scans to investigate each new case. Where interferences cannot be avoided completely, the size of the correction may be predicted. Stored experimental continuum spectra enable corrections for background nonlinearity to be estimated (also for slope where background is measured on one side only, owing to the presence of an unavoidable interference on the other side). Further, analysis conditions (counting times, etc.) required to achieve specified detection limits can be determined in advance.

Key words: background corrections, interference corrections, wavelength-dispersive spectrometry.

In quantitative EPMA using WD spectrometers, peak intensity measurements have to be corrected for the underlying 'background', which is attributable mainly to the continuous X-ray spectrum. Interpolation from measurements made on each side of the peak is generally satisfactory. However, neighbouring X-ray lines sometimes affect these measurements, and become increasingly important for low concentrations. (At or near the detection limit the peak intensity may be only a small fraction of background, making the correction very prone to errors of this sort.)

The presence of potential interferences can be deduced from wavelength tables but their significance is difficult to assess from such tables alone because it depends on the height and width of the interfering peaks. Furthermore, to carry out spectrometer scans of relevant regions of the spectrum is extremely time-consuming if statistical precision commensurate with that of the analytical measurements is to be obtained (especially when long counting times are used in order to obtain a low detection limit). However, since line positions and profiles are invariant (to a sufficient approximation) the need to record wavelength scans experimentally for each

* To whom correspondence should be addressed

new type of sample can be avoided by using computer-generated spectra. Modelling peak profiles by means of mathematical expressions is difficult because of the variation in the peak-shape function with Bragg angle and the presence of satellite lines [1, 2]. Also, values for relative line intensities given in tables are somewhat inadequate, especially for minor lines. Experimental spectra have therefore been used, in order to avoid these problems.

'Virtual' Spectrum Display

A program has been developed which enables 'virtual' WD spectra derived from stored peak profiles to be displayed on the computer screen. Initially a periodic table is displayed, in which elements whose K, L or M lines lie within the wavelength range covered by the specified crystal (TAP, PET, or LiF) are indicated. The analysed element is selected by 'clicking' with the 'mouse' (either α or β lines may be chosen). The display then shows which other elements may affect peak or background determinations for the analysed element, i.e. have a peak (or any part of a peak) within a predetermined wavelength range on each side of the analytical peak. (The presence of high-order lines satisfying this criterion can also be indicated if required.) The user may select from these potentially interfering elements those relevant to the case in hand by 'clicking' on the periodic table, causing the spectra of analytical and interfering elements to be displayed simultaneously (Figs. 1–3).

The spectra are represented in digital form with $\sin\theta$ increments of 0.0001 (θ = Bragg angle). The stored pure-element K, L and M spectra are normalised so that the maximum of the Kα, Lα, or Mα peak is equal to a constant value. The

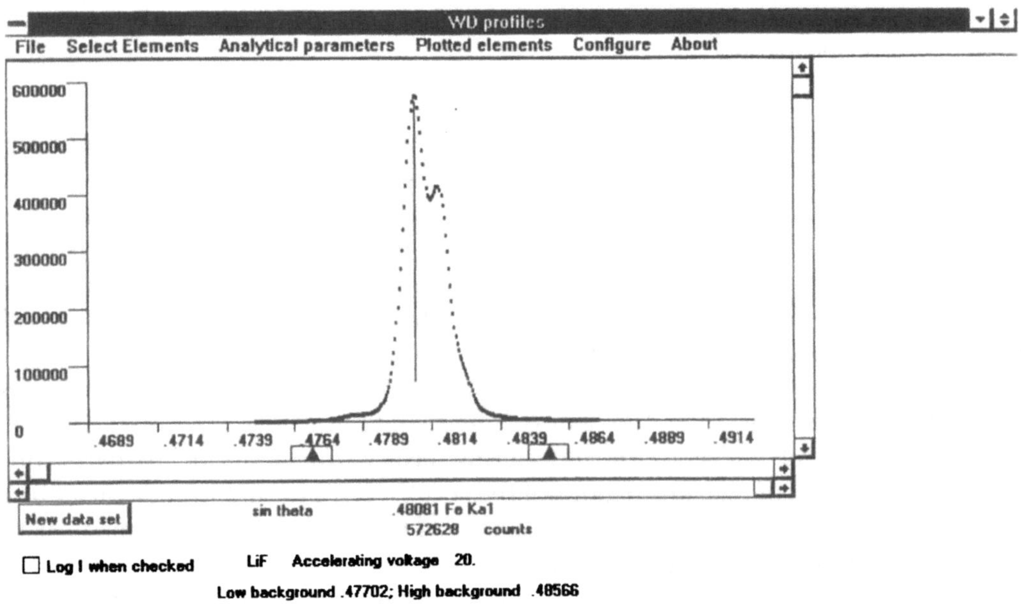

Fig. 1. On-screen display of FeKα peak (for LiF crystal): vertical scale – counts/s/μA, horizontal scale – sine θ; 'sine theta' and 'counts' values given below are for the point indicated by the arrow on the screen (not shown here); background markers below sine θ axis can be moved with the 'mouse'

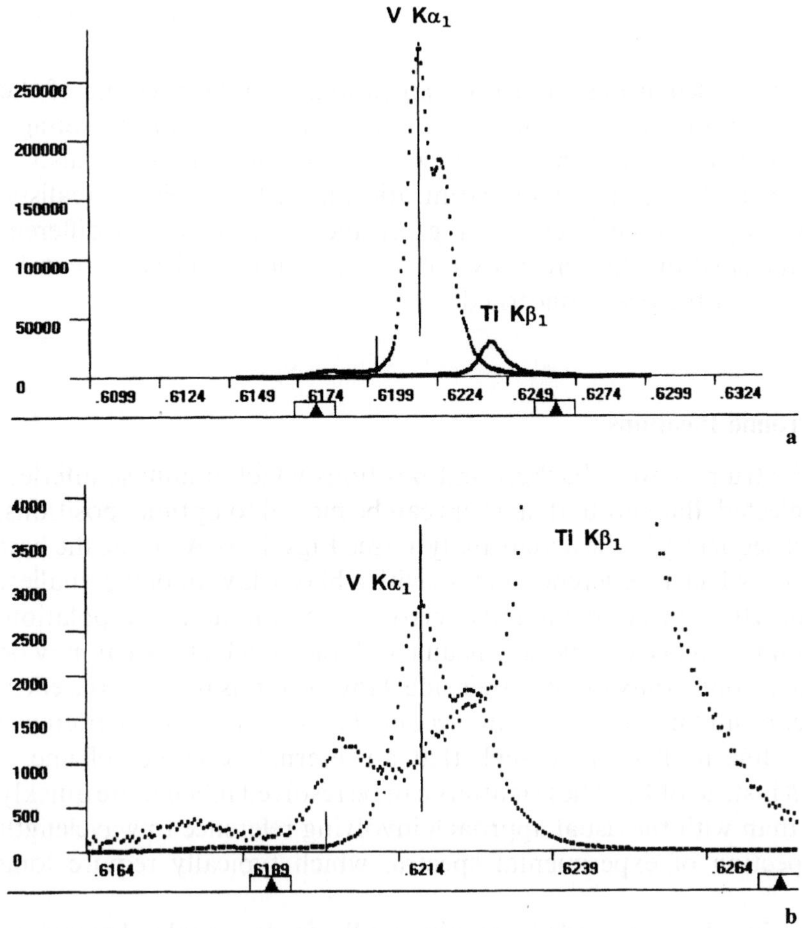

Fig. 2. Example of overlapping peaks: analytical element – V (Kα), interfering line Ti Kβ (LiF crystal): **a** concentrations of both elements 100%; **b** V peak scaled in proportion to concentration of 1% (Ti unaltered) and display expanded both horizontally and vertically. The short vertical line above the sine θ axis marks the position of the Ti K absorption edge. In this case care is needed in choosing background positions. Also, there is significant overlap on the peak, a correction for which can be derived from the 'virtual' spectrum

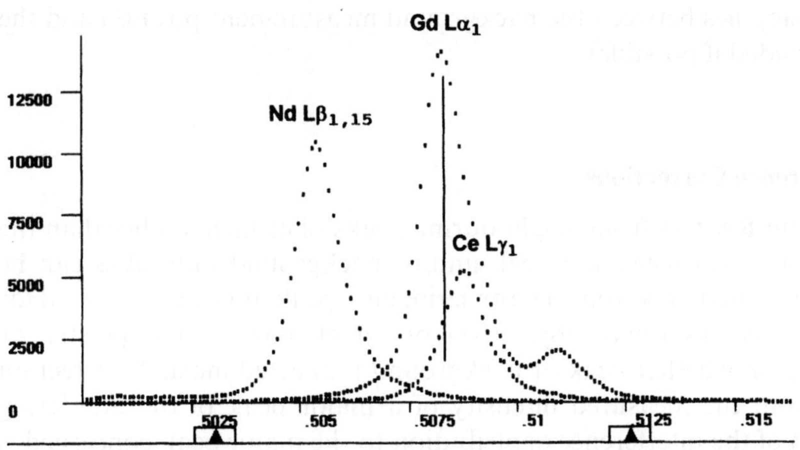

Fig. 3. Another example of peak overlap: analytical element – Gd (Lα), interfering elements – Ce and Nd; concentrations: Gd – 5%, Ce – 40%, Nd – 15%). The importance of choosing suitable background positions and the need for overlap corrections are obvious

displayed spectra are scaled using empirical expressions to take account of the dependence of X-generation efficiency on atomic number and accelerating voltage, and the variation in spectrometer efficiency with wavelength. Intensities may also be scaled in proportion to elemental mass concentration in order to obtain realistic peak intensities for compound specimens (see Figs. 2a and 3). The spectra of different elements are distinguished on the screen by colour, and individual peaks can be identified by 'pointing' at the peak concerned.

Selection of Background Positions

The synthesised spectrum enables background positions which minimise interference effects to be selected. Background markers can be moved to optimal positions, these being used subsequently in the actual analysis (see Figs. 1–3). As a rule the best positions are those for which interference is least, with a bias in favour of the smallest offset from the analytical peak to minimise errors in continuum interpolation. Sometimes serious interference on one side is unavoidable, in which case it may be best to measure background only on the other side. However, this requires the effect of the slope of the continuum to be taken into account (see below). Also, interference with the analytical line itself may be such that an alternative choice of line is appropriate, e.g. $L\beta$ instead of $L\alpha$. These matters can be resolved much more quickly by 'Virtual WDS' than with the usual approach involving reference to wavelength tables and/or inspection of experimental spectra, which typically require long recording times.

The continuum has been recorded experimentally from samples having no emission lines within the wavelength range of the spectrometer crystal concerned. The intensity may be scaled in proportion to the mean atomic number of the specimen, as derived from the assumed composition, in order to make the continuum consistent with the scaled peak intensities. The presence of steps in the continuum due to absorption edges, whose positions are displayed on the screen (see Fig. 2), should also be taken into account when selecting background positions (the situation where a step lies between the background measurement position and the peak should be avoided if possible).

Estimating Interference Corrections

The effect which interferences from neighbouring peaks of elements other than the analysed element have on measured peak and/or background intensities can be estimated from the 'virtual' spectrum. The experimental peak intensity corrected for background in the usual fashion is subject to errors which may be either positive or negative depending on whether peak or background is affected most. A correction may be derived from the measured intensity of a major peak of the interfering element if the ratio of the interference contribution to the major peak concerned is known (Fig. 2). Estimating this ratio from the virtual spectrum makes it possible to decide whether the effect is important in a particular case. If required, the size of the correction may then be estimated and applied to 'real' analytical data.

Background Shape and Curvature

The conventional procedure of measuring background on each side of a peak and interpolating linearly is satisfactory provided curvature is insignificant. The correction required for such nonlinearity can be estimated from the 'virtual' continuum, derived from the recorded peak-free spectrum for the crystal concerned, scaled according to atomic number. Similarly, in cases where background is measured on only one side owing to unavoidable interference on the other, the effect of slope can be estimated.

Predicting Detection Limits

The minimum detectable concentration is that for which the peak is just visible above background: that is, the intensity measured at the peak position minus the background intensity is equal to the statistical uncertainty in the latter, which may be defined as three standard deviations (i.e. three times the square root of the number of background counts). Using empirical data on peak and continuum count-rates (for the appropriate accelerating voltage), detection limits may be calculated for user-specified probe current and counting times. Analytical conditions appropriate to particular requirements can thus be selected in advance, which is more efficient than the usual 'trial and error' approach.

References

[1] G. Remond, P. Coutures, C. Gilles, D. Massiot, *Scann. Microsc.* **1989**, *3*, 1059.

[2] R. L. Myklebust, in: X-ray Spectrometry in Electron Beam Instruments (D. B. Williams, J. I. Goldstein, D. E. Newbury, eds.), Plenum, New York, 1995, p. 275.

Mikrochim. Acta [Suppl.] 13, 485–492 (1996)

Monte Carlo Simulation Program with a Free Configuration of Specimen and Detector Geometries

Ludwig Reimer*, Manfred Kässens, and Lutz Wiese

Physikalisches Institut, Universität Münster, Wilhelm-Klemm-Str. 10, D-48149 Münster, Federal Republic of Germany

Abstract. A FORTRAN 77 program package has been developed for the Monte Carlo simulation of electron trajectories and of secondary electrons (SE), backscattered electrons (BSE) and X-ray linescans for scanning electron microscopy and microanalysis. A data base contains the relativistic elastic Mott cross-sections of all elements, the X-ray absorption coefficients, and the most relevant ionisation and characteristic X-ray energies.

Five different compounds and their specimen geometry or a structure of five multilayers can be configurated by a simple graphic program and the specimen structure is stored in a 350 × 600 matrix. Plots of electron trajectories at different x-positions give an overview of the relation between the specimen geometry and the electron diffusion. For the calculation of linescans, four different BSE detector configurations, with varying take-off directions and range of azimuth angles, consisting of scintillator, semiconductor, BSE/SE converter or channel plate detectors, and for scanning transmission electron microscopy (STEM) conditions can be configurated and calculated simultaneously. Further options are the electron energy, angular and spatial distributions, depth functions of ionization and X-ray generation with and without an absorption correction.

Key words: Monte Carlo simulation, secondary electrons, backscattered electrons, X-ray depth ionisation function, k-factors.

Though Monte Carlo simulations found increasing interest for solving the influence of electron diffusion in scanning electron microscopy (SEM), scanning transmission electron microscopy (STEM), electron-probe microanalysis (EPMA) and Auger electron spectroscopy (AES) due to the increasing speed and storage of PCs, the ability of solving a particular problem with a given algorithm is limited because the specimen geometry description and the chemical composition of the different components are parts of the algorithm. Therefore, we developed a program package

---

* To whom correspondence should be addressed

which allows a fast change in the configuration of specimen geometry and detector strategy without writing a new program.

The Program Package

The program package, written in FORTRAN 77, contains the program described in [1–3] and consists of the following routines.

Configuration of Specimen Geometry

The routine CONF can configurate the specimen geometry and calculates the necessary elastic scattering parameters. It asks at first for the chemical composition of up to 5 different compounds with up to 6 different elements and calculates, with a data base, arrays containing the small-angle ($\theta \leqslant 10°$) and large-angle ($\theta \geqslant 10°$) scattering data for electron energies in the 0.1–50 keV range. The data base contains small-angle and large-angle elastic Mott cross-sections for all elements obtained by solving the Pauli-Dirac equation with the partial-wave method [4]. These cross-sections can be shown by the routine MOTTXABS (see Fig. 1 in [3]) which also can show the energy-dependence of X-ray absorption.

The specimen geometry can be configurated by a graphic program drawing a polygonal surface profile and trapeziform and circular areas in a different colour for each compound. This geometry and composition is stored in a 350 × 600 matrix with pixel values 0 (vacuum), 1 (matrix) and 2–5 (2nd-5th compound), respectively. This specimen structure, configurated in the x-z plane can either extend in the y direction normal to the screen or can show rotational symmetry around the z axis. Fig. 1 and following figures show as an example the geometry of a semiconductor structure consisting of an etched Si substrate, a SiO_2 film, a PMMA resin layer and a Cr conductive path. Such a geometry together with the calculated data for elastic scattering can be stored in an external file and used in the routines PLOT, STEREO, SCAN and XLINE.

Plotting of Electron Trajectories

The routine PLOT can be used to look at plots of trajectories to see the penetration and diffusion of primary and backscattered electrons. Input parameters are the electron energy $E = 0.2$–0.5 keV and the angle of electron incidence. The data for the Bethe continuous-slowing down approximation with a Rao-Sahib-Wittry correction for low energies are generated for the selected electron energy. 10–100 electron trajectories can be plotted at different specimen positions (Figs. 1a, b) to get an overview of how the electrons penetrate and are backscattered at different parts of the specimen for primary electron energies of a) 3 keV and b) 5 keV.

The routine STEREO produces a stereo pair of electron trajectories in red and green with the coordinates

$$x_{1,2} = x \cos\gamma \pm y \sin\gamma \qquad (1)$$

($\gamma = 2$–$5°$) to get an impression about the three-dimensional propagation of electrons by red-green spectacles.

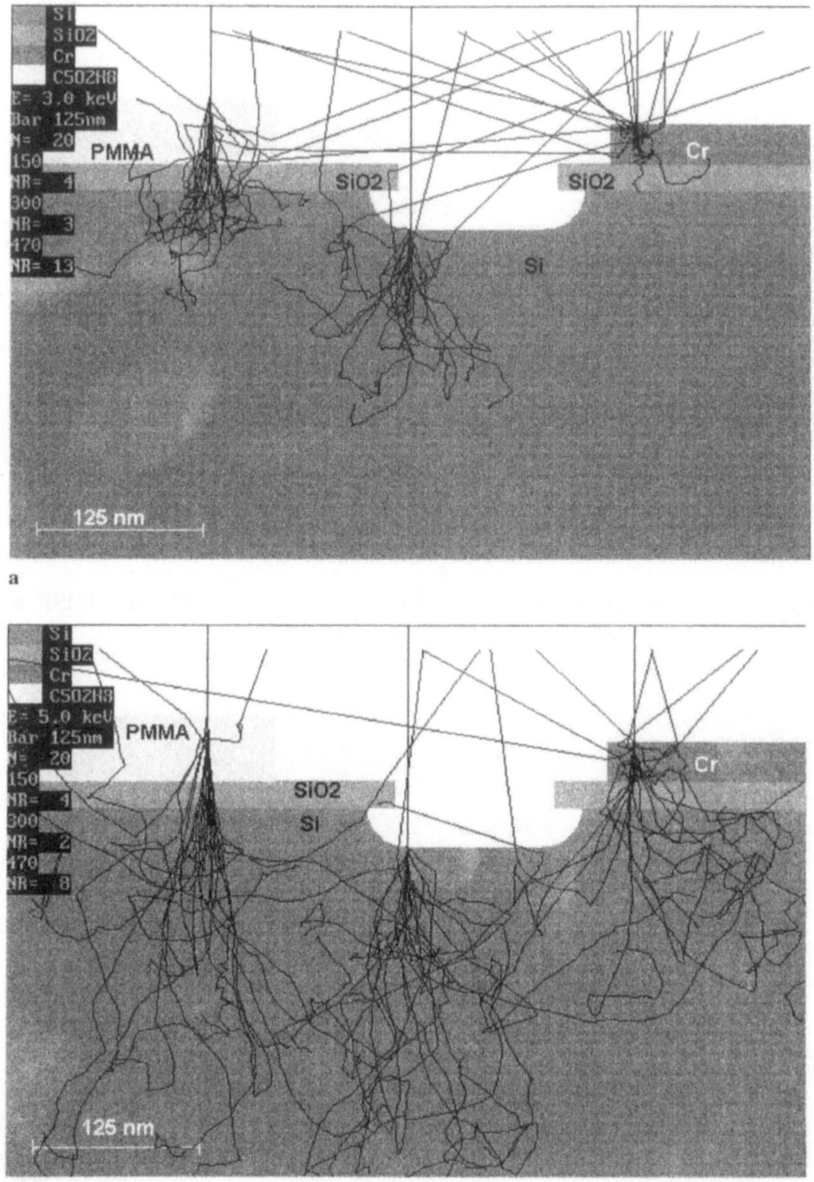

Fig. 1. Simulated electron trajectories for electron energies, **a** $E = 3\,\mathrm{keV}$ and **b** $5\,\mathrm{keV}$ by the routine PLOT entering an etched silicon substrate with SiO_2, Cr and PMMA layers configurated by CONF

Linescans of BSE and SE Signals

The routine SCAN calculates linescans of BSE and SE signals with a configurated detector strategy superposed on the image of the specimen geometry (Fig. 2). After loading the specimen geometry from a file generated in CONF and selecting the electron energy and specimen tilt, four different detector strategies can be configurated for BSE using all BSE, an annular top detector or a cylindrical ring detector

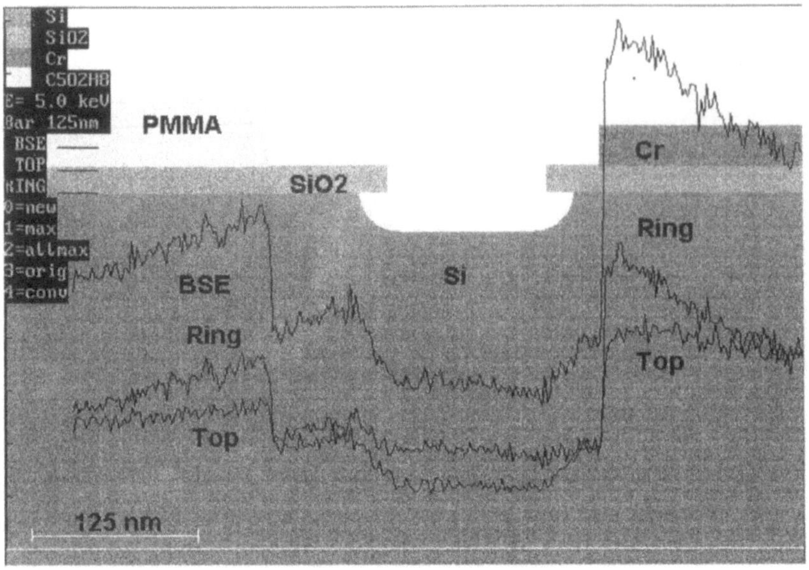

Fig. 2. Linescans of BSE signals across the semiconductor structure of Fig. 1 for all emitted BSE, an annular top detector and a cylindrical ring detector as shown in Fig. 3

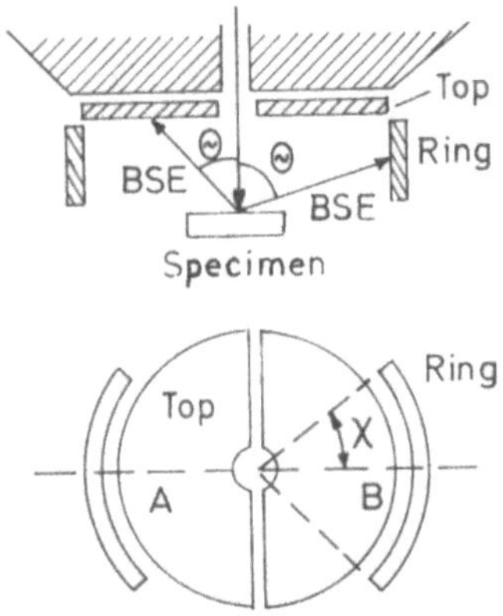

Fig. 3. Geometry of a semi-annular top and ring scintillators coupled per light-pipe to a photomultiplier for detecting BSE

surrounding the specimen (Fig. 3). The scan in Fig. 2 with all BSE and with the ring detector shows an increase of the signals at the PMMA and Cr edges, whereas such an edge effect is not simulated in the signal of the top detector. The program asks for the dimension of the detector in mm and calculates the recorded range of BSE take-off angles. A range of azimuth angles can be selected to compose also semi-annular BSE detectors A and B, for example.

The SE emission is calculated by using a mean energy loss of the order of 10–25 eV for creating a secondary electron and an exponential attenuation $\propto \exp(-s/\Lambda_{SE})$ with a mean exit depth Λ_{SE} of SE of the order of 0.3–10 nm. However, the choice of these two parameters is problematic especially when the SE signals of two different compounds are compared. When an electron passes a surface layer of $3\Lambda_{SE}$, probabilities of SE generation proportional to the Bethe loss are calculated with a random emission angle. The attenuation is a function of the calculated path length s to the surface.

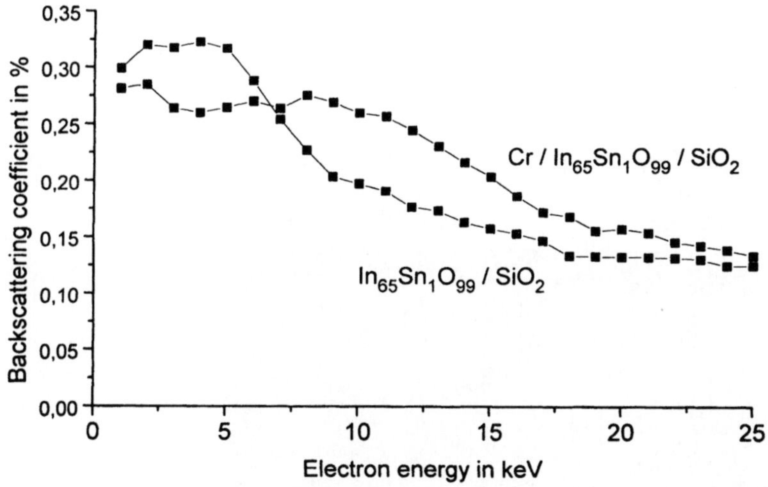

Fig. 4. Calculated energy dependence of the backscattering coefficient η for a 50 nm indium-tin-oxide layer on quartz (SiO_2) with and without a 20 nm Cr layer

Furthermore, a semiconductor or scintillation-detector (signal $\propto E_{BSE} - E_{th}$, E_{th} = threshold energy) inclusively a post-acceleration of low energy BSE, a channel plate or BSE-SE conversion at a metal plate ($\propto \delta$) can be selected for each of the selected detector geometries.

Backscattering and SE Emission of Layered Structures

The routine GRAF calculates the BSE coefficients η and secondary electron yields δ as a function of electron energy and/or surface tilt for a multi-layer system with up to 5 layers parallel to the surface. Reading 0 (vacuum) for the substrate results in a transparent film. As an example Fig. 4 shows the calculated energy dependence of the backscattering coefficient η for a 50 nm indium-tin-oxide film on quartz (SiO_2) with and without a 20 nm Cr layer. The contrast reversal of the BSE signal at about 7 keV is in agreement with experiments [5].

The routine DISTR allows to calculate energy, angular, and exit distributions of BSE and to plot distributions of dissipated energy density which will be of interest for electron beam lithography, cathodoluminescence and the electron-beam induced current (EBIC) mode.

X-Ray Emission

XPHI simultaneously calculates the depth distribution functions $\Phi(\rho z)$ for up to 4 different X-ray lines and the absorption corrected $\Phi_{abs}(\rho z)$ for a selected X-ray take-off angle and a multilayer system of up to 5 layers. Gryzinski cross-sections are used for calculating a probability that a characteristic X-ray quantum is generated at a free-path length. A number of 1000–10000 electrons (20–200 s computation time) is sufficient for good statistics.

When multiplying with the local concentration of elements and integrating $\Phi_{abs}(\rho z)$ over the depth z below the surface, a comparison with a calculation for a pure element of composite standard directly results in the k_{ZA} correction coefficient. An example of calculated $\Phi(\rho z)$-curves is shown in [3].

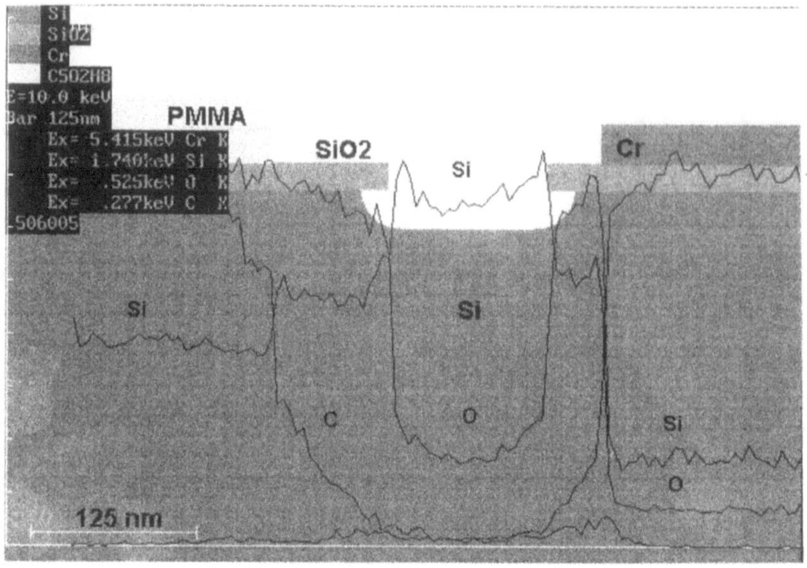

Fig. 5. Linescan of X-ray signals across the semiconductor structure of Fig. 1 excited by 10 keV electrons for Cr, Si, O and C K X-ray quanta across the semiconductor structure (C K emission from the PMMA layer)

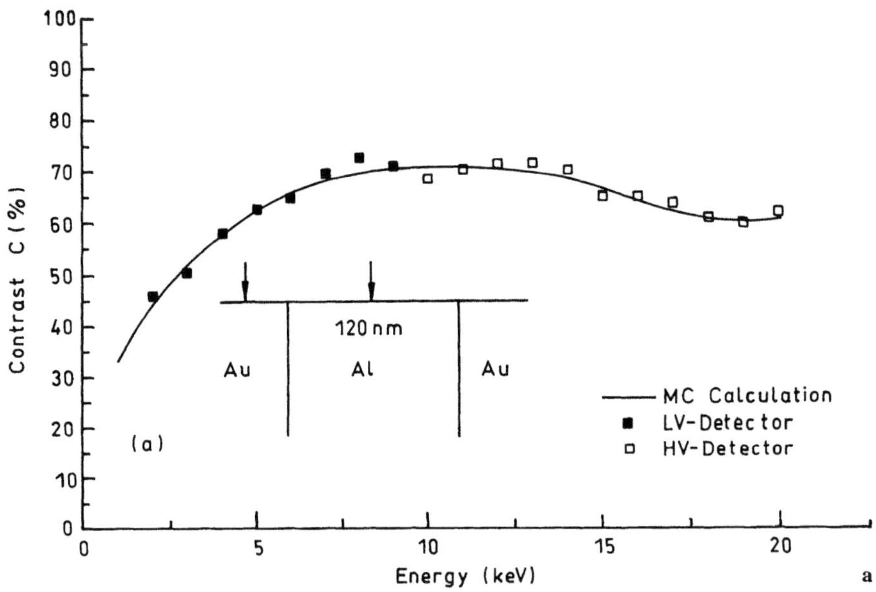

Fig. 6a

For linescans of X-ray signals with characteristic X-ray quanta and for seeing the influence of a more complex specimen composition and geometry on the X-ray signal for quantitative analysis, the routine XLINE can simulate a linescan as shown for an example in [3] and in Fig. 5 for the configurated semiconductor structure. Especially the linescans for Cr Kα ($E_x = 5.415$ keV) and C Kα ($E_x = 0.277$ keV) show that these X-ray quanta are excited also for larger distances from the Cr or PMMA layer.

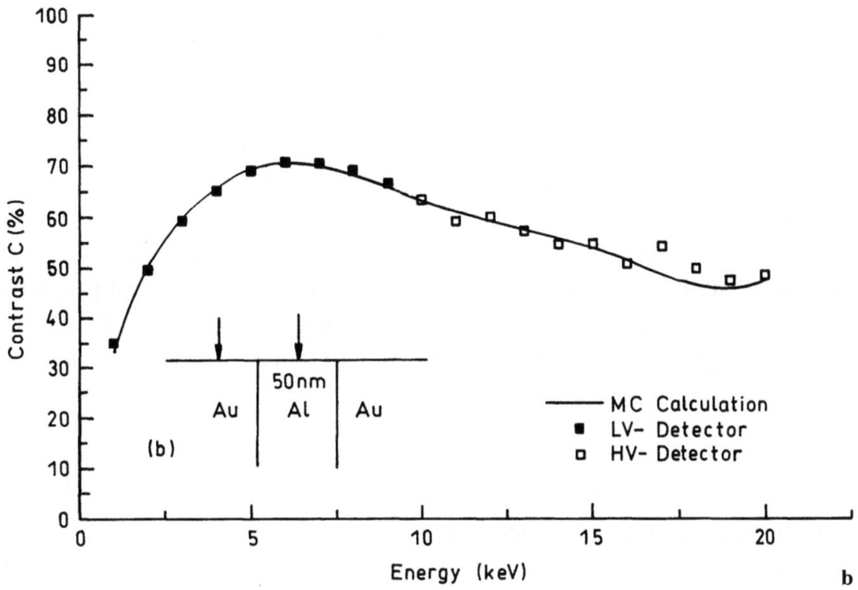

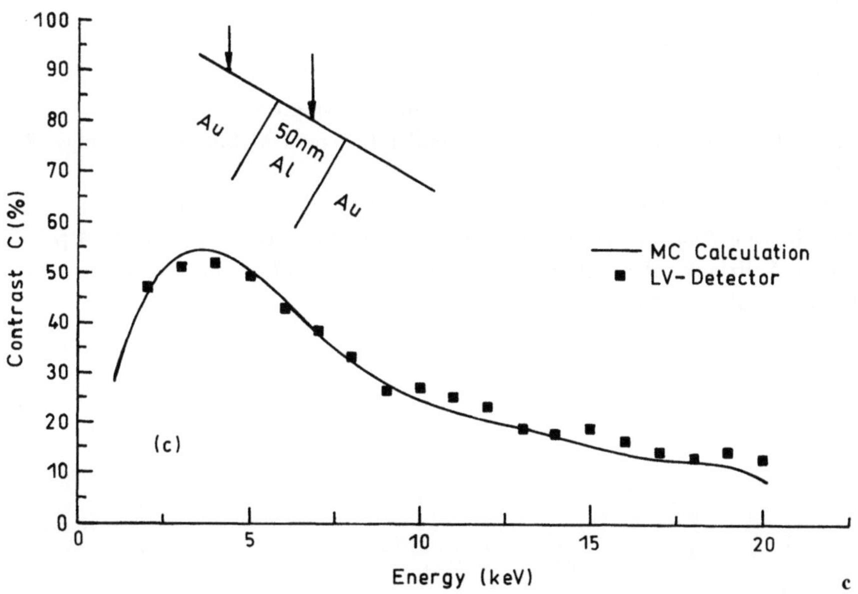

Fig. 6. Measured and simulated contrast C of the BSE signal as a function of electron energy of **a** a 120 nm and **b** a 50 nm aluminium layer in gold viewed head-on a cross-section at normal incidence and **c** the 50 nm layer tilted by 30°

Comparison with Experiments

Monte Carlo simulations are useless when they are not quantitatively compared with experiments. Only this can give the confidence to use MCS for explaining contrast and searching in advances for optimum conditions of energy, tilt and/or detector strategy. We already tested our MCS with measurements of the backscattering coefficient [1–3, 6, 7] and for colloidal gold particles on silicon [8], for example.

As a recent result of the application of this program package, Fig. 6 shows the contrast

$$C = (C_{Au} - C_{Al})/(C_{Au} + C_{Al}) \quad (2)$$

of (a) a 120 nm and (b) a 50 nm Al layer in gold viewed head-on when the electron hits either the Al layer or gold matrix at normal incidence. Calculations and experiments agree rather well and show that the contrast will be largest at about 10 and 7 keV, respectively, when the electrons which hits the centre of the Al layer are backscattered at the Au sidewalls which decreases the backscattering (see also calculations in [9]). The same phenomena of good contrast has been observed for GaAs/GaAlAs layers [10]. Tilting the specimen by 30° (Fig. 6c) results in a strong decrease of contrast.

References

[1] L. Reimer, E. R. Krefting *NBS Spec. Publ.* **1976**, *460*, 45.
[2] L. Reimer, D. Stelter *Scanning* **1986**, *8*, 265.
[3] L. Reimer, *Microchim. Acta*, [*Suppl.*] **1996**, *13*, 1.
[4] F. Salvat, R. Mayol, *Comp. Phys. Comm.* **1993**, *74*, 358.
[5] B. H. Bracher, *Personal Communication.*
[6] L. Reimer, *Image Formation in Low-Voltage Scanning Electron Microscopy*, Tutorial Texts in Opt. Eng. Vol. TT 12, SPIE Opt. Eng. Press, Bellingham, Washington, 1993.
[7] R. Böngeler, U. Golla, M. Kässens, L. Reimer, B. Schindler, R. Senkel, M. Spranck, *Scanning* **1993** *15*, 1.
[8] P. Hirsch, M. Kässens, L. Reimer, R. Senkel, M. Spranck, *Ultramicroscopy* **1993** *50*, 263.
[9] L. Reimer, P. Wahlbring, *Proc. 13th Int. Congr. on Electron Microscopy, Vol. 2*, Paris, 1994, p. 355.
[10] K. Ogura, A. Ono, S. Franchi, P. G. Merli, A. Migliori, *Proc. XIIth Int. Congr. on Electr. Micr., San Francisco, Vol. 1*, 1990, p. 404.

Mikrochim. Acta [Suppl.] 13, 493–500 (1996)

Barriers to Energy Dispersive Spectrometry with Low Energy X-Rays

David G. Rickerby

Institute for Advanced Materials, Joint Research Centre, European Commission, I-21020 Ispra (VA), Italy

Abstract. Application of low energy X-rays for quantitative energy dispersive analysis is limited by absorption in the carbon contamination layer on the specimen surface and the ice layer on the detector crystal. Appropriate methods to correct for these effects are discussed with reference to the analysis of a low atomic number (BN) film on a silicon substrate. Carbon contamination can either be reduced to an insignificant level by acquiring spectra in the scanning mode or treated as an additional layer with a thickness that increases with time. It is shown that correction for the ice layer using the standard technique of estimating the mass thickness by means of the NiL/NiK ratio may be inadequate for the analysis of nitrogen due to the strong absorption of NKα by hydrocarbon impurities in the ice.

Key words: EDS analysis, phi(rhoZ), contamination, thin films, light elements.

The use of low excitation voltages in energy dispersive X-ray analysis of light elements and thin films necessitates increased awareness of the influence of specimen and detector contamination on the apparent peak intensities. Significant absorption of low energy X-rays may occur in the carbonaceous contaminant layer on the specimen surface or an ice layer on the detector crystal. Crystal icing represents a particularly insidious problem for a windowless detector since it leads to a gradual modification of the low energy region of the detection efficiency curve with time. This effect is compounded by the generally larger uncertainties associated with the mass absorption coefficients for low energy X-rays.

Hren [1] has provided a comprehensive review of the specimen contamination mechanisms relevant to X-ray microanalysis in the electron microscope. Surface migration of adsorbed hydrocarbon molecules has been demonstrated [2, 3] to play a determinant role in the build-up of contamination under irradiation with a focused electron beam. Contamination rates have been measured by electron transmission [4] and electron energy loss [5] techniques in TEM, by elastic scattering methods [6] in STEM and by using the backscattered electron signal in SEM [7]. The shape of the contamination spot is related to the current density distribution [7, 8] and can be explained by the theoretical model developed by Müller [9].

The influence of crystal icing on detector performance has been discussed by Musket [10], Müller et al. [11] and Lowe [12]. Ice layers of thickness $\sim 1\ \mu m$ were typically observed under normal conditions of usage. A consequent reduction in spectral efficiency was evident for X-ray energies less than 1 kV. While an ice layer of moderate thickness does not present a problem for analysis with standards, it would be a source of potential error in standardless analysis, particularly of light elements. Decrease in the low energy sensitivity of an ultra thin window (UTW) detector, due to the condensation of hydrocarbon molecules on the window, has also been reported [13].

This paper deals with the diagnosis and correction of contamination and icing effects in energy dispersive analysis of a low atomic number thin film on a silicon substrate. It will be demonstrated that assessment of the extent of detector icing by the commonly recommended method of determining the NiL/NiK integrated peak ratio may not be adequate for analysis of ultra light elements.

Experimental

A c-BN film of approximately 100 nm thickness was deposited on a pure silicon substrate by r.f. sputtering from a boron nitride target. Details of the deposition technique have been published elsewhere [14]. This film was of nearly stoichiometric composition, with a few atomic percent of oxygen the major impurity.

Energy dispersive X-ray spectra were acquired from this specimen using a Link Pentafet Si(Li) windowless detector, with a take-off angle of 35 degrees, mounted on a JEOL 6400F field emission SEM. The acquisition live time was 100 s (dead time 20–30%). Pure boron, pure carbon, silicon nitride and aluminium oxide were employed as standards. An accelerating voltage of 5 kV was used, with a beam current of ~ 1 nA. The vacuum system of the microscope was equipped with an activated alumina filter and a liquid nitrogen foreline trap, allowing attainment of an ultimate specimen chamber pressure of $2 \cdot 10^{-5}$ Pa. To reduce the effects of specimen contamination, spectra were acquired either at TV scanning rate or by displacing the specimen in discrete steps at intervals of ~ 1 s in spot mode. While these two methods are not strictly comparable, because the synchronization delay may lead to a non-uniform dose over the scanned area, they were found to be equally effective in restricting contamination to negligible levels.

In order to perform quantitative analysis with a cold field emission source some form of compensation is required for the inherent instability of the beam current. This was achieved by continuously monitoring the current incident on the condenser lens sperture. The existing circuit for correction of the intensity level of the secondary electron image generates an output signal proportional to this current at all accelerating voltages. This signal can be used, after voltage/frequency conversion, to drive the external clock of the analyser system, thereby automatically adjusting the live time in inverse proportion to the integrated beam current. The correction was found to exhibit good linearity over the entire range of usable beam currents.

Thin film analysis was carried out using the STRATA software [15, 16] based on the XPP model of Pouchou and Pichoir [17]. This model employs an X-ray depth distribution function of the form

$$\Phi(\rho z) = \mathrm{A}\exp(-\mathrm{a}\rho z) + [\mathrm{B}\rho z + \Phi(0) - \mathrm{A}]\exp(-\mathrm{b}\rho z) \qquad (1)$$

where $\Phi(0)$ is the surface ionization and A, B, a, b are coefficients describing the shape of the function. Calculations are performed iteratively to minimize the deviation between experimental data and the theoretical X-ray intensity curves.

Results and Discussion

Crystal Ice Layer

The efficiency of a Si(Li) detector is affected by X-ray absorption in the gold surface layer, the silicon dead layer (more properly termed the region of incomplete charge conversion) and the active region of the crystal [18, 19]. For detectors with either a beryllium or ultra thin window, the absorption in the window material must additionally be considered. In the case of a windowless detector the formation of an ice layer on the crystal will effectively degrade performance. The spectral response in the low energy range is dependent on the time of exposure of the crystal and the quality of the microscope vacuum. The modified detection efficiency ε' is related to the efficiency ε in the absence of an ice layer by the expression

$$\varepsilon' = \varepsilon \exp[-(\mu/\rho)^{A}_{H_2O} \cdot \rho t] \qquad (2)$$

where $(\mu/\rho)^{A}_{H_2O}$ is the mass absorption coefficient for X-rays of element A and ρt is the mass thickness of the ice layer.

Table 1 lists the relative detector efficiencies ε'/ε for the Kα lines of the elements of interest and NiLα for several values of ice layer thickness. An ice layer as thin as ~100 nm will lead to a substantial reduction in the detection efficiency for BKα, while decreasing that for NiLα by only a few percent. The standard method for assessing low energy detector performance by measuring the NiL/NiK peak ratio is therefore not sufficiently informative for quantitative light element analysis with a windowless detector. An alternative method, originally proposed by Musket [10], is to measure the CKα/FKα peak ratios obtained from a Teflon specimen. The intensity of FKα is extremely sensitive to the thickness of the ice layer due to its high mass absorption coefficient (12400 $cm^2\,g^{-1}$) in oxygen.

The influence of crystal icing on the light element efficiency of the Link detector is illustrated in Fig. 1, which shows spectra acquired immediately after conditioning the detector, to remove the ice layer, and following 160 days of intermittent operation in windowless mode. These spectra have been scaled to the number of counts in the highest SiKα channel to facilitate comparison. The NiL/NiK ratio fell from an initial value of 1.26 to 0.83 during this period, corresponding to the formation of an ice layer of mass thickness 0.69 $\mu g\,cm^{-1}$. This would imply relative detection efficiencies ε'/ε of 0.357 for BKα, 0.855 for NKα and 0.928 for OKα. The respective experimentally determined values were 0.355, 0.736 and 0.903.

Such a large discrepancy between the theoretical and experimental detection efficiencies for NKα radiation can be accounted for by assuming that a small weight

Table 1. Relative detection efficiency ε'/ε for low energy X-ray lines with various ice layer thicknesses

t(nm)	BKα	CKα	NKα	OKα	SiKα	NiLα
100	0.8620	0.9473	0.9776	0.9893	0.9909	0.9414
500	0.4760	0.7628	0.8930	0.9478	0.9552	0.7395
1000	0.2266	0.5818	0.7974	0.8983	0.9124	0.5469

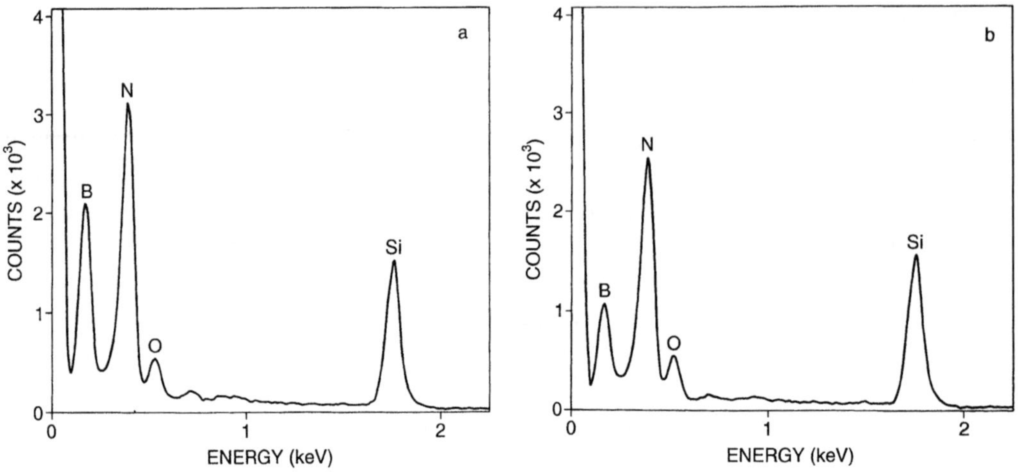

Fig. 1. EDS spectra obtained from a BN film on silicon, **a** without an ice layer on the detector crystal and **b** with an ice layer of approximately 700 nm thickness

fraction, x, of carbon, in the form of hydrocarbons, is incorporated in the ice layer. Then, by solving the simultaneous equations

$$[x(\mu/\rho)_C^{Ni} + (1-x)(\mu/\rho)_{H_2O}^{Ni}]\rho t = \ln(\varepsilon'/\varepsilon)_{Ni} \qquad (3a)$$

$$[x(\mu/\rho)_C^{N} + (1-x)(\mu/\rho)_{H_2O}^{N}]\rho t = \ln(\varepsilon'/\varepsilon)_{N} \qquad (3b)$$

one can obtain a value for x. This resulted in an estimated carbon content of 8.7 wt% (4.55 at%) in the present case. The corresponding mass thickness of the ice + carbon layer is 7.2 μg cm^{-2}, which would give detection efficiencies of 0.361 for BKα, 0.735 for NKα and 0.925 for OKα. The large decrease in the detection efficiency for NKα, relative to that with a layer of pure ice, is attributable to the high mass absorption coefficient (25500 cm^2 g^{-1}) of this line in carbon.

Specimen Contamination

Surface contamination can be regarded as an additional layer of carbon whose thickness varies with time. The correction method is similar to that used to allow for the effect of a deliberately applied coating [13]. An approximate analysis is initially made neglecting the existence of the carbon layer. This first approximation is subsequently used as the starting point for a second calculation in which the carbon layer is now included. If the increase in the thickness of the contamination layer and the resulting decrease in the emitted X-ray intensities are linear functions of time, which will be very nearly true for a carbon layer of several tens of nanometres thickness, the time averaged X-ray intensities can be treated as constants for the purpose of calculation.

Theoretical calculations for the effect of an overlying carbon layer of 10 nm thickness on the relative X-ray intensities (k-ratios) of pure elements at different accelerating voltages, E_0, are summarized in Table 2. Intensities are reduced as a result of electron deceleration and X-ray absorption in the carbon layer. The mass

Table 2. Effect of a 10 nm carbon layer on the k-ratios for pure elements

E_0(kV)	BKα	NKα	OKα	SiKα	NiLα
3	0.931	0.843	0.879	0.832	0.882
5	0.963	0.885	0.928	0.951	0.953
10	0.977	0.902	0.948	0.988	0.983
20	0.979	0.906	0.953	0.996	0.989

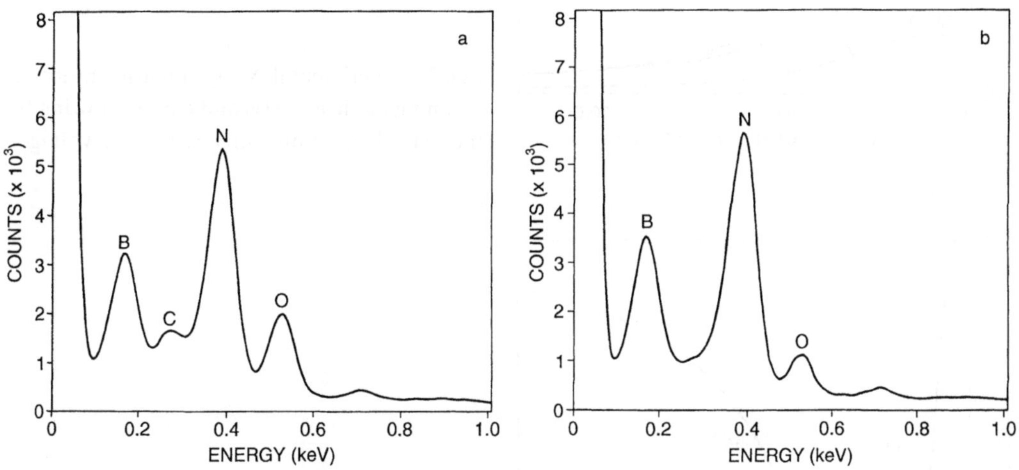

Fig. 2. EDS spectra from the BN film acquired in **a** stationary spot mode and **b** TV scanning mode

absorption coefficients employed in these and all other calculations were those recommended by Pouchou and Pichoir [17], derived mainly from the tables of Heinrich [20] for $Z > 10$ and Henke et al. [21] for $Z < 10$.

EDS spectra obtained from the BN film are shown in Fig. 2. A CKα peak is discernible in the spectrum acquired in stationary spot mode, whereas it is virtually absent in that aquired in TV scanning mode. The increase in the height of the OKα peak in the spot mode spectrum is presumably due to oxidation of the specimen under the action of the electron beam [22].

Experimental X-ray intensities for analysis in the scanning mode are compared in Fig. 3 with theoretical curves indicating the variation with accelerating voltage predicted by the XPP model. Oxygen was assumed to be uniformly distributed throughout the film thickness in these calculations. The computed thickness of the BN film was 103.9 nm with a B:N ratio of 1.09 and an approximately 11 at% oxygen content. Introduction of an oxide layer did not significantly improve the fit. Under the assumption that all the oxygen was concentrated in a surface layer of ideal B_2O_3 composition, the calculated thickness of the oxide layer was 24.6 nm while that of the BN layer became 80.4 nm with a B:N ratio of 0.96. These represent limiting cases, since in reality the oxygen concentration can be expected to vary continuously with depth. Differences in the measured X-ray intensities would, however, only become

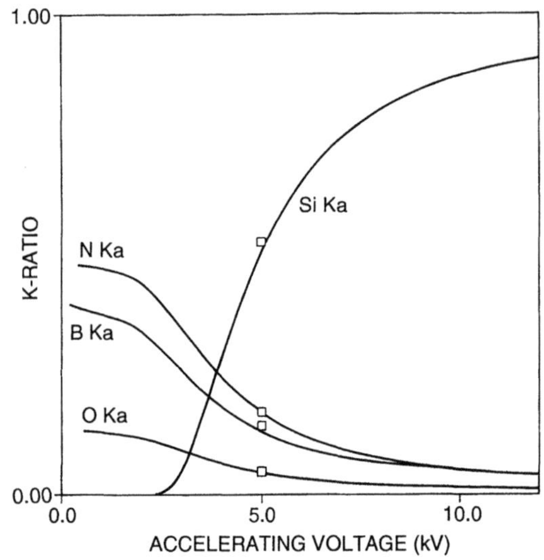

Fig. 3. Experimental X-ray intensity ratios for scanning mode analysis and curves showing the theoretical variation with accelerating voltage

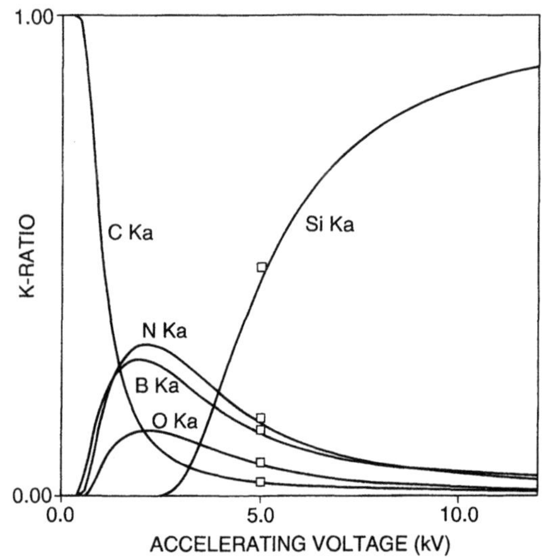

Fig. 4. Experimental X-ray intensity ratios for spot mode analysis and curves showing the theoretical variation with accelerating voltage

apparent at accelerating voltages below 3 kV [23]. Hence, although EDS analysis is able to detect and measure uniform oxide layers of a few nanometres thickness [24], its capability for performing depth composition profiling is restricted [25].

The X-ray intensities obtained for spot mode analysis are plotted in Fig. 4 with theoretical curves again calculated assuming a uniform oxygen distribution. The estimated thickness of the BN film was 106.0 nm with a B:N ratio of 1.04, in reasonable agreement with the previous results for scanning mode analysis, while the (time averaged) thickness of the carbon layer was 10.2 nm. Neglecting the existence of this contamination layer the calculated BN film thickness was 111.6 nm, with a B:N ratio of 1.13. In both instances the oxygen concentration was found to be approximately 16 at%. The assumption of a stochiometric B_2O_3 layer resulted in a thickness of 33.7 nm for the oxide and 75.6 nm for the underlying BN layer, with a B:N ratio of 0.86.

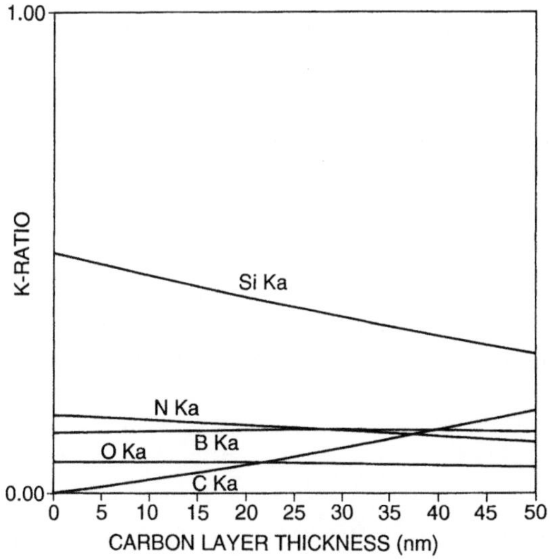

Fig. 5. Predicted variation of X-ray intensities with thickness of the carbon contamination layer

Table 3. Experimentally determined specimen contamination rates

Ref.	E_0(kV)	J($A\,m^{-2}$)	P(Pa)	dz_c/dt($nm\,s^{-1}$)
[4]	100	$4 \cdot 10^5$	10^{-3}–10^{-6}	0.2–1
[5]	80	$5 \cdot 10^8$	$>10^{-3}$	~0.1
[6]	43	$4 \cdot 10^6$	10^{-7}	0.04
[7]	1	$\sim 10^5$	10^{-5}	0.05

Figure 5 shows the theoretical variation of X-ray intensities with the thickness of the contamination layer. The behaviour is practically linear for a carbon layer thickness up to 50 nm. Provided that the contamination rate is also linear, the final thickness of the carbon layer will be twice its time averaged value, i.e., ~20 nm. This corresponds to a contamination rate of $\sim 0.2\,nm\,s^{-1}$ for a mean current density of $4 \cdot 10^6\,A\,m^{-2}$ in the focused spot mode.

Most reported measurements of the contamination rate, dz_c/dt, have been made at accelerating voltages considerably higher than that used in these experiments. Some typical data for different values of accelerating voltage, E_0, current density, J, and vacuum pressure, P, are listed in Table 3. The only available measurements at low voltage in a field emission SEM [7] yielded a contamination rate of $0.05\,nm\,s^{-1}$ at 1 kV with a current density of $\sim 10^5\,A\,m^{-2}$. The greater rate observed in the present work is consistent with an expected increase of contamination with current density.

Conclusions

The intensities of low energy X-rays can be significantly modified by the effect of specimen contamination and the presence of an ice layer on the detector crystal. The detection efficiency for NKα is extremely sensitive to the carbon concentration in the ice layer. This is not revealed by the conventional test of detector performance by measurement of the NiL/NiK ratio.

Specimen contamination effects can be virtually eliminated by acquiring spectra in the scanning mode. Agreement between scanning and spot mode analyses is reasonably close provided that the carbon contamination layer is included in the quantification. Oxidation of reactive specimens under the focused electron beam may lead to overestimation of the oxygen content.

Acknowledgements. The author is grateful to D. Crouch and I. Pearce of Oxford Instruments for their assistance in setting up the beam current compensation system and W. Gissler and J. Haupt for the thin film deposition.

References

[1] J. J. Hren, in: *Principles of Analytical Electron Microscopy* (D. C. Joy, A. D. Roming, J. I. Goldstein, eds.), Plenum, New York, 1986, p. 353.
[2] J. T. Fourie, *Optik* **1975**, *44*, 111.
[3] J. T. Fourie, *Optik* **1978**, *52*, 91.
[4] R. K. Hart, T. F. Kassner, J. K. Maurin, *Philos. Mag.* **1970**, *21*, 453.
[5] R. F. Egerton, C. J. Rossouw, *J. Phys. D. Appl. Phys.* **1976**, *9*, 659.
[6] M. Isaacson, J. Langmore, in: *SEM 1974. Vol. 1*, (O. Johari, I. Corvin, eds.), ITT Research Institute, Chicago, 1974, p. 19.
[7] P. Hirsch, M. Kässens, M. Püttmann, L. Reimer, *Scanning* **1994**, *16*, 101.
[8] L. Reimer, M. Wächter, *Ultramicroscopy* **1978**, *3*, 169.
[9] K.-H. Müller, *Optik* **1971**, *33*, 296.
[10] R. G. Musket, *Nucl. Instrum. Methods* **1986**, *B15*, 735.
[11] P. Müller, F. Riehle, E. Tegeler, B. Wende, *Nucl. Instrum. Methods* **1986**, *A246*, 569.
[12] B. G. Lowe, *Ultramicroscopy* **1989**, *28*, 150.
[13] J.-L. Pouchou, *Mikrochim. Acta* **1994**, *114/115*, 33.
[14] W. Gissler, J. Haupt, A. Hoffmann, P. N. Gibson, D. G. Rickerby, *Thin Solid Films* **1991**, *199*, 113.
[15] J.-L. Pouchou, F. Pichoir, in: *Electron Microscopy, Vol. 1*, EUREM 92, Granada, 1992, p. 293.
[16] D. G. Rickerby, J.-F. Thiot, *Mikrochim. Acta* **1994**, *114/115*, 421.
[17] J.-L. Pouchou, F. Pichoir, in: *Electron Probe Quantitation* (K. F. J. Heinrich, D. E. Newbury, eds.), Plenum, New York, 1991, p. 31.
[18] J. E. Wood, D. B. Williams, J. I. Goldstein, *J. Microsc.* **1984**, *133*, 255.
[19] J. H. Paterson, J. N. Chapman, W. A. P. Nicholson, J. M. Titchmarsh, *J. Microsc.* **1989**, *154*, 1.
[20] K. F. J. Heinrich, *Proc. 11th International Congress on X-ray Optics and Microanalysis*, University of Western Ontario, London (Ont.), 1986, p. 67.
[21] B. L. Henke, P. Lee, T. J. Tanaka, R. L. Shimabukuro, B. K. Fujikawa, *At. Data Nucl. Data Tables* **1986**, *27*, 1.
[22] G. F. Bastin, H. J. M. Heijligers, *Scanning* **1991**, *13*, 325.
[23] D. G. Rickerby, *Microsc. Microanal. Microstruct.* **1995**, *6*, 621.
[24] R. A. Waldo, M. C. Militello, S. W. Gaarenstroom, *Surf. Interface Anal.* **1993**, *20*, 111.
[25] P. Willich, *Mikrochim. Acta* [*Suppl.*] **1992**, *12*, 1.

Mikrochim. Acta [Suppl.] 13, 501–506 (1996)

Measurements of $Ga_{1-x}Al_xAs$ Layers on GaAs with EDS

Kurt Röhrbacher*, Peter Klein, Michael Andrae, and Johann Wernisch

Institute for Applied and Technical Physics, Technical University of Vienna, Wiedner Hauptstrasse 8–10, A-1040 Wien, Austria

Abstract. GaAlAs is a basic material in semiconductor technology and thus of great interest for industrial research. Several $Ga_{1-x}Al_xAs$ bulk samples of predefined composition as well as layered samples of well defined thicknesses and composition were investigated. The experimental data are analyzed with a computer program, which has been developed at the Institute for Applied and Technical Physics at the Technical University of Vienna. Furthermore, analysis of the bulk samples using the program PAP of Pouchou and Pichoir is carried out. An excellent agreement of both correction procedures with the defined composition values is obtained. The determined layer thicknesses also correspond well with the defined data.

For our calculations the total K-peak of aluminum and the total L-peak of gallium and arsenic were measured. Reducing the L-intensities by the already published relative intensity factor for $L\alpha$ radiation leads to less correspondence of the calculated k-ratios compared to the measured values. By additionally regarding the emerging $L\beta$ intensities with partly much higher mass absorption coefficients than the corresponding values for $L\alpha$ radiation a dramatic improvement is achieved.

Key words: GaAlAs, EDS, thickness determination.

The importance of GaAs in semiconductor technology is increasing dramatically. Although this material is still much more expensive than silicon it offers many advantages in electrotechnical research. GaAlAs and pure GaAs have equivalent atomic lattices and replacing gallium atoms in thin films by aluminum is possible without producing crystal imperfections at the interface, by which the electrons could be scattered inelastically. Thus the mobility of electrons in GaAs is five times higher than in Si leading to less heat generation. The energy gap between the valence and conduction band in differently doped layers is dependent on the atomic fraction of aluminum atoms, which can be used for many applications in electronics.

---

* To whom correspondence should be addressed

The purpose of this work was to assess the applicability of low-energy-EPMA with EDS to the analysis of this semiconductor material. Bulk samples of well-known composition, as well as $Ga_{1-x}Al_xAs$-layers on GaAs with defined thickness and composition, were analyzed to check the accuracy of correction procedures in EPMA (August and Wernisch [1, 2], Pouchou and Pichoir [3]) for this particular material.

Experimental

Measurements were carried out with a JEOL JSM-T330A scanning electron microscope (SEM) equipped with an energy dispersive X-ray spectrometer. The detector has a resolution of 140 eV at 5.9 keV and a beryllium window of 8 μm thickness.
Three different compositions of $Ga_{1-x}Al_xAs$-layers on Ga As were analyzed with nominal values of $x = 0.2$, $x = 0.4$, and $x = 0.6$; the nominal layer thickness ranged from 20 to 120 nm for each composition. Since also very thin (20 nm) GaAlAs-films on GaAs were analyzed, the samples were measured at low acceleration voltage, namely 10 kV, to achieve a higher statistical accuracy for the net intensity of aluminum K-lines emerging from the layer. We used the L-lines of gallium and arsenic to obtain overvoltage ratios greater than three necessary for the application of established correction procedures. The experimental k-ratios were evaluated by least-squares fitting of the recorded spectra of Ga, As and Al standards at the same acceleration voltage. The take-off-angle was 35°.

Results

Since Al-atoms replace Ga-atoms without a modification of the crystal structure, the condition Ga (at%) + Al (at%) = 0.5 can be assumed to be valid with respect to stoichiometry. Hence the determination of the atomic fraction of aluminum, denoted

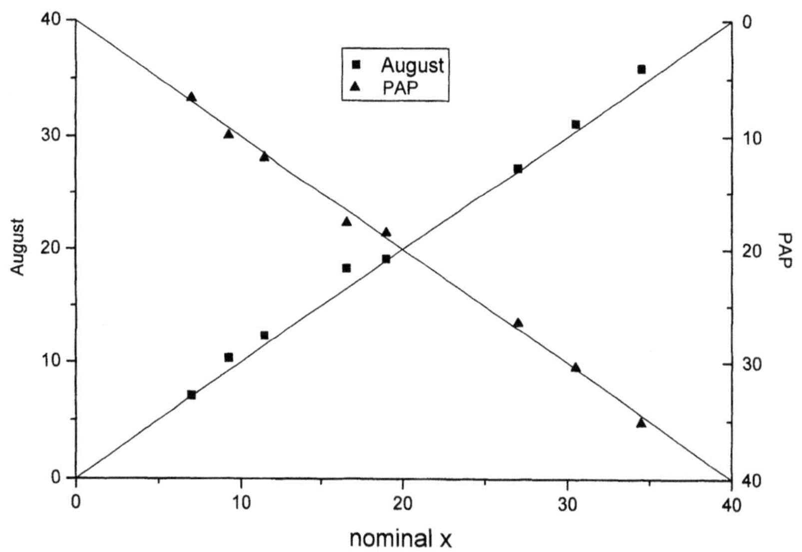

Fig. 1. Atomic fraction of Al in several GaAlAs bulk samples, calculated by using either the electron-scattering model of August and Wernisch [1, 2] or the correction procedure of Pouchou and Pichoir (*PAP*) [3], and plotted versus the nominal values

Table 1. Calculated results of several $Ga_{1-x}Al_xAs$-layers on GaAs. The first two columns are the nominal values for x and the layer thickness. The accurate x-value is given by the mean value of the analytical results for the three particular types of composition. The last column shows the absolute error in at% of aluminum

Nominal x	Nominal thickness	Calculated x	Mean x	Abs. error in at%
0.2	120 nm	0.246	0.250	0.20
0.2	80 nm	0.274	0.250	1.30
0.2	40 nm	0.230	0.250	1.10
0.4	80 nm	0.431	0.445	0.70
0.4	40 nm	0.439	0.445	0.30
0.4	20 nm	0.465	0.445	1.00
0.6	120 nm	0.534	0.565	1.55
0.6	80 nm	0.556	0.565	0.45
0.6	40 nm	0.588	0.565	1.15
0.6	20 nm	0.581	0.565	0.80

as x/2 in this paper, is sufficient to have full information about the composition of the material. The analyses of GaAlAs bulk specimens with the correction models of August and Wernisch [1, 2] and Pouchou and Pichoir [3] lead to excellent agreement concerning the atomic fraction of aluminum with the nominal values. In the first case, we obtained absolute errors less than 1.8 at%, a mean error of 0.774 at% and a standard deviation on the errors of 0.62 at%. For the latter model, we obtained absolute errors less than 1.2 at%, a mean error of 0.524 at% and a standard deviation on the errors of 0.335 at%. The results for both correction procedures are shown in Fig. 1. Excellent results were also obtained for the layered specimens when analyzed with the electron-scattering model developed by August and Wernisch [1, 2]. The specified x-values in the $Ga_{1-x}Al_xAs$-layers had to be corrected from 0.2 to 0.25, from 0.4 to 0.445 and from 0.6 to 0.565. The absolute errors were less than 1.6 at% for aluminum, the mean error was 0.855 at% and the standard deviation on the errors was 0.444 at%. The results for the layered samples are shown in Table 1.

Since our energy dispersive detector unit is not capable of resolving L-lines separately, we used the total L-intensity of gallium and arsenic for the analysis. Hence the $L\alpha$ and $L\beta$ intensities of Ga and As were calculated separately and summed up for the derivation of the theoretical k-ratios, which lead to better agreement with measured *k*-ratios in comparison with calculations of $L\alpha$ intensities only (Fig. 2).

Bulk Measurements

The composition of the bulk specimens was determined with an electron scattering model developed by August and Wernisch [1, 2]. Additionally the model of Pouchou and Pichoir [3] was applied to check the accuracy of the results. The weight fractions of the investigated elements in the samples were altered until

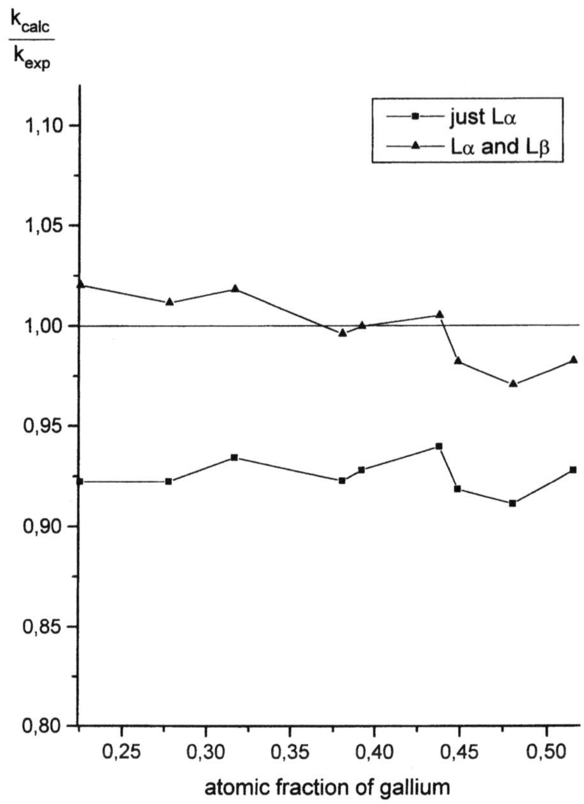

Fig. 2. Theoretical k-ratios of Ga(Lα) and Ga(Lα + Lβ) normalised to the experimental values. An improvement is achieved if Lβ radiation with much higher self-absorption than Ga(Lα)-radiation is also taken into account

a minimum of the total relative deviation between experimental and calculated k-ratios was obtained:

$$\sum_{i=Ga,As,Al} \frac{abs(k_i^{exp} - k_i^{calc})}{k_i^{exp}} \rightarrow \text{minimum} \tag{1}$$

Layered Specimens

The total relative deviation of measured and calculated k-ratios of Ga, As and Al as a function of layer thickness d and atomic fraction of aluminum x/2 had the shape of a long, steep valley – like a folded sheet of paper – shown in Fig. 3. The bottom of this valley has many local minima, which makes the simultaneous matrix and thickness determination impossible. For this reason we calculated the location d_{min} of the minimum as a function of x in the region of interest at equal distances of $\Delta x = 0.05$, as shown in Fig. 4. With known layer thickness the accurate value for the atomic fraction of Al can be easily evaluated by interpolating between the calculated data. The density of the $Ga_{1-x}Al_xAs$-films was calculated according to Vegard's law by

$$\rho = b \cdot \sum_{i=As,Ga,Al} at\%_i \cdot amu_i$$

$$= b/2 \cdot (74{,}922 + (1-x) \cdot 69{,}723 + x \cdot 26{,}982) \text{ in } [g/cm^3] \tag{2}$$

with $b = 0.0742\ g/cm^3$ to obtain the correct density of pure GaAs for $x = 0$.

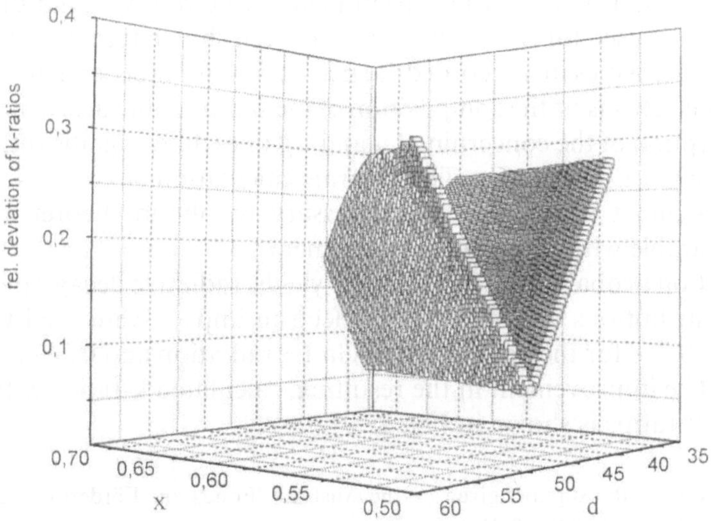

Fig. 3. Sum of relative deviations between calculated and measured k-ratios of Ga, Al and As as a function of x (= 2·at% aluminum) and layer thickness d in nm. The bottom of the resulting, steep valley has many local minima

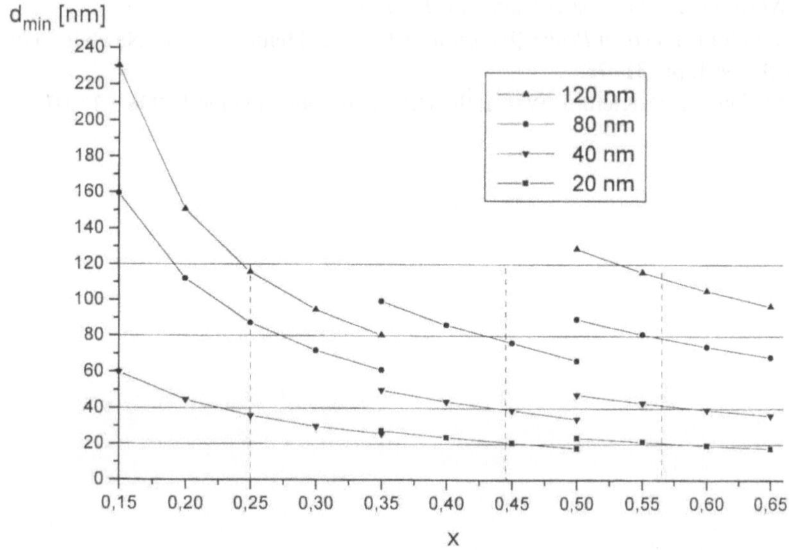

Fig. 4. Location of the minima in Fig. 3 as a function of x. The solid, horizontal lines are the specified layer thicknesses of the investigated samples. The dashed, vertical lines define the accurate values of the atomic fraction of Al in the $Ga_{1-x}Al_xAs$ films

Conclusions

Excellent analytical results with absolute errors less than 2 at% aluminum confirm that an electron microprobe with EDS is an appropriate tool for the destruction-free analysis of semiconductor materials.

Care has to be taken if the L-lines in EDS spectra cannot be resolved and the total, measured L-intensity is used for the analysis. As has been shown recently [4], the ratio of Lα to Lβ line intensities in a recorded L-spectrum is not independent of the energy of the incident electrons or the composition of the sample, but varies even strongly if the mass absorption of the concerning Lα and Lβ lines differ significantly as is the case especially for Ga L-lines in Ga. The separate calculation of Lα and Lβ line intensities and subsequent summation are necessary to obtain theoretical k-ratios which are comparable with those of the experiment.

Coster-Kronig transition probabilities, fluorescence yields, radiative decay rates and primary vacancy distributions in L-shells after electron impact were used to calculate a probability of 0.758 for the generation of Ga Lα radiation and 0.242 for Ga Lβ respectively [4]. The improvement in the resulting, calculated k-ratios with respect to our measured k-ratios is shown in Fig. 2.

Acknowledgement. We acknowledge the support given by the Austrian 'Fonds zur Förderung der wissenschaftlichen Forschung' (Project P10415-PHY).

References

[1] H.-J. August, J. Wernisch, *X-Ray Spectrometry* **1991**, *20*, 131.
[2] H.-J. August, J. Wernisch, *X-Ray Spectrometry* **1991**, *20*, 141.
[3] J.-L. Pouchou, F. Pichoir, *Electron Probe Quantitation* (K. F. J. Heinrich, D. E. Newbury, eds.), Plenum, New York, 1991, pp. 31–76.
[4] K. Röhrbacher, M. Andrae, P. Klein, J. Wernisch, *Mikrochim. Acta* [*Suppl.*] **1996**, *13*, 507.

Mikrochim. Acta [Suppl.] 13, 507–515 (1996)

The Relative Intensity Factor for Lα Radiation Considering the Different Mass Absorption of Lα and Lβ Radiation

Kurt Röhrbacher*, Michael Andrae, Peter Klein, and Johann Wernisch

Institute for Applied and Technical Physics, Technical University of Vienna, Wiedner Hauptstrasse 8-10, A-1040 Wien, Austria

Abstract. The mass attenuation coefficients of Lα and Lβ radiation can differ as much as a factor of five depending on the composition of the sample. A dependence of the relative intensity factor for Lα radiation on the incidence energy is evident. The relative intensity factors are very important not only for standardless EPMA but also for EPMA with standards when an EDS system is used and the whole L-intensity is measured. In this case calculating just Lα intensities leads to acceptable results only if the difference between the mass attenuation coefficients of Lα and Lβ radiation is rather small. If the difference is significant, a much too high mass attenuation coefficient for Lα is observed as can be shown mathematically. A comparison between theoretically calculated and measured intensity factors is carried out and their energy dependence is discussed.

Key words: relative intensity factor, EDS, mass absorption.

Radiative decay rates and effective vacancy distributions have often been used to calculate the probability for generation of Lα radiation with respect to total L-intensities [1–3]. We found that these values must not be compared directly with experimental data, as has been done before, because absorption correction of the emerging X-radiation depending on its generation depth must be taken into account. By replacing the formerly used primary vacancy distribution of L-subshells, namely $N_1 = N_2 = 0.25$ and $N_3 = 0.5$, by more recent, atomic number dependent data [4] we calculated the probability for the generation of Lα X-rays for elements with $26 \leqslant Z \leqslant 55$. Our calculated relative intensity factor (see Fig. 5) is in better agreement with the values measured by Schreiber et al. [2] especially for elements with $Z < 33$ for which self-absorption of the emerging Lβ radiation is more than five times higher than that of the corresponding Lα X-rays (see Fig. 2). Therefore we propose to replace the data fit for Schreiber et al. by a new expression.

---

* To whom correspondence should be addressed

Theory

The characteristic X-ray intensity emerging from a sample at take-off angle ψ produced by incident electrons with energy E_0 can be expressed, according to [5] as:

$$I_L = n^{e-} \cdot (N_0/A) \cdot (\Omega/4\pi) \cdot Q_L(U_0) \cdot \omega_L \cdot \varepsilon \cdot c \cdot \int \cdot \varphi(\rho z) \cdot \exp(-\chi \cdot \rho z) \cdot d\rho z \quad (1)$$

The notations used are:

n^{e-}	number of incident electrons	N_0	Avogadro's number
Q_L	ionization cross section of L-shell	A	atomic mass
U_0	overvoltage ratio E_0/E_c	E_0	energy of incident electrons
E_c	critical energy of the respective shell	Ω	solid angle of detection
c	weight fraction of the considered element	ω_L	fluorescence yield of L-shell
$\varphi(\rho z)$	depth distribution of primary L-ionizations	ε	detector efficiency

The factor $\exp(-\chi \cdot \rho z)$ is the probability for a photon generated at a mass depth ρz to leave the surface in the direction of the detector unit without being absorbed. The parameter χ is given by $\chi = (\mu/\rho) \cdot \sin^{-1} \psi$ with take-off angle ψ between detector axis and surface of the specimen and the mass attenuation coefficient (MAC) for compounds

$$\mu/\rho = \sum_j c_j \cdot (\mu/\rho)_j \quad (2)$$

where c_j represents the weight fraction of element j and $(\mu/\rho)_j$ the mass attenuation of the considered line in element j. If analysis is carried out by measuring the characteristic L-line intensities one has to distinguish between the various emerging L-lines and take the subshell from which they originate as well as their mass attenuation in the sample at each characteristic energy into account. For example the net Lα intensity can be calculated as

$$I_{L\alpha} \propto Q_{L3}(U_0) \cdot (1 + g_{ck}) \cdot \frac{\Gamma_R(L_3 M_{4,5})}{\Gamma_R(L_3)} \cdot \omega_{L3} \cdot \varepsilon(\mathrm{L}\alpha) \cdot c \cdot \int \varphi_\alpha(\rho z) \cdot \exp(-\chi_\alpha \cdot \rho z) \cdot d\rho z \quad (3)$$

The radiative decay rate $\Gamma(L_3 M_{4,5})$ denotes the probability for a vacancy transition from L_3 to M_4 or M_5-subshell and $\Gamma(L_3)$ the probability for all vacancy transitions originating from L_3-subshell respectively. Both values can be derived theoretically [6] by use of relativistic Hartree-Fock wave functions in units of [eV/ħ]. Radiationless transitions can alter the vacancy distribution of the several L-subshells and have to be taken into account by the factor

$$g_{ck} = (f_{13} \cdot N_1 + f_{23} \cdot (N_2 + f_{12} \cdot N_1))/N_3 \quad (4)$$

For the Lβ radiation the situation is even more complicated. The Lβ line results from different vacancy transitions each contributing with its individual intensity but with nearly the same energy. For example the strongest Lβ line, Lβ_1, originates from

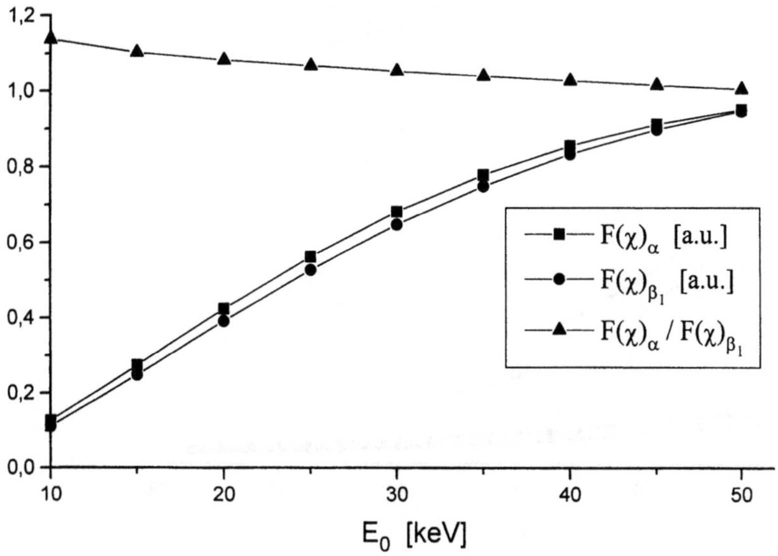

Fig. 1. $F(\chi)$ of silver for various acceleration energies E_0. $F(\chi)_{L\alpha}$ is approximately equal to $F(\chi)_{L\beta_1}$ leading to a ratio $F(\chi)_{L\alpha}/F(\chi)_{L\beta_1}$ of approximately one. MAC(Ag-Lα) in Ag = 526 cm²/g, MAC(Ag-Lβ_1) in Ag = 459 cm²/g

L_2-subshell and its intensity is

$$I_{L\beta_1} \propto Q_{L2}(U_0)\cdot\left(1+f_{12}\cdot\frac{N_1}{N_2}\right)\cdot\frac{\Gamma_R(L_2M_4)}{\Gamma_R(L_2)}\cdot\omega_{L2}\cdot\varepsilon(L\beta_1)\cdot c\cdot\int\varphi_\beta(\rho z)\cdot\exp\cdot(-\chi_\beta\cdot\rho z)\cdot d\rho z \quad (5)$$

In EDS measurements where all emerging Lβ photons of an analyzed element superimpose, giving rise to one total Lβ peak, the intensities of the individual lines have to be calculated by analogy to Eq. (5) and summed afterwards to get the theoretical Lβ intensity. The depth distribution function $\varphi(\rho z)$, utilized in so-called ZAF procedures, represents the amount of ionizations at a mass depth ρz normalized to that of a self-supporting layer of the same composition and under the same experimental conditions. The difference between the critical energies of the several L-subshells is rather small leading to almost identical depth distributions for Lα and Lβ radiation. If self-absorption of Lα and Lβ photons in the investigated sample is of the same order of magnitude the integral $F(\chi)=\int\varphi(\rho z)\cdot\exp(-\chi\cdot\rho z)\cdot d\rho z$, proportional to the emerging intensity, shows the same energy dependence for both Lα and Lβ lines (Fig. 1). The ratio $F(\chi)_{L\alpha}/F(\chi)_{L\beta}$ is then approximately one and the relative intensity factor

$$a_L=\frac{I(L\alpha)}{I(L\ \text{total})}=\frac{I(L\alpha)}{I(L\alpha+L\eta+Ll)+I(L\beta+L\gamma)}\approx\frac{1}{1+\dfrac{I(L\beta)}{I(L\alpha)}} \quad (6)$$

is almost constant. For elements with atomic number less than 33 the characteristic energy of Lβ_1 X-rays, the most intense line of the Lβ series, is slightly higher than the ionization energy of the L_3-subshell of the emitting atom. This results in a very strong self-absorption of the Lβ_1 photons with a strong dependence of the relative weights

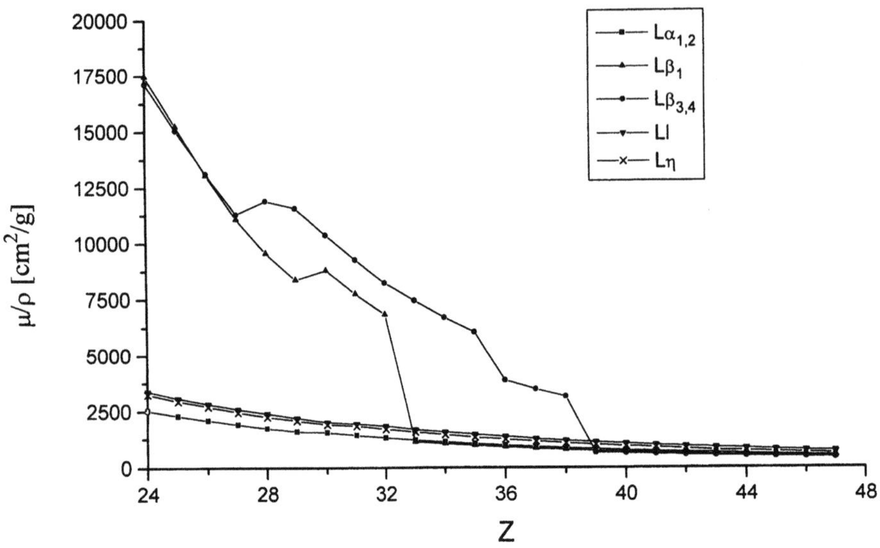

Fig. 2. Self-absorption coefficients of $L\alpha_{1,2}$ and $L\beta_1$ radiation for elements from $Z = 24$ to $Z = 47$ calculated with an algorithm developed by Schmetterer [14]

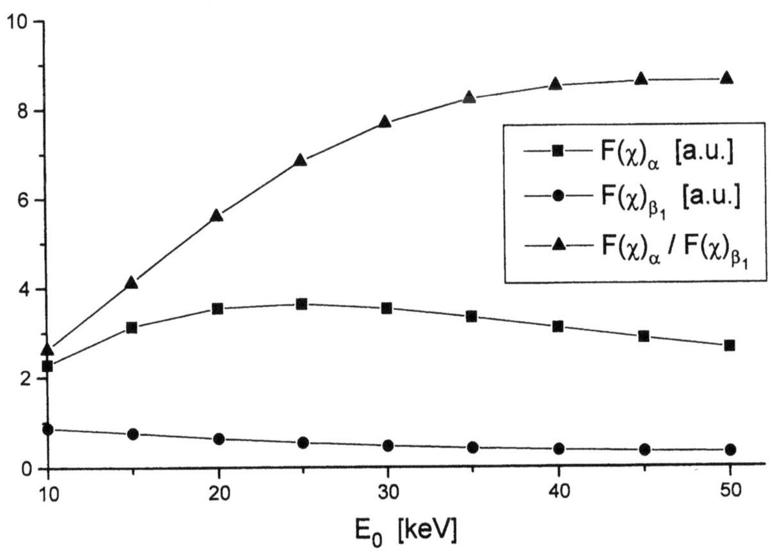

Fig. 3. $F(\chi)$ of copper for various acceleration energies E_0. The significantly different mass absorption of Cu-Lα and Cu-Lβ_1 radiation in Cu results in different energy dependence of these lines. As a consequence the ratio $F(\chi)_{L\alpha}/F(\chi)_{L\beta_1}$ varies strongly with the energy of the incident electrons. MAC(Cu-Lα) in Cu = 1602 cm^2/g, MAC(Cu-Lβ_1) in Cu = 8358 cm^2/g

on the energy of the incident electrons as a consequence. The mass attenuation coefficient for $L\beta_1$ photons is more than five times higher than that of corresponding $L\alpha$ photons (Fig. 2). The ratio $F(\chi)_{L\alpha}/F(\chi)_{L\beta_1}$ of copper for example varies by a factor of 3.25 from 10 to 50 keV (Fig. 3). Furthermore, the relative intensity factor a_L in Eq. (6) is not a constant for a given element anymore but also varies with E_0 (see Fig. 4).

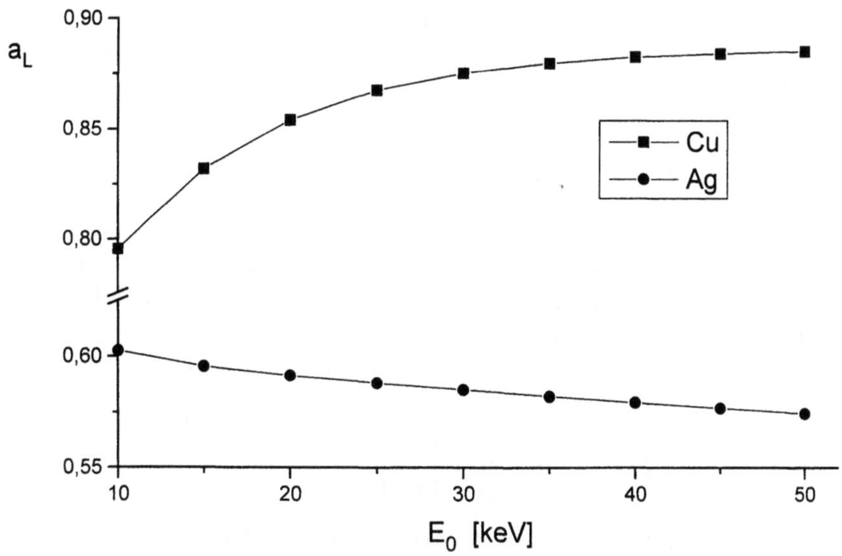

Fig. 4. The relative intensity factor a_L for copper and silver versus acceleration energy. a_L for silver decreases from 10 to 50 keV by 0.027 absolutely (4.4% relatively), while a_L for copper shows a strong energy dependence and increases by 0.082 (10.2% relatively) due to self-absorption effects

The relative line weights are required not only for standardless intensity calculations but also for overlap correction [1]. Hence a lot of work has been done to determine the L-line intensities of pure elements by experiment [2, 3, 7, 8]. Investigating the effect of relative L-line intensity ratios on the accuracy of standardless X-ray microanalysis, Labar [9] found an improvement of the analytical results if the fixed relative line weights in the used computer program were replaced by atomic number dependent line weights proposed by Schreiber and Wims [3].

In measurements with energy dispersive spectrometers Lα and Lβ lines superimpose giving one total L-peak and deconvolution into separate Lα and Lβ peaks is very difficult if the energy difference is less than 50 eV. Furthermore, the unresolved Lα and Lβ peaks form a single peak that is shifted towards higher energies relative to characteristic Lα energy; for example in the spectrum of Cu-L a shift of 10 eV occurs [10]. Especially for elements with low atomic number a separation of Lα and Lβ in EDS spectra is impossible due to the proximity of these two lines (for example: the energy difference of Lα and $L\beta_1$ X-rays of cobalt is 15 eV). When an EDS is used and the relative intensity a_L of the investigated line is known [2] the experimental Lα intensity can be derived to

$$I_{L\alpha} = a_L \cdot I_L \tag{7}$$

where I_L represents the integrated L-peak in the evaluated spectrum. For analysis the net intensity must be calculated according to Eq. (3) and the weight fraction of the considered element is altered until experimental and theoretical intensities are equal. If the difference between the mass absorption of Lα and Lβ radiation is rather small, the relative intensity factor a_L can be assumed to be constant and Schreiber and Wims experimental values [2] measured at 20 kV acceleration voltage give good results if used in Eq. (7). Otherwise a_L depends strongly on the energy of the impinging electrons due to self-absorption of the emerging X-radiation in the investigated specimen. Hence the use of constant a_L-values can lead to remarkable errors: If analysis of a compound is carried out by measurement of characteristic L-lines for example at 10 keV the emerging Lα intensity is

$$(I_{\mathrm{L}\alpha})^{\mathrm{exp}} = a_L(10\,\mathrm{keV})\cdot(I_L)^{\mathrm{exp}} \tag{8}$$

with energy dependent a_L-values. Using Schreiber and Wims data evaluated for $E_0 = 20$ keV in Eq. (7) an experimental Lα signal is obtained which differs from the accurate value given in Eq. (8) by a factor $[a_L(20\,\mathrm{keV})/a_L(10\,\mathrm{keV})]$. Especially with known matrix composition of a sample the mass absorption χ in Eq. (3) has to be replaced by $\bar{\chi}$ to compensate the error caused by the use of wrong relative intensity factors, namely that

$$F(\bar{\chi}) = F(\chi)\cdot\frac{a_L(20\,\mathrm{keV})}{a_L(10\,\mathrm{keV})} \tag{9}$$

is valid. Errors may also occur when standards are used for the calculation of k-ratios. For Lα radiation the experimental k-ratio is

$$k_{\mathrm{L}\alpha}^{\mathrm{exp}} = \frac{(a_L\cdot I_L)_{\mathrm{sample}}^{\mathrm{exp}}}{(a_L\cdot I_L)_{\mathrm{standard}}^{\mathrm{exp}}} \tag{10}$$

and can be set equal to the ratio of the total L-peak of sample to standard

$$k_{\mathrm{L}}^{\mathrm{exp}} = \frac{(I_L)_{\mathrm{sample}}^{\mathrm{exp}}}{(I_L)_{\mathrm{standard}}^{\mathrm{exp}}} \tag{11}$$

only if a_L is supposed to be constant for each element. On the other hand the theoretically derived k-ratio for $L\alpha$ x-rays is

$$k_{\mathrm{L}\alpha}^{\mathrm{calc}} = \frac{(F(\chi)_{\mathrm{L}\alpha})_{\mathrm{sample}}}{(F(\chi)_{\mathrm{L}\alpha})_{\mathrm{standard}}} \tag{12}$$

When elements with atomic numbers less than 33 are analyzed, k_L is much greater than $k_{\mathrm{L}\alpha}$ because $F(\chi)_{\mathrm{L}\beta}$ calculated for the standard is much less than $F(\chi)_{\mathrm{L}\beta}$ of the sample due to the higher self-absorption in the standard. Again if mass absorption is modified to derive theoretical k-ratios (Eq. 12) equal to measured k-ratios (Eq. 10) a much higher mass attenuation coefficient, given by approximately

$$\bar{\chi} = p_{\mathrm{L}\alpha}\cdot\chi_{\mathrm{L}\alpha} + (1 - p_{\mathrm{L}\alpha})\cdot\chi_{\mathrm{L}\beta} \tag{13}$$

is obtained. This can be explained mathematically: the total L-intensity in Eq. (10) must be calculated as

$$I_L = (I_{\mathrm{L}\alpha} + I_{\mathrm{L}\beta}) \propto Q_L\cdot\omega_L\cdot c\cdot\Big(p_{\mathrm{L}\alpha}\cdot\int\phi(\rho z)\cdot\exp(-\chi_{\mathrm{L}\alpha}\cdot\rho z)\cdot\mathrm{d}\rho z$$
$$+ p_{\mathrm{L}\beta}\cdot\int\phi(\rho z)\exp(-\chi_{\mathrm{L}\beta}\cdot\rho z)\cdot\mathrm{d}\rho z\Big) \tag{14}$$

The parameter $p_{\mathrm{L}\alpha}$ is the generation probability of Lα radiation including primary vacancy distributions, Coster-Kronig transitions and fluorescence yields. Since the probability for generation of Lγ is zero or negligibly small for atomic numbers less than 56, $p_{\mathrm{L}\beta} = 1 - p_{\mathrm{L}\alpha}$ is valid (intensities of Lη and Ll are small and can be added to $p_{L\alpha}$ because self-absorption of these lines are of the same order of magnitude). After development of the exponential functions in Eq. (14) the net L-intensity becomes

$$I_L \propto Q_L \cdot \omega_L \cdot c \cdot \int \phi(\rho z) \cdot \mathrm{corr}_{\mathrm{abs}} \cdot d\rho z$$

with absorption correction

$$\mathrm{corr}_{\mathrm{abs}} = 1 - \rho z \cdot (p_{L\alpha} \cdot (\chi_{L\alpha} - \chi_{L\beta}) + \chi_{L\beta}) + 1/2 \cdot (\rho z)^2 \cdot (\chi_{L\beta}^2 + p_{L\alpha} \cdot (\chi_{L\alpha}^2 - \chi_{L\beta}^2)) \\ - 1/6 \cdot (\rho z)^3 \cdot (p_{L\alpha}^2 \cdot (\chi_{L\alpha}^3 - \chi_{L\beta}^3) + \chi_{L\beta}^3) + \cdots \tag{15}$$

ordered by powers of ρz. Eq. (12) is equal to Eq. (10) only if the mass attenuation coefficient of Lα radiation is assumed to be $\bar{\chi} = p_{L\alpha} \cdot \chi_{L\alpha} + (1 - p_{L\alpha}) \cdot \chi_{L\beta}$ because then the development of the exponential function in $F(\chi)_{L\alpha} = \int \varphi(\rho z) \cdot \exp(-\chi_{L\alpha} \cdot \rho z) \cdot d\rho z$ gives

$$\begin{aligned} \exp(-\bar{\chi} \cdot \rho z) &= \exp(-p_{L\alpha} \cdot \chi_{L\alpha} \cdot \rho z - (1 - p_{L\alpha}) \cdot \chi_{L\beta} \cdot \rho z) \\ &= 1 - (p_{L\alpha} \cdot \chi_{L\alpha} + (1 - p_{L\alpha}) \cdot \chi_{L\beta}) \cdot \rho z + 1/2 \cdot (p_{L\alpha} \cdot \chi_{L\alpha} + (1 - p_{L\alpha}) \cdot \chi_{L\beta})^2 \cdot (\rho z)^2 \\ &\quad - 1/6 \cdot (p_{L\alpha} \cdot \chi_{L\alpha} + (1 - p_{L\alpha}) \cdot \chi_{L\beta})^3 \cdot (\rho z)^3 + \cdots \\ &= 1 - \rho z \cdot (p_{L\alpha} \cdot (\chi_{L\alpha} - \chi_{L\beta}) + \chi_{L\beta}) + 1/2 \cdot (\rho z)^2 \cdot (\chi_{L\beta}^2 + p_{L\alpha}^2 \cdot (\chi_{L\alpha}^2 + \chi_{L\beta}^2) \\ &\quad + 2 \cdot p_{L\alpha} \cdot \chi_{L\alpha} \cdot \chi_{L\beta}(1 - p_{L\alpha}) - 2 \cdot p_{L\alpha} \cdot \chi_{L\beta}^2) - 1/6 \cdot (\rho z)^3 \cdot (\cdots) + \cdots \end{aligned} \tag{16}$$

which is at first order equal to the exact net L-intensity in Eq. (15).

In other words, the evaluation of mass attenuation coefficients (MAC) by measurement of Lα intensities according to Eq. (7) is in fact equivalent to the determination of the MAC for one imaginary total L-line, because the calculations are normally carried out in arbitrary units including a constant relative intensity factor. The obtained mass attenuation is *not* that for Lα radiation but for this imaginary total L-line. It is given by the mean value of the MACs of Lα and Lβ photons weighted by their generation probability $p_{L\alpha}$ and $p_{L\beta}$ (see Eq. 13) and is not dependent on the acceleration energy.

For example the summed generation probability for Lα, Ll and Lη radiation in cobalt is 0.78366 according to our calculations. With mass attenuation coefficients for Co-Lα in Co of 1915 cm^2/g and for Co-Lβ_1 in Co of 11059 cm^2/g we get a mean value of 3893 cm^2/g in pure cobalt according to Eq. (13). Recent investigations have shown that the best theoretical net intensities for various electron energies are obtained if a mass absorption of 3847 cm^2/g is assumed [11], confirming our theoretical value for the generation probability of Co-(Lα + Ll + Lη) in cobalt especially.

If evaluation of mass absorption is carried out by fitting theoretical k-ratios to experimental data (see Eq. 11 and 12) we get for example for Co-L in $CoSi_2$ a mean value of 2575 cm^2/g still 34% higher than the accurate value of 1915 cm^2/g for Co-Lα in Co.

Results

We used Scofield's data [6] of radiative decay rates, Krause's table of fluorescence yields and Coster-Kronig transition probabilities [12] and Schiebl and Wernisch

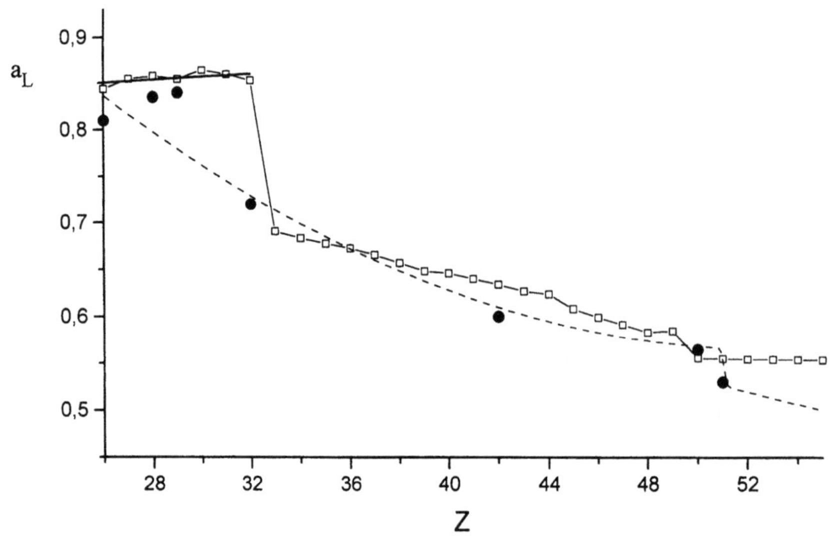

Fig. 5. The relative intensity factor $a_L = I(\mathrm{L}\alpha)/I(L)$ for various elements: dots – measured values by Schreiber and Wims [2]; dashed line – data fit proposed by Schreiber and Wims; squares – our calculation results according to Eq. (3), (5) and (6); solid line – new expression for a_L in the range of $Z < 33$

[4] values of primary vacancy distributions N_i to derive the relative intensity factor shown in Fig. 5. The depth distribution function was calculated according to the model of Pouchou and Pichoir [13]. For a comparison of our theoretical relative intensity factor with experimental data we calculated a_L at 20 keV acceleration energy and 45 degrees take-off angle. Figure 5 shows our results for atomic number $26 \leqslant Z \leqslant 55$ and the values measured by Schreiber and Wims with an electron probe microanalyzer equipped with crystal spectrometers [2]. For atomic numbers less than 26 the characteristic energies lie below 700 eV and a very small detector efficiency leads to a statistical uncertainty which makes measurements with EDS unreliable. For atomic numbers above 55 a deconvolution of the recorded L-spectrum into separate Lα and Lβ peaks can be carried out easily, because the energy difference between these two lines already exceeds 350 eV – for comparison the full width at half maximum (FWHM) of Mn-Kα of usual Si(Li)-detectors is about 140 eV. Furthermore, Schreiber and Wims used a scanning transmission microscope at 200 kV for their measurements of barium and above and their data for $Z > 55$ are therefore not comparable with our calculations.

Conclusion

Our calculated values for the relative intensity of Lα radiation are in good agreement [see Fig. 5] with Schreiber and Wims measured data [2] for a wide range of atomic numbers ($26 \leqslant Z \leqslant 55$). Especially for elements with atomic numbers below 33 the critical energy of the L_3-subshell is lying between the characteristic energies of the respective Lα and Lβ X-rays leading to mass attenuation coefficients of these two lines which differ by a factor five or even more. As a consequence the experimentally obtained relative intensity of Lα radiation is increased by more than 10% at an acceleration energy of 20 keV for elements with $Z < 33$ due to self-absorption (Fig. 5). For this reason we propose to replace the data fit of Schreiber and Wims by the new expression $a_L = (0.00154 \cdot Z + 0.81038)$ in the range of $Z < 33$.

Furthermore, it has been found that in this range the mass attenuation coefficient of Lα radiation must not be determined by measurement of Lα intensities with energy dispersive spectrometers unless the Lα peak can be resolved from the measured L-spectrum. If Lα net intensities are evaluated with constant relative intensity factors according to Eq. (7) a mass attenuation coefficient is obtained which is in fact that of an imaginary total L-line and independent of the energy of the impinging electrons.

Acknowledgement. We acknowledge the support given by the Austrian 'Fonds zur Förderung der wissenschaftlichen Forschung' (Project P10415-PHY).

References

[1]. J. L. Labar, C. E. Fiori, R. L. Myklebust, *Proc. 50th Annual Meeting of the Electron Microscopy Society of America*, San Francisco Press, San Francisco, 1992, pp. 1636–1637.

[2] T. P. Schreiber, A. M. Wims, *X-ray Spectrom.* **1982**, *11*, 42.

[3] T. P. Schreiber, A. M. Wims, in: *Microbeam Analysis*, (K. F. J. Heinrich, ed.), San Francisco Press, San Francisco, 1982, pp. 161–166.

[4] C. Schiebl, J. Wernisch, *Scanning* **1993**, *15*, 203.

[5] J.-L. Pouchou, *Mikrochim. Acta* **1994**, *114/115*, 33.

[6] J. H. Scofield, *Phys. Rev. A* **1974**, *10*(5), 1507.

[7] G. G. Johnson, E. W. White, *ASTM Data Series DS 46*, American Society for Testing and Materials, Philadelphia, 1970.

[8] T. P. Schreiber, A. M. Wims, *Ultramicroscopy* **1981**, *6*, 323.

[9] J. L. Labar, *X-ray Spectrom.* **1987**, *16*, 33.

[10] J. I. Goldstein, D. E. Newbury, P. Echlin, D. C. Joy, A. D. Jr. Romig, C. E. Lyman, C. Fiori, E. Lifshin, *Scanning Electron Microscopy and X-ray Analysis, Second Ed.*, 1992, 353.

[11] C. Schiebl, L. Foster, A. Pfeiffer, A. Testoni, *Microbeam Analysis, Proceedings of the 28th Annual MAS Meeting*, New Orleans, 1994, pp. 199–200.

[12] M. O. Krause, *J. Phys. Ref. Data* **1979**, *8*(2), 307.

[13] J.-L. Pouchou, F. Pichoir, in: *Electron Probe Quantitation* (K. F. J. Heinrich, D. E. Newbury, eds.), Plenum, New York, 1991, pp. 31.

[14] L. Schmetterer, *Thesis*, Technical University Vienna, 1987.

[illegible]

Mikrochim. Acta [Suppl.] 13, 517–523 (1996)

Determination of the Solubility of Cerium in $BaTiO_3$ by Quantitative WDS Electron Probe Microanalysis

Zoran Samardžija*, Miran Čeh, Darko Makovec, and Drago Kolar

Ceramics Department, "Jozef Stefan" Institute, University of Ljubljana, Jamova 39, SLO-61111 Ljubljana, Slovenia

Abstract. Quantitative wavelength dispersive X-ray microanalysis was applied for determination of Ba, Ti and Ce concentrations in Ce-doped $BaTiO_3$ ceramics. There are rare cases in wavelength dispersive analysis where the most intense X-ray lines of the analyzed elements overlap. Such is the case when analyzing matrices containing Ba and Ce, where multiple overlap of $CeL\alpha_{1,2}$ and $BaL\beta_{1,4}$ lines is present. Two independent simplified methods were applied to evaluate this Ce-Ba peak overlap in order to obtain true k-ratios for Ce in $BaTiO_3$: the calibration curve method and the reference spectra fitting method. The experimental conditions of an electron probe microanalyzer were optimized to achieve improved sensitivity and precision of analysis. Pure CeO_2 and $BaTiO_3$ were chosen as standards. K-ratios were corrected by the ZAF matrix correction procedure. Results were calculated for the perovskite ABO_3 formula considering different modes of dopant incorporation and donor charge compensation.

Key words: EPMA, quantitative analysis, solubility of Ce in $BaTiO_3$.

$BaTiO_3$ ceramics modified with donor dopants are extensively studied because of their interesting dielectric and semiconducting properties. It is known that the addition of CeO_2 to $BaTiO_3$ lowers the Curie temperature and that higher values of the dielectric constant can be achieved [1]. A knowledge of the quantity and modality of Ce incorporation into the $BaTiO_3$ perovskite lattice is of fundamental importance in the optimization of the processing of such materials. Since sintered Ce-doped $BaTiO_3$ is generally a multiphase ceramic material, the characterization of such material by using XRD, TGA, DTA and electrical measurements cannot distinguish between the matrix and the secondary phases. These methods give information about solid solubility indirectly by measuring changes in material properties. Therefore, electron probe microanalysis (EPMA) is the most appropriate method for direct analysis of the matrix grains at the micro-level, avoiding the

* To whom correspondence should be addressed

influence of secondary phases [2]. Quantitative wavelength dispersive X-ray spectroscopy (WDS) was applied for the determination of the solubility of Ce in Ce-doped $BaTiO_3$. Good elemental sensitivity and analytical precision allowed accurate quantification after taking into consideration the overlap of the $CeL\alpha_{1,2}$ lines and $BaL\beta_{1,4}$ lines.

Experimental

Ce-doped $BaTiO_3$ ceramic samples were prepared by conventional ceramic technology. $BaTiO_3$, TiO_2 and CeO_2 oxides (analytical p.a. grade) were mixed, pressed into pellets and sintered at 1400 °C in air or in argon for a longer period (5 days) with intermittent cooling, crushing and mixing to promote homogeneity. Starting compositions and sintering conditions for the analyzed samples are given in Table 1. Polished flat surfaces for microanalysis were prepared by conventional metallographic techniques. WDS measurements were performed on a JEOL JXA 840A electron probe microanalyzer equipped with two wavelength dispersive spectrometers (take off angle 40°). Data acquisition and quantitative analysis were performed using a Tracor Series II X-Ray Microanalysis System (TASK software for WDS analysis and ZAF matrix correction procedure). Commercially available version of ZAF (within Tracor Series II Microanalysis System) was used including absorption correction calculated according to the expression of Philibert, modified by Heinrich [3], atomic number correction calculated using the Duncumb – Reed expression [4] and fluorescence correction computed according to the method of Reed [5].

Backscattered electron micrographs of the samples confirmed that the grain size of the matrix was sufficiently large for EPMA. Point analysis was performed on selected Ce-doped $BaTiO_3$ grains, positioning the electron beam (spot) in the center of the grains. Working conditions for WDS analysis were set to 20 kV voltage and 15 nA beam current. A voltage of 20 keV was chosen as optimal accelerating voltage after consideration of the excitation energies of characteristic X-rays for the $BaL\alpha_1$, $TiK\alpha_1$ and $CeL\alpha_1$ lines. Also, at this voltage depth of X-ray generation was found to be less than 2 μm ($\Phi(\rho z)$ curves [6]), which provides sufficient spatial resolution (≈ 1 μm) for analysis of this multiphase material. The measurements of X-ray intensities were carried out using a LiF crystal with sufficiently long counting times to assure 1% standard deviation due to counting statistics.

Table 1. Starting compositions and firing conditions of Ce-doped $BaTiO_3$ samples

Sample	Composition (mol. %) BaO	CeO_2	TiO_2
1	47,4	3,0	49,6
2	52,0	8,0	40,0
3	31,2	15,0	53,8
4	32,0	18,0	50,0

Sample	Firing conditions Temp. (°C)	Atmosphere
1	1400	air
2	1400	argon
3	1400	argon
4	1400	argon

Accurate peak and background positions for the spectral lines analyzed were determined by wavelength scans. Homogeneity of standards and the analyzed phase was checked and verified according to the simplified criterion that all the measured peak intensity counts, N, should fall within $N+/-3(N)^{1/2}$ [7]. During the analysis, a liquid nitrogen cold trap and a cold finger mounted above the samples were used in order to reduce specimen surface contamination. Samples and standards were coated with a thin film of carbon (thickness ≈ 30 nm).

Electron beam current stability was confirmed to be better than 0.5% fo the nominal value. Measured X-ray intensities in the samples were transformed into k-ratios relative to calibrated intensities from known element concentrations in the $BaTiO_3$ and CeO_2 standards. Spectral interference of the $BaL\beta_{1,4}$ and $CeL\alpha_{1,2}$ introduced systematic error in the $CeL\alpha_1$ measured intensity. In order to evaluate the contribution of Ba to the net counts per second (cps) for Ce, two correction methods [8] were applied to the measurements obtained by using $CeL\alpha_1$ line, as follows.

Reference Spectra Fitting Method

The wavelength scans of $CeL\alpha_{1,2}$ lines in CeO_2 and of $BaL\beta_{1,4}$ lines in $BaTiO_3$ were recorded. Background measurements were performed by detuning the wavelength spectrometer slightly above and below the tails of the characteristic peak. The value of the background at the peak position was then established by linear interpolation between the two measured points and subtracted from the peak intensity. After the removal of background, both reference spectra were multiplied by approximate scaling factors, determined from the experimentally obtained spectrum for Ce-doped $BaTiO_3$. Spectra were then added together, processed through the iterative fitting procedure and compared to the average spectrum obtained for Ce-doped $BaTiO_3$ grains with the method of multiple least squares until the best fit (least Chi-squared value) was obtained for certain scaling factors (Fig. 1). The net counts (cps) for Ce in the position of the $CeL\alpha_{1,2}$ line were obtained by subtracting the counts that were contributed by Ba.

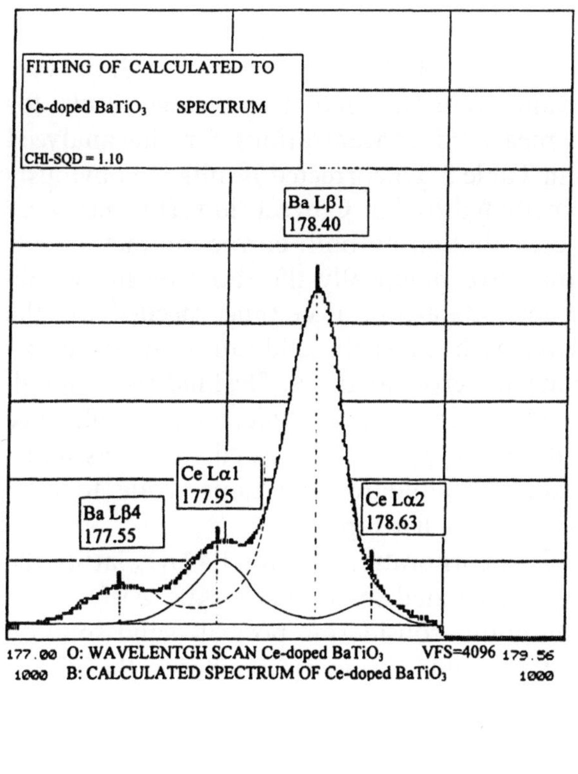

Fig. 1. Fitting of calculated spectrum to average spectrum for Ce-doped $BaTiO_3$

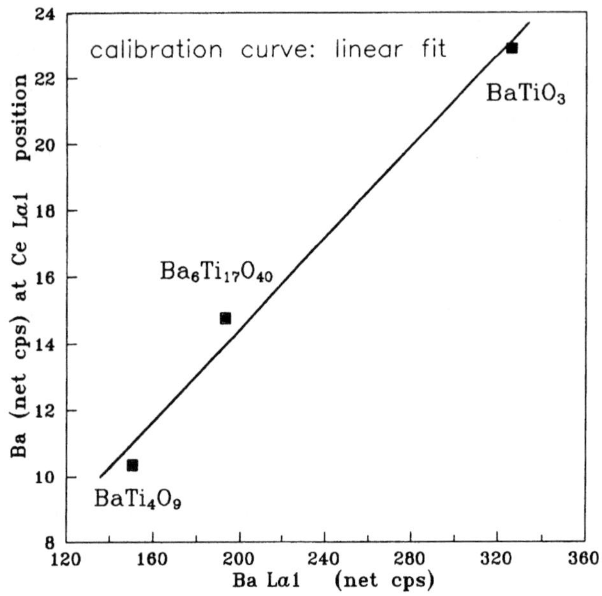

Fig. 2. Calibration curve for the determination of Ba-counts at the position of $CeL\alpha_1$ line

Calibration Curve Method

Net cps for Ba were measured at the positions of the $BaL\alpha_1$ and $CeL\alpha_1$ lines in compounds with different Ba content, using $BaTiO_3$, $Ba_6Ti_{17}O_{40}$ and $BaTi_4O_9$. On the basis of these measurements, the calibration curve of net cps for Ba at the $CeL\alpha_1$ position versus net cps for Ba at the $BaL\alpha_1$ position was plotted. A linear fit between measured points was used in the calculations (Fig. 2). For Ce-doped $BaTiO_3$ grains the net cps for Ce were determined by subtracting counts that were contributed by Ba.

Corrected k-ratios for Ce and measured k-ratios for Ba and Ti were processed through the ZAF quantitative matrix correction procedure to obtain the corresponding element concentrations.

Results and Discussion

Microstructures of the two Ce-doped $BaTiO_3$ samples are shown in Fig. 3. In every sample at least ten doped $BaTiO_3$ grains were analyzed. The average results for $CaL\alpha_1$ k-ratio calculations and Ce measured concentrations for the analyzed Ce-doped $BaTiO_3$ samples are given in Table 2. Uncorrected results are obviously too high due to counts that were contributed by Ba. Correct (lower) values were determined after elimination of this contribution by both correction methods. It should be noted that both procedures gave practically the same results within statistical precision. Because of the long data-collection times needed for the reference spectra fitting procedure (wavelength scans), the calibration curve method is faster and more suitable for performing numerous analyses. The final results for all analyzed elements are given in Table 3. Oxygen content was calculated by difference from 100%. In the following calculations oxygen was adjusted to the assumed valence state of the elements. The quoted standard deviations relate to data measured from n ($n \geqslant 10$) grains analyzed in each sample.

In order to determine the mode of Ce incorporation into the $BaTiO_3$ lattice the measured element concentrations were calculated to the perovskite ABO_3 solid solution formula. Some justified assumptions considering possible solid solution formulae should be taken into account. Ce could be incorporated into $BaTiO_3$ as

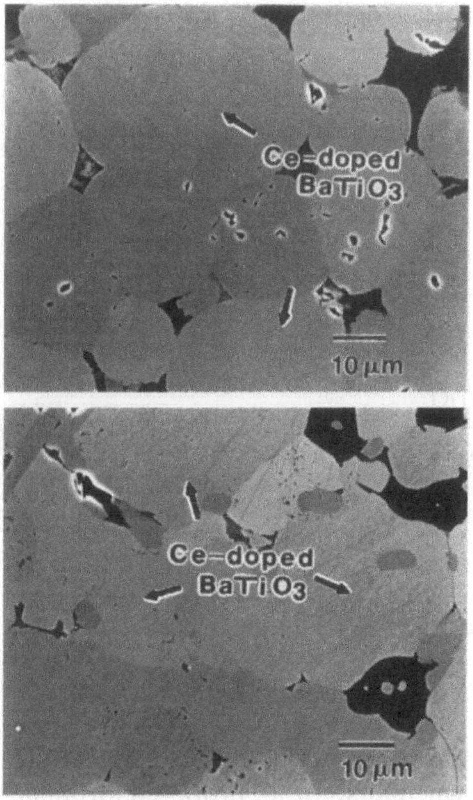

Fig. 3. Backscattered electron micrographs of the two polished Ce-doped $BaTiO_3$ samples (sample 1 and sample 3)

Table 2. Average Ce concentration in Ce-doped $BaTiO_3$ samples

Sample	Line analyzed	k-ratio	Ce (wt.%)
1	$CeL\alpha_1$	0.0743 +/−0.0009	6.02 +/−0.17
	$CeL\alpha_1$[a]	0.0423 +/−0.0011	3.47 +/−0.10
	$CeL\alpha_1$[b]	0.0431 +/−0.0011	3.51 +/−0.11
2	$CeL\alpha_1$	0.1421 +/−0.0027	11.55 +/−0.32
	$CeL\alpha_1$[a]	0.1132 +/−0.0028	9.30 +/−0.24
	$CeL\alpha_1$[b]	0.1166 +/−0.0031	9.43 +/−0.25
3	$CeL\alpha_1$	0.2501 +/−0.0035	20.51 +/−0.37
	$CeL\alpha_1$[a]	0.2269 +/−0.0037	18.69 +/−0.30
	$CeL\alpha_1$[b]	0.2274 +/−0.0041	18.82 +/−0.33
4	$CeL\alpha_1$	0.2773 +/−0.0026	22.56 +/−0.27
	$CeL\alpha_1$[a]	0.2550 +/−0.0026	20.99 +/−0.22
	$CeL\alpha_1$[b]	0.2563 +/−0.0027	21.08 +/−0.23

[a] Calibration curve method.
[b] Reference spectra fitting method.

Ce^{4+} substitutionally at Ti^{4+} [9] sites, as well as Ce^{3+} at Ba^{2+} sites [1, 10]. When entering at Ba sites, Ce acts as a donor. It is assumed that during sintering in air donor charge is compensated by creation of ionized (V''''_{Ti}) vacancies [1, 10], whereas during sintering in an atmosphere with lower oxygen partial pressure donor charge is additionally compensated by reduction of Ti^{4+} to Ti^{3+} [11].

Table 3. Results of WDS microanalysis of Ce-doped $BaTiO_3$ samples

Sample	Ba (wt. %)	Ti (wt. %)	Ce (wt. %)[a]	O (wt. %)[b]
1	54.53 ± 0.64	20.07 ± 0.14	3.47 ± 0.10	21.93 ± 0.57
2	49.29 ± 0.59	19.86 ± 0.23	9.30 ± 0.24	21.55 ± 0.53
3	39.51 ± 0.49	20.39 ± 0.22	18.69 ± 0.30	21.41 ± 0.52
4	37.80 ± 0.43	20.30 ± 0.20	20.99 ± 0.22	20.91 ± 0.56

[a] Calibration curve method.
[b] Calculated by difference.

Table 4. Proposed formulas for Ce solid solution in $BaTiO_3$

Sample	Composition of Ce-doped $BaTiO_3$ phase[a]
1	$Ba_{0.94}Ce_{0.06}Ti_{0.99}(V''''_{Ti})_{0.01}O_3$
2	$Ba_{0.84}Ce_{0.16}Ti_{0.98}(V''''_{Ti})_{0.02}O_3$
3	$Ba_{0.68}Ce_{0.32}Ti_{1.01}O_3$
4	$Ba_{0.65}Ce_{0.35}Ti_{1.00}O_3$

[a] $Ti = Ti^{3+} + Ti^{4+}$.

Considering the above listed assumptions, five different theoretical formulae are possible [12]. Elemental concentrations were calculated for all proposed formulae with normalization to three oxygen atoms per formula unit taking into account the valence of the individual ion. Within the standard deviation of WDS analysis, only one of the proposed formulae fits to the measured concentrations (Table 4). It should be noted that the presented formulae were obtained not only on the basis of results of microanalysis. Electrical conductivity measurements also provide data which enable the distinction between donor charge compensation with Ti^{3+} or (V''''_{Ti}) and were considered in the formulae calculations.

It can be concluded that in the analyzed samples Ce is incorporated as Ce^{3+} at Ba^{2+} sites. Reduction of Ce^{4+} to Ce^{3+} was confirmed by thermogravimetric analysis [1, 10]. Donor charge is compensated by (V''''_{Ti}) vacancies (sample 1), by reduction of Ti^{4+} (samples 3 and 4), and by a combination of both mechanisms (sample 2). Detailed consideration of the calculation of WDS results to the solid solution formulas exceeds the scope of this paper and was published elsewhere [12]. The applied correction methods for Ba-Ce spectral interference allowed correct elemental sample analysis, even for those with a low Ce concentration. Without correction, the Ba to Ce peak overlap introduces a large error in the overall quantitative analysis. WDS quantitative microanalysis was successfully applied for the investigation of Ce-doped $BaTiO_3$. The optimized EPMA procedure allowed determination of the amount of solid solubility of Ce, as well as the investigation of the mode of dopant incorporation in $BaTiO_3$.

Acknowledgements. This publication is based on work sponsored by the U.S.–Slovene Science and Technology Joint Fund in cooperation with the Ministry of Science and Technology of Slovenia under Project Number NIST 95–351.

References

[1] D. F. K. Hennings, B. Schreinemacher, H. Schreinemacher, *J. Europ. Cer. Soc.* **1994**, *13*, 81.

[2] J. S. B. Reed, *Electron Microprobe Analysis*, Cambridge University Press, 1975.

[3] J. Philibert, *Proc. 34rd Intl. Symp. X-Ray Optics and X-Ray Microanalysis*, Academic Press, Stanford University, New York, 1963, p. 379.

[4] P. Duncumb, J. S. B. Reed, in: *The Calculation of Stopping Power and Backscatter Effects in the Electron Probe Microanalysis, Quantitative Electron Probe Microanalysis* (K. F. J. Heinrich, ed.), NBS Special Publication 298, 1968, p. 133.

[5] J. S. B. Reed, *Br. J. Appl. Phys.* **1965**, *16*, 913.

[6] K. F. J. Heinrich, D. E. Newbury, *Electron Probe Quantitation*, Plenum, New York, 1991, p. 19.

[7] J. I. Goldstein, D. E. Newbury, P. Echlin, D. C. Joy, A. D. Romig, C. E. Lyman, C. Fiori, E. Lifshin, *Scanning Electron Microscopy and X-Ray Microanalysis*, Plenum, New York, 1992, p. 493.

[8] M. Čeh, Z. Samardžija, D. Makovec, *Scanning [Suppl. III]* **1994**, *15*, 94.

[9] J. P. Guha, D. Kolar, *J. Am. Cer. Soc.* **1973**, *56*, 5.

[10] D. Makovec, Z. Samardžija, D. Kolar, *Third Euro-Ceramics Proc.* **1993**, *1*, 961.

[11] W. D. Johnston, D. Sestrich, *J. Inorg. Nucl. Chem.* **1961**, *20*, 32.

[12] D. Makovec, Z. Samardžija, D. Kolar, *J. Sol. State Chem.* **1996**, *123*, 30.

Mikrochim. Acta [Suppl.] 13, 525–531 (1996)

Simulation of X-Ray Diffraction Profiles of Gradually Relaxed Epilayers

Alfredo Sanz-Hervás[1,*], Evaristo J. Abril[2], Miguel Aguilar[1], Gabriel de Benito[2], Carlos Llorente[2], and Miguel López[2]

[1] Dpto. Tecnología Electrónica, E.T.S.I. Telecomunicación-U.P.M., Ciudad Universitaria s/n, E-28040 Madrid, Spain

[2] Dpto. Teoría de la Señal y Comunicaciones e Ingeniería Telemática E.T.S.I.T.-U. Valladolid, Real de Burgos, E-47011 Valladolid, Spain

Abstract. A generalisation of the Hornstra-Bartels elasticity theory is proposed in this paper. This extended model allows the calculation of the strain and misorientation of gradually relaxed layers relative to the substrate. These data can be readily included in a X-ray diffraction model based on the Takagi-Taupin formalism. Some examples of gradually relaxed $In_{0.15}Ga_{0.85}As$ layers on (001) GaAs have been studied. It is demonstrated that X-ray diffraction is sensitive to the average relaxation in the layer. However, the actual relaxation profile cannot be readily determined from experimental scans due to the loss of crystalline perfection in the relaxed material.

Key words: X-ray diffraction, strain, relaxation.

Most of the research work on the growth and structural characterisation of epitaxial semiconducting structures makes use of high resolution X-ray diffractometry (HRXRD) as a basic tool for understanding the problems associated with such materials. HRXRD supplies accurate data on the strain, thickness and relaxation of the layers. Nevertheless, the interpretation of the measurements is often not straightforward and the profiles must be simulated in order to obtain the maximum information. Takagi [1] and Taupin [2] developed a dynamical theory of X-ray diffraction applicable to slightly strained crystals such as those found in epitaxial materials. Many researchers have used this model for the calculation of diffraction profiles and some programs are commercially available [3]. Nonetheless, they do not simulate asymmetric reflections for non (001) substrate orientations in spite of the current interest these have. In a previous work we presented a model that calculates any reflection regardless of the substrate orientation [4]. The elastic distortion of the material was obtained through the anisotropic elasticity theory for semiconductors developed by Hornstra and Bartels [5]. In this paper we present a generalisation of this elasticity model that allows the inclusion of relaxed layers.

* To whom correspondence should be addressed

Description of the Elasticity Model

During the growth of an epitaxial layer on a rigid thick substrate the layer material is elastically strained so that the in-plane lattice parameter of the layer a_{II} fits that of the substrate a_S.This effect is called coherent or pseudomorphic growth. Once the layer is thicker than a certain critical thickness, the elastic strain starts being relieved by the generation of defects as misfit dislocations and the material is plastically relaxed. In a HRXRD experiment the angular splitting $\Delta\theta$ between the substrate and layer peaks is due to two contributions, the different interplanar distances of substrate and layer ($\Delta d/d$) and the misorientation between the diffracting planes of substrate and layer ($\Delta\varphi$). Both $\Delta d/d$ and $\Delta\varphi$ are related to the degree of strain in the material and it is necessary to know the elastic distortion in the layer in order to calculate the diffraction pattern.

Uniformly Relaxed Layers of Uniform Composition

Hornstra and Bartels [5] divided the straining process into two steps, as illustrated in Fig. 1: In the first step the relaxed layer is strained in all directions by an amount $\varepsilon_{II} = (a_S - a_L)/a_S$ where a_S and a_L are the substrate and layer bulk lattice parameters respectively. After the first step the unit cell of the layer has the same dimensions as the unit cell of the substrate. In the second step the layer is coherently fixed to the substrate and the stresses are removed. This way, the in-plane strain in the layer is still ε_{II}, but the strain perpendicular to the interface is now $\varepsilon_{\perp}$. The expression that links the strain $(\Delta d/d)_{\mathrm{m}}$ in the direction **m** normal to the reflecting planes with the in-plain misfit strain ε_{II} is

$$-\varepsilon_{II} = \frac{a_L - a_S}{a_S} = P\frac{d_L^* - d_S}{d_S} = P\left(\frac{\Delta d}{d}\right)_{\mathrm{m}} \tag{1}$$

where d_L^* is the interplanar distance of the strained layer and d_S is the interplanar distance of the substrate, both measured in the direction **m** after the second step, and

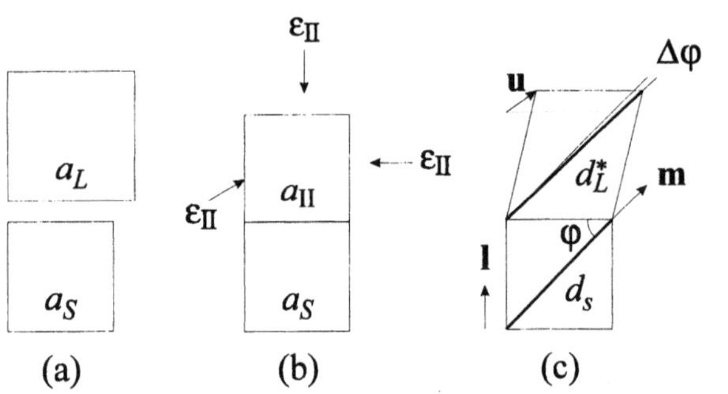

Fig. 1. Steps in the description of the straining process in an epitaxial material. Some important quantities referred to in the text are shown: **a** Relaxed materials in their bulk state; **b** In the first step the layer unit cell is hydrostatically strained to exactly fit the substrate unit cell; **c** Strained coherent layer after the second step. The interplanar distance d_L^* of the strained layer and the angular misorientation $\Delta\varphi$ between the diffracting planes are shown for an arbitrary asymmetric reflection

P is a correction factor calculated as

$$P = \frac{-1}{\sum_i m_i u_i \cos\varphi} \tag{2}$$

where m_i are the components of the unit vector **m**, u_i are the components of the displacement vector **u** of each lattice point in the second step, and φ is the angle between **m** and the unit vector **l** normal to the substrate. The calculation of **u** has been described elsewhere [4, 5]. The strained interplanar distance of the layer d_L^* can be readily calculated from (1) if it is taken into account that before the first step the unit cells are not distorted and all distances d are proportional to the magnitudes a, so that the following relation holds

$$\frac{a_L - a_S}{a_S} = \frac{d_L - d_S}{d_S} \tag{3}$$

With help of Eqs. (1) and (3) the strained interplanar distance can be expressed as

$$d_L^* = d_S\left(1 - \frac{1}{P}\right) + \frac{d_L}{P} \tag{4}$$

The other contribution to the angular splitting in the diffraction profile is the relative misorientation $\Delta\varphi$, which is calculated according to

$$\Delta\varphi = \sum_i m_i u_i \sin\varphi \tag{5}$$

Until this point it has been assumed that the layer was coherently grown on the substrate. This model can be extended to the case where some relaxation occurs in the interface so that the in-plane lattice parameter of the layer a_{II} differs from a_S. This can be macroscopically described by means of a relaxation coefficient R defined as

$$R = \frac{a_S - a_{\text{II}}}{a_S - a_L} \tag{6}$$

From (6) it is clear that $R = 0$ if the layer is coherent ($a_{\text{II}} = a_S$) and $R = 1$ if the layer is totally relaxed ($a_{\text{II}} = a_L$). It is assumed that the relaxation is parallel to the interface and uniform in both in-plane coordinates. We can now repeat the two step procedure for describing the straining process, but in the first step the layer unit cell is strained in all directions to fit a 'virtual' substrate of lattice parameter a_{II} instead of a_S. The expression for the in-plane strain in this case is

$$\varepsilon_{\text{II}} = \frac{a_{\text{II}} - a_L}{a_{\text{II}}} = (1 - R)\frac{a_S - a_L}{a_S} \tag{7}$$

The second step will again consist in relieving all the stresses. Equations (1) and (4) can be modified by substituting a_S for a_{II}. The interplanar distance d_L^* is now

$$d_L^* = d_{\text{II}}\left(1 - \frac{1}{P}\right) + \frac{d_L}{P} \tag{8}$$

Gradually Relaxed Layers of Uniform Composition

In the previous subsection it has been supposed that all the relaxation occurred at the interface. Nevertheless, it has been shown [6] that the presence of dislocations in a layer of uniform composition is not limited to the interface with the substrate and some gradual relaxation takes place throughout the layer. From the HRXRD simulation point of view, gradual relaxation can be implemented by dividing the single layer into a certain number of lamellae parallel to the interface. The state of relaxation of the ith lamella must now be defined in relation to the substrate as follows

$$R_i = \frac{a_S - a_{\text{II}i}}{a_S - a_{Li}} \tag{9}$$

where the relaxation coefficient R_i is now a function of distance from the interface. The strained interplanar distance of the ith layer becomes

$$d^*_{Li} = d_{\text{II}i}\left(1 - \frac{1}{P_i}\right) + \frac{d_{Li}}{P_i} \tag{10}$$

The calculation procedure of a gradually relaxed layer will firstly consist in defining the function $R_i = \mathbf{R}(t_i)$, where t_i is the distance from the interface of the ith lamella. The in-plane lattice parameter $a_{\text{II}i}$ is then calculated from (9), and the strained interplanar distance d^*_{Li} is finally obtained according to (10).

It has been shown experimentally [7] that a significant change in the in-plane strain due to relaxation does not occur until a relaxation critical thickness t_{cr} is reached, which is related to the misfit strain ε_{II} through

$$t_{cr} = \frac{k}{\varepsilon_{\text{II}}} \tag{11}$$

where the experimental parameter k, obtained from HRXRD and TEM results, is 0.83 nm for InGaAs on (001) GaAs [8]. For layers of thickness t greater than t_{cr} the residual average strain is also determined by the same parameter

$$\varepsilon(t) = \frac{k}{t} \tag{12}$$

Equations (11) and (12) predict the final relaxation state of an InGaAs single layer but they give no information regarding the dislocation distribution throughout the layer. TEM results [6] have shown that most of the dislocation density is distributed close to the interface and decreases rapidly as it approaches the upper surface. This would suggest the existence of a region close to the interface where some gradual relaxation occurs. Since HRXRD is very sensitive to strain variations, it is interesting to study the applicability of the technique to the investigation of the actual relaxation profile.

Results and Discussion

We will consider a single InGaAs layer on (001) GaAs with an In content of 15% and a thickness $t_L = 200$ nm. The misfit strain is $\varepsilon_{\text{II}} = 0.01075$, which gives, according to

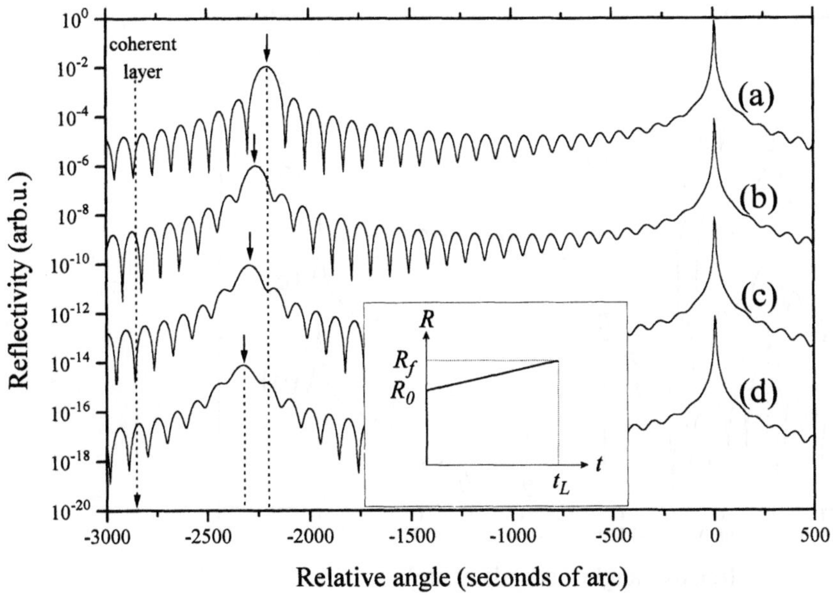

Fig. 2. X-ray diffraction profiles of a gradually relaxed $In_{0.15}Ga_{0.85}As$ layer on (001) GaAs. Inset shows the relaxation profile throughout the layer. The final relaxation is $R_f = 0.386$ in all curves. The interfacial relaxation R_0 in each curve is: *a* 0.386; *b* 0.3; *c* 0.25; *d* 0.2. Small arrows show the angular position of single layers with a uniform relaxation $\langle R \rangle$ equal to the average relaxation in the gradually relaxed layers. The average relaxation $\langle R \rangle$ in each curve is: *a* 0.386; *b* 0.343; *c* 0.318; *d* 0.293

equation (11), a relaxation critical thickness $t_{cr} = 77.2$ nm. The residual strain from (12) is $\varepsilon(t) = 4.15 \cdot 10^{-3}$. This results in a final relaxation coefficient $R_f = 0.386$. Insets in Figs. 2 and 3 show two possible relaxation behaviours. In Fig. 2, R is defined as a linear relaxation throughout the layer of slope

$$\Delta R = \frac{R_f - R_0}{t_L} \tag{13}$$

where R_f is the final relaxation coefficient, R_0 is the initial relaxation coefficient at the interface, and t_L is the total layer thickness. Inset in Fig. 3 shows a behaviour closer to TEM conclusions in which the uppermost material is almost free of misfit dislocations while a high density appears in the vicinity of the interface. This can be described by a region of gradual relaxation of thickness t_f and slope $\Delta R = (R_f - R_0)/t_f$ followed by a region of constant relaxation equal to the value R_f of the inflexion.

Figure 2 shows the 004 reflections for the first relaxation behaviour with different slopes. R_f is kept constant to a value of 0.386, while the interfacial relaxation R_0 differs from curve to curve (see caption of Fig. 2). In all the diffraction profiles pendellösung fringes are clearly visible at both sides of the main layer peak (left). These fringes are due to the layer thickness, while the splitting of the layer peak and the substrate peak (right) is related to the magnitudes $(\Delta d/d)_m$ and $\Delta\varphi$, as explained in the previous section. The angular position of the coherent layer case is marked in the figure. Peak splitting is decreased in all examples due to relaxation. Curve (a) is

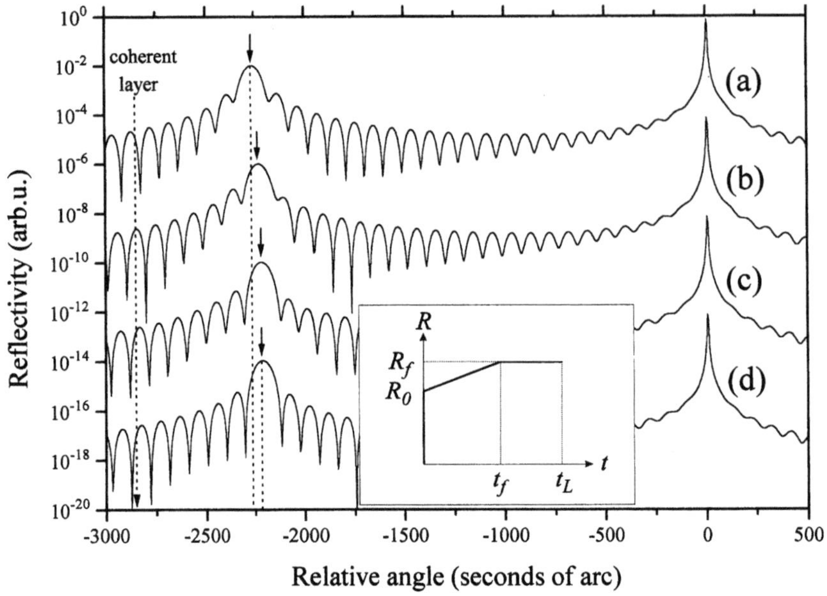

Fig. 3. X-ray diffraction profiles of the same layer in Fig. 2 but with a different relaxation behaviour, as can be seen in the inset. The interfacial relaxation is $R_0 = 0.3$ and the final relaxation is $R_f = 0.386$ in all curves. The thickness t_f of the linearly relaxed region in each curve is: *a* 200 nm; *b* 100 nm; *c* = 50 nm; *d* 25 nm. Small arrows show the angular position of single layers with a uniform relaxation $\langle R \rangle$ equal to the average relaxation in the gradually relaxed layers. The average relaxation $\langle R \rangle$ in each curve is: *a* 0.343; *b* 0.3645; *c* 0.3752; *d* 0.3806

calculated for $\Delta R = 0$, i.e., it assumes an abrupt relaxation behaviour. In curves (b), (c), and (d) the slope ΔR is increased by lowering R_0. The diffraction profiles show some unusual fringes near the layer peak as the slope increases. Simultaneously, the peak splitting decreases. We can now define the average relaxation in a layer as follows

$$\langle R \rangle = \frac{1}{t_L} \int_0^{t_L} R(t)dt \tag{14}$$

and calculate the angular splitting of a layer of uniform relaxation equal to the average relaxation $\langle R \rangle$ of each example in Fig. 2. Plotting the peak positions of the uniformly relaxed layers (small arrows), it can be clearly seen that they exactly fit the peak positions of the gradually relaxed layers. This result indicates that the angular splitting in a gradually relaxed layer is related to the average relaxation in the layer and not to the maximum or minimum relaxation values.

Figure 3 shows the diffraction profiles for a different relaxation behaviour closer to TEM observations [6]. In this case, the upper sublayer, of thickness $t_L - t_f$, is uniformly relaxed with $R_f = 0.386$, while the lower sublayer, of thickness t_f, is linearly relaxed from $R_0 = 0.3$ to $R_f = 0.386$. The thickness t_f of the linearly relaxed sublayer decreases from upper to lower curves. The odd fringes in the vicinity of the layer peak and the peak separation also decrease from upper to lower curves. Small arrows mark the layer peak position for the average relaxation, and an exact agreement with the peak positions is seen. This confirms the result obtained when discussing Fig. 2, according to which, the peak splitting is related to the average relaxation in the layer. This implies that any study of partially relaxed layers by HRXRD [7] yields information about the average relaxation in the layers.

The observation of the unusual pendellösung fringes in the diffraction profiles of Figs. 2 and 3 would suggest a method to study the relaxation profile of the layer as well as its average relaxation. Unfortunately, it must be remembered that the Takagi-Taupin formalism is not valid if the dislocation density is high. The loss of crystallinity and surface quality accompanying the formation of misfit dislocations yields broad and weak diffraction peaks, and any minute detail in the pendellösung oscillations is smeared out by diffuse scattering.

Conclusion

A model for the calculation of X-ray diffraction profiles of gradually relaxed layers has been presented. From this model, the interplanar distance of the strained material and the misorientation between the diffracting planes can be calculated taking into account the state of relaxation throughout the layer. The relaxation model has been integrated in a X-ray diffraction simulation program. The diffraction patterns of an $In_{0.15}Ga_{0.85}As$ layer on a (001) GaAs substrate with different relaxation profiles have been studied. It has been shown that the diffraction profiles are sensitive to the average relaxation in the layer, but the determination of the true relaxation profile is limited due to diffuse scattering that appears in the experimental scans of relaxed layers.

Acknowledgements. This work has been financially supported by the Spanish CICYT through the projects TIC93-0934 and TIC93-0912. We are grateful to A. Sacedón and J. F. Valtueña for some helpful discussions and critical reading of the manuscript. Thanks are due to R. Arias-Bejarano for the correction of the English text.

References

[1] S. Takagi, *J. Phys. Soc. Jpn.* **1969**, *26*, 1239.
[2] D. Taupin, *Bull. Soc. Franç. Minér. Crist.* **1964**, *87*, 469.
[3] RADS Program from Bede Scientific Instruments, Durham, U.K., and PC-HRS Program from Philips Analytical, Almelo, The Netherlands.
[4] A. Sanz-Hervás, E. J. Abril, D. I. Paz, G. de Benito, C. Llorente, M. Aguilar, M. López, *Mat. Sci Technol.* **1995**, *11*, 72.
[5] J. Hornstra, W. J. Bartels, *J. Crystal Growth* **1978**, *44*, 513.
[6] R. Beanland, A. Sacedón, E. Calleja, M. Aindow, *J. Appl. Phys.* **1996**, in press.
[7] D. J. Dunstan, P. Kidd, P. F. Fewster, N. L. Andrew, R. Grey, J. P. R. David, L. González, Y. González, A. Sacedón, F. González-Sanz, *Appl. Phys. Lett.* **1994**, *65*, 839.
[8] A. Sacedón, F. Calle, A. L. Álvarez, E. Muñoz, R. Beanland, P. Goodhow, *Appl. Phys. Lett.* **1994**, *65*, 3212.

Mikrochim. Acta [Suppl.] 13, 533–544 (1996)

Monte Carlo Simulation of Electron Scattering for Arbitrary 2D Structures Using a Modified Quadtree Geometry Discretization

Christian O. Schiebl[1,*,**], Arnold Pfeiffer[1], Horst W. Wagner[1], Wolfgang S. M. Werner[2], and H. Stippel[3]

[1] Digital Equipment Österreich AG, Campus-based Engineering Center, Favoritenstr. 7, A-1040 Wien, Austria

[2] Institute for Applied and Technical Physics, Technical University of Vienna, Wiedner Hauptstrasse 8-10, A-1040 Wien, Austria

[3] National Semiconductor Corporation, P.O. Box 58090, Santa Clara, CA 95052-8090, U.S.A.

Abstract. A quasi 3-dimensional Monte Carlo simulation program, called SESAME, has been developed for the simulation of electron-matter interactions in totally arbitrary 2-dimensional structures. SESAME is designed for application in different analysis techniques, like scanning electron microscopy (SEM), electron probe microanalysis (EPMA), energy and wavelength dispersive X-ray spectroscopy (EDS/WDS), Auger electron spectroscopy (AES), transmission electron microscopy (TEM), electron energy loss spectroscopy (EELS), etc.. The program is extremely flexible in terms of simulation geometries as well as in terms of possessing the ability to handle the majority of electron-matter interaction processes. A modified quadtree discretization for geometry handling is used. Three different scattering models have been implemented to account for the great variety of possible applications.

Key words: Monte Carlo simulation, electron scattering, characteristic X-ray generation, electron probe microanalysis, Auger electron spectroscopy.

Due to the decreasing dimensions of the features on semiconductor wafers, e.g. thicknesses, minimum lateral dimensions, critical defect sizes, etc., it is no longer possible to intuitively predict the excitation and emission volume of X-rays or the area from which Auger electrons are emitted during analysis. The incident electron beam cannot be contained within a single layer and the effects of vertical, horizontal or arbitrarily running interfaces are becoming more pronounced. As a result, the spectroscopist makes many measurements under many different conditions, so that

* Present address: Kremstalstr. 95/2/4, A-3500 Krems, Austria

** To whom correspondence should be addressed

the optimum experimental conditions for the final data collection can be determined. This is not efficient and has caused the average amount of time required to obtain data from 'average' EDS or Auger analyses to increase with each advance in CMOS technology.

In addition, it is becoming more frequently necessary for samples to be cross-sectioned – either mechanically or through the use of a focused ion beam – prior to analysis in order to expose a buried feature for analysis. Experimental set-up and interpretation of spectra and images from these samples is difficult without an understanding of the true analysis volume.

Additionally, in AES the area from which Auger electrons are emitted has been classically assumed to be just the area irradiated by the incident electron beam. This is a good approximation for low to moderate resolution analyses and broad (>1 μm diameter) electron beams but is no longer adequate for high spatial resolution work or when high incident electron-beam energies are employed. Signals excited, for instance, by high energy, subsurface scattered electrons, can contribute substantially to the spectrum even though they are generated far away from the incident electron-beam area. This may confuse the interpretation of spectra, linescans and images and limit quantitative data analysis. Having access to a powerful simulator allows achieving a better experiment planning, a better qualitative and semi-quantitative interpretation of results, and more efficient teaching and data explanation. Therefore, the importance of computer aided materials analysis (CAMA) and hence Monte Carlo simulation for industrial laboratories will definitely grow in future.

Furthermore, Monte Carlo simulation is considered to become a practical method even for the calculation of correction procedures in quantitative EPMA/EDS. Monte Carlo calculations, according to Mackenzie [4], seem to be the only viable means of obtaining accurate ionization depth distribution functions which is the prerequisite for such correction procedures. Nassiopoulos and Valamontes [6], for instance, have already studied thin films on substrates by modelling all correction factors, including fluorescence, using Monte Carlo techniques.

Although considerable success has been reported with analytical approaches for quantitative analysis on the basis of the transport theory by Werner [11] it must be stated here that these results were derived for situations where the boundary conditions used had to be idealized. In contrast, one of the great advantages of the Monte Carlo technique is the generally straightforward treatment of arbitrary geometries. Therefore it can be concluded that in specific cases simulation codes which can handle the most general boundary conditions are of pronounced importance.

The geometries which can be handled by SESAME are quasi 3-dimensional, i.e. they must be 2-dimensional but are infinitely expanded in the direction of the third coordinate-axis perpendicular to the defined geometry. This is the only restriction concerning the shape of the simulation structure. Directions of the electron beam and to the detectors may be chosen totally arbitrarily in 3-dimensional space.

SESAME is written in FORTRAN and C and runs on Open VMS VAX, Open VMS AXP and Digital Unix platforms. Currently, the program is also ported to the PC and will run under Linux there.

The Modified Quadtree

Originally, the quadtree method comes from graphical image processing [2, 8]. It was adopted for our purposes to discretize a given two-dimensional geometrical object in order to allow an easy and fast determination in which area a point is located (point location) and thus check for the region in which an electron is currently moving.

At the beginning of the quadtree discretization the whole geometry is included in one square, the rootsquare. This square is then subdivided into four subsquares, if it is not entirely composed of the same material. This procedure is recursively carried out for every subsquare until either a predetermined accuracy is reached – which is measured by the side-length of the square – or no intersection of this square with the polygons defining the geometry is found.

For the data representation of such a quadtree, a tree is used starting at the root node, which represents the rootsquare. Every subsquare is represented as a child of its predecessor's node. The final squares, which are assumed to be entirely composed of the same material, are the leaves of the tree. The determination of the point location of an electron is thus reduced to a simple search procedure in a tree structure.

This original quadtree allows a very fast and efficient point location, but has some disadvantages. Firstly, it does not represent the geometry exactly because each square is assumed to entirely lie in one material. Secondly, by using only squares, it is possible that even a simple rectangle cannot be discretized exactly. Finally, to obtain a suitable accuracy of the discretization the memory requirements are very high.

To avoid such problems the original quadtree has been modified. In SESAME rectangles instead of squares are used. This results in an improvement for simple geometries such as rectangles. Furthermore, the algorithm of subdividing the initial geometry into rectangles stops already if one of the following three criteria is met: 1) a rectangle lies entirely within a region, or 2) only one point of the geometry is located in the rectangle, or 3) exactly one line of the initial geometry intersects this rectangle. The result of such a geometry discretization is shown in Fig. 1.

This modified method leads to some slightly more complicated leaf areas. But it is faster, because the tree structure is not as complicated as for the original quadtree method. The discretized areas are much bigger. Therefore an electron moving around remains in the same leaf for a longer time. This can be used to further speed up the algorithm. Finally, the memory requirements could be reduced tremendously, although the geometry is represented correctly – there is no discretization error.

Besides the advantage of a drastic reduction of CPU time and memory compared to other, more complicated and numerically more sensitive methods which calculate intersections with sample interfaces and boundaries, the expansion of the quadtree method to fully 3-dimensional geometries is straightforward by means of using an octree for the discretization, as already shown by Stippel and Selberherr [10].

To demonstrate the efficiency of the implemented geometry handling routines, two simulations under the same conditions (sample composition, acceleration energy, number of electrons, etc.) with only a single exception were performed: in one case the most simple quadtree (a single rectangle) and in the other case the quadtree

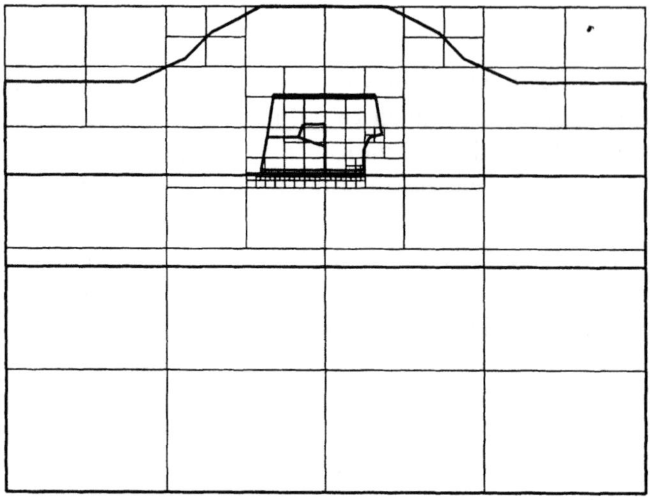

Fig. 1. Geometry discretization using a modified quadtree

shown in Fig. 1 were applied. The first case describes the 'simple bulk case' (no searching in the tree at all) and corresponds to the simple Monte Carlo codes which can only handle bulk structures. Although extensive geometry checking is necessary for the second case, only ca. 8% more CPU-time is needed.

Implemented Scattering Models

Three different electron scattering models have been implemented into SESAME, a single scattering model, a hybrid model and a direct simulation model using dielectric functions.

Elastic scattering, i.e. scattering at the atomic nucleus, is calculated using tabulated Mott cross-sections. The partial wave approach of Mott [5] considers the wave properties of electrons and is therefore based on quantum mechanics. This approach yields better results compared to the frequently used screened Rutherford formula which presents a poor approximation for heavy elements, particularly in the low-energy region.

Since the computation of the differential Mott cross-sections causes considerable computational effort, the cross-sections have been tabulated in SESAME for almost all the elements in the periodic table. The calculation of the tabulated differential Mott cross-sections is based on the Yates [12] method.

For the single scattering model the electrons' energy losses caused by inelastic scattering is considered by a continuous slowing-down approximation using the Bethe [1] stopping-power formula. The Bethe stopping power accounts for energy losses due to core electron ionization, valence and conduction electron excitation, and plasmon generation.

There are two basic problems using the Bethe stopping power formula. First, it is not applicable to energies lower than the mean ionization energies, because it becomes negative, and second, the continuous slowing-down approximation does not allow to describe secondary cascade generation and simple excitation processes.

To eliminate the first drawback, the empirical approach suggested by Rao-Sahib and Wittry [9], which is an extrapolation of Bethe's formula down to zero energy, has been implemented. This correction prevents the stopping power from becoming negative but does not account for the failure of the Bethe approximation at low energies. The second restriction is simply circumvented by supplying a variety of models, each more appropriate for certain tasks than for others.

Inner-shell ionizations are computed by applying the ionization cross-section formula according to Gryzinski [3] along the trajectory of the electron. The probability of the ionization of a certain atomic shell along the path between two adjacent elastic scattering events is calculated and assigned to a randomly determined location along this path segment. This single scattering model is appropriate for the calculation of characteristic X-rays, since it is very fast and already yields acceptable statistics from a relatively small number of simulated electrons.

The second model which has been implemented is a so-called hybrid model. This is a combination of a continuous slowing-down approach and a direct simulation model. Inner-shell ionizations are individually simulated, again applying the Gryzinski formulae for the differential ionization cross-sections. The energy loss between two individual scattering events, which might be elastic scattering or an inner-shell ionization, is caused by valence electron excitations. These are considered by the continuous slowing-down approach using a stopping power which is determined to be the difference of the Bethe stopping power and the stopping power from Gryzinski, taking inner-shell ionizations into account. This hybrid model needs the simulation of many more electron paths to achieve the same statistical accuracy for the generation of ionizations, and therefore characteristic X-rays, as the single scattering model, but has the advantage that also secondary electron generation can be calculated.

The third, and last, model implemented so far is the so-called dielectric function model. This model is based on the dielectric function for the description of inelastic scattering events. The big advantage in comparison with the other models is that the valence electron excitation as well as the inner-shell electron excitation are included exactly according to their oscillator strengths or equivalent energy loss functions. The simulation can be done strictly taking into account the individual inelastic collisions instead of the application of the continuous slowing-down approximation. A further advantage is that this model is applicable to low-energy as well as to high-energy electrons.

The dielectric function needed for the calculation of the differential inelastic mean free path is deduced from experimental optical dielectric data using a method proposed by Penn [7]. Unfortunately, data is available for only a few elements and compounds, thus the simulation is limited to regions composed of elements and compounds for which data is stored in SESAME.

This restriction is partially circumvented in such a way that the simulation model may be specified by the user depending on the regions of the simulation geometry. Thus it is possible to compute interaction processes which need the dielectric function model to get accurate results in one region of interest and follow the electrons in other regions by applying a different model there. That way electrons which would be scattered (back) into the region of interest are not lost.

The applied boundary conditions, i.e. the behavior of the electron crossing an interface, are implemented so that each electron reaching an interface is stopped there. At this location no individual scattering event is calculated, only the energy loss in the continuous slowing-down approximation is calculated for the path between the last scattering event and the current location at the interface. Thereafter a mean free path for the region to be entered is computed. The electron path is continued in the new region without any change of direction at the interface. It can be shown that this procedure exactly fullfills the condition of Poisson distributed mean free paths.

Simulation of Characteristic X-Rays

The computation of characteristic X-ray intensities requires the consideration of the absorption of the generated X-ray quanta in the sample on their way to the X-ray detector. Since, especially in case of more complicated geometries, the calculation of the absorption factor every time an X-ray is generated would be far too time-consuming, a procedure has been implemented into SESAME which prepares appropriate values of the absorption factors depending on 2-dimensional space coordinates for each X-ray line of interest.

Discrete space-dependent values are calculated once, at the beginning of the simulation, and are stored on a rectangular, not necessarily equidistant, grid. This grid is either generated automatically by SESAME according to an estimated excitation volume or may be supplied by the user in a grid file.

The X-ray distributions generated during the simulation are stored on the same grid and the absorption is computed at the end of the simulation.

Simulation of Auger Electrons

To get acceptable statistics in a reasonable period of time it is necessary to implement the generation of Auger electrons in a very efficient way. The generation of Auger electrons is a relatively rare event since the probability that an inner-shell ionization process occurs is much smaller than the probability of elastic scattering. Furthermore, although generated throughout the electron range, only Auger electrons generated in the near surface area (some monolayers) have the possibility of leaving the sample without energy loss, which is the pre-requisite needed to be distinguished from the backscattered and secondary electron background. Additionally, the Auger electron emitted from the sample must reach the detector, i.e. this electron must leave the sample surface in a well defined direction.

To avoid simulating Auger electrons generated further inside the sample, thus wasting valuable CPU time, only Auger electrons which are generated at a depth from the sample surface not exceeding a certain threshold are taken into account. But this alone would not be enough to decrease the CPU time to acceptable values. Therefore, an approach similar to the computation of characteristic X-rays in the single scattering model is used. This means, that at each scattering event taking place within the depth threshold the probability of the generation of an Auger electron is calculated. All such generated Auger electrons are, together with their generation probability, kept in a queue until the simulation of the current primary electron is terminated.

Thereafter all Auger electrons held in the queue are simulated. If an Auger electron is to be detected then only a fraction of it, according to the probability of generation, is counted. This method yields acceptable statistics in a reasonable period of time for all Auger output options provided by SESAME.

Results and Discussion

SESAME supplies a very great variety of output information which may be graphically visualized on the screen or written to files for postprocessing. The output of specific results can be specified prior to simulation. The supplied output options include information of electron (primary and secondary) trajectories (Fig. 2), 1D, 2D, and even 3D distributions of ionizations, generated and detected characteristic X-rays, 2D distributions of generated and detected Auger electrons, backscattered, secondary and Auger electron scans, characteristic X-ray scans, energy and angle distributions of all kind of electrons (backscattered, transmitted, secondary, Auger), etc. Only a few of all these possible features can be demonstrated here.

The geometry of the example under investigation is shown in Fig. 3. The simulation structure represents an aluminum interconnect. To support the characterization of such interconnects, partially the analysis of failed ones, it is necessary to detect aluminum voids and copper precipitates. The samples for analysis come both with and without SiO_2 passivation. The aim of this work is to evaluate what signal at what measurement conditions could give appropriate answers to questions concerning not only the detectability of such voids and precipitates but also concerning their sizes and or locations (depth in the line).

All kinds of different signals have been recorded during the simulation at different energies to find the optimum measurement conditions. One example from

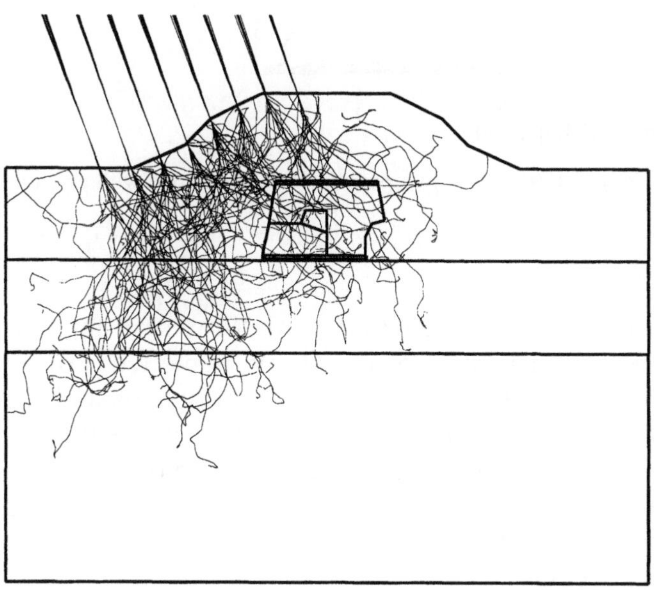

Fig. 2. Trajectories of primary electrons

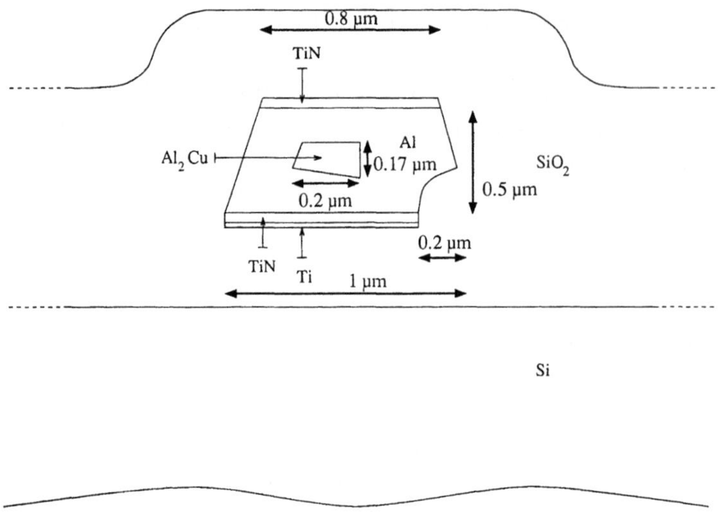

Fig. 3. Representation of the aluminum interconnect structure

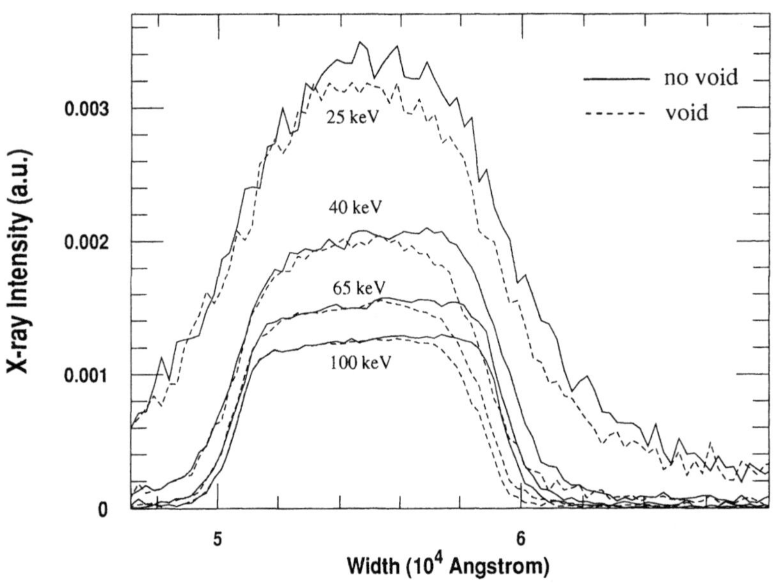

Fig. 4. Al K_α X-ray scans for 25, 50, 75 and 100 keV for two passivated samples, one with and one without an Al void

these simulations is shown in Fig. 4 where the detected Al X-ray scan for four different electron beam energies, ranging from 25 to 100 keV, is plotted. For all examples 50000 electrons were simulated. The tilt angle is zero degrees and the take-off angle is 37.5 degrees. The X-ray scans were performed once with and without a void in the Al line. At 25 keV one would get the highest intensities but the shape and the width of the Al line and the void cannot be determined with the desired accuracy. For higher energies the signal gets weaker but the location of the void can be determined with much higher accuracy.

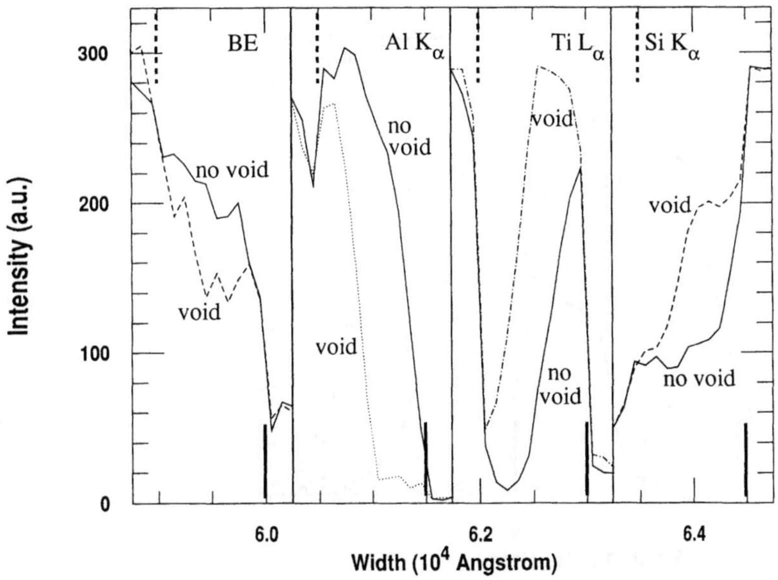

Fig. 5. Scans of backscattered electrons, Al K_α, Ti L_α, and Si K_α X-rays for 6 keV acceleration energy for two samples without passivation, one with and one without an Al void. The Al K_α curves are shifted by 0.15 microns, the TI L_α by 0.3 microns and the Si K_α by 0.45 microns. Vertical solid markers indicate the right base edge and vertical dashed ones the right top edge positions of the Al interconnect line

For the sample without the passivation, the backscattered electron scan and the AlK_α, TiL_α and SiK_α X-ray intensity scans are plotted in Fig. 5 for an acceleration energy of 6 keV. From a combination of these data the Al void not only can be detected but also the lateral dimension may be deduced and information obtained as to whether the TiN/Ti layer underneath the Al line is etched back in the region of the void (the AlK_α curves are shifted by 0.15 μm, the TiL_α curves by 0.3 μm and the SiK_α curves by 0.45 μm for a clearer visualization).

The situation is more complicated in case of a passivated sample. The backscattered electrons and the SiK_α signal do not give a significant contrast anymore. Only the combination of the AlK_α, the TiK_α and TiL_α signal is useful to gain the required information as can be seen from Figs. 6 (the TiK_α curves are shifted by 0.3 microns and the TiL_α curves are shifted by 0.6 microns for the sake of clarity).

The TiL_α intensity originates mainly from the upper TiN layer because of the heavy absorption of these X-rays. Thus, the TiL_α signal can be used to identify the location and lateral dimension of the Al interconnect even if the TiN/Ti layer beneath the Al line is partly etched back. Additionally, the AlK_α intensity is used to identify Al voids and the TiK_α signal delivers information concerning whether this TiN/Ti layer beneath the Al line is etched back.

For the detection of Al_2Cu precipitates and the determination of their size and vertical locations in the Al interconnect the Cu signal delivers the most accurate and complete information. Figure 7 shows the CuK_α X-ray scan for five different electron beam energies in case of the passivated sample. It can be seen that precipitates of the assumed size can already be detected for relatively moderate acceleration energies

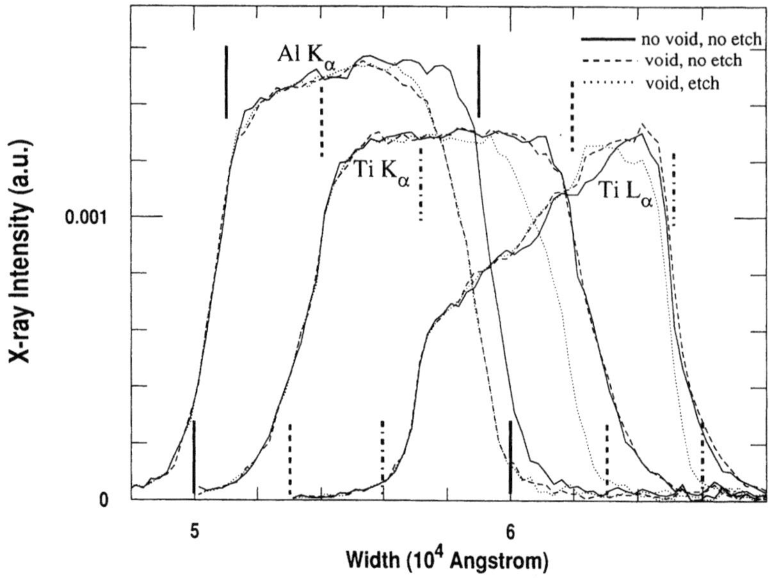

Fig. 6. Scans of Al K_α, Ti K_α, and Ti L_α X-rays for 75 keV and for three samples with passivation, one without two with an Al void. Additionally, for one sample with an Al void the TiN/Ti layer is etched back beneath the void. The Ti K_α curves are shifted by 0.3 microns and the TI L_α curves are shifted by 0.6 microns. Vertical markers indicate the right base and top edge positions of the Al interconnect line

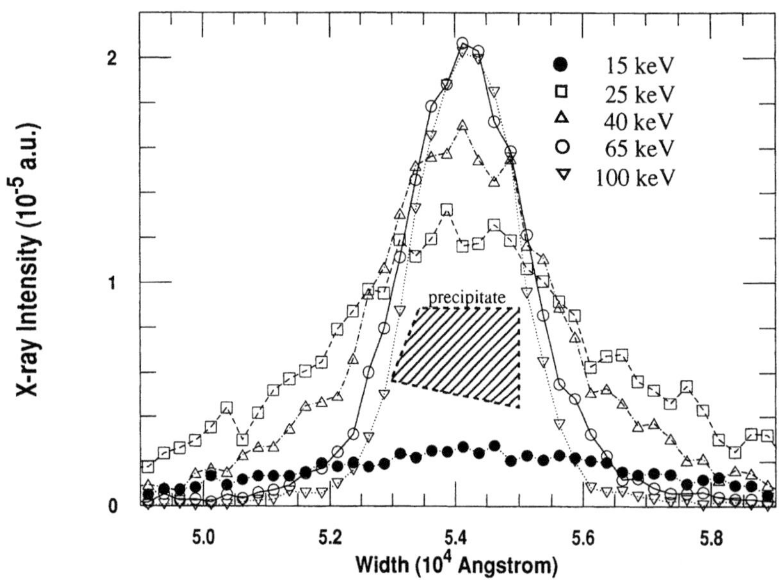

Fig. 7. Cu K_α X-ray scans for 15, 25, 40, 65, and 100 keV acceleration energy for a passivated sample with an Al_2Cu precipitate. The location of the precipitate is indicated by the dashed area

(below 25 keV) – but for the determination of their sizes it is definitely necessary to significantly increase the operating voltage.

The determination of the size and the exact location in the line may be determined by tilting the sample and repeating the measurement. In Fig. 8 the CuK_α X-ray scans for zero and 45-degree sample tilt are plotted. Whereas the scans for zero

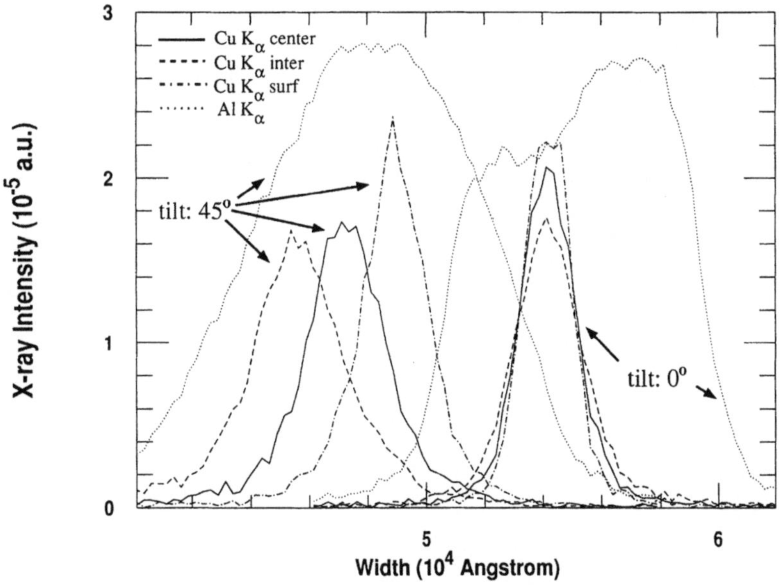

Fig. 8. Cu K_{α} X-ray scans for 65 keV acceleration energy and zero and 45 degree sample tilt for three passivated samples with an Al_2Cu precipitate located at three different depths in the Al line, namely at the center, at the top (surf) and at the bottom (inter). The dashed lines represent the Al K_{α} scans which are needed to determine the location of the precipitate within the Al line

tilt deliver information on the lateral location and the lateral dimension of the precipitate the measurements at 45-degree tilt additonally supply information on the vertical location and the vertical dimension of the precipitate. Both enable a fairly good estimation of the location and size of the precipitate.

The example presented here demonstrates only a very small portion of possible applications of computer aided materials analysis (CAMA). As the analyzed structures are getting more complicated with each CMOS generation there is a growing demand for tools supporting the spectroscopists in analysis planning and interpretation. Since measurement tools are limited and very expensive in acquisition and operation there is a growing need for higher analysis throughput which can be achieved by better experiment planning. Contrary, CPU power is getting cheaper and cheaper and therefore it is expected that the need for appropriate CAMA tools will significantly grow in future.

Acknowledgement. We would like to thank Mrs. Llanda Richardson and Mrs. Anne Testoni, UOG Hudson, Digital Semiconductor, Digital Equipment Corporation, for their continuous support of this work.

References

[1] H. Bethe, *Ann. Phys. Leipzig* **1930**, *5*, 325.
[2] W. D. Fellner, *Computergrafik*, Spektrum Akademischer Verlag, Heidelberg, 1992.
[3] M. Gryzinski, *Phys. Rev.* **1965**, *138A*, 305.
[4] A. P. Mackenzie, *Rep. Prog. Phys.* **1993**, *56*, 557.
[5] N. F. Mott, *Proc. R. Soc.* **1929**, *A124*, 425.

[6] A. G. Nassiopoulos, E. Valamontes, *Surf. Interface Anal.* **1990**, *15*, 405.
[7] D. R. Penn, *Phys. Rev. B* **1987**, *35*(*2*), 482.
[8] F. P. Preparata, *Computational Geometry, An Introduction*, Springer, Berlin Heidelberg New York Tokyo, 1985.
[9] T. S. Rao-Sahib, D. B. Wittry, *J. Appl. Phys.* **1974**, *45*, 5060.
[10] H. Stippel, S. Selberherr, *ICICE Trans. Electron* **1994**, *E77-C*, *2*, 118.
[11] W. S. M. Werner, *Mikrochim. Acta* [*Suppl.*] **1996**, *13*, 13.
[12] A. C. Yates, *Comp. Phys. Comm.* **1971**, *2*, 1975.

Mikrochim. Acta [Suppl.] 13, 545–552 (1996)

Chemical-Bond Characterization of Nanostructures by EELS

Reinhard Schneider[1,*], Jörg Woltersdorf[1], and Andreas Röder[2]

[1] Max-Planck-Institut für Mikrostrukturphysik Halle, Weinberg 2, D-06120 Halle, Federal Republic of Germany

[2] Martin-Luther-Universität Halle-Wittenberg, Fachbereich Physik, Friedemann-Bach-Platz 6, D-06108 Halle, Federal Republic of Germany

Abstract. The paper describes examples of the chemical-bond characterization by ELNES analysis taken from AEM investigations of a SiC-fibre reinforced borosilicate glass exhibiting a carbon-rich reaction layer between fibre and matrix. Chemical bonding is examined across the fibre/matrix interlayer at nanometre resolution, particularly by analysing the Si-L_{23} ELNES. The bonding state of silicon is imaged either along a line by recording series of EEL spectra at high energy resolution, or two-dimensionally by using ELNES details for energy-filtered scanning transmission electron microscopy.

Key words: analytical electron microscopy, electron energy loss spectroscopy, energy loss near edge structure, chemical-bond characterization.

Analytical electron microscopy (AEM), which is the combination of imaging and spectroscopic electron microscope techniques, allows the elucidation of the micro-chemistry of nanometre-sized regions. In general, energy-dispersive X-ray spectroscopy (EDXS) and electron energy loss spectroscopy (EELS) are used as the preferential AEM methods of characterizing the chemical composition of thin specimens (thickness in the order of 50 nm and thinner). Besides qualitative and quantitative element analyses, achievable by both techniques, EELS provides additional information on the chemical bonding of the elements detected. This information is gained from details of the relevant ionization edge, i.e. the energy loss near edge structure (ELNES), which can be observed owing to the high energy resolution (0.5 to 1 eV) of EELS. Near edge fine structures arise because electrons are excited from inner-core levels into unoccupied states above the Fermi energy. A further criterion for the chemical-bond state is the onset energy of the ionization edge, which can be influenced by the specific bond partners (chemical shift). Besides the chemical bonding ELNES analyses also yield data on the

* To whom correspondence should be addressed

coordination of the elements and on the electronic structure of the conduction band (see e.g., [1, 2]).

Fibre-reinforced composites with ceramic and glass matrices are of increasing interest because under mechanical load they may be damage-tolerant and quasi-ductile. The properties of these materials are mainly controlled by the interfacial zone between fibre and matrix. Thus, both the microstructure and the microchemistry of this region have to be known to understand and to optimize the composite behaviour.

In composites of SiC-containing fibres and silicate glass matrices, carbon-rich interlayers arise from chemical reactions during processing (cf. e.g. [3, 4]). These layers of specific phenomena may strongly influence the composite properties.

For SiC(Nicalon®)-fibre reinforced Duran® glasses (B_2O_3-Na_2O-SiO_2), carbon-rich interlayers are generated by chemical reactions during hot-pressing. It is generally assumed that several processes within the fibre and at the fibre/matrix interface generate these layers. For this materials system the dependence of layer formation (microstructure and microchemistry) on the processing conditions (temperature and period of hot-pressing) was studied in detail (cf. e.g. [5, 6]). A model has been developed on the basis of the obtained results and by thermodynamical treatments. This model proves the individual chemical reactions causing carbon-rich interlayers, also in cases of complex morphological configurations [7]. In the model, the main process of C-layer formation is the oxidation of SiC in the periphery of the fibre, taking place step by step, probably via intermediate oxide phases. Thereby, first the SiC of the fibre is oxidized via the chemical reaction

$$2SiC + O_2 \rightarrow 2C + 2SiO$$

and second, owing to a higher O_2 partial pressure near the matrix side, the silicon monoxide is further oxidized according to the reaction

$$2SiO + O_2 \rightarrow 2SiO_2.$$

The microchemical investigations presented in the following imply a selection of studies carried out on Nicalon-fibre reinforced Duran glasses, with the particular aim to identify the chemical bonding of the elements involved in the C-layer generation.

Experimental

The results discussed were obtained for a SiC(Nicalon® NLM202)-fibre reinforced Duran® glass (B_2O_3-Na_2O-SiO_2), which was produced via a sol-gel-supported process developed by *Schott Glaswerke Mainz* [8]. Nicalon is assumed to have a nanocrystalline structure mainly composed of tetrahedra of two types, viz. SiC_4 and SiC_xO_y ($x + y = 4$), surrounded by free aromatic carbon aggregates [9]. After glass impregnation of the fibres, using a SiO_2 generating sol-gel, the composite was hot-pressed at about 1100 °C for 5 min under argon atmosphere. More details of the composite processing procedure are given in [8].

Specimens were prepared in the usual way for transmission electron microscopy examination including the cutting of thin (< 200 µm) slices of 3 mm in diameter, mechanical grinding and dimpling down to a thickness of about 10 µm and final ion milling (Ar, 5 kV).

Analytical electron microscopy was carried out in a *Philips* CM 20 FEG in combination with a parallel EELS (PEELS model 666 of *Gatan*). The microscope was run in the STEM mode at a primary electron energy of 200 keV. Due to the intrinsic energy width of the thermally assisted field emission (FE) cathode the energy resolution is mainly limited to about 0.8 to 1 eV measured at the half width of the zero-loss peak. Series of EEL spectra across the interlayer under investigation were recorded using the PEELS with the STEM probe of about 2 nm in diameter digitally scanned point by point by the *Gatan* Digiscan model 688. Experimental parameters of spectrum recording, i.e. energy loss per channel, integration time and number of read-outs, are given in the text.

In addition, a dedicated STEM of the type *Vacuum Generators* HB 501 UX was used equipped with a serial recording EEL spectrometer (VG ELS 501). In the analytical mode, the cold FE source and the illuminating electron optics offer an electron probe of about 1 nm in diameter with a current of about 1 nA. The instrument was operated at an accelerating voltage of 100 kV. Because of the ultra-high vacuum of some 10^{-7} Pa in the specimen chamber contamination effects are not so critical as for the CM 20 FEG. Owing to the relatively low energy spread of the cold FE gun in the order of about 0.3 eV, the energy resolution amounts to about 0.5 to 1 eV, depending on the stability of the high voltage and on disturbing effects by electrical and magnetic stray fields.

Applying the energy analyser of the EELS system to the STEM signal for energy filtering enabled element-specific or even bond-specific imaging, respectively, depending on the slit width chosen for a specific signal (e.g. integral edge signal or ELNES detail). Images were digitally recorded by means of an image acquisition system (*Kevex* Delta class 5). The digital image data have a resolution of 512 × 256 pixels at 8bit signal depth. Different values of point averaging in the range from 8 to 32 times per image point were chosen to fit the corresponding signal to the signal-to-noise ratio.

Results and Discussion

Chemical-Bond Characterization by EELS Line Profile Analysis

The morphology of the SiC(Nicalon)-fibre reinforced Duran glass (hot-pressed at 1102 °C, 5 min) was previously investigated in much more detail (see e.g. [5]). Figure 1 shows microstructural phenomena: between fibre and matrix there is a carbon-rich reaction layer of about 40 to 50 nm in thickness the structure of which is turbostratic carbon. The interface of the interlayer to the matrix is distinct, but in contrast to that, there is a relatively continuous transition from the carbon layer into the fibre periphery. Earlier EDXS examinations showed rather large amounts of silicon and oxygen inside the reaction layer.

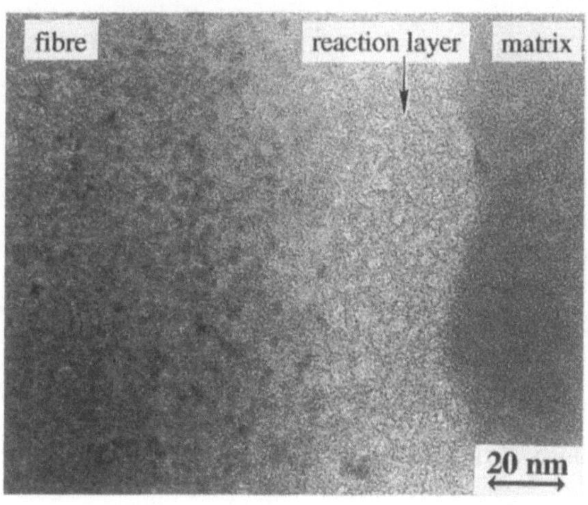

Fig. 1. HREM micrograph of the fibre/matrix interface region

To follow the chemical-bond state of silicon across this interface configuration, a series of EEL spectra were measured in the range of the Si-L_{23} edge at a relatively high energy resolution (about 1 eV) along a line of about 460 nm in extension, as introduced by Coliex et al. [10]. Each spectrum of the EELS line profile analysis visible in Fig. 2 was recorded using an integration time of 4 s at 0.2 eV per channel. The spectra start from the fibre region (lower curves) moving into the glass matrix, thus crossing the zone of the reaction layer. There is a position-dependent change of the chemical bonding of silicon from predominantly carbidic (fibre) to mainly oxidic revealed by analysing the edge onset and ELNES features. In general, the ELNES of SiO_2 has an edge onset at about 105 eV; it is indicated by two peaks distinctly split at about 108 and 115 eV, respectively, with the first exhibiting a small pre-peak (cf. e.g. [11]). Unlike SiO_2, the threshold energy of the Si-L ELNES of SiC amounts to about 101 eV, with only one pronounced peak at about 104 eV. For Nicalon this dominant peak is slightly broadened and, moreover, shifted to higher energies by the presence of oxycarbide phases inside. Besides the general carbidic/oxidic variation during the transition from fibre into matrix it becomes obvious furthermore that already in the outermost zone of the fibre the oxidation state is increasing (cf. Fig. 2).

This is confirmed by single EEL spectra of the line profile enabling the detailed characterization of the bond character as Fig. 3 shows. In this drawing the very top and bottom curves represent the net Si-L ELNES (background subtracted) of matrix and fibre, respectively, corresponding to lines 1 and 40 in Fig. 2. Line spectrum 17 was taken at about the middle of the reaction layer indicating a Si-L ELNES resembling that known from the dioxide. The next lower ones (lines 18 and 19) were recorded from the outermost part of the fibre. They exhibit small residual peaks at

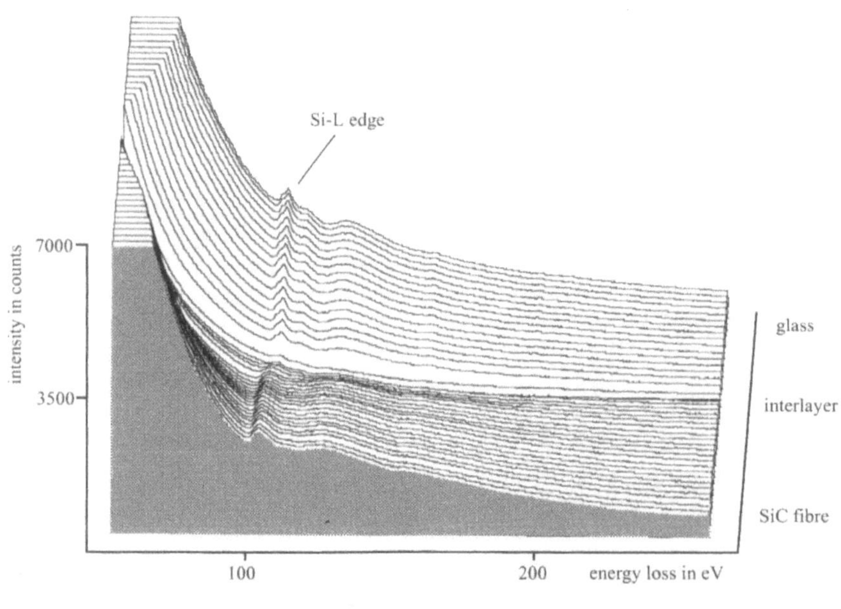

Fig. 2. Line profile analysis exhibiting the change of the Si-L ELNES by crossing the interlayer region

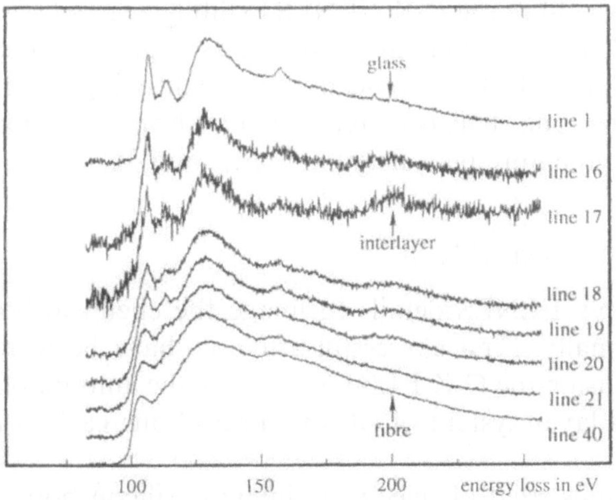

Fig. 3. Single EEL spectra of the line profile allowing the detailed analysis of the bond character of silicon

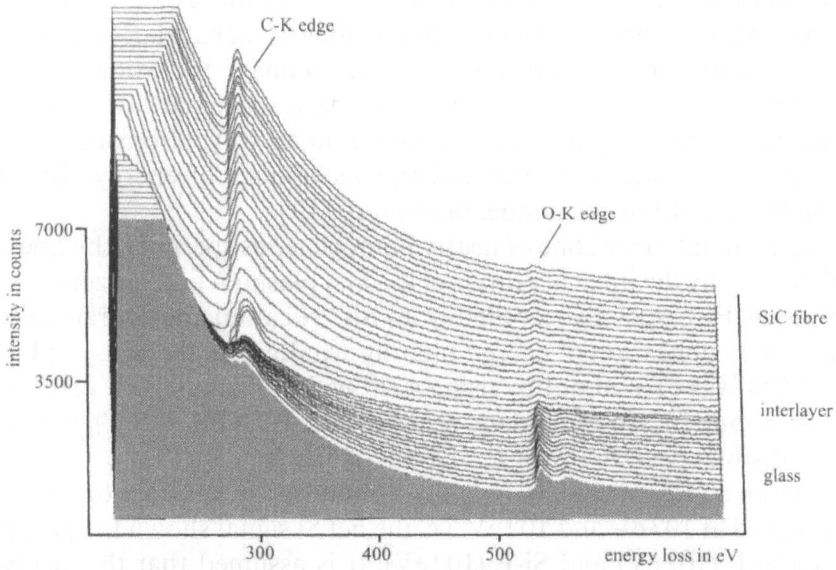

Fig. 4. Series of EEL spectra in the energy loss range of the C-K and O-K edges taken across the fibre/matrix interface region

about 115 eV, which can be attributed to Si-O-Si bonds [11]. Moreover, the onset of the feature, which for SiO_2 originally peaked at about 108 eV, is slightly shifted to that of the Si-L ELNES of the fibre (line 40). Furthermore, a damping of the amplitudes of the peaks at 108 and 115 eV, with respect to the spectrum of the glass, and a peak broadening also occur, both known from silicon oxycarbide phases.

Figure 4 additionally presents a series of EEL spectra over the energy loss range from about 200 to 700 eV to demonstrate the behaviour of the C-K and O-K edges while crossing the same fibre/matrix interface region as in the line profile of Fig. 2. For a better view, especially inside the C-K edge, the direction is here reversed, i.e. the lowest spectrum is from the glass. The carbon-rich interlayer region is indicated by C-K edges showing an ELNES which is typical of partially graphitic-bound carbon, whereas in almost all spectra of the fibre the C-K ELNES exhibits a shape known from silicon carbide. The small carbon content in the glass region is due to contamination. The jump ratios of the O-K edges seem to be slightly higher in the outer fibre region than in the inner one. This is in agreement with the detection of silicon of more oxidic bonding in the fibre periphery (cf. Figs. 2 and 3).

Chemical-Bond Mapping of Oxidic-Bound Silicon

A straightforward idea is to apply ELNES details to image the chemical bonding of an element two-dimensionally. First successful attempts have been made in this field by Muller et al. [12] using the C-K ELNES to image the amount of sp^2 and sp^3 hybridized carbon in a layer system of silicon, amorphous carbon and diamond.

In the following results are presented of selectively imaging silicon bound to oxygen in the interfacial zone of the fibre-reinforced glass described above. Details of the Si-L_{23} ELNES in Figs. 2 and 3 revealed that silicon shows a more oxidic bonding in the outer fibre periphery rather than in the inner regions (cf. particularly the EEL spectra of lines 18 and 40 in Fig. 3). Thus, it was tried to image the oxidic-bound silicon by differences in the ELNES of oxide and carbide, respectively. This is only briefly discussed here, but more systematic studies are under way, also concerning methodical problems associated with the different techniques of energy-filtered imaging (two- and three-window technique, ratio map) [13].

In the experiment, an interface zone of nearly homogeneous specimen thickness was investigated, proven by the intensity ratios of the first plasmon and the zero-loss peaks. Figure 5 shows EEL spectra recorded from SiC(Nicalon) and Duran glass regions illustrating the application of energy filtering windows to the Si-L_{23} edge. Energy-filtered STEM was carried out by using a slit width of about 3 eV at 95 eV (background B), at about 102 eV (dominant peak in Si-L ELNES of SiC), and at about 108 eV (first distinct peak in the Si-L ELNES of SiO_2).

The energy-filtered STEM image of Fig. 6a is obtained by subtraction of the partial images recorded at 107 eV and 102 eV, i.e. the net Si signal shown here is the difference between Si-L (107 eV) and Si-L (102 eV). It is assumed that this net Si

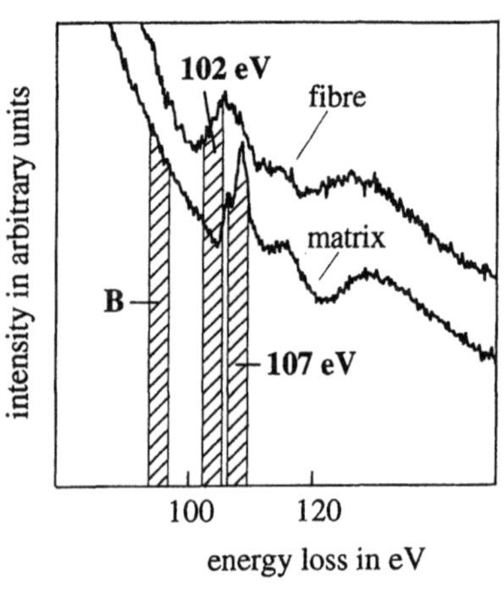

Fig. 5. Energy windows chosen for chemical-bond mapping using details of the Si-L_{23} ELNES

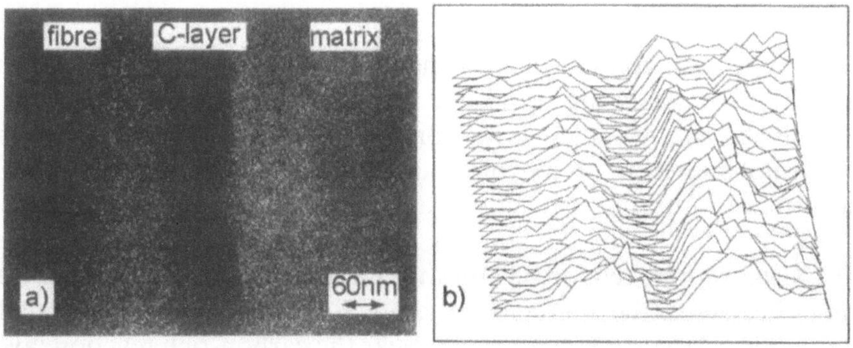

Fig. 6. Map of oxidic-bound silicon obtained by energy-filtered STEM imaging, net Si signal as **a** energy-filtered STEM image, and **b** pseudo-3d plot

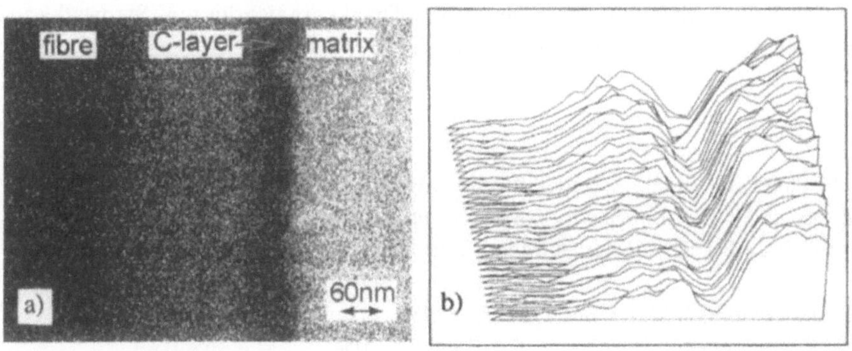

Fig. 7. Distribution of oxygen in the fibre/matrix region, **a** energy-filtered STEM image, and **b** pseudo-3d plot

signal exhibits the silicon which is more strongly oxidic-bound because not only a background is subtracted from the Si-L (107 eV) signal but also a signal corresponding to the carbidic silicon. Therefore, the contrast is mainly due to the signal differences resulting from the chemical shift of the edge onsets and from ELNES. For a better understanding in Fig. 6b the image data are additionally presented in a pseudo-three-dimensional (3d) form. On the fibre side of the carbon reaction layer (indicated by the valley in the middle) there is a ridge of silicon of oxidic bonding. Owing to the presence of SiO_2 inside the glass the mean signal level in the matrix region (right) is higher than that in the fibre.

For comparison, the oxygen distribution in the interface region was conventionally imaged (slit width of about 10 eV), with the individual images taken at 520 eV (background) and 540 eV (O-K edge), respectively. The corresponding net O image is shown in Fig. 7a. In general, there is a good agreement between the tendencies of the O-K edge signal and the net Si-L_{23} edge signal (cf. particularly Figs. 6b and 7b). As the Si-L_{23} signal follows the oxygen one the selective imaging of the oxidic-bound silicon demonstrated above is confirmed.

Conclusions

It is shown that the chemical bonding of an element can laterally be characterized at nanometre resolution by analysing near-edge fine structures and chemical shift effects in a transmission electron microscope combined with EELS. Such investigations are of increasing interest to obtain information on, e.g. the presence of certain phases, the occurrence of chemical reactions and diffusion processes. Respective examples are presented to identify the bonding of silicon in fibre/matrix interlayer regions of a SiC(Nicalon)-fibre reinforced Duran glass. The chemical-bond state was elucidated along a line across the interlayer structure. It was also two-dimensionally imaged by energy-filtered electron microscopy. Near the carbon-rich reaction layer the outer fibre region is enriched with oxidic-bound silicon, which was detected by studying the Si-L_{23} ELNES indicating the oxidation of SiC of the fibre, followed by the C-layer formation.

Acknowledgements. The authors are grateful to Dr. W. Pannhorst, Schott Glaswerke Mainz as well as to Prof. P. Greil and Dr. E. Zschech for providing the samples. Thanks are also due to F. Syrowatka and J. Neumann-Zdralek for their assistance in operating the STEM VG HB 501.

References

[1] R. Brydson, H. Sauer, W. Engel, E. Zeitler, *Microsc. Microanal. Microstruct.* **1991**, *2*, 159.
[2] P. Rez, in: *Transmission Electron Energy Loss Spectrometry in Materials Science* (M. M. Disko, C. C. Ahn, B. Fultz, eds.), The Minerals, Metals and Materials Society, Warrendale, 1992, pp. 107–130.
[3] R. F. Cooper, K. Chyung, *J. Mater. Sci.* **1987**, *22*, 3148.
[4] L. A. Bonney, R. F. Cooper, *J. Am. Ceram. Soc.* **1990**, *73*, 2054.
[5] A. Hähnel, E. Pippel, J. Woltersdorf, *J. Microsc.* **1995**, *177*, 264.
[6] R. Schneider, J. Woltersdorf, *Surf. Interf. Anal.* **1994**, *22*, 263.
[7] A. Hähnel, *PhD Thesis*, Bergakademie Freiberg (TU), 1995.
[8] W. Pannhorst, M. Spallek, R. Brückner, H. Hegeler, C. Reich, G. Grathwohl, B. Meier, D. Spelmann, *Ceram. Eng. Sci. Proc.* **1990**, *11*, 947.
[9] C. Laffon, A. M. Flank, P. Lagarde, M. Laridjani, R. Hagege, P. Orly, J. Cotteret, J. Dixmier, J. L. Miquel, H. Hommel, A. P. Legrand, *J. Mater. Sci.* **1989**, *24*, 1503.
[10] C. Colliex, M. Tencé, E. Lefèvre, C. Mory, H. Gu, D. Bouchet, C. Jeanguillaume, *Mikrochim. Acta* **1994**, *114/115*, 71.
[11] P. E. Batson, K. L. Kavanagh, C. Y. Wong, J. M. Woodall, *Ultramicroscopy* **1987**, *22*, 89.
[12] D. A. Muller, Y. Tzou, R. Raj, J. Silcox, *Nature* **1993**, *366*, 725.
[13] R. Schneider, J. Neumann-Zdralek, J. Woltersdorf, A. Röder, in preparation.

Mikrochim. Acta [Suppl.] 13, 553–558 (1996)

Local Determination of Carbon by Combining Beta-Autoradiography and Electron Microprobe Analysis

Valery G. Senin* and Svetlana N. Shilobreeva

Vernadsky Institute of Geochemistry and Analytical Chemistry, Russian Academy of Sciences, Kosygin Street 19, 117975 Moscow, Russia

Abstract. The methods of β-autoradiography and electron microprobe analysis were used for the determination of low carbon concentrations in synthetic silicate glasses produced at high pressure and temperature. Before the experiments the β-particle emitting radioactive isotope ^{14}C was introduced into the samples. A reference sample of basaltic composition containing 19 µg/g of ^{14}C was prepared. P-type Russian nuclear emulsion plates (10–20 µm thickness) were used. Electron microprobe was employed for measurement of the Ag Lα line intensity. Silver is generated as a result of exposure of the nuclear emulsion to ^{14}C-decay β-particles and subsequent development. The detection limit of this technique is 0.2 µg/g for a scanned area of $50 \times 50\ \mu m^2$ and 3 µg/g for a scanned area of $5 \times 5\ \mu m^2$.

Key words: electron microprobe analysis, β-autoradiography, carbon, silicates.

The precise determination of trace carbon concentrations in silicates at the microscopic level is of considerable interest in various scientific applications. In experimental geochemistry and mineralogy, the major goal of such determinations is to understand the mechanism of carbon entry in minerals and transport diffusion involved in magmatic processes. The samples to be analysed usually have a small weight (less than 100 mg), a complicated phase and chemical composition and small size of phases (Fig. 1). Determination of carbon in these samples may be carried out using local techniques. However the determination by electron microprobe of carbon presents numerous problems resulting from the low cross-sections for ionization, high absorption of the low-energy X-rays, spectral interferences by higher order X-ray lines of heavier elements, and organic pollution.

The technique of autoradiography has been used extensively to provide a map of the distribution of β-particles emitting radioactive isotopes of different elements. Autoradiography ensures cumulative registration of a signal, intensification of it and

* To whom correspondence should be addressed

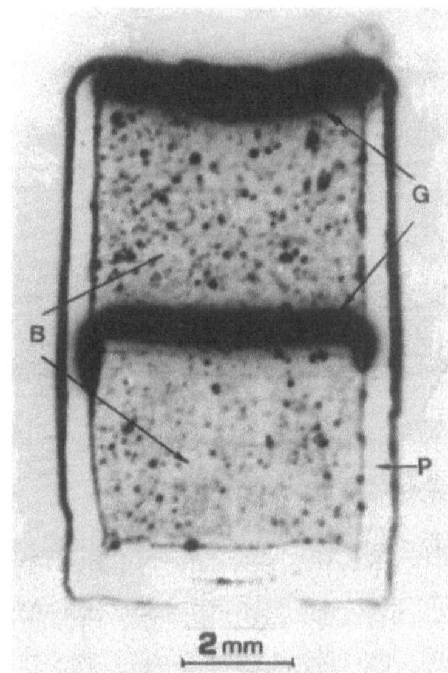

Fig. 1. Beta-track map resulting from exposure of a nuclear emulsion plate to a sample from an experiment in the system: graphite – basaltic melt, 15 kbar, 1350 °C, platinum container (*P*). The top and middle area (*G*) that are totally blackened by beta tracks represent the graphite sources. *B* basaltic melt. The dark spots within the basaltic melt are graphite particles

therefore has been applied in determining low concentrations of elements [1–7]. The technique involves the following steps: 1) introduction of the radioactive isotope (for carbon: ^{14}C) into the sample; 2) distribution of the nuclide between phases in the sample during the experimental process; 3) preparation of a planar surface on the sample; 4) exposure of a nuclear photographic emulsion to the planar sample; 5) development of the emulsion; 6) determination of the density of β-particle tracks in the developed emulsion. The techniques for analytical treatment of autoradiographs include visual counting of silver grains, visual β-track counting, photometric measurements in reflected light, optical density measurements, image analysis by TV scanning systems, grain counting with a scanning electron microscope [8] and measurement of the silver concentration in developed emulsion using an electron X-ray microprobe [9–12]. The latter technique relies on the measurement of the intensity of the characteristic $L\alpha$ X-ray line of silver generated by bombarding the developed emuslion with electrons. This intensity is proportional to the number of β-tracks per unit area emulsion. As the emulsions are disperse systems (silver grains within the volatile organic constituent), the measurements must be carried out with a wide or a scanning beam.

In the present work we have studied the possibilities of a combined technique "Autoradiography – electron microprobe analysis" for the determination of low carbon concentration in silicates.

Experimental

Preparation of Reference Sample

A reference sample of basaltic composition containing 19 µg/g of ^{14}C has been prepared at a pressure of 3 kbar and a temperature of 1250 °C from basalt powder by introduction of ^{14}C in the form of

$Ca^{14}CO_3$. The composition of the basalt is (in wt. %): SiO_2-53.80; CaO-7.80; Al_2O_3-16.20; MgO-5.60; Fe_2O_3-4.18; FeO-5.69; Na_2O-3.50; TiO_2-1.04; K_2O-1.08; H_2O-0.14.

The distribution of carbon in the reference sample was checked for homogeneity using the criterion given by Boyd et al. [13] whereby the sample is considered to be homogeneous if the ratio (homogeneity index) of observed standard deviation of the counts to the standard deviation predicted from counting statistics alone does not exceed 3. These measurements were made using emulsions exposed to the reference sample. Calculated Boyd's index lies within the limit of 3 and we suppose the homogeneous distribution of carbon in the reference sample.

Instrumentation

All nuclear emulsions plates used in this study were of the P type (Russia) with a 10–15 μm thick emulsion layer. The average size of the silver crystals is 0.28 μm in the emulsion and the average size of the developed crystals is 0.50 μm. The content of the silver halide is 82 wt. %. The emulsions were developed using Kodak D-19 developer.

The analytical treatment of autoradiographs was performed on a Camebax MICROBEAM[R] electron microprobe equipped with WD X-ray spectrometers. All exposed emulsions were coated with a carbon layer prior to analysis.

Results and Discussion

Accelerating Voltage of the Electrons

The best sensitivity of microanalysis is achieved at an electron accelerating voltage at which the product $I \cdot K$ has a maximum value [14], where I is the intensity of the analytical line and K is the contrast of this line. We have chosen an accelerating voltage of 15 kV because at this condition the value $I \cdot K$ for the Ag Lα line is maximum. The results are given in Table 1.

Background Correction

The X-ray background for this technique has three sources in the region of the Ag Lα line: 1) silver of the autoradiography image; 2) silver of the emulsion's veil; 3) gelatine of the emulsion. The first component of the background increases with increasing atomic number of the sample which is defined by the content of silver in the

Table 1. Values of $I \cdot K^a$ for Ag Lα line as a function of accelerating voltage of electrons

Accelerating voltage (kV)	$I \cdot K$ (counts per second)
10	2500
15	3700
20	2100

[a] I – the intensity of the analytical line and K – the contrast of this line.

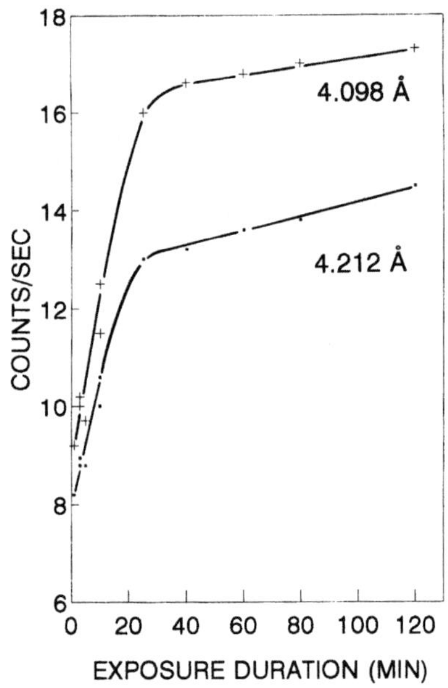

Fig. 2

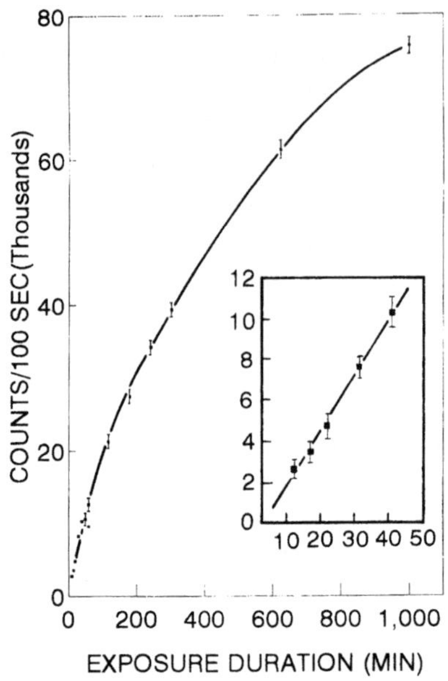

Fig. 3

Fig. 2. Background from the short-wavelength (4.098 Å) and long-wavelength (4.212 Å) sides of the Ag Lα maximum (4.155 Å) as a function of exposure duration

Fig. 3. Ag Lα counting rate as a function of exposure duration. Inset shows the initial range of this plot

developed emulsion. The dependencies of the intensity of the background on the duration of the exposure for the developed emulsion from the short-wavelength and long-wavelength sides of the Ag Lα maximum are presented in Fig. 2. Extrapolation of these dependencies to zero exposure presents the background component value due to the gelatine only. The second (viel) component of the background must be determined from measurements at the Ag Lα peak position on areas of emulsion which were not exposed directly to the ^{14}C containing phases.

Calibration Technique

The relationship between the measured Ag Lα counting rate and exposure duration was examined using the reference sample. An emulsion was exposed to the reference sample for periods ranging from 10 to 1000 min. Results for background-corrected counts are presented in Fig. 3. The measurements were made under the following conditions: 15 kV accelerating voltage; 10 nA beam current; $50 \times 50\,\mu m^2$ scanned area; 100 s counting times. It can be seen that the measured counting rates are proportional to exposure duration up to 40 min.

Determination of Carbon

If the measurements are carried out in the linear region of Fig. 3, the concentration C of carbon in samples can be calculated from [1]:

$$C = i \cdot C_R \cdot k \cdot (t_R/t_U) \cdot (N_U/N_R),$$

Table 2. Values of background standard deviation s_B as a function of scanned area of emulsion

Scanned area (μm^2)	s_B (counts per 100 s)
50 × 50	54
25 × 25	71
13 × 13	97
5 × 5	385

where i is the ratio of the concentration of stable carbon isotopes to the concentration of ^{14}C in the initial material; C_R is the concentration of ^{14}C in the reference sample; k is a correction factor taking into account the difference in the absorption of β-particles for the reference sample and the unknown sample (we used a value $k = d_U/d_R$, where d_U and d_R are the densities of the unknown sample and the reference sample respectively); t_R and t_U are the durations of exposure of the reference sample and unknown sample respectively; N_R and N_U are the measured numbers of pulses for the reference sample and the unknown sample respectively.

Limit of Detection

The limit of detection C_L of carbon by this technique in basalt samples may be calculated according to [15]:

$$C_L = [C_R/(N_R - N_{RB})] \cdot \sqrt{2} \cdot t \cdot s_B/\sqrt{n},$$

where N_{RB} is the number of counted background pulses for the reference sample; t is the Student factor; s_B is the standard deviation of the background measurement; n is the number of measurements. The value s_B increases when decreasing the scanned area on the emulsion (Table 2). For $n = 5$ and a scanned area of $5 \times 5\,\mu m^2$ the calculated value of $C_L = 3\,\mu g/g$, and for a scanned area of $50 \times 50\,\mu m^2$ $C_L = 0.2\,\mu g/g$.

Accuracy

In this work the accuracy of this technique was estimated on an example of investigation of carbon dioxide's solubility in a basaltic melt. Experiments were carried out at a pressure of 3 kbar and a temperature of 1250 °C. The level of saturation is 0.375 wt. % at these parameters [16]. A mixture of magnesite $MgCO_3$ and viterite $Ba^{14}CO_3$ was used as a source of CO_2. The saturation was carried out by diffusion technique. Run durations were 5 hours. Introduced concentrations of CO_2 were: 0.200; 0.320; 0.397 and 0.500 wt. %. Measured and introduced concentrations are in good agreement (Fig. 4).

Applications

The above described method has been used for studies of carbon behaviour in the systems: graphite – basaltic melt, 15–25 kbar, 1400–1800 °C; graphite – basaltic

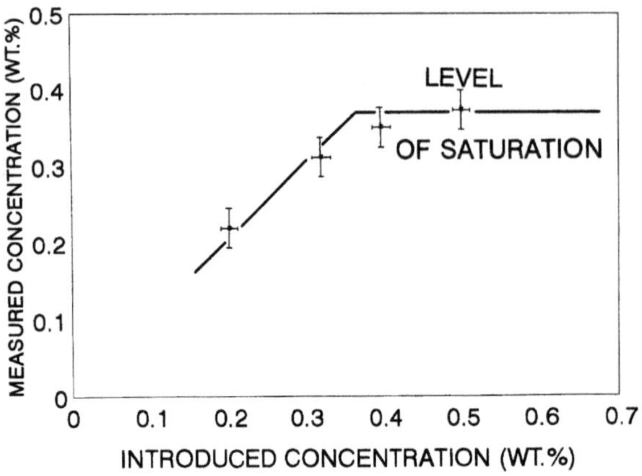

Fig. 4. Comparison of introduced CO_2 concentrations with measured concentrations

melt – crystals of pyroxenes, 30 kbar, 1400 °C; graphite – CO—CO_2 – basaltic melt, 1–5 kbar, 1250–1270 °C; graphite – forsterite, 25–50 kbar, 1700–1800 °C; graphite – CO—CO_2 – forsterite-diopside melt – forsterite, 30 kbar, 1625 °C. The measured carbon concentrations ranged from 0.007 wt. % to 0.770 wt. %.

Conclusion

Combined beta-autoradiography – electron microprobe analysis shows considerable potential as a technique for local determination of low carbon concentration in heterogeneous samples.

Acknowledgement. The authors thank Prof. Arnold Kadik for supplying the experimental samples and for helpful discussions.

References

[1] B. O. Mysen, M. G. Seitz, *J. Geophys. Res.* **1975**, *80*, 2627.
[2] A. W. Hofmann, *Carnegie Inst. Wash. Yearbook* **1975**, *74*, 183.
[3] B. O. Mysen, *Carnegie Inst. Wash. Yearbook* **1976**, *75*, 656.
[4] B. O. Mysen, D. H. Eggler, M. G. Seitz, J. R. Holloway, *Am. J. Sci.* **1976**, *276*, 455.
[5] B. O. Mysen, *Carnegie Inst. Wash. Yearbook* **1977**, *76*, 588.
[6] B. O. Mysen, *Carnegie Inst. Wash. Yearbook* **1977**, *76*, 557.
[7] A. W. Rogers, *Techniques of Autoradiography, 2nd Ed.* Elsevier, Amsterdam, 1973, p. 372.
[8] J. R. Holloway, M. J. Drake, *Geochim. Cosmochim. Acta* **1977**, *41*, 1395.
[9] T. M. Benjamin, N. J. Arndt, J. R. Holloway, *Carnegie Inst. Wash. Yearbook* **1977**, *76*, 658.
[10] J. H. Jones, D. S. Burnett, *Nucl. Instr. Methods* **1981**, *180*, 625.
[11] T. N. Tingle, *Geochim. Cosmochim. Acta* **1987**, *51*, 2479.
[12] V. G. Senin, S. N. Shilobreeva, A. A. Kadik, L. F. Pavlova, *Nukleonika* **1991**, *36 (1–12)*, 45.
[13] F. R. Boyd, L. W. Finger, F. Chaines, *Carnegie Inst. Wash. Yearbook* **1968**, *67*, 210.
[14] T. O. Ziebold, *Anal. Chem.* **1967**, *39*, 858.
[15] R. Theisen, *Quantitative Electron Microprobe Analysis*, Springer, Berlin Heidelberg New York, 1965, p. 158.
[16] S. N. Shilobreeva, *Ph. D. Thesis*, Vernadsky Institute, Moscow, 1984, p. 209.

Mikrochim. Acta [Suppl.] 13, 559–572 (1996)

The Check of the Elastic Scattering Model in Monte-Carlo Simulation

Vladimir Stary

Department of Material Science, Czech Technical University, Karlovo nam. 13, 121 35 Prague, Czech Republic

Abstract. The coefficient of electron backscattering, the angular and energy distribution of transmitted electrons and of backscattered electrons have been calculated by the Monte-Carlo method using several models of electron elastic scattering. The calculations were carried out for C, Al, Cu, Au, in the 10–100 keV energy range and in the 20–1000 nm film thickness range. Our findings were compared with published experimental results.

The cross-section of elastic scattering was calculated either by Rutherford's formula (the screened atomic potential was assumed) or by Mott's theory including electron spin, or was taken from data tables computed from Hartree-Fock atomic wave functions. This provided an opportunity to decide which model of electron-atom elastic interaction is the best for simulation.

The results showed that the backscattering coefficient, and mainly its dependence on film thickness and electron energy, is a quantity which is strongly sensitive to the model type. The best agreement between calculated and experimental data was reached, in the full range of primary electron energies and film thicknesses, for Au and Al by using tables with rigorous partial wave calculations and for Cu by using Rutherford's formula, even though the differences among results for Cu are relatively small.

Key words: Monte-Carlo simulation, electron scattering by matter, backscattering coefficient of thin films, electron probe, X-ray microanalysis.

A Monte-Carlo method (MC) is currently used in electron probe X-ray microanalysis (EPMA) for calculation of electron scattering [1, 2] and for calculation of the intensity of X-ray emission [3–5]; an extensive review is given in [6]. In our recent paper [7] we modified our model of Monte-Carlo calculation for the multicomponent target like [8]. The dependences of the peak-to-background ratio of detected X-rays (P/B) on electron energy and on the thickness of a sample of known composition were calculated and compared with measured ratios. The measurements were carried out using an electron microscope on thin resin standard for biological microanalysis. Hall correction procedure was applied to the measured values. Simultaneously, the source distribution of emitted X-ray photons was calculated in order to make a simple estimate of the interaction volume diameter. Finally, the accuracy of the computational model for biological microanalysis was tested.

MC makes it possible to calculate and estimate the non- or hardly measurable parameters influencing the results of analyses (e.g., the size of interaction volume, the X-ray lateral and depth distribution, etc). Nevertheless, the validity of the model and the accuracy of calculations of some parameters, at least those easily measurable (e.g., the backscattering coefficient), should be tested by comparison with experiments. The algorithm of MC is well known, widely used and accepted without doubts; however, several models for elastic and also inelastic scattering are usually applied. Sometimes their effect on the reliability of calculated results and on their agreement with experiments is not fully checked, even though it is clear that at least some results of MC will depend not only on a suitable algorithm, but also very strongly on the models used (and thus on the accuracy of this models). This can be caused by the fact that one of the most important criteria to apply to the model in question concerns speed of calculation. Thus serious checking is necessary.

The backscattering coefficient and its dependence on film thickness, the angular and energy distributions of backscattered electrons and also the transmission of electrons through thin films was used for checking the MC model in [9]. As was mentioned there, the backscattering coefficient alone is not very sensitive to the model type. The calculation of the cross-section according to Mott's theory of scattering agreed very closely with Seidel's experiments (according to [10]) in the $10 \div 100$ keV energy range for Al and Au. Apart from true backscattered electrons, fast secondary electrons were also used to calculate the backscattering coefficient in [9].

The main purpose of this work was to compare calculated and experimentally measured values of some suitable parameters characterizing the interaction of an electron beam with matter, without taking into account the speed of calculation. In the literature, there is much experimental data on the interaction between electrons and a single element target and even with a target in the form of a self-supporting thin film, for the medium energy range ($10 \div 100$ keV). Most of the data covers the electron energy and film thickness dependences of the backscattering coefficient. From the large number of available experimental works we chose some for our comparison (Seidel [10], Niedrig and Sieber [11], Neubert and Rogaschewski [12] and Cosslett and Thomas [13])). We also used the unpublished work of Dolezel [14].

Electron backscattering from thin films was reviewed by Niedrig [15]. From comparison between theory and experiments he suggested that the coefficient is composed of two components: the first due to direct (Rutherford) large angle scattering ($>90°$) and the second due to electrons diffused to the escape surface after a great number of collisions. This supposition can also be directly compared with the results of MC.

Moreover, there are some experimental data on the dependences of the angular and energy distribution of backscattered and transmitted electrons on the same parameters. These were measured by Kanter [16] (as cited in [9]) and also by Cosslett and Thomas [13]. Measurements of the dependence of the polar diagram of transmitted and backscattered electrons on the material and on the angle of incidence of primary electrons are mentioned in [15]. Unfortunately, the use of these results with our calculations raises some complications.

Models Used for MC-Elastic Scattering

In the first model, Rutherford's formula, corrected for screened potential, was used for calculation of the differential cross-section of elastic scattering. As the differential cross-section diverges for $\vartheta \to 0$, it is necessary to suppose a steeper decrease of potential V at the boundary of an atom. The screening of an atomic nucleus by an electron cloud was taken into account by Wentzel [17]. Thus, the atomic potential V is given by the formula

$$V(r) = \frac{e^2 Z}{4\pi\varepsilon_0 r} \exp\left(-\frac{r}{R}\right), \tag{1}$$

where e is the electron charge, ε_0 is the dielectric constant, Z is the atomic number, r is the distance from the center of an atom and $R = a_H Z^{-1/3}$ is the atomic radius (a_H is the Bohr radius of a hydrogen atom).

The differential cross-section $\frac{d\sigma}{d\Omega}$ is calculated according to formula [18]

$$\frac{d\sigma}{d\Omega} = \frac{4Z^2}{a_H^2} \frac{(1 + E/E_0)^2}{\left[\left(\frac{4\pi}{\lambda_e}\sin\frac{\vartheta}{2}\right)^2 + \frac{1}{R^2}\right]^2}, \tag{2}$$

where λ_e is the wavelength of the electron ($\lambda_e = h/\sqrt{2m_0 E}$, m_0 is the electron mass, E is its energy), E_0 is the rest energy $E_0 = m_0 c^2 = 511\,\mathrm{keV}$. The total cross section is calculated by integration over full solid angle Ω; there is $d\Omega = \sin\vartheta\, d\vartheta\, d\varphi$; ϑ is a polar angle ($0 < \vartheta < \pi$), φ is an azimuthal angle ($0 < \varphi < 2\pi$). For the total cross-section of elastic collisions σ_{tel} we have

$$\sigma_{\mathrm{tel}} = \int_\Omega \frac{d\sigma}{d\Omega} d\Omega = 2\pi \int_0^\pi \frac{d\sigma}{d\Omega} \sin\vartheta\, d\vartheta = \frac{16\pi Z^{4/3}}{\left(\frac{4\pi}{\lambda_e}\right)^2 + \frac{1}{R^2}} \left(1 + \frac{E}{E_0}\right)^2. \tag{3}$$

In the case of nonrelativistic run, the term $(1 + E/E_0)$ is taken as unity. In the relativistic case, the electron wavelength is corrected to be

$$\lambda_e = h/\sqrt{2m_0 E(1 + E/2E_0)}.$$

In the second model, Mott's differential cross-section of elastic scattering was taken from [19], where the ratio of Mott's and Rutherford's cross-section was tabulated. On the basis of these values and the values calculated by the unscreened Rutherford formula with relativistic correction, the values of Mott's cross-section can be deduced up to energy 100 keV in the angle range 10° to 180°. For lower angles, cross-sections by means of *WKB* approximation were tabulated and used for calculation. Unfortunately, data is available only for some elements.

The third model uses the electron-atom elastic cross-sections tabulated in [20] for the whole angle range ($0 \div 90°$) and for the 1–256 keV energy range. This reference gives us real values of the cross-sections for some elements and the tables of their polynomial data fit for all the elements, calculated using Hartree-Fock atomic wave functions. Thus, the differential cross-section is calculated from the simple formula according to this polynomial fit as

$$\frac{d\sigma}{d\Omega} = K \sum_{m=1}^{4} A_m(1 - \cos\vartheta + 2B)^{-m} + \sum_{m=1}^{6} C_n P_n(\cos\vartheta), \tag{4}$$

where ϑ is the angle of scattering, the coefficients A_m, C_n and B are tabulated, P_n are Legendre polynomials and K is the universal constant given as $K = (Z\alpha^2/\gamma\beta^2)^2$, where α is the fine structure constant 1/137, $\gamma = 1 + E/511$ (E in keV) the dimensionless total energy, and $\beta = 1 - \gamma^{-2}$.

The probability density of scattering into angle ϑ is

$$p(\vartheta) = \frac{2\pi \dfrac{d\sigma}{d\Omega} \sin\vartheta}{2\pi \displaystyle\int_0^\pi \frac{d\sigma}{d\Omega} \sin\vartheta \, d\vartheta}, \tag{5}$$

thus, in our MC, the deflection angle ϑ is given by

$$R = \int_0^\vartheta p(\vartheta')\, d\vartheta', \tag{6}$$

where R is the random number.

Models Used for MC-Inelastic Scattering

The modified Bethe model of electrons continuous slowing down [1, 21, 22] has often been used for evaluation of inelastic collisions. We used a similar model. Instead of differential cross-section σ_{in}, the mean free path (mfp) for inelastic scattering λ_{in} was successfully used for calculation. The relation between λ_{in} (in mass per area unit) and inelastic cross-section σ_{in} is given by the well known formula

$$\frac{1}{\lambda_{\text{in}}} = \frac{N_A \sigma_{\text{in}}}{A}, \tag{7}$$

where N_A is the Avogadro number and A the relative atomic weight.

According to [21, 22], the mfp can be calculated by means of formula

$$\lambda_{\text{in}} = \frac{W_{\text{av}}}{S} = \frac{1}{S} \int_{W_{\text{min}}}^{W_{\text{max}}} W F(W)\, dW, \tag{8}$$

where W_{av} is the average energy loss, the $F(W)$ is the normalized probability of losing energy W in an inelastic collision, W_{max} and W_{min} are the limits of electron energy loss in matter (W_{min} usually taken as adjustable parameter) and S is the stopping power which is given in the relativistic case by [23]

$$S = \frac{dE}{ds} = K_0 \frac{Z}{AE} \frac{(E + E_0)^2}{E + 2E_0} \ln \frac{1.166E}{J}, \tag{9}$$

K_0 is the constant, E is the actual energy of the electron, s is the length of trajectory in mass per unit area, J is the mean ionization potential, which is calculated according the Berger and Seltzer formula [24]. Instead of W_{av}, the mean ionization potential J from Bethe's theory is currently used in (9), for example in [25]. In our calculation, according to [21], we supposed that $F(W) = k_1 W^{-2}$ in the interval $W_{max} > W > W_{min}$. Due to normalization $k_1 = W_{max} W_{min}/(W_{max} - W_{min})$ is valid. This gives $W_{av} = k_1 \ln(W_{max}/W_{min})$. Next, we use the relation $W_{max} = E$ (the actual energy of an electron). Moreover we suppose the same shape of $F(W)$ and the same values of $W_{min} = 10\,\mathrm{eV}$ for all the elements. Because the stopping power diverges at energy decreasing to zero, the extrapolation proceeds in the program for a monotonic decrease at low energies according to [26].

By a way similar to that for the angle of deflection, the loss of energy is calculated in the case of an inelastic event, because the normalized probability density is known. It can be seen that our model of inelastic collisions is not fully equivalent to the usual so called continuous slowing down model, but can be taken more exactly as the single scattering MC (i.e. an MC calculation, taking into account each collision as individual). We suppose that the value of scattering angle ϑ is given by $\sin^2\vartheta = W/E$, where W is the actual energy loss.

Subsequent MC Program Details

Using the first model of elastic scattering, the values of integrals can be calculated directly. In the second and the third elastic models, the values of ϑ and energy are given only at defined points of angle and energy range. We tried at first to make the dependences of the cross-section on angle and energy by cubic splines interpolation, which caused the time of calculation to be enormously long. Thus, for a given random number R we calculated the upper integral of (6) by linear interpolation between tabulated values.

The minimal (cut-off) energy for stopping the history of the current electron is put at 1 keV. Below this energy, the electron is regarded as absorbed. The other details are those usually used in MC codes.

Results and Discussion

The calculation was carried out for 3000 to 30000 impinging electrons or for 300 to 1000 backscattered electrons, film thicknesses approximately 5 to 150 μg/cm^2 (i.e. for C it was 25 to 500 nm and for Al it was 19 to 550 nm) or approximately 20 to 1000 μg/cm^2 (i.e. for Cu it was 22 to 1100 nm and for Au it was 26 to 517 nm).

First, we compared the calculated angular distribution of transmitted electrons for Cu at 20 keV and thickness between 19.3 to 172 μg/cm^2 with experimental values [13]. The result for the thickest film is illustrated in Fig. 1. It can be seen that all models of elastic collisions gave very similar curves, and simultaneous results were obtained for lower thicknesses. Thus, it can scarcely be decided which model is the

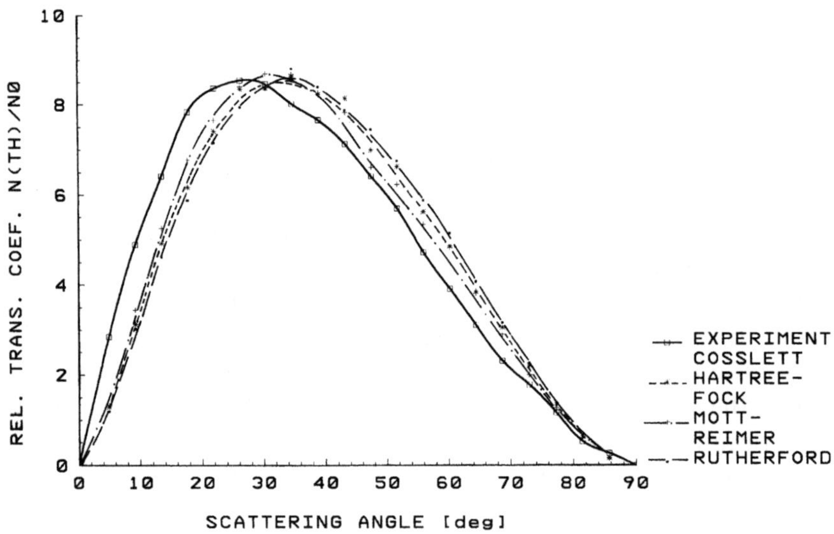

Fig. 1. Angular distribution of electrons transmitted through a thin Cu film, at electron energy 20 keV, thin film thickness 192.2 nm, i.e. 172 μg/cm^2. Full line – experiment [13], dashed lines – MC simulations

best. The most probable scattering angle (irrespective of the model) is slightly shifted from the experimental curve, and the shift is similar for all thicknesses.

Figures 2 and 3 document the comparison between the calculated backscattering coefficient and the experiments of Niedrig and Sieber [11]. We can deduce the excellent agreement for Au at 50 keV (Fig. 2a) using Mott's theory as applied by Reimer [19]. At 20 keV the agreement is only reasonable (Fig. 2b). For the other elements no such good agreement was found at 50 keV (Fig. 3). Moreover the model of the best agreement can be fully changed.

It is clear that the experimental data are subject to systematic errors. In order to minimize these errors, we used the different quantity derived from the backscattering coefficient in [15]. This quantity C_N is given as

$$C_N = \eta (N_1 Z^2 d_{1/2}), \tag{10}$$

where η is the backscattering coefficient, N_1 number of atoms in a volume unit ($N_1 = N_A \rho / A$), Z atomic number and $d_{1/2}$ thickness of the film, where the magnitude of the backscattering coefficient reaches half of the bulk value. We will call this coefficient the Niedrig coefficient. It characterizes mainly the slope of the backscattering coefficient vs. film thickness dependence. In other words, the dependence of the backscattering coefficient of the thin film on the film thickness is defined by only one point in the dependence log C_N vs. log E_p, where E_p is the energy of primary electrons. Theoretically, according to [15] the effect of atomic number Z is eliminated and the points belonging to all the elements could be in one line (in the plot log C_N vs. log E_p). In this type of plot, unfortunately, the agreement of the single models can hardly be tested, so we prepared these plots individually for three elements. (Regrettably, there was not enough data available for carbon).

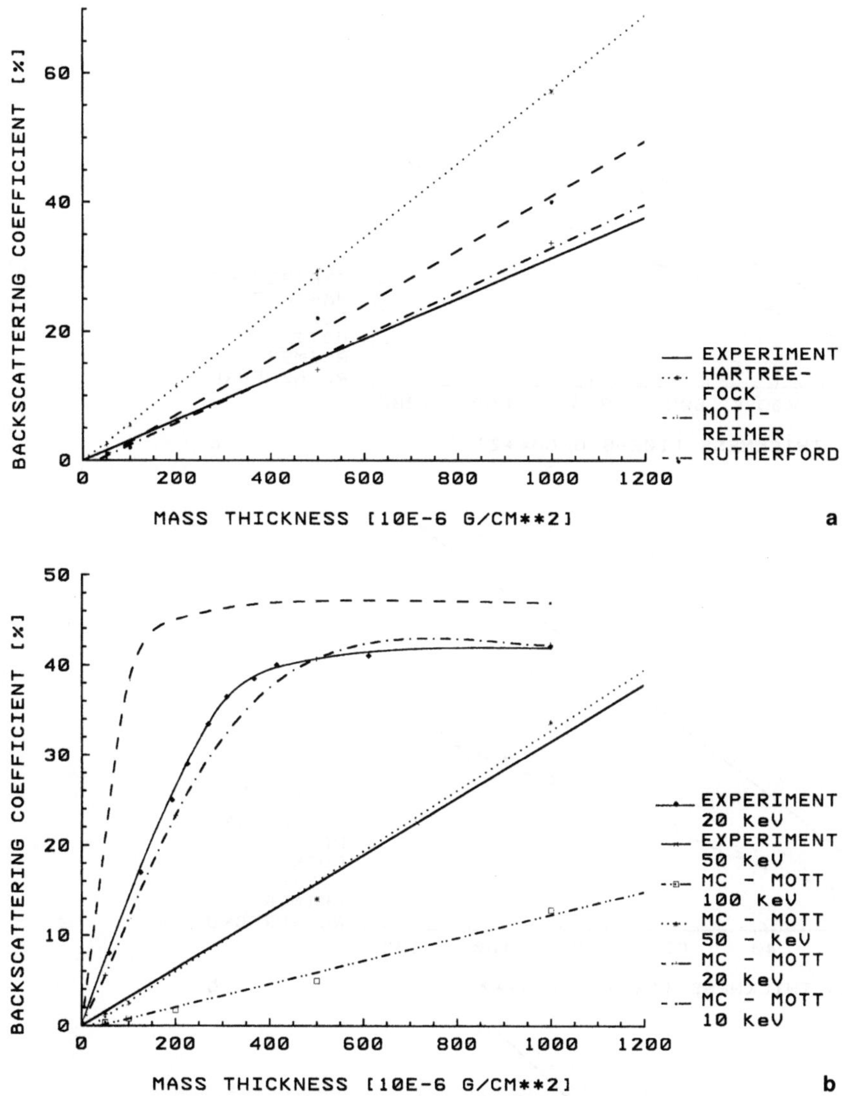

Fig. 2. **a** The dependence of the backscattering coefficient on film thickness for the various models of elastic scattering, for gold at 50 keV. Full line–experiment [11], dashed lines–MC simulations. No experimental values (points), but only line slope was taken from experiment. **b** The dependence of the backscattering coefficient on film thickness for energy of electrons at 10, 20, 50 and 100 keV, respectively, and Mott's theory. Full line–experiments [11], dashed lines–MC simulations

The results are plotted in Fig. 4 for Al, Cu, Au, where we include the experimental results [10–14] and the values calculated by all the models used (screened Rutherford, [19, 20]. Both experimental and MC values of η were recalculated to C_N. All dependences can be fitted by the following functions

$$C_N = AE_p^{-B}. \tag{11}$$

These functions are the straight lines in the log C_N vs. log E_p plot, which describes the

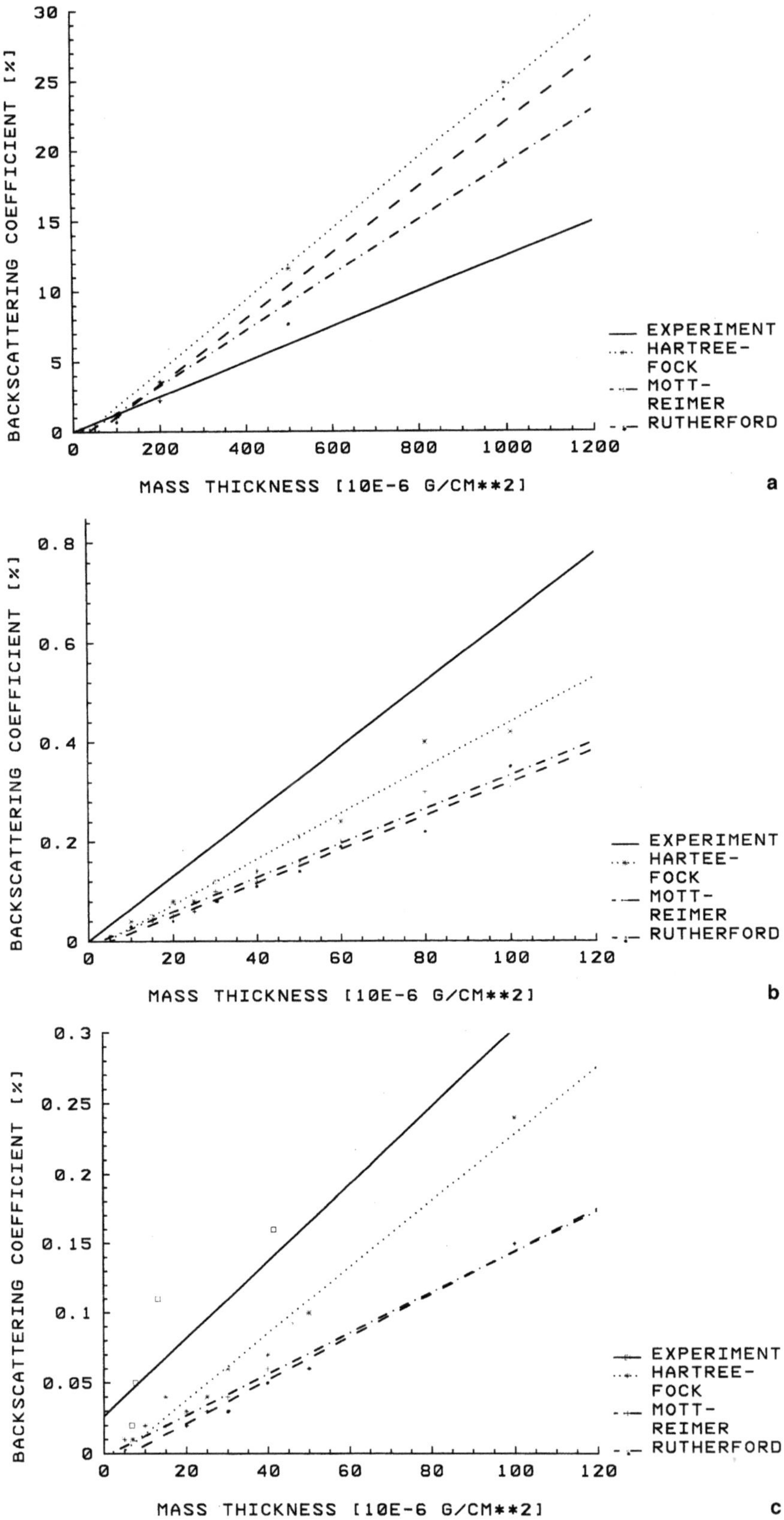

Fig. 3. The dependence of the backscattering coefficient on film thickness for energy of electrons at 50 keV, all models and **a** – Cu, **b** – Al, **c** – C. Full lines – experiments (**a**, **b** [11], **c** [14]), dashed lines – MC simulations. For **a**, **b** again only line slopes were taken from the experiment, for C the experimental values are shown – it is seen that this experiment must be taken with care

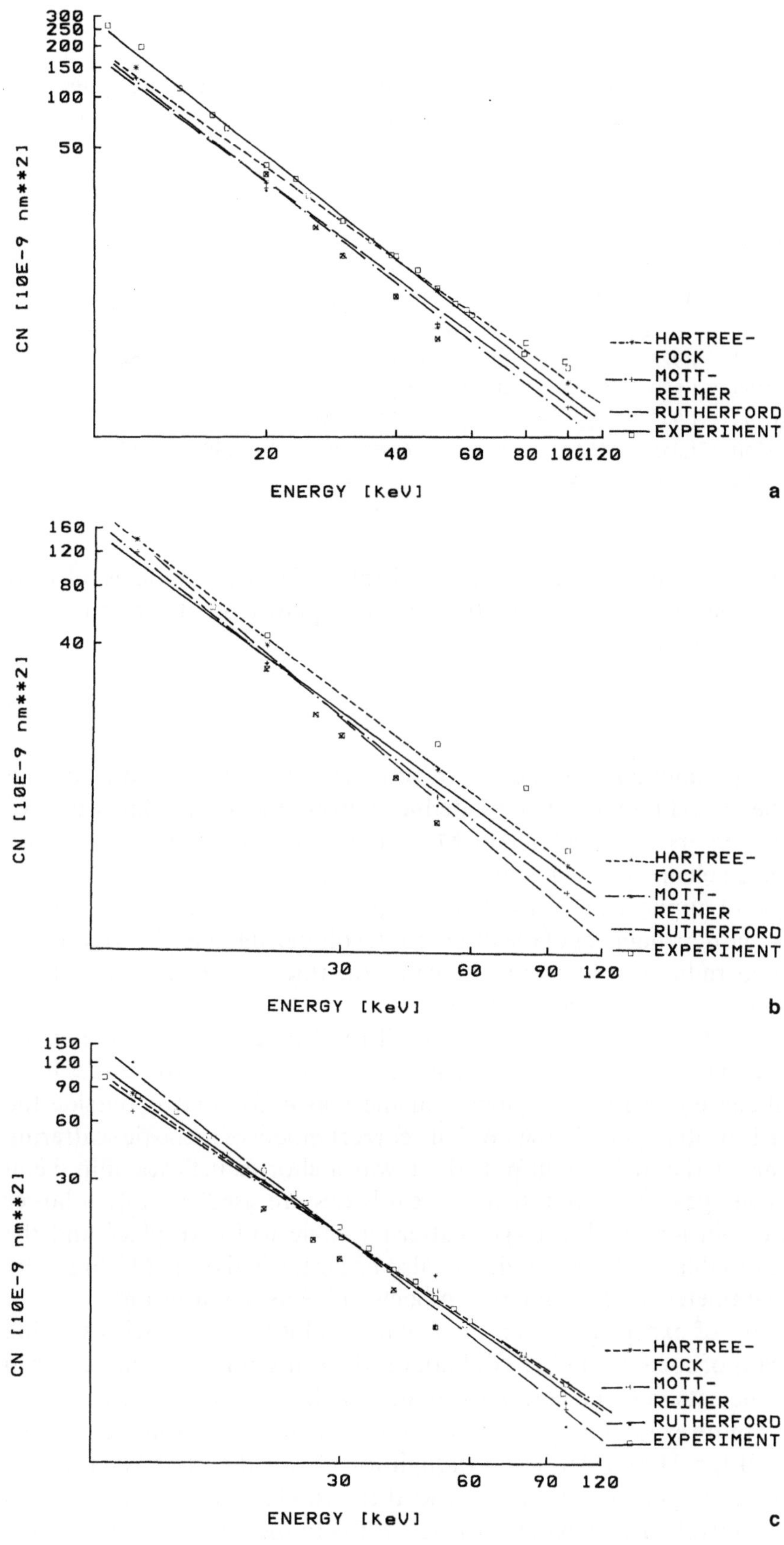

Fig. 4. The dependence of coefficient C_N on electron energy (Niedrig's plot) for **a** – Al, **b** – Cu, **c** – Au. Niedrig's measurements are denoted by crossed squares, the others by empty squares. Full lines – experiments [10–14], dashed lines – MC simulations

Table 1. The calculated fit of all the experimental and calculated values of parameter C_N by means of function $C_N = AE^{-B}$ and the average sum of residuals R_s, giving the agreement between the experimental values and those calculated by MC using the particular model of elastic scattering

Model for calculation	Elements Al			Cu			Au		
	A	B	R_S	A	B	R_S	A	B	R_S
Experiment	12913	1.922		2035	1.392		2949	1.549	
Rutherford	8629	1.880	839	10524	1.906	23	9513	1.847	254
Mott-Reimer	10598	1.947	722	6482	1.754	26	2871	1.592	30
Hartree-Fock	8860	1.816	464	6369	1.674	42	3180	1.555	12

data in the best way. The results of the fits are in Table 1. The best model is selected by comparison of the average value of the sum of squares of the residuals R_s, calculated by the formula [27]

$$R_s = \frac{1}{n}\sum_{i=1}^{n}(C_i - AE_i^{-B})^2, \tag{12}$$

where C_i are the experimental values of C_N at primary energy E_i, A and B are the coefficients of the best fit of C_N values deduced from the particular value of η according to (10), the η being calculated by MC using the particular model of elastic scattering for a given element. The results are also in Table 1.

As an example of application of the MC program for microscopy and microanalysis, we show the dependences of the interaction volume diameter D90 (containing 90% of excited radiation) for some elements on thickness (Fig. 5a) and the diameter D90 for Fe on electron energy (Fig. 5b).

For angular distribution, the agreement of all the MC curves shows that the results do not depend on the model used for simulation. The shift of MC curves from the experimental curve (see Fig. 1) signals that the reason for that is outside the influence studied. It probably lies in the not fully correct model of inelastic scattering which underestimates the inelastic mfp. If there was a shorter inelastic mfp (i.e. if there is a lower average energy loss then the real losses are used, see (8)), a larger number of elastic collisions with a large scattering angle will take place and the calculated most probable scattering angle will also be larger. This could be found by trying different parameters and/or different models of inelastic scattering.

The comparison of experimental and calculated values of the backscattering coefficient by means of coefficient $C_N(E)$ did not produce anytimes big differences in calculated fit coefficients, but the comparison among the models by means of the sum of residuals is more exact. The exact cross-section calculations tabulated in [20] provide the best fit for Al and Au. Even though for Cu all the models gave very similar results, the best agreement is given by Rutherford's formula. For Au and Cu the agreement is relatively better than for Al (according to the absolute values of the best fit R_s). This means that the MC simulation for Al as a whole is less exact. Because the cross-section of inelastic scattering is relatively larger for light elements than for

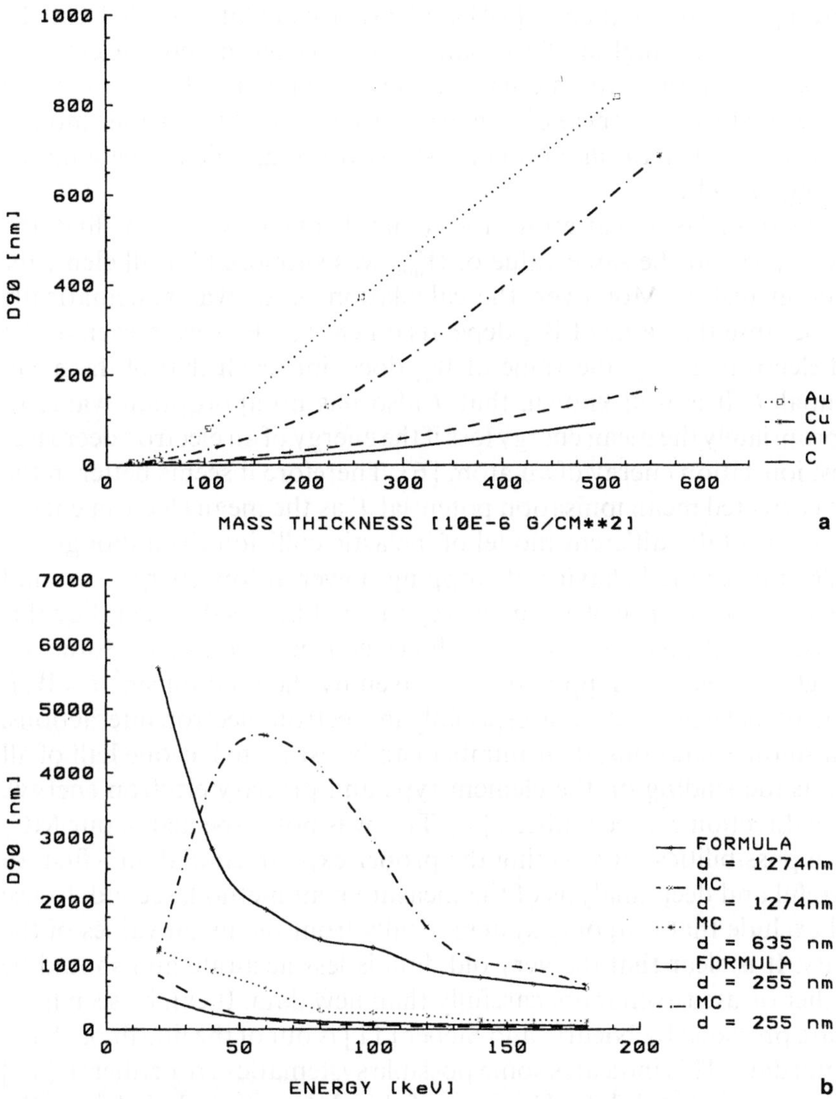

Fig. 5. **a** Dependence of the diameter of interaction volume D90 of C, Al, Cu, Au on film thickness, at 50 keV. **b** Dependence of the diameter of interaction volume D90 of Fe on electron energy for the film thickness *d*. The full lines were calculated by approximate formula [29], dashed lines – MC simulations

heavier elements, we can again probably conclude that the model of inelastic scattering is less appropriate and needs to be optimized.

Nevertheless it may be surprising that the differences of η calculated by means of the Rutherford and Mott models are not so high at primary energy 50 keV, mainly for lighter elements (C, Al, Cu – Fig. 3). It is well known that Rutherford's model crashes for heavy elements (for example Au) and low electron energies (several units of keV and less) [6, 9]. The comparison of absolute magnitudes of differential cross-sections for light elements, which we carried out, shows that the difference between the Rutherford and Mott models and exact cross-sections calculations [20] are very low. On the other hand, the differences between the values of the cross-sections calculated according to Mott's theory [19] and exact calculations [20] can be (mainly for large scattering angles) of the same order as the difference between Rutherford's cross-section and Mott's cross-section (calculated in [19]) for Au at $E_p = 50$ keV (Fig. 2a). Moreover, these differences of η calculated by various models should not depend strongly on the model of inelastic scattering, which is the same for all elastic scattering models.

As to the model of inelastic scattering, a very simple one was used. At first, the same dependence $F(W)$ and the same value of $W_{\min}$ was supposed for all elements, which is not true in reality. Moreover, the calculation of λ_{in} was systematically underestimated, because the value of W_{av} depends on energy. However, even at the highest value of electron energy, the value of W_{av} does not reach that of the mean ionization potential J. It is well known, that J also has no appropriate value to characterize appropriately the mean energy loss if the energy of an electron decreases below the highest ionisation energy of an atom [6]. Therefore it seems better in the future to use the corrected mean ionisation potential J' as the mean electron energy loss [28], or to use a totally different model of inelastic collision.Even though the extrapolation [26] makes the behavior of stopping power at low energies around 1 keV better, the real dependence of λ_{in} on energy under 1 keV is different. For this reason, we selected the cut-off energy of a traveling electron to be 1 keV. Moreover, the scattering angle ϑ is usually supposed to be given by the relation $\sin^2\vartheta = W/E$ only in some kind of inelastic scattering, especially in electron-electron interactions. In the case of plasmon excitations, their number can be estimated as one half of all the inelastic events (depending on the element type and primary electron energy), and the change of direction can be omitted [1]. This was not respected in our MC.

There are two possibilities for selecting the proper experimental data – first, to perform very careful and deep analysis of the measurement method, second, to use all the data and exclude that outgoing systematically from the mean values of the observed quantity. It is clear that the very old data is less accurate and should be excluded altogether or used even more carefully than new data. It can be seen from Fig. 4 that the data provided by Niedrig and Sieber [11] is out of the line fit of all the other experimental data. This indicates some possible systematic error either in [11] or in all the other experimental data. However, if this data was excluded from the calculation of R_s, the results of comparison of particular models did not change. Thus, our approach of using the large set of data from different authors shows that possible errors are eliminated in the comparison.

For only single scattering, the coefficient B of the fit (11) should, according to [15], be equal to 2 and the single backscattered electrons constitute at energy 20 keV about 10–12% of all the backscattered electrons. Instead of this number, in our MC, we calculated the number of all electrons with given number of collisions altogether, not only that of backscattered electrons. In spite of that, in our simulations there were usually almost no single scattered electrons escaping the film at energy 100 or 50 keV. However, their number increases if the energy of primary electrons decreases and it reaches value of about 0.1% of all backscattered electrons at 20 keV on Cu. The reason for this big difference is not clear and a more precise analysis would be desirable.

The error of calculated values of the number of backscattered electrons (and simultaneously of the number of electrons which have lost some energy and/or are scattered into some solid angle, i.e. the number N of electrons with a given property) can be estimated, if we assume a binomial distribution of the resulting numbers of electrons. In this case, the standard deviation can be estimated as $s = \sqrt{N_o \hat{p}(1-\hat{p})}$, where N_0 is the full number of electrons and $\hat{p}$ is the estimated probability of this phenomenon, $\hat{p} = N/N_0$. The upper limit of this estimation is $s_{max} = \sqrt{N_0/4}$. At high number N_0 and at low value of probability p this distribution can be approximated by Poisson's distribution and the standard deviation is about $\sqrt{N}$).

Conclusions

We compared the experimental results with those calculated by means of several theoretical models of elastic scattering. The comparison of the calculated values versus the measured values has shown that

- the angular distribution of transmitted electrons for several film thicknesses is nearly the same for all models;
- the values of the backscattering coefficient η from single element films and especialy its dependence on the film thickness depend on the model; in the energy range 10 to 100 keV and thickness range 20 to 1000 nm for Al and Au the value of cross-sections calculated by means of Hartree-Fock atomic functions give the best agreement; while for heavier elements (Au), the agreement of experiments with the best model is very good, for lighter element (Al) the agreement is worse; for Cu the difference between the models is relatively low.

Simultaneously, we calculated other physical quantities, such as the source distribution of the emitted X-ray photons. These calculations enable us to make a simple estimate of the interaction volume diameter and, in this way, to determine the spatial resolution of electron probe X-ray microanalysis in the range of electron energy, film thickness and element atomic number, and to calculate the element concentrations in some geometrically nonregular objects and particles, respectively.

It is necessary to check all the models used in simulation; the checking of inelastic scattering models by this method is therefore now under preparation.

Acknowledgements. For critical reading of this manuscript and for helpful discussions, I would like to thank Dr. K. Jurek and Dr. B. Smola, for advice in statistical evaluation I should like to thank Dr. M. Huskova and Dr. K. Zvàra, for English review R. A. Healey, M. A. is gratefully acknowledged. The work was supported by GA-CR grant No. 103/94/1708.

References

[1] R. Shimizu, Y. Kataoka, T. Ikuta, T. Koshikawa, H. Hashimoto, *J. Phys. D. Appl. Phys.* **1976**, *9*, 101.

[2] L. Reichelt, A. Engel, *Ultramicroscopy* **1984**, *13*, 279.

[3] G. Love, M. G. C. Cox, V. D. Scott, *J. Phys. D. Appl. Phys.* **1977**, *10*, 7.

[4] R. Rosa, A. Armigliato, *X-Ray Spectrom.* **1989**, *18*, 19.

[5] P. Karduck et al., *Microbeam Analysis* **1990** (J. R. Michael, P. Ingram, eds.), San Francisco Press, San Francisco, 1990, p. 21; N. Amman, P. Karduck, *Microbeam Analysis 1990* (J. R. Michael, P. Ingram, eds.), San Francisco Press, San Francisco, 1990, p. 150.

[6] R. Shimizu, Z.-J. Ding, *Rep. Progr. Phys.* **1992**, *55*, 487.

[7] V. Stary, *Microchimica Acta* **1994**, *114/115*, 463.

[8] D. F. Kyser, in: *Introduction to Analytical Electron Microscopy* (J. J. Hren, J. I. Goldstein, D. C. Joy, eds.), Plenum, New York, 1989, p. 199.

[9] L. Reimer, E. R. Krefting, in: *Use of Monte-Carlo Calculations in Electron Probe Microanalysis and Scanning Electron Microscopy* (K. F. J. Heinrich et al., eds.), NBS Special Publication 460, U.S. Dept. of Comm., Washington, DC 1976, p. 45.

[10] H. Seidel, cited in: L. Reimer, G. Pfefferkorn, *Rasterelektronenmikroskopie*, Springer, Berlin Heidelberg New York Tokyo, 1983, p. 38.

[11] H. Niedrig, P. Sieber, *Z. Angew. Phys.* **1971**, *31*, 27.

[12] G. Neubert, S. Rogaschewski, *J. Phys. D. Appl. Phys.* **1984**, *17*, 2439.

[13] V. E. Cosslett, R. N. Thomas, *Br. J. Appl. Phys.* **1965**, *16*, 779. V. E. Cosslett, R. N. Thomas, *Br. J. Appl. Phys.* **1964**, *15*, 883, 1283.

[14] T. Dolezel, *Diploma Thesis*, Charles University, Prague, 1982.

[15] H. Niedrig, *J. Appl. Phys.* **1982**, *53*, R15.

[16] H. Kanter, *Ann. Phys.* **1957**, *20*, 144.

[17] G. Wentzel, *Ann. Phys.* **1922**, *69*, 335.

[18] L. Reimer, *Transmission Electron Microscopy*, Springer, Berlin Heidelberg New York Tokyo, 1984, p. 138.

[19] L. Reimer, B. Lodding, *Scanning* **1984**, *6*, 128.

[20] M. E. Riley, C. J. MacCallum, F. Biggs, *At. Data Nucl. Data Tables* **1975**, *15*, 443.

[21] D. Liljequist, *J. Phys. D. Appl. Phys.* **1978**, *11*, 839.

[22] A. Spalek, *Nucl. Instr. Meth.* **1982**, *198*, 399.

[23] K. Siegbahn, in: *Alpha-, Beta- and Gamma-Ray Spectroscopy* (K. Siegbahn, ed.), North Holland, Amsterdam, 1965, p. 12.

[24] M. J. Berger, S. M. Seltzer *Nuclear Science Series Report No. 39, NAS-NRC Publication No. 1133*, National Academy of Science, Washington DC, 1964, p. 205.

[25] R. Shimizu, Y. Kataoka, T. Matsukawa, T. Ikuta, K. Murata, H. Hashimoto, *J. Phys. D. Appl. Phys.* **1975**, *8*, 820.

[26] T. S. Rao-Sahib, D. B. Wittry, *J. Appl. Phys.* **1974**, *45*, 5060.

[27] N. R. Draper, H. Smith, *Applied Regression Analysis*, Wiley, New York, 1966.

[28] D. C. Joy, S. Luo, *Scanning* **1989**, *11*, 176.

[29] J. I. Goldstein, in: Introduction to Analytical Electron Microscopy (J. J. Hren, J. I. Goldstein, D. C. Joy, eds.), Plenum, New York, 1989, p. 101.

Mikrochim. Acta [Suppl.] 13, 573–579 (1996)

True Colour X-Ray Vision for Electron Microscopy and Microanalysis

Peter J. Statham

Oxford Instruments, Microanalysis Group, Halifax Road, High Wycombe, Bucks HP 12 3SE, England, U.K.

Abstract. Secondary and backscattered electron signals in the SEM provide high resolution images and features of interest are selected for subsequent microanalysis based on recognisable topography and morphology. However the images are intrinsically monochrome so contrast is the only differentiator. Colour aids recognition of familiar objects and can provide warning of unusual situations so a true colour image would provide many more cues to guide the observer.

The X-ray spectrum detected by EDX can be used to construct a true colour response that would be obtained if we offset the human visual sensitivity to the electromagnetic spectrum into the X-ray wavelength region (<5 nm). This colour input can then be used to augment a conventional electron image just like the chrominance signal is used to augment the luminance to give colour television.

While this coloured image still retains the detail of the original electron signal image, it also portrays the underlying elemental composition because the spectrum from each compound gives it a characteristic colour. Furthermore, the colour does not change with beam current and is not subjective as it would be with pseudo-coloured BSED images or combinations of X-ray maps from user-defined energy windows. Changes in topography and shadowing affect the colour in the same way as visible objects so the effect is still intuitive. Thus, topographic and compositional information from all elements is compressed into a single view. This true colour "television" image gives the analyst a useful first look to guide further microanalysis.

Key words: X-ray electron microscope colour imaging.

Perhaps the most endearing feature of the scanning electron microscope (SEM) is that it produces magnified images of small objects which are instantly recognisable. Human beings have not developed any sense for electron intensity and the physics of image formation is totally different from that for objects illuminated by visual light, yet the image of an insects head on a flat SEM viewing screen is undoubtedly perceived as a real illuminated object, just as if it were broadcast on a monochrome television set. Unfortunately, this intuitive view of the sample is often lost when microanalysis is required because specimen preparation usually destroys the raw topography and limits the analyst to look at flat prepared sections of material. Once a sample has been modified in this way, the analyst is left to choose the right tools and parameters to measure the signals emanating from the sample and achieve an objective result. Instead of a magnifying glass for exploration of the unknown, the SEM becomes more of an oscilloscope for monitoring the expected.

The advent of low cost colour television display systems brought with it new opportunities for introducing colour into the displayed SEM image. Rather than attempt to reference work in this area, this author recommends a review by Saparin and Obyden [1] which contains an extremely comprehensive bibliography of most SEM colour enhancement techniques published prior to 1988. Colour coding is useful because the eye is capable of discriminating many more colour tints than grey scale intensity levels. Each level of image intensity is displayed as a specific colour, rather than as a grey tone. The choice of colour scale depends on whether the aim is to enhance contrast or to clarify where the signal level falls in certain ranges. For example, if the SEM image of a flat sample is acquired using a backscattered electron detector, then signal strength correlates with mean atomic number of the specimen so that specific colours can be used to denote areas in the sample likely to have the same elemental composition. This sort of encoding is often referred to as "false colour" because the colour is not implicit in the signal so it is artificial and added under the control of the operator. When there are other signals available which give additional information about the specimen, several images can be acquired from the same field of view. For example, these could be X-ray maps, each map corresponding to the characteristic energy range of a different element, or electron images from quadrant backscatter detectors or separate secondary electron detectors. In this case, colour encoding can be useful to expose where certain combinations of signal strengths occur. Specific methods for combining 3 or 4 X-ray maps to reveal chemical phases are demonstrated in reference [2], and there are many more examples in [1] including several plates showing applications using cathodoluminescence (CL) and electron beam induced current (EBIC) signals. In the case of CL, where the sample does emit visible light, the displayed colour can be related directly to the dominant wavelength of emission so that the image appears in "true colour", as would be seen in an optical microscope. However, in most other cases, the colour is still artificial and intended for highlighting areas which conform to required ratios and signal combinations so the operator still has to decide how best to expose what is expected to occur. Furthermore, with artificial colour enhancement, images do not seem to come from the "real world" so sample investigation is planned rather than intuitive.

A recent invention by Scharf [3] introduces colour based on the various responses from several wide energy bandwidth secondary electron detectors yet preserves the "real world" intuitive nature of images. Because of their positioning, each detector has different sensitivity to topographic and compositional information and by controlling signal mixtures and colours, the operator effectively has access to a powerful "studio" with flexible illumination for generating images which are both aesthetically pleasing and can enhance certain types of information particularly from biological specimens. Commercial systems have also been produced which give the operator a console with a variety of controls to mix and assign colour to image signals from secondary and backscatter electron detectors. Although the resulting image is intuitive, such systems are subjective as success depends on the skill of the operator in generating the desired special effects and results are unlikely to be reproducible day-to-day or from one operator to another.

There are many instances where we have a sample, such as a fracture surface or a collection of particulates, which cannot be subjected to any extreme specimen preparation techniques such as grinding and polishing. If the sample has topography and fine image detail which is particularly relevant yet we are searching for compositional variations which might explain exceptional behaviour, we need a reproducible, objective technique to display both the topography and composition which gives the operator a clue where to perform further more detailed analysis within the field of view.

To see how this can be achieved, it is necessary to consider how colour appears to us in the natural world. Objects exhibit colour because of a variety of mechanisms, but perhaps the most common is selective absorption of certain wavelengths of white light due to the molecular structure of the material being illuminated. The human eye perceives colour through the stimulation of visual pigments in the cones of the retina which behave as sensors for three different overlapping bands of wavelengths. The relative levels of excitation of these sensors determines the perceived colour. If all sensors are excited equally, the perception is that of white light but if they are imbalanced, the sensation is that of colour. Because the spectral sensitivity ranges overlap, the same colour sensation can be produced from monochromatic radiation of a particular wavelength or from a mixture of wavelengths. For example, the almost monochromatic light from a sodium arc lamp will give a similar "yellow" sensation to a mixture of 700 nm ("red") and 520 nm ("green") wavelength radiations. Thus, although a given spectral distribution will produce a single colour sensation, the perceived colour does not correspond to a unique spectral distribution of light entering the eye. Nevertheless, the overal sensation gives us an indication of the composition of the object we are observing and this helps us distinguish otherwise similar objects. For example, a green apple is distinguishable from a red tomato, the yellow of gold metal distinguishes it from the red tinge of a fresh surface of copper but oxidised copper is characteristically green. An object with pronounced topography may exhibit quite a large range of colour sensations to the eye because of the uneven illumination. However, once the brain has made sense of the shape and illumination direction, such colour variations due to topography are distinguishable from real variations in composition.

In the SEM, we use electrons rather than visible light to illuminate the sample. The detected signal is eventually converted to visible white light of varying brightness using the phosphor of a television picture tube. When we are interested in elemental composition, we know that following excitation, atoms from different elements emit radiation with wavelength characteristic of the element concerned. Therefore, to give the specimen a colour characteristic of composition, we need to introduce colour in a consistent way based on the emitted X-ray energy spectrum. Using the X-ray spectrum detected by EDX, we can construct a true colour response that would be obtained if we could offset the human visual response to the electromagnetic spectrum into the X-ray wavelength region (< 5 nm). This colour input can then be used to augment a conventional electron image just like the chrominance signal is used to augment the luminance to give colour television.

Experimental

Figure 1 shows the normal human visual response to light with wavelengths from about 400 to 700 nm. To offset the visual response into the X-ray region, an energy dispersive X-ray detector and analysis system (Link ISIS) is used to assign colour to photons based on their effective wavelengths using a rainbow colour scale. For example, with a 300 eV X-ray assigned red and a 5 keV photon assigned blue, the offset visual response appears as in the lower graph in Fig. 1. Since most elements emit a characteristic line in this energy band, any major changes in concentration for any element should cause a change in colour sensation so this is a useful default band of sensitivity. While the electron beam is resting on a point on the sample, after each photon is detected, its energy is measured and converted to a colour (r, g, b) with the condition $r + g + b = 1$ and the results added to three colour accumulators which maintain totals R, G, B. At the end of the acquisition period, R, G, B effectively represents the colour sensation to the emitted X-ray spectrum which can be reproduced by display on a colour monitor. This process is repeated for all the pixels in a rectangular grid over the field of view to build up a complete full colour 2D image in approximately 1 minute; if necessary the acquisition can be extended for further frames to improve counting statistics. The resultant colour X-ray image effectively provides a low bandwidth chrominance signal which is combined with a high resolution monochrome electron image previously acquired from the same field of view using the same beam hardware. As in conventional broadcast television, the combination is designed to maintain the same luminance as the electron image and thus retain high resolution image detail and topographic appearance. More details are provided in reference [4].

In the example shown in Fig. 2b, the bottom edge of the picture represents the energy range from 0 to 20 keV and shows the colour scale look up table for energies from 0.2 keV to 5 keV. This is a conventional "rainbow" scale with smooth transitions from red to yellow to green to violet to blue. Photons with energy outside this range are ignored. The smooth change in hue is essential to exploit the eye/brain's capability to interpret subtle colour changes in the context of local topography. The break points in the scale are not critical but have been chosen to maximise the likelihood that characteristic lines from different elements will produce a different hue. The constraint, $r + g + b = 1$, is imposed so that the statistical significance of a single photon is not distorted by the look up table and remains independent of energy.

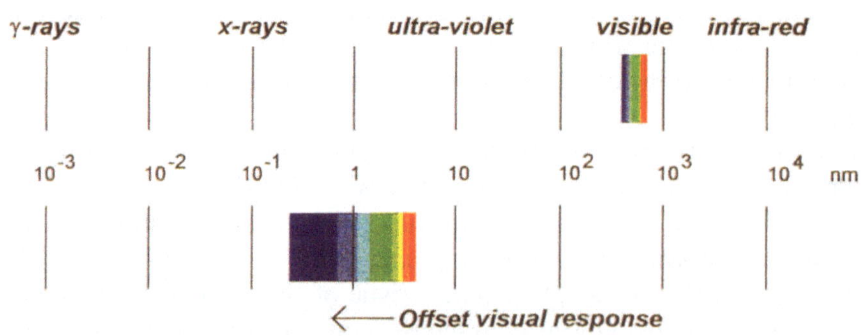

Fig. 1. Wavelengths in nm for electromagnetic spectrum showing region where human visual sensors normally operate to give the sensation of colour. Offsetting this visual response to the low energy X-ray region would allow us to see characteristic X-ray emissions in colour

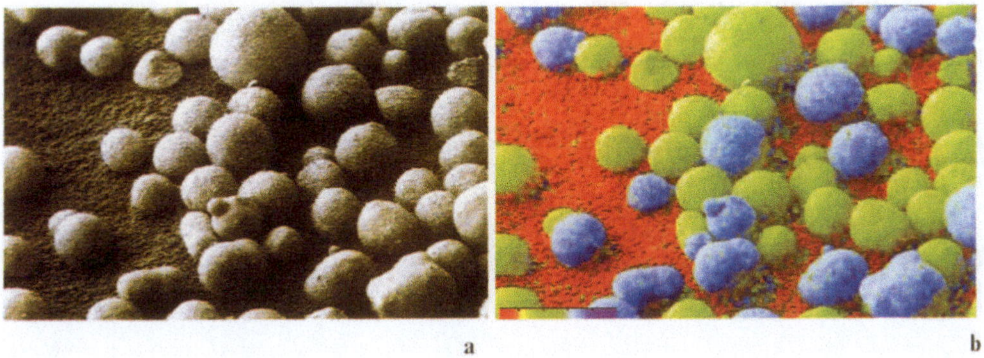

Fig. 2. SEM images of metal balls taken at 20 keV, field width approximately 1000 μm. **a** Secondary electron image 512 × 336. **b** Colour augmented image using a visual response offset to the 0.2 keV–5 keV X-ray region, acquisition time 3 mins, total spectrum count rate approx. 10 kcps. In this case, the colours are easy to understand because the W L lines fall in the green and the Ti K lines fall in the blue region of the offset visual response

Results and Discussion

In Fig. 2a, the topography of the small balls is clearly apparent but they are only differentiated based on size. In Fig. 2b the colour augmented image still maintains the topography and fine detail yet clearly differentiates the tungsten-rich from the titanium-rich balls and shows the substrate (carbon cement) is generally much lower atomic number. Note that the green ball at top centre shows some blue colouration which appears to be a reflection of the adjacent blue ball; although the physics of electron scattering would clearly show that analysis of the green ball in such areas is impractical, the presentation of X-ray information together with topographical detail gives an immediate intuitive view which prevents the analyst from choosing such a region for acquiring a representative spectrum. In Fig. 3a the metal turnings clearly show topographic detail but there is no indication of the types of material present. The colour augmented Fig. 3b maintains resolution and topography while segmenting the field into blue, yellow, pink and green particles on a reddish brown substrate. The green particle lower left of centre would be difficult to spot in the monochrome electron Fig. 3a. Furthermore, there are several particles with similar grey level in Fig. 3a which clearly have different composition as shown by Fig. 3b. By positioning the beam on representative particles, presence of the following elements can be confirmed: blue = Fe + Cr + Ni, yellow = Al, pink = Cu + Zn and green = Sn while brownish substrate is primarily carbon cement. In Fig. 4a the backscattered electron image in compositional mode indicates that there is indeed some material contrast. Near the centre of the image is a mid grey feature which looks similar in size and brightness to two other features on the right hand side. However, in the colour augmented Fig. 4b, it is clear that the orange central feature is quite different in composition from the green features on the right. Subsequent analysis of X-ray spectra acquired from representative points shows that white/blue = Ba + S + O, orange = O + Fe, green = Si, yellow = F + Zn + Al and

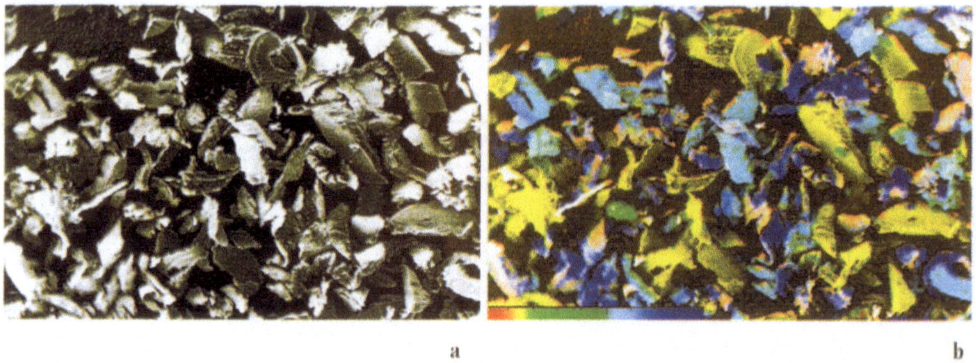

Fig. 3. SEM images of metal turnings taken at 20 kV, field width approximately 2400 μm. **a** Secondary electron image 512 × 336. **b** Colour augmented image using a visual response offset to the 0.2 keV–9 keV X-ray region, acquisition time 2 mins, total spectrum count rate approx. 10 kcps

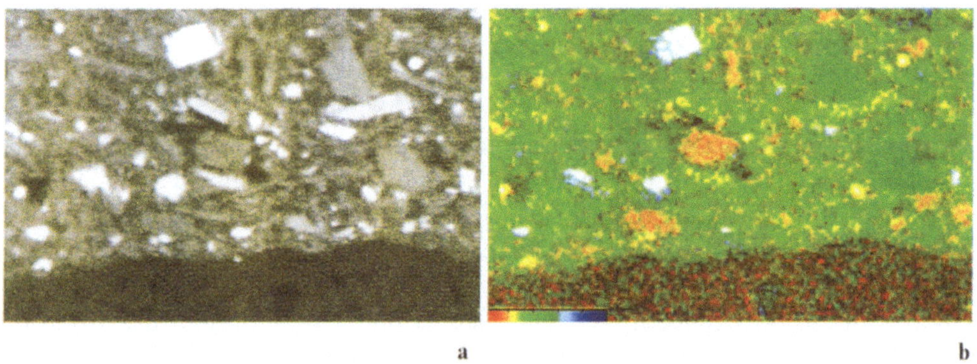

Fig. 4. SEM images of cross-section of a paint flake taken at 20 kV, field width approximately 95 μm. **a** Backscattered (all quadrants) electron image 128 × 84. **b** Colour augmented image using a visual response offset to the 0.2 keV–5 keV X-ray region, acquisition time 1 minute, total spectrum count rate approx. 10 kcps

the dark coloured band at the bottom of the image is carbon cement used to mount the cross-section.

The strength of this approach is in providing consistent visual cues for the eye/brain to interpret so that the viewer has an intuitive understanding of sample chemistry and morphology. The colour will change wherever the relative intensities of characteristic peaks and background are substantially altered. Although this may not reveal changes in trace element concentration, it will delineate phases where there are complex relationships between major components. The mapping of the visual response into an X-ray energy band is different for Figs. 2 and 3. Obviously, such "tuning" of the response will affect the colours that are seen. If colours are to be learnt so that viewers can instantly recognise colours for specific compositions, changes in the colour look up table have to be avoided wherever practical. So far, experiments have shown that the 0.2 to 5 keV scale is a useful default for a wide range of materials and this could be used as a standard if such X-ray colour images were to become as recognisable as characteristic cathodoluminescence emission colours for example. However, some laboratories dealing with a limited class of materials have found it useful to choose a different scale to amplify colour contrast; the facility to change scales allows them to establish a local standard more suited to a particular range of applications. Of course, if images from different laboratories are to be compared, it is imperative to know what visual offset has been used and this is why the colour scale is routinely imprinted along the base of each image as shown in Figs. 2, 3, and 4.

The examples demonstrate how micrographs can be augmented by colour generated by offsetting the visual response in the electromagnetic spectrum. Provided the same energy response band is used, the colour will be reproduced on any microscope under similar conditions so compounds will appear in memorable colours which can be learnt as for CL microscopy. Change in beam voltage produces a bias to the colour rather like inserting a colour filter in front of an optical image. Unlike arbitrary pseudo-colour encoding of backscattered electron signal intensity or other signal mixing techniques, the colour is independent of beam current, insensitive to local geometry and no user intervention is required to calibrate thresholds. Unlike conventional X-ray mapping techniques, there is no need to choose specific elements and set appropriate windows. The augmented image is intuitive and easy to interpret yet contains a considerable amount of encoded information equivalent to a multitude of X-ray images. A single image can be acquired in about one minute with sufficient statistics to give an informative "first look" at a sample. Faced with a complex analytical problem, this true colour vision gives the analyst a useful cue for positioning the probe for subsequent more detailed analysis and a way of compressing topographical and compositional information into a single image for archiving.

Acknowledgements. I am grateful to my colleagues, particularly Neville Cox for software implementation and Judith Root, Mohinder Dosanjh and Pierre Rolland for experimental work.

References

[1] G. V. Saparin, S. K. Obyden, *Scanning* **1988**, *10*, 87.
[2] P. J. Statham, M. Jones, *Scanning* **1980**, *3*, 168.
[3] D. Scharf, *U.S. Patent 5212383*, May 18, 1993.
[4] P. Statham, *U.S. Patent 5357110*, October 18, 1994.

Mikrochim. Acta [Suppl.] 13, 581–586 (1996)

Determination of the Oxidation States of Nb by Auger Electron Spectroscopy

Zoltán Tass* and Gabor Horvath

Department of Atomic Physics, Technical University of Budapest, Budapest Budafoki ut 8, H-1111 Budapest, Hungary

Abstract. The oxidation state of niobium is investigated by Auger electron spectroscopy. An empirical function is determined, which establishes a direct relation between the Nb(132), Nb(142) normalised peak-to-peak heights and the oxidation number of Nb. The oxidation state determination applying this function has proved to be more beneficial from many aspects than "classical" lineshape and chemical shift analysis. Auger peaks with similar behaviour are found in other elements as well, which implies that the method can be extended to several cases. However, the explanation of the behaviour of the observed transitions does not seem to be straightforward for those elements where no valance electrons are involved.

Key words: Nb oxides, chemical state analysis, Auger transition, spin splitting.

It is known that besides qualitative and quantitative analysis, Auger electron spectroscopy can also provide information on the chemical state of the investigated material, most cases the determination of the chemical states takes place via Auger lineshape and/or chemical shift analysis. However, we have shown previously [1, 2] that particular intensities of a series of XYZ Auger transitions can also depend on the chemical state of the particular element. We have proved that the oxidation states of Ni and Ti can be characterised on the basis of Ni (663), Ni (676) and Ti (354), Ti (364) peaks. In the case of Ti, we have introduced a characteristic number Ω, which can be directly related to the oxidation state of Ti Ω was defined as follows [2]:

$$\Omega = \frac{h_2 + h_3}{h_1} \tag{1}$$

where the meaning of h_1, h_2, and h_3 are illustrated in Fig. 1 which shows the Ti(354) and Ti(364) differentiated peaks of pure (1a) and oxidised (1b) Ti. The usage of Ω can be more beneficial in several cases than that of "classical" lineshape and chemical shift analysis. Compared with lineshape analysis, the method based on Ω can deliver

* To whom correspondence should be addressed

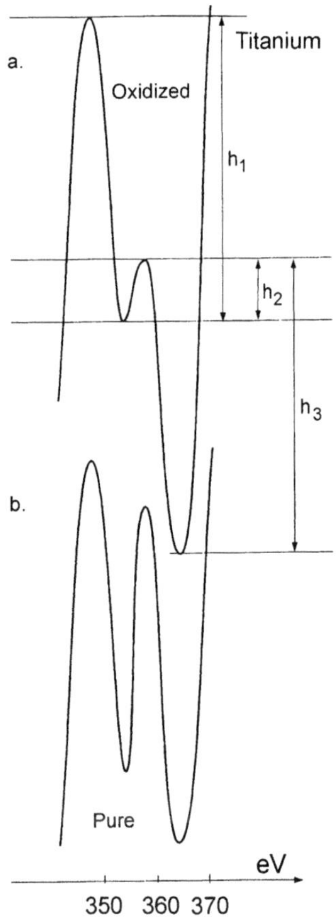

Fig. 1. The Ti(354) and Ti(365) peaks obtained from the oxidised sample (*a*) and the pure metal (*b*). The definitions of h_i values are also shown

more definite and more precise results in a very simple way. On the other hand, the energy resolution of the widely used single-pass cylindrical mirror analyser may not be good enough to detect the subtle energy shifts due to different chemical environments. Moreover, this method also allows the traditional differentiated spectra rather than direct ones only to be employed.

In this paper we are attempting to extend the above-mentioned method for the case of niobium. A similar approach is applied as for Ti in ref. [1], namely, first in situ study of the oxidation process of Nb is carried out, then standard samples are used to determine the Ω-oxidation number function empirically. As a representative of other Nb compounds, $LiNbO_3$ is employed to check the validity of this function.

Experimental

The measurements were carried out by a PHI 545 SAM equipment at a background pressure of 8×10^{-8} Pa. The Auger spectra were recorded with a single-pass CMA in differentiated mode with 4 V peak modulation. The applied primary electron beam had a 5 keV energy with about 20 μm spot diameter, whereas the electron current was varied from sample to sample in order to obtain the best signal-to-noise ratio. Nevertheless, a typical primary current was about 2 μA.

Prior to the measurements each sample was cleaned by Ar^+ sputtering of 3.8 keV energy and 0.5 μA current. During the investigation of the Nb oxidation process the oxygen partial pressure was of 1×10^{-4} Pa, while the sample temperature was kept at 350 °C.

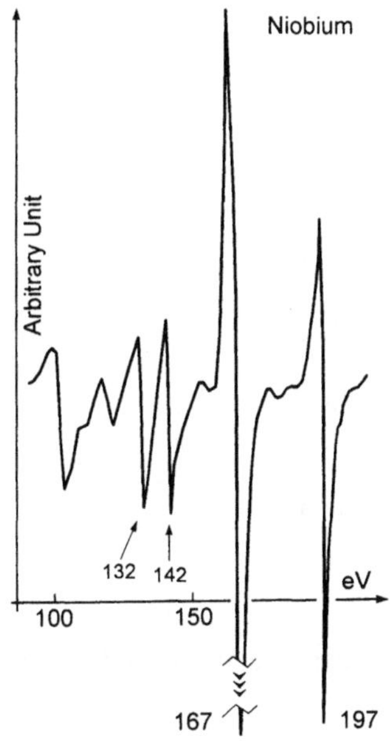

Fig. 2. Auger spectrum of pure Nb in the range of 80–220 eV

Standard NbO samples were prepared by the method described by Oechsner and coworkers [3]. Bulk Nb_2O_5 samples were prepared by heating of Nb powder in furnace at 800 °C, in air for 1 hour. The stoichiometry and the structure of the obtained materials was checked by X-ray diffraction method.

Results and Discussion

In Fig. 2 we show a spectrum taken from a pure niobium sample, where the referred peaks are indicated by the actual energies in eV.

It was observed that the Nb(132) and Nb(142) peaks behave similarly during the oxidation as the Ti(354), Ti(364) and the Ni(663), Ni(676) ones, therefore the Ω characteristic number can be defined for these peaks as well. The development of the oxidation process with time is illustrated in Fig. 3, where Ω, and the $O(KL_2L_2)/Nb(M_{4,5}N_{2,3}N_{4,5})$ peak-to-peak height ratios are shown as functions of time. Similarly to the case of Ti [2], the Ω characteristic number changes markedly as the oxidation process is proceeding. The relatively large scatter in the Ω values is due to the fact that as time passes the heights of h_2 and h_3 become smaller leading to decreased signal-to-noise ratio. The O/Nb peak-to-peak height ratio can serve as a basis for a rough quantification. Following the example of Szalkowski and Somorjai [4] we used stoichiometric NbO and Nb_2O_5 samples as references, and assumed the O/Nb peak-to-peak height ratio to be a linear function of the Nb oxidation number. In this way we can plot Ω against the Nb oxidation number as shown in Fig. 4. The oxidation number can be interpreted in this case as the average oxidation number of that volume of the sample, where the detected electrons come from. To see whether the Ω is an unambigous function of the oxidation number we can determine the Ω values for standard samples with known stoichiometry. The Ω values obtained from standard NbO and Nb_2O_5 samples are also shown in Fig. 4. During the Auger the measurements of the standard samples the peak intensities did not change in time. It can be seen that the Ω values reasonably fit the function

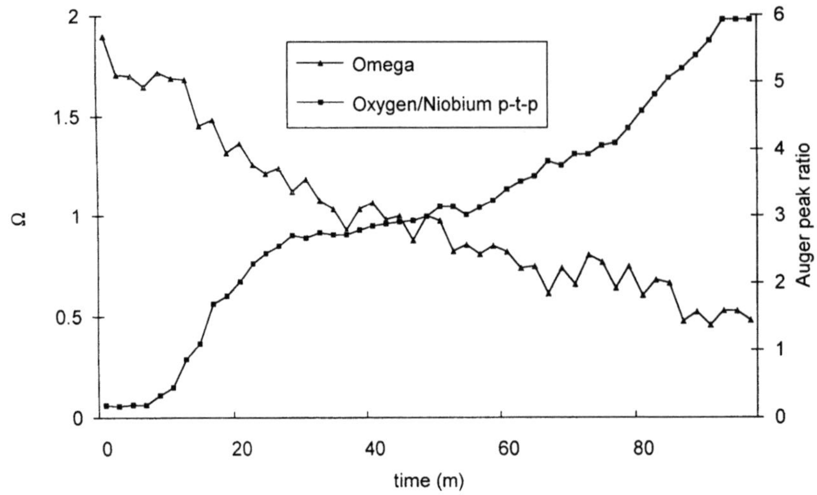

Fig. 3. The $O(KL_2L_2)/Nb(M_{4,5}N_{2,3}N_{4,5})$ peak-to-peak ratio and Ω as functions of the oxidation time. The oxidation was carried out at elevated temperature

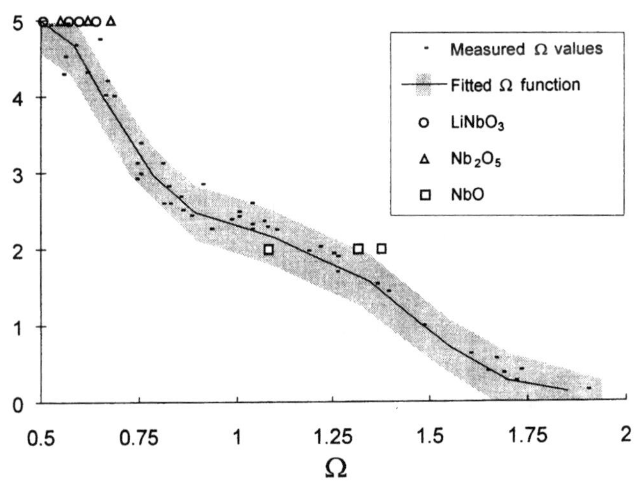

Fig. 4. The oxidation number of niobium versus Ω. The solid line is fitted to the measured points. The Ω values obtained from standards (NbO, Nb_2O_5) and $LiNbO_3$ are also plotted

determined on the basis of the oxidation procedure. This implies that the empirical function of Fig. 4 is valid in general, irrespective of whether it is a result of a time-dependent oxidation based, or a time-independent standard based calibration.

The validity of the Ω-oxidation number function was also examined in the case of more complex niobium-oxides. As a representative, $LiNbO_3$ powder samples were chosen. The obtained result, plotted in Fig. 4 as well, shows that the Ω values for Nb_2O_5 and $LiNbO_3$ are very close to each other, which agrees well with the expectation based on formal oxidation numbers. Thus, the determined, empirical Ω-oxidation number function seems to be valid for $LiNbO_3$ as well.

Table 1. The elements, for which the method based on Ω predicted to be suitable, with the appropriate Auger energies

Element	Auger	Energies (eV) [5]
Sc	310	319
Ti	354	364
V	399	411
Cr	447	459
Mn	498	510
Fe	550	562
Co	606	618
Ni	663	676
Cu	718	731
Zn	773	789
Ga	830	847
Ge	891	906
As	951	970
Zr	116	124
Nb	132	142
Mo	148	161

According to these findings, the method proposed to determine the oxidation states of Ti and Ni is quite promising for Nb as well. The common feature of these cases is that the higher energy peaks of the Ni(663–676), Ti(354–364) and Nb(132–142) doublets are remarkable sensitive to the chemical states of the actual element. After investigating the spectra of elements published in the Auger Handbook [5], similar doublet peaks can be found in several other cases as well. These particular peaks are shown in Table 1.

On the basis of the Table above, and the observed behaviour of these peaks for Ni, Ti and Nb, one may predict that these doublets can be suitable for chemical state characterization for the listed elements. However, we should treat separately the elements from Sc to As and from Zr to Mo. In the former case the studied doublets are ascribed to $L_3M_1M_{23}$ (^1P) and $L_3M_1M_{23}$ (^3P) transitions due to the spin exchange splitting of the $L_3M_1M_{23}$ peak [6, 7]. Regarding these assignations, further investigation is needed to explain why the ^3P final state is less probable in the oxidised state than in the elemental state. As far as Zr, Nb and Mo are concerned, the explanation seems to be more straightforward. In these elements the Auger peaks in question are assigned to the $M_{45}N_{23}N_{23}$ and the $M_{45}N_1N_{45}$ ([7]) transitions. That is, the intensity of the higher energy peak is related to the population of the valence state levels. If the electrons of these levels are partially "removed", due to the high electronegativity of oxygen, the transition involving these electrons is expected to be less probable.

Conclusion

The oxidation state of Nb was determined by Auger electron spectroscopy, applying the method previously proposed for Ti. However, in the case of Nb the features of the studied peaks ($M_{45}N_{23}N_{23}$ and $M_{45}N_1N_{45}$) can be explained by a simple argument, since the valence electrons are also involved in the latter transition, which is not the case for Ti. For Ti and Ni further work should be done to show, why the $L_3M_1M_{23}$ (^1P, ^3P) peaks behave differently in distinct chemical environments. A suggestion has been made that the transitions above might be suitable for determining the oxidation states of other elements as well.

Acknowledgement. The authors are thankful to A. Nagy (Technical University of Budapest) for her valuable help in preparing the samples, to F. Hange (GE Lighting) for the XRD measurements and to the OTKA for supporting this project. G. Horvath is grateful to the Foundation "Magyar Tudomanyert".

References

[1] Z. Tass, V. K. Josepovits, *Mikrochim Acta* **1994**, *114/115*, 481.
[2] Z. Tass, G. Horvath, V. K. Josepovits, *Surf. Sci.* **1995**, *331–333*, 272–276.
[3] H. Oechsner, J. Giber, H. J. Fusser, A. Darlinski, *Thin Film Solids* **1985**, *124*, pp. 199.
[4] F. J. Szalkowski, G. A. Somorjai, *J. Chem. Phys.* **1971**, *56(12)*, 6097.
[5] L. E. Davies, N. C. MacDonald, P. W. Palmberg, G. E. Riach, R. E. Weber, *Handbook of Auger Spectroscopy, 2. Ed*, PHI Eden, Praire, 1976.
[6] J. P. Coad, *Phys. Lett.* **1971**, *37A(5)*, 437.
[7] D. Briggs, M. P. Seah (eds.), *Practical Surface Analysis, 2nd Ed.*, Wiley, Chichester, 1990, p. 646.

Mikrochim. Acta [Suppl.] 13, 587–595 (1996)

Study by SIMS of the ^{54}Cr and ^{18}O Diffusion in Cr_2O_3 and in Cr_2O_3 Scales

S. C. Tsai[1,*], A. M. Huntz[1], C. Dolin[2], and C. Monty[3]

[1] Laboratoire de Métallurgie Structurale, CNRS URA 1107, Bat 413, Université Paris XI-Orsay, 91405 Orsay Cedex, France

[2] Laboratoire de Physique du Solide, Ecole Supérieure de Physique et de Chimie Industrielles de la Ville de Paris, 75231 Paris Cedex 05, France

[3] Institut de Science et de Génie des Matériaux et Procédés, B. P. 5 Odeillo, 66125 Font-Romeu Cedex, France

Abstract. The diffusivities, in lattice and along grain boundary, of ^{54}Cr and ^{18}O were determined at 900 °C and $pO_2 = 10^4$ Pa in Cr_2O_3 (single and polycrystals) and in Cr_2O_3 scales formed on a $Ni_{70}Cr_{30}$ alloy by oxidation. All the concentration profiles were established by SIMS which allows an accurate determination of the diffusion profiles in materials such as oxides which are characterized by a very low diffusivity, even with an important surface roughness as in oxide scales. With our analysis for the concentration profiles and the f value, fraction of atomic sites associated with grain boundaries modified by a ridge model, it appears that the lattice diffusion coefficients of Cr and O determined in Cr_2O_3 scales are close to those determined in Cr_2O_3 single or polycrystals. However, the grain boundary diffusion coefficients of Cr and O determined in polycrystals are smaller than those determined in scales. The lattice diffusion coefficients of Cr and O are close to each other. But, the chromium grain boundary diffusion coefficient is slightly greater than that of oxygen.

Key words: chromium, oxygen, lattice diffusion, grain boundary diffusion, Cr_2O_3, Cr_2O_3 scales.

Cr_2O_3 scale is generally regarded as a barrier to protect metallic alloys at high temperatures. As a consequence, the transport properties of Cr_2O_3 are of considerable interest. Many attempts have been made to measure the diffusion coefficients of chromium and oxygen in Cr_2O_3, and to determine from diffusion data the rates at which Cr_2O_3 scales grow on chromia-forming materials.

From early studies of the self-diffusion of oxygen and chromium summarized by Kofstad [1], it was shown that oxygen diffusion was lower than chromium diffusion

---

* To whom correspondence should be addressed

by about 3 orders of magnitude. Later, from oxidation studies, it was concluded that Cr_2O_3 scales grow on chromium metal by counter-current diffusion of chromium and oxygen, mainly via grain boundaries [2, 3]. Nevertheless, Kofstad [4] and Atkinson [5] suggested that outward cationic diffusion is more important for the growth of undoped Cr_2O_3 scales. On the contrary, Sabioni et al. [6–8] recently determined the lattice and grain boundary self-diffusion coefficients of oxygen and chromium in single crystals and in polycrystals of Cr_2O_3 in the range of temperatures 1100–1450 °C and showed that the oxygen diffusion coefficients are greater than those of chromium. This agrees with results obtained by King and Park [9]. But the oxygen and chromium diffusion coefficients of Sabioni are too small to justify Cr_2O_3 scales growth using Wagner theory.

In the literature, it appears that the values of oxygen and chromium diffusion coefficients determined in single crystals are some orders of magnitude smaller than data obtained on polycrystalline materials or on scales. According to our earlier work [10–12], in the case of polycrystalline Cr_2O_3, the first part of the penetration profiles corresponds to an apparent diffusion, and the second part of the profiles is related to grain boundary diffusion. Moreover, considering the surface roughness of Cr_2O_3 scales, a ridge model was proposed to modify the f-value, fraction of atomic sites associated with a grain boundary. The modified f-value becomes smaller than that generally calculated with a classical formula $f = 3\delta/\Phi$ (δ: grain boundary width and Φ: grain size) which is applied for smooth surface samples, for instance polycrystals of Cr_2O_3. With such an analysis and a modified f-value, oxygen lattice diffusion coefficients determined in Cr_2O_3 scales are in very good agreement with those determined in massive Cr_2O_3, either in single crystals or in massive polycrystals.

Due to the discrepancy between the values of oxygen and chromium diffusion coeffcients determined by other workers, the main objective of this work was, with the same analysis of the penetration profiles as mentioned above and the ridge model, to simultaneously determine lattice (D_L) and grain boundary (D_{gb}) diffusion

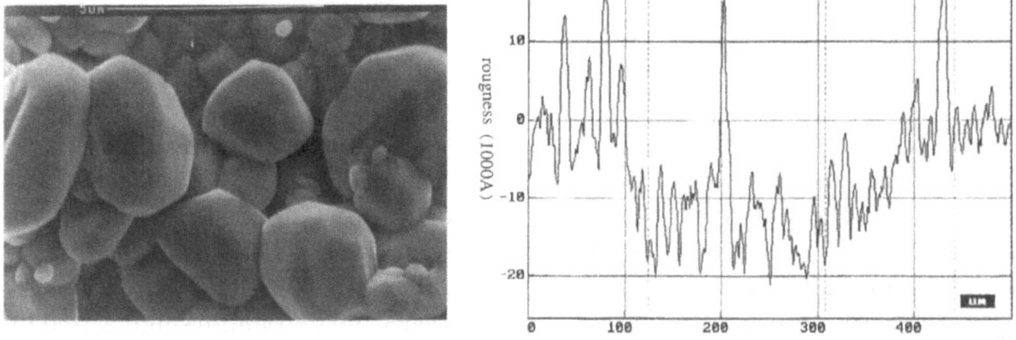

a b

Fig. 1. a SEM micrograph of a Cr_2O_3 scale surface formed on the $Ni_{70}Cr_{30}$ alloy at 900 °C and $pO_2 = 1 \times 10^5$ Pa during 165 hours. **b** Crater obtained on an oxide scale surface showing that the roughness and the penetration are of the same order of magnitude coefficients of oxygen and chromium in Cr_2O_3 single crystals and polycrystals, as well as in Cr_2O_3 scales obtained by oxidation of a $Ni_{70}Cr_{30}$ alloy to allow a direct comparison between oxygen and chromium diffusion. All the diffusion profiles were measured by secondary ion mass spectrometry (SIMS, CNRS in Bellevue) which is probably the most accurate modern technique for measuring diffusion profiles in the case of the materials tested here. Indeed, as will be seen later, the penetration depth is too small to allow other more classical techniques, and the SIMS method allows to work with stable isotopes. Besides, this method is the only one which allows a sputtering parallel to the surface even if this surface is rough which is the case of oxide scales as shown in Fig. 1a and b. Thus, it is possible to establish diffusion profiles on a depth which is on the order of magnitude of the roughness. Moreover, the measurement of the crater depth with a profilometer (Fig. 1b), after SIMS analysis, allows to determine the surface roughness and to calculate the f-value in the case of the ridge model [12].

Experimental

Materials

Single crystals and polycrystals of Cr_2O_3: The preparation of these two high-purity materials has been described earlier [12]. The polycrystals of Cr_2O_3 are made of Cr_2O_3 powder by the hot pressing method. The average grain size for polycrystals is 9 μm. All samples were polished to 3 μm diamond, and were ultrasonically cleaned in aceton, then washed with ethanol.

Cr_2O_3 scales: The $Ni_{70}Cr_{30}$ alloy was melted by CECM (Vitry, France). The impurities in the alloy are 0.2 wt% Si, 500 μg/g Fe, 500 μg/g Ca and 150 μg/g Al. All samples were cold rolled (40%), preannealed for 15 h at 900 °C in argon, and then polished to 1200 mesh grit SiC. The scales were then developed at $p^{16}O_2 = 10^5$ Pa and at 900 °C during 165 hours. The scales consist of compact Cr_2O_3 grains with small amounts of Ni and Al. The average grain size of the scales is about 2 μm (Fig. 1a) and the roughness is about 1 μm (Fig. 1b). ^{54}Cr tracer was deposited on the surface of the Cr_2O_3 scales by evaporation at a total pressure of about 10^{-4} Pa.

^{18}O and ^{54}Cr tracer: High purity $^{18}O_2$ gas and ^{54}Cr used as stable tracers were obtained from the Commissariat à l'Energie Atomique (France). The average thickness of the ^{54}Cr tracer deposited is about 15 nm as determined by SIMS. According to our diffusion data and conditions, the average diffusion length ($\sqrt{D_L t}$) of chromium is about 600 nm. So, the ^{54}Cr tracer on surface of the sample is considered as a thin film.

Experiments

Experimental conditions: All diffusion experimental conditions of the sample are given in Tables 1 and 2. The oxygen and chromium diffusion was studied by means of ^{18}O-^{16}O and ^{54}Cr-^{52}Cr isotopic exchange, respectively.

At first, it was tried to simultaneously determine the oxygen and chromium diffusion coefficient under the same experimental condition, as shown in Table 1. But, the oxygen diffusion profile obtained in this case was not always satisfying. So, the diffusion time was increased, as shown in Table 2, and then the oxygen diffusion profiles were obtained.

Table 1. Experimental conditions for ^{54}Cr at 900 °C

Sample	Material	Experimental conditions
Cr1	single crystal	^{54}Cr layer + ^{18}O, 10^4 Pa, 2 h
Cr2	massive polyc.	^{54}Cr layer + ^{18}O, 10^4 Pa, 2 h
Cr3	scales	^{16}O, 10^5 Pa, 165 h + cooling (40 °C/h) + ^{54}Cr layer + ^{18}O, 10^4 Pa, 2 h

Table 2. Experimental conditions for ^{18}O at 900 °C

Sample	Material	Experimental conditions
O1	single crystal	^{54}Cr layer + ^{18}O, 10^4 Pa, 4 h
O2	massive polyc.	^{54}Cr layer + ^{18}O, 10^4 Pa, 4 h
O3	scales	^{16}O, 10^5 Pa, 165 h + cooling (40 °C/h) + ^{54}Cr layer + ^{18}O, 10^4 Pa, 4 h

Analysis of the diffusion of ^{18}O and ^{54}Cr: For all experiments, the concentration profiles were established by SIMS (CAMECA 4F), using a 10 keV Cs^+ ion source for oxygen and a 10 keV O^+ ion source for chromium. The scanned area was 250 × 250 μm and the analyzed zone was 62 μm in diameter. The sputtering rate was determined by measuring the crater depth with a profilometer.

Determination of the diffusion coefficients: To determine the oxygen diffusion coefficient, the Fick solution given for constant surface concentration was used,

$$[C(x) - C_S]/[C_0 - C_S] = \operatorname{erf}(x/2(Dt)^{1/2}) \quad (1)$$

with C_0 = the tracer natural concentration in the sample, equal to 0.2%
C_S = the tracer constant surface concentration
t = the diffusion time

A thin film method solution was used to determine chromium diffusion coefficients:

$$C(x) = C_0 \exp(-x^2/4Dt) \quad (2)$$

In case of single crystals, the D_L for oxygen and chromium can be determined by Eqs. 1 and 2, respectively.

In case of polycrystals of Cr_2O_3 and Cr_2O_3 scales, as it was considered that diffusion of oxygen and chromium occurs in a B regime, i.e. $\delta \ll (D_L t)^{1/2} \ll \Phi$ [13], the first part of the oxygen and chromium penetration profiles corresponds to apparent diffusion. So, D in Eqs. (1) and (2) is considered as the apparent diffusion coefficient (D_{app}), defined by Hart [13]:

$$D_{app} = (1-f)\, D_L + f\, D_{gb} \quad (3)$$

In the second part of the profiles corresponding to grain boundary diffusion, as the values of $\beta\,(\approx D_{gb}\,\delta/D_L\, 2(D_L t)^{1/2})$ are greater than 10, D_{gb} can be determined by the following equation proposed by Le Claire [14]:

$$D_{gb} \cdot \delta = 0.661 \cdot (-P_{gb})^{-5/3} \cdot (4D_L/t)^{1/2} \quad (4)$$

where P_{gb} is the slope of the curve $\ln C = f(x^{6/5})$ plotted in the second part of the experimental profiles.

Taking $\delta = 10^{-7}$ cm and combining Eqs. (3) and (4), a two order degree equation with an unknown D_L is obtained:

$$(1-f)D_L + 1.332\cdot 10^7\cdot f\cdot(-P_{gb})^{-5/3}\cdot(t)^{-1/2}\cdot(D_L)^{1/2} - D_{app} = 0 \quad (5)$$

With Eq. (5), D_L can be calculated, then D_{gb} can be determined with Eq. (4).

Results and Discussion

Some oxygen and chromium penetration profiles are shown in Figs. 2 to 4. All profiles have the same general shape. The results of the oxygen and chromium diffusion coefficients in single crystals and polycrystals of Cr_2O_3 and in Cr_2O_3 scales are presented in Tables 3 and 4.

Single crystals: Our oxygen lattice diffusion coefficient is $5.7 \times 10^{-17}\,cm^2\,s^{-1}$. This oxygen D_L value is slightly greater than those determined in single crystals without ^{54}Cr layer in our earlier work [12] ($D_L^0 \approx 10^{-17}\,cm^2\,s^{-1}$, see Table 3). It is possible that there was a thin polycrystalline film of $^{54}Cr_2\,^{18}O_3$ formed on the surface of the single crystal with ^{54}Cr layer which leads to a greater oxygen "D_L" value. Sockel

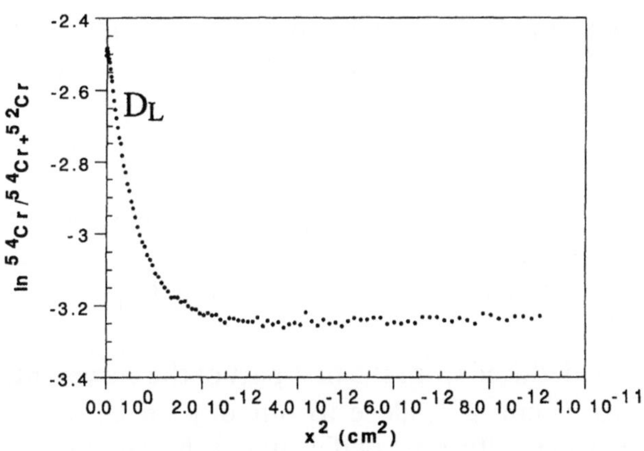

Fig. 2. ^{54}Cr concentration profile plot against x^2 in single crystal (sample Cr1)

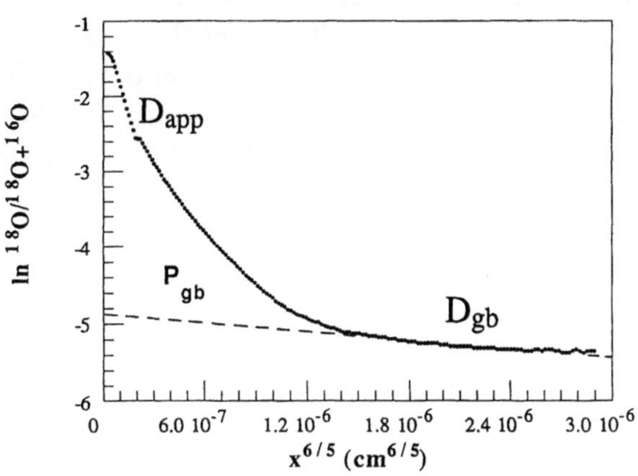

Fig. 3. ^{18}O concentration profile plot against $x^{6/5}$ in polycrystals (sample O2)

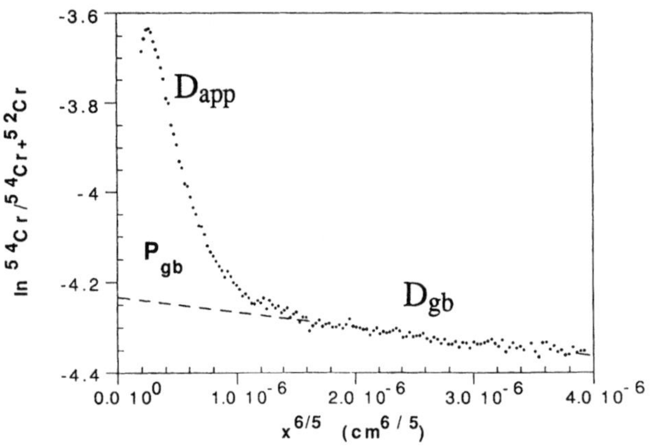

Fig. 4. ^{54}Cr concentration profile plot against $x^{6/5}$ in Cr_2O_3 scales (sample Cr3)

Table 3. D_{app}, $\mathbf{D}_L$ and $\mathbf{D}_{gb}$ ($cm^2\,s^{-1}$) of ^{18}O determined in single crystals, in massive polycrystals and in scales with and without (*) [ref. 12] ^{54}Cr layer at 900 °C and $pO_2 = 10^4$ Pa

Sample	Material	D_{app}	D_L	D_{gb}
O1	single crystal		5.7×10^{-17}	
O2	massive polyc.	1.8×10^{-16}	1.4×10^{-17}	4.9×10^{-13}
O3	scales	1.0×10^{-15}	1.1×10^{-17}	1.6×10^{-12}
O1*	single crystal		1.3×10^{-17}	
O2*	massive polyc.	6.0×10^{-17}	8.1×10^{-18}	1.6×10^{-13}
O3*	scales	2.9×10^{-15}	5.1×10^{-18}	4.8×10^{-12}

et al. [15] studied oxygen diffusion in polycrystals of Cr_2O_3 between 1050–850 °C. In their work at 900 °C D_L^0 is about $10^{-17}\,cm^2\,s^{-1}$, close to our value [12], as is also the extrapolated D_L value determined in single crystal by Sabioni et al. [7] which is about $3 \times 10^{-18}\,cm^2\,s^{-1}$.

The chromium diffusion coefficient, deduced from the first part of the curves lnC versus x^2 (Eq. (2), Fig. 2), is $4.9 \times 10^{-17}\,cm^2\,s^{-1}$, close to our oxygen D_L value. Sabioni et al. [6] studied chromium lattice diffusion coefficients in the same single crystal as ours at $pO_2 = 5$ Pa over the temperature range 1200–1450 °C. The D_L value extrapolated at 900°C is about $2 \times 10^{-21}\,cm^2 s^{-1}$ much smaller than ours. An extrapolated D_L value at 900°C and $pO_2 = 0.2 - 1$ Pa from Atkinson and Taylor's work [5] is about $10^{-20}\,cm^2\,s^{-1}$, also smaller than our chromium D_L value. Another extrapolated chromium D_L value, determined from Hagel and Seybolt [16] in a hot-pressed compact Cr_2O_3 and in a nitrogen atmosphere, is about $7 \times 10^{-13}\,cm^2\,s^{-1}$. It appears that, for chromium diffusion coefficients, the data are very scattered due to different experimental conditions and samples.

Polycrystals of Cr_2O_3: The curves lnC versus $x^{6/5}$ (Fig. 3), clearly indicate that there are two domains in the penetration profiles, the first one related to apparent diffusion and the second to grain boundary diffusion. Taking the average grain size,

Table 4. D_{app}, D_L and D_{gb} ($cm^2\,s^{-1}$) of ^{54}Cr determined in single crystal, in massive polycrystals and in scales at 900 °C and $pO_2 = 10^4$ Pa

Sample	Material	D_{app}	D_L	D_{gb}
Cr1	single crystal		4.9×10^{-17}	
Cr2	massive polyc.	1.0×10^{-15}	2.1×10^{-17}	2.9×10^{-12}
Cr3	scales	4.7×10^{-15}	2.0×10^{-17}	9.3×10^{-12}

$\Phi = 9\,\mu m$ and the grain boundary width, $\delta = 1$ nm, a f-value was obtained: $f = 0.00034$ ($= 3\delta/\Phi$). With Eqs. 4 and 5, the lattice (D_L) and grain boundary (D_{gb}) diffusion coefficients of oxygen and chromium were determined. From Table 3 and 4, it appears that: (1) D_L values for oxygen and chromium determined in this case are not far from those in single crystals. (2) D_L, D_{app} and D_{gb} values of oxygen are smaller than those of chromium.

Comparing our oxygen D_{gb} value with that of Sockel et al. [15], it was found that their oxygen D_{gb} value is about $10^{-13}\,cm^2\,s^{-1}$ at 900 °C, close to ours ($\approx 4.9 \times 10^{-13}\,cm^2\,s^{-1}$). But our D_{gb} values are greater than those obtained by Sabioni et al. at higher temperature (their D_{gb} value $\approx 10^{-13}\,cm^2\,s^{-1}$ at 1100 °C and $pO_2 = 2 \times 10^4$ Pa) [8].

Our chromium D_{gb} value is $2.9 \times 10^{-12}\,cm^2\,s^{-1}$. From Sabioni's work [8], chromium extrapolated D_{gb} value at 900 °C is about $10^{-21}\,cm^2\,s^{-1}$ ($pO_2 = 5$ Pa), very close to his extrapolated chromium D_L value ($\approx 4 \times 10^{-22}\,cm^2\,s^{-1}$) [8]. A D_L value should not be so close to D_{gb} value. An extrapolated D_{gb} value of Atkinson and Taylor [17] at 900 °C and $pO_2 \approx 0.2 - 1$ Pa is about $10^{-18}\,cm^2\,s^{-1}$, much smaller than ours and greater than that of Sabioni.

There is a lot of difference between our oxygen and chromium D_L and D_{gb} values and some extrapolated values determined by other workers. This is firstly due to the analysis of the first part of the diffusion profiles: whereas apparent diffusion is considered in our case, lattice diffusion was only taken into account by other authors. Secondly, the extrapolated D_{gb} values were obtained on the basis of few points, which may lead to greater uncertainty. Finally, the microchemistry of the grain boundaries can vary with the temperature and the material purity.

In Cr_2O_3 scales: As for polycrystals, the plots lnC against $x^{6/5}$ show two parts (Fig. 4). According to our earlier work, the f-value was modified by a ridge model taking into account the surface roughness [12]. The lattice and grain boundary diffusion coefficients of oxygen and chromium are shown in Tables 3 and 4. Some remarks can be made: (1) D_L values of oxygen and chromium determined in scales are close to those determined in single crystals and polycrystals of Cr_2O_3. (2) D_{app} and D_{gb} values of oxygen and chromium determined in Cr_2O_3 scales are greater than those determined in polycrystals of Cr_2O_3. (3) D_L values of oxygen and chromium are close to each other, but D_{app} and D_{gb} values of oxygen are smaller than those of chromium (4) D_L, D_{app} and D_{gb} values of oxygen are not modified by the ^{54}Cr

layer deposited on the samples (see the value for O1, O2, O3 and O1*, O2*, O3* in Table 3).

Graham et al. [18] determined oxygen diffusion coefficients in Cr_2O_3 scales formed on single crystals and polycrystals of Cr metal in the temperature range 550–900 °C. Their oxygen D_{gb} value is about $10^{-12}\,cm^2\,s^{-1}$ at 900 °C, in good agreement with our oxygen D_{gb} values.

Lobnig et al. [19] determined chromium lattice diffusion coefficients whose order of magnitude is about $10^{-15}\,cm^2\,s^{-1}$ at 900 °C. This is close to our D_{app} values (Table 3) and bigger than our D_L values, possibly because they did not consider the effect of apparent diffusion in the first part of their oxygen concentration profiles.

Conclusions

At 900 °C and $pO_2 = 10^4$ Pa, the lattice and grain boundary diffusion coefficients of oxygen and chromium were studied in single and polycrystals of Cr_2O_3 and in Cr_2O_3 scales by SIMS, a technique which allows to plot penetration profiles over a small depth and even in the presence of rough surface. It can be concluded that:

(1) The lattice diffusions of oxygen and chromium determined in single crystal are close to each other.
(2) In the case of Cr_2O_3 scales, the oxygen and chromium lattice diffusion coefficients are in good agreement with those determined in polycrystals of Cr_2O_3. But grain boundary diffusion coefficients of oxygen and chromium are greater than those determined in polycrystals.
(3) Chromium diffusivities, especially for apparent and grain boundary diffusion, are greater than those of oxygen.
(4) In spite of the roughness of oxide scales like Cr_2O_3 scales, the SIMS technique allows the diffusion coefficients in scales to be determined.

References

[1] P. Kofstad, *Non-Stoichiometry, Diffusion and Electrical Conductivity in Binary Metal Oxide*, Wiley, New York, 1972.
[2] P. Kofstad, K. P. Lillerud, *J. Electrochem. Soc.* **1980**, *127*, 2410.
[3] K. P. Lillerud, P. Kofstad, *J. Electrochem. Soc.* **1980**, *127*, 2397.
[4] P. Kofstad, *High Temperature Corrosion*, Barking, Essex, Elsevier Applied Science, 1988.
[5] A. Atkinson, *Rev. Mod. Phys.* **1985**, *57*, 437.
[6] A. C. S. Sabioni, B. Lesage, A. M. Huntz, J. C. Pivin, C. Monty, *Phil. Mag. A* **1992**, *66*, 333.
[7] A. C. S. Sabioni, A. M. Huntz, F. Millot, C. Monty, *Phil. Mag. A* **1992**, *66*, 351.
[8] A. C. S. Sabioni, A. M. Huntz, F. Millot, C. Monty, *Phil. Mag. A* **1992**, *66*, 361.
[9] W. E. King, J. H. Park, *Proceedings of the Materials Research Society Spring Meeting*, 1988.
[10] A. M. Huntz, S. C. Tsai, *Mat. Sci. Letter* **1994**, *13*, 821.
[11] S. C. Tsai, A. M. Huntz, C. Dolin, C. Monty, *Proceedings of the 7th Europhysical Conference on Defects in Insulating Materials 1994, Radiation Effects and Defects in Solids* **1995**, *137*, 285.
[12] S. C. Tsai, A. M. Huntz, C. Dolin, *Oxid. Met.* **1995**, *43(5/6)*, 581.

[13] E. W. Hart, *Acta Metall.* **1957**, *5*, 597.
[14] A. D. Le Claire, *J. Appl. Phys.* **1963**, *14*, 351.
[15] H. G. Sockel, B. Saal, M. Heilmaier, *Surface Interface Anal.* **1988**, *12*, 531.
[16] W. C. Hagel, A. U. Seybolt, *J. Electrochem. Soc.* **1961**, *108*, 1146.
[17] A. Atkinson, R. I. Taylor, *NATO Adv. Study Inst. Ser. B* **1984**, *129*, 285.
[18] M. J. Graham, J. I. Elridge, D. F. Mitchell, R. J. Hussey, *Material Science Forum, Vol. 43*, Trans Tech Publications, Switzerland, 1989, p. 207.
[19] R. E. Lobnig, H. P. Schmidt, K. Hennesen, H. J. Grabke, *Oxid. Met.* **1992**, *37(1/2)*, 81.

Mikrochim. Acta [Suppl.] 13, 597–603 (1996)

Comparison of Back-Foil Scanning X-Ray Microfluorescence and Electron Probe X-Ray Microanalysis for the Elemental Characterisation of Thin Coatings

Evangelos Valamontes and Androula G. Nassiopoulos*

Institute of Microelectronics, NRCPS "Demokritos", P.O. Box 60228, 153 10 Aghia Paraskevi, Attiki, Greece

Abstract. Back-foil scanning X-ray microfluorescence, developed in a scanning electron microscope and applied for the analysis of very thin coatings is compared with electron probe X-ray microanalysis (EPMA). Both experimental results and Monte-Carlo calculations are used in this respect. The signal to background ratio as a function of the primary electron beam energy and angle of incidence and for different film thicknesses is obtained for both techniques and a comparative study of sensitivity is made. Back-foil scanning X-ray microfluorescence (SXRF) used in optimised experimental conditions, is found to be more sensitive than EPMA, especially in the case of film thicknesses below ≈ 100 nm. The resolving power of back-foil SXRF is also calculated for the anode used by Monte-Carlo simulations.

Key words: X-ray microanalysis, X-ray microfluorescence, thin coatings

X-ray fluorescence (XRF) and X-ray microanalysis (EPMA) are two analytical techniques with quite different characteristics. EPMA uses electron excitation in an electron microscope while XRF uses an X-ray excitation source, which offers several advantages over the electron excitation: a) relative detection limits of the order of 1 ppm compared to ≈ 1000 ppm by weight in energy dispersive EPMA, b) possibility of analysing insulators without depositing a conducting layer on the top of them, c) possibility of nondestructive analysis of beam sensitive biological samples. On the other hand, EPMA offers the advantage of a much better lateral resolution, in the μm range, which allows imaging of the analysed element of the sample.

Several attempts have been made to use the electron source of the electron microscope in order to obtain an X-ray source inside the microscope. Middleman and Geller [1] fixed an attachment with a 25 μm thick molybdenum foil, used as an

* To whom correspondence should be addressed

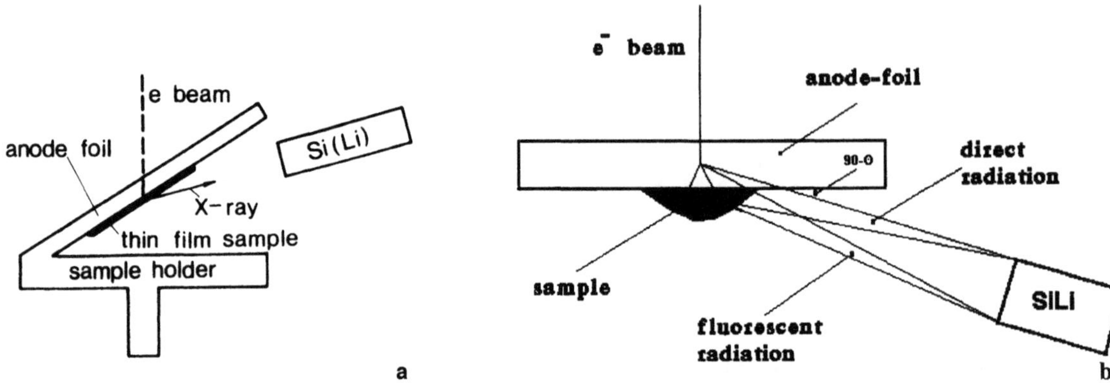

Fig. 1. a Experimental setup used for back-foil XRF, the anode is composed of a nickel foil, 5 μm thick. **b** Illustration of the trajectories of primary and fluorescence X-rays in back-foil XRF. The background is reduced by absorption of the continuous radiation, created by the electron beam, within the anode

X-ray source, generated by a 30 keV electron beam. Analogous attachments were proposed by Linnemann and Reimer [2], Wendt [3, 4] and Weiss [5], all of them mounting their transmission X-ray sources far from the specimen to be analysed. Little different, but always with the X-ray source far from the specimen were the attachments proposed by Eckert (both reflection [6] and transmission [7, 8]) and Pozsgai (transmission mode, but with the analysed specimen in a closed space, protected from unwanted X-rays [9]). Other works, dealing with analogous SEM attachments, are nicely described in a review article by Pozsgai [10].

A somewhat different solution was proposed by Cazaux [11] and implemented first in a surface analysis instrument [12] and later in a scanning electron microscope [13]. It consists of using an anode in thin film form (X-ray source) in close contact to the sample, also in thin film form. The X-rays are created within the anode and ionise the atoms of the sample on their way out. The corresponding SEM attachment [13] is illustrated in Fig. 1. Compared to all the other proposed solutions, this experimental arrangement offers the ultimate resolution, which is in the μm range when the anode is sufficiently thin (of the order of the electron range). In this paper, this last solution is used (back-foil scanning X-ray microfluorescence) and the sensitivity of the technique for the analysis of very thin overlayers, where EPMA reaches its detection limits, is examined. Both experimental results and Monte-Carlo simulations are used in this respect. Direct comparison with EPMA is also made. The radial distribution of the obtained fluorescence signal is also calculated by Monte-Carlo simulations.

Experimental Setup

The experimental setup of Fig. 1a was used for back-foil XRF. The anode was composed of a nickel foil, 5 μm thick. The anode was chosen sufficiently thick in order to prevent primary beam electrons to reach the Ti film on the back side of the anode. On the other hand, a limited thickness is necessary in order to minimize selfabsorption of the generated X-rays. Titanium films of different thicknesses were deposited on one side by electron gun evaporation in high vacuum (10^{-3} Pa). A Si(Li) detector and an EDAX system for quantitative analysis was used for X-ray signal acquisition. The X-ray signal, integrated under the peak and the integrated background, also under the peak and in the same energy window, were used. Standard EPMA, using the same experimental conditions of beam current and electron beam focussing was applied to the same samples.

Monte-Carlo Calculations

The basic computational model for the calculation of electron trajectories was described in detail elsewhere [14]. It is based on the following assumptions:

- elastic scattering is described by screened Rutherford cross-section;
- angle deviation is considered to be only due to elastic scattering, this is a good approximation, as inelastic scattering occurs through very small angles;
- between two scattering points, electrons are considered to lose energy continuously.

The energy loss is described by the Bethe equation for energies $E > 6.4\,J$, where J is the mean ionisation potential. For lower energies, the Rao-Sahib and Wittry's expression is used.

- Each step length for elastic scattering is considered to be constant and equal to 1/100 of the total electron range (plural scattering model). The total length of the electron trajectory within the sample is taken to be the Bethe range. With this approximation, the calculation time is reduced significantly while little error is introduced. In order to evaluate the X-ray signal induced by electrons, each step is divided into smaller steps. At each small step, the energy and position of the electron is known, so the probability for X-ray generation is given by the corresponding cross-section.

The X-ray emission is considered to be isotropic, two random numbers are used for the direction of X-ray emission. Part of them enter into the film after possible absorption within the anode.

The X-ray fluorescence signal is calculated by dividing the film in 10 layers and considering the probability that the primary X-ray entering the layer are absorbed by the layer on their way out of the film. Absorption correction of the fluorescent X-ray signal created within the film is also considered. The absorption coefficients used are taken from the X-ray cross-section compilation by the Kaman Science Corporation [16].

Results and Discussion

Figure 1b illustrates the trajectory of the direct and fluorescence X-rays in back-foil XRF. The background is reduced by absorption of the continuous radiation, created by the electron beam, within the anode.

Figure 2 indicates the experimental results of the signal to background ratio (S/B) in the case of back-foil XRF for different film thicknesses as a function of the

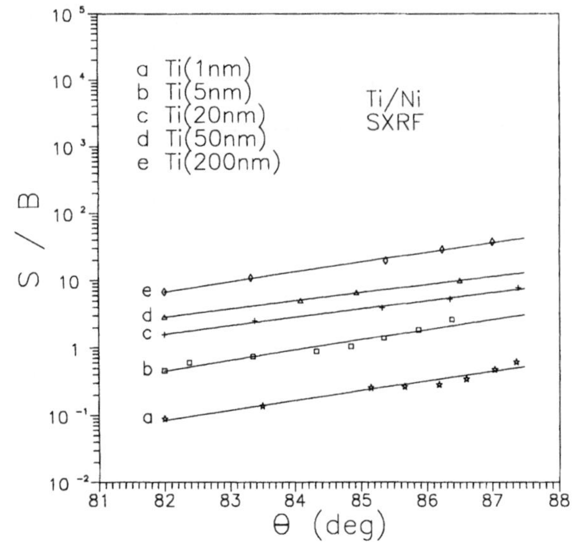

Fig. 2. Experimental values of the signal to background ratio (S/B) in the case of back-foil XRF for different film thicknesses as a function of the detection angle Θ

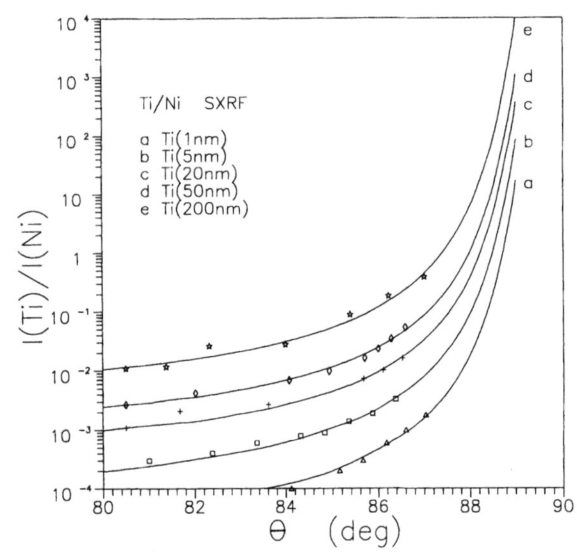

Fig. 3. Ratio of the Ti X-ray signal from the film to the Ni X-ray signal from the anode. Points correspond to experimental results and full lines to Monte-Carlo calculations

detection angle Θ. The acquisition time of the whole spectrum was equal to 600 s. When the angle is increased, the S/B ratio is improved, due to further reduction of the background and an increase of the analysed film area.

Figure 3 shows the ratio of the analysed Ti X-ray signal from the film to the Ni signal from the anode. Points correspond to experimental results and full lines to Monte-Carlo calculations. Good agreement is observed between experiment and calculations. The Ti signal from the film is increased with respect to the signal from the anode as the detection angle Θ is increased. This is due to an increase of the selfabsorption of the Ni X-ray line and to an increased analysed area.

EPMA was performed on the same sample for comparison. The signal from a thin film in EPMA is improved at low primary beam energies and high angles of incidence Figs. 4a, 4b and 4c illustrate experimental results of the signal to

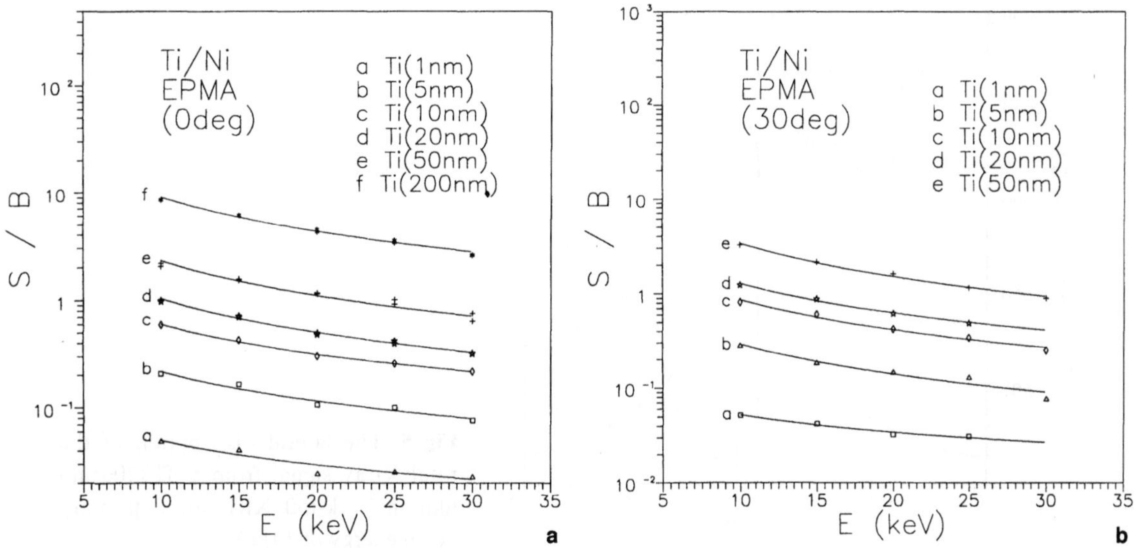

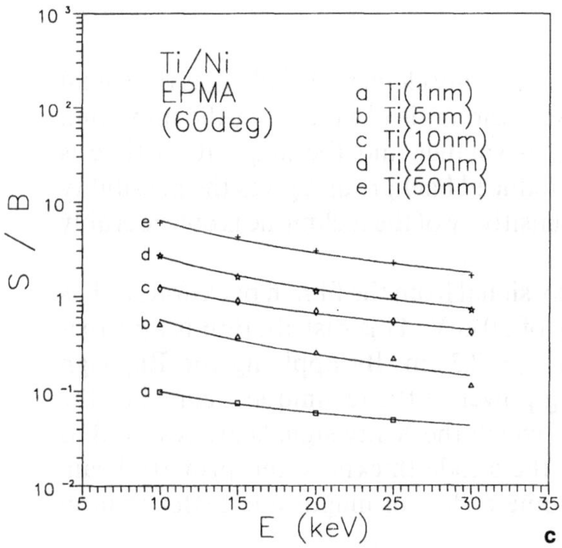

Fig. 4. a-c Experimental values (EPMA) of the signal to background ratio (S/B) for three different angles of incidence (0, 30, 60 deg)

background ratio (S/B) for three different angles of incidence. If we compare Fig. 2 with Figs. 4a, 4b and 4c we see that in back-foil XRF the signal to background ratio is 2 or 3 times higher than in EPMA. The same acquisition time was used for both EPMA and back-foil XRF spectra (600 sec). But we have to note that the reduced background in the case of back-foil XRF permits to increase the acquisition time and improve sensitivity. Indeed the minimum detectable concentration (or thickness) is calculated by considering the Rose criterion, according to which a signal is detectable if the corresponding peak is at least 3 times the standard deviation of the background. By taking that into account the minimum detectable concentration X_m of an element A in a matrix may be expressed as:

$$X_m = \frac{3}{I_s}\sqrt{\frac{I_{BG}}{t}}$$

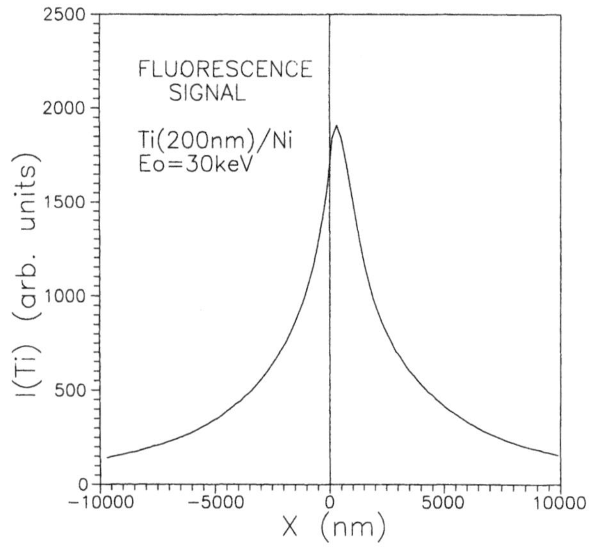

Fig. 5. The lateral distribution of the total X-ray signal from a Ti (200 nm) film in back-foil XRF for a primary beam energy of 30 keV

where I_s is the signal from a standard containing only element A and I_{BG} is the signal under the background. From the above expression it is obvious that the minimum detectable concentration is reduced if I_{BG} is smaller and the acquisition time is increased. In the case of back-foil XRF the reduced background gives the possibility to increase the measuring time so that the sensitivity of the technique is considerably improved compared to that of EPMA.

The lateral distribution of the total X-ray signal from the film in back-foil XRF is given in Fig. 5 for a primary beam energy of 30 keV. This distribution is approximately gaussian with standard deviation $\sigma_0 = 2.3\ \mu m$. By applying the Rayleigh criterion as in reference [15], the resolving power of the technique, defined as the minimum distance between two points for which the X-ray signals are resolved, is calculated. A value of 3.7 μm is found for the anode thickness and primary beam energy used. This resolving power is by some orders of magnitude better than in standard XRF.

Conclusion

Back-foil XRF and EPMA are both applied to the analysis of very thin films. The sensitivity of back-foil XRF is better than that of EPMA in the case of very small film thicknesses (up to a few tens of nm). The obtained experimental results are verified by Monte-Carlo calculations. The resolving power of back-foil XRF is of the order of some micrometers. It is much better than in conventional XRF and of the same order of magnitude as in EPMA.

References

[1] L. M. Middleman, J. D. Geller, *Scanning Electron Microscopy, Part I*, IITRI, Chicago, 1976, pp. 171–177.

[2] B. Linnemann, L. Reimer, *Scanning* **1978**, *1*, 109.

[3] M. Wendt, *Patentschrift DD-WP139*, 1978, p. 304.
[4] M. Wendt, Th. Krajewski, in: *Proc. of 5 Tagung Microsonde*, Physikalische der DDR, Leipzig, 1981, pp. 125–129.
[5] R. M. Weiss, *Beitr. Elektronenmikroskop. Direktabl. Oberfl.* **1979**, *2*, 209.
[6] R. Eckert, in: *Scanning Electron Microscopy, Vol. IV*, SEM inc. AMF O' Hare, Chicago, 1983, pp. 1535–1545.
[7] W. Plannet, *Leaflet Röntgenbox*, Plano, Marburg, 1983.
[8] R. Eckert, *Scanning* **1986**, *8*, 232.
[9] I. Pozsgai, *Inst. Phys. Conf. Ser.* **1985**, *78*, 201.
[10] I. Pozsgai, *X-ray Spectrometry* **1991**, *20*, 215.
[11] J. Cazaux, *Rev. Phys. Appl.* **1975**, *10*, 263.
[12] D. Mouze, X. Thomas, J. Cazaux, in: *Proceed. of 11th Internat. Congr. on X-ray Optics and Microanalysis* (J. D. Brown, R. H. Packwood, ed.), London, Canada, 1986, pp. 63–66.
[13] E. Valamontes, A. G. Nassiopoulos, in: *Proceedings of III Balkan Congress on Electron Microscopy, Athens*, 1989, pp. 267–268.
[14] A. G. Nassiopoulos, E. Valamontes, *Surf. Interf. Anal.* **1990**, *15*, 405.
[15] E. Valamontes, A. G. Nassiopoulos, N. Glezos, *Surf. Interf. Anal.* **1992**, *19*, 419.
[16] *X-ray Cross-section Compilations from 0.1 keV to 1 MeV*, Kaman Science Corporation, Colorado Springs, Colorado.

Mikrochim. Acta [Suppl.] 13, 605–610 (1996)

Electron Probe X-Ray Microanalysis of Coatings

Evangelos Valamontes and Androula G. Nassiopoulos*

Institute of Microelectronics, NRCPS "Demokritos", P.O. Box 60228, 153 10 Aghia Paraskevi, Attiki, Greece

Abstract. Monte-Carlo simulations were applied to electron probe X-ray microanalysis of thin coatings on a bulk material. The X-ray signal from the film and its lateral extent as a function of film thickness, primary beam energy and angle of incidence were calculated. All contributions to the measured total X-ray signal were taken into account. These are: a) The signal induced by the primary electron beam. b) The signal induced by backscattering, considered both within the film and the substrate. c) The signal induced by characteristic and continuous X-rays, created within the substrate by incident and backscattered electrons, which ionise the atoms of the film in their way out of it.

Key words: X-ray microanalysis, thin coatings, Monte-Carlo simulations.

Electron probe X-ray microanalysis (EPMA) is a well known and established technique for the quantitative analysis of bulk materials [1–4]. The formalism used is applied to samples which are homogeneous in composition in the whole analysed volume. But when the analysed elements are not homogeneously distributed in the sample, as is the case of small inclusions in a different matrix or stratified materials, the formalism used for bulk materials is no longer valid and different approaches are needed. The first attempts to apply EPMA to the analysis of stratified materials were based on the determination of the $\varphi(\rho z)$-function giving the distribution of the generated X-ray signal as a function of depth (Philibert and Brown [5], Reuter [6] and Brown [7]). This approach is quite complicated and several assumptions were needed. Monte-Carlo simulations were also used (Bishop and Poole [8], Kyser and Murata [9]. Foulu and Henoc [10]), but also with important simplifications, in order to reduce computer time. A more elaborated model was that of Pouchou and Pichoir [11–13] which constitutes an analytical model established by considering both general laws and Monte-Carlo simulations in order to obtain the $\varphi(\rho z)$-curve of stratified materials as a function of primary beam energy. The chemical composition and film thickness of the analysed samples are obtained by measuring experimentally the ratio of intensities of the analysed element from the sample and from a standard of known and homogeneous composition, containing the same element.

* To whom correspondence should be addressed

A simplified model, applied in the case of coatings which are sufficiently thin compared to the corresponding electron range (thickness $t \leqslant R_e/10$, R_e = electron range) was that of Cazaux et al. [14–15], based on a formalism analogous to that applied in Auger electron spectroscopy. In this formalism, the X-ray signal emitted by element A of a film of thickness t on a substrate B is written as:

$$I_{A/B} = C_A \cdot \rho \cdot \left(\frac{\mathrm{N}}{\mathrm{A}}\right) \cdot I_p \cdot \omega_{ij} \cdot \sigma_X^A(E_0) \cdot (1 + T_{A/B}) \cdot t \cdot T,$$

where C_A is the concentration of element A in the film, ρ is the volume density and A the atomic mass, N is Avogadro's number. I_p (in e/sec) is the primary beam current, ω_{ij} the fluorescence yield, $\sigma_X^A(E_0)$ the ionisation cross-section by electrons of energy, E_0, T is the collection efficiency and $T_{A/B}$ is a total correction factor taking into account all ionisation within the substrate (backscattered electrons and X-rays created within the substrate, which ionise the atoms of the film). Only characteristic X-rays from the substrate are in general taken into account. Nassiopoulos and Valamontes [16–18] and Valamontes et al. [19] made more exact calculations of the correction factor $T_{A/B}$ by including also fluorescence from continuous X-rays created within the substrate. They used Monte-Carlo simulations for these calculations and they found that in some cases (high primary beam energy and substrate of a high Z material) the contribution from continuous X-rays is very important and not negligible. The factor $T_{A/B}$ was also measured experimentally by this group [17] and by Cazaux et al. [14], by measuring under the same conditions, the X-ray signal of element A from the film on a substrate B and the signal from a thin unsupported film of the same thickness, and by calculating the ratio of these two signals. The signal from the thin unsupported film is equal to the signal due to the primary beam.

In the present work, the above calculations are extended to the case where the film is not thin, compared to the electron range. Backscattering within the film is in this case taken into account and the change of the ratio I_{total}/I_{film} with film thickness as a function of primary beam energy and angle of incidence will be discussed.

The radial distribution of the X-ray signal from the film is also calculated in this work for different film thicknesses and angles of incidence. The corresponding Monte-Carlo programme is quite general and takes into account all contributions to the total signal, as described above. But even though the Monte-Carlo programme is quite complicated, it can be run on a personal computer.

Results and Discussion

The Monte-Carlo programme has been described in detail elsewhere [16]. The incident electron beam is normally distributed (gaussian curve) around the point of incidence, with a standard deviation $\sigma_0 = 4\,\mathrm{nm}$. For each incident electron the electron trajectory within the sample (including both the film and the substrate) is first calculated. Backscattering within both the film and the substrate is considered. The X-ray signal from the film is then calculated for the part of electron trajectory situated within the film. Fluorescence from characteristic X-rays is considered in the following way: the created characteristic X-rays along the electron path which have enough energy, ionise the analysed atoms from the film with a probability given by the X-ray ionisation cross-section at the corresponding energy. A random number

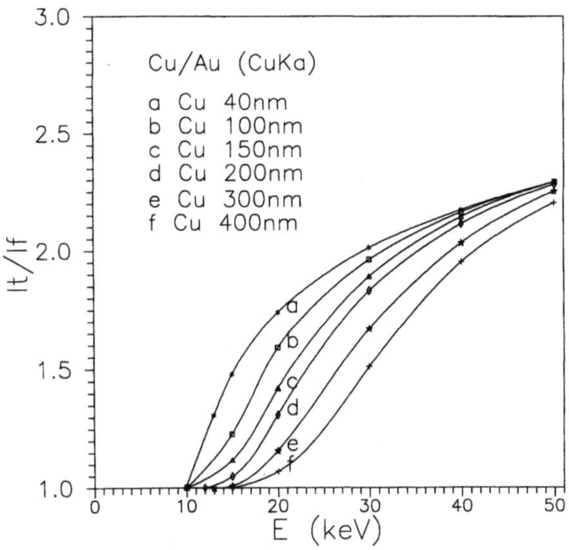

Fig. 1. Variation of the total X-ray signal I_t from an overlayer of Cu on a gold substrate normalised to the signal I_f from a thin unsupported film of the same thickness as a function of energy, for different overlayer thicknesses

determines the emission angle of these X-rays which produce the signal of interest on their way out of the sample. Fluorescence due to continuous X-rays is calculated as described in reference [16]. Only continuous X-rays from the substrate are taken into account.

An example of application of the Monte-Carlo programme is given in Fig. 1 which shows the total X-ray signal from a copper film on a gold substrate as a function of primary beam energies, in the range 10–50 keV. The film thicknesses are between 40 nm and 400 nm. The analysed line is CuK_α. The total signal is normalised to the signal from an unsupported film of the same thickness, calculated separately. The contribution of the substrate is thus separated from the signal generated directly within the film by the incident beam. As expected, this contribution is more important when the film is thin. Experiments were also performed in order to verify the Monte-Carlo results. Figures 2 and 3 show experimental results obtained from a series of Cu films on various substrates for film thicknesses of 40 to 150 nm. Full lines indicate results from Monte-Carlo calculations and the points show the experimental results [17].

Figure 4 shows a second series of experiments in comparison to our calculated results from reference [14]. The total normalised signal I_t/I_f is given. Al films, 40 nm thick, on various substrates are used. The agreement between calculations and experimental results is quite good in all cases within the limits of experimental and calculation errors.

The extent of the total X-ray signal in an x-y axis is calculated the following way: A line is considered, which is defined by the intersection of the plane defined by the sample surface and that defined by the perpendicular to this surface and the direction of the incident beam (see Fig. 5). The lateral extent of the X-ray signal is greater in this direction than in a direction perpendicular to that one. So the more restrictive case was considered. We then consider planes perpendicular to the above line and the calculated signal at each point of these planes in the interaction volume is attributed to one point of this line.

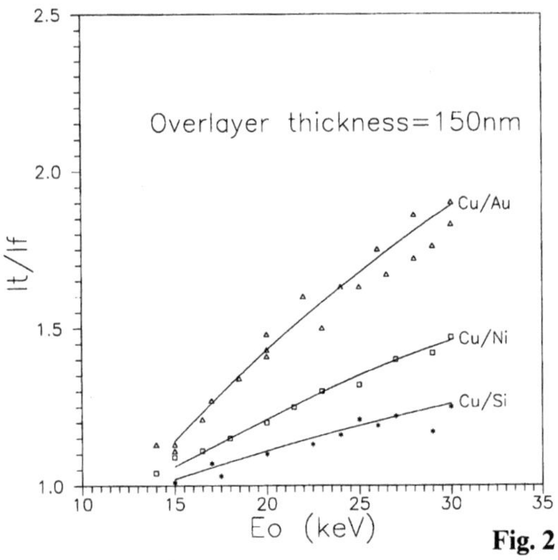

Fig. 2

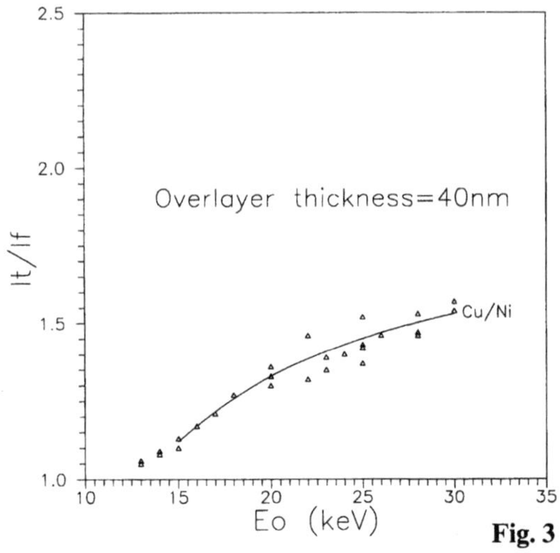

Fig. 3

Figs. 2, 3. Experimental results obtained from a series of Cu films on various substrates for film thicknesses of 40 to 150 nm. Full lines indicate results from Monte-Carlo calculations and the points show the experimental results

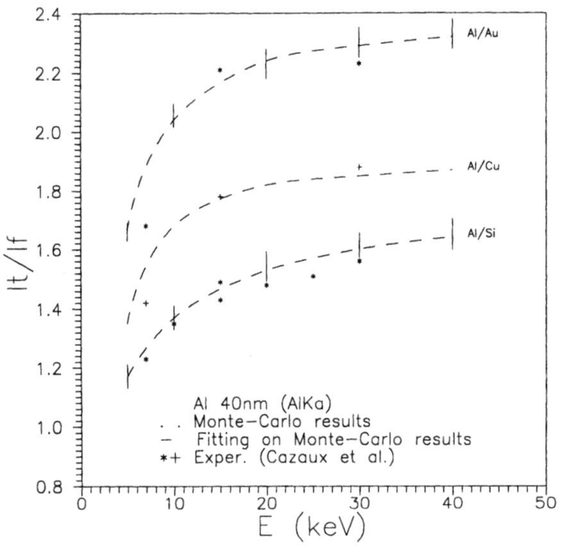

Fig. 4. A second series of experiments in comparison to our calculated results. The total normalised signal I_t/I_f is given. Al films, 40 nm thick, on various substrates are used

Figure 6 shows the distribution of the total X-ray signal as defined above, for the case of a Cu film on Au (analysed line: CuK_α). The film thickness was 40 nm and 3 different angles of incidence were considered (30° (a), 60° (b) and 85° (c)) for different primary beam energies. When $\Theta = 30°$ and below and for high primary beam energies, the obtained distribution is quite similar to that at normal incidence (normally distributed around the point of incidence). The main contribution is that due to the primary beam and the other contributions constitute a low background, distributed over a very large area. At large angles the extent of the signal is larger and it is several hundreds of nm at 85°, being even larger at higher primary beam energies. When the film thickness is increased, the extent of the X-ray signal is increased and it is larger even at small angles of incidence (30°) (see Fig. 6a–c).

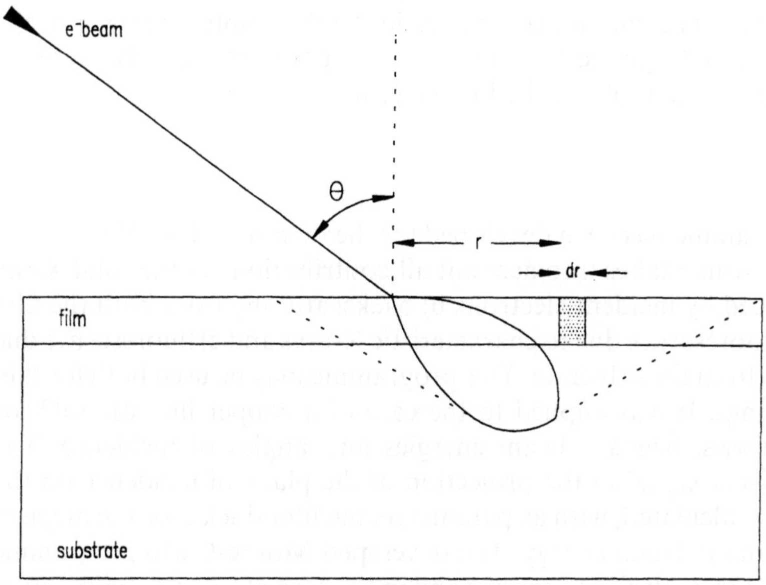

Fig. 5. The extent of the total X-ray signal is considered by projecting the calculated 3-D signal to a line defined by the plane of incidence and the sample surface

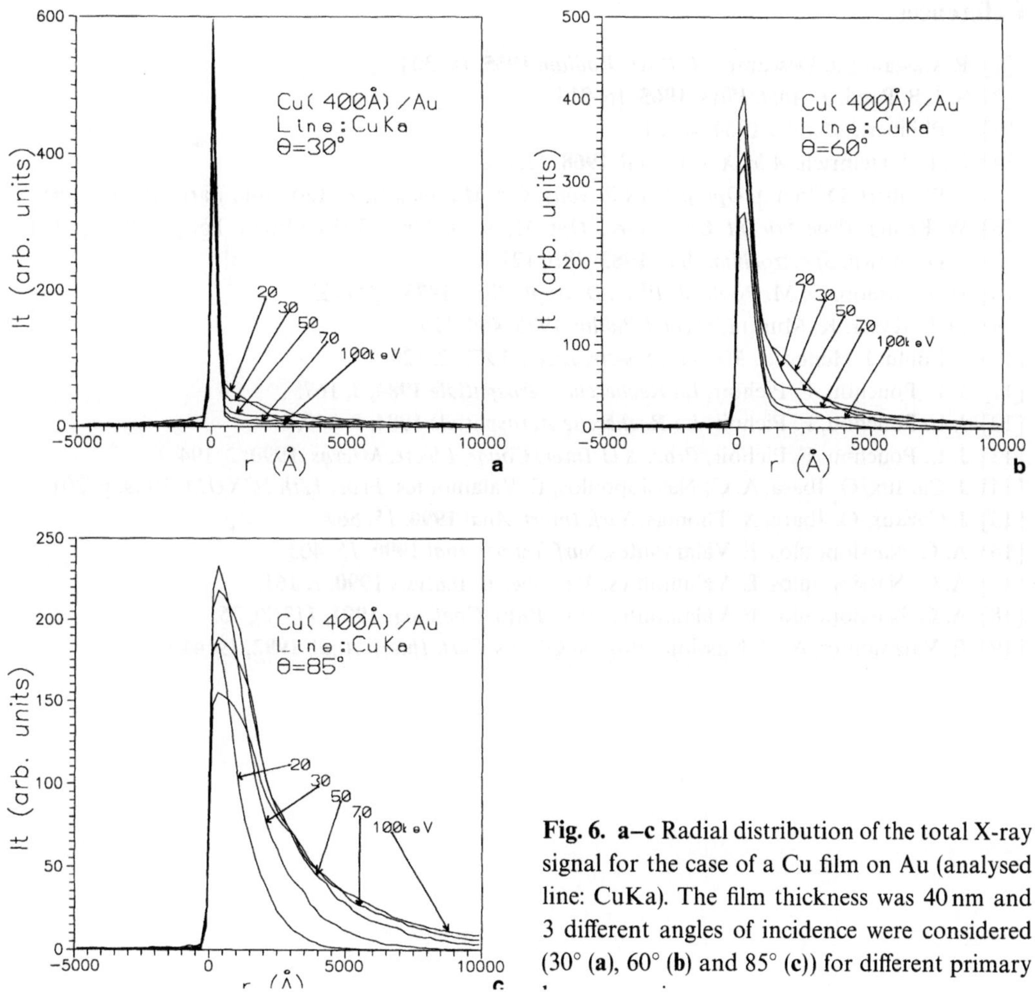

Fig. 6. a–c Radial distribution of the total X-ray signal for the case of a Cu film on Au (analysed line: CuKa). The film thickness was 40 nm and 3 different angles of incidence were considered (30° (**a**), 60° (**b**) and 85° (**c**)) for different primary

From the obtained curves we can roughly estimate the resolving power of the technique at the corresponding conditions by considering the obtained distributions as "distorted gaussians" and applying the known criteria [18, 19].

Conclusion

A Monte-Carlo programme has been developed for the determination of the X-ray signal from coatings, which takes into account all contributions of the total X-ray signal: a) signal induced by incident electrons b) backscattering from both the film and the substrate c) fluorescence due to characteristic X-rays and c) fluorescence due to continuous X-rays from the substrate. This programme may be used both for thin and for thicker coatings. It was applied to the case of a copper film on gold for different film thicknesses, primary beam energies and angles of incidence. The distribution of the X-ray signal in the projection of the plane of incidence on the sample surface is also calculated, with as parameters the film thickness, the angle of incidence and the primary beam energy. The developed Monte-Carlo programme may be used both for the analysis of coatings and in the case of known composition for film thickness determination.

References

[1] R. Castaing, J. Descamps, *J. Phys. Radium* **1955**, *16*, 304.
[2] S. J. B. Reed, *J. Appl. Phys.* **1965**, *16*, 913.
[3] J. Philibert, *Metaux* **1964**, *465*, 157.
[4] K. F. J. Heinrich, *Adv. X-ray Anal.* **1968**, *11*, 40.
[5] J. Philibert, D. Penot, *Optique des Rayons X et Microanalyse*, Hermann, Paris, 1966, p. 365.
[6] W. Reuter, *Proc. 6th Int. Conf. X-ray Opt. Microan.*, Univ. Tokyo Press, Tokyo, 1972, p. 121.
[7] J. D. Brown, *Spectrochim. Acta* **1983**, *38B*, 121.
[8] H. E. Bishop, D. M. Poole, *J. Phys. D. Appl. Phys.* **1973**, *6*, 1142.
[9] D. F. Kyser, K. Murata, *Special Public.* **1976**, *460*, 129.
[10] J. Foulu, J. Henoc, *J. Microsc. Spectr. Electr* **1977**, *2*, 129.
[11] J. L. Pouchou, F. Pichoir, *La Recherche Aerospatiale* **1984**, *3*, 167.
[12] J. L. Pouchou, F. Pichoir, *La Recherche Aerospatiale* **1984**, *5*, 349.
[13] J. L. Pouchou, F. Pichoir, *Proc. XII Inter. Congr. Electr. Micros.* **1990**, *2*, 194.
[14] J. Cazaux, O. Jbara, A. G. Nassiopoulos, E. Valamontes, *Proc. 12th ICXOM*, 1989, p. 201.
[15] J. Cazaux, O. Jbara, X. Thomas, *Surf. Interf. Anal* **1990**, *15*, 567.
[16] A. G. Nassiopoulos, E. Valamontes, *Surf. Interf. Anal* **1990**, *15*, 405.
[17] A. G. Nassiopoulos, E. Valamontes, *Microbeam Analysis* **1990**, *1*, 161.
[18] A. G. Nassiopoulos, E. Valamontes, *Inst. Phys. Conf. Ser.* **1991**, *117*(2), 75.
[19] E. Valamontes, A. G. Nassiopoulos, N. Glezos, *Surf. Interf. Anal.* **1992**, *19*, 419.

Mikrochim. Acta [Suppl.] 13, 611–622 (1996)

Analysis of Layers: X-Ray Maps of Change in Thickness Obtained by Electron Macroprobe

D. Viale* and G. Petitgand

IRSID, Voie Romaine – B.P. 320, F-57214 Maizieres les Metz, France

Abstract. Surface coatings on steels can significantly change the properties of this material, thus the characterization of such thin films is an important objective of the steelmakers. The aim of the present work is to investigate a method of obtaining elemental compositions and thickness profiles of coatings using X-ray mapping. The analyses are carried out by electron probe micro analysis using a technique based on the phi-rho-Z model of the depth distribution of X-ray emission.

Thickness X-ray maps of multilayered coatings can be built using suitable computer software combined with an accurate calibration method. The software used in STRATA from SAMx (France). The analyser used is a Camebax SXmacro from CAMECA. It is shown that, for the simple case of a chromium film (100 to 600 nm thick) on a steel substrate, the analysis can be performed at a single excitation voltage of 20 keV.

Of course, in a more complicated case, such as multilayers containing many elements it is necessary to work with several accelerating voltages. In a second example, a thin film containing Fe, Cr, Mo, Si and Ti was deposited on the previous chromium layer. With Strata computer software and our calibration procedure it is possible to obtain the composition and the coating thickness variation in the regions where the film was not too thick for analysis.

Key words: thickness, X-ray maps, thin films, multilayers.

At the present time, steels are often coated with thin films to improve their properties (to protect them against corrosion, to promote the contact between a tool and the material or to put a decoration) and increase their value. In this context, it is absolutely necessary to have some techniques to measure precisely both composition and thickness of these coatings which change the properties of the modern steels. When the layers are very thin, electron and ion microbeam methods are commonly used for these analyses. The previous workshop of the European

* To whom correspondence should be addressed

Microbeam Analysis Society on "Modern developments and applications in microbeam analysis" (EMAS '93) discussed the advantages of various techniques [1–2] (electron microprobe and ion microbeam methods like SIMS, RBS, channeling or NRA).

The electron microprobe is a very good tool, and some computer software capable of analyzing multilayered structures have been developed. The basic idea is to change the accelerating voltage of the electron beam to vary the depth of the analyzed volume. For example, low voltages (5 keV) are applied to the analysis of superficial thin layers whilst higher beam energies are used to investigate the thickest coatings or sub-surface layers. Using the STRATA software specially developed for thin layers analysis, combined with a specific work method, we can construct maps of composition and thickness of coatings on the surface of steels. These maps also allow us to verify the homogeneity of such films. We will demonstrate the feasibility of this method using two examples.

Short Theoretical Reminders

The depth distribution of ionisation events in a sample is influenced by the accelerating voltage of the electron probe. Thus, by varying this voltage it is possible to investigate elemental distributions as a function of depth. However, accurate interpretation of experimental data acquired from layered or inhomogeneous specimens requires prior knowledge of the structure, particularly when modelling the parameter $\Phi(\rho z)$ which represents the distribution in depth of the primary ionisation, according to the definition of Castaing [3]. The first $\Phi(\rho z)$ curves were originally determined by experiment. Monte-Carlo calculations [4] have produced $\Phi(\rho z)$ curves in good agreement with experimental data and have formed the basis for the development of convenient analytical models for $\Phi(\rho z)$. These models can be easily parameterised by using "simple" expressions [5]. The STRATA software utilised in this work uses both parabolic (PAP) [6–7] and exponential (XPP) [8–9] models to simulate the form of $\Phi(\rho z)$ curves.

If the knowledge of the distribution in depth of the primary ionisation is indispensable for the thin layer analysis, it is because this function $\Phi(\rho z)$ is the important parameter which affects the intensity of the X-ray emerging from the sample analysed. Indeed, the X-ray intensity measured by the microprobe derives directly from the number of ionisation of the considered element, while taking into account the number of photons that will be absorbed by the layer in which they are produced (equation (1)).

In the case of a surface coating of thickness ρz_c the characteristic X-ray intensity I_A produced by element A can be expressed as :

$$I_A \propto C_A \int_0^{\rho z^c} \Phi_A(\rho z) \exp(-\chi_A \rho z) d\rho z \tag{1}$$

Where C_A is the mass concentration of element A, $\Phi(\rho z)$ is defined such that: $\int_0^{\rho z^c} \Phi_A(\rho z) d\rho z$: total intensity generated in the film by atoms of element A, and χ_A is the absorption factor for A radiation in the film.

Similarly in a standard the measured intensity is proportional to:

$$I_A \propto C'_A \int_0^{\infty} \Phi'_A(\rho z) \exp(-\chi'_A \rho z) d\rho z \tag{2}$$

It is worth noting that the distribution $\Phi_A(\rho z)$ and $\Phi'_A(\rho z)$ in the sample and the standard are different, and similarly for the absorption factors. Moreover the concentration C'_A is equal to 1 for a pure standard.

In practice, a more useful quantity is the relative intensity K_A obtained by taking a bulk standard as a reference (note that factors of proportionality omitted in Eqs. (1) and (2) eliminate in the calculation of K_A).

$$K_A = \frac{I_A}{I_{\text{std}}} = \frac{C_A \int_0^{\rho z} \Phi_A(\rho z) \exp(-\chi_A \rho z) d\rho z}{C'_A \int_0^{\infty} \Phi'_A(\rho z) \exp(-\chi'_A \rho z) d\rho z} \tag{3}$$

The analysis of layered samples requires a modification of the equation (1), in order to include absorption of the photons emitted in a buried layer by upper layers In this simplest case of a two layered system, the emergent intensity produced by element A in the second layer is proportional to:

$$I_A^2 \propto C_A^2 \exp(-\chi_A^1 \rho Z_1) \int_{\rho Z_1}^{\rho Z_2} \Phi_A(\rho Z) \exp[-\chi_A^2(\rho Z - \rho Z_2)] d\rho z \tag{4}$$

where C_A^2 is the mass concentration of element A in layer 2
χ_A^1 is the absorption factor for A radiation by layer 1
χ_A^2 is the absorption factor for A radiation by layer 2
ρZ_1 is the mass thickness of layer 1 (superficial layer)
$\rho(Z_2 - Z_1)$ is the mass thickness of layer 2

Equation (4) can be written as:

$$I_A^2 \propto C_A^2 \exp[\rho z_1(\chi_A^2 - \chi_A^1)] \int_{\rho Z_1}^{\rho Z_2} \Phi_A(\rho z) \exp(-\chi_A^2 \rho z) d\rho z \tag{5}$$

More generally for a buried layer i of mass thickness $\rho(Z_{i+1} - Z_i)$:

$$I_A^i = C_A^i T_A^i \int_{\rho Z_i}^{\rho Z_{i+1}} \Phi_A(\rho z) \exp(-\chi_A^i \rho z) d\rho z$$

where

$$T_A^i = \prod_{k=1}^{i-1} \exp[\Delta\rho Zk(\chi_A^i - \chi_A^k)] \tag{6}$$

where $\Delta\rho zk$ represents the mass thickness of the layer K, χ_A^i and χ_A^k are factors of absorption of the photons coming from the element A, by the layers i and k respectively.

It emerges therefore from preceding expressions that the measured intensity depends both on the concentration C_A of the element, the thickness of the film in which the photons are emitted (by the integration of $\Phi(\rho z)$) and the absorption of the possible superior layers; the problem is therefore to separate these two

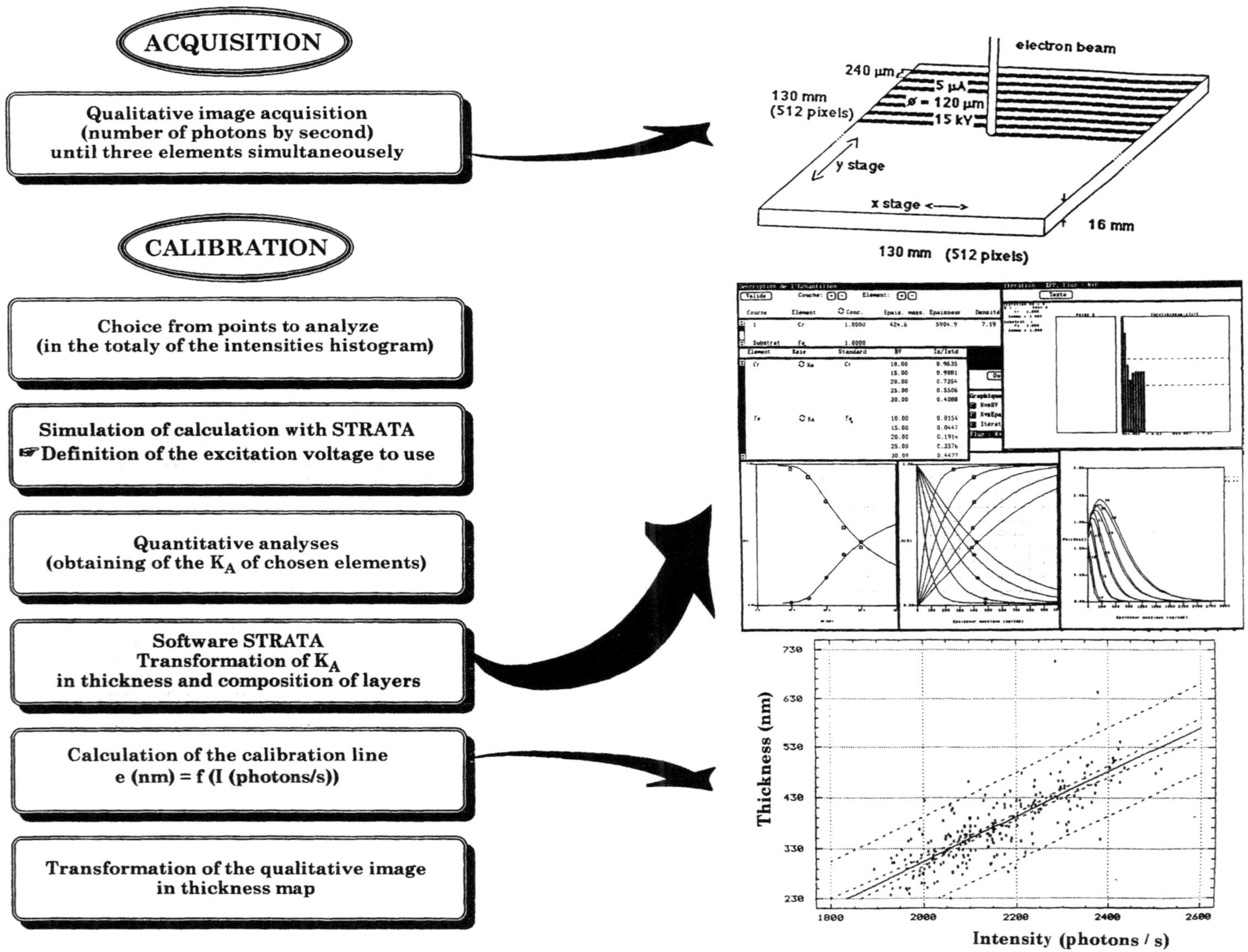

Fig. 1. Method of work to obtain a cartography in thicknesses

contributions. This is possible due to the fact that $\Phi(\rho z)$ varies with the electron accelerating voltage, what would impose theoretically to work with two different voltages to determine both composition and thickness of coatings.

Experimental

The experimental procedure used to obtain a cartography in thickness is rather similar to that developed for quantitative analysis of the segregation of elements on large surfaces [10–11].

The first step consists in acquiring qualitative images representing the intensity measured in number of photons per second. These images should be acquired at an accelerating voltage great enough so that the ionisation depth is greater than the coating thickness. The defocus beam mode [11] allows to obtain maps from areas of a few mm^2 in size up to approximately $130 \times 130\,mm^2$.

The next steps involve transforming these qualitative images into thickness maps. In this procedure of calibration summarized in Fig. 1, it is necessary to underline the two specific steps to the realization of thickness maps:

- firstly, before beginning the quantitative analysis to obtain the k_A values as defined in Eq. (3), it is necessary to determine the optimum accelerating voltages for the calibration measurements. This requires some prior knowledge of the approximate thickness and composition of the specimen. Given this information, the STRATA software calculates theoretical $K_A = f(E_0)$ curves. From such curves it is possible to choose the optimum analysis voltages.
- secondly, the specific step to obtain thickness maps consists in transforming k_A values (obtained by a quantitative analysis of a set of points previously chosen) into thickness and local composition of layers. The software allows this transformation by an iterative calculation based on theoretical data summarized previously.

Lastly, a thickness–intensity ($th = f(I)$) calibration line is plotted. This calibration is used to convert the original qualitative image into a thickness map.

Results and Discussion

The previously described methodology has been applied to the analysis of steel samples coated with various films. Results obtained on two examples will be presented: first the relatively simple case of a monoelement layer with a thickness variation, then the more complex case of multielement multilayer. In this latter case, some elements are common to several layers.

Thicknesses Map of Chromium Layers on the Surface of a Steel Sample

The purpose of this first study was simply to verify the feasibility of thickness mapping by electron microprobe. A polished substrate of steel (without Cr) was coated with a magnetron sputtered chromium layer (with this method of deposition, the knowledge of evaporation time allows to estimate the thickness of the coating). In a second time, a mask has been put on the sample and a second layer of the same metal deposited on the first. Figure 2 represents the sample thus realized with 2 thicknesses x and y to be determined. Qualitative X-ray map of iron and chromium signals (accelerating voltage: 15 keV, current intensity: 3 μA, diameter of probe: 40 μm and image of 512 × 512 pixels (262144 analysis X points)) are shown in Fig. 3.

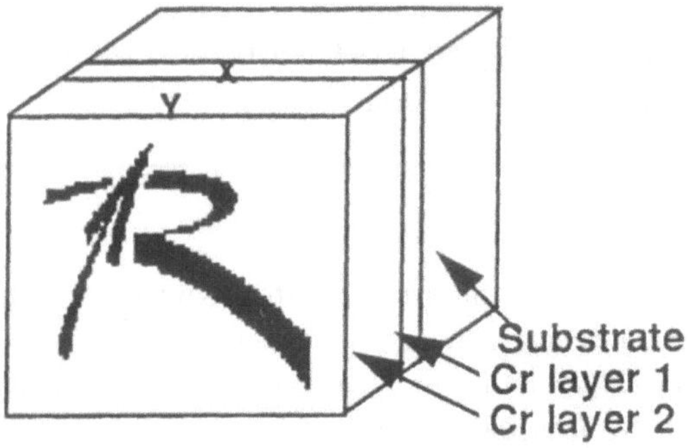

Fig. 2. Studied sample

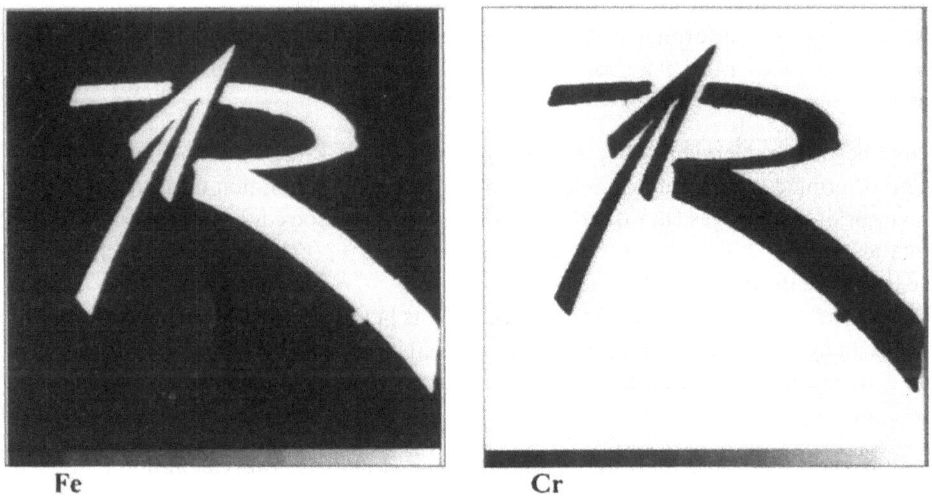

Fig. 3. Qualitative X-ray maps of iron and chromium

They contain some interesting information:

- first of all, the variations in the Fe and Cr signals clearly indicate the difference in thickness between the pattern and its contour,
- these images show that the "R" pattern is in hollow (only the thickness x covers the plate of steel, while a thickness x + y is deposed on the contour outside the "R"),
- the two images are well complementary; when the thickness of the coating increases the X intensity measured for the chromium of the layer increases, while the reverse is observed for the iron of the substrate (the intensity increase with the yellow colour on the maps).

The first preparatory step in the quantitative analysis is to determine the optimum voltage conditions. A simulation has therefore been realized with the software of layers analyses. It is necessary to define the elemental concentrations in the layer and the substrate, and then estimate the possible thickness of the film. Figure 4 presents theoretical $K = f(E_0)$ curves produced by STRATA for the chromium and the iron in the two extreme hypotheses where the sample would be covered 100 Å or 1 μm. In the case of 100 Å Cr layer it is impossible to undertake the analysis using a high accelerating voltage because above 20 keV such a layer is

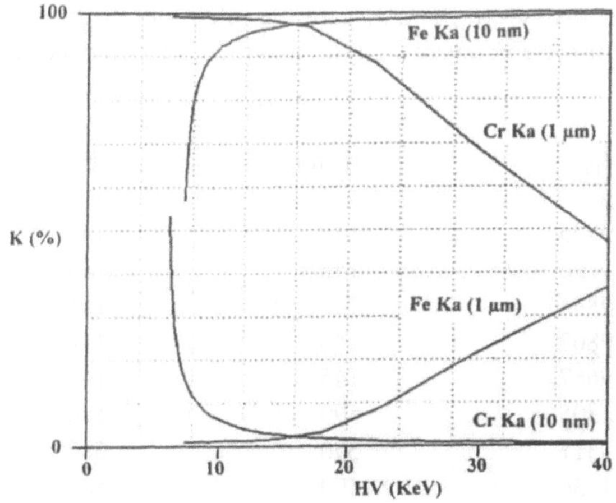

Fig. 4. Diagram $K = f(E_0)$ obtained by simulation

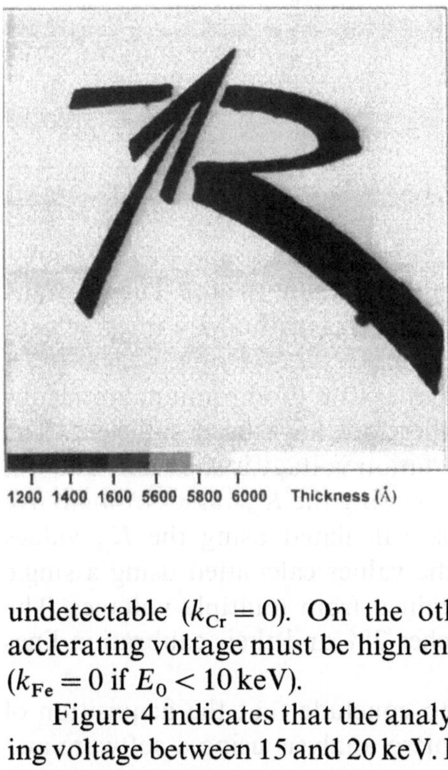

Fig. 5. Thickness map of the chromium layer

undetectable ($k_{Cr} = 0$). On the other hand, if the film becomes too thick, the acclerating voltage must be high enough to produce X-rays from the iron substrate ($k_{Fe} = 0$ if $E_0 < 10$ keV).

Figure 4 indicates that the analysis could be carried out using a single accelerating voltage between 15 and 20 keV. In fact, quantitative analyses were undertaken to 5 different voltage (10, 15, 20, 25 and 30 keV) in order to determine the accuracy of the analysis under optimum conditions. For each analysis point (up to 150), the experimental data (k_{Cr} and k_{Fe}) is input into the STRATA software which calculates the thickness. With the thickness determined from each point of analysis it is possible to construct a thickness (th) versus intensity (I) calibration line for each characteristic signal. For this example, the correlation coefficients of the two calibration lines ($th = f(I_{Fe})$ and $th = f(I_{Cr})$) are larger than 0.99 (no dispersion). The resulting thickness map is shown in Fig. 5. It indicates that the two superposed Cr layers are homogeneous in thickness; the first thickness (x) being 145 nm and the second (y) of 445 nm for a total thickness of 590 nm seen on the map (the difference of thickness

Table 1. Examples of results obtained to some analyses. I_{Cr} and I_{Fe} are the intensities noticed on qualitative maps, th is the thickness calculated for these points of analysis, and th20 is the thicknesses calculated from a single measurement at 20 keV

N	I_{Cr}	I_{Fe}	th (nm)	th20 (nm)
1	8034	9905	145	144
2	8024	9796	146	144
3	7942	9854	146	145
4	8051	9862	146	144
5	8071	9667	145	144
11	12344	1305	590	586
12	12373	1311	590	587
13	12319	1336	590	587
14	12411	1309	591	587
15	12405	1265	591	587
21	12359	1340	586	582
22	12429	1267	591	588
23	12494	1267	583	580
24	8078	9768	146	144
25	8117	9852	146	144

determined from each analysis point is never greater than 10 nm). These results seem very satisfactory since thicknesses measured by this method are in agreement with those expected according to the deposition times.

Thickness maps made from qualitative images of iron or chromium are identical; the acquisition of a single image would therefore have been sufficient. The possibility this map by operating at a single excitation voltage is shown in Table 1, where the thicknesses calculated at various points using the K_A values from all five accelerating voltages are compared with those calculated using the K_A values acquired at 20 keV only. It can be seen that the values calculated using a single voltage are almost identical to the results obtained from multiple voltages. The difference between these two results never reaches 2% and their calibration lines would be comparable.

In conclusion, this first experience allows to conclude that the acquisition of thickness maps is possible by electron probe micro analysis using a software and a methodology adapted.

Thickness Map of Complex Superposed Coatings

The problem approached in this second example is more complex since onto the preceding sample (Fig. 2), it has been deposited a layer containing 5 unknown elements. As previously, a part of the sample has been covered with a mask during of the deposition (the mask covered the name IRSID). The purpose of this study is to determine both composition and thickness of the superficial layer. This analysis is very interesting because the different elements of the target do not evaporate with the

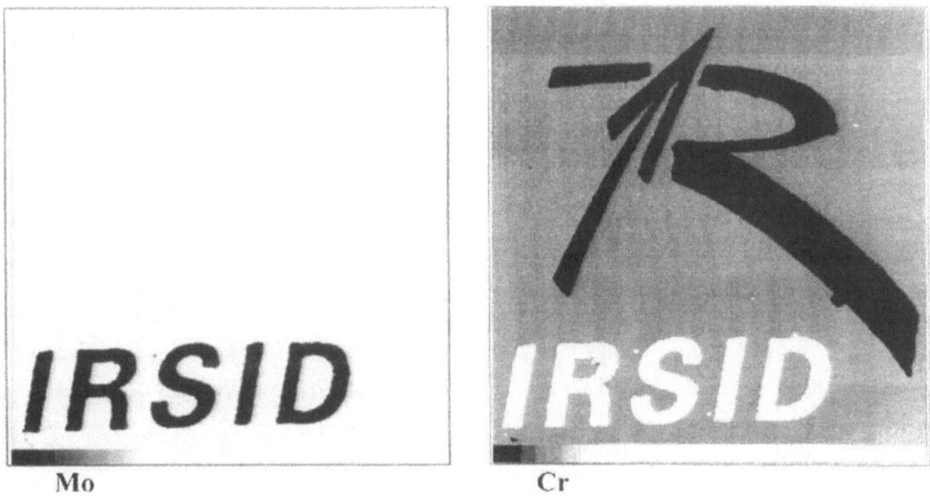

Fig. 6. Qualitative X-ray maps of molybdenum and chromium

same speed; it suggests therefore to verify if the deposition is homogeneous. The first step is to identify the elements present in the superficial layer. A global spectrum of every peak detected by the spectrometers was traced revealing that the elements Fe (intensity higher in the layer than outside), Mo, Ti, Si and Cr are present in the layer. Quantitative images of all five elements were acquired at 15 keV, (with the other parameters unchanged). The images of the Mo, Si, Ti are identical. They are easily interpretable since molybdenum, titanium and silicon are only present outside the "IRSID" name. Figure 6 contains one of these images. The colouring is uniform in all three cases, which indicates that the superficial layer is homogeneous in composition and in thickness (the probability that the concentrations and thickness of the coating varing in such a way as to produce qualitative images of uniform intensity is extremely small). Further, the "*R*" pattern in the original chromium layers does not appear in Fig. 6; the thickness of the multielement layer into and out of the pattern of the original chromium layer is equal.

Images of iron and chromium are more complex since both elements are present at more than one level in the sample. The intensity measured for iron comes from both the multielement surface layer and the substrate. It is at a minimum inside the masked region ("IRSID" name) of the multielement coating, and it is at a maximum value where the original Cr layer is thinnest (inside the "R" pattern). In contrast, the chromium intensity comes both from the multielement layer and the original Cr layers. The largest contribution comes from the chromium underlayers, hence the signal is strong when the original Cr layer is thick and weak then it is thinner (inside "*R*" pattern). The intensity is at a maximum inside the "IRSID" pattern where the chromium layers are exposed (Fig. 6).

Analysis and calculation steps involved in determining the deposit thickness are similar to those described previously, except that the coating composition is determined at each step of the iterative calculation. The presence of chromium in both the original and multielement layers requires that the results of the first analysis are input into the second calculation.

Table 2. Composition and thickness of the superficial coating

I_{Fe}	Fe%	I_{Mo}	Mo%	I_{Si}	Si%	I_{Cr}	Cr%	I_{Ti}	Ti%	th(nm)
12979	45	7894	29.5	10014	3	7615	16.5	11665	6	410
13015	45	8021	29	9860	3	7469	17	11799	6	415
13040	44.5	8026	29.5	9967	3	7585	17	11742	6	413
13021	45	7930	29.5	9879	3	7478	16.5	11561	6	405
13104	45.4	8075	29.2	9954	3	7381	16.7	11742	5.8	416
12942	45	8026	29.2	9989	3	7592	16.9	11736	5.9	418
12822	45.1	7934	29.4	9840	3	7550	16.7	11487	5.9	413
12868	45	8091	29.6	9887	3	7301	16.3	11834	6	409
12947	44.6	8215	29.2	10132	3	7302	17.2	11962	6	418
13122	45.3	8086	29.2	10116	3	7410	16.5	11705	6	413
12272		7892		10192		8504		11666		409
12239		7996		9970		8578		11726		406
12148		8115		10086		8584		11955		410
12317		8044		10055		8657		11837		409
12338		8059		9959		8588		11630		412
12260		8130		10183		8493		11675		413
12220		7955		10014		8732		11862		410
12312		7939		10190		8668		11851		407
12233		8069		10129		8623		12131		410
12271		7949		10217		8737		11947		410

An example of results obtained for 20 analysis points are given in Table 2. The upper part gives values for the characteristic intensities recorded at points inside the "*R*" pattern. For each point the calculated composition (weight %) and layer thickness are given. The calculated values for the compositions are identical to those given by the manufacturer for the magnetron target. The average thickness for this region is 410 nm giving a total thickness of 500 nm including the original chromium layers.

The results in the lower part of Table 2 come from points where the total thickness of the deposited layers exceed 1 μm. In this case it is not possible to determine both composition and thickness of the surface film with the available accelerating voltages. To achieve the required penetration depth for an accurate determination of $\Phi(\rho z)$ would require working at a higher voltage.

In addition to this limitation because of the maximum thickness, the second problem revealed by this study is the impossibility of determining simultaneously the thickness of the two superposed layers. It was necessary to know the results of the first analysis (previous example) to be able to calculate the composition and the thickness of the multielement surface coating. This difficulty comes from the fact that, in the present case, the single constituent (Cr) of one layer is also present into another layer. This conclusion is totally in agreement with that of [12] who observed that "both thickness and composition of superposed layers can be determined provided that the same element is not present in more than one layer. Only the thickness is obtained if layers are composed of same elements".

Fig. 7. Thickness map of the superficial multielement layer

Figure 7 shows the cartography of thickness variation of the Mo, Cr, Fe, Si and Ti surface layer. It allows us to conclude that this coating has an homogeneous thickness of 410 nm (standard deviation below 30).

Conclusions

These different studies show that it is possible to obtain coating thickness distribution maps on relatively large samples using an electron macroprobe in conjunction with suitable work method and the appropriate analysis software.

In simple cases (a single homogeneous surface layer with one or two constituents) this cartography can be obtained by working with a single accelerating voltage. In the more complex cases (multilayers containing common elements ···) the use of multiple voltages is absolutely necessary.

In these complex cases, the method has shown of its limits:

- maximum thickness of the different deposits,
- impossible to determine simultaneously the thickness and composition of two superposed layers where all the elements contained in one are also present in the other; thicknesses can be determined provided the composition of the different films is accurately known [12].

There remains several different situations to examine, for example the possibility of obtaining simultaneously thickness maps of several superposed layers, provided that each of them has at least one element that is not contained in the others. One of the other points that will soon be approached is the determination of the detection limit achievable by this technique. Lastly, the ultimate purpose of this study will be to analyze layers which are inhomogeneous either in thickness or composition. Coatings maps of variation in thickness and in true concentration distribution will then be realized.

References

[1] 3rd European Workshop on "Modern developments and applications in microbeam analysis" EMAS'93, 9–13 May 1993, Rimini.
[2] A. Armigliato, L. Van't dack, H. W. Werner, J. Wernisch (eds.), *Mikrochim. Acta* **1994**, *114/115*.
[3] R. Castaing, *Thèse*, Université de Paris, 1951.
[4] K. Murata, M. Kotera, K. Nagami, *J. Appl. Phys.* **1993**, *54*, 1110.
[5] P. Duncumb, *Mikrochim. Acta* **1994**, *114/115*, 3.
[6] J. L. Pouchou, F. Pichoir, *Electron Probe Quantitation*, Plenum, 1991, p. 31.
[7] J. L. Pouchou, F. Pichoir, *J. Phys.* **1984**, *2*, 45, C2.
[8] J. L. Pouchou, F. Pichoir, *Microbeam Analysis*, San Francisco Press, San Francisco, 1988, p. 315.
[9] J. L. Pouchou, F. Pichoir, *J. Microsc. Spectrosc. Electron* **1985**, *10*, 279.
[10] C. Daly, D. Jeulin, D. Benoit, G. Auclair, *ISIJ International* **1990**, *30*, 7, 529.
[11] G. Auclair, D. Benoit, *Proc. XIIth EMSA*, Seattle, 1990, p. 260.
[12] D. G. Rickerby, J. F. Thiot, *Mikrochim. Acta* **1994**, *114/115*, 421.

Mikrochim. Acta [Suppl.] 13, 623–629 (1996)

Comparison of Simulated and Experimental Auger Intensities of Au, Pt, Ni and Si in Absolute Units

Horst W. Wagner[1,2,*], Christian O. Schiebl[1], and Wolfgang S. M. Werner[2]

[1] Campusbased Engineering Center (CEC), Digital Equipment Österreich AG, Favoritenstrasse 7, A-1040 Vienna, Austria

[2] Institute for Applied Physics, Vienna University of Technology, Wiedner Hauptstrasse 8–10, A-1040 Vienna, Austria

Abstract. A comprehensive tool for the simulation of electron transport in solids with an arbitrary geometry has recently been developed at the CEC Vienna. To test the physical model on which the software package SESAME (Simulation of Electron transport Supporting Analysis of MatErials) is based, spectra of backscattered electrons and Auger electrons for primary energies in the range of 1–8 keV for elemental Si, Ni, Pt and Au have been measured in absolute units. The spectrometer was calibrated employing elastic peak spectroscopy. The simulation results are in agreement with these experimental values within the experimental error. Systematic deviations were observed for Si. Evidence has been found that the simulation model used is essentially valid for scattering in the bulk whereas surface excitations are known to effect the emission process.

Key words: Auger yield, electron transport, Monte Carlo simulation, backscattering factor.

As the precision of most electron microbeam techniques approaches 1%, and in some cases even better, increasingly accurate quantification schemes are necessary to ensure that the accuracy of analysis is of the same magnitude. This implies that the signal generation process needs to be quantitatively understood. In particular, fundamental quantities like the (differential) inelastic mean free path [1], the differential and total elastic cross-section [2], linear range [3], ionization cross-section [4], fluorescence yield [5], photo-ionization cross-section [6], and X-ray absorption length [7] need to be known with high accuracy.

In the present paper a simple method to measure Auger spectra in absolute units is outlined and applied to polycrystalline Au, Ni, Pt specimens and an amorphized Si-wafer. The method is based on experimental elastic peak intensities allowing

* To whom correspondence should be addressed

assessment of the apparatus transmission using the quantitative knowledge about the elastic backscattering process.

After calibration of the spectrometer, absolute Auger intensities were obtained for the investigated samples. These were compared with Monte Carlo calculations simulating the entire Auger generation process from incidence of the primaries to emission of the secondaries. These simulations were performed with the comprehensive SESAME code, recently developed at the CEC Vienna [8]. The simulated intensities agree with the measured values, within experimental error, for all investigated samples and for primary energies ranging from 1 to 8 keV.

Measurements

The samples were cleaned, polished and mounted in an UHV system. Targets were cleaned *In situ* with 2 keV Kr^+ bombardment until no traces of contamination could be detected in the Auger Spectra. All spectra were recorded at normal incidence of the primary beam using a CMA in constant $\Delta E/E$ mode with an angle of acceptance of $42° \pm 3°$. A Faraday cup integrated in the sample holder and a Keithley picoamperemeter were used to monitor the primary current.

To obtain absolute Auger yields the energy dependent effective transmission, $T_{eff}(E)$, of the apparatus has to be known. This quantity is defined as to take into account all processes between emission of an electron from the surface in the direction of the solid angle of acceptance of the CMA and its detection. This includes effects due to stray magnetic fields, the true transmission of the analyzer, efficiency of the electron multiplier, etc. The measured spectrum, $N_{\text{exp}}(E)$, is then related to the true energy/angular distribution of emitted electrons, $Y_{\text{abs}}(E, \Omega)$, as:

$$N_{\text{exp}}(E) = I_0 \Delta t\ T_{eff}(E) \int_{\Delta\Omega_{CMA}} Y_{abs}(E, \Omega) d\Omega \tag{1}$$

where I_0 denotes the primary beam current and Δt the sampling time. To experimentally obtain T_{eff}, the elastic reflection coefficients of the samples were measured for primary energies, E_p, ranging from 100 to 2000 eV. For a given primary current, the effective transmission can then be directly derived if the integral on the right-hand side of Eq. (1) is known. In the present work, this integral corresponding to the true elastic reflection coefficient, was evaluated by means of Monte Carlo simulations. The result of this procedure is shown in Fig. 1.

The values for T_{eff} obtained in this way are plotted for all four investigated elements. The solid line denotes a polynomial least square fit based on the results from all samples. At energies above 1200 eV the evaluation of the elastic peak was complicated due to problems connected with the electron optics of the system. Systematic deviations are seen especially in the case of the Si sample. For evaluation of absolute Auger yields (see below) the polynomial fit for T_{eff} shown in the Fig. 1 was used.

Auger spectra of all samples were recorded at primary energies ranging from 1 to 8 keV. The procedure to evaluate the Auger intensities from the spectra is illustrated in Fig. 2. First the data were normalized to the primary current, I_0, sampling time, Δt, and the acceptance angle of the CMA, $\Delta\Omega_{CMA}$, to obtain values of the spectra in absolute units (curve a). Subsequently the background of inelastically scattered primary electrons was subtracted using a polynomial fit (curve b). Then the spectra were corrected for the effective transmission using the solid curve in Fig. 1 (curve c). Finally, the background of inelastically scattered Auger electrons was subtracted using the deconvolution method described in ref. [9] (curve d). This approach is based on analysis of the kinetic equation for electron transport in the quasi-elastic regime, resulting in the following expression:

$$N(E) = y(\mu, E) - \kappa(\mu) \int_E^\infty dE' y(\mu, E') \omega(E' - E) \tag{2}$$

where ω denotes the normalized differential inelastic mean free path, $y(\mu, E)$ the measured yield at

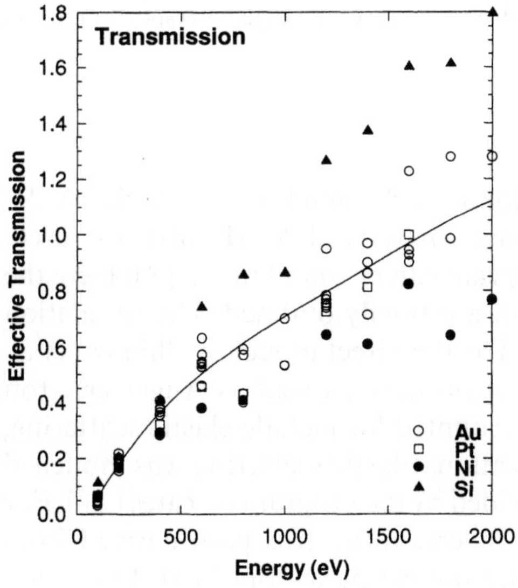

Fig. 1. Results of measurement of the effective transmission employing elastic peak spectroscopy. The solid line denotes the polynomial fit used to convert the measurements of Auger yields into absolute units

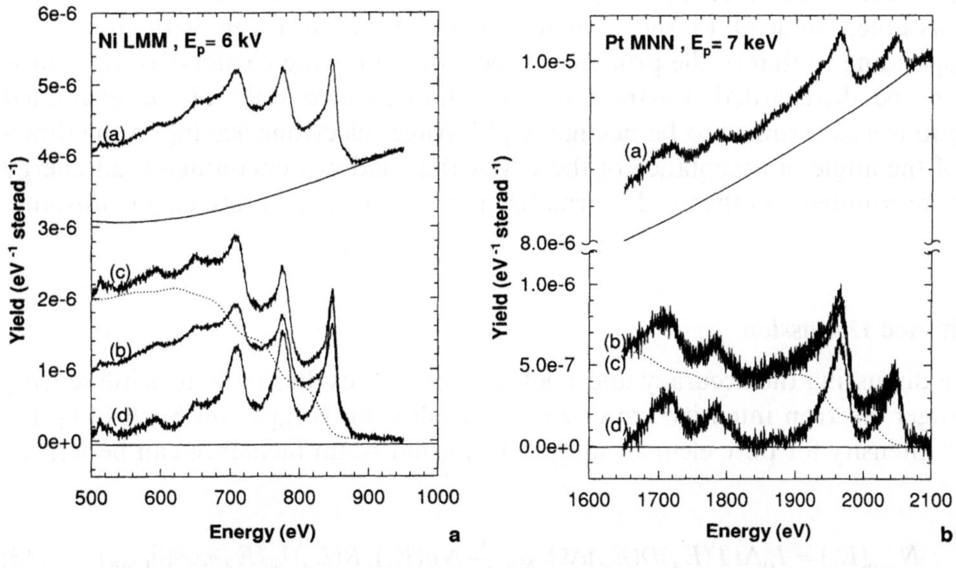

Fig. 2. a, b Evaluation procedure of Auger peaks for Ni and Pt. *a* denotes the original normalized spectrum, *b* after subtraction of inelastic primaries, *c* after correction due to effective transmission, *d* after application of the background procedure. The dotted line represents the background of inelastically scattered Auger electrons

energy E, the emission angle $\theta = \arccos(\mu)$ defined relative to the outward surface normal, and N(E) the deconvoluted result. The factor κ is a function of emission angle, transport mean free path [9], and inelastic mean free path (IMFP). Typical values of κ for Auger electron spectroscopy (AES) are in the range 0.8–0.95. The uncertainty in κ due to the lack of precise knowledge of the mean free path lengths does not critically influence the performance of this method. However, complete neglect of elastic scattering ($\kappa = 1$) leads to an overestimation of the background.

The deconvoluted Auger spectrum was integrated over the energy range corresponding to all relevant transitions of the ionized shell.

Simulation Model

For the simulations the newly developed Monte Carlo simulation code SESAME (Simulation of Electron transport Supporting Analysis of MatErials) was used. A detailed description of this simulation program can be found in ref. [8]. Here the basic features of interest for the present work are briefly outlined. The simulation model used corresponds to the one described as the direct model by Shimizu et al. [10] and was applied for both simulation of primary as well as Auger electron trajectories. The fundamental interactions accounted for include elastic scattering, inelastic scattering and electron impact ionization. Elastic scattering was modelled using the relativistic Mott cross-sections provided by the program of Yates [11]. For inelastic scattering the differential inelastic inverse mean free path derived from dielectric theory [12] has been applied using optical data from [13]. Ionization cross-sections have been calculated with the expression of Gryzinski [4]. Different energies of subshells were explicitly considered in the simulation but all ionizations were assumed to lead to the emission of an Auger electron with an energy corresponding to that of the principal Auger peak allowing Coster-Kronig transitions to be disregarded. Furthermore, the fluorescence yield of the examined transitions was assumed to be negligible [5]. Auger electrons leaving in the direction of the angle of acceptance of the CMA that had not encountered an energy loss were counted and the result normalized to give the Auger intensity in absolute units.

Results and Discussion

Before discussing the accuracy and relevance of our results, all factors influencing the Auger electron intensity are described. Following Briggs and Seah [14], the Auger intensity for pure element targets at normal beam incidence can be written as:

$$N_{\exp}(E_p) = I_0 \Delta t T(E_A) D(E_A) \Delta\Omega_{\mathrm{CMA}} \frac{1}{4\pi} N\sigma(E_p)\gamma R(E_p)\lambda_m(E_A)\cos(\theta_{\mathrm{CMA}}) \tag{3}$$

The product of analyzer transmission, T, and detector efficiency, D, in this equation corresponds to the effective transmission, T_{eff}, of this work. N denotes the atomic density of the target, σ the ionization cross-section of the examined shell, γ the Auger transition probability, R the backscattering factor which accounts for ionizations due to backscattered primary electrons. The term $\lambda_m(E_A)\cos(\theta_{\mathrm{CMA}})$ describes the escape probability in the direction of the acceptance angle of the analyzer.

The first five factors on the right-hand side of Eq. (3) describe the experimental factors influencing the intensity, the remaining terms account solely for the elemental interaction processes. The experimental error in the presented data is almost entirely determined by the uncertainty in T_{eff} which is estimated to be $\approx 30\%$ accounting for errors in the experimental determination and simulation of the elastic peak area

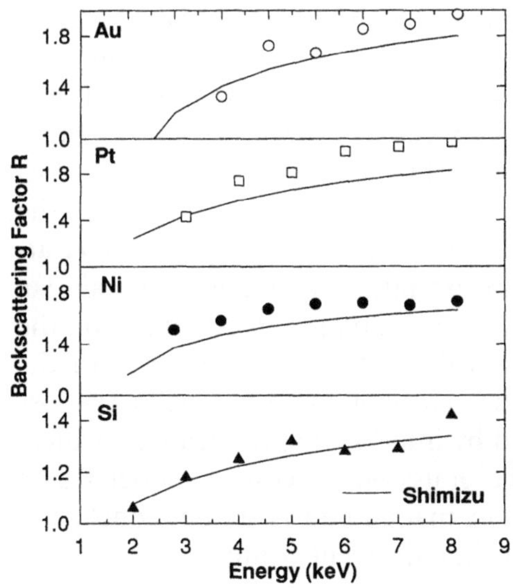

Fig. 3. Simulation results of the backscattering factor as a function of primary energy compared with the expression given by Shimizu (solid line) from similar model calculations

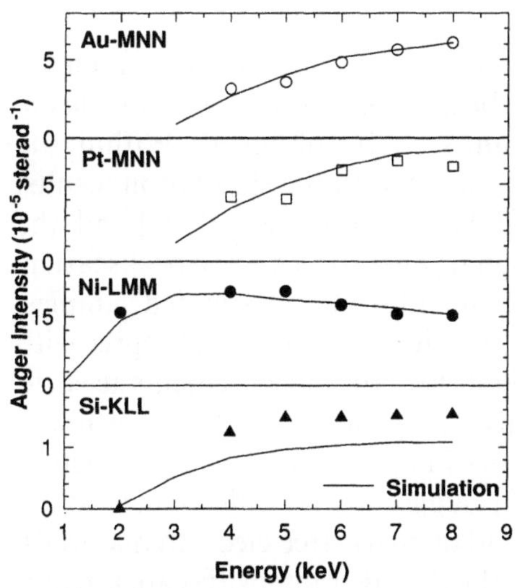

Fig. 4. Comparison of simulated and measured Auger yields. Note that values are both given in absolute units

as well as the statistical scatter of the data around the fit. The error in the simulation results is solely determined by the accuracy of the basic physical quantities. A comparison of the backscattering factor R calculated with our model with the expression given by Shimizu [15] is shown in Fig. 3. Although this author used the continuous slowing down approximation to describe inelastic scattering, the results agree well with the present model employing the differential mean free path based on optical data. On the other hand, the electron emission probabilities from our model agree within $\leqslant 3\%$ with transport theory [9]. Thus, it can be concluded that the uncertainty in the simulation is governed by the accuracy of σ and λ_m. The simulation results for the Auger electron intensities are compared with the experimental data in Fig. 4. In view of the rather large estimated error in our results, the agreement between simulation and measurement in surprisingly good. The only exceptions are the experimental results for Si which are consistently higher than the Monte-Carlo data. The relative intensities of all four samples at 5 keV agree reasonably well with the peak-to-peak sensitivities reported by Davis et al. [16]. Inconsistencies for Si were not only found for the Auger yields, but also for the elastic reflection probability as can be deduced from Fig. 1. A similar deviation of the elastic peak intensity of Al was found in previous experiments [17] using a different experimental setup. On the basis of REEL (reflection electron energy loss) measurement [18], this was attributed to differences in the mechanics of surface excitations between the metals Au, Ni and Pt on the one hand and the nearly free electron materials Al and Si on the other hand. The reason for these differences is likely to be connected with the plasmon governed loss function of the nearly free electron materials, while the loss functions of the other materials are determined by interband and intraband transitions. These findings suggest that, also in the emission process, a model which accounts only for bulk-like excitation is generally inadequate. A similar conclusion was drawn by Yoshikawa and co-workers [19] in a different context.

Summary and Conclusions

Auger intensities of Si, Ni, Pt and Au have been both measured and simulated for primary energies ranging from 1 to 8 keV. A calibration procedure based on elastic peak intensities allowed results in absolute units to be obtained. Within the experimental error, these results show good agreement with the simulation results. However, results for Si disagree with the simulations. On the other hand the simulation of the main quantity describing the signal generation, i.e. the backscattering factor, agrees well with earlier results by Shimizu for all investigated samples. Moreover, the simulation of the emission process was found to accurately agree with results obtained from transport theory [9]. The reflection measurements used to calibrate the spectrometer exhibit a similar inconsistency for Si as earlier experimental data for Al obtained on a different apparatus [17]. Analysis of the reflection energy loss spectra of these earlier data [18] suggests that this is caused by a difference in the surface excitations of metals and of nearly free electron materials. These considerations indicate that deviations in the Si Auger intensities are attributable to the model which accounts for bulk-like plasmon excitations only. Summarizing, it can be concluded that the simulation model used in SESAME is a powerful approach to the simulation of electron beam analysis. To improve the accuracy for surface studies (AES/XPS) the results indicate that a more sophisticated emission model is required.

References

[1] S. Tanuma, C. Powell, D. Penn, *Surf. Interface Anal.* **1988**, *11*, 577.
[2] N. F. Mott, *Proc. R. Soc. A* **1929**, *124*, 425.
[3] H. Bethe, *Ann. Phys.* **1930**, *5*, 325.
[4] M. Gryzinski, *Phys. Rev.* **1965**, *138 A*, 336.
[5] W. Bambynek, B. Crasemann, R. W. Fink, H. U. Freund, H. Mark, C. D. Swift, R. E. Price, P. Rao, *Rev. Mod. Phys.* **1972**, *44*, 716.
[6] J. H. Scofield, *J. Electron. Spectros.* **1976**, *8*, 129.

[7] K. Heinrich, *The Electron Microprobe*, Wiley, New York, 1966.
[8] C. Schiebl, A. Pfeiffer, H. W. Wagner, W. S. M. Werner, H. Stippel, *Microchim. Acta* [*Suppl.*] **1996**, *13*, 533.
[9] I. Tilinin, W. S. M. Werner, *Surface. Sci.* **1993**, *290*, 119.
[10] R. Shimizu, D. Ze-Jun, *Rep. Prog. Phys.* **1992**, 487.
[11] A. C. Yates, *Comp. Phys. Commun.* **1971**, *2*, 175.
[12] D. Penn, *Phys. Rev. B* **1987**, *35*, 482.
[13] E. Palik (ed.), *Handbook of Optical Constants of Solids*, Academic Press, San Diego, 1985.
[14] D. Briggs, M. P. Seah (eds.), *Practical Surface Analysis, Vol. 1*, Wiley, New York, 1990.
[15] R. Shimizu, *J. Appl. Phys.* **1983**, *22*, 1631.
[16] L. E. Davis, N. C. MacDonald, P. W. Palmberg, G. E. Riach, R. E. Weber, *Handbook of Auger Spectroscopy*, Physical Electronics Industries Inc., Minnesota, 1976.
[17] H. Beilschmidt, I. S. Tilinin, W. S. M. Werner, *Surf. Interface. Anal.* **1994**, *22*, p. 120.
[18] W. S. M. Werner, M. Hayek, *Surf. Interface Anal.* **1994**, *22*, 79.
[19] H. Yoshikawa, R. Shimizu, Z. Ding, *Surface Sci.* **1992**, *261*, 403.

Mikrochim. Acta [Suppl.] 13, 631–638 (1996)

Practical Aspects and Applications of EPMA at Low Electron Energies

Peter Willich* and Reinhold Bethke

Fraunhofer-Institut für Schicht- und Oberflächentechnik, Bienroder Weg 54E, D-38108 Braunschweig, Federal Republic of Germany

Abstract. Electron probe microanalysis is carried out at electron energies of 3–5 keV to reduce the volume of analysis of dimensions of 0.1 μm (depth) or 0.3 μm (lateral resolution). At these conditions the minimum detectable concentration for most elements is of the order of 0.1 wt% by WDS or 1 wt% by EDS. The accuracy of quantitative analysis is estimated to be at least 5% (WDS). Significantly higher (> 10%) analytical errors may occur in cases of strong chemical shift and due to the influence of surface contamination. The practical use of EPMA at low electron energies is demonstrated by examples of stratified materials.

Key words: EPMA, low voltage, quantitative analysis, stratified specimens, microsections.

Electron probe microanalysis (EPMA) can be performed by wavelength dispersive X-ray spectrometry (WDS) or by energy dispersive X-ray spectrometry (EDS) in a scanning electron microscope. Typical electron energies (E_0) are in the range of 15–25 keV. The depth of analysis is given by the ultimate depth of X-ray emission(R_e), which in cases of not too strong X-ray absorption is almost identical with the maximum depth of X-ray generation (R_g). R_g can be estimated by Castaing's formula [1]

$$R_g[\mu m] = 0.033/\rho\,(E_0^{1.7} - E_c^{1.7})A/Z, \tag{1}$$

where ρ [g/cm^3] is the density of the target, A and Z are the mean atomic weight and the mean atomic number of the material. E_c [keV] is the critical excitation energy of the X-ray line. For $E_0 = 15$–25 keV the depth of analysis ($\sim R_g$) is typically in the range of 1–3 μm. In the case of strong X-ray absorption, e.g., for the low energy X-rays of the ultralight elements (*B-O*), R_e, which is in fact the depth of analysis, may be significantly smaller than E_g. As an example, for CKα in SiC at $E_0 = 10$ keV, values of $R_g = 1.1$ μm and of $R_e = 0.5$ μm can be derived from the calculation of $\Phi(\rho z)$

* To whom correspondence should be addressed

depth distribution functions [2]. The X-ray lateral spatial resolution or X-ray source size is equal to the sum of R_g and the size of the electron beam (D_e). In an electron microprobe D_e ($<0.2\,\mu$m) is usually much smaller than R_g ($>1\,\mu$m), consequently, at electron energies of 15–25 keV, the lateral spatial resolution is limited to dimensions of about 1 μm.

In principle, it should be possible to reduce the depth and lateral resolution of EPMA to dimensions of $<0.3\,\mu$m (Eq. 1) by operating the instrument at $E_0 < 5$ keV in combination with the use of soft X-ray lines ($E_c < 3$ keV). Under these conditions it becomes possible to treat surface coatings, with thicknesses as small as 0.15 μm, as bulk samples [2, 3]. For such applications, the use of a WDS instrument is preferable with respect to sensitivity and quantitative analysis of ultralight elements. For applications under conditions of an ultimate lateral resolution, the combination SEM/EDS may be of interest. A lateral resolution of 0.1 μm was demonstrated [4]. Experiments were carried out by a field emission gun SEM, over a range of beam energies from 2–5 keV. In general, EPMA at low voltages to reduce the volume of microanalysis to dimensions of $<0.3\,\mu$m seems to be practicable. In the following, practical aspects and problems of low voltage EPMA will be discussed. This concerns the analytical sensitivity and the accuracy of quantitative analysis. Moreover, submicron lateral resolution by WDS-EPMA and special applications to thin film characterization are presented.

Experimental

WDS-EPMA was carried out in a CAMECA SX 50 microprobe using TAP and PET crystals for X-rays with E_c in the range of 0.6–2.6 keV. Multilayer monochromators were used to determine the ultralight elements B–O. At $E_0 = 4$ keV the instrument provided a stable electron beam of 100 nA with a minimum beam size of 2 μm. For a beam current of 5 nA the minimum beam size was about 0.1 μm. Before and during data acquisition for quantitative analysis, an oxygen gas-jet device was activated. Experiments by EDS were performed by an EDAX (Be-window) detector attached to a PHILIPS 505 (LaB_6) scanning electron microscope.

Analytical Sensitivity

The sensitivity of electron probe microanalysis may be expressed by the minimum detectable concentration c_{MDL}. The equation for c_{MDL} can be written as

$$c_{\mathrm{MDL}}[\mathrm{wt\%}] = 3\, c_S\, (I_B t)^{1/2}\, (I_{P-B} t)^{-1}, \tag{2}$$

where c_S is the concentration of the element in the specimen, I_B and I_{P-B} are the count rate (in c/s) of the background and the net count rate of the X-ray line, t (in s) is the counting time. Figure 1 shows the minimum detectable concentration of Si in Fe as a function of the ultimate depth of X-ray generation (R_g). An electron energy of 4 keV is correlated with an R_g of only 0.1 μm, as compared with about 1 μm at $E_0 = 20$ keV. The increase of c_{MDL}, when going from 20 to 4 keV, is about one order of magnitude. However, at $E_0 = 4$ keV, WDS-EPMA still provides a sensitivity of $c_{\mathrm{MDL}} = 0.05$ wt%, which is sufficient for many practical applications. The situation with respect to EDS, where c_{MDL} is about 1 wt%, is more critical. In practise, it may be difficult at $E_0 < 5$ keV to determine concentrations <10 wt% with a precision of

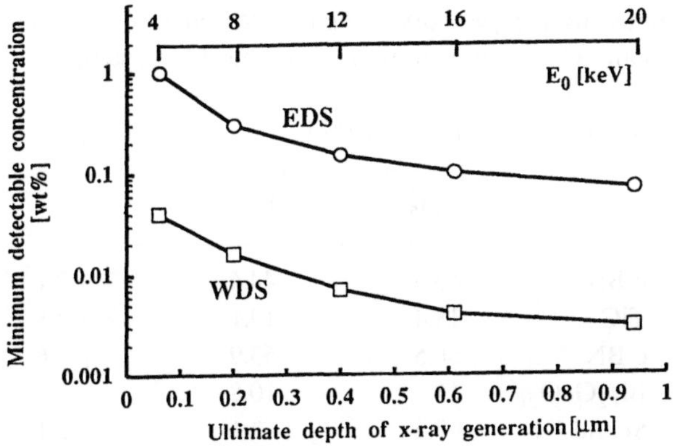

Fig. 1. Minimum detectable concentration for Si as a function of the ultimate depth of X-ray generation. Sample of 3.2 wt% Si in Fe. EDS $I_0 = 1$ nA, $t = 200$ s. WDS $I_0 = 100$ nA, $t = 100$ s (peak), 100 s (background)

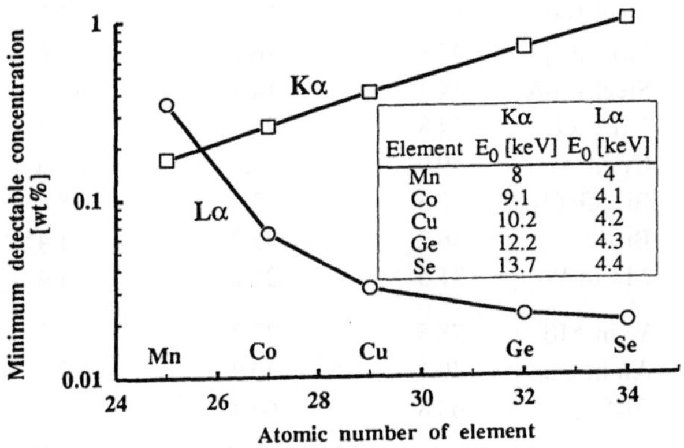

Fig. 2. Minimum detectable concentration at electron energies corresponding to 75 μg/cm^2 (~0.1 μm) for the depth of analysis. WDS $I_0 = 100$ nA, $t = 100$ s (peak), 100 s (background)

better than 20% relative by EDS, whereas WDS enables a precision of <5% for counting times of the order of 100 s.

As an alternative to experiments at low electron energies to reduce the depth of X-ray generation, one may consider the possibility of using relatively "hard" X-rays (E_c = 5–10 keV) in combination with electron energies of 7–15 keV. According to Eq. (1), R_g, as a consequence of the low overvoltage E_0/E_c, is also reduced to dimensions of 0.2 μm or even less. In Fig. 2 the analytical sensitivity of some transition elements determined by K-lines at electron energies of 8–14 keV is compared with the sensitivity of the corresponding L-lines determined at E_0 = 4–5 keV. EPMA was carried out on the pure elements at a value of E_0, corresponding to R_g = 75 μg/cm^2 (0.11 μm at a density of 7 g/cm^3) as a fixed depth of analysis. As an example, for Cu Lα at E_0 = 4.2 keV, the minimum detectable concentration is about 0.03 wt% (WDS) in comparison with 0.4 wt% for Cu Kα at 10.2 keV (Fig. 2). For elements with $Z < 26$ (Fe) the use of K-lines is preferable. For elements with $Z > 73$ (Ta), the minimum detectable concentration using M-lines is on the order of 0.1 wt% for electron energies of 4–6 keV. The density of alloys containing elements with $Z > 73$ may be high (> 10 g/cm^3), consequently, R_g is reduced to dimensions of 50 nm or even less.

Table 1. Quantitative analysis of reference materials at $E_0 = 4\,\mathrm{keV}$ using the PAP matrix correction model. Experimental: peak area k-ratios for B and C, gas-jet operation on standards and samples

Element	Line	Standard	Sample	c_{sample} [wt%]		Δ[%]
				Bulk	EPMA	
B	Kα	B	c-BN	43.5	44.4	+2.1
C	Kα	vitreous C	TiC	18.4	19.1	+3.8
N	Kα	Cr_2N	c-BN	56.5	53.9	−4.6
O	Kα	Fe_2O_3	$Bi_{12}GeO_{20}$	11	10.8	−1.8
Si	Kα	Si	Si and Fe	3.2	3.1	−3.1
S	Kα	FeS_2	Sb_2S_3	28.5	29.4	+3.2
Cr	Lα	Cr	Steel 160A	18.7	17.9	−4.5
Mn	Lα	Mn	Steel 160A	1.6	1.3	−19
Fe	Lα	Fe	Steel 160A	61.6	62.4	+1.3
		Fe_2O_3	$NiFe_2O_4$	47.8	46.1	−3.6
Ni	Lα	Ni	Steel 160A	14.1	14.8	+5
		NiO	$NiFe_2O_4$	24.8	25.9	+4.4
Cu	Lα	Cu	Cu in Au	59.9	61.9	+3.3
Ge	Lα	Ge	$Bi_{12}GeO_{20}$	2.5	2.3	−8
Se	Lα	Se	Bi_2Se_3	36.2	35.7	−1.4
Mo	Lα	Mo	Mo in W	21.5	21.2	−1.4
W	Mα	W	W in Mo	78.5	77.2	−1.7
Au	Mα	Au	Au in Cu	40.1	40.9	+2
Bi	Mα	Bi	Bi_2Se_3	63.8	64.4	+1

Quantitative Analysis

The accuracy of quantification at $E_0 = 4\,\mathrm{keV}$ was tested by analysis of the reference materials in Table 1. Matrix correction was carried out using the PAP model [5]. Table 1 includes EPMA based on K-lines (elements B-S), L-lines (elements Cr-Mo) and M-lines (elements W-Bi). The accuracy for the determination by K-lines, including the ultralight elements (B-O), is reasonable. Also EPMA by L- and M-lines with $E_c > 0.9\,\mathrm{keV}$ (Cu-Mo and W-Bi) causes no problems, the result for Ge (concentration 2.5 wt%) is probably influenced by poor counting statistics. The most critical situation proved to be the use of L-lines to perform quantitative analysis of Cr, Mn, Fe and Ni. Serious problems of chemical shift may occur, the determination of peak-area intensities leads to results even worse than those listed in Table 1, which are based on peak-height intensities. Moreover, it is known that some transition elements exhibit significant changes in their mass absorption coefficient, depending on chemical bonding [6]. To keep the effects of chemical bonding as low as possible, it is strictly recommended to use "matched" standards, for which the nature of the chemical bond is similar to that of the sample, e.g. oxide standards for the analysis of oxide materials. Following this rule, the examples of Table 1 (Cr, Mn, Fe, Ni in steel, $NiFe_2O_4$) show an accuracy of better than 5% relative for the use of L-lines, apart from Mn in steel, where the concentration is very low (1.3 wt%).

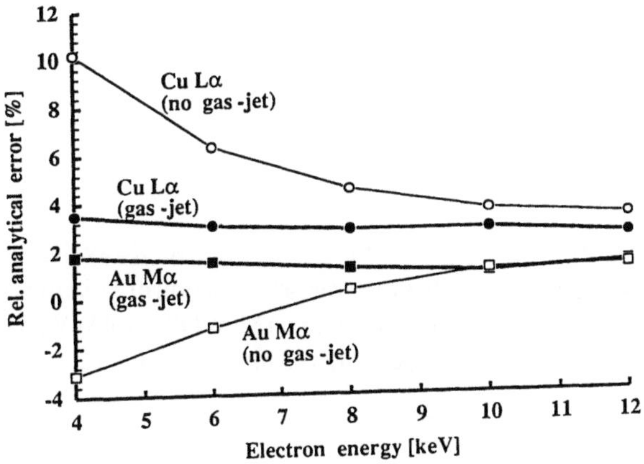

Fig. 3. Influence of surface contamination on the accuracy of quantitative analysis. Gas-jet operation on sample and standards before and during data acquisition

A serious experimental problem of quantitative analysis at low electron energies concerns the strong influence of surface contamination. This is demonstrated by the example of an Au-Cu alloy (Fig. 3). At first instance, analysis was carried out on not specially cleaned standards (pure elements) and sample. For low electron energies, e.g. 4 keV, the analytical error proved to be as high as 10% (Cu Lα). The intensity of C Kα was also determined, the values can be used [2] to estimate the thickness of carbonaceous deposits on the surface: 4 nm on Au-standard, 16 nm on Cu-standard, 9 nm on Au-Cu alloy. Consequently, analysis was repeated using the oxygen gas-jet device of the instrument before (120 s) and during the intensity measurements, which enables one to remove the carbonaceous deposits almost completely (<2 nm). As shown in Fig. 3, the results of quantitative analysis, as a consequence of gas-jet operation, are now more consistent, the error for Cu Lα at 4 keV (3.8%) is acceptable. This example shows, that in the case of operation at low voltages, the contamination on samples and standards has a strong influence on the accuracy of quantification. Deceleration of the low energy primary electrons, due to surface contamination, is relatively high. As a consequence, less X-rays are generated in the underlying material. Surface contamination also causes X-ray absorption, although this effect is less dominating for Cu Lα and Au Mα. Moreover, X-ray absorption is independent of primary electron energy.

Submicron Lateral Resolution

For EPMA at low electron energies, the ultimate depth of X-ray generation (R_g) may be 0.3 μm or even less. The X-ray source size or lateral spatial resolution $D_x = R_g + D_e$ may have the same order of magnitude, provided that the size of the electron beam (D_e) is reduced to 0.1 μm or less. In the case of WDS-EPMA, the minimum beam current is about 5 nA, the corresponding X-ray count rates at, e.g., $E_0 = 4$ keV, are typically the order of 500–5000 cps (pure elements) for the X-ray lines listed in Table 1. The size of the electron beam with a beam current of 5 nA is on the order of 0.05–0.1 μm, as roughly estimated from the corresponding SEM-image. Figure 4 shows the example of line scans (beam scanning with 0.1 μm step width) over the cross-section of a CVD-diamond coating deposited on Si. Experiments were carried out at 4 and 10 keV with a beam current of 5 nA. The experimental X-ray

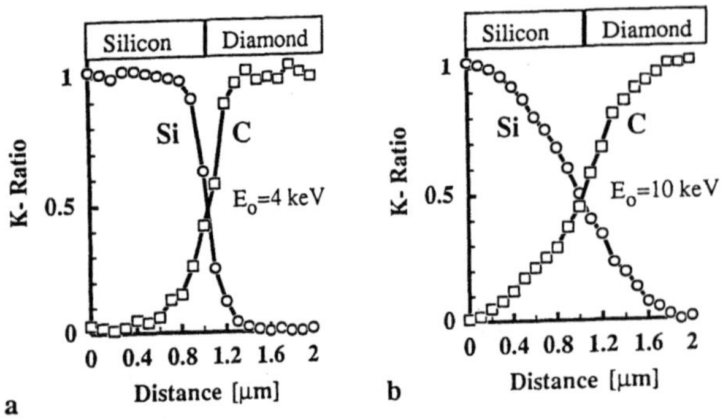

Fig. 4. Profile of relative X-ray intensities over the cross-section of a diamond film deposited on Si. WDS at **a** 4 keV and **b** 10 keV, $I_0 = 5$ nA

source size (D_{xe}) may be defined as the region in which the signal intensity at an infinitely steep concentration step changes from 16% to 84% of the step height. The D_{xe}-values at 4 keV (Fig. 4a) are 0.3 μm (Si Kα) and 0.4 μm (C Kα) compared with 0.8 μm (Si Kα) and 0.9 μm (C Kα) at 10 keV (Fig. 4b). These values are in agreement with $D_x = R_g + D_e$, when calculating R_g by Eq. 1 and assuming an electron beam size of $D_e = 0.1$ μm. A further attempt was made to apply an electron energy of only 2.5 keV. However, the D_{xe}-values became worse (of the order of 0.5 μm). This indicates the inability of current microprobe instruments (tungsten hairpin filament) to provide a sufficiently intense (beam current 5–10 nA) and well-focused beam at electron energies below 3 keV.

Thin Films and Stratified Specimens

Surface coatings with film thicknesses as small as 0.1–0.3 μm (depending on the density of the material) can be treated as bulk samples [2, 3]. A favourable situation is represented by the determination of ultralight elements (B-O). At low electron energies, e.g., 4 keV, the sensitivity is almost optimal, particularly when EPMA is carried out on materials exhibiting strong matrix absorption [2]. The shallow generation of the low-energy X-rays also reduces the magnitude of the matrix (absorption) correction, which may improve the accuracy of quantitative analysis [2].

The low depth of analysis at $E_0 < 5$ keV can be used to perform "depth profiling" on stratified specimens. For this application a microsection (angle lapping) of the sample is prepared, e.g., by the technique of grinding a flat spherical crater into the surface using a steel ball covered with diamond paste ("ball cratering"). An example is given in Fig. 5, the step width of the line scan was 10 μm, the size of the electron beam (100 nA) was about 2 μm at $E_0 = 5$ keV. The "depth resolution" of about 0.2 μm at $E_0 = 5$ keV is only determined by the ultimate depth of X-ray emission, in contrast to the technique of perpendicular line scans on cross-sections (Fig. 4), where the "depth resolution" is also dependent on the size of the electron beam.

EPMA can be applied to the characterization of very thin films where the film thickness is only a fraction of the ultimate depth of X-ray generation [3, 6]. The X-ray intensities are related to the film (mass) thickness and composition, which can be determined simultaneously. In the case of film thicknesses of 20 nm or less the optimal sensitivity is obtained at electron energies in the range of 5 keV. As

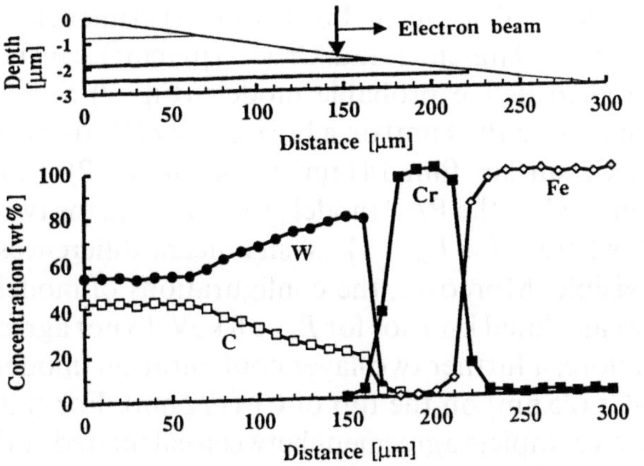

Fig. 5. Depth profile of W-C/Cr/steel layered structure. Lateral scan over a microsection (angle lapping) using WDS at $E_0 = 5\,\mathrm{keV}$

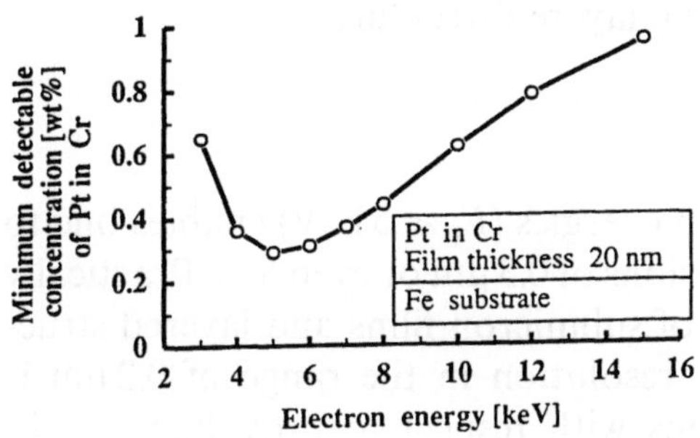

Fig. 6. Minimum detectable concentration of Pt (Mα-line) in a thin Cr-film as a function of E_0. WDS: $I_0 = 100\,\mathrm{nA}$, $t = 100\,\mathrm{s}$ (peak), 100 s (background)

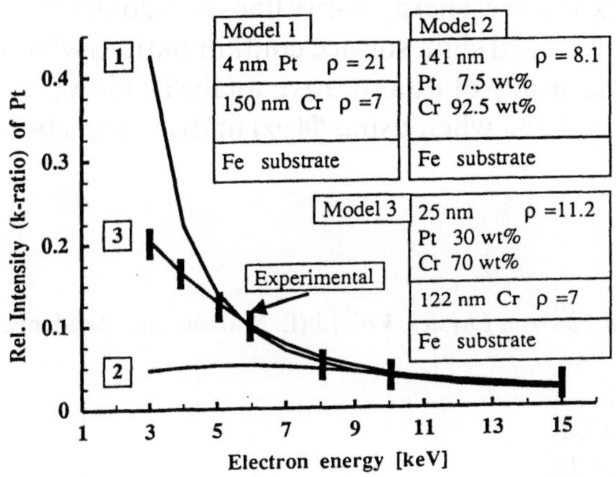

Fig. 7. Characterization of a Pt/Cr stratified film on Fe. K-ratio vs. E_0 data calculated by the PAP model for different configurations of the layered structure

illustrated in Fig. 6, EPMA enables one to detect a concentration of 0.3 wt% Pt in a Cr- film of 20 nm. The detection limit corresponds to $0.04\,\mu\mathrm{g/cm^2}$ for the mass thickness of Pt, which, in terms of a "pure" Pt- film, represents a "theoretical" film thickness of only 0.02 nm.

The evaluation of EPMA experiments at different electron energies enables one to describe stratified targets in terms of the parameters film (mass) thickness and composition [3, 6]. An example is given in Fig. 7. A surface coating containing the elements Pt and Cr was deposited on a substrate of Fe. From experiments at $E_0 = 15$ keV the mass thicknesses of Pt (8.4 μg/cm²) and Cr (105 μg/cm²) can be determined. These values can be used to define probable models (Fig. 7) of the near-surface configuration, e.g., a film of pure Pt (4 nm) on a buried layer (150 nm) of pure Cr (model 1) or a "mixed" Pt-Cr surface film 141 nm thick (model 2). For $E_0 > 8$ keV the k-ratios of Pt Mα calculated by the PAP model [6], assuming the two configurations, are almost identical, whereas for $E_0 < 8$ keV significant differences between the predicted k-ratios are visible. Moreover, the configurations of model 1 and model 2 can be excluded, as the calculated k-ratios for $E_0 < 8$ keV do not agree with the experimental k-ratios. Therefore, a further two-layer configuration (model 3) was established, consisting of Pt-Cr (25 nm) on the top of Cr (122 nm). For this "most probable" structure, Fig. 7 shows complete agreement between calculated and experimental k-ratios at electron energies of 3–15 keV. The example of Fig. 7 demonstrates that experiments at low electron energies ($E_0 < 6$ keV) are essential to decide between probable configurations of submicron layered structures.

Conclusions

Electron probe microanalysis at low electron energies ($E_0 < 5$ keV) enables one to reduce the volume of microanalysis to dimensions of 0.3 μm or even less. Practically useful applications are the characterization of submicron films and layered structures. The principal possibility of a lateral resolution in the range of 0.2 μm is influenced by actual instrumental restrictions with respect to providing a well-focused (< 0.1 μm) but sufficiently intense electron beam at $E_0 < 5$ keV. In practise, the determination of low concentrations (< 5 wt% by EDS, < 0.5 wt by WDS) is difficult. The use of low electron energies and low energy X-ray lines is accompanied by certain experimental problems (e.g., chemical shift, surface contamination) which may have a negative influence on the accuracy of quantitative analysis. However, typical analytical errors should not exceed 5% when using $\Phi(\rho z)$ matrix correction models.

References

[1] R. Castaing, in: *Advances in Electronics and Electron Physics, Vol. 13* (L. Marton, ed.), Academic Press, New York, 1960, p. 317.
[2] P. Willich, *Microbeam Analysis* **1993**, *2*, 45.
[3] P. Willich, *Mikrochim Acta* [*Suppl.*] **1992**, *12*, 1.
[4] C. E. Nockolds, *Microbeam Analysis* **1994**, *3*, 185.
[5] J. L. Pouchou, F. Pichoir, in: *Electron Probe Quantitation*, Plenum, 1991, pp. 31–75.
[6] J. L. Pouchou, F. Pichoir, *Scanning Microsc.* [*Suppl*] **1993**, *7*, 167.

Mikrochim. Acta [Suppl.] 13, 639–643 (1996)

Oxidation and Reduction Processes of Be/BeO Induced by Electrons

Peter Zanel and Hildegard Hammer*

Institut für Angewandte Physik, Heinrich-Heine-Universität, D-40225 Düsseldorf, Federal Republic of Germany

Abstract. The electron induced oxidation and reduction processes have been studied by Auger electron spectroscopy (AES) and scanning Auger electron microscopy (SAM). The latter experiments enabled us to confirm an equilibrium reaction for the oxidation and reduction depending on the oxygen pressure and the dose of incident electrons. We found an intensified oxidation by exposure to X-rays too. Therefore, we are interpreting the oxidation and reduction process as Auger electron induced desorption.

Key words: beryllium, electron induced desorption, AES/SAM.

Be and BeO have often been studied by Auger electron spectroscopy (AES) and by diffraction of low energy electrons (LEED). On the one hand, the oxidation process [1] and the growth of the oxide film [2] has been studied depending on the oxygen partial pressure. On the other hand, a reduction process mainly induced by exposure to high doses of incident electrons has been mentioned in the past [3]. The main interpretation for the latter effect was the presence of electrons being responsible for the reduction process and no further systematic experiments have been done. Our AES-experiments [4] under UHV-conditions and at low sample currents have always shown an oxidation rather than a reduction process. Our longtime AES and SAM experiments have given the hint to an equilibrium reaction.

Experimental

The experimental arrangement was a scanning Auger electron microscope MICROLAB 305 (FISONS/VG Instruments). The residual gas pressure was usually in the 10^{-9} Pa region. To study the oxidation process depending on oxygen partial pressure we opened the valve to the transfer chamber. Then the pressure was about 10^{-7} Pa. With an external power supply it was possible to study the oxidation at sample temperatures between 300 K and about 600 K. The electron gun has a LaB_6 filament and the electron energy is variable between 1 keV and 10 keV. The sample current has been regulated between 0.05 nA and 1 μA. For this high sample current, the objective aperture has been

* To whom correspondence should be addressed

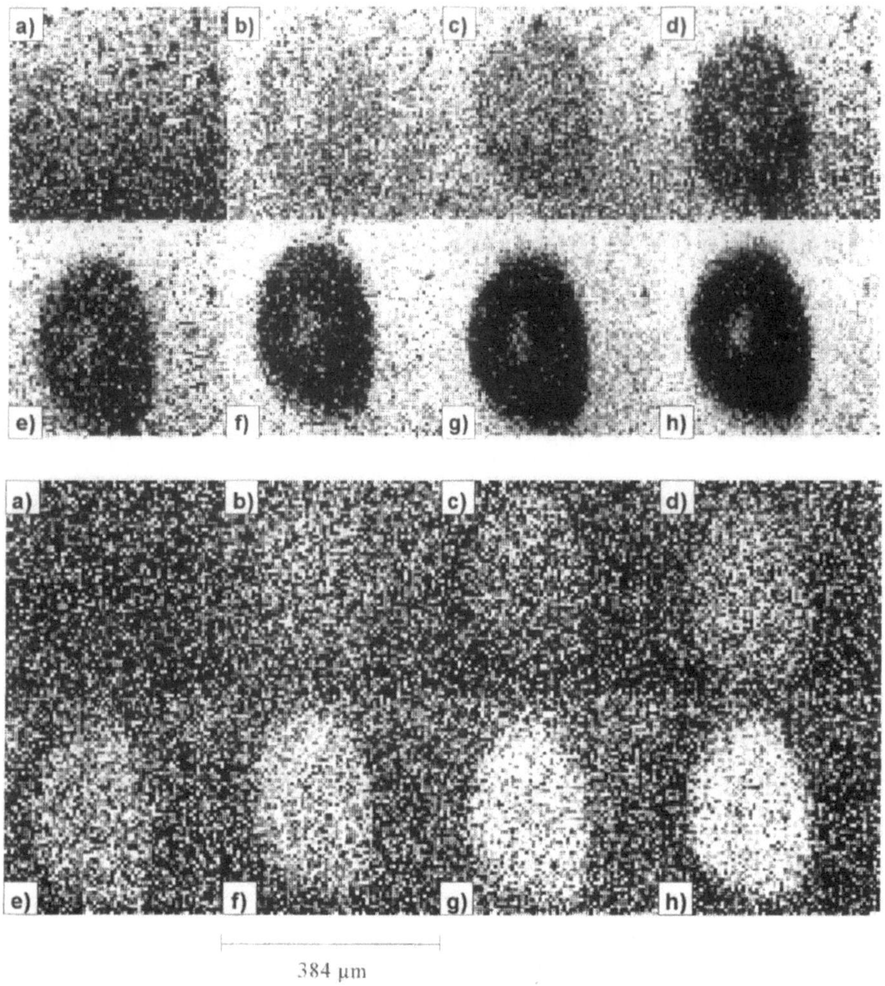

Fig. 1. Demonstration of artefacts induced by the examination itself by means of SAM-images of Be 104 eV (top) and O 507 eV (bottom); the increment of exposure to electrons was 20 min. from **a** to **h**; magnification: 625-times

removed. The result was a large region of non-collimated electrons reaching the sample. These electrons caused, additionally to the primary beam electrons, a well defined large oxidized region in the SAM images.

Figure 1 shows the formation of the oxidized region and the less oxidized or reduced central area for Be-SAM (top) and O-SAM (bottom) From a) to h) there have been twenty minutes of exposure

---

Fig. 2. **a** Demonstration of the electron induced oxidation and reduction process in SAM-images of Be 104 eV (top), Be(ox.) 98 eV (middle) and O 507 eV (bottom); magnification 625-times; left column: end of oxidation by exposure to electrons; middle column: partly reduced (after some hours), right column: further reduced Be surface by exposure to electrons (after 1 day). **b** AES survey spectra belonging to the SAM images of Fig. 2a taken at magnification 10.000-times; *1* spectrum of a clean Be surface at the beginning of the experiment; *2* at the end of exposure to electrons; *3* new position of the sample for the reduction experiment; *4* at the end of reduction by the exposure to electrons. The low energy region belongs to the beryllium AES lines, the high energy region to the oxygen AES lines

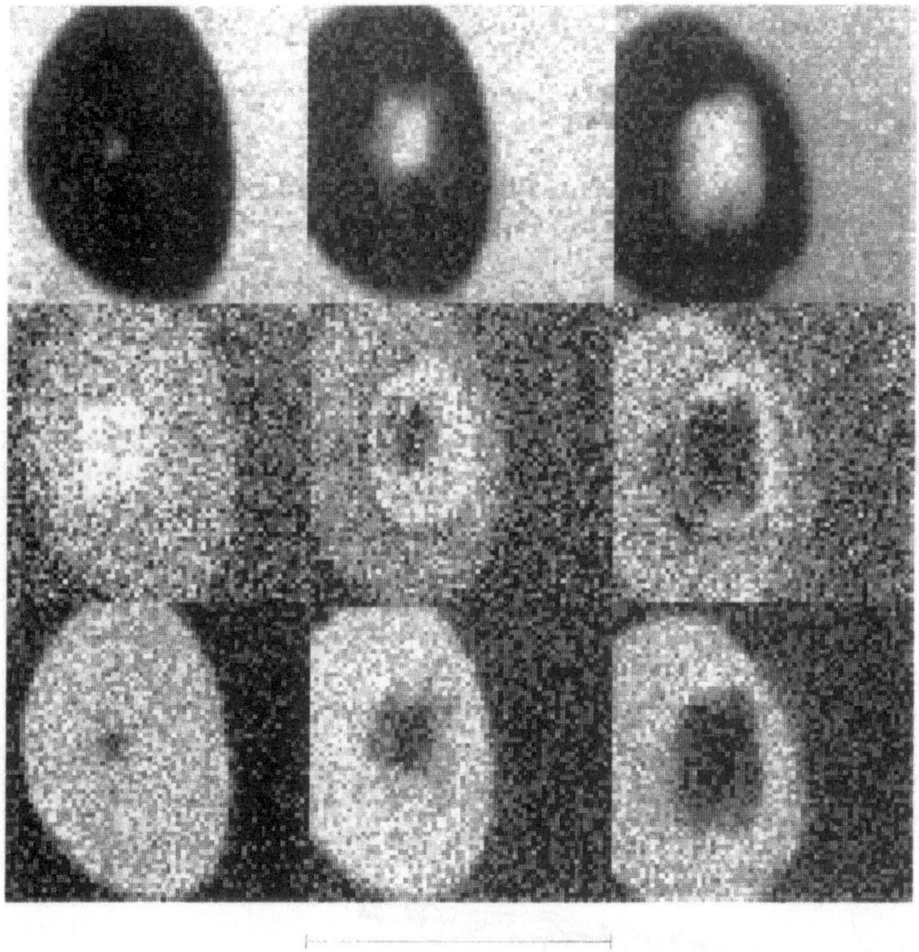
384 µm

a

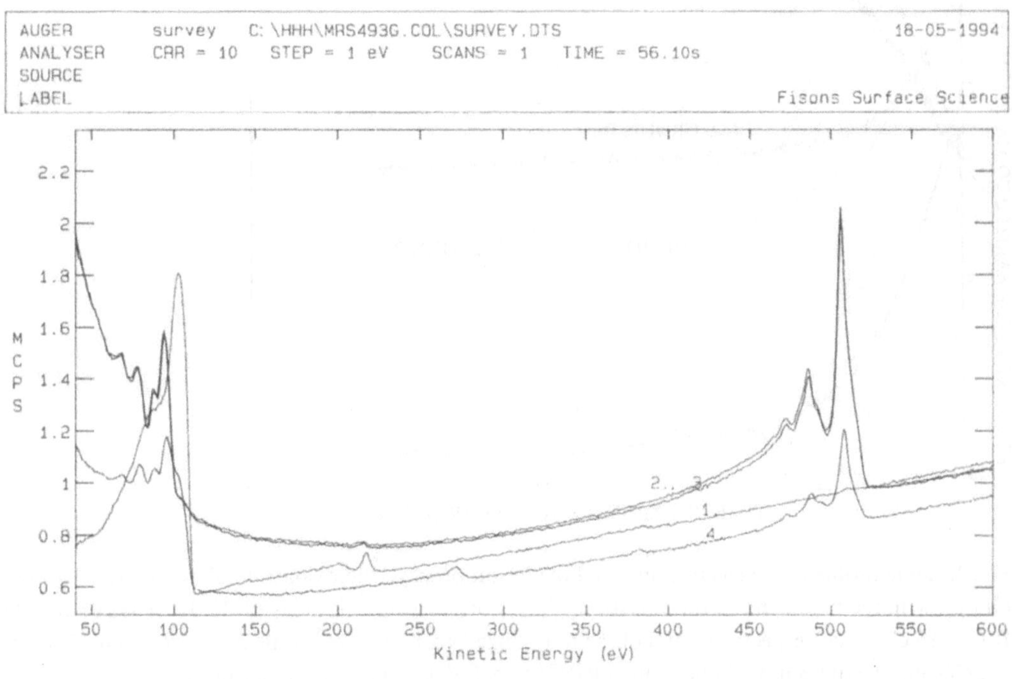
AUGER survey C:\HHH\MRS493G.COL\SURVEY.DTS 18-05-1994
ANALYSER CRR = 10 STEP = 1 eV SCANS = 1 TIME = 56.10s
SOURCE
LABEL
Fisons Surface Science
MCPS
2.2
2
1.8
1.6
1.4
1.2
1
0.8
0.6
50
100
150
200
250
300
350
400
450
500
550
600
Kinetic Energy (eV)
1
2
3
4
C:\HHH\MRS492D.COL\SURVEY.DTS C:\HHH\MRS493A.COL\SURVEY.DTS C:\HHH\MRS493G.COL\SURVEY.DTS

b

to a high dose of electrons between the registration of the SAM-images at a low dose of incident electrons. Figure 2a demonstrates the reduction process for Be-SAM (top), Be(ox)-SAM (middle) and O-SAM (bottom) and Fig. 2b shows the AES spectra of the clean Be surface just at the beginning of the exposure to electrons, the spectra of the oxidized and of the partly reduced surface, respectively. The procedure was as follows: first (left row in Fig. 2a) the sample has been oxidized as in Fig. 1. Then the sample has been shifted so that the scan was centered on the oxidized region. Some hours later (middle row) the reduced area appeared and increased during one day of exposure (right row).

Results and Discussion

The energy of the incident electrons had no measurable influence on the oxidation/reduction process. The sample temperature accelerated the oxidation process, but the transition region between the electron induced oxidized area and the unoxidized area remained unchanged. This region is mainly determined by the interaction volume of the electrons in beryllium depending on the energy and the dose of the incident electrons.

It is known [1, 2] that the oxidation depends on the oxygen partial pressure. The experiments [2] where the oxygen dose has been varied have shown, that during oxidation the beryllium oxide is at first growing more or less in islands. Figure 3 shows the results of a time profile for oxidation. The changing slopes of the curves hint at a layer by layer growth of the beryllium oxide film [5]. Our pressure

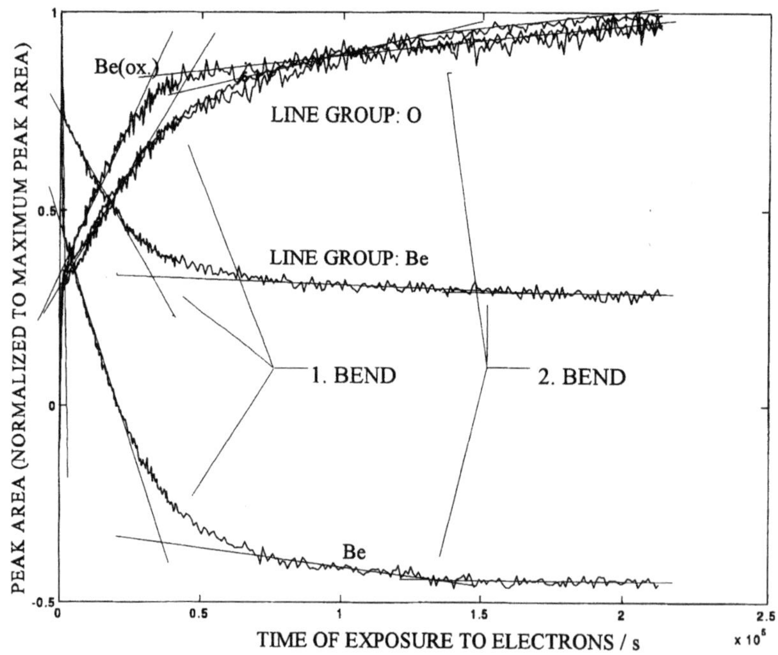

Fig. 3. Time profile of the electron induced oxidation process: the oxidation follows a layer by layer film growth; the figure shows the variation of the normalized peak areas of the Be (104 eV) and the Be(ox) (98 eV) lines, respectively, and of the normalized peak areas of the Be AES line group (52 eV to 112 eV) and the oxygen AES line group (460 eV to 520 eV), respectively (see Fig. 2b)

AUGER ELECTRON INDUCED DESORPTION

Reduction:	$2\,O^{2-}$	$\leftarrow$	$O_2 + 4\,e^-$
	Be	$\leftarrow$	$Be^{2+} + 2\,e^-$
Oxidation:	$(O^{2-} + Be^{2+}) + e^-$	$\rightarrow$	$O + Be^{2+} + 3\,e^-$
	$Be + e^-$	$\rightarrow$	$Be^{2+} + 3\,e^-$
Result:	$Be + O_2$	$\rightleftarrows$	2 BeO

REACTION OF EQUILIBRIUM
DEPENDING ON OXYGEN PRESSURE AND
DOSE OF INCIDENT ELECTRONS

Fig. 4. Reaction equations for the oxidation, the reduction and the equilibrium reaction by Auger electron induced desorption

dependent experiments have shown [4] that there is a certain final state of oxidation for residual gas pressures in the 10^{-9} Pa region and a higher one in the 10^{-7} Pa region. Lowering the pressure to 10^{-9} Pa again, the sample is reduced and the first state of oxidation is reached again. This means that there is an equilibrium reaction (Fig. 4) depending on the oxygen dose and on the incident electrons. The equilibrium is shifted to the oxidation part of the equilibrium for low doses of incident electrons, shifted to the reduction part at higher doses. The main process for the oxidation and reduction is the so called Auger electron reduction [6].

Acknowledgement. The authors would like to thank Prof. Kisker for some financial support.

References

[1] D. L. Seymour, E. B. Pattinson, *Surf. Sci.* **1987**, *189/190*, 1003.
[2] D. E. Fowler, J. M. Blakely, *Surf. Sci.* **1984**, *148*, 265.
[3] S. Thomas, *J. Appl. Phys.* **1974**, *45*, 161.
[4] P. Zanel, *Thesis*, Düsseldorf, 1995.
[5] C. Argile, G. E. Rhead, *Surf. Sci. Rep.* **1989**, *10*, 277.
[6] P. J. Feibelman, M. L. Knotek, *Phys. Rev.* **1978**, *B18*, 6531.

MIX
Papier aus verantwortungsvollen Quellen
Paper from responsible sources
FSC® C105338

If you have any concerns about our products,
you can contact us on
ProductSafety@springernature.com

In case Publisher is established outside the EU,
the EU authorized representative is:
Springer Nature Customer Service Center GmbH
Europaplatz 3, 69115 Heidelberg, Germany

Printed by Libri Plureos GmbH
in Hamburg, Germany